21ST CENTURY
ASTRONOMY

 W · W · NORTON & COMPANY
NEW YORK · LONDON

21ST CENTURY
ASTRONOMY

JEFF HESTER
Arizona State University

DAVID BURSTEIN
Arizona State University

GEORGE BLUMENTHAL
University of California—Santa Cruz

RONALD GREELEY
Arizona State University

BRADFORD SMITH
University of Hawaii—Manoa

HOWARD VOSS
Arizona State University

GARY WEGNER
Dartmouth College

The text of this book is composed in Bembo, with the display set in Charlemagne.
Composition by Techbooks.
Manufacturing by Courier, Kendallville.
Cover photograph: Detail of the Trifid Nebula taken from the NASA *Hubble Space Telescope*.
Photograph by Jeff Hester (Arizona State University) and NASA.

Editor: John Byram
Director of Manufacturing—College: Roy Tedoff
Copy Editor: Susan Middleton
Project Editor: Nan Sinauer
Photo Editor: Neil Ryder Hoos
Photo Assistant: Penni Zivian
Editorial Assistant: Garrett Michaels
Book Designer: Rubina Yeh
Illustrations: J/B Woolsey Associates
Layout Artist: Cathy Lombardi

Library of Congress Cataloging-in-Publication Data

21st century astronomy / Jeff Hester . . . [et al.].
    p. cm.
  Includes index.
  ISBN 0-393-97400-6
    1. Astronomy.   I. Title: Twenty first century astronomy.   II. Hester, John Jeffrey.

  QB45.2 .A14 2002
  520—dc21
                                                        2001044886

W. W. Norton & Company, Inc., 500 Fifth Avenue, New York, N. Y. 10110
www.wwnorton.com

W. W. Norton & Company Ltd., Castle House, 75/76 Wells Street, London W1T 3QT

1 2 3 4 5 6 7 8 9 0

David Burstein wishes to thank Gail, Jon, and Liz for their love and help.

George Blumenthal gratefully thanks his wife, Kelly Weisberg, and his children, Aaron and Sarah Blumenthal, for their support during this project.

Ronald Greeley dedicates this book to his wife, Cindy.

Bradford Smith dedicates this book to Diane and Don.

Howard Voss dedicates this book to Helen Ann.

Gary Wegner wishes to thank Cynthia Kay, Emma, Belle, Nicky, and Stella for their constant support and encouragement during the course of this work.

Jeff Hester dedicates this book to Vicki to whom he is still married, even after "The Book"; to his children—may they learn about rushing in where the wise fear to tread; to his colleagues and graduate students, who too often were left to pick up the slack as this project became all consuming; and finally to the many students, past and future, who make it all worthwhile.

CONTENTS

PART IV   GALAXIES, THE UNIVERSE, AND COSMOLOGY

PREFACE

Why, you might ask, would a group of accomplished scientists choose to get together and write yet another introductory astronomy textbook for the college and university market? It certainly could not be a response to a desperate shortage of books for such a course. Nor could it be that we have more time on our hands than we know what to do with. Perhaps it's the money? Count the number of authors and guess again.

In truth, our reasons for becoming involved in this project are almost certainly as varied as we are, but there is one reason that we all share. Having taught many sections of introductory science courses, we were unhappy with the available offerings for the introductory astronomy market. The titles sitting on our shelves did not match our idea of what a book for an introductory non-majors' astronomy course should be. They did not convey our sense of what our science is about or how we think and feel about our chosen field.

It means little to discuss the merits of an astronomy text without first answering a more basic question: What purpose does an introductory course in astronomy serve on today's college and university campuses? There are those intrepid students who would boldly answer this question, "To satisfy my life-long curiosity about astronomy, physics, and planetary science!" Thank goodness that such students exist. It is their intent faces and interested questions that get us through those occasional class periods when we think to ourselves, "Boy, am I glad I didn't have to *sit through* the lecture I just gave!"

Unfortunately, such students are the exception rather than the rule. Many students take "Stars for Po-

ets" to satisfy a science requirement. When faced with the choices at hand—biology (too squishy), physics (turn the catalog page quickly), chemistry (too smelly), geology (what's geology?)—many students opt for astronomy as the least of the available evils. Looking overhead at night they see the stars and think, "Big Dipper, Little Dipper, North Star . . . sure, I can do stars!" This is not meant as a disparaging remark, but simply as an acknowledgment that many of the bright faces we see in class on that first day have little idea about what astronomy really is. It is, of course, ironic that astronomy turns out to be a mixture of physics, chemistry, and geology, with a bit of math and biology thrown in for good measure.

The point is this: The students who take introductory astronomy typically come from very diverse backgrounds and, as often as not, their introductory astronomy course is the only formal exposure to science that they will have in their lives! When viewed in this way, teaching Introductory Astronomy seems like a big responsibility, and so it is. As instructors of introductory astronomy courses, everything we do should turn on the answer to basic questions such as: What should a student carry away from this course that will still be with them 20 years from now? How should this course change the students who take it? And how can we, as scientists and educators, facilitate those changes within our students?

A thoughtful response to these questions does *not* lead to a course that is primarily about memorization of facts. At best, such facts will go into a student's short-term memory for long enough to be regurgitated on an exam. At worst, a student might come to think that science really is nothing more than memorizing information. It is hard to imagine doing students a greater disservice than misrepresenting science in this way, especially since this is just the preconception that many students bring to class with them on the first day.

Granted, facts are an essential part of any course. It is impossible to reason or to understand apart from a basis in fact. The problem is that many students arrive at our doors with the idea that *learning* and *memorization* are synonymous. They have been taught in course after course to think of their brains as sponges, there to soak up information and then be wrung out into the final exam booklet, leaving them empty for the next term. They confuse *knowing* with the ability to call an object by name or recite a few relevant pieces of information about it, as if stars once gathered into constellations and named were of no further interest. If an introductory science course does nothing else, it should relieve students of any such notion. It should encourage students to think of their brains as they would a muscle that grows strong only with exercise and training. An introduction to science should be an introduction to what it means to think deeply about the Universe, finding within the world patterns and relationships that go beyond the specifics of a particular object or setting, and applying those patterns and relationships broadly. An introductory science course should first and foremost be about understanding the world through the eyes of a scientist.

One of the other authors of this text, Howard Voss, enjoys telling of a time when he was teaching an introductory physics course for nonscientists. One afternoon after a class discussing Newton's laws, one of his students came up to him in tears. He braced for the worst, but instead of a complaint about the difficulty of the course, he was met by a triumphant "I get it!" The student's tears were tears of joy. She had come to see the *sense* of Newton's laws. Not only did she know what they say, but she had come to understand something of what they mean and of how they can be used to make sense of the world.

In that moment of insight into Newton's laws, Howard's student caught a glimpse of what knowledge really is—of what it means to understand something rather than just know about it—and her discovery of that world of understanding was an emotional experience. I recall a time when a history major sprinted up to me at the end of a class to tell me, "If the Sun were more massive than it is, then it would *have* to be more luminous!" The magic in the moment was not a student successfully repeating some memorized fact but rather a student seeing why increasing the mass of a main sequence star *must* cause the star's luminosity to increase. Granted, 20 years from now that student will not remember the details of stellar structure, but he *will* remember what it feels like to see and understand the world through the eyes of science. If that student goes on to search for that same quality of understanding in the broader context of his life, then I will have done my job. This is the essence of an introductory science course.

An introductory astronomy course should also show students something of the epistemology of science: the way that knowledge is arrived at by constantly challenging and testing our ideas; the faith of the scientist that we live in a Universe that is not capricious but behaves according to consistent, explicable rules; and finally the great strength inherent in the fact that all scientific knowledge is provisional. How can it be that in science we never really prove things to be true, but instead only fail to show that they are false? Because that is how we give what is *really* true precedence over what we would *like* to be true. It is precisely because science acknowledges lack of certainty that the pinnacle of human knowledge is a well-tested, well-corroborated scientific theory. Here is the key to the difference between science and the myriad pseudoscientific pretenders that would borrow the credibility of science by adopting its trappings but forego its intellectual rigor. If a student comes to know and recognize the difference between science and pseudoscience, and if 20 years from now a former student still chuckles inwardly at the naiveté of the phrase, "it's only a theory," then we will have succeeded.

There is a strong practical side to what we teach. How can one claim to be educated in today's world without a basic understanding, not just of the products of science, but of the process of science? Thinking about the world as a scientist means learning to sort the wheat from the chaff. Physical scientists tend not to buy lottery tickets, call psychic help lines, or consume the many other forms of modern-day snake oil. Our students are served well if we pass on a healthy dose of our practical skepticism. A democratic society, which in the long run can fare no better than its citizens can think, is served also.

Yet while a well-taught astronomy course offers students meaningful tools for thinking about the world, the real reason why it matters that students come to see the world through the eyes of a scientist is because of the sheer beauty of the view. There can be no doubt of the majesty of the Universe in which we live or of the staggering implications of what we have come to know about the Universe and our place in it. Simply to stare at a deep image of distant galaxies or the glow of a newly formed star or a picture of Jupiter's swirling

Great Red Spot and try to comprehend what we see can be a mind-boggling experience. Add to that a dose of quantum mechanics or the idea of the tortured fabric of space-time around a black hole, and we are into territory that would awe Aristotle. Science—this marvelous expression of human curiosity and passion and reason—has shattered the mental cage in which we have lived since our ancestors first looked to the sky and wondered about what they saw there. To face these vistas surrounded, not by speculation, but by the meticulously constructed edifice of science is extraordinary. Finally, to live during that moment in history when we first looked at the stars and recognized our own origins in the Big Bang that occurred some 14 billion years ago—*that* is something to write home about!

Science is not the passionless, tedious, snore that many students believe it is. Rather, it is a vital, exciting, ongoing expression of our humanity. While an introductory astronomy book should teach students about what science is, showing them what science has revealed and giving them powerful tools for thinking about the world, it should also share with them something of the passion that scientists feel for the endeavor to which we have devoted our lives.

The book that meets these goals would have to be very different from most textbooks. Rather than presenting a blocked-out set of facts and descriptions, that book would need to tell a story. On a small scale, it would need to tell the story of specific ideas. Understanding does not come from a recitation of facts. Understanding comes from thinking carefully and critically about things and how they work. Helping a student understand a concept as a scientist means guiding that student through the concept, making heavy use of examples and analogies, and tying the concept back to everyday phenomena and experiences to which the student can relate. That is what the text we envisioned would have to do.

So rather than the expository style of traditional textbooks, this book tries for the explanatory style of a good teacher. Rather than stating that "the altitude of the north celestial pole above the horizon equals the observer's latitude" and leaving it at that, we talk about what the sky looks like over the course of the day as viewed from the North Pole and why, then invite the student to follow along as we head south, seeing how things change along the way. Rather than talking strictly about the equilibrium between heating by sunlight and cooling by infrared radiation that determines the temperatures of terrestrial planets, we begin instead by talking about water flowing into a can with a hole in the bottom and how the water level tends toward that

point where water flows in and out of the can at the same rate.

On a larger scale, *21st Century Astronomy* tells a web of different stories. Some of these stories have to do with what science is and how it is done. Why did Newton choose the form that he did for his universal law of gravitation? What are the fundamental differences between Kepler's empirical "laws" and Newton's theoretical derivation of the same relationships? And if Einstein was "right," why wasn't Newton "wrong"? Other stories explore the human side of science, such as how science has led us to think beyond the box that evolution built for us. Representing science fairly and honestly even means telling those stories that might make some students uncomfortable, like the story of why our understanding of the evolving Universe and our place in it is science, while "creation science" is not.

Of course, the stories that make up most of this work are the stories of the ideas themselves. And stories they are: the story of motion, the story of light, the story of matter and energy, the story of Earth and our planetary system, the story of the Sun and stars, the story of life, and the story of the Universe. These are told as stories, with one idea flowing into the next and each idea connected to the whole, because that is the way humans learn. Knowledge and understanding are constructed, with each idea and insight given meaning by how it fits into the whole. None of the stories in *21st Century Astronomy* exist in isolation. They are threads in a grand intellectual tapestry, woven from the recurrent themes of the physics of matter, energy, radiation, and motion.

Science is not about moldy ideas found on dusty shelves in the back of long-forgotten libraries. Science is about the excitement of our ongoing discovery of the Universe. The story of science is not light entertainment, but it is most definitely entertaining. We have adopted a writing style for *21st Century Astronomy* that fits this vision of science. The writing you will find here is far less formal than that found in most textbooks. We allow ourselves to be conversational, to pursue the occasional digression, to be irreverent or provocative or idiosyncratic when it suits our taste, and to allow the sense of passion and wonder that we feel for the subject to show through from time to time. As the principal author of the textbook, it was my goal to write in a style more characteristic of popular nonfiction than of textbooks. My hope was to catch the feel of a pleasant, stimulating conversation with a student during office hours.

Which brings us to the first of many tensions that were with us throughout the writing and editing of

21st Century Astronomy—the question of who is sitting across the desk, listening to what we have to say. There is a very real pull in the market in the direction of simplification. Books filled cover to cover with the latest images from NASA missions and simple, factual descriptions look pretty and are doubtless much easier to teach out of than more conceptually challenging treatments of the material. The problem is that such texts are so accessible that what they contain sometimes seems hardly worth accessing at all. For the most part, students pass through such courses untouched, the course forgotten the moment the student walks out the door.

There are plenty of books around that can be adapted to a course aimed at the lowest common denominator. That is not the slice of the market we chose to target. Instead, we hope that *21st Century Astronomy* will appeal to instructors who, like us, have found that a serious nonscience student in today's colleges and universities is capable of thinking more deeply about more conceptually challenging material than is often assumed. Frankly, with the story that *21st Century Astronomy* has to tell—a story that deals head on with some of the most fundamental, fascinating, and far-reaching questions humans have ever asked themselves—few students can resist being drawn in if approached in the right way.

If level of presentation is the most fundamental tension in this book, length and scope was certainly the most difficult tension upon which to reach a final compromise. It was impossible to cover everything we would have liked to cover, cover it in the depth and style we prefer, and stay within the number of pages customary for a book in this market. The authors might have been content to let the book run long and leave it to the instructor to decide what to discuss in depth. However, writing a textbook that we like but that no one adopts because it is too long seemed a Pyrrhic victory. So compromises were made and made and made. The number of chapters in the book was reduced, and material was cut. Discussions and examples were shortened or eliminated. Topics were excised altogether or moved off onto the CD-ROM and web page being developed by others in parallel with the text. In the end, trimming the book down to size was a painful process. While we are comfortable with the suite of topics that remain, we know that some of you will look through the table of contents and discover that your favorite topic is missing. If so, you might take some small comfort in knowing that the pages you seek probably exist somewhere in the heap on the cutting-room floor.

Another significant tension in the book is the tension between a traditional organization of topics that may be more comfortable for instructors and an organization that more accurately represents the way that scientists think about topics at the opening of the 21st century. True to our purpose, we went with an approach that reflects the state of our science. (The students who will read this book have no emotional attachment to the way an author first laid out an introductory astronomy text 30 years ago. Why should we?) For example, we organized the Solar System section around a theme of comparative planetology, rather than the more traditional "this is Tuesday so it must be Mars" approach. We pulled the discussion of tidal interactions, orbital resonances, chaos, and similar phenomena into a separate chapter (Chapter 9) entitled "Gravity Is More than Kepler's Laws." We posited the existence of a rotating, collapsing interstellar cloud and a rotating protoplanetary disk and discussed the formation of the Solar System in Chapter 5 *before* discussing the planets themselves. This allowed us to look at the planets from the outset within the context of how they formed. Enough examples—you get the idea.

Even so, there is flexibility in the text. For example, we feel the flow of material through the early chapters discussing physical principles works well. We have tried to include enough discussion of the human and social aspects of science to break up the treatment of conceptually difficult material. On the other hand, some instructors may like to show their students a bit more astronomy and planetary science before hitting the physics. Such instructors might want to start with Chapter 5 on the formation of the Solar System and then pull in material on orbits and radiation as they are needed in what follows. Another approach is to start with stars in Chapter 12, pull in radiation and orbits as needed to understand the properties of stars, then insert the Solar System as an extended excursion on the topic of the formation of low-mass stars. While we wrote the book to tell a complete, well-integrated story, we tried to allow for different paths tailored to the needs and tastes of individual instructors.

So far this preface has tried to give you some sense of what we set out to accomplish. We have done what we have done, and you hold the result in your hands. It is up to you, the reader, to decide for yourself whether we have been successful and whether our vision of the "better mousetrap" of astronomy textbooks conforms to your own.

However, in closing I should say a few words about just who we are. Among the authors of this text you will find a group of accomplished and well-known scientists who have been at the forefront of many of the most exciting and significant events in astronomy and planetary science in the second half of the 20th century. The authors of this text include members of scientific teams that built three of the cameras that have flown on the *Hubble Space Telescope*, team members and leaders of many of the major planetary missions including the *Voyager* and *Galileo* missions to the giant planets and the *Mars Pathfinder* mission, and a former president of the American Association of Physics Teachers. Among us, we have been involved in significant fundamental research on topics ranging from planetary geology to the origin and evolution of the Solar System to the formation and evolution of stars to the structure and dynamics of the interstellar medium to the nature of galaxies to the origin of structure in the Universe. There are remarkably few fields discussed in this book to which one or more of the authors have not contributed in some important way over the years. For us, 21st century astronomy is not a textbook. Rather, it is the life that we live. We hope that you find value in our attempt to share with you what *we* see when *we* look up at the sky at night.

Jeff Hester
Tempe, Arizona
December 2001

ACKNOWLEDGMENTS

Production of a book like this is a far larger enterprise than any of us imagined when we jumped on board. We would particularly like to thank the group of teachers who reviewed portions of the manuscript along the way: Jeff Adams, William Andersen, Barbara Anthony-Twarog, Allen Armstrong, Keith Ashman, Edward Baron, David Baum, Dwight Beery, William Bittle, John Blake, Anita Corn, John Cummins, Kathy Eastwood, Terry Ellis, Randy Emmons, Thomas English, Eric Feigelson, Simonetta Frittelli, Martin Gaskell, Billy Graves, Kim Griest, Erick Guerra, Javier Hasbrun, Paul Heckert, Roger Hewins, Paul Hinds, Eric Hintz, David Hufnagel, Andrew Ingersoll, Adam Johnston, Khondkar Karim, Frank Kowalski, Claud Lacy, John Laird, Kenneth LaSota, Irene Little-Marenin, Bruce Margon, Bradley Matson, George McGill, Kenneth McLaughlin, Karen Meech, Zdzislaw Musielak, Robert O'Connell, Aileen O'Donoghue, Charles Peterson, Cynthia Peterson, Randy Phelps, Carlton Pryor, George Rosensteel, Anthony Russo, Carl Rutledge, Stephen Schneider, William Schoenfeld, Michael Sitko, Tim Slater, Larry Smith, Larry Sromovsky, Thomas Statler, Christine Staver, Curtis Struck, Donald Terndrup, Suzanne Willis, Louis Winkler, and George Wolf.

There are many at W. W. Norton & Company without whom this project would not have come together. The authors would like to thank our two editors, Stephen Mosberg, who first looked across the table at us and said, "We will find a way," and John Byram, who stepped into the project midstream and held it together through difficult times. Roby Harrington gave us enough rope to hang ourselves, then helped us out when we tried to do just that. Susan Middleton is a first-rate copy editor and also an interested and critical reader. Among those in the trenches are Garrett Michaels, Editorial Assistant; Neil Hoos, Manager of Photo Permissions; Penni Zivian, Photo Assistant; Jane Carter, Associate Managing Editor—College Books; Roy Tedoff, Director of Manufacturing—College; Rubina Yeh, Associate Art Director and designer of this book; and Nan Sinauer, Project Editor. Thanks to Paul Scowen, who assisted with art development and whose door was always open for a sanity check, and who is also developing the student CD-ROM. Thanks also to Karen Vanlandingham, who is writing the study guide to accompany the text. Finally, we would like to thank John Woolsey and Craig Durant of J/B Woolsey Associates, who became true collaborators on the project, rather than simply artists drawing to spec.

INTRODUCTION TO ASTRONOMY

The most beautiful thing we can experience is the mysterious.
It is the source of all true art and all science.
He to whom this emotion is a stranger,
who can no longer pause to wonder and stand rapt in awe,
is as good as dead: his eyes are closed.

ALBERT EINSTEIN (1879–1955)

WHY LEARN ASTRONOMY?

1.1 STARTING WITH A SPARK OF INTEREST

Not everyone has a general fascination for science, but almost everyone harbors a spark of interest in astronomy. Since you are reading this book, you probably share this spark as well. The spark may have been struck when you were a child looking at the sky and found yourself wondering about what you saw there. What are the Sun and Moon made of? How far away are they? What are the stars? How do they work? Do they have anything to do with me? The prominence of the Sun, Moon, and stars in cave paintings and rock drawings (such as those in **Figure 1.1**) dating back thousands of years tells us that these questions have long occupied the human imagination. Your initial spark of interest in astronomy may have grown over the years as you saw or read news reports about spectacular discoveries made in your lifetime. Some of these discoveries may have sounded so amazing that it was difficult to draw the line between science fact and science fiction.

If you nurture your spark of interest in astronomy as you continue through this book, you may be surprised to find that spark growing into a flame. The title of this book—*21st Century Astronomy*—was chosen to emphasize that this is the most fascinating time in history to be studying this most ancient of sciences. This book will take you to places you never imagined going, and

KEY CONCEPTS

Before traveling through unfamiliar terrain, it helps to have some idea of where you are going, what you might see along the way, and what you should pack for the journey. In *21st Century Astronomy*, we will learn not only about the wonders of the Universe, but also about what it means to look at the world through the eyes of science. In this chapter's overview of what is to come, we will find that:

* The Universe is vast beyond all human experience, yet it is governed by the same physical laws that shape our daily lives;
* We are a product of that Universe; the very atoms of which we are made were formed in stars that died long before the Sun and Earth were formed;
* Science is a creative human activity like art, literature, and music, and is also a remarkably powerful, successful, and aesthetically beautiful way of viewing the world;
* Understanding comes from thinking carefully and deeply about patterns in the world, and not simply from memorization of facts; and
* Like climbing a mountain, the journey we are about to make requires effort, but the view from the top is amazing to behold.

Figure 1.1 *Ancient petroglyphs often include depictions of the Sun, Moon, and stars.*

will lead you to insights and understandings you never imagined having. To those of you who are reading this book for a course in astronomy at your college or university, we have a special note. The authors of this text have taught many sections of introductory astronomy over the years. We recognize that you may be in this course primarily because you need a science credit to graduate. As you flipped through your course catalog, perhaps you were reminded of your spark of interest in astronomy, and that led you to choose astronomy over your other options. (Or perhaps you simply considered astronomy to be the least of the available evils!) Whatever your expectations, the story in *21st Century Astronomy* can fascinate you if you open your mind to it.

The journey of discovery on which we are embarking is not always an easy one, but few worthwhile journeys are. A hike in the mountains can at times be an easy stroll and at other times a more strenuous climb, but when you arrive, the view from the top is hard to beat. In much the same way, this book will ask you to exercise your mental muscles in different, possibly unaccustomed ways. But, as with the hike in the mountains, we feel certain that you will find the rewards worth the investment.

GETTING A FEEL FOR THE NEIGHBORHOOD

If you are like many people, your conception of astronomy may not go much beyond learning about the constellations and the names of the stars in them. The word **astronomy** *means* "naming the stars." But modern astronomy—the astronomy we will talk about in this book—has become far more than looking at the sky and cataloging what is visible there. It may seem something of a contradiction, but a great deal of front-line astronomy is now carried out in physics laboratories like the one shown in **Figure 1.2.** Today astronomers work along with their colleagues in related fields such as physics, chemistry, geology, and planetary science to sharpen our understanding of the physical laws that govern the behavior of **matter** and **energy,** and to use this understanding to make sense of our observations of the cosmos.

We are confident in our answers to many of the questions that you may have asked yourself as a child when you looked at the sky. We all live on a planet called Earth, which is orbiting under the influence of gravity about a star called the Sun. The Sun is an ordinary, middle-aged star, more massive and luminous than some stars but less massive and luminous than others. The Sun is extraordinary only because of its impor- **Earth exists in the context of the Universe.** tance within our own Solar System. The Sun is located about half way out from the center in a flattened collection of approximately 100 billion stars referred to as the Milky Way Galaxy. The Milky Way in turn is a member of a small collection of a few dozen galaxies called the Local Group, which is part of a vastly larger collection of thousands of galaxies called a supercluster. But even this vast structure is part of the *local* Universe. The part of the Universe that we can see extends outward for the distance that light travels in 14 billion years or so, and in

Figure 1.2 *This laboratory, where physicists are studying the properties of atoms, might seem an unlikely place to be doing astronomy. But laboratory astrophysics, studying astronomically important physical processes in a laboratory, has become an important part of astronomy.*

this volume it is estimated that there are about 100 billion galaxies—roughly as many galaxies as there are stars in the Milky Way!

One of the first conceptual hurdles that we face as we begin to think about the Universe is its sheer size. If a hill is big, then a mountain is really big. If a mountain is really big, then Earth is enormous. But where do we go from there? We quickly run out of superlatives as the scale of what we are talking about comes to dwarf our human existence. One technique that can help us develop a sense for the size of things in the Universe is to use a little slight of hand and move from a discussion of distance to talking instead about time. If you are driving down the highway at 60 miles per hour, a mile is how far you go in a minute. Sixty miles is how far you go in an hour. Six hundred miles is how far you go in a 10-hour day. So to get a feeling for the difference in size between 600 miles and 1 mile, you can think about the difference between how long a 10-hour day is compared with how long a single minute is.

We can play this same game in astronomy, but the speed of a car on the highway is far too low to be useful. Instead, we will use the greatest speed in the Universe—the speed of light. Light travels at a speed of 300,000 kilometers per second. At that speed, light is able to circle Earth (a distance of 40,000 kilometers,

Light travel time helps in understanding size.

a very long way) in just under $\frac{1}{7}$ of a second, or about the time that it takes you to snap your fingers. Fix that comparison in your mind. The size of Earth is like— *snap!*—a snap of your fingers. Follow along in **Figure 1.3** as we move outward into the Universe. We next encounter the Moon, 384,000 kilometers away, or a bit over $1\frac{1}{4}$ seconds when moving at the speed of light. So, if the size of Earth is a snap of your fingers, the distance to the Moon is about the time that it takes to turn a page in this book. Continuing on, we find that at this speed the Sun is $8\frac{1}{3}$ minutes away, or the length of a hurried lunch at the student union. To cross from one side of the orbit of Pluto, the outermost planet in our Solar System, to the other takes about 11 hours. Think about that for a minute. Let it sink in. Comparing the size of Pluto's orbit to the circumference of Earth is like comparing the time between sunrise and sunset to a single snap of your fingers.

Yet in crossing Pluto's orbit we have only just begun our journey. Many steps remain. It takes us a bit over four years to cover the distance from Earth to the nearest star (other than the Sun), or as much time as you spent in high school. At this point even our analogy using light travel time can no longer bring astronomical distance to a human scale. For light to travel from the center of our Galaxy to its edge takes about 100,000 years, or the time that *Homo sapiens* has walked the surface of Earth. To reach the nearest large galaxies beyond our own takes several million years, or the time since our australopithecine ancestors appeared on the scene. To reach the limits of the currently observable Universe takes light roughly 14 billion years—the age of the Universe, or about three times the age of Earth.

Look at that comparison again. The size of Earth is to the vast expanse of the Universe as a single snap of your fingers is to three times the amount of time that has passed since the Sun and Earth were formed! Here is something to ponder the next time you look up at a star-filled summer sky.

As you go through this text, you will occasionally see material set aside in boxes. Material has been boxed either because it is somewhat out of the main stream of our journey, or because we wish to highlight it. In particular, Foundations boxes deal with material that is central to our physical understanding of the Universe. Tools boxes discuss the technology and techniques that astronomers and planetary scientists use. Excursions boxes are short but interesting side trips. Finally, Connections boxes draw attention to recurring themes—bridges between different parts of our journey.

GLIMPSING OUR PLACE IN THE UNIVERSE

While seeking knowledge about the **Universe** and how it works, modern astronomy and physics have repeatedly come face to face with a number of age-old questions long thought to be forever the domain of philosophers. Issues as seemingly metaphysical as the origin and fate of the Universe and the nature of space and time themselves have become the subjects of rigorous scientific investigation. The answers we are finding to these questions are often far more wondrous than our predecessors could have dreamed. They are changing not only our view of the cosmos, but our view of ourselves as well.

Figure 1.4 envisions a scholar raising the veil of the heavens to see what wonders lie there. Throughout most of history, philosophers looked at the Universe and saw it as remote and different from Earth—disconnected from our terrestrial existence. When modern astronomers look at the Universe, they see instead a network of ongoing processes that we are a part of. Astronomy begins by looking out at the Universe, but increasingly that outward gaze turns to introspection

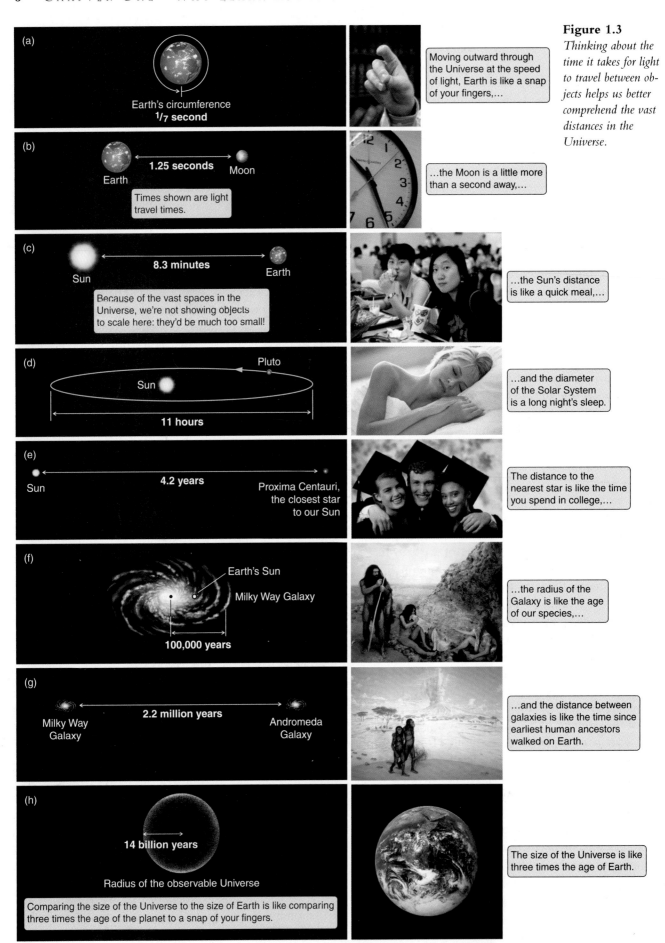

Figure 1.3
Thinking about the time it takes for light to travel between objects helps us better comprehend the vast distances in the Universe.

(a) Earth's circumference **1/7 second**

Moving outward through the Universe at the speed of light, Earth is like a snap of your fingers,…

(b) Earth **1.25 seconds** Moon

Times shown are light travel times.

…the Moon is a little more than a second away,…

(c) Sun **8.3 minutes** Earth

Because of the vast spaces in the Universe, we're not showing objects to scale here: they'd be much too small!

…the Sun's distance is like a quick meal,…

(d) Pluto Sun **11 hours**

…and the diameter of the Solar System is a long night's sleep.

(e) Sun **4.2 years** Proxima Centauri, the closest star to our Sun

The distance to the nearest star is like the time you spend in college,…

(f) Earth's Sun Milky Way Galaxy **100,000 years**

…the radius of the Galaxy is like the age of our species,…

(g) Milky Way Galaxy **2.2 million years** Andromeda Galaxy

…and the distance between galaxies is like the time since earliest human ancestors walked on Earth.

(h) **14 billion years** Radius of the observable Universe

The size of the Universe is like three times the age of Earth.

Comparing the size of the Universe to the size of Earth is like comparing three times the age of the planet to a snap of your fingers.

Figure 1.4 *Throughout most of history humans conceived of the rest of the Universe as a place apart from us. Here a traveler raises the curtain of the firmament to get a glimpse at what lies beyond.*

as we come to appreciate that our very existence is a consequence of those same processes.

The study of the chemical evolution of the Universe is a case in point. As a result of both observation and theoretical work, we now understand that when the Universe was young (14 billion years ago, or so), the only chemical elements to be found in abundance were hydrogen and helium, plus tiny amounts of lithium, beryllium, and boron. Yet we are not made exclusively of these lightest of elements. Our bodies are built of carbon, nitrogen, oxygen, sodium, phosphorus, and a host of other chemical elements. We live on a planet with a core consisting mostly of iron and nickel, surrounded by a mantle made up of rocks containing large amounts of silicon and other elements. If these more massive elements were not present in the early Universe, where did they come from?

We are stardust.

To answer this question we have but to look at the lights in the night sky. The energy to power stars comes from nuclear fusion reactions that occur deep within their interiors. Fusion reactions in stars take less massive atoms like hydrogen and combine them, forming more massive atoms, accomplishing the alchemist's dream of transforming one element into another. When a star exhausts its nuclear fuel and nears the end of its life, it often loses much of its mass—including some of the new atoms formed in its interior—blasting it back into interstellar space. We will talk later about the life and death of stars. For now it is enough to note that our Sun and Solar System formed from a cloud of interstellar gas and dust that had been "polluted" by the chemi-

cal effluent from earlier generations of stars. This chemical legacy supplies the building blocks for the interesting chemical processes that go on around us—chemical processes such as life. **Figure 1.5** symbolizes this intimate relationship between the world around us and our heritage in the stars. Look around you. The atoms that make up everything that you see were formed in the hearts of stars. Poets sometimes say that "we are stardust," but this is not poetry. It is literal truth.

WE LIVE IN AN AGE OF EXPLORATION AND DISCOVERY

Another reason this is a fascinating time to be learning about astronomy is that we live in an age of exploration. The 1957 launch of *Sputnik*, the first human-made satellite, occurred just one year before the birth of the youngest of the authors of this book. Little more than four decades later, as we begin the 21st century, we have seen humans walk on the Moon **(Figure 1.6),** and have seen unmanned probes visit all of the planets except Pluto. Spacecraft have flown by asteroids, comets, and even the Sun. Our inventions have landed on Mars and Venus and have plunged into the atmosphere of

Space exploration has expanded our view of the Universe.

Figure 1.5 *You and everything around you, including beautiful waterfalls, are composed of atoms that were forged in the interior of stars that lived and died before the Sun and Earth were formed. The left panel shows a cloud of chemically enriched material that has been ejected from the star Eta Carinae.*

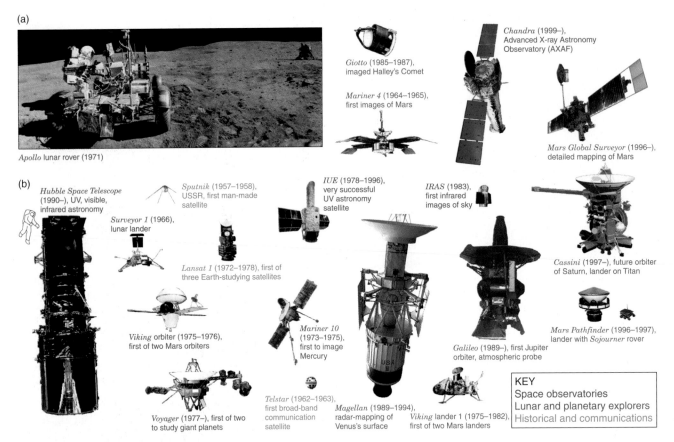

Figure 1.6 (a) Apollo 15 *astronaut James B. Irwin stands by the lunar rover during an excursion to explore and collect samples from the Moon.* (b) *Artificial satellites and space probes have progressed a long way since the 1957 launch of* Sputnik I. *These spacecraft are all shown to the same scale. Some are astronomical observatories that view space from Earth's orbit. Others are interplanetary explorers sent to investigate other worlds within our Solar System.*

Jupiter. Most of what we know of the Solar System has been learned during the last part of the 20th century as a result of this burst of exploration.

Satellite observatories in orbit around Earth have also provided us with many new perspectives on the Universe. The same atmosphere that shields us from harmful radiation also blinds us to much of what is going on around us. Space astronomy continues to show us vistas hidden from the gaze of so-called ground-based astronomers by the protective but obscuring blanket of our atmosphere. Satellites capable of detecting radiation—ranging from gamma rays and X-rays, to ultraviolet radiation, to infrared radiation to microwaves—have brought surprising discovery after surprising discovery. Each has forever altered our perception of the Universe, further expanding the domain of the human mind. At the same time, the closing years of the 20th century have also witnessed a renewed vigor in astronomical observations from the surface of Earth. The view of the sky seen by radio telescopes shown in **Figure 1.7** illustrates

Figure 1.7 *In the 20th century, new tools opened new windows on the Universe. This is the sky as we would see it if our eyes were sensitive to radio waves, shown as a backdrop to the National Radio Astronomy Observatory site in Green Bank, West Virginia.*

the new perspectives that have been opened up by our growing technological prowess.

Astronomy has also benefited enormously from the computer revolution. The 21st-century astronomer spends far more time peering at a computer screen than peering through the eyepiece of a telescope. Computers are used to do everything from collecting and analyzing data from telescopes, to calculating physical models of the conditions that exist in the hearts of stars, to preparing and disseminating the results of our work.

We truly live in a golden age of exploration and discovery. When we look back at the Renaissance, we recall little of the concerns that dominated the day-to-day existence of those alive at that time. Instead, we remember the Renaissance as a time of great art, literature, and music. We remember it especially as a time when the spirit of inquiry was reawakened and when much of what we think of as science was born. What might a historian 300 years in the future consider to be of lasting significance about *our* time? It is doubtful that our hypothetical historian of the future will care much about who won the Super Bowl, or which performer was at the top of the pop charts, or which brand of toothpaste tasted best. The media spectacles that often seem to dominate public consciousness will merit little more than an occasional footnote.[1]

> **A growing understanding of the Universe is a hallmark of our time.**

Much of what seems so significant to us today will be dust in the wind 300 years from now. However, we can be certain that our future historian *will* remember ours as the time when humankind first stepped beyond the world of our birth, and we began to reach out with our minds and our science to touch the fabric of the Universe itself. It is probably safe to say that few things will have a more lasting impact on our culture than this revolution in our understanding of the Universe and our place in it. No history book will ever again be complete without the headline in **Figure 1.8.** What has yet to be determined is whether our future historian will remember us for reaching out to touch the Universe and embracing what we found, or whether we will instead be remembered for a loss of spirit—for stepping back and turning away from the frontier of exploration and discovery. The direction we take from here is a decision in which you will play a part.

---

[1]That footnote will probably comment on a civilization whose technical ability to spread information had far outstripped its judgment about what information was worth spreading.

Figure 1.8 *There is no doubt that history will remember ours as the time when humankind first stepped beyond our home world and reached out with our minds to embrace the Universe.*

1.2 SCIENCE IS A WAY OF VIEWING THE WORLD

As we look at the Universe through the eyes of astronomers, we will also learn something of how **science** works. It is almost impossible to overstate the importance of science in our civilization. One obvious manifestation of science is the fact that almost everything that you use in your everyday life is a product of our scientific understanding of the world. The "simple life" is often romanticized, but it would be hard for any of us today even to begin to imagine what a truly pretechnological existence was like. Even a "primitive" trip into the backcountry is often accomplished today by eating freeze-dried food, sheltering under ultralight, ultrastrong synthetic fabrics, using a cellular phone to stay minutes away from emergency medical assistance, and following maps using a handheld Global Positioning System

(GPS) receiver. Given the visible importance of technology in our lives, you might even be tempted to say that science *is* technology.

It is true that science forms the basis of technology and that a mutually supportive relationship exists between science and technology, in which each enables advances in the other. Yet science is much more than technology. Science can no more be reduced to its practical technological application than the accomplishment of an Olympic athlete can be reduced to the utilization of the athlete's fame to market shoes and soda pop.

If science is more than technology, you might instead suggest that science is the **scientific method.** When you took science in high school you probably had the scientific method drilled into you—hypothesis and theory, followed by prediction, followed by experiments to test those predictions. There is good reason for placing emphasis on the scientific method. For all practical purposes, it defines what we mean when we use the verb *to know*. It is sometimes said that the scientific method is how scientists prove things to be true, but actually it is a way of proving things to be *false*.

Figure 1.9 *The scientific worldview is as aesthetically pleasing as that of art, music, or literature, but unlike art, nature has the final say about what scientific theories have lasting value.*

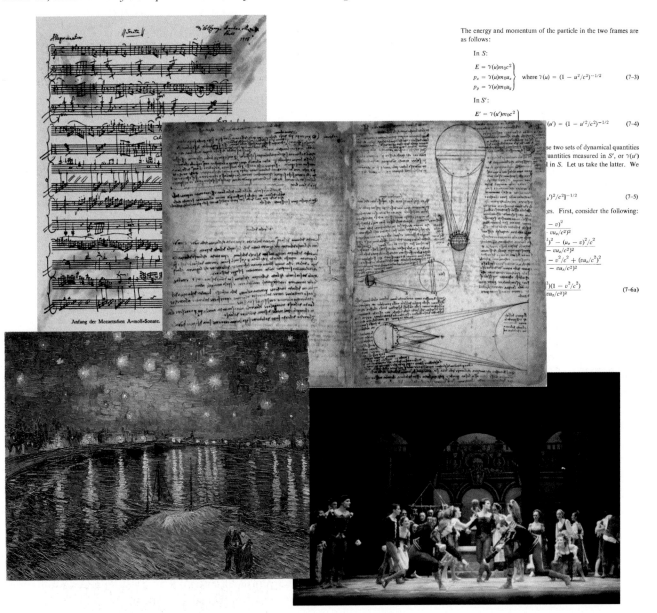

Before scientists accept something as true, they work very hard to show that it is false. Only after repeated attempts to disprove an idea have failed do scientists begin to accept its likely validity.

But again, science can no more be said to *be* the scientific method than music can be said to *be* the rules for writing down a musical score. The scientific method provides the rules for asking **nature** whether an idea is false, but it offers no insight into where the idea came from in the first place, or how an experiment was designed. If you were to listen to a group of scientists discussing their work, you might be surprised to hear them using words such as *insight, intuition,* and *creativity.* Scientists speak of a beautiful theory in the same way that an artist speaks of a beautiful painting or a musician speaks of a beautiful performance. Yet science is not the same as art or music in one very important respect. Art and music are judged by a human jury alone, while in science it is nature (through the application of the scientific method) that provides the final decisions about what theories can be kept and what theories must be discarded. Nature is totally unconcerned about what we *want* to be true. In the history of science many a "beautiful" and dearly beloved theory has been abandoned by the wayside. At the same time, however, **Figure 1.9** makes the point that there is an aesthetic to science that is as human and as profound as any found in the arts.

Nature is the arbiter of science.

It is also incorrect to say that science is a body of facts. We do not pretend to have all of the answers, and are constantly having to refine our ideas in response to new data and new insights. (This is, after all, what it means to learn.) The vulnerability of knowledge that is implicit in the scientific method may seem like a weakness at first. "Gee, you really don't know anything," the cynical student might say. But this vulnerability is actually science's great strength. It is what keeps us honest. Once an idea is declared to be "truth," then all progress stops. In contrast, in science even our most cherished ideas about nature remain fair game, subject to challenge by evidence according to the rules of the scientific method. Many of history's best scientists earned their place in the forward march of knowledge by successfully goring a sacred cow.

All scientific knowledge is provisional.

Scientists spend most of their time working within a framework of understanding, extending, and refining that framework and testing its boundaries. Occasionally, however, major shifts occur in the framework of some scientific field itself. Many books have been written about how science progresses, perhaps the most influential being *The Structure of Scientific Revolutions* by Thomas Kuhn. In this work, Kuhn emphasizes the constant tension between the scientist's human need to construct a system of beliefs within which to interpret the world, and the occasional (and likely painful) need to drastically overhaul that system of beliefs.

Scientific revolutions are major shifts in understanding.

A scientific revolution is not a trivial thing. We cannot just wish that the Universe were one way or another and then expect the Universe to oblige. A new theory or way of viewing the world must be able to explain everything that the previous theory could, while extending this understanding to new territory into which the earlier theory could not go. If one face can be said to symbolize modern science, it is that of Albert Einstein (1879–1955; **Figure 1.10**). Einstein's theories of special and general relativity replaced the 300-year-old edifice of Newtonian mechanics not by proving Newton wrong, but by showing that Newton's mechanics was a special case of a far more general and powerful set of physical laws. Einstein's new ideas unified our concepts of mass and energy and destroyed our conventional notion of space and time as separate things. Yet scientific revolutions are seldom comfortable for those who live through them, and even the greatest of scientists can be left behind as bastions of the old guard. Albert Einstein actually helped to start two scientific revolutions. He saw the first of these—relativity—through, and embraced the world that it opened up. Yet Einstein was unable to accept the implications of the second

Figure 1.10 *Albert Einstein, perhaps the most famous scientist of the 20th century, and* Time *magazine's selection as Person of the Century. Einstein helped to usher in two different scientific revolutions, one of which he himself was never able to accept.*

revolution he helped start—quantum mechanics—and went to his grave unwilling to embrace the view of the world it offered.

Science is not simply a body of facts. Science is not simply technology. Science is not simply the scientific method. Perhaps more than anything, science is a way of thinking about the world. It is a way of relating to nature. It is a search for the relationships that make our world what it is. It is a belief that nature is not capricious, but instead operates by consistent, explicable, inviolate rules. It is a collection of ideas about how the Universe works, coupled with an acceptance of the fact that what is known today may be superseded tomorrow. The scientist's faith is that there is an order in the Universe, and that the human mind is capable of grasping the essence of the rules underlying that order, or at least of inventing ever better approximations to those rules. The scientist's creed is that nature, through observation and experiment, is the final arbiter of the only thing worthy of the term *objective truth*. Science is an exquisite blend of aesthetics and practicality. And in the final analysis, science has found such a central place in our civilization because *science works*.

> **More than anything, science is a way of thinking about the world.**

It is beyond the scope of this book for us to try to provide you with a detailed justification for all that we will say. However, we will try to offer some idea of where an idea comes from and why we believe it to be valid. We will not present something as fact unless there is a compelling reason to believe it. We will try to be honest when we are on uncertain, speculative ground, and will admit it when the honest answer is, "we really do not know." This book is not a compendium of revealed truth or a font of accepted wisdom. Rather, it is an introduction to a body of knowledge and understanding that was painstakingly built (and sometimes torn down and rebuilt) brick by brick.

For those of us who grew up in a world transformed by science, the scientific worldview might seem anything but subversive. However, for much of history, knowledge was sought in the pronouncements of "authority" rather than through observation of nature. This worldview slowed the advance of knowledge throughout Western Europe for the millennium prior to the European renaissance. (It was largely the Chinese and Arab cultures that kept the spark of inquiry alive during this time.) The greatest scientific revolution of them all was the revolution which overthrew "authority" and replaced

> **Pronouncements of "authority" do not matter in science.**

it with rational inquiry and the scientific method. Science is not *just* one of many possible worldviews. Science is the most successful worldview in the history of our species. It is worth noting that so far science itself has passed its own test. The foundations of the scientific worldview have withstood centuries of fine minds trying to prove them false.

THE COSMOLOGICAL PRINCIPLE

The scientific revolution brought about a dramatic shift in our ideas about what knowledge is and how it is sought, but this change alone was not enough to open the Universe to our probing gaze. It is likely that every civilization had some system of beliefs about the relationship between the heavens and Earth. Prior to the Renaissance most of these included two fundamental tenets. The first was the belief that Earth occupies a special, unique place, usually at the center of the Universe. The second was the belief that objects in the heavens are made of a different type of substance than Earth, and behave according to their own rules. The key to our understanding of the Universe turned out to be the literal and total negation of both of these beliefs.

At the heart of modern astronomy is a fundamental idea called the **cosmological principle.** The cosmological principle asserts that there is nothing special or unique about Earth, either in our place in the Universe

> **There is nothing special about our place in the Universe.**

or in the rules which govern the behavior of matter here. The cosmological principle states that we are *a part of* the Universe, rather than *apart from* the Universe. The cosmological principle has two very important aspects. The first is that when we look out around us, what we see is representative of what the Universe is generally like. Our location in the Universe is what it happens to be by chance—nothing more, nothing less. In a *deep image* (an image showing very faint objects), even a piece of apparently empty sky is filled with distant galaxies. There are as many galaxies in the observable Universe as there are stars in our own Milky Way. The cosmological principle says that nothing sets our own Milky Way apart from this group of galaxies. The impression that we get of the Universe from our vantage point is representative of the whole. The second aspect of the cosmological principle is the premise that matter and energy obey the same physical laws throughout space and time as they do today on Earth. This means that the same physical laws we learn about in terrestrial laboratories can be used to understand what goes on in the centers of stars or in the hearts of distant galaxies.

The cosmological principle is a theory, and like all scientific theories it is subject to whatever tests our ingenuity and the scientific method can bring to bear. We will not visit a distant star or galaxy in our lifetimes, nor relive the early days of the Universe. Yet we can analyze the light that reaches us from events distant in time and space, and ask whether those events follow the same laws that apply today on Earth. Each new success that comes from applying the cosmological principle to observations of the Universe around us—each new theory that succeeds in explaining or predicting patterns and relationships among celestial objects—adds to our confidence in the validity of this cornerstone of our worldview.

1.3 PATTERNS MAKE OUR LIVES AND SCIENCE POSSIBLE

Imagine what life would be like if sometimes when you let go of an object it fell up instead of down. What if one day apples were essential nutrition, but when you took a bite out of an apple the next day you discovered that it was instead deadly poison? What if, unpredictably, one day the Sun rose at noon and set at 1:00 P.M., the next day it rose at 6:00 A.M. and set at 10:00 P.M., and the next day the Sun did not rise at all? In fact, objects always fall toward the ground. Our biochemistry remains stable. The Sun rises, sets, then rises again. Spring turns into summer, summer turns into autumn, autumn turns into winter, and winter turns into spring. The rhythms of nature produce patterns in our lives, and we count on these patterns being there. If nature did not behave according to regular patterns, then our lives—indeed, life itself—would not be possible.

The same patterns that make our lives possible also make science possible. The goal of science is to identify and characterize these patterns and to use them to understand the world around us. Some of the most regular and easily identified patterns in nature are the patterns that we see in the sky. What in the sky will look different or the same a week from now? A month from now? A year from now? Most of us today lead an indoor and in-town existence, removed from an everyday awareness of the patterns in the sky. However, away from the smog and glare of our cities, the patterns and rhythms of the sky are as easy to see today as they were in an-

cient times. Patterns in the sky mark the changing of the seasons **(Figure 1.11)**, the coming of the rains, the movement of the herds, and the planting and harvesting of the crops. Patterns in the sky share the rhythms of our lives. It is no surprise that astronomy, which is the expression of our human need to understand these patterns, is the oldest of all sciences.

Patterns in our lives echo patterns in the sky.

MATHEMATICS IS THE LANGUAGE OF SCIENCE

There are many kinds of mathematics, most of which deal with more than just numbers. Arithmetic is about counting things. Algebra is about manipulating symbols and the relationships between things. Geometry is about shapes. Calculus is about change. Other types of mathematics deal with topics such as the properties of surfaces, or groups of objects and their relationships. What do all of these have in common? Why do we consider all of these to be a part of a single discipline called "mathematics"? All of these share one thing—they all deal with patterns. The best working definition of **mathematics** is that "mathematics is the science and language of patterns."

Mathematics is the science and language of patterns.

We have seen that science is about patterns—patterns in relationships, patterns in behaviors, and patterns in characteristics. Astronomy is that part of science that concerns itself with patterns related to celestial objects. If patterns are the heart of science, and mathematics is the language of patterns, it should come as no surprise that *mathematics is the language of science*. Trying to study science while totally avoiding mathematics is the practical equivalent of trying to study Shakespeare while totally avoiding the written or spoken word. It quite simply cannot be done, or at least cannot be done meaningfully.

On the other hand, as the authors of this book we understand (as humorously pointed out by **Figure 1.12**) that for many of you, *math* is not *a* four-letter word—it is *the* four-letter word. Many people decide early on in their education that they cannot "do" math, and from that day forward the mere mention of the word causes their eyes to glaze over and their palms to sweat. A distaste for mathematics is one of the most common obstacles standing between a nonscientist and an appreciation of the beauty and elegance of the world as seen through the eyes of a scientist. To move beyond this obstacle, scientist and nonscientist alike need to find common ground.

Part of the responsibility for moving beyond this obstacle lies with us, the authors. It is our job to take on the role of translators, using words to express as many concepts as possible, even when these concepts are more concisely and accurately expressed mathematically. When we do make use of mathematics, we will explain in everyday language what the equations mean, and try to show you how equations express concepts that you can connect to the world. We will also limit the mathematics to a few basic tools that all college students should have been exposed to. Scientific notation is needed because of the vast range of sizes of the objects involved. Units are needed for distinguishing between a

time and a distance and a mass and an energy. A bit of geometry is necessary for understanding the distances and sizes and shapes and volumes of things. Finally, some algebra—mostly a few ratios and proportionality—will provide a way of expressing the patterns that relate one physical quantity to another. "Basic" does not necessarily mean easy, but it does mean that we will use the tools that are most accessible to you and that will make our journey of discovery as comfortable and informative as possible.

Your responsibility is to accept the challenge and make an honest effort to think through the mathematical concepts that we use. Do not become your own worst enemy by conceding defeat while still in the starting blocks. It is very likely that you know what it means to square a number, or to take its square root, or to raise

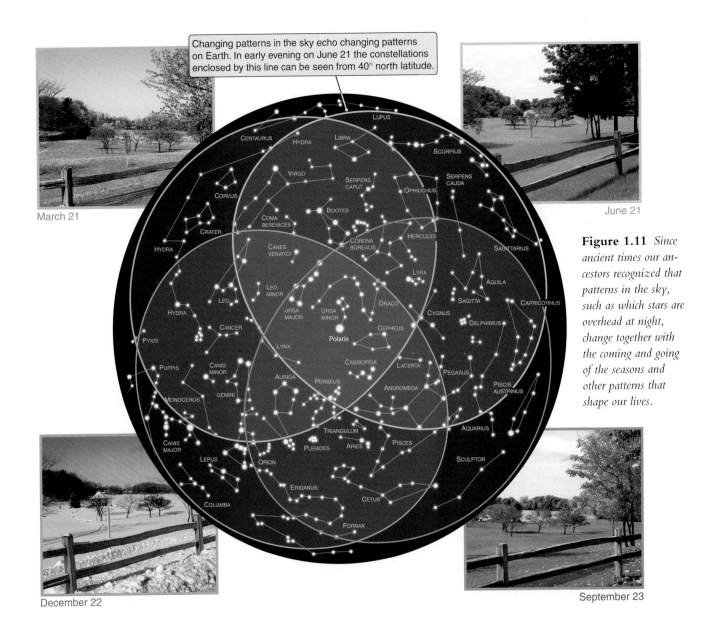

Figure 1.11 *Since ancient times our ancestors recognized that patterns in the sky, such as which stars are overhead at night, change together with the coming and going of the seasons and other patterns that shape our lives.*

NON SEQUITUR   by WILEY

Figure 1.12 *Mathematics is the science of patterns, which makes mathematics the language of science.*

it to the third power. The mathematics in this book is on a par with what it takes to balance a checkbook, build a bookshelf that stands up straight, check your gas mileage, estimate how long it will take you to drive to another city, figure your taxes, or buy enough food to feed an extra guest or two at dinner. Foundations 1.1 describes several of the basic mathematical tools we will use throughout this book. (Also see Appendix 1.)

FOUNDATIONS 1.1

MATHEMATICAL TOOLS

Mathematics provides scientists with the tools they need to understand the patterns they see, and to communicate that understanding to others. As the authors of this text, we are aware that mathematics is not a friend to many of you taking this course, and so have worked to keep the math in this text to a minimum. Even so, there are a few tools that we will need:

Scientific notation Scientific notation is the way that scientists deal with numbers of vastly different sizes. Rather than writing out 7,540,000,000,000,000,000,000, we write 7.54×10^{21}. Rather than writing out 0.00000000000005, we write 5×10^{-12}.

Ratios Ratios are the most common way that astronomers use to compare things. A star may be 10 times as massive as the Sun or 10,000 times as luminous as the Sun. These are ratios.

Geometry To describe and understand objects in astronomy and physics, we use concepts such as distance, shape, area, and volume. Apparent separations between objects in the sky are expressed as *angles*. Earth's orbit is an *ellipse* with the Sun at one *focus*. The planets in the Solar System lie close to a *plane*. Geometry provides the tools for working with these concepts.

Algebra Algebra provides a way of using and manipulating symbols that represent numbers or quantities. We will use algebra to express relationships that are valid not just for a single case, but for many cases. Algebra lets us conveniently express ideas such as "the distance that you travel is equal to the speed at which you are moving times the length of time you go that speed" (in other words, $d = s \times t$, where d is distance, s is speed, and t is time). Algebra also lets us combine these ideas with other ideas to arrive at new relationships.

Proportionality Often, understanding a concept amounts to understanding the *sense* of the relationships that it predicts or describes. "If you have twice as far to go, it will take you twice as long to get here." "If you have half as much money, you will only be able to buy half as much gas." These are examples of proportionality. If you are traveling at a constant speed, then time is proportional to distance. We write $t \propto d$, where $\propto$ means "is proportional to." Proportionalities often involve quantities raised to some power. A circle of radius r has an area A equal to πr^2, so we say that area is proportional to the square of the radius, and write $A \propto r^2$. This means that if you make the radius of a circle three times as large, its area will grow by a factor of 3^2, or 9.

1.4 BENDING YOUR BRAIN INTO SHAPE

This book will likely ask you to think in ways that are different from the ways in which you are accustomed to thinking, and to learn to view the world from new and unfamiliar perspectives. Knowledge and understanding have nothing to do with shoving facts into short-term memory so that they can be regurgitated on an exam, and that is certainly not what astronomy is about. Changing the way you think about things takes more effort. It also means studying in ways that may be different from your normal habits. Here are a few practical suggestions for how you might better study this text:

Building understanding is an active process.

* **Read the text actively.** Think about each section after you have read it. What were the major concepts discussed in the section? How are they related to what you have read so far? Have you run into similar concepts elsewhere? Why are the contents of that section important enough to be included in the book? Briefly summarize the section and your thoughts about it in your class notes.

* **Draw a picture.** Many physical and mathematical concepts are most easily understood if they are visualized. If you understand a concept well enough to draw a picture that expresses it, then you probably understand the concept fairly well. Trying to sketch a picture will also help you better identify what things you understand and what things you do not.

* When trying to understand a concept, **ask yourself, "What if?"** What if Earth were more massive? How would that affect Earth's gravity? What if the Sun were hotter? How would that affect the color of the light from the Sun or the amount of energy that the Sun radiates? If you cannot "what if" a concept, then you probably do not really understand it yet.

* It is often said that you do not really understand something until you try to **teach it.** At the end of a reading assignment, talk about the material with a friend or family member. Try to find a partner or a group in your class whom you can meet with, and *take turns teaching each other the material.* Each of you should read the assigned text, then divide up responsibility for who will present which sections. Ask each other lots of questions—the harder the better! (Make a game of playing "stump the instructor.") Even explaining a concept out loud to yourself can help.

* **Share your ideas and insights** with your discussion group. If a particular concept really "clicks" for you—if you really think it is neat—share both your understanding and your enthusiasm with your group.

* **Be honest with yourself** about what things you understand and what things you do not, and try not to avoid concepts you find difficult. Personal growth comes from real accomplishment. Getting the most from this journey will come from facing challenging concepts.

* **Focus your discussion on concepts, relationships, and connections.** You have to know the facts, but the facts are the starting point rather than the end. Make use of the key concepts, study questions, and other study aids to identify and concentrate your effort on the most important ideas.

* **Do not let discomfort with math keep you from succeeding.** Mathematical formulas are not magical incantations. Rather, they are expressions of logical ideas. We will always present a plain-English discussion of the idea behind any mathematics that we use. Begin by focusing on this discussion. After you have a grasp of the idea, then look at the math. Try to see how the relationships between the quantities in the equation embody the idea. (If you need help with basic math skills, work through the appendices and other aids available to you on the CD-ROM or the Internet. Your instructor is also there to help.)

* **Above all, remember that building understanding is always an *active* process, never passive!**

Our best suggestion for a successful journey through *21st Century Astronomy* is to nurture your spark of interest in astronomy until it grows enough to draw you in. When, as Einstein said in the opening quote, you "pause to wonder and stand rapt in awe"—*then* you will have learned the secret!

1.5 LET THE JOURNEY BEGIN

The journey we are about to begin is sometimes entertaining, sometimes enlightening, often surprising, usually challenging, and always remarkable. We will explore the wonders of the Universe, and along the way take a look at what science is, how it is done, and what it means to look at the world as a scientist does. It is a journey that will

* teach us as much about ourselves as it will about what is "out there";
* open our minds to new ways of thinking about and experiencing the world;
* and pay tribute to what the human mind and the human spirit are capable of achieving.

Reading the book and taking the journey with us are two different things. This is only a guidebook. It can lead you to the trailhead and tell you something of what you might find along the way, but *you* have to walk the path! If you become an active participant in this adventure, rather than a passive spectator, then what you gain from the journey will remain a part of you long after the final exam is forgotten. And if along the way you find yourself applying your understanding to new situations and new information, and if you learn to combine your understandings and arrive at new insights that are greater than the sum of their parts, then you will have learned something far more than just astronomy.

SEEING THE FOREST THROUGH THE TREES

At the end of each chapter of *21st Century Astronomy* you will find a brief narrative about the content of that chapter. The purpose of "Seeing the Forest Through the Trees" is not to rehash the entire contents of the chapter, but rather to pick out a few of the high points and put them into a broader context. Building on the analogy of a hike in the mountains, we will spend a lot of time looking in detail at the rocks and the trees, but every so often we need to step back and look around at the forest as a whole.

We live in a world that has been profoundly shaped by the scientific revolution that took hold of Western thought during the Renaissance. That revolution fundamentally altered our way of thinking about the world, as well as our view of the relationship between ourselves and the Universe of which we are a part. A new spirit of rational inquiry was turned on the heavens, dislodging Earth and humankind from the center of the cosmos. Observation, experiment, and rigorously applied reason came to replace dogma and authority as the arbiter of knowledge. The heavens became a realm not of mysticism and magic, but instead of physical law—the same physical law that governs the behavior of matter and energy in laboratories here on Earth.

The closing years of the 20th century saw our knowledge of the Universe charge ahead at an ever-accelerating pace. This progress comes courtesy of a great many advances both in our technology and in the sophistication of our physical understanding of matter and energy and of space and time themselves. We have seen many fundamental questions about the origin and fate of the Universe and the threads that tie our existence to the cosmos move from the realm of philosophical speculation into the realm of hard scientific inquiry. The insights that this age of exploration and discovery have brought are often far more profound and startling than dreamed of even a few decades ago. There can be little doubt that this time will be looked on as one of the more significant moments in the intellectual and cultural history of our species.

Like a hike through the mountains, *21st Century Astronomy* will not be an effortless journey. Many travelers will find that they have to flex a few mental muscles in ways they are not used to, and may even have to face an old adversary or two on the trail. But muscles that are sore after the first day of a hike grow comfortable and strong with time, and adversaries can become the best of friends.

In Chapter 2 we begin the journey in earnest, and as with most journeys our starting point is home. What patterns do we see in the skies of our planet Earth, and how do those patterns come to be? This is not an easy or gentle slope on which to begin our trek, but our vistas will change rapidly as we climb.

STUDENT QUESTIONS

THINKING ABOUT THE CONCEPTS

1. List patterns in your own life that repeat regularly. How do these patterns affect you? Which patterns are of your own making, which are set by others, and which are determined by nature?

2. When you are in a section of town you have not previously seen, or a new city, or a new country, what patterns do you look for that help you find your way around?

3. List and discuss some of the ways in which you may have already noticed that the cosmological principle works.

4. What experiment might you do that, if it turned out a certain way, would falsify the cosmological principle? (Your experiment might or might not be practical.)

5. Astrology makes testable predictions. For example, it predicts that the horoscope for your star sign on any day should fit you better than horoscopes for other star signs. Read each of the horoscopes in yesterday's paper without regard to your own sign. How many of them might fit the day that you had yesterday? Repeat the experiment every day for a week and keep records. Was your horoscope consistently the best description of your experiences?

6. A scientist on television states that it is a known fact that life does not exist beyond Earth. Would you consider this scientist to be reputable? Why or why not?

7. Some astrologers use elaborate mathematical formulas and procedures to predict the future. Does this show that astrology is a science? Why or why not?

8. You run across an old newspaper with the headline "EINSTEIN PROVES NEWTON WRONG!" Did the newspaper get this story right? Explain.

9. Hydrogen and Oxygen are two common elements that combine to form water, an essential ingredient of life. A friend tells you that some galaxies contain no water at all. How would you respond to your friend's statement?

APPLYING THE CONCEPTS

10. The surface area of a sphere is proportional to the square of its radius. If the Moon has a radius only one-quarter that of Earth, how does the surface area of the Moon compare with that of Earth?

11. If it takes about 8 minutes for light to travel from the Sun to Earth, and Pluto is 40 times this distance from us, how long does it take light to reach Earth from Pluto? Radio waves travel at the speed of light. What does this imply about the problems you would have if you tried to conduct a two-way conversation between Earth and a spacecraft orbiting Pluto?

12. Civilization on Earth has existed for about 10,000 years. If we were to launch a spacecraft that travels at 10^9 km/h (roughly one-tenth the speed of light), how far could it explore within our Galaxy over the next 10,000 years?

13. Describe an experiment that would determine whether the distance something travels at a constant speed is proportional to the time or to the square of the time it takes to complete the trip.

14. If you understand proportionality, then you understand most of the math you need to follow this text. Make a list of at least five different proportionalities from your everyday life. (For example, the price of a bag of apples is proportional to the weight of the bag of apples.) What does each tell you about your world? For each proportionality, identify the constant of proportionality (such as, the price per pound of apples). How are these constants determined?

15. Imagine the Sun to be the size of a grain of sand and Earth a tiny speck 83 mm away. (On this scale, each light-minute of distance equals 10 mm.) How far would it be from Earth to the Moon on this scale? From the Sun to Pluto? From Earth to the nearest star? To the nearest large galaxies? (Note that 1 m = 10^3 mm and that 1 km = 10^3 m.) At what point do you lose your "feeling" for these distances?

16. The circumference of a circle is given by $C = 2\pi r$.
    a. Calculate the approximate circumference of Earth's orbit around the Sun, assuming that the orbit is a circle with a radius of 1.5×10^8 km. You can also approximate π as being about 3.
    b. About how fast, in kilometers per hour, does Earth move in its orbit?
    c. How far along its orbit does Earth move in one day?

. . . . marking the conclave of all the night's stars,
those potentates blazing in the heavens
that bring winter and summer to mortal men,
the constellations, when they wane, when they rise.

AESCHYLUS (525–456 B.C.)

Patterns in the Sky—
Motions of Earth

2.1 A View from Long Ago

The herds have reached the high meadows where they spend the warm season, and for a time the life of the tribe has settled in as well. The weather is pleasing and comfortable, and the days are long, bringing none of the hardships that accompany the time of cold and snow and long, dark nights. It is a time of plenty and a time for telling the age-old stories of the tribe around a fire that guards against the chill of the gathering night. You feel a sense of contentment, and are thankful to the gods for this time when life is good. As the embers die down, you turn your gaze toward the familiar canopy of stars overhead, and as you often do in such moments, you wonder about what you see.

To survive, you must learn the subtle patterns of your world. You must know the ways of the herds, and recognize the gathering of clouds that heralds a coming storm. When you turn your keen eye toward the heavens, you find subtle and changing patterns there as well—patterns that somehow echo those of your life. The spirits of the great animals dwell in the sky; as a child you learned to recognize their pictures there. Above you now are the stars that rule the time of the short nights. These are the stars that bring summer and lead the herds to this pleasant place.

Some of the spirits of the sky can be difficult to please. The mischievous planets wander from place

KEY CONCEPTS

In this chapter we begin our journey in earnest, starting out as our ancestors did when they first gazed at the Sun, Moon, and stars, and tried to understand what they saw. With the benefit of knowledge hard-won over the course of centuries, we will look at patterns present both in the sky and on Earth, and then look beyond appearances to the underlying motions that cause those patterns. Here we will discover:

* How the stars appear to move through the sky as Earth rotates on its axis, and how those motions differ as seen from different latitudes on Earth;
* The fundamental concept of a frame of reference, and how Earth's rotating frame of reference affects weather patterns and other terrestrial phenomena;
* How Earth's motion around the Sun and the tilt of Earth's axis relative to the plane of its orbit combine to determine which stars we see at night, and the seasons we feel through the year; and
* The motion of the Moon on its orbit about Earth, and how that motion, together with the motion of Earth and the Moon around the Sun, shape the phases of the Moon and the spectacle of eclipses.

EXCURSIONS 2.1

WHERE ARE THE CONSTELLATIONS?

Where are the **constellations?** The answer may seem obvious: "The constellations are overhead in the sky, for all to see." Yet if you look at the sky, no pictures of winged horses or dragons or chained maidens are painted there. Instead, there is only the random pattern of stars—about 6,000 of them visible to the naked eye—spread out across the sky. Constellations exist in one place and one place only: within the imagination of the human mind. Constellations are the ideas and pictures that humans imposed on the lights in the sky in an effort to connect our lives on Earth with the workings of the heavens.

As illustrated in **Figure 2.1,** there have been as many different sets of constellations and stories to go with them as there have been cultural traditions in our history. Modern constellations visible from the Northern Hemisphere draw heavily from the list compiled 2,000 years ago by the Alexandrian astronomer Ptolemy. Constellations in the southern sky are drawn from the lists put together by European explorers visiting the Southern Hemisphere during the 17th and 18th centuries. Today astronomers use an officially sanctioned set of constellations as a kind of road map of the sky. The entire sky is broken into 88 different constellations, much as continental landmasses are divided into countries by invisible lines. Every star in the sky lies within the borders of a single constellation, and the names of constellations are used in naming the stars that lie within their boundaries. For example, Sirius, the brightest star in the sky, lies within the boundaries of the constellation Canis Major (meaning the "big dog"). Sirius's official name is therefore Alpha *Canis Majoris,* indicating that it is the brightest star in that constellation, and earning its nickname, the Dog Star.

to place, using their fearsome powers to sow chaos through the heavens. The Moon sometimes turns blood red, and the Sun is consumed by an ominous beast. But as long as the tribe remembers them, the gods and spirits of the sky will continue to bring the seasons and send the stars to guide the tribe. This is as your elders taught you when you were young, and this is as you teach the young ones today. So it has always been, and so it shall always be.

Our ancestors lived their lives attuned to the ebb and flow of nature, and the patterns in the sky were a part of that ebb and flow. The coming of night and day, the changing of the seasons, the rising and falling of the tides, the movement of the herds—all of these march in lockstep with the changes that we see in the sky. The repeating patterns of the Sun, Moon, and stars echo the rhythms that have defined the lives of humans since before the beginning of recorded history. By watching the patterns in the sky our ancestors found that they could predict when the seasons would change and the rains would come and the herds would move. Knowledge of the sky offered knowledge of the world, and knowledge of the world was power. It was but a small step from here to thinking of the unreachable, untouchable stars as not only a reflection of the patterns in the world, but also the very *cause* of those patterns. The stars

Patterns in the sky have always been important to our species.

found a special place in legend and mythology as the realm of gods and goddesses, holding sway over the lives of humankind. As writing came to replace oral traditions and legends, mythologies of the sky became more elaborate as well. And as humans invented numbers and mathematics to describe and predict and account for things in the world, predictions of the motions of the stars and planets were among their greatest successes. Some of our ancestors came to look upon the orderly and predictable patterns of the sky as the *true* patterns of the world, and our own lives as but imperfect reflections of this heavenly reality. They looked for ways to use their knowledge of the sky to find order in the seeming chaos of their everyday lives, and astrology was born.

Elements of this same basic history played themselves out many times over and in every corner of our globe. From Africa to Asia, from Europe to Central America, from North America to the British Isles, the archeological record holds evidence of early humans who projected ideas from their own cultures onto what they saw in the sky (see Excursions 2.1). The connection between the patterns in the sky and the patterns in their lives was simply too compelling to be missed. The idea of the sky as a realm of mysticism and magic is deeply rooted in the traditions and beliefs and history of our species. There is no mystery about the currents of mind that led our ancestors to their belief in astrology and other celestial mythologies. At a time when the causes of things were

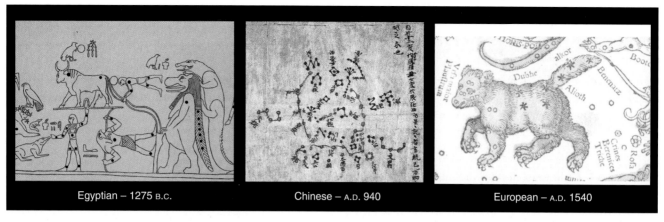

Egyptian – 1275 B.C.   Chinese – A.D. 940   European – A.D. 1540

Figure 2.1 *The region around what we now call the Big Dipper (Ursa Major, or the "Great Bear") as viewed by three different civilizations. Constellations exist only in the human mind.*

unknown, and humans existed at the seeming whim of forces they could not comprehend, the sky seemed to offer a window into a mystical and powerful world of spirits and gods and devils and angels.

What an unwelcome shock it must have been when a few remarkable individuals, with names like Copernicus and Kepler and Galileo and Newton, tugged at the threads of this comfortable and familiar tapestry, *only to discover that it fell apart in their hands!* The stars and other heavenly bodies do not rotate about Earth each day, as humans have thought since they first took notice of the sky. Rather, it is *Earth* that spins on its axis, giving the stars, planets, Sun, and Moon the appearance of following daily paths through the heavens. Nor is Earth at the center of all existence, as befits the home of humankind, the pinnacle of all Creation. Earth is but one of many planets orbiting about the Sun. The subtle complexity of the changing patterns we see in the sky results from the motions of planets and moons as they step through their gravitational dance with the Sun. Even the Sun itself, whose radiant energy makes our world what it is, is but one of countless stars, adrift in a Universe whose full extent is unknown even today.

The magic of astrology properly belongs to a time long-dead, when in the minds of humans Earth rode on the back of a giant sea turtle, and with each passing month the Sun moved from one stellar "house" to the next. Today it is a matter of experimentally verifiable fact that the imaginary patterns seen in the stars hold no more influence over our lives than the random patterns of leaves blowing down the street on an autumn day. The astrologers' quest for deep connections between our lives and the patterns in the sky was both understandable and well placed, but knowledge of the true

> **Science shattered ancient mystical views of the heavens.**

nature of those connections had to wait for the birth of modern science. Today the sky has become a window of knowledge on the *physical* world. This knowledge has proven worth the wait.

Today saw a long hard climb before you reached the meadow by the river where you now camp, and tomorrow's trek promises to be just as demanding. Even so, the evening is pleasant, and you are content. The embers of your campfire have almost died away when the distant sound of a jetliner interrupts your reverie. Without really thinking you look up to catch sight of the plane high overhead, and are caught off guard by the blazing spectacle of the summer Milky Way and the thousands of pinpoints of light, which seem so close that you can almost reach out and touch them. For a moment the thousands of years separating you from a long-dead tribal nomad vanish as you share the same sense of wonder and awe that has always defined humankind's experience of the Universe.

It is here that we begin the journey of *21st Century Astronomy*—with the changing patterns in the sky that captured the attention and imagination of that long-ago nomad and that still beacon overhead on a dark, cloudless night. Yet unlike that nomad, we look on those changing patterns with the perspective of centuries of hard-won knowledge. We will find that patterns of change in the sky are often the understandable and even unavoidable consequences of the daily rotation of Earth about its axis and Earth's annual trip about the Sun. This is an example of science at its best—the discovery of wonderful variety arising from simple and elegant underlying causes. And just as happened over the course of the history of our species, curiosity about the changing patterns in the sky will show us the way outward into a Universe far more vast and awesome than our distant ancestor could have imagined.

2.2 EARTH SPINS ON ITS AXIS

Despite the apocryphal stories you may have learned in grade school, Columbus did not discover that the world is round. Long before his famous (or possibly infamous) journey to the New World, anyone who had read Aristotle or the other Greek philosophers (as had Columbus) knew that Earth is a ball. Far more difficult to accept was the idea that the changes occurring in the sky from day to day and month to month are the result of the motion of Earth rather than the motion of the Sun and stars. The most apparent of these motions is Earth's rotation on its axis, which sets the very rhythm of life on Earth— the passage of day and night. When our remote ancestors first noticed the sky with something approaching human awareness, it was doubtless the daily motion of the Sun in the sky that drew their attention.

As viewed from above Earth's North Pole, Earth rotates in a counterclockwise direction **(Figure 2.2)**, completing one rotation in a 24-hour period. As the rotating Earth carries us from west to east, objects in the sky *appear* to move in the other direction, from east to west. The path a celestial body makes across the sky as seen from Earth is called its *apparent daily motion.* The Sun is one such object. When we say "noon," we mean the time of day when our location on Earth faces most directly toward the Sun. By convention, astronomers divide the sky evenly into eastern and western halves along an imaginary north-south arc called the **meridian.** The meridian runs from due north to due south, passing through the point directly overhead in the sky, called the **zenith.** True *local noon* occurs when the Sun crosses the meridian at our location. Half a day later our spot on Earth comes closest to facing directly away from the Sun. This is *local midnight.*

> **Earth's rotation is counterclockwise when viewed from above the North Pole.**

THE VIEW FROM THE POLES

The apparent daily motions of the stars and the Sun witnessed by ancient nomadic tribes would have depended on where on the surface of the planet they happened to live. The apparent daily motions of celestial objects in Northern Europe, for example, are quite different from the apparent daily motions seen from a tropical island. (Differences in apparent motions of the Sun are responsible for many of the differences in culture between people living in these two regions.) The daily motions of the stars are easiest to understand when viewed from a place where humans did not set foot until 1909—Earth's North Pole.

Imagine you are standing on the North Pole watching the sky, as shown in **Figure 2.3.** (Ignore the Sun for the moment, and suppose that you can always see stars in the sky.) You are standing on Earth's axis of rotation, which is much the same as standing at the center of a rotating carousel. As Earth rotates, the spot directly above you remains fixed while everything else in the sky appears to revolve in a counterclockwise direction around this spot. (If you are having trouble visualizing this, find a globe and, as you spin it, imagine standing at the pole of the globe.) The direction in which Earth's axis of rotation

> **Standing on the North Pole is like standing at the center of a rotating carousel.**

Figure 2.2 *The rotation of Earth and the Moon, the revolution of Earth and the planets about the Sun, and the orbit of the Moon about Earth, are counterclockwise as viewed from above Earth's North Pole.*

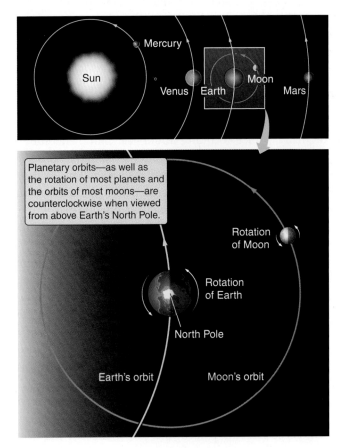

Sun
Mercury
Venus Earth Moon Mars

Planetary orbits—as well as the rotation of most planets and the orbits of most moons—are counterclockwise when viewed from above Earth's North Pole.

Rotation of Moon

Rotation of Earth

North Pole

Earth's orbit   Moon's orbit

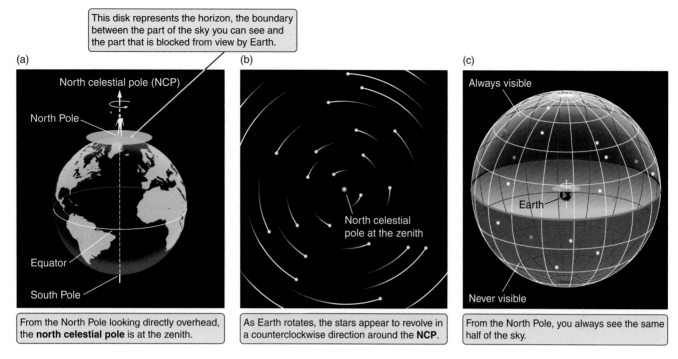

(a)

This disk represents the horizon, the boundary between the part of the sky you can see and the part that is blocked from view by Earth.

North celestial pole (NCP)

North Pole

Equator

South Pole

From the North Pole looking directly overhead, the **north celestial pole** is at the zenith.

(b)

North celestial pole at the zenith

As Earth rotates, the stars appear to revolve in a counterclockwise direction around the **NCP**.

(c)

Always visible

Earth

Never visible

From the North Pole, you always see the same half of the sky.

Figure 2.3 (a) *As viewed from Earth's North Pole* (b) *stars move over the course of the night on counterclockwise circular paths about the zenith.* (c) *The same half of the sky is always visible from the North Pole.*

points, and about which the stars appear to revolve as Earth turns, is called the **north celestial pole.** The greater the angular distance between an object and the north celestial pole, the larger the circular path the object appears to follow. Objects close to the pole appear to follow small circles, while the largest circles are followed by objects nearest to the horizon.

You can never see more than half of the sky at any one time, regardless of where you are on the surface of our planet. The other half of the sky is blocked from view by Earth. The boundary between the part of the sky you can see and the part that is blocked by Earth is called the **horizon.** From most locations on Earth, the half of the sky that we can see above the horizon is constantly changing as Earth rotates. (The direction in space in which our zenith points at the moment is different now from what it was 12 hours ago, or even 12 seconds ago.) In contrast, Earth's North Pole points in the *same* direction, hour after hour and day after day. From the North Pole we always see the *same* half of the sky. Nothing rises or sets as Earth turns beneath you. If you look off toward the horizon, you will see that the objects visible there follow circular paths that keep them always the same distance above the horizon.

> The same half of the sky is always visible from the North Pole.

The view from Earth's South Pole is much the same, but with two major differences. First, the South Pole is on the opposite side of Earth from the North Pole, so the half of the sky you see overhead is precisely the half that is hidden from view from the North Pole. The direction in space that is at the zenith at the South Pole, and about which everything appears to spin as Earth rotates, is now the **south celestial pole.** The second difference is that, instead of appearing to move counterclockwise around the sky, stars appear to move *clockwise* around the south celestial pole. (To see this, sit in a swivel chair and spin it around from right to left. As you look at the ceiling things appear to move in a counterclockwise direction, but as you look at the floor it will appear to be moving clockwise.)

> From the South Pole, the other half of the sky is visible, and stars circle in a clockwise direction.

AWAY FROM THE POLES THE PART OF THE SKY WE SEE IS CONSTANTLY CHANGING

Latitude is a measure of how far north or south we are on the face of Earth. Imagine a line from the center of Earth to your location on the surface of the planet. Now imagine a second line from the center of Earth to the point on the equator closest to you. (Refer to **Figure 2.4** for help imagining these lines.) The angle between these two lines is your latitude. The latitude of any point on the equator is 0°. The latitude of the North Pole is 90° north latitude, while the South Pole is at 90° south latitude. The latitude of Phoenix, Arizona, is 33.5° north.

Imagine what we see as we leave the North Pole and travel south to lower latitudes. As we follow the curve of Earth, our horizon tilts and our zenith moves away from the north celestial pole. By the time we reach a latitude of 60° north (as shown in Figure 2.4b), the north celestial pole has

> North latitude is the same as the angle of the north celestial pole above the northern horizon.

dropped to 60° above the northern horizon. This equality between north latitude and the height of the north celestial pole above the northern horizon holds everywhere. When we reach a latitude of 30° north (as in Figure 2.4c), the direction of the north celestial pole has dropped to 30° above the horizon. (A more accurate way to say it is this: The north celestial pole lies in the same direction regardless of where we are on Earth. It is *our horizon* that is tilted at 30° from the north celestial pole when we are at a latitude of 30° north.)

In Figure 2.4(d) we have reached Earth's equator, at a latitude of 0°. The north celestial pole is now sitting on the northern horizon. At the same time we get our first look at the south celestial pole, which is sitting opposite the north celestial pole on the southern horizon. Continuing on into the Southern Hemisphere,

Figure 2.4 *Our perspective on the sky depends on our location on Earth. Here we see how the locations of the celestial poles and celestial equator depend on an observer's latitude.*

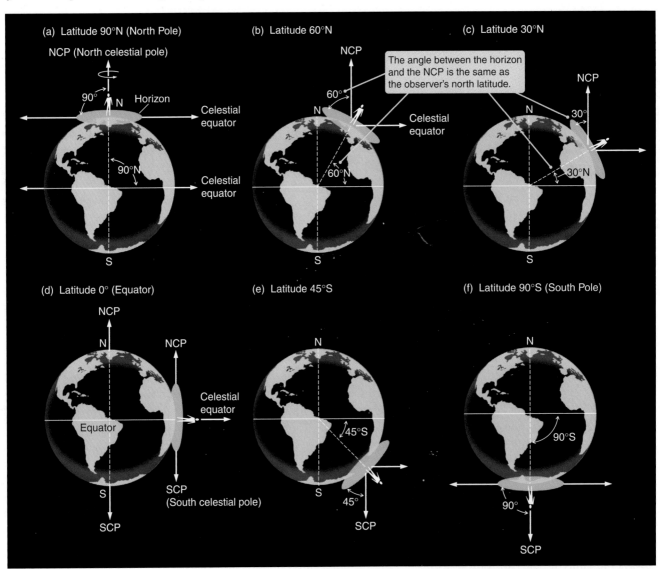

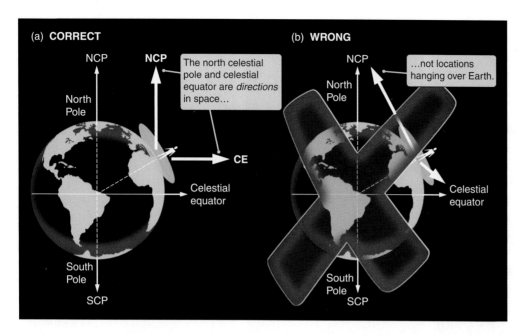

Figure 2.5 (a) *The celestial poles and the celestial equator are directions in space, not locations hanging above Earth. Do not make the mistake shown in (b).*

the south celestial pole is now visible above the southern horizon, while the north celestial pole is hidden from view by the northern horizon. At a latitude of 45° south (Figure 2.4e), the south celestial pole lies 45° above the southern horizon. At the South Pole (90° south latitude—Figure 2.4f), the south celestial pole is at the zenith, 90° above the horizon.

In the Southern Hemisphere the south celestial pole is above the horizon.

Probably the best way to cement your understanding of how the view of the sky changes from one latitude to another is to draw pictures like those in **Figure 2.5(a)** for different latitudes. If you can draw a picture like this for any latitude—filling in the values for each of the angles in the drawing and imagining what the sky looks like from that location—then you will be well on your way to developing a working knowledge of the appearance of the sky. That knowledge will prove useful later when we discuss a variety of phenomena such as the changing of the seasons. When practicing your sketch, however, take care not to make the common mistake shown in **Figure 2.5(b).** The *north celestial pole is not a location* in space, hovering over Earth's North Pole. Instead, *it is a direction* in space—the direction parallel to Earth's axis of rotation.

For many centuries, travelers, including sailors at sea, have used the stars for navigation. Perhaps the simplest of the navigator's techniques is to use the equality between latitude and the altitude of the north (or south) celestial pole. The north or south celestial poles can be found by recognizing the stars that surround them. In the Northern Hemisphere it happens by chance that a moderately bright star is seen within about $\frac{3}{4}°$ of the north celestial pole. This star is called **Polaris,** or more commonly the **north star.** If you can find Polaris in the sky and measure the angle between the north celestial pole and the horizon, then you know your latitude. If you are in Phoenix, Arizona, for example (latitude 33.5° north), you will find the north celestial pole 33.5° above your northern horizon. On the other hand, if you are studying astronomy in Murmansk, Russia (latitude 68.6° north), Polaris sits much higher overhead, 68.6° above the horizon in the north.

Currently Polaris just happens to be near the north celestial pole.

The location of the north celestial pole in the sky can be used to measure the size of Earth. Suppose we start out in Phoenix, Arizona, and head north. By the time we reach the Grand Canyon, about 290 km (kilometers) later, we notice that the north celestial pole has risen from 33.5° to about 36° above the horizon. This change—2.5°—is 1/144 of the way around a circle. (A circle is 360°, and 2.5°/360° = 1/144.) This means that we must have traveled 1/144 of the way around circumference of Earth, so the circumference of Earth must be about 144 × 290 km, or about 42,000 km. The actual circumference of Earth is just a shade over 40,000 km, so our simple measurement was not too bad, given our sloppy measurements of angles and distances. The radius of Earth is this circumference divided by 2π, or about 6,400 km.

The size of Earth was first measured using differences in the appearance of the sky.

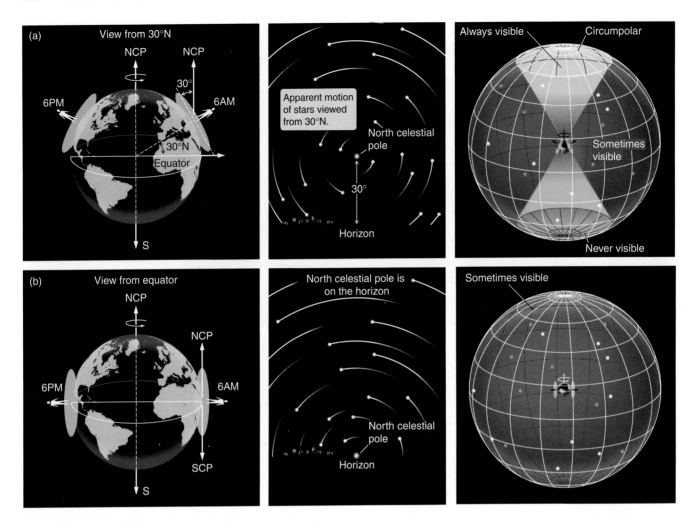

Figure 2.6 (a) *As viewed from 30° north latitude the north celestial pole is 30° above the northern horizon. Stars appear to move on counterclockwise paths around this point. At this latitude some parts of the sky are always visible, while others are never visible.* (b) *From the equator the north and south celestial poles are seen on the horizon, and the entire sky is visible over the course of 24 hours.*

It was in much this way that the Greek astronomer Eratosthenes made the first accurate measurements of the size of Earth around 230 B.C. (*well before Columbus's time*).

The apparent motions of the stars about the celestial poles also differ from latitude to latitude. **Figure 2.6(a)** shows an observer at a point in the Northern Hemisphere other than the North Pole. As Earth rotates, the part of the sky visible to this observer is constantly changing. Of course, from the perspective of the observer it is the horizon that seems to remain fixed, while the stars appear to move past overhead. If we focus our attention on the north celestial pole, from this perspective we still see much the same thing we saw

The part of the sky that we see changes throughout the day.

from Earth's North Pole. The north celestial pole remains fixed in the sky, and all of the stars appear to move in daily counterclockwise circular paths around that point. But since the north celestial pole is no longer directly overhead, the apparent circular paths of the stars are now tipped relative to the horizon. (Or more correctly, our horizon is now tipped relative to the apparent circular paths of the stars.)

Stars located close enough to the north celestial pole can still be seen 24 hours a day as they complete their apparent paths around the pole (see **Figures** 2.6(a) and 2.7). This always-visible region of our sky is referred to as being **circumpolar,** which means "around the pole." There also remains a part of the sky that can *never*

Circumpolar stars are always above the horizon.

be seen from this latitude. This is the part of the sky near the *south* celestial pole that never rises above your horizon. And between this region and the always-visible circumpolar region lies a portion of the sky that can be seen for *part but not all* of each day. Stars in this intermediate region appear to rise above and set below Earth's shifting horizon as Earth turns. The only place on Earth where you can see the entire sky over the course of 24 hours is the equator. From the equator **(Figure 2.6b)** the north and south celestial poles sit on the northern and southern horizons, respectively, and the whole of the heavens passes through the sky each day.

THE CELESTIAL SPHERE IS A USEFUL FICTION

It is sometimes useful to think about the sky as if it were a huge sphere with the stars on its surface and Earth at its center. (Our ancient tribesman probably thought this really was the case.) Astronomers refer to this imaginary sphere as the **celestial sphere.** The celestial sphere is a useful concept because it is easy to draw and to visualize, but never forget that it is imaginary! Each point on the celestial sphere actually corresponds to a *direction* in space. **Figure 2.8** shows the celestial sphere as seen by observers at different places on Earth.

We divide the celestial sphere into a northern half and a southern half with an imaginary circle called the **celestial equator.** Just as the north celestial pole is the projection of the direction of Earth's North Pole into the sky, the celestial equator is the projection of the plane of Earth's equator into the sky. If you are on Earth's equator (Figure 2.8c), then the celestial equator runs east to west, passing directly overhead through the zenith. As you move north away from Earth's equator, the celestial equator tips toward the southern horizon by the same amount that the north celestial pole appears to rise above the northern horizon. Just as Earth's North Pole is 90° away from Earth's equator, the north celestial pole is always 90° away from the celestial equator. If you point one arm at a point on the celestial equator and one arm toward the north celestial pole, your arms will always form a right angle. If you are in the Southern Hemisphere, the same holds true there. The angle between the celestial equator and the south celestial pole is 90° as well.

> **The celestial equator is the projection of Earth's equator into space.**

Look at the location of the celestial equator in Figure 2.8. The points where the celestial equator intersects the horizon are always due east and due west. (The only exception to this is at the poles, where the celestial equator is coincident with the horizon.) An object on the celestial equator rises due east and sets due west. Objects that are north of the celestial equator rise north of east and

> **The celestial equator intersects your horizon due east and due west of where you stand.**

From a location in the Canadian woods, the north celestial pole appears high in the sky…

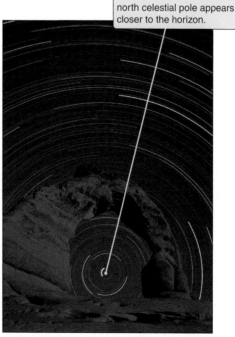

…but at lower latitudes the north celestial pole appears closer to the horizon.

Figure 2.7 *Time exposures of the sky showing the apparent motions of stars through the night. Note the difference in the circumpolar portion of the sky as seen from the two different latitudes.*

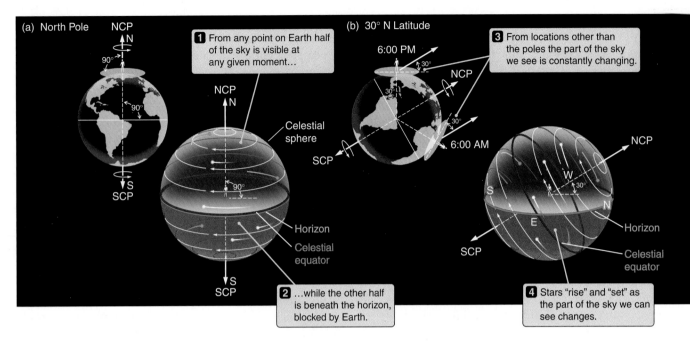

Figure 2.8 *The celestial sphere is a useful fiction for thinking about the appearance and apparent motion of the stars in the sky. Here the celestial sphere is shown as viewed by observers at four different latitudes. At most latitudes, stars rise and set as the part of the celestial sphere that we see changes during the day.*

set north of west. Objects that are south of the celestial equator rise south of east and set south of west.

Also note from Figure 2.8 that regardless of where you are on Earth, half of the celestial equator is always visible above the horizon (again with the exception of the poles). Since half of the celestial equator is always visible, it follows that you can see any object that lies in the direction of the celestial equator half of the time. An object that is in the direction of the celestial equator rises due east, is above the horizon for exactly 12 hours, and sets due west. This is not true for objects that are not on the celestial equator. A look at Figure 2.8(b) shows that from the Northern Hemisphere, you can see more than half of the apparent circular path of any star that is north of the celestial equator. And if you can see more than half of a star's path, then the star is above the horizon for more than half of the time.

As seen from the Northern Hemisphere, stars north of the celestial equator remain above the horizon for more than 12 hours each day. The farther north the star is, the longer it stays up. The circumpolar stars near the north celestial pole are the extreme example of this. They are up 24 hours a day. In contrast, objects south of the celestial equator are above the horizon for less than 12 hours a day, and the farther south you look, the less time a star is visible. Near the south celestial pole there are stars that never rise above our horizon.

Different parts of the sky are above the horizon for different amounts of time.

If you were an observer in the Southern Hemisphere (Figure 2.8d), the reverse of the preceding discussion would be true: Objects on the celestial equator would still be up for 12 hours a day, but now it is objects south of the celestial equator that would be up longer than 12 hours and objects north of the celestial equator that would be up less than 12 hours.

A SWINGING PENDULUM, A FLYING CANNONBALL, AND A SWIRLING STORM FEEL THE EFFECT OF EARTH'S ROTATION

One reason the ancients did not believe that Earth rotates is that they could not perceive the spinning motion of Earth. Put yourself in their place. As a result of Earth's rotation, the surface of Earth is moving along at a very respectable speed—1,674 km/h (kilometers per hour) at the equator (calculated by dividing the circumference of Earth by the period of its rotation). Even so, we do not feel that motion any more than we would "feel" the speed of a car with a perfectly smooth ride cruising down a straight highway. However, Earth's rotation *does* have a number of measurable effects on objects riding along on its surface.

Jean-Bernard-Léon Foucault (pronounced "Foo-coe"; 1819–1868) was the first to carry out an experiment demonstrating that Earth rotates. In 1851,

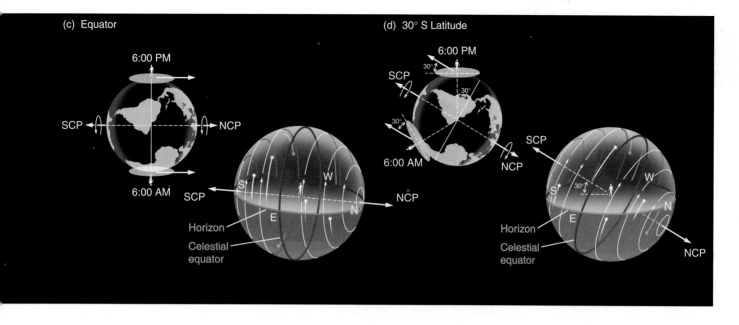

(c) Equator

6:00 PM

SCP — NCP

6:00 AM   SCP

Horizon

Celestial equator

S   W

E

N

NCP

(d) 30° S Latitude

6:00 PM

SCP

6:00 AM

SCP

NCP

Horizon

Celestial equator

S   W

E

N

NCP

Foucault constructed a **pendulum** by suspending a 28 kg (kilogram) weight at the end of a steel wire 67 meters long from the dome of the Pantheón in Paris, and set the pendulum swinging. A pendulum swings back and forth because of the combined effects of Earth's gravity and the tension in the wire. A child swinging on a rope is a good example of a pendulum. He falls toward Earth, picking up speed, then slows to a stop as he climbs on the other side of his swing. He then repeats the processes in reverse, returning to its starting point. Left on its own, the laws of physics say that a pendulum will keep swinging back and forth, its motion remaining in the same plane. Yet when Foucault observed his pendulum over the course of several hours, he found that the plane in which the pendulum swings gradually *changes,* rotating in a clockwise direction as viewed from above!

It is easiest to understand the behavior of a **Foucault pendulum** when it is placed at Earth's North or South Pole, as shown in **Figure 2.9.** The pendulum swings back and forth, staying in the same plane, *but Earth is rotating underneath it.* This is obvious to an observer watching from space (Figure 2.9a), yet to us riding on the rotating Earth (Figure 2.9b), the direction of the pendulum's swing appears to change. After 6 hours, Earth will have rotated 90° on its axis. The pendulum, which is still swinging in the same plane as it always was,

Foucault's pendulum provided the first experimental demonstration of Earth's rotation.

appears to us to have changed the direction of its swing by 90° in the opposite direction. (We stress that the back-and-forth motion of any pendulum is due to Earth's gravity and the tension in the wire, and *not* to anything having to do with Earth's rotation. It is only the *apparent* change in the *direction* of the back-and-forth swing relative to the ground that is caused by Earth's rotation.)

The plane in which a Foucault pendulum at the North Pole swings appears to make one complete rotation in 24 hours. In contrast, a Foucault pendulum on Earth's equator behaves very differently. Here imagine that the pendulum is set swinging back and forth along the equator (Figure 2.9c). As Earth rotates, the pendulum keeps swinging in the same plane, and so from the standpoint of an observer riding along on Earth, the pendulum shows no change in direction. Earth is no longer spinning underneath the pendulum. Instead, the pendulum is riding around on a big circle with the surface of Earth.

At the equator the plane of the pendulum's swing does not appear to rotate.

The case for latitudes between the pole and the equator is more difficult to understand in detail, but you can probably guess the answer. If it takes exactly one day for the plane of the swing of a pendulum at the pole to appear to rotate once, and if it takes forever for a pendulum on the equator to change the plane of its swing, then for latitudes between the two it probably

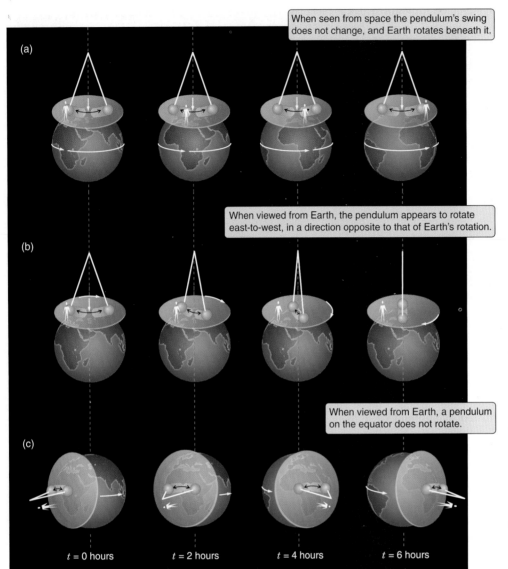

When seen from space the pendulum's swing does not change, and Earth rotates beneath it.

(a)

When viewed from Earth, the pendulum appears to rotate east-to-west, in a direction opposite to that of Earth's rotation.

(b)

When viewed from Earth, a pendulum on the equator does not rotate.

(c)

$t = 0$ hours $t = 2$ hours $t = 4$ hours $t = 6$ hours

Figure 2.9 (a) *A Foucault pendulum at Earth's North Pole swings back and forth in the same plane while Earth rotates underneath it. (b) To an observer on the rotating Earth, however, it is the swing of the pendulum that seems to change directions. (c) A Foucault pendulum on the equator does not appear to rotate.*

takes longer than a day but less than forever to go around once. This guess would be correct. Foucault's original pendulum, located in Paris at a latitude of 49°, took about 32 hours to complete one rotation. (It is often true in science that we start by working out the "easy" or "limiting" cases—the pendulum at the pole or the equator, for example—and then use these to help us think about what happens in more complicated situations.)

DIFFERENCES IN SPEED BETWEEN DIFFERENT LATITUDES CAUSE THE CORIOLIS EFFECT Foucault may have been the first to conduct an experiment showing that Earth rotates, but he was not the first to experience effects of this rotation. Earth's rotation actually influences things as diverse as the motion of weather patterns on Earth, to

how an artillery gunner must aim at a distant target. Any object sitting on the surface of Earth follows a circle each day as Earth rotates on its axis. This circle is larger for objects near Earth's equator, and smaller for objects nearer to one of Earth's poles, but because Earth is a solid body, all objects must complete their circular motion in one day. Since an object nearer to the equator has farther to go each day than an object nearer a pole, the object nearer the equator must be moving *faster* than the object at a greater latitude. If an object starts out at one latitude and then moves to another, its apparent motion over the surface of Earth is influenced by this difference in speed.

Imagine that you are riding in a car traveling down a straight section of highway at a constant speed. The

> **Objects closer to the equator move faster than objects farther from the equator.**

(a) Frame of reference: Viewer on the street

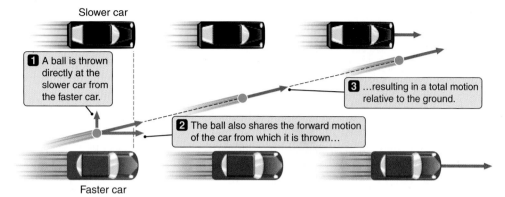

Figure 2.10 *The motion of an object depends on the frame of reference of the observer.*

Slower car

1 A ball is thrown directly at the slower car from the faster car.

2 The ball also shares the forward motion of the car from which it is thrown…

3 …resulting in a total motion relative to the ground.

Faster car

(b) Frame of reference: Viewer in faster car

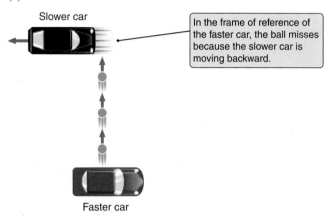

Slower car

In the frame of reference of the faster car, the ball misses because the slower car is moving backward.

Faster car

(c) Frame of reference: Viewer in slower car

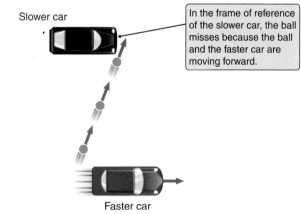

Slower car

In the frame of reference of the slower car, the ball misses because the ball and the faster car are moving forward.

Faster car

windows are blacked over so that you cannot see the scenery go by. How would you tell the difference between this and sitting completely still in your driveway? The fact is that, apart from the vibration due to the roughness of the road, you could not. You might try throwing a ball around in the car. You toss it up and it falls right back in

Only relative motion counts.

your lap. You can play catch with the person sitting next to you. Even though the car is speeding down the road, the relative motions between the objects in the car are small, and it is only the **relative motions** that count.

Now imagine that two cars are driving down the road at *different* speeds, as shown in **Figure 2.10.** Ignoring for the moment any real-world complications like wind resistance, if you were to throw a ball from the faster moving car directly at the slower moving car as the two cars pass, you would miss. The ball shares the forward motion of the faster car, so the ball outruns the forward motion of the slower car. From your perspective in the faster car (Figure 2.10b), the slower car lagged behind the ball. From the slower car's perspective (Figure 2.10c), your car and the ball sped on ahead.

Now do the same experiment, but instead of two cars, think about two locations at different latitudes. Suppose you fire a cannon directly north from a point in the Northern Hemisphere, as shown in **Figure 2.11(b).** Because the cannon is located nearer to the equator than its target is, the cannon itself is moving toward the east faster than its target. Even though the cannonball is fired toward the north, it shares in the eastward velocity of the cannon itself. This means that the cannonball is *also* moving toward the east faster than its target! Recall how the ball thrown from the faster car outpaced the slower-moving car. Similarly, as the cannonball flies north, it finds itself moving toward the east faster than the ground underneath it is. To an observer on the ground, it looks like the cannonball begins to curve toward the east as it outruns the eastward motion of the ground it is crossing. The farther north the cannonball flies, the greater the difference between its eastward velocity and the eastward velocity of the ground. As a result the cannonball follows a path that appears to curve more and more to the east the farther north it goes. If you are located in the Northern Hemisphere and fire a cannonball *south* toward the equator **(Figure 2.11c)**, just the opposite effect will occur. Now the cannon is moving toward the east more slowly than its target. As the cannonball flies toward the south, its eastward motion lags behind that of the ground underneath it, and the cannonball appears to curve toward the west.

This effect of Earth's rotation is called the **Coriolis effect.** In the Northern Hemisphere the Coriolis effect causes a cannonball fired north to drift to the east as seen from the surface of Earth. In other words, the cannonball appears to curve to the right. A cannonball fired south appears to curve to the west, which also gives it the appearance of curving to the right. In the Northern Hemisphere the Coriolis effect seems to deflect things to the *right*. If you think through this example for the Southern Hemisphere, you will see that south of the equator the Coriolis effect seems to deflect things to the *left*. In between these two, at the equator itself, the Coriolis effect vanishes.

> **Deflection caused by the Coriolis effect is to the right in the Northern Hemisphere.**

These differences in the Coriolis effect between the Northern and Southern Hemisphere are seen in the rotation of weather systems. As air is pushed from regions of higher pressure toward regions of lower pressure, these motions are influenced by the Coriolis effect. Think about a low-pressure region in the Northern Hemisphere. When air is pushed toward this region of low pressure from the south, the Coriolis effect deflects this flow of air toward the east. Similarly, air moving toward the region of low pressure from the north is deflected to the west by the Coriolis effect. The net effect is that as air moves toward a region of low pressure in the Northern Hemisphere, the Coriolis effect deflects it into a counterclockwise circulation **(Figure 2.12)**. (Think about this carefully. The Coriolis effect deflects objects toward the right in the Northern Hemisphere, which results in weather patterns that rotate toward the *left*.) The next time you see a television weather map with a low-pressure region, look at the direction of the winds around the region and you will see this counterclockwise flow. The most spectacular example of this **cyclonic motion** is the swirl of wind and clouds

> **The Coriolis effect causes the counterclockwise rotation of northern hurricanes.**

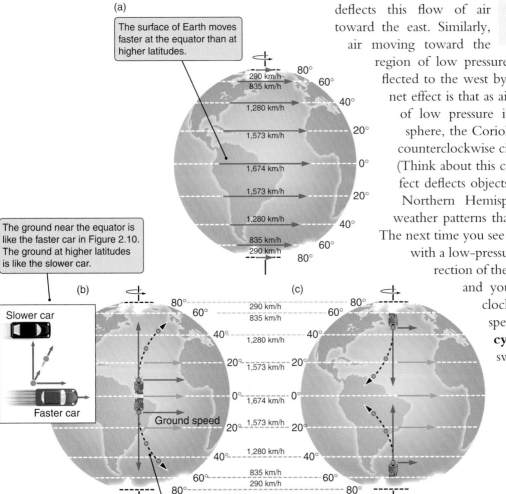

(a)

The surface of Earth moves faster at the equator than at higher latitudes.

The ground near the equator is like the faster car in Figure 2.10. The ground at higher latitudes is like the slower car.

(b)

Slower car

Faster car

Ground speed

A cannonball fired away from the equator outruns the ground it flies over, so its ground track curves to the east.

(c)

A cannonball fired toward the equator lags behind the eastward motion of the ground, so its ground track curves to the west.

Figure 2.11 *The Coriolis effect causes objects to appear to be deflected as they move across the surface of Earth.*

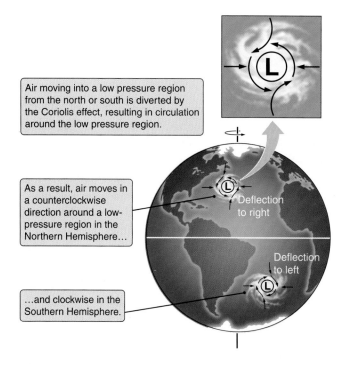

Air moving into a low pressure region from the north or south is diverted by the Coriolis effect, resulting in circulation around the low pressure region.

As a result, air moves in a counterclockwise direction around a low-pressure region in the Northern Hemisphere…

…and clockwise in the Southern Hemisphere.

Deflection to right

Deflection to left

Figure 2.12 *As a result of the Coriolis efffect, air circulates around regions of low pressure on the rotating Earth.*

around the deep low-pressure area at the eye of a hurricane or typhoon. As the air moves in closer and closer to the central region of low pressure, it rotates faster and faster, giving rise to the winds of hundreds of kilometers per hour that make hurricanes so destructive.

With what you now know about the Coriolis effect, you can show that hurricanes in the Southern Hemisphere rotate in the opposite direction from hurricanes in the Northern Hemisphere. Here instead of curving to the right, air moving into a region of low-pressure curves to the left, causing a clockwise rotation around the low-pressure region. In crossing from the Northern Hemisphere to the Southern, the Coriolis effect goes away at the equator. The difference in the direction of rotation between the hemispheres and the weakness of the Coriolis effect near the equator mean that a northern hurricane would literally collapse if it tried to cross into the Southern Hemisphere. (Imagine throwing your car into reverse while traveling down the road at 150 km/h.) This is why countries around Earth's equator do not experience the ravages of hurricanes.

Southern hurricanes rotate clockwise.

As discussed in Foundations 2.1, the Coriolis effect is only one example of the type of effect associated with relative motions. On a humorous note, the Coriolis effect has nothing to do with the direction that water swirls in a toilet bowl, as many people imagine. The difference in the speed of Earth's motion on one side of a toilet bowl is not enough different from the speed of Earth's motion on the other side of the toilet bowl to matter much. Other effects, such as the direction in which the water flows into the bowl, are much larger. However, the Coriolis effect is enough to deflect a fly ball hit into deep left field in a stadium in the northern United States by about a half a centimeter. At some time or other the Coriolis effect has probably determined the outcome of a ball game.

2.3 REVOLUTION ABOUT THE SUN LEADS TO CHANGES DURING THE YEAR

The second motion we will discuss is the motion of Earth about the Sun. Earth revolves around the Sun in the same direction Earth spins about its axis—counterclockwise as viewed from above Earth's North Pole. A **year** is the time it takes for Earth to complete one revolution around the Sun. The motion of Earth around the Sun is responsible for many of the patterns of change we see in the sky and on Earth, including changes in which stars we see at night. When you look overhead at midnight, you are looking away from the Sun. As Earth moves around the Sun, this direction changes. Six months from now, Earth will be on the other side of the Sun, and the stars that we see overhead at midnight will be in nearly the opposite direction from the stars we see near overhead at midnight tonight. The stars that were overhead at midnight six months ago are the stars that are overhead today at noon, but we cannot see those stars today because of the glare of the Sun.

If you make note of the position of the Sun relative to the stars each day for a year, you will find it traces

Earth's orbital motion is counterclockwise as viewed from above Earth's North Pole.

We see different stars as we view the Universe in different directions throughout the year.

FOUNDATIONS 2.1

RELATIVE MOTIONS AND FALSE FORCES

Aside from looking out of the window or feeling the vibrations from the road, there is no experiment that you could easily do to tell the difference between riding in a car down a straight section of highway and sitting in the car while it is parked in your driveway. Because everything in the car is moving together, the relative motions between objects in the car are all that count. In fact, the only reason you can feel the roughness of the road is that it causes the motion of the car to change slightly. It is these brief accelerations that you feel as the car's vibration.

The idea that only relative motions count occurs again and again in astronomy and physics. There are numerous examples in this chapter alone. For example, even though Earth is spinning on its axis and flying through space on its orbit about the Sun, the resulting relative motions between objects that are near each other on Earth are small—so small that for most of history humans assumed that Earth was sitting still. Newton's realization that motions are only meaningful when tied to the **frame of reference** of some observer is also at the heart of Einstein's theories of relativity. These theories, which we will return to later, wound up changing the way we think about space and time.

The Coriolis effect, discussed in this chapter, is an example of what can happen if motion is viewed from a frame of reference whose velocity is changing. The Coriolis effect is sometimes *wrongly* called the Coriolis force, because to someone on the ground it looks like a force has acted on the cannonball, pushing it to the side. In fact, no additional force is acting on the cannonball. It is flying true. The ball *appears* to curve because of the shifting frame of reference of the ground over which it travels. The Coriolis effect and other false forces can be seen in many contexts. When you turn a sharp corner in a car, it seems as though you are thrown against the door by some Coriolis-like force. Actually, no force is throwing you against the door at all. Instead, your body is trying to continue on in a straight line while the frame of reference of the car is changing. For a playground example of the Coriolis effect and such false forces, return to the same carousel we used to understand the apparent motions of the stars. Sit near the center of the spinning carousel, and try to play catch with someone sitting near the edge. This exercise, if carried out physically and not just in your mind, will help you develop a feeling for how the Coriolis effect works. (It might be good for a few laughs or a sick stomach as well.)

As we continue on our journey, keep your eyes open for more examples of relative motions. Relative motions among planets, relative motions among stars, relative motions among galaxies, relative motions among atoms—they will all be there.

---

out a **great circle** against the background of the stars **(Figure 2.13).** On September 1, the Sun appears to be in the direction of the constellation of Leo. Six months later, on March 1, Earth is on the other side of the Sun, and the Sun appears to be in the direction of the constellation of Aquarius. The apparent path that the Sun follows against the background of the stars is called the **ecliptic.** The constellations that lie along the ecliptic and through which the Sun appears to move are called the constellations of the **zodiac.** This is why ancient astrologers assigned special mystical significance to these stars. Actually, the constellations of the zodiac are nothing more than random patterns of distant stars that happen by chance to lie near the plane of Earth's orbit about the Sun.

The Sun's apparent yearly path against the background of stars is the ecliptic.

EARTH'S MOTION THROUGH SPACE IS MEASURED USING ABERRATION OF STARLIGHT

Just as it is difficult to "feel" the effects of Earth's rotation on its axis, it is even harder to sense the motion of Earth around the Sun. Through most of the history of our species, humans believed that Earth remains stationary while the Sun, the Moon, and the heavens revolve around us. The history of modern astronomy,

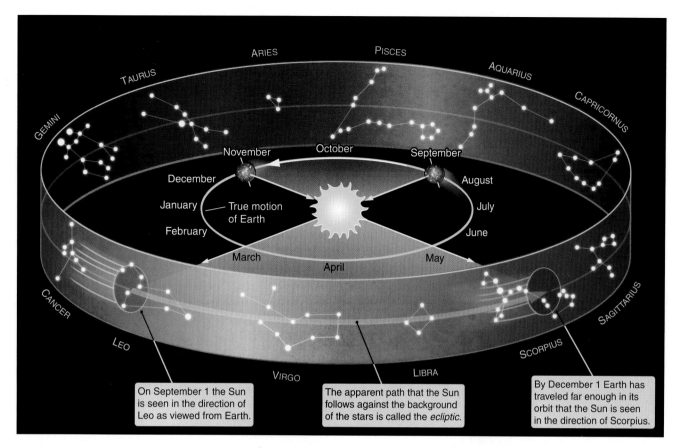

ARIES   PISCES

TAURUS   AQUARIUS

GEMINI   CAPRICORNUS

November   October   September

December   August

January — True motion of Earth   July

February   June

March   May

April

CANCER   SAGITTARIUS

LEO   SCORPIUS

VIRGO   LIBRA

On September 1 the Sun is seen in the direction of Leo as viewed from Earth.

The apparent path that the Sun follows against the background of the stars is called the *ecliptic*.

By December 1 Earth has traveled far enough in its orbit that the Sun is seen in the direction of Scorpius.

Figure 2.13 *As Earth orbits about the Sun, the Sun's apparent position against the background of stars changes. The imaginary circle apparently traced by the Sun is called the ecliptic. Constellations along the ecliptic form the zodiac.*

and to some degree the story of the rise of modern science, can be told as the story of how this view was overthrown during the 17th and 18th centuries. However, the first direct measurement of the effect of Earth's motion did not come until the 18th century. To understand how this measurement was made, we return to our automotive example of a moving frame of reference.

Imagine you are sitting in a car in a windless rainstorm, as shown in **Figure 2.14.** If the car is sitting still and the rain is falling vertically, then when you look out of your side window you see raindrops falling straight down. That is, if you were to hold a vertical tube out of the window, raindrops would fall straight through the tube. When the car is moving forward, however, the situation is different. Between the time a raindrop appears below the top of your window and the time it disappears beneath the bottom of your window, the car has moved forward. The raindrop disappears beneath the window *behind* the point at which it appeared, which means the raindrop *looks as if* it falls at an angle, even though in reality it is falling straight down. For raindrops to fall directly through the tube

Figure 2.14 *The direction in which rain falls depends on the reference frame in which it is viewed. From a stationary car, rain is seen to fall vertically downward. From a moving car, rain is seen to fall at an angle determined by the speed and direction of the car's motion.*

From the vantage point of a person outside, the rain falls vertically, even when the car is moving.

Car stationary   Car moving

From the frame of reference of a person inside the car...

...the rain falls vertically if the car is stationary...

...but at an angle if the car is moving.

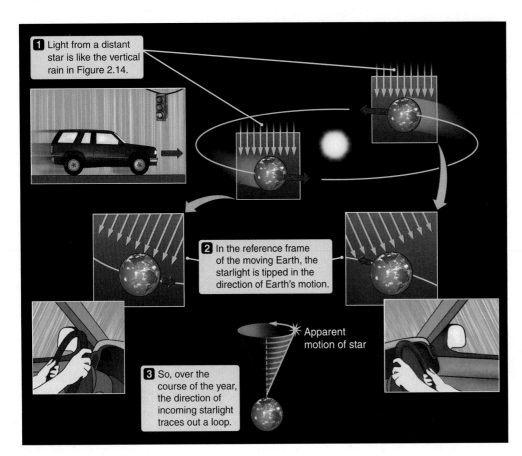

1 Light from a distant star is like the vertical rain in Figure 2.14.

2 In the reference frame of the moving Earth, the starlight is tipped in the direction of Earth's motion.

Apparent motion of star

3 So, over the course of the year, the direction of incoming starlight traces out a loop.

Figure 2.15 *The apparent positions of stars are deflected slightly toward the direction in which Earth is moving. As Earth orbits the Sun, stars appear to trace out small ellipses in the sky. This effect is called aberration of starlight.*

you are holding out the window now, you would have to tilt the top of the tube forward. The faster you go, the greater the apparent front-to-back motion of the raindrops, and the more tilted their apparent paths become. An observer by the side of the road would say the raindrops are coming from directly overhead, but to you in the moving car they are coming from in front of the car.

This same phenomenon occurs with starlight, as shown in **Figure 2.15.** The light from a distant star arrives at Earth from the direction to this star. Because Earth moves, however, to an observer on Earth the starlight appears to be coming from a slightly different direction, just as the raindrops appeared to be coming from in front of the car.[1] As the direction of Earth's motion around the Sun continuously changes during the year, the position of a star in the sky moves in a small loop. This shift in apparent position is what we refer to as the **aberration of starlight.**

The apparent direction of stars depends on Earth's motion.

The speed and direction of Earth's motion and the part of the sky a star is in determine the exact shift in the apparent position of a star. The apparent deflection of a star located at a right angle to Earth's motion is about half of what the typical unaided human eye can detect. This much deflection is easy to measure with a telescope. Aberration of starlight was first detected in the 1720s by two English astronomers, Samuel Molyneux and James Bradley. Measurement of the aberration of starlight shows that Earth is moving on a roughly (but not exactly) circular path about the Sun with an average speed of just under 30 km/s (kilometers per second). Since distance equals speed times time, the distance around this near-circle—its circumference—is just the speed of Earth (29.8 km/s) times the length of one year (3.16×10^7 s). The circumference of Earth's orbit is then

Earth's orbital motion around the Sun was first measured from the aberration of starlight.

$$\text{Distance} = \text{speed} \times \text{time}$$

$$= (29.8 \tfrac{\text{km}}{\text{s}}) \times (3.16 \times 10^7 \text{ s})$$

$$= 9.42 \times 10^8 \text{ km.}$$

---

[1]Actually, in the frame of reference of Earth, the starlight *is* coming from a different direction, and in the frame of reference of the moving car, the raindrops *are* coming from in front of the car.

The radius of Earth's nearly circular orbit is this circumference divided by 2π, or 1.50×10^8 km, or 150 million kilometers. Astronomers refer to this distance—the average distance between the center of the Sun and the center of Earth[2]—as one **astronomical unit,** abbreviated AU. The astronomical unit is a good unit for measuring distances within the Solar System. Modern measurements of the size of the astronomical

The astronomical unit (AU) is the average distance between the Sun and Earth.

unit are made in very different ways, such as bouncing radar signals off Venus. However, the aberration of starlight provided a simple and compelling demonstration that Earth orbits about the Sun—and a pretty good value for the size of Earth's orbit as well.

SEASONS ARE DUE TO THE TILT OF EARTH'S AXIS

If you ask most people why it is cold in the winter and warm in the summer, they are likely to tell you that it is because Earth is closer to the Sun in the summer and farther away in the winter. This is a common (and commonsense) idea that *does* have something to do with the seasons on Mars, but has virtually *nothing* to do with the seasons on Earth! Earth's orbit around the Sun is very nearly a circle centered on the Sun[3], and so the distance from the Sun changes very little during the year. In fact, Earth is slightly closer to the Sun during the northern winter than it is during the northern summer. So far we have discussed the rotation of Earth on its axis and the revolution of Earth about the Sun, and the consequences of each. To understand the changing of the seasons, we need to consider the combined effects of these two motions.

If Earth's spin axis were exactly perpendicular to the plane of Earth's orbit (the ecliptic plane), then the Sun would always appear to lie on the celestial equator. Since the position of the celestial equator is fixed in our sky, the Sun would follow the same path through the sky day after day, rising due east each morning and setting due west each evening. If the Sun were always

on the celestial equator, it would be above the horizon for exactly half the time, and days and nights would always be exactly 12 hours long. In short, if Earth's axis were exactly perpendicular to the plane of Earth's orbit, each day would be just like the last, and there would be no seasons.

However, Earth's axis of rotation is *not* exactly perpendicular to the plane of the ecliptic. Instead, it is tilted by 23.5° from the perpendicular. As Earth moves around the Sun, its axis points in almost exactly the same direction through the year and from one year to the next. As a result, sometimes Earth's North Pole is tilted more toward the Sun, and at other times it is pointed more away from the Sun. When Earth's North Pole is tilted toward the Sun, an observer on Earth sees the

Seasons result from the 23.5° tilt of Earth's axis with respect to a line perpendicular to the plane of its orbit.

Sun as lying *north* of the celestial equator. Six months later, when Earth's axis is tilted away from the Sun, the Sun is seen as lying *south* of the celestial equator. If we look at the circle of the Sun's apparent path through the stars—the ecliptic—we see that it is tilted by 23.5° with respect to the celestial equator.

To understand the effect that this has on Earth, begin by looking at **Figure 2.16(a).** This shows the situation on June 21, the day that Earth's North Pole is tilted most directly toward the Sun.[4] Note first that the Sun is north of the celestial equator. We found earlier in the chapter that, from the perspective of an observer in the Northern Hemisphere, an object north of the celestial equator can be seen above the horizon for more than half the time. This is true for the Sun as well as for any other celestial object. Saying that the Sun is above the horizon for more than half the time is just another way of saying

The length of day and night at a given latitude depends on where Earth is in its orbit.

that the days are longer than 12 hours. You can see this directly in Figure 2.16(a) by noting that when Earth's North Pole is tilted toward the Sun, over half of Earth's Northern Hemisphere is illuminated by sunlight. These are the long days of the northern summer. Six months later, on December 22, the situation is very different. On December 22 **(Figure 2.16b)** Earth's

---

[2]Distances between celestial objects are almost always taken to be between their centers.

[3]Planetary orbits are actually elliptical, as will be discussed in Chapter 3.

[4]We are a bit sloppy with language here. Earth's North Pole tilts in the same direction year-round. It is just that on the first day of the northern summer, Earth is on the side of the Sun where the tilt of the North Pole is toward the Sun. On the first day of the northern winter, Earth is on the opposite side of the Sun, so the tilt of the North Pole is away from the Sun.

2.2
SEASONS

North Pole is tilted away from the Sun, so the Sun appears in the sky south of the celestial equator. Someone in the Northern Hemisphere will see the Sun for less than 12 hours each day. Less than half of the Northern Hemisphere is illuminated by the Sun. It is winter in the north.

Over the course of the year the length of the day changes, courtesy of the 23.5° tilt of Earth's axis relative to the perpendicular to the plane of its orbit. In the preceding paragraph we were very careful to specify the length of the day in the *Northern* Hemisphere, because things are very different in the Southern Hemisphere. In fact, things in the Southern Hemisphere are exactly reversed from what is going on in the north. Look again at Figure 2.16. On June 21, while the Northern Hemisphere is enjoying long days and short nights, Earth's South Pole is tilted in the direction away from the Sun. Less than half of

Seasons in the Southern Hemisphere are the reverse of those in the Northern Hemisphere.

the Southern Hemisphere is illuminated by the Sun, and days are shorter than 12 hours. Similarly, on December 22 Earth's South Pole is tilted toward the Sun, and the southern days are long.

The differing length of days through the year is part of the explanation for the changing seasons, but there is another important effect that we need to consider. The Sun appears higher in the sky during the summer than it does during the winter, so sunlight strikes the ground *more directly* during the summer than during the winter. To see why this is important, hold a piece of cardboard toward the Sun and look at the size of its shadow. If the cardboard is held so that it is directly face-on to the Sun, then its shadow is large. However, as you turn the cardboard more edge-on, the size of its shadow shrinks. The size of the cardboard's shadow tells you that the cardboard catches

The angle of sunlight to the ground is closer to perpendicular in summer than in winter, so there is more heating per unit area.

Figure 2.16 (a) *On the first day of the northern summer (June 21, the summer solstice), the northern end of Earth's axis is tilted most nearly toward the Sun, while the Southern Hemisphere is tipped away. Seasons are opposite in the Northern and Southern Hemispheres.* (b) *Six months later, on the first day of the northern winter, the situation is reversed.*

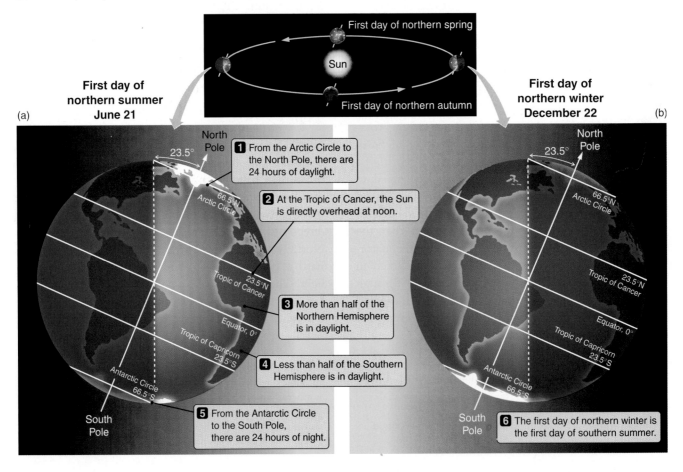

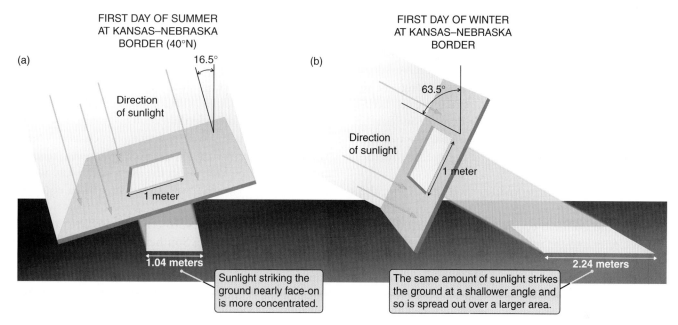

Figure 2.17 *Noon at the Kansas–Nebraska border (40° north latitude). (a) On the first day of northern summer, sunlight strikes the ground almost face-on. (b) At noon on the first day of northern winter, sunlight strikes the ground more obliquely, and less than half as much sunlight falls on each square meter of ground each second.*

less energy from the Sun each second when it is tilted relative to the Sun than it does when it is face-on to the Sun. This is exactly what happens with the changing seasons. During the summer, Earth's surface is more nearly face-on to the incoming sunlight, so more energy falls on each square meter of ground each second. During the winter the surface of Earth is more inclined with respect to the sunlight, so less energy falls on each square meter of the ground each second. That is the main reason why it is hotter in the summer and colder in the winter.

Look at **Figure 2.17,** which shows the direction of incoming sunlight striking Earth at the border between Kansas and Nebraska (40° north latitude). At noon on the first day of summer the Sun is high in the sky—73.5° above the horizon, and only 16.5° away from the zenith. Sunlight strikes the ground almost face on. In contrast, at noon on the first day of winter, the Sun is only 26.5° above the horizon, or 63.5° from the zenith. Sunlight strikes the ground at a very shallow angle. The difference between the two cases is great. At the Kansas–Nebraska border, over twice as much solar energy falls on each square meter of ground per second at noon on June 21 as falls there at noon on December 22. Together these two effects—the directness of sunlight and the differing length of the day—mean that during the summer there is more heating from the Sun and during the winter there is less heating from the Sun.

We do not have to wait for the seasons to change to see the effect that the height of the Sun in the sky has on terrestrial climate. We only have to compare the climates found at different latitudes on Earth. Near the equator the Sun passes high overhead every day, regardless of the season. As a result, the climate is warm throughout the year. At high latitudes, however, the Sun is *never* very high in the sky, and the climate can be cold and harsh even during the summer.

Climates are different at different latitudes.

FOUR SPECIAL DAYS MARK THE PASSAGE OF THE SEASONS

The apparent path that the Sun follows through the stars each year (the ecliptic) is a great circle that is tilted 23.5° with respect to the celestial equator. Follow along in **Figure 2.18** as we note the four special points on this path that mark the passage of the seasons. Begin with the point in March when Earth's axis is perpendicular to the direction to the Sun. This is the point where the Sun's apparent motion along the ecliptic crosses the celestial equator moving from the south to the north. This direction in the sky, located in the constellation Pisces, is called the **vernal equinox.** The term *vernal equinox* also refers to the day—about March 21—when the Sun appears at this location. Since the Sun lies on the celestial equator on the vernal equinox, days are 12 hours long. (The term *equinox* means literally "equal night," because, everywhere on Earth, night and day are the same length on the days of the equinoxes.) In the Northern Hemisphere, this is the first day of spring.

The changing seasons are marked by equinoxes and solstices.

As Earth continues its journey around the Sun, the direction to the Sun moves into closer alignment with the tilt of the northern end of Earth's axis, and the Sun climbs higher into the northern sky. The Sun is lined up with the tilt of Earth's axis about three months after the vernal equinox. When this happens, the Sun reaches its northern-most point in the sky, located in the constellation Taurus, near its border with Gemini. This day, which marks the longest day of the year and the beginning of summer in the Northern Hemisphere, occurs around June 21. This is the **summer solstice.** (*Solstice* literally means "sun standing still," because the Sun's north-south motion stops as it reverses its direction.) Note that on this same day, the southern end of Earth's axis is tipped directly away from the Sun. This is the shortest day of the year in the Southern Hemisphere, marking the beginning of the southern winter.

Three months later, around September 23, the Sun is again crossing the celestial equator. This point on the Sun's apparent path, located in the constellation Virgo, is called the **autumnal equinox.** *Autumnal equinox* refers both to the location in the sky and the date when this happens. Since the Sun is on the celestial equator, days and nights are exactly 12 hours long. It is the first day of autumn in the Northern Hemisphere, and the first day of spring in the Southern Hemisphere.

Around December 22 the Sun reaches its most southern point in the sky as its apparent path takes it through the constellation Sagittarius. This day is called the **winter solstice.** Earth's North Pole is tipped most directly away from the Sun on this day. In the Northern Hemisphere this is the shortest day of the year— the first day of winter. As the Sun passes the winter solstice and moves on toward the vernal equinox, the northern days begin growing longer again. Almost all cultural traditions in the Northern Hemisphere include some major celebration late in the month of December **(Figure 2.19).** Christmas, for example, is celebrated just four days after the winter solstice. These winter festivals have many different meanings to the different celebrants, but they all share one thing: They celebrate the return of the source of Earth's light and warmth. The days have stopped growing shorter and are beginning to get longer. It is a "new year," and once again the Sun will hold sway over night. Spring will come again.

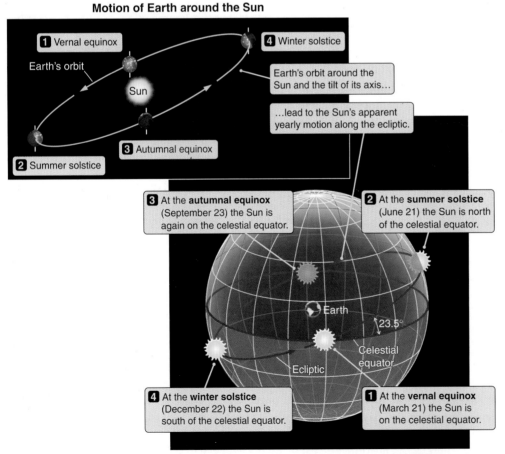

Motion of Earth around the Sun

1 Vernal equinox

Earth's orbit

Sun

4 Winter solstice

Earth's orbit around the Sun and the tilt of its axis...

...lead to the Sun's apparent yearly motion along the ecliptic.

3 Autumnal equinox

2 Summer solstice

3 At the **autumnal equinox** (September 23) the Sun is again on the celestial equator.

2 At the **summer solstice** (June 21) the Sun is north of the celestial equator.

Earth

23.5°

Celestial equator

Ecliptic

4 At the **winter solstice** (December 22) the Sun is south of the celestial equator.

1 At the **vernal equinox** (March 21) the Sun is on the celestial equator.

Apparent motion of the Sun seen from Earth

Figure 2.18 *The motion of Earth about the Sun as seen from the frame of reference of the Sun* (upper panel), *and Earth* (lower panel).

Figure 2.19 *Most cultural traditions in the Northern Hemisphere include a major celebration in late December, around the time when days begin to grow longer.*

As an aside, it is interesting to note that there is more to the seasons we feel than the amount of energy we are receiving from the Sun. Just as it takes a pot of water on a stove time to heat up when the burner is turned up and time to cool off when the burner is turned down, it takes Earth time to respond to changes in heating from the Sun. The hottest months of the summer are usually July and August, which come *after* the summer solstice, when the days are growing shorter. Similarly, the coldest months of winter are usually January and February, which occur *after* the winter solstice, when the days are growing longer. The climatic seasons on Earth lag behind changes in the amount of heating we are receiving from the Sun.

> **Seasonal temperatures lag behind changes in the directness of sunlight.**

Our picture of the seasons must be modified somewhat near Earth's poles. At latitudes north of 66.5° north latitude and south of 66.5° south latitude, the Sun is circumpolar for a part of the year surrounding the first day of summer. These lines of latitude are called the **Arctic Circle** and the **Antarctic Circle,** respectively. When the Sun is circumpolar, it is above the horizon 24 hours a day, earning the arctic regions the nickname "land of the midnight Sun." The Arctic and Antarctic regions pay for these long days,

> **The regions around the poles are called the "land of the midnight Sun."**

however, with an equally long period surrounding the first day of winter when the Sun never rises and the nights are 24 hours long. The Sun never rises very high in the Arctic or Antarctic sky, which means that sunlight is never very direct. This is why even with the long days at the height of summer, the Arctic and Antarctic regions remain relatively cool.

The seasons are also different near the equator. Recall from Figure 2.8(c) that for an observer on the equator, *all* stars are above the horizon 12 hours a day, and the Sun is no exception. On the equator, days and nights are 12 hours long throughout the year. Changes in the directness of sunlight through the year are also different on the equator. Here the Sun passes directly overhead on the first day of spring and the first day of autumn, because these are the days when the Sun is on the celestial equator. Sunlight is most direct on the equator on these days. At the summer solstice the Sun is at its northern-most point along the ecliptic. It is on this day, and on the winter solstice, that the Sun is *farthest* from the zenith at noon, and therefore sunlight is *least* direct. Strictly speaking, the equator experiences eight seasons a year instead of four: two summers, two winters, two springs, and two falls. However, none of these seasons is very different from the others. The Sun is always up for 12 hours a day, and the Sun is always so close to being overhead at noon that the directness of sunlight changes by only 8% throughout the year.

If you live between the latitudes of 23.5° south and 23.5° north, twice during the year the Sun will be directly overhead at noon. This band is called the **Tropics.** The northern limit of this region is called the tropic of Cancer. The southern limit is called the tropic of Capricorn. (As a challenge, think about what the seasons are like at different locations within the Tropics.)

Seasons behave differently in the Tropics.

EARTH'S AXIS WOBBLES, AND THE SEASONS SHIFT THROUGH THE YEAR

After our discussion of the apparent motion of the Sun along the ecliptic, you might wonder about the names given to the Tropics. The summer solstice is located in the constellation Taurus, so why do we call the northern tropic the tropic of Cancer rather than the tropic of Taurus? Similarly, why is the southern tropic not referred to as the tropic of Sagittarius, since the Sun is seen in that constellation on the winter solstice? The answer has to do with the fact that Earth's axis wobbles like the axis of a spinning top. The wobble is very slow, taking about 26,000 years to complete one cycle. During this time the north celestial pole makes one trip around a large circle through the stars. Polaris is a modern name for the star we see near the north celestial pole. If you could time-travel several thousand

Earth's axis wobbles like the axis of a top.

years into the past or future, you would find that the point about which the northern sky appears to rotate is no longer near Polaris.

Recall that the celestial equator is the set of directions in the sky that are perpendicular to Earth's axis. As Earth's axis wobbles, then, so must the celestial equator. And as the celestial equator wobbles, the locations where it crosses the ecliptic—the equinoxes—change as well. Over the course of each 26,000-year wobble of Earth's axis, the locations of the equinoxes make one complete circuit around the celestial equator. This phenomenon is called the **precession of the equinoxes.** Ptolemy and his cohorts were formalizing their knowledge of the positions and motions of objects in the sky two thousand years ago. At that time, the Sun *was* in the constellation of Cancer on the first day of northern summer, and it *was* in the constellation of Capricorn on the first day of northern winter—hence the names of the Tropics.[5]

A 26,000-year wobble causes the position of the equinoxes to slowly precess.

The tendency of the seasons to shift through the year from century to century has played havoc with human efforts to construct reliable calendars, and history is full of interesting anecdotes related to this difficulty. For example, for much of the 16th, 17th, and 18th centuries the calendars in Protestant Europe lagged behind the calendars in Catholic Europe by first 10, and later 11, days. It was not until 1752 that England and her colonies, including those in America, finally dropped 11 days from their calendars to bring them into line. This progressive step was met by riots among the people, who somehow felt that these 11 days had been stolen from them. Today's modern calendar (which was not adopted in Russia until the 1917 Bolshevik Revolution!) is based on the **tropical year,** which is 365.242199 solar days long. The tropical year measures the time from one vernal equinox to the next—from the start of spring to the start of spring. Notice that the tropical year is not an integer number of days long. An elaborate system of leap years is used in our calendar to make up for the extra fraction of a day, preventing the seasons from slowly "sliding through" the year (i.e., getting increasingly out of synch with the months): winter in December one year, in August in another. Excursions 2.2 gives further interesting details on this topic.

---

[5] In fact, the location of the summer solstice just passed from Gemini into Taurus in 1990. In about 600 years the vernal equinox will pass due to precession from Pisces into Aquarius, marking the beginning of the "Age of Aquarius."

WHY IS IT SURPRISING THAT A.D. 2000 WAS A LEAP YEAR?

Everyone knows that years that are divisible by 4 are leap years, and A.D. 2000 was no exception to that rule. Yet A.D. 2000 *was* a special case. To understand why, we need to take a look at the way our calendar is constructed, and the purpose that leap years serve.

It takes Earth 365.242199 days to travel from one vernal equinox to the next. This tropical year is *not* exactly equal to the 365 days we normally think of as making up a year. (And, thanks to the precession of the equinoxes, is it also not equal to the time—365.256366 days—it takes for Earth to complete one orbit about the Sun!) For reasons of convenience we count years as starting at midnight on the morning of January 1, and ending 365 days later at midnight on the night of December 31. But what about that extra fraction of a day (0.24 plus a bit) that we have not accounted for? Here is where leap years come in. A true tropical year is 0.242199 days—or about $\frac{1}{4}$ day—longer than a 365-day year, which means that after 4 years these extra parts have added up to make about 1 extra day. So every 4 years we add the extra day back in, and February gets a 29th day. If we did not add in this extra day, our calendar would slip by $\frac{1}{4}$ day each year, which means that the seasons would shift by not quite a month each century. If we did not correct for leap years, the first day of summer would come in July, then in August, then in September . . .

Even after we add a leap year, the calendar still has problems. A 365-day year is short by 0.242199 days, which is a bit *less* than $\frac{1}{4}$. If we keep adding an extra day *every* 4 years, then over time we would accumulate an average of 0.007801 extra days per year. In 400 years the calendar would have slipped by about 3 days. To fix this it is necessary to get rid of 3 days every 400 years. This is accomplished by making century years into common 365-day years, *except* for those century years which are divisible by 400—such as A.D. 2000—which remain leap years. With one slight further revision—making years divisible by 4,000 back into common 365-day years—the modern *Gregorian calendar* now slips by only about one day in 20,000 years.

2.4 THE MOTIONS AND PHASES OF THE MOON

The second most prominent object in the sky after the Sun is the Moon. Just as Earth orbits about the Sun, the Moon orbits around Earth. (Actually, Earth and the Moon orbit around each other and together they orbit the Sun, as we will see in Chapter 3.) In some respects the appearance of the Moon is constantly changing, but we begin our discussion of the motion of the Moon by talking about an aspect of the Moon's appearance that does *not* change.

WE ALWAYS SEE THE SAME FACE OF THE MOON

The Moon constantly changes its lighted shape and position in the sky, but one thing that does *not* change is the face of the Moon that we see. If we were to go outside next week or next month or 20 years from now or 20,000 *centuries* from now, we would still see the same side of the Moon as tonight. Because of this fact, there is a common misconception that the Moon does not rotate. The Moon *does* rotate on its axis—exactly once for each revolution that it makes about Earth.

The Moon rotates on its axis once each orbit around Earth.

Imagine walking around the Washington Monument while keeping your face toward the monument at all times. (A reasonable thing to do. You want to get a good look at it.) By the time you complete one circle around the monument, your head has turned completely around once. (When you were south of the monument, you were facing north; when you were west of the monument, you were facing east; and so on.) But someone looking at you from the monument would never have seen anything other than your face. The Moon does exactly the same thing, rotating on its axis once per revolution around

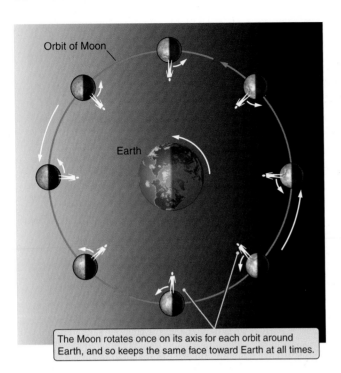

The Moon rotates once on its axis for each orbit around Earth, and so keeps the same face toward Earth at all times.

Figure 2.20 *The Moon rotates once on its axis for each orbit around Earth, an effect called synchronous rotation.*

Earth, always keeping the same face toward Earth, as shown in **Figure 2.20.** This phenomenon is referred to as the Moon's **synchronous rotation.** (The Moon's synchronous rotation is not an accident. In Chapter 9 we will find that its cause is related to why we have tides on Earth.)

THE CHANGING PHASES OF THE MOON

Sometimes the Moon appears as a circular disk in the sky. At other times it is as nothing more than a thin

sliver. At still others, its face is all dark. The Moon has no light source of its own. Like the planets, including Earth, it shines by reflected sunlight. Like Earth, half of the Moon is always in bright daylight, and half of the Moon is always in darkness. The different **phases** of the Moon result from the fact that the illuminated portion of the Moon is constantly changing. Sometimes (during a new Moon), the side facing away from us is illuminated, and sometimes (during a full Moon), the side facing toward us is illuminated. The rest of the time, only part of the illuminated portion can be seen from Earth (see Excursions 2.3).

The Moon shines by reflected sunlight.

To help you visualize the changing phases of the Moon, go outside at night and have a friend hold up a soccer ball so that it is illuminated from one side by a nearby streetlight (representing the Sun). Have your friend walk around you in a circle, and watch the changes in the ball's appearance. When you are between the ball and the streetlight, the face of the ball that is toward you is fully illuminated. The ball appears to be a bright circular disk. As the ball moves around its circle, you will see a progression of shapes, depending on how much of the bright side of the ball you can see and how much of the dark side of the ball you can see. This progression of shapes exactly mimics the changing phases of the Moon.

The phase of the Moon is determined by how much of the bright side you can see.

Figure 2.21 shows the changing phases of the Moon. When the Moon is between Earth and the Sun, the illuminated side of the Moon faces away from us, and we see only its dark side. This is called a **new Moon.** Study Figure 2.21 to see that a new Moon can only be "seen" from the illuminated side of Earth. It appears close to the Sun in the sky, and so it rises in the east at sunrise, crosses the meridian near noon, and sets

EXCURSIONS 2.3

THE DARK SIDE OF THE MOON

Popular culture often refers to the side of the Moon away from Earth as the "dark side of the Moon." This is even the title of one of the most popular pieces of music from the 20th century. But in fact, there is no dark side of the Moon. At any given time, half of the Moon is in sunlight and half of the Moon is in darkness just as at any given time, half of

Earth is in sunlight and half of Earth is in darkness. The side of the Moon that faces away from Earth spends just as much time in sunlight as the side of the Moon that faces toward Earth.

On the other hand, "I'll see you on the backside of the Moon," might not have been so wildly successful as a song lyric.

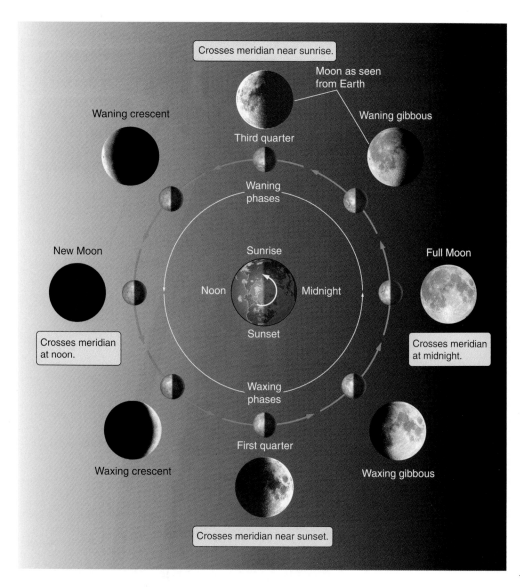

Figure 2.21 *The inner circle of images shows the Moon as it orbits Earth, as seen by an observer far above Earth's North Pole. The outer ring of images shows the corresponding phases of the Moon as seen from Earth.*

in the west near sunset. A new Moon is never visible in the nighttime sky.

As the Moon continues on its orbit around Earth, a small part of the portion being illuminated becomes visible. This shape is called a **crescent.** Because the Moon appears to be "filling out" from night to night at this time, the full name for this phase of the Moon is a waxing crescent Moon. (**Waxing** here means "growing in size and brilliance.") From our perspective the Moon has also moved away from the Sun in the sky. Because the Moon travels around Earth in the same direction in which Earth rotates, we now see the Moon located to the east of the Sun. A waxing crescent Moon is visible in the western sky in the evening, near the setting Sun but remaining above

A waxing Moon is growing more full each night.

the horizon after the Sun sets. The "horns" of the crescent always point away from the Sun.

As the Moon moves farther along in its orbit, more and more of its illuminated side becomes visible each night, so the crescent continues to fill out. At the same time the angular separation in the sky between the Moon and the Sun grows. After about a week the Moon has moved a quarter of the way around Earth. We now see half the Moon in daylight and half the Moon as dark—a phase that we call **first quarter Moon.** A look at Figure 2.21 shows that the first quarter Moon rises at noon, crosses the meridian at sunset, and sets at midnight. Note that *first quarter* refers not to how much of the face of the Moon that we see illuminated, but rather to the fact that the Moon has completed the first quarter of its cycle from new Moon to new Moon.

2.3
PHASES

As the Moon moves beyond first quarter, we are able to see more than half of its bright side. This phase is called a waxing gibbous Moon. The gibbous Moon continues nightly to "grow" until finally Earth is between the Sun and the Moon and we see the entire bright side of the Moon—a **full Moon.** The Sun and the Moon now appear opposite each other in the sky. The full Moon rises as the Sun sets, crosses the meridian at midnight, and sets in the morning as the Sun rises.

The second half of the Moon's orbit proceeds just like the first half but in reverse. The Moon continues on its orbit, again appearing gibbous but now becoming smaller each night. This

A waning Moon is growing less full each night.

phase is called a waning gibbous Moon. (**Waning** means "becoming smaller.")

A **third quarter Moon** occurs when we once again see half of the illuminated part of the Moon and half of the dark part of the Moon. A third quarter Moon rises at midnight, crosses the meridian near sunrise, and sets at noon. The Moon continues on, visible now as a waning crescent Moon in the morning sky, until the Moon again appears as nothing but a dark circle rising and setting with the Sun, and the cycle begins again.

It takes the Moon 27.32 days to complete one revolution about Earth; this is called its **sidereal** period. However, because of the changing relationship between Earth, the Moon, and the Sun due to Earth's orbital motion, it takes 29.53 days to go from one full Moon to the next; this is called its **synodic**

The time from one new Moon to the next is not the same as the time it takes the Moon to orbit Earth.

period. (**Figure 2.22** shows how this works.) You can always tell a waxing from a waning Moon because the side that is illuminated is always the side facing the Sun. When the Moon is waxing, it appears in the evening sky, so it is its western side that is illuminated (the right side as viewed from the Northern Hemisphere). Conversely, when the Moon is waning, it is the eastern side (the left side as viewed from the Northern Hemisphere) that appears bright.

Do not try to memorize all of the possible combinations of where the Moon is in the sky at what phase and at what time of day. You do not have to. Instead, work on *understanding* the motion and phases of the Moon, then use your understanding to figure out the specifics of any given case. As a way to study the phases of the Moon, draw a picture like Figure 2.21,

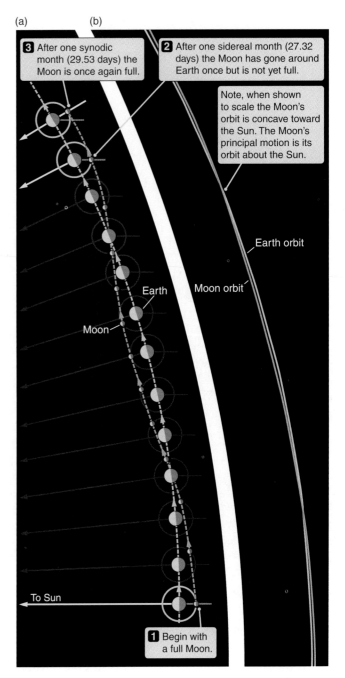

Figure 2.22 (a) *The Moon completes one sidereal orbit in 27.32 days, but the synodic period (the period between phases seen from Earth) from one full Moon to the next is 29.53 days. (The orange line to the right of the Moon indicates a fixed direction in space.)* (b) *The orbits of Earth and the Moon are shown here to scale.*

and use it to follow the Moon around its orbit. Figure out from the drawing what phase you would see and where it would appear in the sky at a given time of day. You might also enjoy thinking about what phase someone on the Moon would see when looking back at Earth.

2.5 ECLIPSES: PASSING THROUGH A SHADOW

Put yourself in the place of our nomadic friend from the opening section of this chapter. You are finely attuned to the patterns of the sky, and you view these patterns not as the inexorable consequences of physical law but as visible signs from the gods. Can you imagine any celestial event that would strike more terror in your heart than to look up and see the Sun, giver of light and warmth, being eaten away as if by a giant dragon? There is archeological evidence that our ancestors put great effort into trying to find the pattern of eclipses and thereby bring them into the orderly scheme of the heavens. For example, the ancient and massive stone artifact in the English countryside called Stonehenge, pictured in **Figure 2.23,** may have allowed its builders to predict when eclipses might occur. The lives and motivations of the builders of Stonehenge, 4,000 years dead, may be lost in antiquity. Yet how can we doubt their desire to exert some control over the terror of eclipses by learning a few of their secrets—and in the process assuring themselves that an eclipse did not mean that all was lost?

Figure 2.23 *Stonehenge is an ancient artifact in the English countryside, used 4,000 years ago to keep track of celestial events.*

VARIETIES OF ECLIPSES

The type of eclipse described above, in which Earth moves through the shadow of the Moon, is called a **solar eclipse.** Three different types of solar eclipses are possible. To see why, begin by looking at the structure of the shadow of the Sun cast by a round object such as the Moon, as shown in **Figure 2.24.** An observer at point A could see no part of the surface of the Sun. This darkest, inner part of the shadow is called the **umbra.** If a point on Earth passes through the Moon's umbra, the Sun's light is totally blocked by the Moon. This is called a **total solar eclipse.** Now look instead at points B and C in Figure 2.24. From these points an observer can see one side of the disk of the Sun but not the other. This outer region, which is only partially in shadow, is the **penumbra.** If a point on the surface of Earth passes through the Moon's penumbra, the result is a **partial solar eclipse,** in which the disk of the Moon blocks the light from a portion of the Sun's disk.

There are three different types of solar eclipses.

(Figure 2.4 ECLIPSE I)

There is a third type of eclipse, called an **annular solar eclipse,** in which the Sun appears as a bright ring surrounding the dark disk of the Moon. An observer at point D is far enough from the Moon that the Moon appears to be smaller than the Sun. You may be wondering, how can one eclipse be total and another

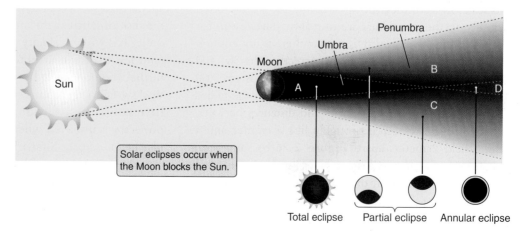

Figure 2.24 *Different parts of the Sun are blocked at different places within the Moon's shadow. An observer in the umbra (A) sees a total solar eclipse, observers in the penumbra (B and C) see a partial eclipse, while observers in region D see an annular eclipse.*

Solar eclipses occur when the Moon blocks the Sun.

Total eclipse   Partial eclipse   Annular eclipse

(a)

(b)

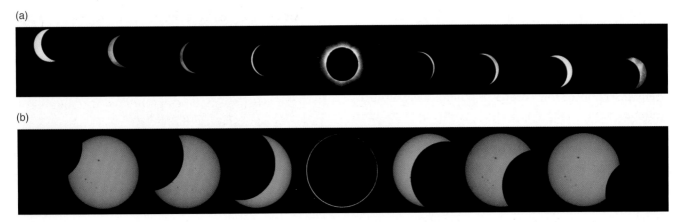

Figure 2.25 *Time sequences of images of the Sun taken (a) during a total solar eclipse, and (b) during an annular solar eclipse.*

annular? Two things make this possible. One is a fluke of nature: the diameter of the Sun is about 400 times the diameter of the Moon, and the Sun is about 400 times farther away from Earth than the Moon is. As a result, the Moon and Sun have almost exactly the same apparent size in the sky. The other factor is that the Moon's orbit is not a perfect circle. So when the Moon and Earth are a bit closer together than average, the Moon appears larger in the sky than the Sun. An eclipse occurring at that time will be total. When the Moon and Earth are farther apart than average, the Moon appears smaller than the Sun, so eclipses occurring during this time will be annular. Pictures of total and annular solar eclipses are shown in **Figure 2.25.** Roughly one-third of solar eclipses are total, one-third are annular, and one-third are seen only as partial eclipses.

> **The Moon's apparent size in the sky is almost exactly the same as the Sun's.**

Figure 2.26(a) shows a drawing of the geometry of a solar eclipse, with the Moon's shadow falling on the surface of Earth. It is important to realize that figures like this (or like Figures 2.20 or 2.21) are seldom drawn to scale. Instead, they show Earth and the Moon much closer together than they are in reality. The reason for distorting figures this way is simple: there is not enough room on the page to draw them correctly. The 384,400 km distance between Earth and the Moon is over 60 times the radius of Earth, and over 220 times the radius of the Moon. The relative sizes and distances between Earth and the Moon are roughly like a quarter and a dime held 2 meters apart. **Figure 2.26(b)** shows the geometry of a solar eclipse with Earth, the Moon, and the separation between them drawn to scale. Compare this to Figure 2.26(a), and you will understand why artistic license is normally taken in drawings of Earth

and the Moon. If the Sun were drawn to scale in Figure 2.26(b), it would be six-tenths of a meter across and located almost 64 meters off of the left side of the page.

The Moon's penumbra is quite large. In fact, with a bit of thought and a pencil and paper you can convince yourself that the Moon's penumbra, where it hits Earth, must have about twice the diameter of the Moon itself, or almost 7,000 km. This part of the shadow is large enough to cover a substantial fraction of Earth, so partial solar eclipses are often seen from much of the planet. In contrast, the path along which a total solar eclipse can be seen **(Figure 2.27)** covers only a tiny fraction of the surface of Earth. Earth is so close to the tip of the Moon's umbra that, even when the distance between Earth and the Moon is at a minimum, the umbra is only 269 km wide at the surface of Earth. As the Moon moves along in its orbit, this tiny shadow sweeps across the face of Earth at breakneck speed. The Moon moves on its orbit around Earth at a speed of about 3,400 km/h, and its shadow sweeps across the disk of Earth at the same rate. Earth is also rotating on its axis with a velocity of 1,670 km/h at the equator (and less than that at other latitudes). The situation is further complicated by the fact that the Moon's shadow falls on the curved surface of Earth. You may have noticed that the image projected by an overhead projector is distorted when the beam is not perpendicular to the screen. Similarly, the curvature of Earth often causes the region shaded by the Moon during a solar eclipse to be elongated by differing amounts. The curvature can even cause an eclipse that started out as annular to become total.

> **Partial solar eclipses are seen over large regions.**

> **Total solar eclipses are localized and short-lived.**

When all of these effects are considered, the result is that a total solar eclipse can never last longer than $7\frac{1}{2}$ minutes, and is usually significantly shorter. Even so, it is one of the most amazing and awesome sights in nature. People flock from the world over to the most remote corners of Earth to witness the fleeting spectacle of the bright disk of the Sun blotted out of the daytime sky, leaving behind the eerie glow of the Sun's outer atmosphere.

Lunar eclipses are very different in character from solar eclipses. The geometry of a lunar eclipse is shown in **Figure 2.26(c)** (and is shown drawn to scale in **Figure 2.26d**). Because Earth is much larger than the Moon, the dark umbra of Earth's shadow at the distance of the Moon is about 9,200 km in diameter, or over 2.5 times the diameter of the Moon. A **total lunar eclipse** is a much more leisurely affair than a total solar eclipse, with the Moon spending as long as 1 hour and 40 minutes in the umbra of Earth's shadow.

Lunar eclipses last much longer than solar eclipses.

A **penumbral lunar eclipse** occurs when the Moon passes through the penumbra of Earth's shadow. A penumbral eclipse can be anything but spectacular, because its appearance from Earth is nothing more than a fading in the brightness of the full Moon. While the penumbra of Earth is 16,000 km across at the distance of the Moon—over four times the diameter of the Moon—a penumbral eclipse is only noticeable when the Moon passes within about 1,000 km of the umbra.

2.5
ECLIPSE
II

ECLIPSE SEASONS OCCUR ROUGHLY TWICE EVERY 11 MONTHS

If the Moon's orbit were in exactly the same plane as the orbit of Earth (imagine Earth, the Moon, and the Sun all sitting on the same flat tabletop), then the

Figure 2.26 *A solar eclipse occurs when the shadow of the Moon falls on the surface of Earth. A lunar eclipse occurs when the Moon passes through Earth's shadow. Note that (b) and (d) are drawn to proper scale.*

(a) Solar eclipse geometry (not to scale)

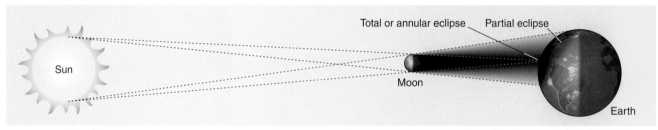

(b) Solar eclipse to scale

(c) Lunar eclipse geometry (not to scale)

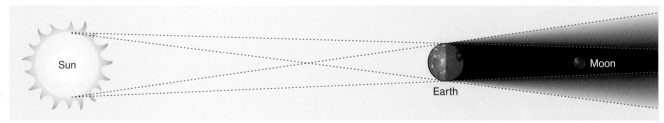

(d) Lunar eclipse to scale

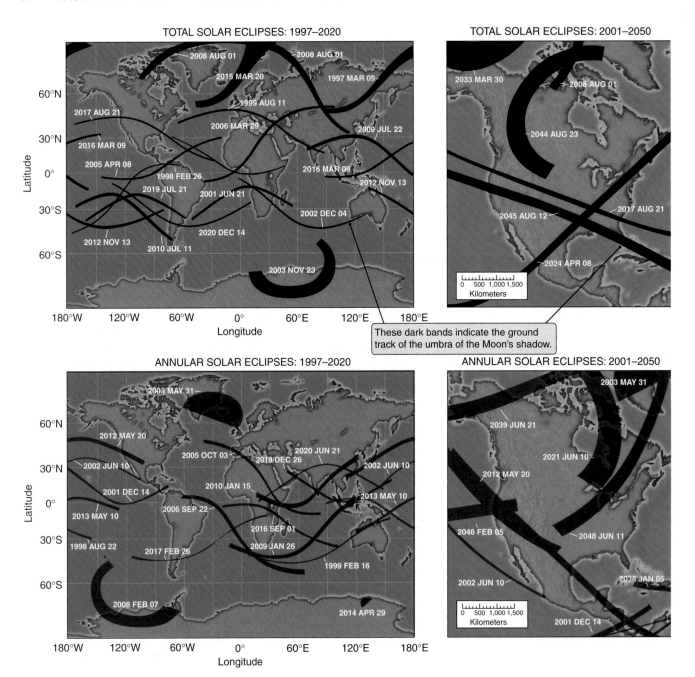

Figure 2.27 *The paths of total and annular solar eclipses predicted for the late 20th and early 21st centuries.*

Moon would pass directly between Earth and the Sun at every new Moon. The Moon's shadow would pass across the face of Earth, and we would see a solar eclipse. Similarly, Earth would pass directly between the Sun and the Moon every synodic month, and each full Moon would be marked by a lunar eclipse.

Solar and lunar eclipses do *not* happen every month because the Moon's orbit does not lie in exactly the same plane as the orbit of Earth. Look at **Figure 2.28** to see how this works. The plane of the Moon's orbit about Earth is inclined by

The Moon's orbit is tilted . . .

about 5.2° with respect to the plane of Earth's orbit about the Sun. The line along which the two orbital planes intersect is called the **line of nodes.** For part of the year, the line of nodes passes close to the Sun. During these times, called **eclipse seasons,** a new Moon passes between the Sun and Earth, casting its shadow on Earth's surface and causing a solar eclipse. Similarly, a full Moon occuring during an eclipse season passes through Earth's shadow, and a lunar eclipse results. An eclipse season lasts for only 38 days. That is how long the Sun is close enough to the line of nodes for eclipses to occur. Most of the time, the line of nodes points farther away from the

Sun, and Earth, Moon, and Sun cannot line up closely enough for an eclipse to occur. A solar eclipse cannot be

. . . so most months, Earth and the Moon miss each other's shadows.

seen because the shadow of a new Moon passes "above" or "below" Earth. Similarly, no lunar eclipse can be seen because a full Moon passes "above" or "below" the shadow of Earth.

If the plane of the Moon's orbit always had the same orientation, then eclipse seasons would occur twice a year, as suggested by the drawing in Figure 2.28. In actuality, eclipse seasons occur about every 5 months and 20 days. The roughly 10-day difference is due to the fact that the plane of the Moon's orbit slowly wobbles, much like the wobble of a spinning plate balanced on the end of a circus performer's stick. As it does so, the line of nodes changes direction. This wobble rotates in the direction opposite the direction of the motion of the Moon in its orbit. (That is, the line of nodes moves clockwise as viewed from the north.) It takes the Moon's orbit 18.6 years to complete one "wobble" of 360°, so we say that the line of nodes *regresses* by about 360°/18.6 years, or 19.4° per year. This amounts to about a 20-day regression each year. If January 1 marks the middle

of an eclipse season, the next eclipse season would be centered around May 20, and the one after that, December 10.

We have come far since our nomadic ancestor looked at the sky and saw there a mystical reflection of the patterns and events that marked the life of the tribe. Yet when we look at the sky today our sense of awe and majesty is no less than that experienced by that long-ago tribesman. Our distant ancestor had to look for patterns in the world in order to survive. It is this same human impulse to seek out patterns that led the astrologers to look for connections between ourselves and the heavens, and later to seek out patterns that gave birth to science. We now know that the patterns of the sky are connected to us much more directly than any mystical link invented by an astrologer. The patterns and changes that we see in the sky are caused by the same forces of nature that bind us to our planet and that cause the wind to blow and the rain to fall. They are the same forces that push the blood through our veins and carry the electric impulses of thought through our brains. So far in our look at the changing patterns of the sky, we have only glimpsed these connections, so it is to these underlying causes that we now turn our attention.

Figure 2.28 *Eclipses are only possible when the Sun, Moon, and Earth lie along a line. When the Sun does not lie along the line of nodes, Earth passes under or over the shadow of a new Moon, and a full Moon passes under or over the shadow of Earth.*

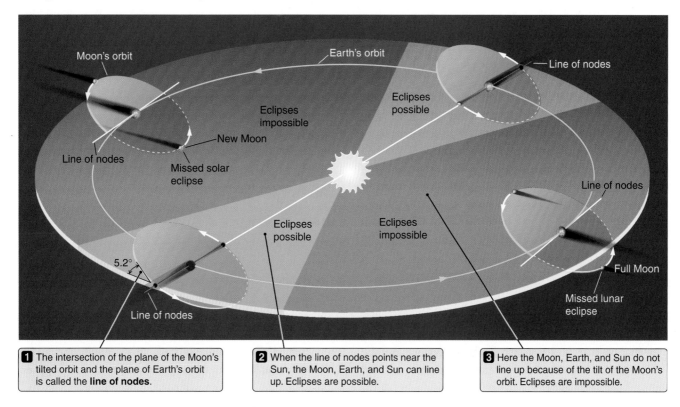

1 The intersection of the plane of the Moon's tilted orbit and the plane of Earth's orbit is called the **line of nodes**.

2 When the line of nodes points near the Sun, the Moon, Earth, and Sun can line up. Eclipses are possible.

3 Here the Moon, Earth, and Sun do not line up because of the tilt of the Moon's orbit. Eclipses are impossible.

SEEING THE FOREST THROUGH THE TREES

Patterns in the sky change in lockstep with the cycles of life on Earth. It is no surprise that our ancestors sought some link between the stars and the events in their lives, but the astrologers were ultimately limited by their misconceptions about the Universe. Earth does not reside at the center of Creation, nor does the Sun move from "house" to "house" along the ecliptic. The eerie spectacle of a solar eclipse blotting out the Sun is no more mystical than the shadow you cast on the sidewalk on a sunny afternoon. The heavens are not a place of magic but of physical law, knowable though observation and experiment, and subject to test by the scientific method. The signs of the zodiac that grace the entertainment section of your local newspaper and the covers of supermarket tabloids are anachronisms—pictures born from the human imagination and painted onto the random splash of stars across the night sky.

Yet, while astrologers were off the mark in their conclusions, their quest was well motivated. The motions of Earth and the properties of the Sun do, indeed, give rise to the most basic of all of the patterns faced by life on Earth. Earth's rotation is responsible for the coming of night and day. Earth's passage around the Sun brings the changing of the seasons. Changes in the direction in which sunlight falls on Earth cause dramatic differences in climate from the equator to the poles. These patterns set the stage for the evolution of life on Earth, and they remain with us today, buried deep within the genetic code of our species. Even as humans begin to venture beyond Earth and into space, we carry these patterns with us in everything from the length of our cycle of waking and sleeping, to the temperature we prefer, to the amount of light we need in order to see. Much of human culture also has its roots in the apparent motions of celestial objects. Many of our legends and traditions arose in the ancient view of the sky as a place of gods and spirits. At the same time, patterns of objects in the sky spurred the development of mathematics and, as we will see in Chapter 3, the development of a physical understanding of the world around us. Much of who and what we are as a thinking species has its origins in our experience of the larger Universe.

The errors in perception that shaped our views of the Universe throughout most of human history are both understandable and forgivable. It is very difficult to directly sense the effects of Earth's motion. As you read this, you are probably (depending on your latitude) moving at more than 1,000 km/h on a circular path around Earth's axis, while Earth itself is moving at over 100,000 km/h in its orbit about the Sun. (And the Sun itself is moving through our Galaxy at almost 1 million km/h.) Yet you feel none of this motion. Objects around you share your motion, and so to you are motionless. The telltale signs of your motion are far too subtle to perceive directly. As you watch the Sun and stars cross the sky, it certainly seems as if it is you about which the cosmos revolves. In fact, without the benefit of discussions such as ours, it would be difficult to avoid that impression.

When we replace simple perception with careful experiment and reason, however, this commonsense notion evaporates before our eyes. Minute changes in the direction of starlight through the year provide proof of Earth's motion in its orbit, and allow us to measure our path around the Sun. The behavior of a simple pendulum hung from the ceiling of a dome in Paris, when carefully considered, provides direct evidence of Earth's rotation about its axis. And as our physical understanding of Earth's motion improves, even the blowing of the wind and the need for a fire on a cold, dark winter night become side effects of Earth's motions through space.

Moving along in our journey we will again encounter many of the motions discussed in this chapter, but from a rather different perspective. So far we have concentrated on *describing* the motion of Earth about the Sun and the Moon about Earth. From here we will take the step that separates post-Renaissance science from all that came before by taking up the question of *why* these motions are as they are. The search for understanding will lead us to a discovery that changed our perception of the Universe and our place in it—that the same natural forces that dictate the path of a well-hit baseball also govern the clockwork motions of Earth, the Moon, and planets.

STUDENT QUESTIONS

THINKING ABOUT THE CONCEPTS

1. Draw a picture of Earth showing the location of an observer at 20° north latitude. Where are the north and south celestial poles and the celestial equator located relative to the horizon from this observer's perspective? How long would stars in different parts of the sky remain above this observer's horizon each day?

2. Assume that the Moon's orbit is circular. Suppose

you are standing on the side of the Moon that faces Earth. How would Earth appear to move in the sky as the Moon made one revolution around Earth? How would the "phases of Earth" appear to you, as compared to the phases of the Moon as seen from Earth?

3. You are way out in open ocean sailing from the Carolinas to Bermuda (that is, heading due east), and you know there is a hurricane nearby. A strong wind is blowing straight out of the south. Do you continue on to Bermuda or head back? To arrive at your answer, draw a diagram of the Coriolis effect acting on air flowing into the region of low-pressure at the center of a hurricane.

4. Many U.S. cities have main streets laid out in east-west, north-south alignments. Why are there frequent traffic jams on east-west streets during both morning and evening rush hours within a few days of the equinoxes? On this basis, if you work in the city during the day, would you rather live east or west of the city?

5. The vernal equinox is now in the zodiacal constellation of Pisces. Wobbling of Earth's axis will eventually cause the vernal equinox to move into Aquarius, beginning the legendary, long-awaited "Age of Aquarius." How long, on average, does the vernal equinox spend in each of the 12 zodiacal constellations?

6. Sometimes artists paint the horns of the crescent moon pointing toward the horizon. Is this realistic? Explain.

7. What is the advantage of launching satellites from spaceports located near the equator? Why are satellites never launched in a westerly direction?

8. Does the occurence of solar and lunar eclipses disprove the notion that the Sun and the Moon both orbit around Earth? Explain your reasoning.

9. What is the approximate time of day when you see the full Moon near the meridian? At what time is the first quarter (waxing) Moon on the eastern horizon? Use a sketch to help explain your answers.

10. A soldier fires a cannon directly at a distant target toward the east and makes a perfect hit. She then fires a shot directly at another distant target toward the north and finds that her shot has hit to the west of the target. Was the soldier in Australia or Canada? Would her marksmanship have improved if she had been in Ecuador, on the equator? Explain.

11. The tilt of Jupiter's rotational axis is 3°. If Earth's axis had this tilt, explain how it would affect our seasons.

APPLYING THE CONCEPTS

12. Assume that rain is falling at speed of 5 meters per second (m/s) and you are driving along in the rain at a leisurely 5 m/s (or 18 km/h). Estimate the angle from the vertical at which the rain appears to be falling.

13. The Moon's orbit is tilted by about 5° relative to Earth's orbit around the Sun. What is the highest altitude in the sky that the Moon can reach as seen in Philadelphia (latitude 40°)?

14. Assume that you and your sister can both throw a baseball at a speed of 100 km/h. If you are in an airplane traveling at 800 km/h and play catch with your sister who is near the front of the plane (you being toward the rear), how fast would the ball be traveling as seen by an observer on the ground when thrown (a) by you and (b) by your sister? How fast does it appear to move as seen by you and your sister?

15. Solar (and lunar) eclipses can occur only when the Moon is near one of its nodes and is new (or full) during its orbit around Earth. The time between successive passages of the Moon through the same node is 27.2122 days. The period between successive new Moons is 29.5306 days. Show that 223 successive new Moons (223 lunar months) is the same interval (to within one part in 250,000) as 242 successive crossings of the same node, and that this is very close to 18 years in length. It seems likely that the builders of Stonehenge knew about this cycle some 4,000 to 5,000 years ago.

16. Using an atlas, determine the latitude where you live. Draw and label a diagram showing that your latitude is the same as
   a. the altitude of the north celestial pole; and
   b. the angle (along the meridian) between the celestial equator and your local zenith.

17. What is the noontime altitude of the Sun as seen from your home at the times of winter solstice and summer solstice?

18. Suppose the tilt of Earth's equator relative to its orbit were 10° instead of 23.5°. At what latitudes would the Arctic and Antarctic Circles and the Tropics be located?

19. Assume Earth is a perfect sphere with a radius of 6,400 km. What is the distance at Earth's equator that corresponds to 1° of longitude? What would be the distance corresponding to 1° of latitude? Estimate the distance corresponding to 1° of longitude at latitudes of 30° and 60°. What would be the distance of 1° of latitude at these same latitudes?

The Newtonian principle of gravitation is now more firmly established, on the basis of reason, than it would be were the government to step in, and to make it an article of necessary faith. Reason and experiment have been indulged, and error has fled before them.

THOMAS JEFFERSON (1743–1826)

GRAVITY AND ORBITS—
A CELESTIAL BALLET

3.1 GRAVITY!

Today it is a rare person who does not know that the planets, including Earth, orbit around the Sun. Yet it was not always so. Only about 500 years have passed since a soft-spoken Polish monk named **Nicholas Copernicus** (1473–1543) started a revolution when he revived the idea, discarded by the Greeks two thousand years earlier, that it is the Sun rather than Earth that lies at the center of Creation. At the time, this suggestion seemed outlandish. To think that Copernicus would have us believe that humankind—the pinnacle of Creation—resides anywhere but at the center of all things! To imagine that we occupy but one of several planets circling the Sun's central fire, that we are nothing but a pebble among all the pebbles on the beach—absurd!

It would be easy from our "modern" perspective to chuckle at the naiveté of our ancestors and their Earth-centered view of the Universe, but to do so would be unfair. The previous chapter showed that hard evidence of the motions of Earth was remarkably hard to come by. To an ancient scholar, educated in Greek and Roman philosophy, Copernicus's view of the Universe simply made no sense. Copernicus knew quite well that his ideas flew in the face of authority, and would not be welcomed by the powers that be. Wishing to avoid the controversy that his theory would certainly cause,

KEY CONCEPTS

In this chapter we will follow the story of the birth of modern science, as humans discovered regular patterns in the motions of the planets, then went on to explain those patterns in terms of fundamental physical laws. Along the way we will explore:

✳ Empirical rules discovered by Kepler that describe the elliptical orbits of planets around the Sun;

✳ Theoretical physical laws discovered by Newton and Galileo that govern the motion of all objects;

✳ The use of proportionality to describe patterns and relationships in nature;

✳ The nature of scientific theories, the roles of empirical and theoretical science, and the difference between science and pseudoscience;

✳ How Newton's laws of motion and an inverse square law of gravitation combine to explain an orbit as one body falling freely around another;

✳ The cosmological principle that grew from the startling realization that the heavens and Earth are governed by the same physical laws; and

✳ How theories lead to new knowledge, such as the way Newton's derivation of Kepler's third law is used to measure the masses of objects from observations of orbital motions.

Copernicus chose not to publish his ideas until late in his life. His great work, *De revolutionibus orbium coelestrium* (On the Revolution of the Celestial Spheres), did not appear until the year of his death. It was this work that pointed the way toward our modern cosmological principle. Copernicus knew that his ideas would not be popular, but he had no way of guessing the consequences of what he had started. In retrospect we see now that he knocked the bottom out from under the house of cards representing humankind's view of the world around them, and the following centuries would see that house of cards slowly fall apart. The repercussions of Copernicus's insight would come not only to shape our understanding of the Universe around us, but would change the direction of the progress of human civilization, itself.

Copernicus knocked humankind from the center of the Universe.

Even today we need to avoid too much complacency about what we think we know. People often confuse knowing the name of something with actually knowing *about* something. For example, we happily talk about spacecraft in orbit about Earth, or planets in orbit about the Sun, but remarkably few people understand what those words mean. When asked why astronauts float about the cabin of a spacecraft, most educated adults answer, "The spacecraft has escaped Earth's gravity." Yet the gravitational force acting on a space shuttle orbiting Earth is only slightly weaker than when the shuttle is sitting on the launch pad. In fact, were it not for Earth's gravity, the spacecraft would not orbit about Earth at all!

Orbiting spacecraft do not escape Earth's gravity.

Copernicus knew nothing of gravity, but his ideas inevitably led to it. As physicists and astronomers have come to better understand gravity, they have realized that in most respects it is gravity that holds the Universe together. Our Solar System is a gravitational symphony. The Sun's gravity shapes the motions of the planets and every other object in its vicinity. These motions range from the almost circular orbits of a number of the planets, to the extremely elongated orbits of comets. A comet's orbit may carry it from tens of thousands of astronomical units[1] from the Sun, to somewhere inside the orbit of Mercury, and back out again. Within this grand symphony, subthemes arise. A chorus of particles orbiting the giant planets gives rise

to majestic systems of rings which, in turn, play counterpoint to the gravitational ballet of the planets and their moons. The analogy between the motions of objects in the Solar System and the patterns of music is not new. **Johannes Kepler** (1571–1630), who will figure prominently in our story, entitled his great work of 1619 *Harmonice mundi,* or "Harmonies of the World."

As our grasp of the Universe has expanded, we have come to realize that our Solar System is but one gravitational opus in a far larger opera. Gravity binds stars into the colossal groups we call galaxies, and works to slow the expansion of the Universe. It is gravity that holds the planets and stars together, and keeps the thin blanket of air we breathe close to the surface of the planet we live on. It is gravity that caused a vast interstellar cloud of gas and dust to collapse 4.5 billion years ago to form our Sun and Solar System. It is gravity that gives space and time their very shape. As we continue our journey outward through the cosmos, we will come to each of these ideas in turn, and time and again find gravity at the center of our growing understanding.

3.2 AN EMPIRICAL BEGINNING: KEPLER'S LAWS DESCRIBE THE OBSERVED MOTIONS OF THE PLANETS

Now that we have extolled the wonders of gravity and the role it plays in the Universe, you might expect us immediately to turn to a discussion of what gravity is and how it works. But the great minds that brought us to our modern understanding did not have the benefit of our 20-20 hindsight. All they could do was watch the motions of the planets over the course of months and years, and puzzle over what they saw. With no way even to judge the distances to the planets, it is no wonder that it took humans thousands of years to begin to see the reality behind the patterns before their eyes.

In Chapter 1 we painted a picture of science as a worldview in which nature is governed by physical laws, and in which mathematical descriptions of these physical laws are used to explain natural phenomena. But how do scientists go about *discovering* these physical laws? When faced with phenomena as complex and puzzling as the motions of the planets in the heavens, where can we find a toehold? The wise sailor settles for

---

[1]Recall from Chapter 2 that an astronomical unit (abbreviated AU), is the average distance between the Sun and Earth.

any port in a storm, and the wise scientist knows that when faced with a complex and poorly understood phenomenon there may be little choice but to turn directly to the information our senses provide. We carefully observe the phenomenon under study, systematically recording as much information as we can, as accurately as possible. Then, as our observations start piling up, we look for patterns in those observations and start trying to formulate rules that seem to do a good job of *describing* what we have seen.

Imagine that you are a scientist from another world, setting foot for the first time on planet Earth. You notice right away that there are many interesting structures sticking out of the ground on this new planet. As you record your observations, you find that some of these structures are large, some are small, some spread out over the ground, some stick up in the air, and so on. But after a time you realize that almost all of these structures are covered with some kind of appendage, and that those appendages are green in color. So you form a descriptive rule about these objects (call them "plants"): most have green appendages. You decide that "green appendages" must be fundamental to the nature of plants, so you begin to study what makes these appendages green. After a time you discover that the green appearance always comes from the same chemical substance, and when you study that chemical substance you find that it is capable of absorbing light and turning water and carbon dioxide into more complex organic molecules. You have discovered photosynthesis, the process responsible for powering the majority of life on Earth. *But you did not start out to discover photosynthesis. You started out noting that plants have green leaves.*

The quest to first note and then accurately describe patterns in nature is called **empirical** science. Empirical science often involves a great deal of creativity and not a small amount of pure (but educated) guesswork. Copernicus's theory that Earth and the planets move on circular orbits about the Sun is an example of empirical science. Copernicus did not understand *why* the planets move about the Sun, but he did realize that his Sun-centered picture provided a much *simpler* description of the observed motions of planets than a model with Earth at its center did. Copernicus's work was made great by the fact that he was able to see beyond the prejudice of his time and to think the unthinkable—that perhaps Earth is "merely" one planet among many.

Copernicus's work paved the way for another great empiricist, Johannes Kepler. Science has often benefited

Empirical science involves description of observations.

from unlikely and chancy collaborations, and Kepler's is one such story. Kepler, a mathematician who had studied the ideas of Copernicus, worked in 1600 as an assistant to **Tycho Brahe** (1546–1601), a firm believer in an Earth-centered Universe. While Tycho is described as anything but a pleasant individual, he was also one of the greatest observational astronomers of all time. Toiling away through long nights with what we would now consider unimaginably primitive equipment, Tycho amassed a wealth of remarkably accurate observations of the positions of the planets over the course of decades. Kepler, using Tycho's data, took the next major step toward understanding the motions of the planets. Working first primarily with Tycho's observations of Mars, Kepler was able to deduce three empirical rules that elegantly and accurately describe the motions of the planets. These three rules are now almost universally referred to as **Kepler's laws.**

Kepler used Tycho's observations to refine Copernicus's empirical picture.

KEPLER'S FIRST LAW: PLANETS MOVE ON ELLIPTICAL ORBITS WITH THE SUN AT ONE FOCUS

When Kepler used Copernicus's model to calculate where in the sky a planet should be at a particular time, he found quite a lot of disagreement between these predictions and Tycho's data. He was not the first to notice such discrepancies. Rather than discarding Copernicus's ideas, however, Kepler played with Copernicus's Sun-centered model. Kepler discovered that if he replaced Copernicus' circular orbits with elongated elliptical orbits, and displaced these orbits to one side so that the Sun was no longer at the center, then his calculations and Tycho's observations fell into almost perfect agreement.

Planetary orbits are ellipses.

You probably think of an ellipse as an oval shape, but to make sense of Kepler's discovery we need to be a bit more precise about what an ellipse is. The most concrete way to define an **ellipse** is to call it the shape that results when you attach the two ends of a piece of string to a piece of paper, stretch the string tight with the tip of a pencil, and then draw around those two points keeping the string taut (see **Figure 3.1**). Each of the points at which the string is attached is called a **focus** of the ellipse. The closer the two *foci* (plural of focus) are to each other, the more nearly circular the ellipse is. In fact, a circle is just an ellipse with the two foci at the same place. (To see this, just think about the shape you would draw if the two ends of the string were attached at the same spot. In this case, each half of the string would become a radius of the circle.) As the two foci are moved farther apart, however, the ellipse becomes more and more elongated. Kepler found that the orbit of each planet is an ellipse with the Sun located at one focus. This result is now known as **Kepler's first law** of planetary motion. (You might ask, "If the Sun is located at *one* focus, what is at the other focus?" The answer is, "nothing but empty space.")

The Sun is at one focus of a planet's elliptical orbit.

Figure 3.2 shows a drawing of an ellipse. Half of the length of the long axis of the ellipse is called the **semimajor axis** of the ellipse, often denoted by the letter A. The semimajor axis of an orbit turns out to be a handy way to describe the orbit because, apart from being half the longer dimension of the ellipse, it is also the average distance between one focus and the ellipse itself. The average distance between the Sun and Earth, for example, is equal to the semimajor axis of Earth's orbit. The same is true for the orbits of all the planets.

In the case of a circular orbit, the semimajor axis is just the radius of the circle. Some ellipses, on the other hand, are very elongated. When describing the shape of an ellipse, we speak of its **eccentricity.** The eccentricity of an ellipse is defined as the separation between the two foci divided by the length of the long axis of

Eccentricity gives the shape of an ellipse.

Kepler's First Law

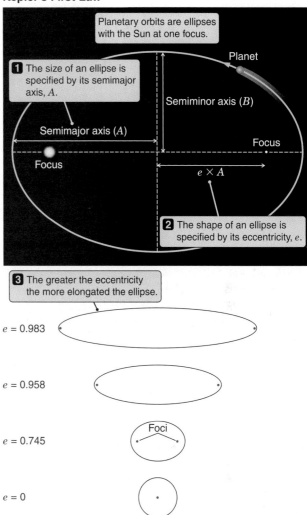

Figure 3.2 *Planets move on elliptical orbits with the Sun at one focus. Ellipses range from circles to elongated eccentric shapes.*

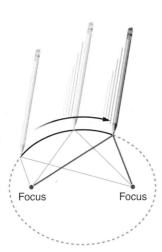

Figure 3.1 *An ellipse can be drawn by attaching a length of string to a piece of paper at two points (called foci), then pulling the string around as shown.*

the ellipse.[2] A circle has an eccentricity of 0. The more elongated the ellipse becomes, the closer the eccentricity gets to 1 (see Figure 3.2). Most planets have nearly circular orbits with eccentricities close to 0. The eccentricity of Earth's orbit, for example, is 0.017, which means that the distance between the Sun and Earth departs from its average value by only 1.7%. The distance between the two bodies varies from about 0.983 AU to 1.017 AU. It is hard to tell the difference between the

---

[2]This book tends to use operational definitions. An ellipse, for example, is defined according to how it is drawn. Eccentricity is defined according to how you would calculate its value. Definitions do not mean much in science unless they are connected to how something is actually measured or calculated.

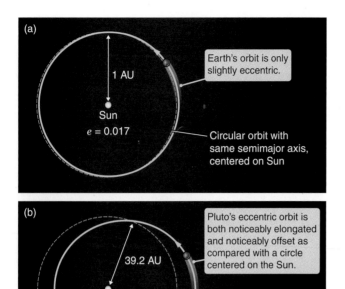

Figure 3.3 panel (a):

(a)

1 AU

Sun
e = 0.017

Earth's orbit is only slightly eccentric.

Circular orbit with same semimajor axis, centered on Sun

Figure 3.3 panel (b):

(b)

39.2 AU

Sun
e = 0.244

Pluto's eccentric orbit is both noticeably elongated and noticeably offset as compared with a circle centered on the Sun.

Circular orbit with same semimajor axis, centered on Sun

Figure 3.3 *The shapes of the orbits of Pluto and Earth as compared with circles centered on the Sun.*

orbit of Earth and a circle centered on the Sun **(Figure 3.3a).** The planet with the most eccentric orbit is Pluto. With an eccentricity of 0.244, the distance between the Sun and Pluto varies by 24.4% from its average value, ranging from 75.6% of average to 124.4% of average. Pluto's orbit is both noticeably oblong and noticeably displaced from the Sun **(Figure 3.3b).**

KEPLER'S SECOND LAW: PLANETS SWEEP OUT EQUAL AREAS IN EQUAL TIMES

The next empirical rule that Kepler found has to do with how fast planets move at different places on their orbits. A planet moves most

Planets move fastest when they are closest to the Sun.

rapidly when it is closest to the Sun, and is at its slowest when it is farthest from the Sun. The average speed of Earth in its orbit about the Sun is 29.8 km/s. When Earth is closest to the Sun, it is traveling at 30.3 km/s. When it is farthest from the Sun, it is traveling at 29.3 km/s.

Kepler found an elegant way to describe the changing speed of a planet in its orbit about the Sun. Look at

Figure 3.4, which shows one planet at six different points in its revolution about the Sun. Imagine there is a straight line connecting the Sun with this planet. We can think of this line as "sweeping out" an area as it moves with the planet from one point to another. Area A is swept out between times t_1 and t_2, area B is swept out between times t_3 and t_4, and area C is swept out between times t_5 and t_6. When the planet is closest to the Sun, it is moving rapidly, but the distance between the planet and the Sun is small (area A in Figure 3.4). Kepler realized that changes in the distance between the Sun and a planet and changes in the speed of a planet exactly cancel each other out. As a result, the area swept out by a planet in the same amount of time is always the same, regardless of the location of the planet in its orbit. In Figure 3.4, this means that if the three time intervals are equal ($t_1 \rightarrow t_2 = t_3 \rightarrow t_4 = t_5 \rightarrow t_6$) then the three areas A, B, and C will be equal as well.

3.2
K II

This is **Kepler's second law,** which is also referred to as Kepler's **Law of Equal Areas.** It states that the line connecting a planet to the Sun sweeps out equal areas in equal times, regardless of where the planet is in its or-

A planet "sweeps out" equal areas in equal times.

bit. (Note that this law applies to only one planet at a time. The area swept out by Earth in a given time is always the same. Likewise, the area swept out by Mars in a

Figure 3.4 *An imaginary line between a planet and the Sun sweeps out an area as the planet orbits. Kepler's second law states that if the three intervals of time shown are equal, then the three areas A, B, and C will be the same.*

Kepler's Second Law

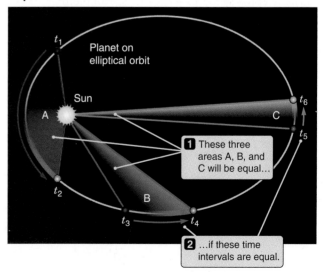

Planet on elliptical orbit

t_1

Sun

A

t_2

t_3

B

t_4

C

t_5

t_6

1 These three areas A, B, and C will be equal...

2 ...if these time intervals are equal.

given time is always the same. But the area swept out by Earth and the area swept out by Mars in a given time are *not* the same.)

KEPLER'S THIRD LAW: THE HARMONY OF THE WORLDS

Kepler's first law describes the shapes of planetary orbits, and Kepler's second law describes how the speed of a planet *changes* as it goes around its orbit. But neither of these laws tells us how long it takes for a planet to complete one orbit about the Sun (referred to as the **period** of the orbit). Nor do they tell us how this time depends on the distance between the Sun and a planet.

Planets that are closer to the Sun do not have as far to go to complete one orbit as do planets that are farther from the Sun. Jupiter, for example, has an average distance of 5.2 AU from the Sun—5.2 times as far from the Sun as Earth is. That means Jupiter has 5.2 times farther to go on its orbit about the Sun than Earth does. We might guess, then, that if the two planets travel at about the same speed in their orbits, Jupiter would complete one orbit in 5.2 years. But such a guess would be wrong. Jupiter takes almost 12 years to complete one orbit. Clearly, Jupiter not only has farther to go in its orbit, but must be *moving more slowly than Earth* as well. This trend holds true for all of the planets. As we go farther out from the Sun, the circumferences of planetary orbits are longer and longer, while the speed at which a planet travels becomes less and less. Mercury, at an average distance of 0.387 AU from the Sun, whizzes around its short orbit at an average speed of 47.9 km/s, completing one revolution in only 88 days. At an average distance of 39.2 AU from the Sun, Pluto lumbers along at an average speed of 4.72 km/s, and takes 247.7 years to make it once around the Sun.

Kepler discovered a simple mathematical relationship between the period of a planet's orbit and its distance from the Sun. **Kepler's third law** states that the square of the period of a planet's orbit, measured in years, is equal to the cube of the semimajor axis of the planet's orbit, measured in astronomical units. Written as an equation, this says that

Outer planets have farther to go and move more slowly on their orbits about the Sun.

The square of a planet's orbital period equals the cube of the orbit's semimajor axis.

$$(P_{\text{years}})^2 = (A_{\text{AU}})^3,$$

TABLE 3.1

KEPLER'S THIRD LAW: $P^2 = A^3$

The orbital properties of the planets.

Planet	Period P years	Semimajor axis A (AU)	$\dfrac{P^2}{A^3}$
Mercury	0.241	0.387	$\dfrac{0.241^2}{0.387^3} = 1.00$
Venus	0.615	0.723	$\dfrac{0.615^2}{0.723^3} = 1.00$
Earth	1.000	1.000	$\dfrac{1.000^2}{1.000^3} = 1.00$
Mars	1.881	1.524	$\dfrac{1.881^2}{1.524^3} = 1.00$
Jupiter	11.86	5.204	$\dfrac{11.86^2}{5.204^3} = 1.00$
Saturn	29.46	9.582	$\dfrac{29.46^2}{9.582^3} = 0.99^1$
Uranus	84.01	19.201	$\dfrac{84.01^2}{19.201^3} = 1.00$
Neptune	164.79	30.047	$\dfrac{164.79^2}{30.047^3} = 1.00$
Pluto	247.68	39.236	$\dfrac{247.68^2}{39.236^3} = 1.02^1$

[1]These are not exactly 1.00, due to slight perturbations from the gravity of other planets.

where P_{years} represents the period of the orbit divided by 1 year, and A_{AU} the semimajor axis of the orbit divided by 1 AU.

This is a case where astronomers use nonstandard units as a matter of convenience. Years are handy units for measuring the periods of orbits, AU are handy units for measuring the sizes of orbits, and when we use years and AU as our units, we get the simple relationship shown above. It is important to realize, however, that *our choice of units in no way changes the physical relationship* we are studying. If we instead stayed with standard metric units, this relationship would read $(P_{\text{seconds}})^2 = 3 \times 10^{-19}(A_{\text{meters}})^3$. **Table 3.1** lists the periods and semimajor axes of the orbits of all nine planets, together with the value of the ratio P^2 divided by A^3.

Judge for yourself how well Kepler's third law works. These data are also plotted in **Figure 3.5**. This relationship was so beautiful to Kepler that he referred to it as his harmonic law or, more poetically, as the "Harmony of the Worlds."

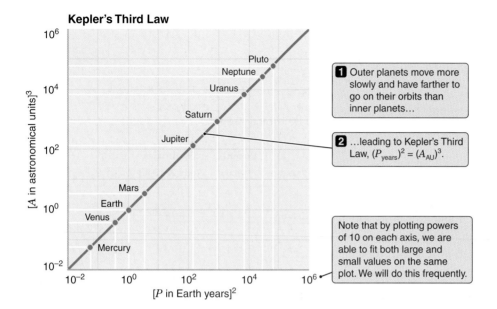

Figure 3.5 *A plot of* A^3 *versus* P^2 *for the nine planets in our Solar System shows that they obey Kepler's third law.*

3.3 THE RISE OF SCIENTIFIC THEORY: NEWTON'S LAWS GOVERN THE MOTION OF ALL OBJECTS

When investigating a newly discovered or poorly understood phenomenon, an empirical approach is often the only available way to proceed. Such was certainly the case with Kepler. The development of Kepler's laws of planetary motion was an intellectual accomplishment with few peers. Yet to a modern scientist the sort of empirical rules that Kepler spent his life pursuing is only the first step in the study of a phenomenon. Such empirical laws *describe* some phenomenon, and are even useful in predicting what will happen in the future, but they do little to *explain* that behavior. Taken at face value, empirical rules offer little insight into the more fundamental laws describing nature. Kepler was able to characterize the orbits of planets as ellipses, but he did not understand *why* they should be so.

Empirical laws describe but do not explain.

Once the empirical rules that describe some phenomenon have been discovered, a modern scientist will next try to understand those empirical rules in terms of more general physical principles or laws. Beginning with basic physical principles and using the tools of mathematics, the scientist works to *derive* the empirically

determined rules. At other times a scientist may start with physical laws and predict relationships which are then verified empirically. This technique is sometimes referred to as the **theoretical** approach to science. In practice, if the relevant physical laws are already understood, this process is often short-circuited. A scientist may make a theoretical prediction about the behavior of a system, then compare the prediction with experimental data directly to see how well they fit.

Theoretical science uses physical laws to explain empirical rules.

Today a great deal of science is done without ever trying to invent an empirical rule. This only works if the relevant physical laws are known ahead of time. If the relevant physical laws are *not* known—as was the case for planetary motions—the empirical rules become a way of *discovering* the physical laws themselves. Can we invent hypothetical physical laws that will allow us to derive the empirical rules? If so, what other predictions might we make on the basis of these hypothetical laws? Are these predictions also borne out by experiment and observation? If so, then we may have discovered something more fundamental about the way the Universe works. This is how physical laws are discovered and tested.

Hypothetical laws that work become accepted as physical laws.

One of the earliest great advances in theoretical science was also arguably one of the greatest intellectual accomplishments in the history of our species. In many ways, the work of **Sir Isaac Newton** (1642–1723) on the nature of motion set the standard for what we now refer to as *scientific theory* and *physical law*. Building on the work of Kepler and others, Newton proposed three laws that he believed to govern the motions of all objects in the heavens and on Earth. Today **Newton's laws** remain the basis for all of what is known as **classical mechanics.** By the time physicists and astronomers have completed their formal education, they will have taken many hours of course work studying the wondrously subtle and complex consequences of Newton's three laws of motion. (Physics professors pride themselves on their ability to invent truly nasty problems to challenge their graduate students' understanding of Newton's laws.) Even so, Newton's laws themselves are beautifully elegant, and the relationships they describe between such everyday concepts as force, velocity, acceleration, and mass are accessible to all.

Newton's laws of motion are the basis of classical mechanics.

However fascinating Newton's laws of motion may be, you might reasonably ask why they constitute an essential stop on our journey through *21st Century Astronomy.* "After all," you might say, "this is a book on astronomy, not physics." Yet in a very real sense it is with Newton's laws, published in 1687, that truly modern astronomy got its start. It was these laws that allowed Newton to look at the motion of a shot fired from a cannon, and see instead the motions of the planets on their orbits around the Sun. It was with Newton's laws that the chasm between our thinking about the heavens and about Earth was banished once and for all, and Earth took its true place in the Universe.

Newton's laws removed the barrier between terrestrial and heavenly phenomena.

NEWTON'S FIRST LAW: OBJECTS AT REST STAY AT REST, OBJECTS IN MOTION STAY IN MOTION

In one of the strange quirks of history, the physical law almost invariably referred to today as *Newton's first law of motion* did not originate with Newton at all. It was the brainchild of a contemporary of Kepler's by the name of **Galileo Galilei** (1564–1642). Galileo is probably best known in the mind of the public as the first person to use a telescope to make significant discoveries about the heavens and to report those discoveries. In the history of science, however, Galileo's work on the motion of objects is at least as fundamental a contribution as his astronomical observations.

By Galileo's day, Copernicus and others had begun to turn toward the view that knowledge comes from observing nature, rather than only from reading the works of the classical Greek and Roman philosophers. Yet even in the 16th and 17th centuries the works of one of the greatest of these philosophers, Aristotle, who lived almost 2,000 years earlier, still carried the weight of authority. Aristotle believed that the natural state of all objects was to be at rest, and that an object in motion would tend toward this natural state. This seemed to be a good empirical rule about the way objects in the world around us behaved, and Aristotle had elevated this observation into a fundamental tenet of his philosophy of nature. Aristotle was an extremely sharp individual, and this idea was not easy to refute. A cart rolling down the street coasts to a stop when it is no longer being pulled. A bouncing ball eventually settles to the ground. Even an arrow shot from a bow loses much of its speed before striking its target.

As discussed in Excursions 3.1, Galileo was no stranger to controversy. In his writings on motion, Galileo challenged Aristotle's authority by proposing that this apparent tendency of objects to come to rest was a mirage. Galileo argued that in all of the cases above—indeed in *every* such case—there are hidden reasons why objects come to rest. There is friction as the axle of the cart rubs against its bearing, resisting the motion and eventually bringing it to a halt. Every time a ball bounces, its shape is distorted, and what we might think of as "internal friction" within the ball causes it to bounce less high each time. The resistance of the air, which the arrow must push out of the way and which drags against the arrow's shaft, slows the arrow's progress.

Galileo agreed with Aristotle that an object at rest remains at rest unless something causes it to move. But Galileo disagreed with Aristotle by asserting that, *left on its own, an object in motion will remain in motion.* Specifically, Galileo said that *an object in motion will continue moving along a straight line with a constant velocity until an* **unbalanced force** *acts on it to change its state of motion.* Galileo referred to the resistance of an object to changes in its state of motion as **inertia.** Galileo's great insight formed the starting point for Newton's tour de force that was to come. The work of great scientists is always built on the foundation of the great scientists who came before.[3] It is a tribute to Galileo that his law of inertia became the cornerstone of physics as **Newton's first law of motion.**

Galileo found that an object left in motion remains in motion.

---

[3]Newton himself is credited with the famous quote, "If I have seen further [than you] it is by standing upon the shoulders of giants."

EXCURSIONS 3.1

CHALLENGING AUTHORITY

It is seldom safe to challenge the entrenched wisdom of your day, and that was especially true at a time when intellectual and religious authority and political power resided in the same hands. In 1600 Giordano Bruno fell victim to the Inquisition and was burned at the stake for his beliefs. These included his support of Copernicus, his belief that the Universe is infinite, and his suggestion that Earth is but one of many habitable planets. In 1632 Galileo actually invited the ire of the powers that be when he published his great work, *Dialogo sopra i due massimi sistemi del mondo* (Dialogue on the Two Great World Systems). In the *Dialogo* the champion of the Copernican (Sun-centered) view of the Universe is a brilliant, witty, and erudite philosopher named Salviati. The *Dialogo*'s defender of Aristotelian authority is named Simplicio, and is as much an ignorant buffoon as the name might imply. In Galileo's story the "neutral" but intelligent moderator, Sagredo, is quick to see the truth in Salviati's arguments, and to dismiss

Simplicio's rebuttals as patently absurd. Galileo's prose is lively and entertaining, and he wrote the *Dialogo* in Italian rather than Latin so that it would be easily accessible to the person on the street. When Galileo published the *Dialogo,* he actually thought he had the tacit approval of the Vatican, which held to the Aristotelian view. However, when he placed a number of the Pope's own arguments into the unflattering mouth of Simplicio, he found that the Vatican's tolerance had limits. Fortunately for Galileo he had more friends in high places than Bruno did, and spent the closing years of his life under house arrest rather than the closing moments of his life tied to the stake. To escape harsher sentence, Galileo was forced to publicly recant the Copernican theory that he had supported with such fervor. In one of the great apocryphal stories of the history of astronomy, it is said that as he left the courtroom following his sentencing, he stamped his foot on the ground and muttered, "But it moves!"

The idea of inertia has come a long way since the days of Galileo and Newton. In fact, we have already seen a number of very sophisticated applications of this idea. What Galileo and Newton called inertia can actually be viewed as a consequence of what we discovered in our previous discussions about relative motions and frames of reference. To say that only *relative* motions between objects have meaning is the same as saying that there is *no difference* between an object at rest and an object in uniform motion. What objects are at rest and what objects are in motion, anyway? The object at rest beside you on the front seat of your car as you drive down the highway is moving at 60 miles per hour according to the bystander along the side of the road, and is moving at 120 mph according to the car in oncoming traffic. All of these perspectives are equally valid.

The connection between inertia and the relative nature of motion is so fundamental that a reference frame moving in a straight line at a constant speed is referred to as an **inertial frame of reference.** Motion is only meaningful when measured relative to an inertial frame of reference, and *any* inertial frame of reference is as good as any other. The realization that the laws of physics are the same in *any* inertial frame of reference is one of the deepest in-

There is a deep connection between inertia and the idea of relative motions.

3.4
N I

sights ever made into the nature of the Universe. When thought of in this way, *of course* an object moving in a straight line at a constant speed remains in motion. As illustrated in **Figure 3.6,** in the frame of reference of that object, *it is already at rest.*

NEWTON'S SECOND LAW: MOTION IS CHANGED BY UNBALANCED FORCES

Newton often gets credit for Galileo's insight about inertia because it was Newton who took the crucial next step. If Newton's first law says that in the absence

Figure 3.6 *An object moving in a straight line at a constant speed is at rest in its own inertial reference frame.*

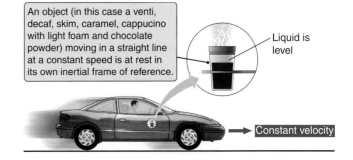

An object (in this case a venti, decaf, skim, caramel, cappucino with light foam and chocolate powder) moving in a straight line at a constant speed is at rest in its own inertial frame of reference.

Liquid is level

Constant velocity

Unbalanced forces cause changes in motion.

of an unbalanced force an object's motion does not change, then **Newton's second law of motion** says that *if there is an unbalanced force acting on an object, then the object's motion does change.* Even more, Newton's second law tells us *how* the object's motion changes in response to that force.

Before going any further, it would be wise to pause to be sure we are all together on this. In the preceding paragraphs we spoke of changes in an object's motion, but what does that phrase really mean? When you are in the driver's seat of a car, a number of controls are at your disposal. On the floor of the car are a gas pedal and a brake. You use these to make the car speed up or slow down. A *change in speed* is one way the motion of an object can change. But remember your hands on the steering wheel: when you are moving down the road and you turn the wheel, your speed does not necessarily change, but the direction of your motion does. A *change in direction* is also a kind of change in motion.

Together, the speed and direction of an object's motion are called the object's **velocity.** A change in velocity is called an **acceleration.** Acceleration actually refers to how rapidly the change in velocity happens. If you go from 0 to 60 mph in 4 seconds, you feel the back of your seat shoving your body forward, causing

you to accelerate along with the car. If you take 2 minutes to get from 0 to 60 mph, on the other hand, the acceleration is so slight that you hardly notice it. To formalize this a bit, your acceleration is given by how much your velocity changes divided by how long it takes that change to happen, or

Acceleration measures how quickly a change in motion takes place.

$$\text{Acceleration} = \frac{\text{how much velocity changes}}{\text{how long the change takes}}.$$

For example, if an object's speed goes from 5 m/s (meters per second) to 15 m/s, then the change in velocity is 10 m/s. If that change happens over the course of 2 seconds, then the acceleration is 10 m/s divided by 2 seconds, which equals 5 meters per second per second. (This is the same as saying "5 meters per second squared," which is written 5 m/s^2, or 5 m s^{-2}.)

Because the gas pedal on a car is often called the accelerator, some people think acceleration means that an object is speeding up. But we need to stress that *any* change in motion is an acceleration. **Figure 3.7** illustrates the point. Slamming on your brakes and going from 60 to 0 mph in 4 seconds is just as much acceleration as going from 0 to 60 mph in 4

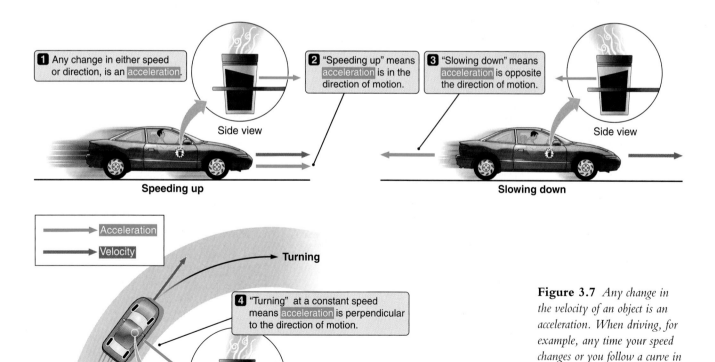

1 Any change in either speed or direction, is an acceleration.

Side view

2 "Speeding up" means acceleration is in the direction of motion.

3 "Slowing down" means acceleration is opposite the direction of motion.

Side view

Speeding up

Slowing down

→ Acceleration
→ Velocity

Turning

4 "Turning" at a constant speed means acceleration is perpendicular to the direction of motion.

Rear view

Figure 3.7 *Any change in the velocity of an object is an acceleration. When driving, for example, any time your speed changes or you follow a curve in the road, you are experiencing an acceleration. (Throughout the text velocity arrows will be shown as red and acceleration arrows will be shown as green.)*

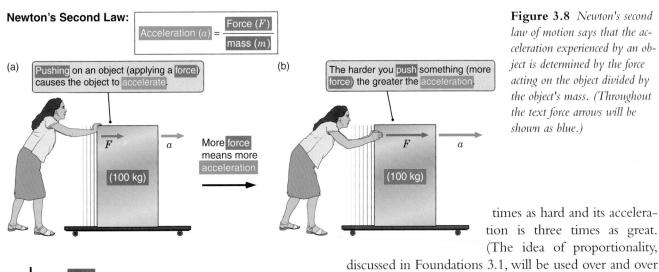

Figure 3.8 *Newton's second law of motion says that the acceleration experienced by an object is determined by the force acting on the object divided by the object's mass. (Throughout the text force arrows will be shown as blue.)*

times as hard and its acceleration is three times as great. (The idea of proportionality, discussed in Foundations 3.1, will be used over and over again throughout our journey.) The resulting change in motion occurs in the direction in which the unbalanced force is imposed. Push something forward and it speeds up. Push it to the left and it veers in that direction.

The acceleration that an object experiences also depends on the degree to which the object resists changes in motion (Figure 3.8c). Some objects—say a baseball—are easily shoved around by humans. A baseball is thrown at great velocity by the pitcher, only to be hit with a bat and have its motion abruptly changed again. The hard-hit line drive comes suddenly to a stop in the glove of the second baseman. A baseball resists changes in its motion—that is, it has inertia—but not *too* much inertia. Other objects are less obliging. A piece of solid lead the size of a baseball would make a poor substitute in the game. A pitcher would be *very* hard pressed to throw such a lead ball hard enough to get it over home plate, and if he did, being catcher would become an even more dangerous job. A baseball and a ball of lead may be the same size, but they are quite different in the degree to which they resist changes in their motion. The property of an object that determines its resistance to changes in motion—the measure of an object's inertia, if you will—is referred to as the object's **mass.**

> **Mass is the property of matter that resists changes in motion.**

You probably knew this answer intuitively. The lead ball is "heftier" and therefore harder to throw than a baseball. However, you may not have thought much about what we really *mean* when we say that a ball of lead "has more mass"—or "is more massive"—than a baseball. You might say that the ball of lead is made up of "more stuff" than the baseball, but again, what is meant by "more stuff?" If you grapple with this question for a time, you may find yourself chasing the question around in circles. When it comes right down

seconds. Similarly, the acceleration you experience as you go through a fast, tight turn at a constant speed is every bit as real as the acceleration you feel when you slam your foot on the gas pedal. Faster, slower, turn left, turn right—*if you are not moving in a straight line at a constant speed, then you are experiencing an acceleration.*

Newton's second law of motion says that changes in motion—accelerations—are caused by unbalanced forces. The acceleration that an object experiences depends on two things, as shown in **Figure 3.8.** First, it depends on the unbalanced force that is acting on the object to change its motion. This is a pretty common-sense idea. The stronger the force, the greater the acceleration. In fact, the acceleration of an object is *proportional* to the unbalanced force applied. Push on something twice as hard (Figure 3.8b), and it experiences twice as much acceleration. Push on something three

> **Greater force means greater acceleration.**

FOUNDATIONS 3.1

PROPORTIONALITY

Often in this text we will say that one quantity is *proportional* to another. Proportionality is a way of getting the gist of how something works—understanding the relationships between things—without having to actually calculate the details of one case after another.

Proportionality

If two quantities are **proportional** to each other, then making one of them larger means making the other quantity larger as well. For example, think about the weight of a bag of apples and how much the bag costs. Double the weight of the bag of apples and you double the cost. Increase the weight of the bag of apples by a factor of 5, and the cost goes up by a factor of 5 as well. *The cost of a bag of apples is proportional to the weight of the bag of apples.* We write this relationship as

$$\text{Cost} \propto \text{weight},$$

where the symbol $\propto$ means "is proportional to." This expression captures the essence of the relationship between the cost and the weight of apples. It tells us that the more apples we buy, the more they will cost us.

Constants of Proportionality

Sometimes it is enough to know that two quantities are proportional to each other, but sometimes it is not. What if you need to know how much money one of those bags of apples will actually set you back? We know that the full relationship between the cost and weight of a bag of apples is that the cost is equal to the price per pound of apples times the weight of the bag. We write

$$\text{Cost} = \text{price per pound} \times \text{weight}.$$

Compare this expression with the previous one. When we say that two quantities are proportional to

each other, what we mean is that one quantity equals some number *times* the other quantity. The number by which one quantity is multiplied to get the other number is called the **constant of proportionality.** In our example, the constant of proportionality is just the price per pound of apples.

Look at the difference between the two expressions. The fact that the cost of a bag of apples is proportional to the weight of the apples is a statement about the *relationship* between things. It is a statement about how things "work." More apples do not cost *less* than fewer apples. More apples cost *more* than fewer apples. The constant of proportionality—here the price per pound of apples—means something very different. Hidden within the price per pound of apples is a great deal of information, such as the cost of growing apples, the cost of transporting them from the orchard, and the profit margin the grocer needs to make to stay in business. The constant of proportionality carries with it information about how this aspect of the world *is*.

Very often, physical laws work in this same fashion. Proportionalities tell us about *relationships,* how two things vary with one another. They let us get a feeling for the "how" in how something works. In this chapter, for example, we find that gravitational force is proportional to an object's mass. Constants of proportionality more precisely tell us about the way the Universe is. The universal gravitational constant G is a constant of proportionality that tells us about the intrinsic strength of gravitational interactions and allows calculation of the numerical value of this force. Constants of proportionality are needed if we are to turn an understanding of relationships into hard numbers.

More mass means less acceleration.

to it, *the property of matter that we refer to as mass is nothing more and nothing less than the degree to which an object resists changes in its motion.* (Mass is measured in units of kilograms. An object with a mass of 2 kg is twice as hard to push around as an object with a mass of 1 kg. An object with a mass of 9 kg is three times as hard to push around as an object with a mass of 3 kg.)

So, if we want to know how an object's motion is changing, we need to know two things: what force is acting on the object, and what is the resistance of the object to that force? We can put this into equation form as follows:

$$\frac{\text{The acceleration experienced by an object}} {} = \frac{\text{the force acting to change the object's motion}}{\text{the object's resistance to that change}} = \frac{\text{force}}{\text{mass}}$$

Instead of spelling it all out in words every time, we can introduce a convenient bit of shorthand—*a* for acceleration, *F* for force, and *m* for mass—and we get

$$a = \frac{F}{m}.$$

Acceleration is force divided by mass.

This is the succinct mathematical statement of Newton's second law of motion.[4] If you are comfortable with mathematics, this elegant expression may speak to you clearly and directly. If not, when you see this equation, remind yourself that Newton's second law is nothing more than the embodiment of three commonsense ideas: (1) when you push on an object, that object accelerates in the direction in which you are pushing; (2) the harder you push on an object, the more the object accelerates; and (3) the more massive the object is, the harder it is to change its state of motion.

NEWTON'S THIRD LAW: WHATEVER IS PUSHED, PUSHES BACK

Imagine you are a child again, sitting in a wagon or standing on a skateboard, and pushing yourself along with your foot. Each shove of your foot against the ground sends you faster along your way. But why does this happen? Your muscles flex and your foot exerts a force on the ground, it is true. (Earth does not respond much to that force because its great mass gives it great inertia.) Yet this does not explain why *you* experience an acceleration. The fact that you accelerate at the same time means that, as you push on the ground, *the ground must be pushing back on you.*

Part of Newton's great genius was his ability to see sublime patterns in such mundane events. Newton realized that *every* time one object exerts a force on another, a matching force is exerted by the second object on the first. That second force is exactly as strong as the first force, but is in exactly the *opposite* direction. The child pushes back on Earth, and Earth pushes the child forward. A canoeist's paddle pushes backward through the water, and the water pushes forward on the paddle, sending the canoe along its

For every force there is an equal and opposite force.

[4]Newton's second law is often written as *F = ma,* giving force as units of mass times units of acceleration, or kg m/s². These units are aptly named "newtons," abbreviated N.

Newton's Third Law:

For every force ...
...there is an equal and opposite force.

Woman pulls on rope
Rope pulls on cage
Rope pulls on woman
Cage pulls on rope
Floor pushes on woman
Woman pushes on floor

Both forces are equal, but the more massive bat experiences less acceleration.
Bat pushes on baseball
Baseball pushes on bat

Acceleration
The unbalanced force of the road on the tire causes the car to accelerate.
Road pushes forward against tire
Tire pushes backwards against road

The upward push of the floor balances the downward pull of gravity. There are no unbalanced forces so there is no acceleration.
Earth's gravity pulls down on Man
Floor pushes up on Man
Man's gravity pulls up on Earth
Man pushes down on floor

Moon pulls on Earth
Earth pulls on Moon
F
a

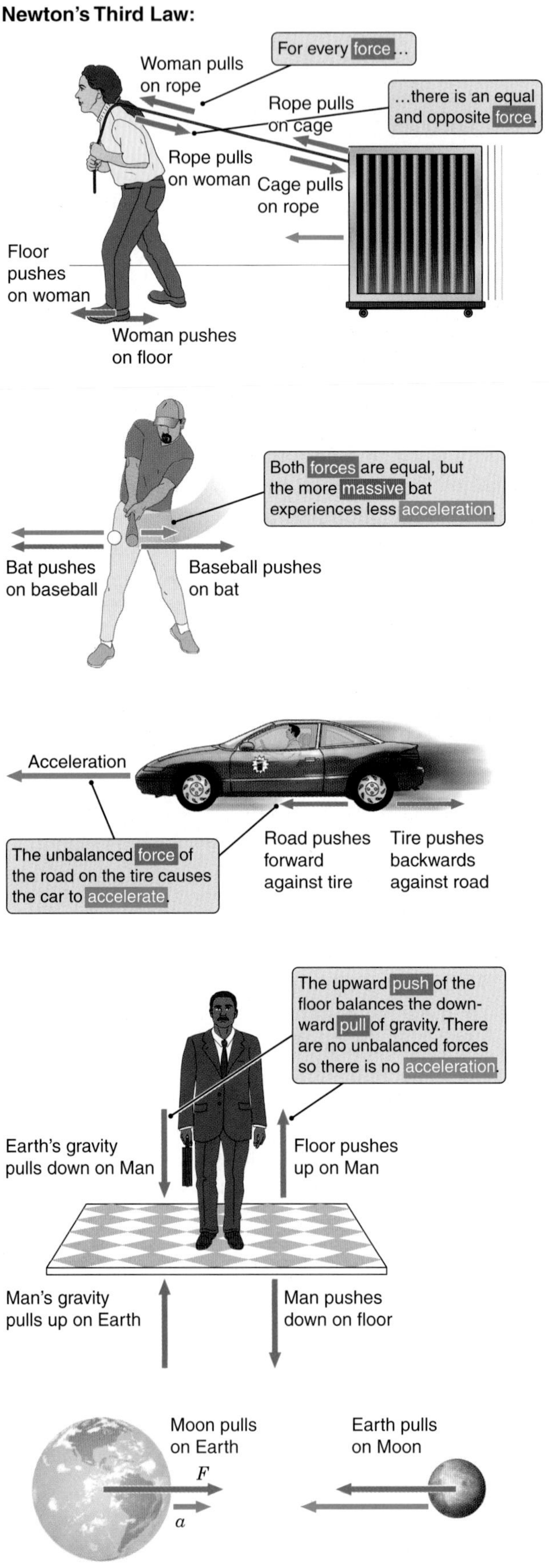

Figure 3.9 *Newton's third law states that for every force there is always an equal and opposite force. These opposing forces always act on the two different objects in the same pair.*

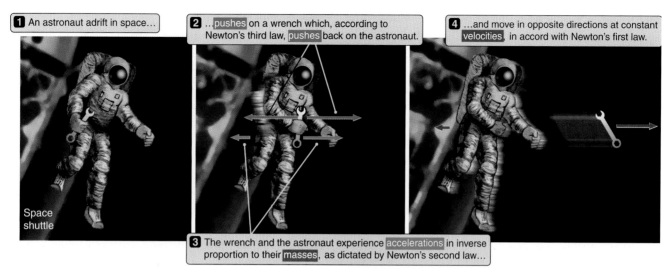

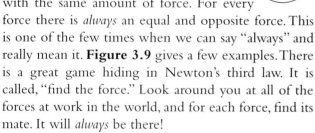

Figure 3.10 *According to Newton's laws, if an astronaut throws a wrench the two will move in opposite directions at speeds that are inversely proportional to their masses. (Acceleration and velocity arrows are not drawn to scale.)*

way. A rocket engine pushes hot gases out of its nozzle, and those hot gases push back on the rocket, propelling it into space.

All of these are examples of **Newton's third law of motion,** which says that *forces always come in pairs, and those pairs are always equal in magnitude but opposite in direction.* The forces in these action/reaction pairs always act on two different objects. Your weight pushes down on the floor, and the floor pushes back on you with the same amount of force. For every force there is *always* an equal and opposite force. This is one of the few times when we can say "always" and really mean it. **Figure 3.9** gives a few examples. There is a great game hiding in Newton's third law. It is called, "find the force." Look around you at all of the forces at work in the world, and for each force, find its mate. It will *always* be there!

To see how Newton's three laws of motion work together, think about the following situation, shown in **Figure 3.10.** An astronaut is adrift in space, motionless with respect to the nearby space shuttle. With no tether to pull on, how can the astronaut get back to his ship? The answer? Throw something. Suppose the 100 kg astronaut throws a 1 kg wrench directly away from the shuttle at a speed of 10 m/s. Newton's second law says that in order to cause the motion of the wrench to change, the astronaut has to apply a force to it in the direction away from the shuttle. Newton's third law says that the wrench must therefore push back on the astronaut with as much force but in the opposite direction. The force of the wrench on the astronaut causes the astronaut to begin drifting toward the shuttle. How fast will the astronaut move? Turn to

Newton's second law again. A force that causes the 1 kg wrench to accelerate to 10 m/s will not have much effect on the 100 kg astronaut. Since acceleration equals force divided by mass, the 100 kg astronaut will experience only 1/100 as much acceleration as the 1 kg wrench. The astronaut will drift toward the shuttle at the leisurely rate of 1/100 × 10 m/s, or 0.1 m/s.

3.4 GRAVITY IS A FORCE BETWEEN ANY TWO OBJECTS DUE TO THEIR MASSES

Drop a ball and the ball falls toward the ground, picking up speed as it goes. It accelerates toward Earth. Newton's second law says that where there is acceleration, there is force. But *where* is the force that causes the ball to accelerate? Many

> Gravity is "force at a distance."

of the forces that we see in everyday life involve "direct contact" between objects.[5] The cue ball slams into the eight ball, knocking it into the pocket. The shoe of the child in the wagon shoves directly onto the surface of the pavement. In cases where there is physical con-

---

[5]Actually the "direct contact" between billiard balls or other "solid" objects is also force at a distance—electric force acting at a distance between the electrons and protons in the atoms of which the objects are made.

tact between two objects, the source of the forces between them is easy to see. But the ball falling toward Earth is an example of a different kind of force, one that acts at a distance across the intervening void of space. The ball falling toward Earth is accelerating in response to the force of **gravity.** We began this chapter with a qualitative discussion of the fundamental role that gravity plays in the Universe. Having explored both Kepler's empirical description of the motions of planets about the Sun and Newton's theory of motion, it is time to return to gravity, for it is gravity that unites these two pillars of empirical and theoretical science.

You probably will not be surprised to learn that once again it is Newton we turn to for a law of gravitation. It is customary at this point in an introductory textbook simply to present Newton's law of gravitation as a "done deal" and go straight to its application, but such a leap misses one of the most interesting aspects of this stretch of our journey. A common misconception about the way science works is the notion that new theories just spring fully formed into the mind of a scientist, as if by magic. This idea is certainly supported by the grade school story of the apple falling on Newton's head, literally knocking the idea of gravity into his brain. One could almost get the idea that scientific theories are arbitrary—that Newton could have invented some *other* law of gravity that would have worked just as well. (The idea that scientific knowledge is a cultural construct is discussed in Excursions 3.2.) While it might seem at first glance that his work was arbitrary, nothing could be further from the truth. Where did

The story of gravity illustrates how science is done.

Newton's ideas of gravity come from? What guided him in his development of those ideas, and how did he turn them into a theory with *testable predictions?* How did he confront that theory in the crucible of experiment and observation? It is by answering these questions, not by making a simple statement of Newton's law of gravitation, that we will gain some insight into what science is.

WHERE DO THEORIES COME FROM? NEWTON REASONS HIS WAY TO A LAW OF GRAVITY

As with inertia, the story of gravity begins with the insight and observation of Galileo. Galileo discovered that all freely falling objects accelerate toward Earth at

the same rate, regardless of their mass. Drop a marble and a cannonball, at the same time and from the

All objects on Earth fall with the same acceleration, g.

same height, and they will hit the ground together. The gravitational acceleration near the surface of Earth is usually written as g, and has a value of 9.8 m/s^2. Whether you drop a marble or a cannonball, after 1 second it will be falling at a speed of 9.8 m/s, after 2 seconds at 19.6 m/s, and after 3 seconds at 29.4 m/s. (These numbers assume that we can neglect air resistance, which is reasonably negligible for relatively dense objects.)

Having worked out the laws governing the motion of objects, Newton saw something deeper in Galileo's findings. Newton realized that if all objects fall with the same acceleration, then the gravitational *force* on an object must be determined by the object's *mass.* To see why, look back at Newton's second law (acceleration equals force divided by mass). The only way for gravitational acceleration to be the same for all objects is if the value of the force divided by the mass is the same for all objects. A greater mass *must,* therefore, be accompanied by a stronger gravitational force. In other words, the gravitational force on an

Greater mass means greater gravitational force.

object on Earth is, according to Newton's second law, the object's mass times the acceleration due to gravity, or $F_{grav} = mg$. Make an object twice as massive, and you double the gravitational force acting on it. Make an object three times as massive, and you triple the gravitational force acting on it.

The gravitational force acting on an object is commonly referred to as the object's **weight.** It is easy to see why people often confuse mass and weight. On the surface of Earth, weight is just mass times the constant g. The situation is not helped by the sloppy way we use language. We often say that an object with a mass of 2 kg "weighs 2 kg," but it is more correct to express a weight in terms of newtons (N): thus an object with a *mass* of 2 kg has a *weight* of 2 kg × 9.8 m/s^2 or 19.6 N.

Newton's next great insight came from applying his third law of motion to gravity. For every force there is an equal and opposite force. If Earth exerts a force of 19.6 newtons on a 2 kg mass sitting on its surface,

Like all forces, gravity is a two-way force.

then that 2 kg mass must exert a force of 19.6 newtons on Earth as well. Drop a 20 kg cannonball, and it falls toward Earth, but at the same time Earth falls toward the 20 kg cannonball! The reason we do not notice the motion of Earth is because Earth is very massive. It has a lot of resistance to a change in its motion. In the time it takes a 20 kg cannonball to fall to the ground from a height of

EXCURSIONS 3.2

SCIENCE AND CULTURE

In recent years, some critics of science have drawn attention to how science is influenced by culture. It is hard to avoid the conclusion that political and cultural considerations strongly influence which scientific research projects are funded. This choice of funding channels the directions in which scientific knowledge advances, and can lead to serious ethical issues. For example, moral judgments about nontraditional lifestyles greatly restricted the funding available for AIDS research during the decade or so after its discovery.

Some critics even carry this a step further, arguing that scientific knowledge itself is an arbitrary cultural construct. Yet, as illustrated by the discussion of Newton's law of gravity, successful scientific theories are *never* arbitrary. Scientific theories must be consistent with all that we know of how nature works, and turning a clever idea into a real theory with testable predictions is a matter of careful thought and effort. One of the most remarkable aspects of scientific knowledge is its *independence* from culture. Scientists are people, and politics and culture enter into the day-to-day practice of science. But in the end, *scientific theories are judged not by cultural norms, but by whether their predictions are borne out by observation and experiment.* As long as the results of experiments are repeatable and do not depend on the culture of the experimenter— that is, as long as there is such a thing as objective physical reality—scientific knowledge cannot be called a cultural construct.

Nor does it seem that the path to knowledge embodied in science is any more arbitrary than the logic it is built on. It is significant that no philosopher critical of science has ever offered a viable alternative for obtaining reliable knowledge of the workings of nature. Had science not arisen when and where it did, something very much like it would likely have arisen at some time and in some location. Furthermore, no other category of human knowledge is subject to standards as rigorous and unforgiving as those of science. For this reason, scientific knowledge is reliable in a way that no other form of knowledge can claim. Whether you want to design a building that will not fall over, or decide on the best treatment for a disease, or calculate the orbit of a spacecraft on its way to the Moon, you had better consult a scientist rather than a psychic—regardless of your cultural heritage.

---

1 km, Earth has "fallen" toward the cannonball by about 3.4×10^{-21} meters, which is only about 1/300 of the diameter of a hydrogen atom!

Newton reasoned that this should work both ways. If doubling the mass of an object doubles the gravitational force between the object and Earth, then doubling the mass of Earth ought to do the same. In short, the gravitational force between Earth and an object must be equal to the product of the two masses times something, or

$$\text{Gravitational force} = \text{something} \times \text{mass of Earth} \times \text{mass of object.}$$

If the mass of the object is three times greater, then the force of gravity will be three times greater. Likewise, if the mass of Earth were three times what it is, the force of gravity would have to be three times greater as well. If *both* the mass of Earth *and* the mass of the object were three times greater, the gravitational force would increase by a factor of 3×3, or nine times. Since objects fall toward the center of Earth, we know that this force is an attractive force acting along a line between the two masses.

"And by the way," reasoned Newton, "why are we restricting our attention to Earth's gravity?" If gravity is a force that depends on mass, then there should be a gravitational force between *any* two masses. Say we have two masses—call them mass 1 and mass 2, or m_1 and m_2 for short—then the gravitational force between them is something times the product of the masses, or

The force of gravity depends on the product of two masses.

$$\text{Gravitational force between two objects} = \text{something} \times m_1 \times m_2.$$

Realize that we have gotten this far just by combining Galileo's observations of falling objects with Newton's laws of motion, and Newton's belief that Earth is a mass just like any other mass. There has been no wiggle room—there is nothing arbitrary in what we have done. But what about that "something" in the expression above? Today we have sensitive enough instruments to allow us to put two masses close to each other in a laboratory, measure the force between them, and determine that something directly. Yet Newton had no such instruments. He had to look elsewhere to go further with his exploration of gravity.

It turns out that Kepler had already thought about this question. He reasoned that since the Sun is the focal point for planetary orbits, the Sun must be responsible for exerting an influence over the motions of the planets. Kepler speculated that whatever this influence is, it must grow weaker with distance from the Sun. (After all, it must surely require a stronger influence to keep Mercury whipping around in its tight, fast orbit than it does to keep the outer planets lumbering along on their paths around the Sun.) Kepler's speculation went even further. While he did not know about forces or inertia or gravity, he did know quite a lot about geometry, and geometry alone suggested how this solar "influence" might change for planets progressively farther from the Sun.

Imagine you have a certain amount of plaster to spread over the surface of a spherical ball. If the sphere is small, then when you spread the plaster over the sphere you get a thick coat. But if the sphere is larger, the plaster has to spread farther, and so the coat you get is thinner. The surface area of a sphere depends on the square of the sphere's radius. Double the radius of a sphere, and the sphere's surface becomes four times what it was. If you plaster this new, larger sphere, the plaster must cover four times the area, and so the thickness of the plaster will only be a fourth of what it was on the smaller sphere. Triple the radius of the sphere, and the sphere's surface is nine times as large, and the thickness of the coat of plaster will be only a ninth as thick.

3.7 GRAVITY I

Kepler thought that the influence that the Sun exerts over the planets might be like the plaster in this example. As the influence of the Sun extends farther and farther into space, it would have to spread out to cover the surface of a larger and larger imaginary sphere centered on the Sun. (We will learn later that light works in exactly this way.) If so, then, like the thickness of the plaster, the influence of the Sun should be proportional to 1 divided by the square of the distance between the Sun and a planet. Double the distance between the Sun and a planet, and this influence declines by a factor of $2 \times 2 = 4$, to $\frac{1}{4}$ of its original strength. Triple the distance, and this influence declines by a factor of $3^2 = 9$, becoming $\frac{1}{9}$ of its initial strength. When something (like the thickness of the plaster or the strength of Kepler's solar "influence") changes in proportion to 1 divided by the square of the distance, we refer to it as an **inverse square law** (see Connections 3.1).

> **Gravity is an inverse square law.**

Kepler had an interesting idea, but not a scientific theory with testable predictions. What Kepler lacked was a good idea of the true source of this influence, or the mathematical tools to calculate how an object would move under such an influence. Newton had both. If gravity is a force between *any* two objects, then there should be a gravitational force between the Sun and each of the planets. Might this gravitational force be the same as Kepler's "influence"? If so, then the something in Newton's expression for gravity might be a term that diminishes according to the square of the distance between two objects. Gravity might be an inverse square law. Newton's expression for gravity now came to look like this:

Gravitational force between two objects

$$= \text{something} \times \frac{m_1 \times m_2}{(\text{distance between objects})^2}.$$

CONNECTIONS 3.1

INVERSE SQUARE LAWS

In this chapter we discover that gravity obeys what is called an *inverse square law*. This means that the force of gravity is proportional to 1 divided by the square of the distance between two objects, or

$$F_{\text{grav}} \propto \frac{1}{r^2}.$$

If two objects are moved so that they are twice as far apart as they were originally, the force of gravity between them becomes only $\frac{1}{4}$ of what it was. If two objects are taken three times as far apart, the force of gravity drops to $\frac{1}{9}$ of its original value.

Gravity is only one of several inverse square laws found in nature. The other important one that we will deal with in this book involves radiation. The intensity of radiation from an object is also proportional to 1 over the square of the distance between the objects. Our discussion of radiation in Chapter 4 will present a clear picture of *why* an inverse square law applies in that case.

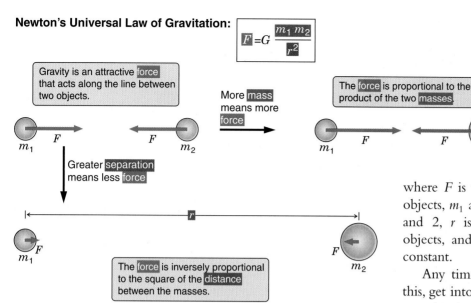

Newton's Universal Law of Gravitation:

$$F = G\,\frac{m_1\,m_2}{r^2}$$

Gravity is an attractive force that acts along the line between two objects.

More mass means more force

The force is proportional to the product of the two masses.

Greater separation means less force

r

The force is inversely proportional to the square of the distance between the masses.

Figure 3.11 *Gravity is an attractive force between two objects. The force of gravity depends on the masses of the objects and the distance between them.*

There is still a "something" left in this expression. Newton guessed that it was a measure of the intrinsic strength of gravitation interactions, and would turn out to be the same for all objects. He named this something the **universal gravitational constant.** This quantity is now written as G.

PUTTING THE PIECES TOGETHER: A UNIVERSAL LAW FOR GRAVITATION

Newton had good reasons every step of the way in his thinking about gravity—reasons directly tied to observations of how things in the world behave. Newton's chain of logic and reason brought him to what has come to be known as Newton's **universal law of gravitation.** This law, illustrated in **Figure 3.11,** states that gravity is a force between any two objects and has these properties:

1. It is an attractive force acting along a straight line between the two objects.
2. It is proportional to the mass of one object times the mass of the other object: $F_{grav} \propto m_1 m_2$.
3. And it decreases in proportion to 1 divided by the square of the distance between the two objects:

$$F_{grav} \propto \frac{1}{r^2}.$$

Written as a mathematical formula, the universal law of gravitation states that

$$F_{grav} = G \times \frac{m_1 \times m_2}{r^2},$$

where F is the force of gravity between two objects, m_1 and m_2 are the masses of objects 1 and 2, r is the distance between the two objects, and G is the universal gravitational constant.

Any time you run across a statement like this, get into the habit of pulling it apart to be sure that it makes sense. First, gravity is an attractive force between two masses that acts along the straight line between the two masses.[6] Regardless of where you stand on the surface of Earth, Earth's gravity pulls you *toward the center of Earth.* Second, the force of gravity depends on the product of the two masses. If you make m_1 twice as large, then the gravitational force between m_1 and m_2 becomes twice as large. Again, this should make sense. Doubling the mass of an object also doubles its weight. A subtle—but in retrospect very important—point lurks in this statement. The mass that appears in the universal law of gravitation is the *same* mass that appears in Newton's laws of motion. *The same property of an object that gives it inertia is the property of the object that makes it interact gravitationally.* (This equivalence between the effect of gravitation and the effect of inertia later became the basis for Einstein's general theory of relativity, in which mass literally warps space and time. We will return to this idea later in our journey.)

Inertial mass is also gravitational mass.

The third part of Newton's universal law of gravitation tells us that the force of gravity is inversely proportional to the *square* of the distance between two objects. Doubling the distance between two objects reduces the strength of gravity to $\frac{1}{2^2}$, or $\frac{1}{4}$, of its original value. Tripling the distance between two objects reduces the strength of gravity to $\frac{1}{3^2}$, or $\frac{1}{9}$, of its original value (see **Figure 3.12**). Gravity is only one of several laws that we will see in which the strength of some effect diminishes in proportion to the square of the distance.

---

[6]This is not obvious from the equation alone. More properly, this equation should be written using *vector notation,* which would include information about the direction of the force.

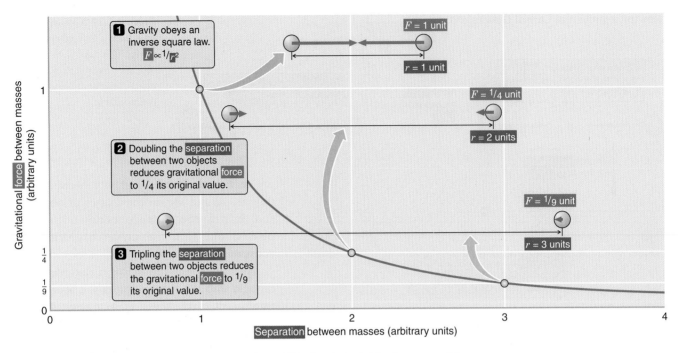

Figure 3.12 *The gravitational force between two objects falls off as 1 divided by the square of the distance between them.*

PLAYING WITH NEWTON'S LAWS OF MOTION AND GRAVITATION

If you have ever watched a child play with blocks, you have probably noticed how the child will put the blocks together in different ways, seeing what she can build. Perhaps if you are an artist, you play with colors and patterns of light and dark as you create new works. If you are a musician, you might play with tones and rhythms as you compose. Writers play with combinations of words. All of these uses of the term *play* mean the same thing. Play is very serious business, because it is by playing that we explore the world. And in exactly the same sense, scientists often play with the equations describing natural laws as they seek new insights into the world around them. We can play a bit with Newton's laws of gravitation and motion and see what interesting things turn up. (If you have difficulty following the math here, do not worry too much. Pay attention instead to the sense of what we are doing.)

The universal gravitational constant *G* has a value of 6.67×10^{-11} Nm²/kg². This makes gravity a *very* weak force. The gravitational force between two 16-pound (7.26 kg) bowling balls sitting 0.3 m (about a foot) apart is only

Play is serious business.

$$F_{grav} = 6.67 \times 10^{-11} \frac{Nm^2}{kg^2}$$
$$\times \frac{7.26 \text{ kg} \times 7.26 \text{ kg}}{(0.3 \text{ m})^2}$$
$$= 3.9 \times 10^{-8} \text{ N},$$

or 0.000000039 newtons. This is about equal to the weight on Earth of a single bacterium! Gravity is such an important force in our everyday lives only because Earth is so very massive.

There are two different ways that we can think about the gravitational force that Earth exerts on an object with mass *m*. The first is to look at gravitational force from the perspective of Newton's second law of motion: gravitational force equals mass times gravitational acceleration, or

$$F_{grav} = mg.$$

The other way to think about the force is from the perspective of the universal law of gravitation, which says that

$$F_{grav} = G \frac{M_\oplus m}{(R_\oplus)^2}.$$

(Here $M_\oplus$ is the mass of Earth and $R_\oplus$ the radius of Earth.) Since the force of gravity on an object is what it is, the two expressions describing this force must be equal to each other. $F_{grav} = F_{grav}$, so

$$mg = G\frac{M_\oplus m}{(R_\oplus)^2}.$$

The mass m is there on both sides of the equation, so we can divide it out. The equation then becomes

$$g = G\frac{M_\oplus}{(R_\oplus)^2}.$$

This is interesting. We started out to calculate the gravitational acceleration experienced by an object of mass m on the surface of Earth. The expression that we arrived at says that this acceleration (g) is determined by the mass of Earth ($M_\oplus$) and by the radius of Earth ($R_\oplus$). But the mass of the object itself (m) appears nowhere in this expression. So, according to this equation, changing m has no effect on the gravitational acceleration experienced by an object on Earth. In other words, our play with Newton's laws has shown us that all objects experience the same gravitational acceleration, regardless of their mass. *This is just what Galileo found in his experiments with falling objects!* We already saw that Galileo's work *shaped* Newton's thinking about gravity. Here we find that Galileo's discoveries about gravity are *contained within* Newton's laws of motion and gravitation.

Galileo's discoveries are contained within Newton's laws.

What else can we discover? If we rearrange that last equation a bit so that the mass of Earth is on the left and everything else is on the right, we get

$$M_\oplus = \frac{g(R_\oplus)^2}{G}.$$

Everything on the right side of this equation is known. Galileo measured a value for g, the acceleration due to gravity on the surface of Earth, almost 400 years ago, and in about 235 B.C. Eratosthenes measured the radius of Earth in the manner described in Chapter 2. The universal gravitational constant G is a bit tougher, but it too can be measured in the laboratory, for example, by measuring the slight gravitational forces between two large metal spheres. With everything on the right side now known, we can calculate the mass of Earth:

We can calculate the mass of Earth.

$$M_\oplus = \frac{gR_\oplus^2}{G}.$$

$$= \frac{\left(9.80\,\frac{m}{s^2}\right) \times (6.38 \times 10^6\,m)^2}{6.67 \times 10^{-11}\,\frac{m^3}{kg\,s^2}}$$

$$= 5.98 \times 10^{24}\,kg$$

You may have wondered how we know the mass of Earth. (After all, we cannot just pick up a planet and set it on a bathroom scale.) Now you know. By playing with theories and equations, much as a child plays with building blocks, scientists discover new relationships between things in the Universe. And from those new relationships come new knowledge.[7]

Playing with relationships leads to new knowledge.

3.5 ORBITS ARE ONE BODY "FALLING AROUND" ANOTHER

If you have been following our discussion closely, you may be about ready to take us to task. Kepler may have speculated about the dependence of the solar "influence" that holds the planets in their orbits, and Newton may have speculated that this influence is gravity, but physical law is *not* a matter of speculation! Newton could not measure the gravitational force between two objects in the laboratory directly, so how did he test his universal law of gravitation? Again it was Kepler who provided what Newton lacked. Newton used his laws of motion and his proposed law of gravity to *calculate* the paths that planets should follow as they move around the Sun. When he did so, his calculations predicted that planetary orbits should be ellipses with the Sun at one focus, that equal areas should be swept out during equal times, and that the square of the period of a planet's orbit should vary as the cube of the semimajor axis of that ellipse. In short, Newton's universal law of gravitation *predicted* that planets should orbit the Sun in just the way that Kepler's empirical laws described. This was the moment when it all came together. By *explaining* Kepler's laws, Newton found important corroboration for his law of gravitation. And in the process he moved the cosmological principle out of the realm of interesting ideas and into the realm of testable scientific theories. To get

Kepler's empirical laws confirm a prediction of Newton's law of gravity.

The cosmological principle became a testable theory.

[7]Newton actually turned this around. He *guessed* at the mass of Earth by assuming it had about the same density as typical rocks. Then he used this mass and the equation above to get a rough idea of the value of G.

a sense for how this happened, we need to look below the surface of how scientists go about connecting their theoretical ideas with events in the real world.

Newton's laws tell us about how an object's motion changes in response to forces and how objects interact with each other through gravity. To go from statements about how an object's motion is *changing* to more practical statements about where an object *is*, we have to carefully "add up" the object's motion over time. We must keep careful track of how the motion changes from one instant to the next, and then ask where that motion has gotten us.

Newton invented calculus to keep track of changing motions.

Doing this requires rather more sophisticated play than we have time for at the moment. Learning how to make the jump from laws of gravitation and motion to calculations of the paths of the planets about the Sun led Newton to become one of the coinventors of the branch of mathematics known as *calculus*. Fortunately we do not need calculus to build a conceptual understanding of such motions. Instead we can begin with a *thought experiment*—the same thought experiment that helped lead Newton to his understanding of planetary motions.

NEWTON FIRES A SHOT AROUND THE WORLD

Drop a cannonball and it falls directly to the ground, just as any mass does. However, if instead we fire the cannonball out of a cannon that is level with the ground, as shown in **Figure 3.13(a),** it behaves differently. The ball still falls to the ground in the same time as before, but while it is falling, it is also traveling *over* the ground, following a curved path that carries it some horizontal distance before it finally lands. The faster the cannonball is fired from the cannon **(Figure 3.13b),** the farther it will go before hitting the ground.

In the real world, the experiment reaches a natural limit. To travel through air, the cannonball must push the air out of its way—an effect we normally refer to as *air resistance*—which slows it down. Air resistance increases very rapidly with increasing speed. (For example, doubling the speed increases air resistance by much more than two times.) But because this is only a thought experiment, we can ignore such real-world complications as air resistance. Let us imagine instead that, having inertia, the cannonball continues along its course until it runs into something. As the cannonball is fired faster,

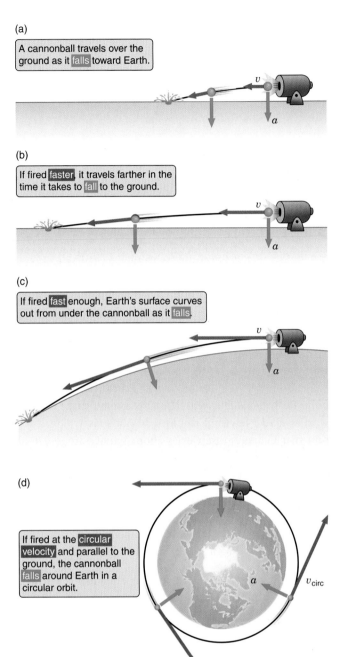

Figure 3.13 *Newton realized that a cannonball fired at the right speed would fall around Earth in a circle.*

and faster, it goes farther and farther before hitting the ground. If the cannonball flies far enough, the curvature of Earth starts to matter. As the cannonball falls toward Earth, Earth's surface "curves out from under it" **(Figure 3.13c).** Eventually we reach a point where the cannonball is flying so fast that the surface of Earth curves away from the cannonball at exactly the same rate as the rate at which the cannonball is falling toward Earth. This is the case shown in **Figure 3.13(d).** At this point, the cannonball, which always falls *toward the center of Earth,* is literally "falling around the world."

In 1957 the Soviet Union used a rocket to lift an object about the size of a basketball high enough above Earth's blanket of air that wind resistance ceased to be a concern, and Newton's thought experiment became a matter of great practical importance.[8] This object, called *Sputnik I,* was moving so fast that it fell around Earth, just as the cannonball did in Newton's mind. *Sputnik I* was the first human-made object to **orbit** Earth. This is what an orbit is: one object freely falling around another.

Orbiting **means falling around the world.**

In Section 3.1 we asked why it is that astronauts float freely about the cabin of a spacecraft. We now see that it is *not* because they have escaped Earth's gravity. It is Earth's gravity that holds them in their orbit. Instead the answer lies in Galileo's early observation that any object falls in just the same way, regardless of its mass. The astronauts and the spacecraft are both moving in the same direction, at the same speed, and are experiencing the same gravitational acceleration, *so they fall around Earth together.* **Figure 3.14** demonstrates the point. The astronaut is orbiting Earth just as the spacecraft is orbiting Earth. On the surface of Earth our bodies try to fall toward the center of Earth, but the ground gets in the way. We experience our weight when we are standing on Earth because the ground pushes on us hard enough to counteract the force of gravity, which is trying to pull us down. In the spacecraft, however, nothing interrupts the astronauts' fall because the spacecraft is falling around Earth on just the same orbit. The astronaut is not truly weightless. Instead, the astronaut is in **free fall.**

Astronauts and spacecraft fall around the world together.

When one object is falling around another, much more massive object, we say that the less massive object is a **satellite** of the more massive object. Planets are satellites of the Sun, and moons are natural satellites of planets. Newton's imaginary cannonball is a satellite. *Sputnik I,* the first artificial satellite (*sputnik* means "satellite" in Russian), was the early forerunner to the spacecraft and the astronauts, which are independent satellites of Earth that conveniently happen to be sharing the same orbit.

[8]Actually, wind resistance did not totally cease to be a concern. Objects in orbit within a few hundred kilometers of Earth are moving through the thin outer part of Earth's atmosphere. Friction with this thin atmosphere will oppose the object's motion and cause its orbit to *decay.*

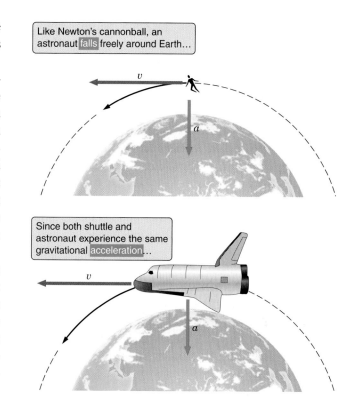

Like Newton's cannonball, an astronaut falls freely around Earth...

Since both shuttle and astronaut experience the same gravitational acceleration...

...both are independent satellites sharing the same orbit.

Figure 3.14 *A "weightless" astronaut has not escaped Earth's gravity. Rather, an astronaut and a spacecraft share the same orbit as they fall around Earth together.*

HOW FAST MUST NEWTON'S CANNONBALL FLY?

If fired fast enough, Newton's cannonball falls around the world, but just how fast is "fast enough"? Newton's orbiting cannonball moves along a circular path at constant speed. This type of motion, referred to as **uniform circular motion,** is discussed in more depth in Appendix 7. There are other examples of uniform

Centripetal forces maintain circular motion.

circular motion that you are probably familiar with. For example, think about a ball whirling about your head on a string, as shown in **Figure 3.15(a).** If you were to let go of the string, the ball would fly off in a straight line in whatever direction it was traveling at the time, just as Newton's first law says. It is the string that keeps this from happening. The string exerts a steady force on the ball, causing it constantly to change the direction of its motion, always bending its flight toward the center of the circle. This central force is called a **centripetal force.** Using a more massive ball, speeding up its motion, or making the circle smaller so the turn is tighter all increase the force needed to keep the ball from being carried off in a straight line by its inertia.

In the case of Newton's cannonball, there is no string to hold the ball in its circular motion. Instead, the centripetal force is provided by gravity, as illustrated in **Figure 3.15(b).** For Newton's thought experiment to work, the force of gravity must be just right to keep the cannonball moving on its circular path. In this case we can say that

> **Gravity provides the centripetal force that holds a satellite in its orbit.**

$$\text{The force needed for uniform circular motion} = \frac{\text{force provided}}{\text{by gravity}}.$$

In Appendix 7 we derive an expression for the centripetal force needed to keep an object moving in a circle at a steady speed. If we put that expression on the left side of the above equation and the universal law of gravitation on the right side (then do some algebra), we arrive at

$$v_{\text{circ}} = \sqrt{\frac{GM}{r}},$$

where M is the mass of the object being orbited and r is the radius of the circular orbit. This value is called the **circular velocity.**

Here is the result we were looking for. If a satellite is in a stable circular orbit, then it *must* be moving at a velocity v_{circ} where v_{circ} is given by the above expression. *If the satellite were moving at any other velocity, then it would not be moving in a circular orbit.* Remember the cannonball. If the cannonball were moving too slowly, it would drop below the circular path and hit the ground. Similarly, if the cannonball were moving too fast, its motion would carry it above the circular orbit. Only a cannonball moving at just the right velocity—the circular velocity—will fall around Earth on a circular path. The circular velocity at Earth's surface is about 8 km/s (see Foundations 3.2).

> **An object in a circular orbit must have a speed of V_{circ}.**

Figure 3.15 (a) *A string provides the centripetal force that keeps a ball moving in a circle. (We have ignored the smaller force of gravity that also acts on the ball.) (b) Similarly, gravity provides the centripetal force that holds a satellite on a circular orbit.*

(a)

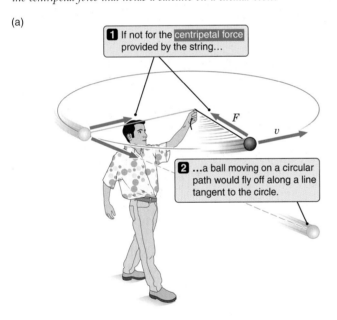

1 If not for the centripetal force provided by the string…

2 …a ball moving on a circular path would fly off along a line tangent to the circle.

(b)

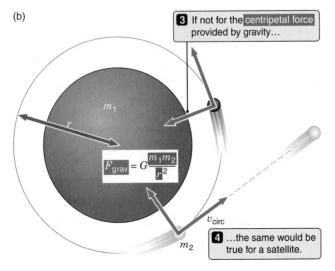

3 If not for the centripetal force provided by gravity…

$$F_{\text{grav}} = G\frac{m_1 m_2}{r^2}$$

v_{circ}

4 …the same would be true for a satellite.

PLANETS ARE JUST LIKE NEWTON'S CANNONBALL

We can take this same idea and apply it to the motion of Earth around the Sun. In the case of Earth's orbit, we already know from our discussion of the aberration of starlight that Earth travels at a speed of 2.98×10^4 m/s

> **We use Earth's orbit to calculate the Sun's mass.**

or (29.8 km/s) on its orbit about the Sun. We also know that the radius of Earth's orbit is 1.50×10^{11} m. So we know everything about the circular orbit except for the mass of the Sun. If we can skip a step or two, a little algebra applied to the equation for v_{circ} gives the mass of the Sun $(M_\odot)$ as

$$
\begin{aligned}
M_\odot &= \frac{(v_{circ})^2 \times r}{G} \\
&= \frac{\left(2.98 \times 10^4 \, \frac{\text{m}}{\text{s}}\right)^2 \times (1.50 \times 10^{11} \, \text{m})}{6.67 \times 10^{-11} \, \frac{\text{m}^3}{\text{kg s}^2}} \\
&= 1.99 \times 10^{30} \, \text{kg}.
\end{aligned}
$$

Once again a bit of play has taken us places we might not have imagined. We began with Newton's thought experiment about a cannonball fired around the world, and ended up knowing the mass of the star our planet orbits about.

As long as we are at it, we can carry our game one step further. Kepler's third law talks about the time for a planet to complete one orbit about the Sun, known as the period P of a planet's orbit. The time it takes an object to make one trip around a circle is just the circumference of the circle $(2\pi r)$ divided by the object's speed. (Time equals distance divided by speed.) If the object is a planet on a circular orbit about the Sun, then its speed must be equal to the circular velocity we calculated above. Bringing this together, we get

$$
\text{Period } P = \frac{\text{circumference of orbit}}{\text{circular velocity}} = \frac{2\pi r}{\sqrt{\dfrac{GM_\odot}{r}}}.
$$

Some algebra gives us

$$
P^2 = \frac{4\pi^2}{GM_\odot} \times r^3.
$$

Once again our play has really gotten us somewhere interesting. The square of the period of an orbit is equal to a constant (i.e., $4\pi^2/GM_\odot$) times the cube of the radius of the orbit. *This is just Kepler's third law applied to circular orbits.* This is how Newton showed, at least in

Kepler's third law emerges from Newton's laws of motion.

the special case of a circular orbit, that Kepler's third law—his beautiful "harmony of the worlds"—is a direct consequence of the way objects move under the force of gravity. A more complete treatment of the problem—the problem for which Newton invented calculus—shows that Newton's laws of motion and gravitation predict *all* of Kepler's empirical laws of planetary motion—for elliptical as well as circular orbits.

You may well be wondering why we are taking such a long and sometimes strenuous excursion through the work of Galileo, Kepler, and Newton. Sometimes when hiking in the mountains it is hard to see the summit from the perspective of the trail. Now that we have arrived at the top of this particular pass, we can look around and appreciate what we have gained. When Newton carried out his calculations, he found that his laws of motion and gravitation predicted elliptical

Kepler's laws provided an empirical test of Newton's laws.

orbits that agree exactly with Kepler's empirical laws. *This is how Newton tested his theory that the planets obey the same laws of motion as cannonballs and how he confirmed that his law of gravitation is correct.* Had Newton used *any* other rule for gravity, he would have predicted something different for the way planets orbit the Sun, and these predictions would have failed Kepler's observational test. If Newton's laws had *failed* to predict motions that agreed with Kepler—if Newton's predictions had *not* been borne out by observation—then Newton would have had to throw them out and go back to the

FOUNDATIONS 3.2

CIRCULAR VELOCITY

It is interesting to put some values into the equation for circular velocity to see how fast Newton's cannonball would really have to travel. The radius of Earth is 6.38×10^6 m, the mass of Earth is 5.98×10^{24} kg, and the gravitational constant is $6.67 \times 10^{-11} \text{m}^3/(\text{kg s}^2)$. (We have seen how each of these values is measured.) Putting these values into the expression for v_{circ} gives

$$
v_{circ} = \sqrt{\frac{\left(6.67 \times 10^{-11} \frac{\text{m}^3}{\text{kg s}^2}\right) \times (5.98 \times 10^{24} \, \text{kg})}{6.38 \times 10^6 \, \text{m}}}
$$

$$
= 7.9 \times 10^3 \, \text{m/s}.
$$

Newton's cannonball would have to be traveling about 8 km/s—over 28,000 km/hr—to stay in its circular orbit. That's well beyond the reach of a typical cannon, but just what we routinely accomplish with rockets.

drawing board! All of his work would have fallen onto that large heap of beautiful ideas that do not pass nature's test.

We promised we would show you *how* science works. Well, this is how science works. Kepler's empirical rules for planetary motion pointed the way for Newton, and provided the crucial observational test for Newton's laws of motion and gravitation. At the same time, Newton's laws of motion and gravitation provided a powerful new understanding of why planets and satellites move as they do. Theory and empirical observation work together hand in hand, and our understanding of the Universe strides forward.

If it does not work this way, it is not science.

Rather than hiding from challenges that might prove their theories wrong, scientists actively seek out and confront nature's judgment. And even if one scientist fails to find and probe the potential weaknesses of her favorite theory, she can rest assured that another scientist will. As discussed in Excursions 3.3, a well-tested scientific theory is a far cry from the "theory" that Elvis was abducted by aliens. A well-tested scientific theory is about as close to certain knowledge as we humans can come. This is the main difference between science and the pseudosciences—astrology, creationism, quack medicine, homeopathy, parapsychology, numerology, and a host of others—which hide slipshod thinking and untested, untestable, or even disproven speculation behind scientific-sounding jargon, and which actively ignore the wealth of evidence against them. Here is the key to the difference between science and what *pretends* to be science. In our discussion of Newton we have focused on the basis of scientific knowledge and on how science is done. If you run across activities that are *not* done this way—if they do not make testable predictions and do not search for and embrace every observation or experiment that might in principle prove them wrong—then they are *not* science, no matter how many "authorities" would have you believe otherwise.

REAL-WORLD ORBITS ARE NOT CIRCLES

So far we have concentrated on circular orbits and simply asserted that everything works out for elliptical orbits as well. As pointed out before, it is much more difficult to carry out these calculations for ellipses than for circles, so we will leave that particular peak unscaled on this journey. Even so, from our current vantage point we can get an idea from afar of what that peak looks like—of how elliptical orbits differ from circular orbits, and why Newton's calculations work out the way they do.

Begin again with a satellite in a circular orbit about Earth. The satellite is traveling at the circular velocity, so it remains the same distance from Earth at all times, neither speeding up nor slowing down in its orbit. But now change the rules a bit. What if the

Figure 3.16 (a) *A ball thrown into the air slows as it climbs away from Earth, then speeds up as it heads back toward Earth.* (b) *A planet on an elliptical orbit around the Sun does the same thing. (While no planet has as eccentric an orbit as that shown, cometary orbits can be far more eccentric.)*

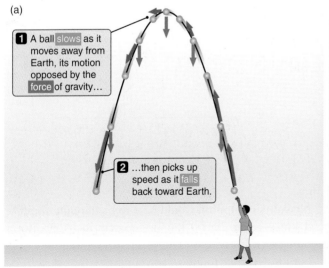

(a)

1 A ball slows as it moves away from Earth, its motion opposed by the force of gravity…

2 …then picks up speed as it falls back toward Earth.

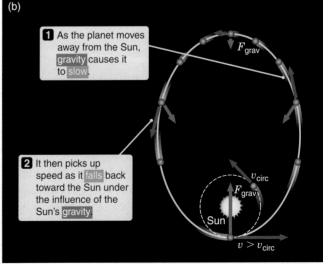

(b)

1 As the planet moves away from the Sun, gravity causes it to slow.

2 It then picks up speed as it falls back toward the Sun under the influence of the Sun's gravity

F_{grav}

v_{circ}

F_{grav}

Sun

$v > v_{circ}$

EXCURSIONS 3.3

"AFTER ALL, IT'S ONLY A THEORY"

Science is sometimes misunderstood because of the special ways that scientists use everyday words. An example is the word *theory*. In everyday language, a theory may be little more than a conjecture or a guess: "Have you any theory about who might have done it?" "My theory is that a third party could win the next election." In everyday parlance, a theory is something worthy of little serious regard. "After all," we say, "it is only a theory."

In stark contrast, a scientific theory is a carefully constructed proposition that takes into account all of the relevant data and all of our understanding of the way the world works, and makes testable predictions about the outcome of future observations and experiments. A theory is a well-developed idea that is ready to be confronted by nature. A well-corroborated theory is a theory that has survived many such tests. Rather than simple speculation, scientific theories represent and summarize bodies of knowledge and understanding that provide our fundamental insights into the world around as. A successful and well-corroborated theory is the pinnacle of human knowledge about the world.

Theories fill a place in a loosely defined hierarchy of scientific knowledge. In science, the word idea has its everyday use. An idea is just a notion about how something might be. A **hypothesis** is an idea that leads to testable predictions. A hypothesis may be the forerunner of a scientific theory, or it may be based on an existing theory, or both. When an idea has been thought about carefully enough, has been tied solidly to existing theoretical and experimental knowledge, and makes testable predictions, then that idea has become a **theory.** Scientists build **theoretical models** that are used to connect theories with the behavior of complex systems. Competing theories are ultimately decided among on the basis of the success of their predictions. Some theories become so well tested and are of such fundamental importance that we come to refer to them as **physical laws.** A scientific **principle** is a general idea or sense about how the Universe is that guides us in the construction of new theories. **Occam's razor,** for example, is a guiding principle in science that says that when faced with two hypotheses that explain some phenomenon equally well, we should adopt the simpler of the two.

Unfortunately, there is room for confusion here because our terminology does not always follow these categories. For example, Newton's laws are physical laws, but Kepler's laws are really empirical rules (based on data and observation, but not on more fundamental principles or laws). Furthermore, science is a living enterprise, and the status of ideas changes in time. For example, principles can themselves become theories. The cosmological principle has been instrumental in shaping countless theories about the Universe. In turn, the success of those theories has effectively become the experimental test of the cosmological principle itself.

Newton's theory of motion illustrates the point of our digression. This theory is the basis of our worldwide technological civilization, and comes about as close to certain knowledge as humankind can ever hope for. Yet to scientists this knowledge remains a theory, subject to observational and experimental tests via the formal and rigorous application of the scientific method.

So think twice the next time you hear someone casually dismiss some body of scientific knowledge as being "only a theory."

If the velocity is not the circular velocity, then the orbit cannot be circular.

satellite were in the same place in its orbit and moving in the same direction, but traveling *faster* than the circular velocity? The pull of Earth is as strong as ever, but because the satellite has a greater speed, its path is not bent by Earth's gravity sharply enough to hold it in a circle. So the satellite begins to climb above a circular orbit.

As the distance between Earth and the satellite begins to increase, an interesting thing starts to happen. Think about a ball thrown into the air, as shown in **Figure 3.16(a).** As the ball climbs higher, the pull of Earth's gravity opposes its motion, slowing the ball down. The ball climbs more and more slowly until its vertical motion stops for an instant, then is reversed and the ball begins to fall back toward Earth, picking up speed along the way. Our satellite does exactly the

A satellite's speed changes in the same way as a ball thrown in the air.

same thing as the ball. As the satellite climbs above a circular orbit and begins to move away from Earth, Earth's gravity opposes the satellite's outward motion, slowing the satellite down. The farther the satellite pulls away from Earth, the more slowly the satellite moves—just as happened with the ball thrown into the air. And just like the ball, the satellite reaches a maximum height on its curving path, then begins falling back toward Earth. Now, as the satellite falls back in toward Earth, Earth's gravity is pulling it along, causing it to pick up more and more speed as it gets closer and closer to Earth.

What is true for a satellite orbiting Earth on an elliptical orbit is also true for any object in an elliptical orbit, including a planet orbiting about the Sun. As we saw earlier, the basic idea of Kepler's Law of Equal Areas is that a planet moves fastest when it is closest to the Sun, and slowest when it is farthest from the Sun. Now we know why. As shown in **Figure 3.16(b)**, planets lose speed as they pull away from the Sun, then gain that speed back as they fall inward toward the Sun.

Insight is given into Kepler's Law of Equal Areas.

Newton's laws do more than explain Kepler's laws. Newton's laws also predict that there are different types of orbits that are beyond Kepler's empirical experience. **Figure 3.17** shows a whole series of satellites, each with the same point of closest approach to Earth, but with different velocities at that point. A look at the figure shows that the greater the speed a satellite has at its closest approach to Earth, the farther the satellite is able to pull away from Earth, and the more eccentric its or-

bit becomes. Yet no matter how eccentric it becomes, as long as it remains elliptical, an orbit will eventually bring a satellite back to the planet that it orbits, and bring a planet back to the Sun.

You might imagine, though, that somewhere in this sequence of faster and faster satellites there comes a point of no return—a point when the satellite is moving so fast that gravity is unable to reverse its outward motion, so the satellite coasts away from Earth, never to return. This indeed is possible. The lowest speed at which this happens is called the **escape velocity.** If we were to work through the calculation, we would find that the escape velocity is a factor of $\sqrt{2}$, or 1.414 . . . , larger than the circular velocity. This can be expressed as

A satellite moving fast enough will escape a planet's gravity.

$$v_{esc} = \sqrt{\frac{2GM}{R}} = \sqrt{2}\, v_{circ}.$$

Look at this equation for a minute to be sure it makes sense to you. The larger the mass (M) of a planet, the stronger its gravity is, and so it stands to reason that a more massive planet would be harder to escape from than a less massive planet. Indeed, the equation says that the more massive the planet, the greater the required escape velocity. It also stands to reason that the closer we are to the planet, the harder it will be to escape from its gravitational attraction. Again, the equation confirms our intuition. As the distance R becomes larger (that is, as we get farther from the planet), v_{esc} becomes smaller (in other words, it is easier to escape from the planet's gravitational pull). For objects at the surface of Earth, the escape veloc-

Figure 3.17 (a) *A range of different orbits that share the same perigee but differ in velocity at that point.* (b) *Perigee velocities for the orbits in (a). An object's velocity determines the orbit shape and whether the orbit is bound.*

(a) Representative orbits

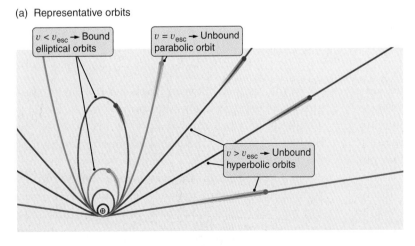

(b) Velocity at closest approach

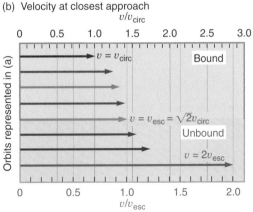

ity is 11.2 km/s, or about 40,000 km/h. Once an object reaches escape velocity, the shape of its orbit is no longer an ellipse. But when Newton solved his equations of motion, he found that ellipses are not the only possible shape an orbit can have.

If a satellite's velocity is less than the escape velocity, its orbit will be elliptical. Elliptical orbits close on themselves. An object following an elliptical orbit is destined to follow the same path, over and over again. For this reason, *elliptical orbits* (orbits whose velocity is less than v_{esc}) are said to be **bound orbits,** since the satellite is bound to the object it is orbiting about. If a satellite has a velocity greater than the escape velocity, then the shape of its orbit is a hyperbola. A hyperbola does not close up like an ellipse but instead keeps opening up forever. A satellite on a hyperbolic orbit makes only a single pass around a planet, then is back off into deep space, never to return. *Hyperbolic orbits* are called **unbound orbits** because an object in hyperbolic orbit is not bound by gravity to the object it is orbiting. *Parabolic orbits,* the third kind of orbit, are the borderline case. In these orbits, the orbiting object has exactly the escape velocity. If the object had any less velocity it would be on an elliptical orbit, any more and its orbit would be hyperbolic. As an object on a parabolic orbit moves away from the planet, its velocity gets closer and closer to zero. Like the hyperbolic orbit, it is never turned around by the planet's gravity, and makes only a single pass by the planet.

> **Bound orbits are ellipses.**

> **Unbound orbits are hyperbolas or parabolas.**

NEWTON'S THEORY IS A POWERFUL TOOL FOR MEASURING MASS

As mentioned earlier, Kepler's empirical laws describe the motion of the planets but do not explain them. On the basis of Kepler's laws alone we might imagine that angels carry the planets around in their orbits, just as many people believed during the 16th century! Newton's derivation of Kepler's laws changed all of that. Newton showed that the *same* physical laws that describe the flight of a cannonball on Earth—or the fall of the apocryphal apple on his head—also describe the motions of the planets through the heavens. In this way Newton shattered the prevailing concept of the heavens and Earth, and at the same time opened up an entirely new way of investigating the Universe. Copernicus may

have dislodged Earth from the center of the Universe and started us on the way toward the cosmological principle, but it was Newton who moved the cosmological principle out of the realm of philosophy and into the realm of testable scientific theory. It was with Newton's work that **astrophysics** was born.

> **Newton's work was the birth of astrophysics.**

Not only is Newton's method more philosophically satisfying than simple empiricism—it is far more powerful as well. We have already seen, for example, how Newton's laws can be used to measure the mass of the Sun and Earth. This could never be done with Kepler's empirical rules. This is especially important when we remember that Newton's laws apply to *all* objects, not just the Sun and Earth. This fact will prove very handy as we continue our journey.

Astronomers often rearrange Newton's form of Kepler's third law to read

$$M = \frac{4\pi^2}{G} \times \frac{A^3}{P^2}.$$

Everything on the right side of this equation is either a constant (such as 4, π, and G) or a quantity we can measure (like the semimajor axis A and period P of an orbit). The left side of the equation is the mass of the object at the focus of the ellipse.

> **Newton's derivation of Kepler's laws allows cosmic mass calculations.**

It is important to note that we cut a couple of corners to get to this point. For one thing, we arrived at this relationship by thinking about circular orbits, then simply asserting that it holds for elliptical orbits as well. Another corner we cut was assuming that a low-mass object such as a cannonball is orbiting a more massive object such as Earth. Earth's gravity has a strong influence on the cannonball, but as we have seen, the cannonball's gravity has very little effect on Earth. For this reason we can imagine that Earth remains motionless while the cannonball follows its elliptical orbit. In the same way it is a good approximation to say that the Sun remains motionless as the planets orbit about it.

This picture changes when two objects are closer to having the same mass. In this case, *both* objects experience significant accelerations in response to their mutual gravitational attraction. We now must think of the two objects as falling around *each other,* with each mass moving on its own elliptical orbit around a point located between the two. Yet even in this most general of cases the equation remains valid. The mass M now refers to the *sum* of the masses of the two objects. So, if we can measure the size and period of an orbit—*any* orbit—then we can use this equation to calculate the mass of the orbiting objects. This is true not only for the masses of Earth and the Sun, but also for the masses of other planets, distant stars, our Galaxy and distant galaxies, and vast clusters of galaxies. In fact, it turns out that *almost all of our knowledge about the masses of astronomical objects comes directly from the application of this one equation.* In a sense, this one equation will even allow us to tackle the question "What is the mass of the Universe itself?"

We have come a long way in this chapter. We have gone from Copernicus's simple picture of planets moving on circles around the Sun, to the pinnacle of Newton's comprehensive laws of motion and gravitation. The work of Copernicus, Galileo, Kepler, and Newton was not just *a* scientific revolution, it was *the* scientific revolution, which changed forever not only our view of the Universe, but also our very notion of what it means "to know."

Before getting too carried away, however, we need to put our accomplishment in perspective. Newton's grand theoretical edifice leaves us with the feeling that we understand the "how come" of Kepler's laws and a great deal more—as if we have lifted up the hood of the Universe and taken a peek underneath. And indeed we have. But remember that we have not talked about *why* acceleration equals force divided by mass, or *why* objects have inertia, or *why* one inertial reference frame is as good as any other. These are good, fundamental laws—so fundamental that they form the basis of our understanding of the world we live in and the foundation of our entire technological civilization. Yet the law equating acceleration with force divided by mass, the inverse square law of gravity, and the law of inertia are still basically empirical facts about how objects are observed to move. These are not articles of faith, but rather scientific hypotheses, the validity of which remain to this day subject to test through experimentation and observation.

SEEING THE FOREST THROUGH THE TREES

The story of planetary motions is also the story of how we as a species learned to do science. No one taught Copernicus or Galileo or Kepler or Newton about the differences between empiricism and theory, or about how the scientific method could be used to test their ideas. They had to figure these lessons out on their own as they went along. Yet the accomplishments they made remain near the top of the all-time intellectual feats of humankind, and the trail they blazed pointed the way for all that was to come. Galileo offered powerful insights into the nature of matter and gravity. Kepler built on Copernicus's revolutionary ideas about planetary motions, and uncovered three empirical rules that showed the orbits of all of the planets to be reflections of the same underlying patterns. Newton combined the two sets of ideas: he built Galileo's insights into a powerful theoretical edifice describing the motions of all objects, and then tested this edifice against Kepler's empirical reality. The culmination of this intellectual campaign was far greater than the sum of its parts. The walls separating the heavens and Earth came down once and for all, and the science of astrophysics was born.

The model for science that emerged during this era remains the template for how science is done to this day. Careful empiricism uncovers patterns in nature in need of explanation. Scientific theory seeks to discover the more fundamental truths underlying all things. And in the meeting of the two—the test of theory against the unforgiving and stalwart challenge of empirical fact—new knowledge and understanding emerge.

Newton's work became the cornerstone of what is often referred to as classical mechanics. All objects have inertia. An object will continue to move in a straight line at a constant speed unless an unbalanced force acts to change its motion. Mass is the property of matter that resists changes in motion. Every force is matched by another force that is equal in size but opposite in direction. Gravity is a force between any two masses, proportional to the product of the two masses and inversely proportional to the square of the distance between them. Putting all of this together, we find that objects "fall around" the Sun and Earth, on elliptical, parabolic, or hyperbolic paths. Orbits are ultimately given their shape by the gravitational attraction of the objects involved, which in turn is a reflection of the mass of these objects. We now understand that it is gravity that holds the Universe together, giving planets, stars, and galaxies their very shapes as well as controlling their motions through space. Using Newton's theoretical insight we look backward along this chain, turning observations of motions of objects throughout the Universe into measurements of the masses of objects that no human has ever, or in most cases will ever, visit.

This brings us to the next stage of our journey. If the ancients could have stepped off Earth and touched the planets, they would never have fallen into the conceptual errors that muddled our thinking for millennia—but they had no such luxury. Today we have sent our robotic surrogates to all of the planets in the Solar System save one, and humans have walked on the surface of the Moon. Even so, we have made only the most cursory visits to our immediate neighborhood. Even the nearest stars remain thousands of times more distant than the most far-flung of our robotic planetary explorers. For the most part we, like the ancients, are left with nothing to base our knowledge of the Universe on but the signals reaching us from across space. Far and away the most common of these signals is electromagnetic radiation, which includes the light by which you are reading this book. Our ability to interpret these signals depends on what we know of light. What is it? How does it originate? How does it interact with matter? What changes does it experience during its journey? We now turn to these questions.

STUDENT QUESTIONS

THINKING ABOUT THE CONCEPTS

1. The orbits of the planets around the Sun and the orbits of satellites around these planets are always ellipses rather than perfect circles. Why is this so?

2. Picture a swinging pendulum. During a single swing, the bob of the pendulum first falls toward Earth and then moves away until the gravitational force between it and Earth finally stops its motion. Would a pendulum swing if it were in orbit? Explain.

3. When riding in a car, we can sense changes in speed or direction through the forces the car applies on us. Do we wear seat belts in cars and airplanes to protect us from speed or from acceleration? Explain.

4. Describe two distinct ways to measure the mass of an object.

5. In 1920, a *New York Times* editor would not publish an article based on rocket pioneer Robert Goddard's paper that predicted spaceflight, saying that "rockets could not work in outer space because they have nothing to push against" (a statement the *Times* did not retract until July 20, 1969, the date of the *Apollo 11* Moon landing). You, of course, know better. What was wrong with the editor's logic?

6. Aristotle taught that the natural state of all objects is to be at rest. Even though this seems consistent with what we observe around us, explain how we know that this conclusion was wrong.

7. Imagine a planet moving in a perfectly circular orbit around the Sun and, because the orbit is circular, the planet is moving at a constant speed. Is this planet experiencing acceleration? Explain.

8. An astronaut standing on Earth could easily lift a wrench having a mass of 1 kg, but not a scientific instrument with a mass of 100 kg. In the International Space Station she is quite capable of manipulating both, although the scientific instrument moves more slowly than the wrench. Explain.

9. Two comets are leaving the vicinity of the Sun, one traveling in an elliptical orbit and the other in a hyperbolic orbit. What can you say about the future of these two comets? Would you expect either of them to eventually return?

10. Even if the Sun were dark and invisible, we could still tell that we are in an elliptical orbit around a massive object and at which of the two foci the Sun is located. Explain.

11. Had Kepler lived on one of a group of planets orbiting a star three times as massive as our Sun, would he have deduced the same empirical laws? If not, how would they have been different. Explain.

12. How do we reconcile Newton's first law of motion with the Coriolis effect, in which the paths of objects seem to curve, even in the absence of forces acting in the direction of that curve.

APPLYING THE CONCEPTS

13. During the latter half of the 19th century, a few astronomers thought there might be a planet circling the Sun inside Mercury's orbit. They even gave it a name, Vulcan. We now know that Vulcan does not exist. If there were such a planet with an orbit $\frac{1}{4}$ the size of Mercury's, what would be its orbital period relative to that of Mercury?

14. Using values given in the Appendices for G and for Earth's radius and mass, show that the acceleration of gravity at the surface of Earth is 9.80 m/s^2.

15. What does an acceleration of 9.80 m/s^2 mean? If you should jump from a stationary balloon, after 1 second you would be falling at a speed of 9.80 m/s. After 2 seconds your speed would be $2 \times 9.80 = 19.6$ m/s or slightly more than 70 km/h! In the absence of any air resistance, how fast would you be falling after 20 seconds?

16. How long does it take Newton's mythical cannon-ball, moving at 7.9 km/s just above Earth's surface, to complete one orbit around Earth?

17. Weight refers to the force of gravity acting on a mass. We often calculate the weight of an object by multiplying its mass by the local acceleration due to gravity. The value of gravitational acceleration on the surface of Mars is 0.39 times that on Earth. Assume your mass is 85 kg. Then your weight on Earth is 833 N (833 N = mg = 85 kg $\times$ 9.8 m/s^2). What would your mass and weight be on Mars?

18. At the surface of Earth, the escape velocity is 11.2 km/s. What would be the escape velocity at the surface of a very small asteroid having a radius 10^{-4} of Earth's and a mass 10^{-12} of Earth's? If you were standing on the asteroid and threw a baseball with a strong pitch, what would happen to it?

Then God said, "Let there be light,"
and there was light.
And God saw that the light was good;
and God separated the light from the darkness.
And God called the light day, and the darkness
*   He called night.*
And there was evening and there was morning, one day.

GENESIS 1:3–5

LIGHT

4.1 THEN GOD SAID, "LET THERE BE LIGHT"

There can be little wonder why the author of the book of Genesis chose to begin the Biblical account of Creation with this moving and powerful prose. Light is a fundamental part of our experience of the world because it is through light that we most directly perceive the world beyond our physical grasp. The symbolic nature of light is everywhere in our language. A close companion may be the "light of our lives." To understand something is to "see it." A "bright" idea is symbolized by a lightbulb going off over our heads. As we leave our ignorance behind, we become "enlightened." A symbol of hope is a "light at the end of the tunnel." Throughout our language and culture, light is a metaphor for knowledge and information. Those who have lost their sight face a challenge greater than most of us can imagine. At the same time, light plays another, even more important role in our lives. It is light from the Sun that warms Earth, drives the wind and the rain, and powers photosynthesis in plants, which lie at the bottom of the terrestrial food chain. The energy that you expend as you move through your day arrived on the planet in the form of light.

KEY CONCEPTS

Unlike the physicist or the chemist, who have control over the conditions in their laboratories, the astronomer must try to glean the secrets of the Universe from the light that reaches us from distant objects. On this leg of our journey we turn our attention to light, a most informative messenger, and find that:

* Light is an electromagnetic wave with a spectrum extending far beyond the colors of the rainbow;
* Light is *also* a stream of particles called photons;
* Reconciling the wave and particle nature of light and matter points beyond Newton's physics, and challenges our everyday ideas about what is "real";
* Measurements of the speed of light also require that we think beyond classical physics and reassess our understanding of time and space;
* The wave/particle nature of light and matter gives different types of atoms unique spectral "fingerprints" that can be used to measure the composition and properties of distant objects;
* Temperature measures the thermal energy of an object, and determines the amount and spectrum of light that a dense object emits; and
* Light is not only a messenger, but is also a way that energy is carried throughout the Universe.

The role of light in astronomy closely parallels the role of light in our everyday existence. Your mental picture of an astronomer is probably of a denizen of the night with head bent to the eyepiece of a telescope, peering at the distant Universe. While this picture is a bit quaint in this modern era of electronic cameras and telescopes orbiting Earth, in many ways it remains on the mark. Whether the telescope is on a mountain-top, or in orbit about Earth, or part of a spacecraft hurtling toward a comet makes little difference. Our knowledge of the Universe beyond Earth comes over-whelmingly from light given off or reflected by astro-nomical objects. Fortunately, light is a *very* informative messenger. It carries with it information about the tem-peratures of objects, what they are made of, the speed they are moving at, and even the nature of the material that the light passed through on its way to Earth.

Light is how we sense the Universe.

Yet light plays a far larger role in astronomy than just being a messenger. Light is one of the main ways that energy is transported from place to place in the Universe. Light carries the energy generated in the heart of a star outward through the star and off into space. From stars to planets to vast clouds of gas and dust filling interstellar space—absorption of light heats objects up while emission of light cools them off.

While light allows us to see the world, we cannot ac-tually "see" light. That is, we do not see light in the same way that we see a tree or we see the stars at night. Light is the messenger, but is itself invisible. It is hard to study something that cannot be seen, so an understanding of light was a long time coming. The property of light that is easiest to *try* to measure is the speed at which light travels. This might seem a straightforward problem, and yet the answer led 19th- and 20th-century physicists to change the way we think about the very fabric of space and time. As we continue our journey we will find that light sets the standard for what we mean by "when," "where," or "how fast," for nothing can travel faster than the speed of light.

4.2 OUR PICTURE OF LIGHT EVOLVED WITH TIME

Suppose you are a scientist living in the 16th or 17th century. How might you go about measuring the speed of light? One obvious way would be to have a friend stand on a hilltop far away. You uncover a lantern, and the instant your friend sees the light from your lantern, your friend uncovers her own lantern. The time it takes from when you uncover your lantern to when you see your friend's light will be the light's round-trip travel time—plus, of course, your friend's reaction time. Galileo mea-sured the speed of sound in just this way, but when he tried this method on light, he failed. He could not mea-sure any delay. Galileo concluded that the speed of light must be very great indeed, possibly even infinite.

If light travels so rapidly, then to measure its speed we will need either very large distances over which to measure its flight, or very good clocks. Galileo had neither at his disposal, but by the end of the 18th century astronomers had both. The great distances were the dis-tances between the planets, while the good clock was provided courtesy of Kepler and Newton. According to Newton's derivation of Kepler's laws, orbital periods should be rock-steady, with each orbit taking exactly as much time as the orbit before. This applies to moons or-biting about planets just as it applies to planets orbiting about the Sun.

Rømer was the first to measure the speed of light.

In the 1670s, **Ole Rømer** was studying the moons of Jupiter, making measurements of the times when each moon disappeared behind the planet. Much to his amaze-ment, Rømer found that, rather than maintaining a regu-lar schedule, the observed times of these events would slowly drift in comparison with predictions. Sometimes the moons disappeared behind Jupiter too soon, while at other times they were seen going behind Jupiter later than expected. Rømer realized that the difference de-pended on where Earth was in its orbit. When he started keeping track of the moons at the point when Earth was closest to Jupiter, then by the time Earth was farthest from Jupiter, the moons were a bit over $16\frac{1}{2}$ minutes "late." But when he waited until Earth was once again at its closest point to Jupiter, the moons "made up" the lost time, and once again passed behind Jupiter at the predicted times.

It is often the case in science that a difference be-tween theoretical predictions and experimental results points the way to new knowledge, and Rømer's work was no exception. Rømer correctly surmised that rather than a failure of Kepler's laws, he was seeing the first clear evidence that light travels at a finite speed. As shown in **Figure 4.1,** the moons appeared "late" when Earth was farther from Jupiter because of the time needed for light to travel the extra distance between the two planets. Over the course of Earth's yearly trip around the Sun, the distance between Earth and Jupiter changes by 2 AU, (astronomical units), which is about 3×10^{11} m. The speed of light equals this distance divided by Rømer's 16.7-minute delay, or about 3×10^8 m/s. The value Rømer actually announced in 1676 was a bit on the low side—2.25×10^8 m/s—because the length of 1 astro-nomical unit was not well known. But Rømer's result

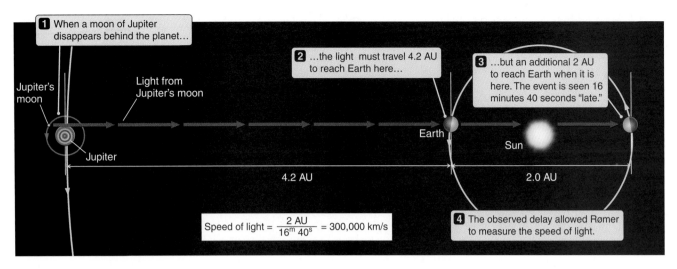

Figure 4.1 *Ole Rømer measured the speed of light by noting that apparent delays in the orbital motions of Jupiter's moons depend on the distance between Earth and Jupiter.*

The speed of light is 300,000 km/s.

was more than adequate to make the point. The speed of light is very great indeed! The orbiting space shuttle moves around Earth at a dazzling speed of about 28,000 km/h (almost 8000 m/s). Light travels almost 40,000 times faster than this. It could circle Earth in only $\frac{1}{7}$ of a second. No wonder Galileo's attempts to measure the speed of light failed!

A good deal of work has been done to improve on Rømer's original result. Modern measurements of the speed of light made with the benefit of high-speed electronics give a value of 2.99792458×10^8 m/s in a vacuum. As of October 1983 the length of a meter is now *defined* as the distance traveled by light in a vacuum in 1/299,792,458 of a second.

LIGHT IS AN ELECTROMAGNETIC WAVE

Since the earliest investigations of light, there has been a good deal of controversy over the question of whether light is composed of particles, as Newton believed, or is instead a wave. (A **wave** is a disturbance that travels from one point to another.) This controversy was seemingly put to rest once and for all in 1873 by the Scottish physicist **James Clerk Maxwell** (1831–1879). One of Maxwell's many accomplishments was the discovery of the fundamental laws that describe electricity and magnetism. Electricity and magnetism are actually two aspects of the same electro-

Maxwell discovered the laws governing electricity and magnetism.

magnetic force. The **electric force** is the push and pull between electrically charged particles. Opposite charges attract and like charges repel. The **magnetic force**, on the other hand, is a force between electrically charged particles arising from their motion.

To describe the electric and magnetic forces Maxwell introduced the concepts of the **electric field** and the **magnetic field**. A charged particle gives rise to an electric field that points toward it or away from it, as shown for a positive charge in **Figure 4.2(a)**. To find out how much force is exerted on a charged particle, we multiply the charge of the particle by the strength of the electric field at its location. Because the electric field points directly away from a positively charged particle (or directly toward a negatively charged particle), the force that a second charge feels is either directly toward or directly away from the first charged particle.

The picture gets a bit more interesting if we quickly move the first charged particle (q_1) by some amount, as shown in **Figure 4.2(b)**. We might expect the force on the second particle (q_2) to change immediately, so that it points away from the new position of the first particle. Yet experiments show that it does not. Immediately after the first charge moves, there is *no* change in the force felt by the second charge at all. Only later does the second particle feel the change in location of the first **(Figure 4.2c)**. The situation is something like what happens if you are holding on to one end of a long piece of rope and a friend is holding on to the other end. When you yank your end of the rope up and down, your friend does not feel the result immediately. Instead your yank starts

Changes in an electric field spread outward at a finite speed.

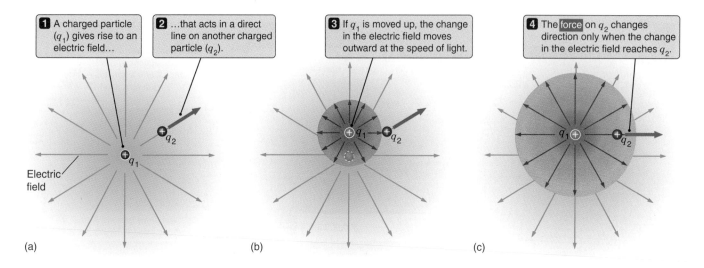

1 A charged particle (q_1) gives rise to an electric field...

2 ...that acts in a direct line on another charged particle (q_2).

3 If q_1 is moved up, the change in the electric field moves outward at the speed of light.

4 The force on q_2 changes direction only when the change in the electric field reaches q_2.

Electric field

(a)   (b)   (c)

Figure 4.2 *When a charged particle accelerates, changes in the electric field move outward at the speed of light. (In (b) the charge is shown moving instantly from one place to another for clarity. In reality, this could not happen.)*

a pulse—a wave—that travels down the rope. Your friend notices the yank only when this wave arrives at his end. Similarly, when you move a charged particle, information about the change travels outward through space as a wave in the electric field. Other charged particles do not know that the first particle has moved until the wave reaches them.

Maxwell summarized the behavior of electric and magnetic fields in four elegant equations. Among other things, these equations say that a changing electric field causes a magnetic field, and that a changing magnetic field causes an electric field. These changes "feed" on themselves. A change in the motion of a charged particle causes a changing electric field, which causes a changing magnetic field, which causes a changing electric field . . . Once you get the process started, a self-sustaining procession of oscillating electric and magnetic fields moves out in all directions through space. Instead of a purely electric wave, an accelerating charged particle gives rise to an **electromagnetic wave.**

Changing electric and magnetic fields lead to a self-sustaining electromagnetic wave.

In addition to predicting that electromagnetic waves should exist, Maxwell's equations also predict how rapidly the disturbance in the electric and magnetic fields should move. In short, Maxwell's equations *predict* the speed at which an electromagnetic wave should travel. When Maxwell carried out this calculation, he discovered that his electromagnetic waves should travel at 3×10^8 m/s—which

Figure 4.3 (a) *The bobbing motion of a floating cork generates waves that move outward across the water's surface.* (b) *In similar fashion, an accelerated electric charge generates electromagnetic waves that move away at the speed of light.*

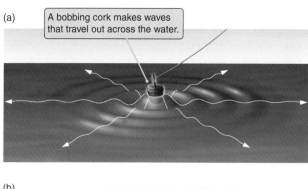

(a) A bobbing cork makes waves that travel out across the water.

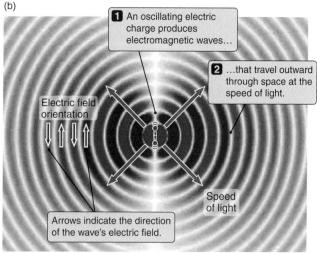

(b)

1 An oscillating electric charge produces electromagnetic waves...

2 ...that travel outward through space at the speed of light.

Electric field orientation

Speed of light

Arrows indicate the direction of the wave's electric field.

Maxwell showed that light is such a wave.

is the speed of light! This agreement could not be simple coincidence. Maxwell had shown that light is an electromagnetic wave.

Maxwell's wave description of light also gives us an idea of where light comes from and how light interacts with matter. Imagine a cork floating on a lake on a perfectly calm day. The surface of the lake is as smooth and flat as a mirror, until a fish tugs on the hook and line dangling beneath the cork. The motion of the cork causes a disturbance that moves outward as a ripple on the surface of the lake (see **Figure 4.3a**). In much the same way, an accelerating electric charge **(Figure 4.3b)** causes a disturbance that moves outward through space as an electromagnetic wave. (An electric charge that is moving at a constant velocity is stationary in its inertial frame of reference, and so does not radiate.) According to Maxwell's equations, *any* time an electrically charged particle is accelerated, the result is an electromagnetic wave. *Accelerating charges are the sources of electromagnetic radiation.*

Accelerating charges cause electromagnetic waves.

WAVES ARE CHARACTERIZED BY WAVELENGTH, FREQUENCY, SPEED, AND AMPLITUDE

Along our journey we will encounter waves of a number of different kinds, ranging from electromagnetic waves crossing the vast expanse of the Universe to seismic waves traveling through Earth. In the most general sense a wave is a disturbance that travels away from its source. If you drop a pebble in a pond, ripples spread out over the surface of the water. This kind of wave is called a **transverse wave** because the wave's displacement is perpendicular, or "transverse", to its direction of travel. The waves on a plucked guitar string are also transverse waves (see **Figure 4.4a**). The reason why such waves travel is because, when the material is disturbed, forces try to even out that disturbance. A portion of the material moves back toward its undisturbed position. But because it is still moving and has inertia, when it reaches that position, it overshoots,

A wave is a disturbance that travels away from a source.

4.1 WAVES

Figure 4.4 *Mechanical waves result from forces that try to even out disturbances. (a) A transverse wave involves oscillations that are perpendicular to the direction in which the wave travels. (b) A longitudinal wave involves oscillations along the direction of travel of the wave.*

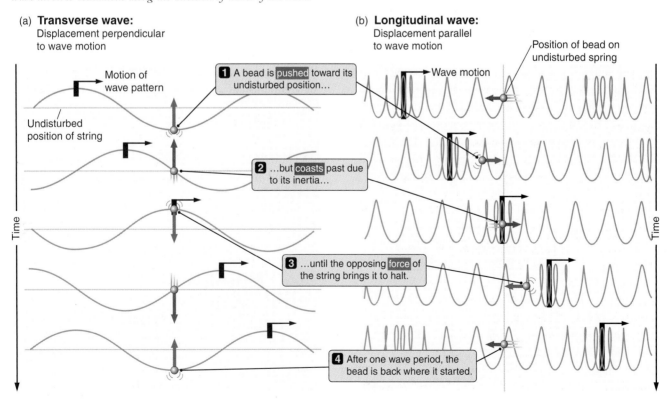

(a) Transverse wave:
Displacement perpendicular to wave motion

Motion of wave pattern

Undisturbed position of string

1 A bead is pushed toward its undisturbed position...

2 ...but coasts past due to its inertia...

3 ...until the opposing force of the string brings it to halt.

4 After one wave period, the bead is back where it started.

Time

(b) Longitudinal wave:
Displacement parallel to wave motion

Position of bead on undisturbed spring

Wave motion

Time

distorting the material in the opposite way from how it was originally distorted. Forces *now* try to push the material back the other direction, but again there is an overshoot. The material oscillates back and forth from one side of its undisturbed position to the other, and the wave moves along. In the case of the guitar string (Figure 4.4a), the force responsible for the wave is the tension which tries to keep the string straight. In the case of water, the forces responsible for creating the wave are gravity, which tries to pull down the crest of the wave, and pressure, which tries to push the trough of the wave up.

Another kind of wave is called a **longitudinal wave.** Sound waves are an example of longitudinal waves, and so are the waves in a spring **(Figure 4.4b).** In a spring, compressed regions try to push out into the stretched-out regions to even the spring out. But when a portion of the spring reaches its undisturbed position, it is still moving, and overshoots. Parts of the spring that were originally compressed are now stretched out, and the parts that were originally stretched out are now compressed. Regions in which the spring is alternately compressed and stretched move along the length of the spring as the cycle at each point repeats itself again and again. In the case of sound waves, it is air pressure that provides the forces that keep the wave moving.

In these examples, the waves result from *mechanical* distortions of the medium the wave travels through (for example, distortion of the surface of the pond, or distortion of the coils in the spring). Mechanical waves involve distortions measured as distances, and media that have mass. Maxwell showed that light waves are a fundamentally different type of wave. Light waves involve no mechanical distortion of a medium, but instead involve periodic changes in the strength of the electric and magnetic fields. Even so, light waves are generally thought of as transverse waves, because the directions of the electric and magnetic fields are perpendicular to the direction in which the wave travels (see **Figure 4.5**).

Waves are generally characterized by four quantities, which are shown in **Figure 4.6.** The **amplitude** of a wave is the maximum excursion from its undisturbed or relaxed position. The wave travels at some speed, which is usually written as v (except in the case of light, where the speed of light is written as c). The number of wave crests passing a point in space each second is called the wave's **frequency,** denoted by f. The unit of frequency is cycles per second, which is generally referred to as **hertz** (abbreviated Hz), after the 19th-century physicist **Heinrich Hertz,** who was the first to experimentally confirm Maxwell's predictions about electromagnetic radiation. The time taken for one complete cycle is called the **period,** P, which is measured in seconds.

> **Waves are characterized by their wavelength, frequency, speed, and amplitude.**

The distance a wave travels during one complete oscillation is called the **wavelength.** This is just the distance from one wave crest to the next, or the distance from one wave trough to the next. The wavelength is usually denoted by the Greek letter λ (pronounced "lambda"). There is a clear relationship between the frequency of a wave and the wavelength of the wave. If the period of a wave is $\frac{1}{2}$ second—that is, if it takes $\frac{1}{2}$ second for one wave to pass by, crest to crest—then two waves will go by in 1 second. So a wave with a period of $\frac{1}{2}$ second per cycle has a frequency of 2 cycles per second. Similarly, if a wave has a period of 1/100 second per cycle, then 100 waves will pass by each second. This wave has a frequency of 100 Hz. More generally, the frequency of a wave is just 1 divided by its period:

> **A long period means a low frequency, and vice versa.**

$$\text{Frequency} = \frac{1}{\text{period}} \quad \text{or} \quad f = \frac{1}{P}$$

There is also a relationship between the period of a wave and its wavelength. The period of a wave is the

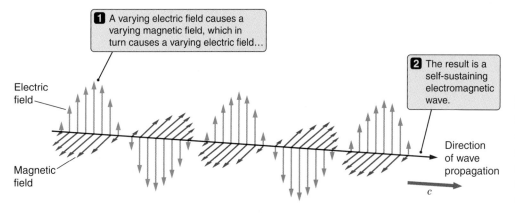

1 A varying electric field causes a varying magnetic field, which in turn causes a varying electric field...

2 The result is a self-sustaining electromagnetic wave.

Electric field

Magnetic field

Direction of wave propagation

c

Figure 4.5 *Far from its source, an electromagnetic wave consists of oscillating electric and magnetic fields that are perpendicular both to each other and to the direction in which the wave travels.*

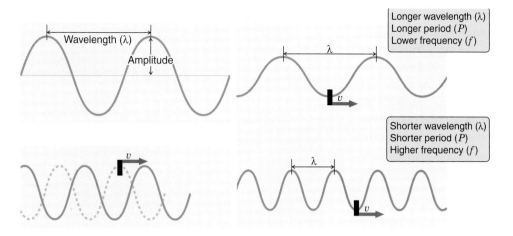

Figure 4.6 *A wave is characterized by the distance over which the wave repeats itself (called the wavelength, λ), the maximum excursion from its undisturbed state (called the amplitude), and the speed (v) at which the wave pattern travels. In an electromagnetic wave, the amplitude is the maximum strength of the electric field, and the speed of light is written as c.*

time between the arrival of one wave crest and the next. During this time the wave travels a distance equal to the separation between the two wave crests, or one wavelength. So far, so good. Now add to the picture the fact that distance traveled equals speed times time taken. Change "distance traveled" to one wavelength and change "time taken" to one period and we find that the wavelength of a wave equals the speed the wave is traveling at times the period of the wave:

$$\text{Wavelength} = \text{speed} \times \text{period}$$

The speed of light is c, so we can say that

$$\lambda = c \times P.$$

Using the relationship between period and frequency given above, we can also write

$$\text{Wavelength} = \frac{\text{speed}}{\text{frequency}}, \quad \text{or} \quad \lambda = \frac{c}{f}.$$

So if the speed of a wave is known, then knowing one of the three properties—its wavelength, period, or frequency—will tell us the other two.

Look at this relationship more closely. The longer the wavelength of a wave, the longer you have to wait between wave crests, and so the frequency of the wave will be lower. A shorter wavelength means less distance between wave crests, which means a shorter wait until the next wave comes along. Therefore, a shorter wavelength means a higher frequency. A tremendous amount of information can be carried by waves—intelligible speech, for example, or complex and beautiful music. As we continue with our study of the Universe, time and again we will find that the information we receive, whether information about the interior of Earth or about a distant star or galaxy, rides in on a wave.

A long wavelength means a low frequency, and vice versa.

Maxwell's equations also describe how electromagnetic waves interact with the matter they encounter. Returning to the analogy of the lake illustrated in Figure 4.3, imagine now that a second cork is afloat on the lake some distance from the first, as in **Figure 4.7(a).**

Figure 4.7 (a) *When waves moving across the surface of water reach a cork they cause the cork to bob up and down.* (b) *Similarly, a passing electromagnetic wave causes an electric charge to wiggle in response to the wave.*

(a)

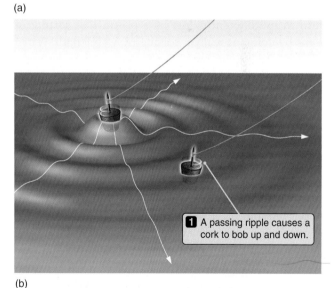

1 A passing ripple causes a cork to bob up and down.

(b)

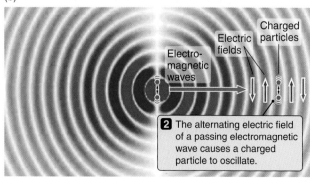

2 The alternating electric field of a passing electromagnetic wave causes a charged particle to oscillate.

Electromagnetic waves cause other charged particles to accelerate.

The second cork remains stationary until the ripple from the first cork reaches it. As the ripple passes by, the rising and falling of the water causes the cork to rise and fall as well. Similarly the oscillating electric field of an electromagnetic wave causes an oscillating force on any charged particle that the wave encounters, and this force causes the particle to move about as well **(Figure 4.7b).** It takes energy to produce an electromagnetic wave, and that energy is carried through space by the wave. Matter far from the source of the wave can absorb this energy. In this way some of the energy lost by the particles generating the electromagnetic wave is transferred to other charged particles. The emission and absorption of light by matter is the result of the interaction of electric and magnetic fields with electrically charged particles.

ELECTROMAGNETIC WAVES OF DIFFERENT WAVELENGTHS MAKE UP THE ELECTROMAGNETIC SPECTRUM

You have almost certainly seen a rainbow like the one in **Figure 4.8** spread out across the sky, or sunlight split into many different colors by a prism. This sorting of light by colors is really a sorting by wavelength. When we talk about light spread out according to wavelength, we refer to the **spectrum** of the light. On the long-wavelength (and therefore low-frequency) end of the visible spectrum is red light. A convenient unit for measuring the wavelength of visible light is the **micrometer,** or **micron.** A micrometer is a millionth $\left(\frac{1}{10^6}\right)$ of a meter. The abbreviation used for a micron is μm (where μ is the Greek letter "mu"). The wavelengths of the light we perceive as red fall between about 0.6 and 0.7 μm. At the other end of the visible spectrum is violet light, which is the bluest of blue light.

The spectrum of visible light is seen as the colors of the rainbow.

The shortest-wavelength violet light that the eye can see has a wavelength of around 0.35 μm. Stretched out between the two, literally in a rainbow, is the rest of the visible spectrum. The colors in the visible spectrum in order of decreasing wavelength can be remembered as a name: "Roy G. Biv," which stands for

Red Orange Yellow Green Blue Indigo Violet.

Figure 4.8 *The visible part of the electromagnetic spectrum is laid out in all its glory in the colors of this rainbow.*

The eye is most sensitive to light in the green to yellow part of the spectrum. This light has a wavelength of around 0.5 to 0.55 μm. Green light with a wavelength of 0.52 μm has a frequency of

$$f = \frac{c}{\lambda} = \frac{3.00 \times 10^8 \frac{\text{m}}{\text{s}}}{0.52 \times 10^{-6}\,\text{m}} = \frac{5.8 \times 10^{14}}{\text{s}}$$
$$\text{or} \quad 5.8 \times 10^{14}\,\text{Hz}.$$

That frequency corresponds to 580 *trillion* wave crests passing by each second!

When we say "visible light," what we mean is "the light which the light-sensitive cells in our eyes respond to." But this is not the whole range of possible wavelengths for electromagnetic radiation. Radiation can have wavelengths that are much shorter or much longer than the eye can perceive. The whole range of different wavelengths of light is collectively referred to as the **electromagnetic spectrum.**

Visible light is only one small segment of the electromagnetic spectrum.

Follow along in **Figure 4.9** as we take a tour of the electromagnetic spectrum, beginning with visible

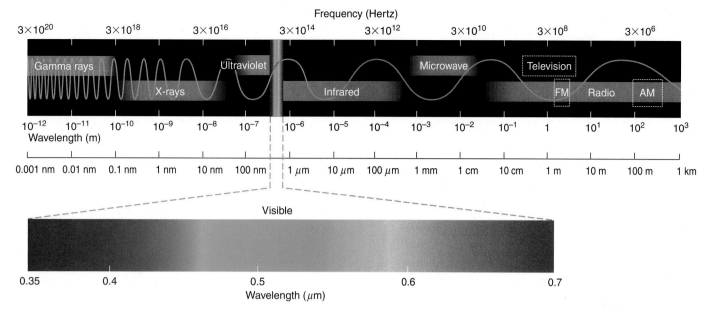

Figure 4.9 *By convention the electromagnetic spectrum is broken into loosely defined regions ranging from gamma rays to radio waves.*

light, and then working our way to shorter and longer wavelengths. We begin with the blue end of the visible spectrum. Beyond this short-wavelength, high-frequency side of the visible spectrum, there is light that is "bluer than blue," or actually, "more violet than violet." This light, with wavelengths between 0.04 and 0.35 μm, is called **ultraviolet (UV) radiation.** You can remember what ultraviolet light is just by looking at the name. The prefix *ultra-* means "extreme," so *ultra*violet light is light that is more "extremely" violet than violet. It is important to remember that ultraviolet light is fundamentally no different from visible light, anymore than high C on a piano is fundamentally different from middle C.

As we go to shorter and shorter wavelengths of light (and so to higher and higher frequencies), we pass through the ultraviolet part of the spectrum. At a wavelength shorter than 0.04 μm, or 4×10^{-8} m, we stop calling light ultraviolet and instead start calling it **X-rays.** This distinction comes for historical reasons. When X-rays were discovered in the last part of the 19th century, they were given the name "X" by their discoverer, **W. C. Roentgen** (1845–1923), to indicate they were "a new kind of ray." As we continue to even shorter wavelengths, we come to another somewhat arbitrary break. Electromagnetic radiation with the very shortest wavelengths (less than about 10^{-10} m) is referred to as **gamma rays.** Again the reasons are histor-

ical. Gamma rays (or γ-rays) were first discovered as one of the types of radiation given off by radioactive material. It was only later that their true nature became known.

So far we have been considering ever shorter wavelengths and higher frequencies. In principle, there is no limit to this process. We can conceive of gamma rays of arbitrarily short wavelengths and arbitrarily high frequencies (even though practical considerations do eventually come into play). We can also go in the other direction. Just as there is light that is more violet than violet, there is also light that is "redder than red." Such light, covering wavelengths longer than about 0.7 μm and shorter than 500 μm (5×10^{-4} m), is referred to as **infrared (IR) radiation.** Again the key to remembering what infrared light is comes from looking at the word itself. *Infra-* is a prefix that means "below." *Infra*red light is light that has a frequency that is lower than (below) that of red light. When the wavelength of light gets longer than this, we start calling it instead **microwave radiation.** The longest-wavelength (and therefore lowest-frequency) electromagnetic radiation, with wavelengths longer than a few centimeters and ranging up to arbitrarily long wavelength, is called **radio waves.** Tools 4.1 discusses the telescopes used by astronomers to capture and analyze electromagnetic radiation.

TOOLS 4.1

The Tools of Astronomy

The tools used by astronomers are remarkably diverse, ranging from atomic physics laboratories to powerful supercomputers to robotic probes sent to cruise the Solar System. Astronomy's tools are not only diverse; they are also rapidly changing. It seems that every few months we are likely to see the commissioning of a new mountaintop telescope, the launch of a satellite observatory, or the arrival of a spacecraft at some remote planetary destination. In fact, in the time it takes between the final changes made by the authors of this volume and its appearance on your bookshelf, much of what we might say about the latest of astronomy's tools will already have become yesterday's news. Rather than give in to instant obsolesence, we will instead make use of the technology of the 21st century to bring you a discussion of (what else?) the technology of the 21st century! On the CD-ROM and web site that accompany this text, you will find sections discussing many of the tools that astronomers use to explore the Universe. These will include Internet links to take you into the worlds of NASA missions, ground-based observatories, university scientists, and wherever else we need to go to find astronomy's technological frontier.

While the details surrounding the tools of astronomy are constantly changing, some fundamentals remain constant. When most people think of astronomy, the mental image that comes to mind is a toy-store telescope pointed at the night sky. The image is fitting. Some form of telescope forms the heart of almost every tool that astronomers have used to directly observe the heavens. In fact, humanity's use of telescopes dates back to far earlier than the first night Galileo turned his telescope skyward. You were born with two telescopes of your own—your eyes. The human eye is a **refracting telescope,** which means that it uses a lens to bend the light passing through it, bringing that light to a sharp focus. Eyes are such amazingly useful things that biologists believe they have evolved *independently* as many as 60 different times during the history of terrestrial life! While the human eye is wonderfully evolved to meet our daily needs, however, it is not so well suited for doing astronomy.

The first of the eye's limitations as an astronomical telescope is **resolution.** When we speak of resolution, we are referring to how close two points of light can be to each other before a telescope is no longer able to split the light into two separate images. Unaided, the human eye can resolve objects separated by an angular distance of about 1 minute of arc, or a 30th of the diameter of the full Moon. This may seem small, and in our daily lives it is, yet when we look at the sky, thousands of stars and galaxies may hide within the smallest area the unaided human eye can resolve. **Figure 4.10(a)** shows the path followed by rays of light from two distant stars as they pass through the lens of a refracting telescope. The distance between the telescope lens and the images formed is referred to as the **focal length** of the telescope. Comparison with **Figure 4.10(b)** illustrates the fact that the longer the focal length, the greater the separation between the images. The focal length of a human eye is typically about 20 mm. In comparison, telescopes used by professional astronomers often have focal lengths of tens or even hundreds of meters. Such telescopes make images that are far larger than those formed by your eye, so contain far more detail.

Focal length explains part of the difference between the resolution of telescopes and the unaided eye. The other difference results from the wave nature of light. As waves of light pass through the lens of a telescope, those waves spread out from the edges of the lens, as illustrated in **Figure 4.11.** The distortion of light as it passes a sharp edge is called **diffraction.** Diffraction "diverts" some of the light from its path, slightly blurring the image made by the telescope. The degree of blurring depends on wavelength of the light in comparison with the diameter of the telescope lens. The larger the lens relative to the waves of light it is focusing, the less of a problem is posed by diffraction. The ultimate limit on the resolution of a telescope, called the **diffraction limit,** is determined by the ratio of the wavelength of light passing through it to the diameter of the lens. The smaller this ratio, the better the resolution of the telescope.

(Tools 4.1 continued on page 96.)

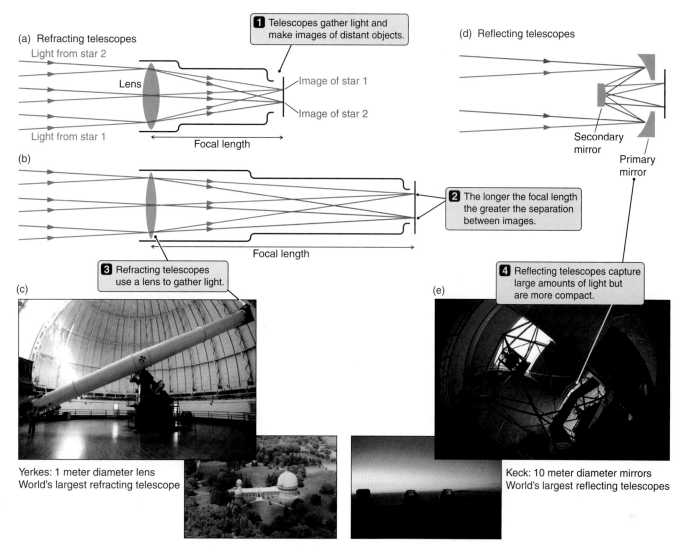

(a) Refracting telescopes

Light from star 2

1 Telescopes gather light and make images of distant objects.

Image of star 1

Lens

Image of star 2

Light from star 1

Focal length

(b)

2 The longer the focal length the greater the separation between images.

Focal length

3 Refracting telescopes use a lens to gather light.

(c)

Yerkes: 1 meter diameter lens
World's largest refracting telescope

(d) Reflecting telescopes

Secondary mirror

Primary mirror

4 Reflecting telescopes capture large amounts of light but are more compact.

(e)

Keck: 10 meter diameter mirrors
World's largest reflecting telescopes

Figure 4.10 (a) *A refracting telescope uses a lens to focus light from two stars into images.* (b) *Longer-focal-length telescopes make larger, more widely separated images.* (c) *The Yerkes 1-m telescope is the world's largest refractor.* (d) *A reflecting telescope uses mirrors to focus light. Note that even though this reflecting telescope is very compact, it makes images that are as widely separated as the much longer refracting telescope in part* b. (e) *The world's largest reflecting telescopes are the twin Keck telescopes.*

Figure 4.11 *Diffraction of incoming waves by the edges of a telescope lens or mirror causes a slight blurring of the image the telescope forms, limiting the telescope's ability to resolve objects. Images at right are diffraction patterns from a source such as a star.*

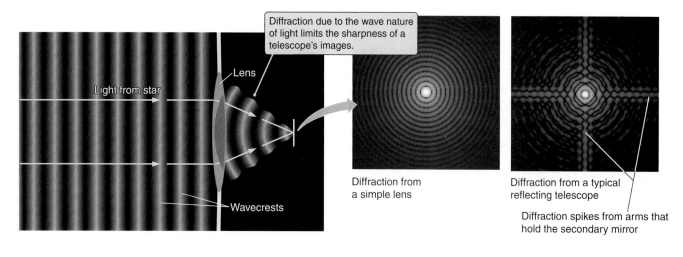

Diffraction due to the wave nature of light limits the sharpness of a telescope's images.

Lens

Light from star

Wavecrests

Diffraction from a simple lens

Diffraction from a typical reflecting telescope

Diffraction spikes from arms that hold the secondary mirror

The diffraction limit means that larger telescopes get better resolution. This is what limits the resolution of the human eye. However, for telescopes with apertures larger than about a meter, Earth's atmosphere stands in the way of better resolution. If you have ever looked out across the desert on a summer day, you have seen the distant horizon shimmer as light from that horizon is constantly bent this way and that by the turbulent atmosphere rising off of the hot desert floor. The problem is less pronounced when looking overhead, but the twinkling of stars in the night sky tells us the problem still exists. The limit on the resolution of a telescope on the surface of Earth caused by this atmospheric distortion is called **astronomical seeing.** One advantage of launching telescopes such as the *Hubble Space Telescope* into orbit around Earth is that from their vantage point above the atmosphere, telescopes get a much clearer view of the Universe, unhampered by seeing. Ground based telescopes have also begun using computer-controlled "adaptive optics" to partially correct for atmospheric distortion.

Now for the second limitation on the human eye as a tool of astronomy. Suppose that you were thirsty, and your only source of water was the rain water that you could collect. You would not be content to just stand there with your mouth open to the sky, catching the water that fell into it. Instead, you would want to use the biggest tub you could find to catch as much rain as possible. The same holds true for telescopes. The light reaching us from distant objects is often like the faintest of mists, so feeble that no human eye alone can capture enough of the light to detect. There are billions of stars filling our skies, but only about 6,000 of these are bright enough to be seen with the unaided human eye. To study the rest we need large telescopes—big "light buckets"—to catch the incoming light. It is next to impossible to build a large light bucket using a lens. **Figure 4.10(c)** shows the ungainly bulk of the Yerkes refractor which, with a lens 1 meter in diameter, is the world's largest refracting telescope. Instead, **reflecting telescopes** use curved mirrors to collect and focus light. **Figure 4.10(d)** shows the path that light follows through the most common type of reflecting telescope in use today. At the dawn of the 21st century the world's largest visible-light telescopes are the twin Keck telescopes that sit atop the Mauna Kea volcano in Hawaii **(Figure 4.10e).** Each of the two Keck telescopes has a mirror with a diameter of 10 meters, giv-

ing it 4,000,000 times the light gathering power of a human eye!

The next limitation on our eyes as astronomical telescopes results not from the lens itself, but from what lies behind it. The lens of your eye focuses incoming light on a mat of light-sensitive cells called a retina. Your retina works fine for the tasks that it evolved to carry out. However, it has shortcomings. For one thing, all that your retina can do is see an image of the world. In astronomy we often use instruments called **spectrographs** to take the light from an object and spread it out into its component wavelengths. Your retina is also hampered by the fact that it has no memory. Today, astronomers use solid-state detectors such as CCDs (similar to the light detectors in home video cameras) to capture and record for hours at a time almost 100% of the radiation collected by a telescope, and turn that light directly into a digital signal that can be processed and analyzed using computers.

The final limitation on the human eye is that it is sensitive only to light in the visible part of the electromagnetic spectrum. (That is, after all, why we call it the "visible" part of the spectrum!) Even though visible light is only a small part of the electromagnetic spectrum, it is anything but happenstance that our eyes work in this range of wavelengths. Our atmosphere is transparent in the visible part of the spectrum, but for most of the spectrum outside this restricted wavelength band, trying to see through our atmosphere is like trying to see through a brick wall. Almost all of the X-ray, ultraviolet, and infrared light arriving at Earth is blocked before it reaches the ground by the layer of atmosphere that surrounds our planet. The visible part of the spectrum is a fairly narrow window through which we can look at the Universe. (There are a few other **atmospheric windows** in the spectrum as well, as shown in **Figure 4.12.**) Just because light does not reach the surface of Earth does not mean that it is uninteresting, however. There are many things that we can learn only by observing the Universe outside of the visible window. While radio observations are also possible from the ground, we owe a large fraction of what we know about the Universe to a host of ultraviolet, X-ray, gamma ray, and infrared telescopes that, beginning in the 1960s, were carried above Earth's atmosphere by rockets. All sorts of fascinating information on space astronomy missions—past, present, and future—is as close as your text's CD-ROM and the nearest Internet connection.

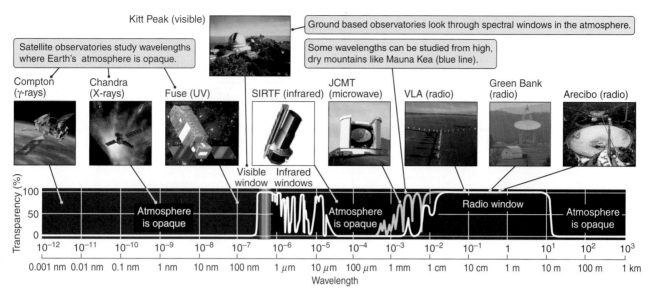

Figure 4.12 *Earth's atmosphere blocks most electromagnetic radiation. Earth-bound telescopes observe the sky through transparent "windows" in the atmosphere. To observe wavelengths outside of these windows requires sending telescopes into space.*

4.3 THE SPEED OF LIGHT IS A VERY SPECIAL VALUE

Maxwell's description of light as a wave was a great success, but without realizing it he had also found a flaw in the armor of Newtonian physics. To understand this flaw, we need to think back to Chapter 2 where we used a moving car as an example of a moving frame of reference. If you are sitting in a car and a ball is sitting there on the seat beside you, then in your frame of reference the ball is at rest. But if the car is moving at 50 mph down the highway, someone sitting by the side of the road will say that the ball is moving at 50 mph as well. To someone in oncoming traffic moving at 50 mph in the other direction, the relative speed of your car and hence of the ball would be 100 mph. There really is *no difference* between these three perspectives. The laws of physics are the same in *any* inertial frame of reference.

As a variant of the "ball in the car" experiment, imagine that as your car moves down the highway at 50 mph you pitch a good, solid fast ball forward at 100 mph (see **Figure 4.13a**).

In our everyday experiences, velocities simply add.

In your frame of reference the ball is moving at 100 mph, but to an observer standing by the side of the road the ball is moving at 150 mph. (The ball has the original speed of 50 mph that the car had, plus the additional 100 mph that you gave it with your throw.) In the frame of reference of a car in oncoming traffic traveling at 50 mph in the other direction, the ball is moving at 200 mph. (This is the 150 mph that the ball is moving relative to the ground, plus the 50 mph motion of the oncoming car.) In our everyday experience velocities simply add. This is also how Newton's laws say the Universe should behave.

Now do exactly the same thought experiment, but with two changes. Instead of a car traveling at 50 mph, imagine you are in a spaceship traveling at half the speed of light, or 0.5*c*, as shown in **Figure 4.13(b)**.

Based on everday experience, the speed of light should depend on the motion of an observer . . .

Instead of throwing a baseball, you shine a beam of light forward. To you, the light is moving at the speed of light, *c*. If we replace "100 mph" with "*c*" in the previous paragraph, we think we know what to expect for other observers. To an observer on a nearby planet, the light should travel by at a speed equal to the speed of your spacecraft plus the speed of light, or 1.5*c*. Similarly, to an observer in an oncoming spacecraft traveling at 0.5*c* in the other direction, the light should travel at a speed of 2*c*.

We do not have the luxury of performing this experiment while traveling through space at half the speed of light, but physicists are ingenious folk. During the closing years of the 19th century and the early years of the 20th century, physicists were conducting laboratory experiments that

. . . yet experiments show that the speed of light is the same for all observers.

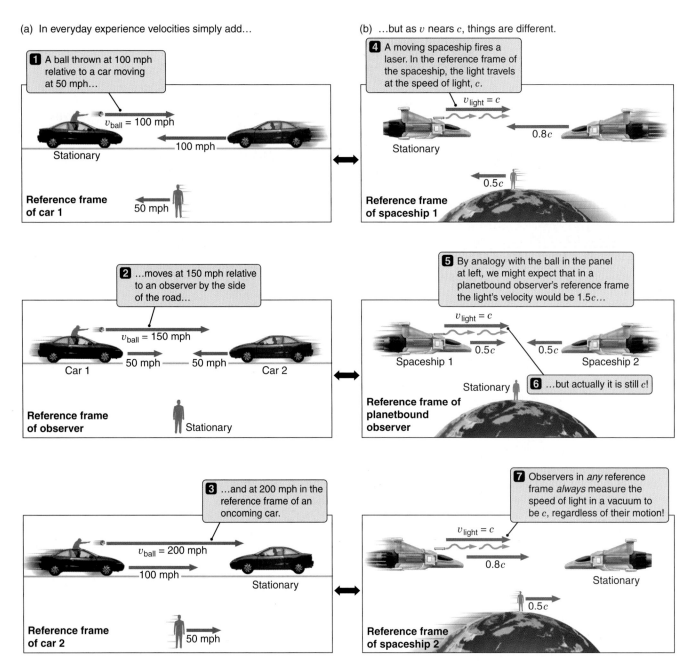

Figure 4.13 *The rules of motion that apply in our daily lives break down when speeds approach the speed of light. The fact that light itself always travels at the same speed for any observer is the basis of special relativity. (Note that relativity also affects the relative speeds of the two spacecraft.)*

were the functional equivalent of the thought experiment in the preceding paragraph. What they were finding had them pulling their hair. Instead of finding that the speed of the beam of light differed from one observer to the next, as expected on the basis of Newton's physics and "common sense," they found instead that *every observer measures exactly the same value for the speed of the beam of light, regardless of their motion!*

As you ride in your rocket ship, you measure the speed of the beam of light to be *c*, or 3×10^8 m/s. That is as expected since you are holding the source of the

light. But the observer on the planet *also* measures the speed of the passing beam of light to be 3×10^8 m/s. Even the passenger in the oncoming spacecraft finds that the beam from your light is traveling at exactly *c* in her own frame of reference. In fact, it turns out that *every observer always finds that light in a vacuum travels at exactly the same speed c, regardless of their own motion or the motion of the source of the light.*

If at this point you have a queasy feeling in your gut and are saying to yourself, "This is very bizarre," then you probably have followed what the last few para-

This experimental result falsified a prediction of Newtonian physics.

graphs have to say. And if this discussion bothers you, imagine the reaction of the poor physicists! Newton's laws of motion had been the bedrock of science for 200 years, facing every experimental challenge that came their way. Now suddenly that bedrock seemed to turn to sand. How could light have the same speed for *all* observers, regardless of their own velocity? Preposterous! And yet that was the inescapable experimental result. Despite all of its spectacular successes, Newtonian physics seemed to be in serious trouble.

Enter a young German-born Swiss patent clerk by the name of **Albert Einstein** (1879–1955). As a 16-year-old schoolboy Einstein had already realized that there was trouble afoot. Light travels in a straight line at a constant speed. Einstein reasoned that, according to Newton's laws of motion, there should then be a perfectly good inertial frame of reference that moves along with the light and in which the light is stationary. That is, you should be able to "catch up" with light, so that you are moving right along with it. But if you could do that, the light would be an oscillating electric and magnetic wave *that does not move.* Such a thing was impossible according to Maxwell's equations for electromagnetic waves. There was a contradiction here. Either Maxwell was wrong in his understanding of electricity and magnetism, or Newtonian physics did not apply at very large velocities. As the experimental results rolled in on measurements of the speed of light, it became clear that it was Newtonian physics that needed revision.

TIME IS A RELATIVE THING

Einstein resolved the difficulty between Maxwell and Newton and ushered in a scientific revolution (see Connections 4.1) with his theory of **special relativity,** which was published in 1905. Special relativity was Einstein's answer to the question, "What must the Universe be like if every observer always measures the same value for the speed of light in a vacuum?" Einstein focused his

CONNECTIONS 4.1

A SCIENTIFIC REVOLUTION

Throughout the first three chapters of this book we interlaced our story of the motions of the sky and the discovery of Newton's laws of motion and gravitation with a discussion of the nature of scientific knowledge and the way science progresses. We stressed that scientific knowledge differs from all other forms of knowledge in that even our most cherished and fundamental knowledge is open to challenge by new observations and experiments. In this chapter we will see this drama play itself out several times over.

By the middle of the 19th century, many physicists felt that our fundamental understanding of physical law was more or less complete. For over a century, Newtonian physics had withstood the scrutiny of scientists the world over. It seemed that little remained but cleanup work—filling in the details. Some even went so far as to pronounce it the "end of science." Yet during the late 19th and early 20th centuries, physics was rocked by a series of scientific revolutions that shook the very foundations of our understanding of the nature of reality. In this chapter we come across several of these revolutions. Einstein's theory of relativity erased the classical dis-

tinction between space and time, and united our concepts of matter and energy. Quantum mechanics forced us to abandon our everyday understanding of "substance," and even to part with the notion that we live in a Universe in which effect follows cause in lockstep. Together these revolutions led to the birth of what has come to be known as **modern physics.** While modern physics *contains* Newtonian physics, the understanding of the Universe offered by modern physics is far more sublime and far more powerful than the earlier understanding which it subsumed.

As we continue on our journey, we will encounter many other discoveries and successful ideas that forced scientists either to abandon their treasured notions, or be left behind, hopelessly locked into a worldview that had ultimately failed the test of observation and experiment. The point is this: In physical science, we are not just paying lip service to a hollow ideal when we say that the rigorous standards of scientific knowledge respect no authorities and take no prisoners. No theory, no matter how central or how strongly held, is immune from the rules.

thinking on pairs of *events*. In relativity, an **event** refers to a particular location in space and a particular time. When you snap your fingers, that is an event. From everyday experience we know that the distance between any two events depends on the frame of reference of the person observing them. Suppose you are sitting in a car that is traveling down the highway in a straight line at a constant 60 mph. You snap your fingers (event 1), and a minute later you snap your fingers again (event 2). In your frame of reference *you* are stationary and the two events happened at exactly the same place. They are separated by a minute in *time,* but there is no separation between the two events in *space*. This is very different from what happens in the frame of reference of an observer sitting by the side of the road. This observer agrees that the second snap of your fingers (event 2) occurred a minute after the first snap of your fingers (event 1), but to this observer the two events were separated from each other in space by a mile. In this everyday, "Newtonian" view, the *distance* between two events depends on the motion of the observer, but the *time* between the two events does not.

> **Special relativity concerns the relationship between events in space and time.**

> **In our everyday Newtonian view, space is relative but time is absolute. . . .**

Einstein questioned why there was such a distinction between the way Newton treated space and the way Newton treated time. Einstein realized that the *only* way the speed of light can be the same for all observers is if *the passage of time is different from one observer to the next!* This is a *very* counterintuitive idea, but it is so central to our modern understanding of the Universe that it is worth wrestling with a bit. Hang on to your hat, while we reconstruct something of the reasoning that led Einstein to this remarkable conclusion.

To measure time, the first thing that we need is a clock. What better way to build a clock than to base it on a value that everyone can agree on—the speed of light. **Figure 4.14(a)** shows just such a clock as seen by observer 1 who is stationary with respect to the clock. At time t_1 a flashlamp gives off a pulse of light. Call this event 1. The light bounces off a mirror a distance l meters away, then heads back toward its source. At time t_2 the light arrives and is recorded by a photodetector. Call this event 2. The time between events 1 and 2 is just the distance the light travels ($2l$ meters), divided by the speed of light, or $t_2 - t_1 = 2l/c$.

So far, so good, but now look at the clock from the perspective of observer 2 in a frame of reference that is moving relative to the clock. In *this* observer's frame of reference he is stationary, and it is the *clock* that is mov-

Figure 4.14 *The "tick" of a light clock as seen in two different reference frames. As Einstein's thought experiment demonstrates, if the speed of light is the same for every observer, then moving clocks must run slow.*

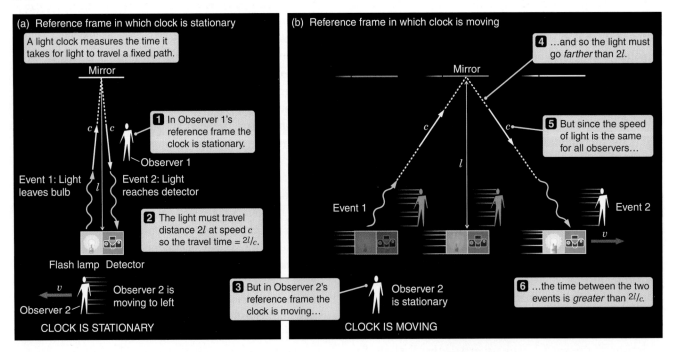

(a) Reference frame in which clock is stationary

A light clock measures the time it takes for light to travel a fixed path.

Mirror

1 In Observer 1's reference frame the clock is stationary.

Observer 1

Event 1: Light leaves bulb Event 2: Light reaches detector

2 The light must travel distance $2l$ at speed c so the travel time = $2l/c$.

Flash lamp Detector

Observer 2 is moving to left

Observer 2

CLOCK IS STATIONARY

(b) Reference frame in which clock is moving

4 ...and so the light must go *farther* than $2l$.

Mirror

5 But since the speed of light is the same for all observers...

Event 1   Event 2

3 But in Observer 2's reference frame the clock is moving...

Observer 2 is stationary

6 ...the time between the two events is *greater* than $2l/c$.

CLOCK IS MOVING

ing at speed *v*, as shown in **Figure 4.14(b).** (Recall that since any inertial frame of reference is as good as any other, this observer's perspective is as valid as the first observer's perspective.) We see the same two events as before; event 1 when the light leaves the flashlamp and event 2 when the light arrives at the detector. There is a difference, however. In this frame of reference the clock *moves* between the two events, and so the light has *farther to go.* (If you do not see this straight away, use a ruler to measure the length of the light path in Figure 4.14(b) and compare it with the length of the light path in Figure 4.14a.) The time between the two events is still the distance traveled divided by the speed of light, but now that distance is *longer* than *2l* meters. Since the speed of light is the same for all observers, the time between the two events must be longer as well!

. . . but if *c* is constant, then time *must* be different for different observers.

Go over that again. The two events are the *same two events,* regardless of the frame of reference from which they are observed. The question is, how much time passed between the two events? Because the speed of light is the same for all observers, there *must* be more time between the two events when viewed from a frame of reference in which the clock is moving. It takes a moving clock more time than a stationary clock to complete one "tick." Moving clocks *must* run slow, and the passage of time *must* depend on an observer's frame of reference.

To Newton, and to us in our everyday lives, the march of time seems immutable and constant. But in reality the only thing that is truly constant is the speed of light, and even time itself "flows" differently for different observers.

Here, in a nutshell, is the heart of Einstein's theory of special relativity. In our everyday Newtonian view of the world, we live in a three-dimensional space, through which time marches steadily onward. Events occur in space, at a certain time. By the time Einstein finished working out the implications of his insight, he had reshaped this three dimensional universe into a four-dimensional **space-time.** Events occur at specific locations within this four-dimensional space-time, but how this space-time translates into what we perceive as "space" and what we perceive as "time" depends on our frame of reference.

Space and time together form a four-dimensional "space-time."

It is very important to state that Einstein did not "throw out" Newtonian physics. We were not wasting our time in Chapter 3 when we studied Newton's laws of motion. Instead, Einstein found that Newtonian physics is *contained within* special relativity. In our everyday experience we never encounter speeds that approach that of light. Even the breakneck speed of the space shuttle is only about 0.000025*c*. When Einstein's special relativity is restricted to cases where velocities are much, much less than the speed of light, then Einstein's equations become the very equations that describe Newtonian physics! In our everyday lives we experience a Newtonian world. It is only when relative velocities approach that of light that things begin to depart from the predictions of Newtonian physics. When great velocities cause something to turn out differently than we would expect based on Newtonian physics, it is referred to as a **relativistic** effect.

Newtonian physics is contained within special relativity.

THE IMPLICATIONS OF RELATIVITY ARE FAR RANGING

The story of special relativity is another case study of the way science really works. Newton's laws had been around for a long time, and had proven to be an extraordinarily powerful way of viewing the world. But as science turned its attention to a different phenomenon—the phenomenon of light—difficulties arose. Newton's theory of motion, Maxwell's theory of electromagnetic radiation, and empirical measurements of the speed of light met head on. Such conflicts are what the scientist lives for, for it is conflicts like this that point the way to new knowledge and new understanding. Einstein was able to step in and reconcile this conflict, and in the process he changed the way we think about the very fabric of the Universe. Einstein's ideas remained controversial well into the 20th century. However, as experiment after experiment confirmed the many strange and counterintuitive predictions of relativity, scientists came to accept its validity. Today, special relativity is an integral and indispensable part of all of physics, shaping our thinking about the motions of the tiniest of subatomic particles as well as the motions of the most distant of galaxies.

Conflicts between theory and experiment point the way to new knowledge.

It would be great fun to linger here for a time and explore. Yet our journey has hardly begun, and there is so much more to see. We hope you will find the time at some point to come back and stay a while, exploring the wonders of the relativistic world in which we live. Puzzling out relativity is time well spent. In the meantime, here are a few of the interesting insights that come from Einstein's work:

1. **What we think of as "mass" and what we think of as "energy" are actually two manifestations of the same thing.** Usually we think of the energy of an object as depending on its speed. The faster it moves, the more energy it has. But Einstein's famous equation, $E = mc^2$, says that even a *stationary* object has an intrinsic "rest" energy that is equal to the mass m of the object times the speed of light, c, squared. The speed of light is a very large number. This relationship between mass and energy says that a single tablespoon of water has a rest energy equal to the energy released in the explosion of over 300,000 tons of TNT! All reactions that produce energy do so by converting some of the mass of the reactants into other forms of energy. But even the most efficient chemical or nuclear reactions release only tiny fractions of the total energy available. Exploding TNT, for example, converts less than a trillionth of its mass into energy. Even the explosion of a hydrogen bomb releases far less than 1% of the energy contained in the mass of the bomb.

   The equivalence between mass and energy points both ways. We found in Chapter 3 that the only sensible definition of mass is that property of matter that resists changes in motion. Does the energy of an object really increase its resistance to changes in motion? Yes. Even adding to the energy of motion of an object increases its inertia. For example, a proton in a high-energy particle accelerator may approach the speed of light so closely that its total energy is 1,000 times greater than its rest energy. Such an energetic proton is, indeed, harder to "push around" (in other words, it has more inertia) than a proton at rest.

2. **The speed of light is the ultimate speed limit.** There are several ways to think about this. We already discussed the insight that led Einstein to relativity in the first place. Were it possible to travel at the speed of light, then in that frame of reference light would cease to be a traveling wave, and all of the laws of physics would come tumbling down around our ears. We can also think about this limit in terms of the equivalence of mass and energy discussed above. As the speed of an object gets closer and closer to the speed of light, its energy becomes greater and greater, and so it becomes more and

more resistant to further changes in its motion. We can continue to push on it all we like, making it go faster and faster, but we face diminishing returns. The situation is like trying to get from 0 to 1 by halving the remainder again and again. The resulting sequence—0, $\frac{1}{2}$, $\frac{3}{4}$, $\frac{7}{8}$, $\frac{15}{16}$, $\frac{31}{32}$, $\frac{63}{64}$, . . . —gets arbitrarily close to 1 but never actually reaches it. In the same way, a continuous force applied to an object will cause its velocity to get closer and closer to the speed of light, but it will never actually reach the speed of light. You just cannot get there. It would take an *infinite* amount of energy to accelerate an object with a nonzero rest mass to the speed of light. In short, all of the energy in the entire Universe is inadequate to accelerate a single electron to the speed of light. We can get the electron arbitrarily close to that number—0.99999999999999999999 . . . × c is no problem, at least in principle—but there is no getting over the hump. Faster-than-light travel may be a mainstay of science fiction, but "Sorry, Jim. I just cannot make her go faster than $c!$" (We leave the Scottish accent up to your imagination. Here is one of those cases where wishing that something is physically possible does not necessarily "make it so.")

3. **Time passes more slowly in a moving reference frame.** This phenomenon is referred to as **time dilation,** since time is "spread out" in the moving reference frame. Were you to compare clocks with an observer moving at $\frac{9}{10}$ the speed of light ($0.9c$), you would find that the other observer's clock was running less than half as fast as your clock (about 0.44 times as fast)[1]. You might guess that to the other observer, your clock would be fast, but actually the other observer would find instead that it is *your* clock that is running slow! A bit of thought shows why it must be this way. To you, the other observer may be moving at $0.9c$, but to the other observer, it is *you* who are moving! Either of your frames of reference is equally valid, so it stands to reason that if a clock in a moving reference frame runs slow, then you would each find the other's clock to be slow!

4. **"At the same time" is a relative concept.** Two events that occur at the *same* time for one observer, may occur at *different* times for a different observer.

---

[1] The factor by which time and space are dilated is given by $\dfrac{1}{\sqrt{1 - \dfrac{v^2}{c^2}}}$. This factor is often referred to as γ.

Hold out your arms and snap the fingers on both hands at the same time. For you, the two snaps were simultaneous. But to an observer moving by you from right to left at close to the speed of light, you snapped the fingers of your left hand first and the fingers of your right hand later.

5. **An object in motion is shorter than it is at rest.** More specifically, moving objects are compressed in the direction of their motion. A meter stick moving at $0.9c$ is only 43.6 cm (centimeters) long.

These different consequences of relativity can be combined in what is often called the *twin paradox*. You head off on a trip to the center of the Milky Way Galaxy,

The twin paradox illustrates many aspects of relativity.

roughly 25,000 light-years distant. Your spectacularly powerful star drive accelerates your ship up to $0.9999999992c$. To you, it is the Galaxy that is moving by at this speed, so the 25,000 light-years to the center of the Galaxy is compressed by a factor of 25,000 to a distance of a single light-year (see number 5 above). At your speed, you cross this distance in a single year. You snap a picture of the gas swirling around the black hole at the center of the Galaxy, then turn around to head home and show it to your twin. Again, the return trip takes only a year. So in 2 years you have traveled to the center of the Galaxy and back again. (Who says that interstellar travel is such a big deal?) The only problem is that when you return, you find that your twin died 50,000 years ago. In the reference frame of Earth, your spacecraft crossed the 25,000 light-year distance to the center of the Galaxy moving at just under the speed of light. The only reason you survived the journey, according to an Earth-bound observer, is because in your moving frames of reference time ran way too slow (number 3 above). Each leg of the two-way journey took 25,000 years to observers on Earth, and your twin just could not wait that long.

You might puzzle over the twin paradox a bit. Both on the way out and on the way back, in your reference frame it is the clocks on Earth that are running slowly, so you are aging *faster* than your twin. Yet when you return, more time has passed for your twin than for you. How can this be? The answer is that, unlike your twin, you *changed reference frames* during your trip. Event 1 is when you left Earth, and event 2 is when you returned to Earth. Your twin went from one event to the other, riding along in Earth's frame of reference. You, on the other hand, changed reference frames when you left Earth, changed again when you stopped at the center of the Galaxy, changed a third time when you left the

galactic center to return home, and changed reference frames one final time when you arrived back at Earth. It happens that the path through space-time that you followed between the two events involved the passage of only 2 years of what you experienced as time, while your twin's path involved 50,000 years of what your twin experienced as time.

A purist would say that the difference between you and your twin is that you experienced acceleration during your trip, while your twin did not. We will return to the subject of accelerated reference frames later in the book when we talk about the connection between Einstein's work and gravity.

4.4 LIGHT IS A WAVE, BUT IT IS *ALSO* A PARTICLE

Maxwell may have achieved a great success with his theory of electromagnetic waves, and may have opened the crack in Newtonian physics that led to the theory of relativity, yet Maxwell's accomplishments themselves would soon need serious revision. As mentioned earlier, from very early on scientists disagreed over whether light consists of waves or of particles. Maxwell's work seemed to put the issue to rest by showing that light is an electromagnetic wave, but before too many years had gone by, the particle description raised its head again. While the electromagnetic wave theory of light has had a great many successes in describing phenomena, there are also many phenomena that it does not de-

Maxwell's wave theory could not account for all of the phenomena associated with light.

scribe well. These range from the presence of sharp bright and dark "lines" at specific wavelengths in the light from some objects, to the shape of the continuous spectrum of light emitted by a lightbulb. Many of these difficulties with the wave model of light have to do with the way light interacts with atoms and molecules.

Scientists working in the late 19th and early 20th centuries discovered that many of the puzzling aspects of light could be better understood if light energy came in discrete packages. (It was for his contributions to this work that Einstein received the Nobel Prize in 1921.) Effectively, these scientists were reintroducing the particle picture of light. In some ways the particle description of light is easier to think about than the wave description of light. In this model we think about light as being made up of particles called **photons** (*phot-*

means "light," as in *photograph*, and *-on* signifies a particle, as in *electron, neutron,* and *proton*). Photons always travel at the speed of light, and they carry energy. (After our discussion of relativity in Section 4.3, you may wonder how a particle can travel at the speed of light. The answer is that a photon has no mass. A massless particle can *only* travel at the speed of light.)

The particle description of light is tied to the wave description of light by a relationship between the energy of a photon and the frequency or wavelength of the wave. The higher the frequency of the electromagnetic wave, the greater the energy carried by each photon. Specifically we write

The energy of a photon is proportional to its frequency.

$$E = hf \quad \text{or} \quad E = \frac{hc}{\lambda}.$$

The *h* in this equation is called **Planck's constant,** and has the value $h = 6.63 \times 10^{-34}$ joule-second. (Planck's constant is named after the German physicist, **Max Planck,** 1858–1947.) According to the particle description of light, the electromagnetic spectrum is a spectrum of photon energies. Photons of shorter wavelength (higher frequency) carry more energy than photons of longer wavelength (lower frequency). For example, photons of blue light carry more energy than photons of longer-wavelength red light. Ultraviolet photons carry more energy than photons of visible light, and X-ray photons carry more energy than ultraviolet photons. The lowest-energy photons are radio-wave photons.

A blue photon has more energy than a red photon, and an X-ray photon has more energy than a radio photon.

The **intensity** of light measures the *total* amount of energy that a beam of the light carries. A beam of red light can be just as intense as a beam of blue light—that is, it can carry just as much energy—but because the energy of a red photon is less than the energy of a blue photon, it will take more red photons to reach that intensity than it would take blue photons. This relationship is a lot like money. A hundred dollars is a hundred dollars, but it takes a lot more pennies (low energy photons) to make up a hundred dollars than it takes fifty-cent pieces (high energy photons).

When physicists speak of the energy of light as broken into discrete packets called photons, they say that the light energy is **quantized.** The word *quantized* has the same root as the word *quantity,* and means that something is subdivided into discrete units. A pho-

Photons are the *quantum mechanical* description of light.

ton is referred to as a **quantum of light.** The branch of physics that deals with the quantization of energy and of other properties of matter is called **quantum mechanics.**

Quantum mechanics, like special relativity, is very counterintuitive for we humans, but its predictions have been confirmed over and over again by experiment. The conflict between everyday, commonsense ideas about the world, and the world as revealed through modern science is discussed in Connections 4.2.

ATOMS CAN ONLY OCCUPY CERTAIN DISCRETE ENERGY STATES

If we want to understand better how light interacts with matter, we need to start by pinning down exactly what we mean by "matter" in the first place. To a physicist, **matter** is anything that occupies space and has mass. Virtually *all* of the matter we have direct experience with is composed of **atoms.** The computer keyboard this is being typed on is made of atoms, and the neurons in your brain that are changing their structure as you read are made of atoms. Atoms are incredibly tiny—so tiny that a single teaspoon of water contains about 10^{23} atoms. (There are more atoms in a single teaspoon of water than there are stars in the observable Universe.) When we talk about the interaction of light with matter, what we are really talking about is the interaction of light with atoms, and the things atoms themselves are composed of. So the next question is, "What are atoms?"

Virtually all matter we encounter is composed of atoms.

Atoms are built from three types of elementary particles as illustrated in **Figure 4.15(a).** Sitting in the center of the atom is the **nucleus,** which is composed of positively charged **protons** and electrically neutral **neutrons.** An atom may have many protons and neutrons in its nucleus. Surrounding the nucleus of the atom are negatively charged **electrons.** For an atom to be electrically neutral, it must have the same number of electrons as protons. Electrons have much less mass than protons or neutrons, and so almost all of the mass of an atom is found in its nucleus. This naturally leads to a mental picture of an atom as a "tiny solar system," with the massive nucleus siting in the center and the smaller electrons orbiting about much like planets orbit about the Sun **(Figure 4.15b).**

Unless you have thought about atoms a great deal, this is probably the conception of an atom that you have

CONNECTIONS 4.2

THINKING OUTSIDE THE BOX

As you read about the combined wave and particle description of light, you will likely find yourself scratching your head in confusion over just exactly what light really is. If you do, then consider yourself in good company. The scientists who invented the seemingly bizarre quantum description of nature had a great deal of trouble thinking about light as well. The wave model of light is clearly the correct description to use in many instances, just as Maxwell has shown. At the same time the particle description of light is also clearly the correct description to use in other cases, as scientists like Planck and Einstein demonstrated. But how can the same thing—light—be both a wave *and* a particle? It is hard for us to imagine a single thing sharing the properties of a wave on the ocean *and* a beach ball, yet light does just that.

Our trouble with thinking of light as both a wave and a particle only begins to scratch the surface of the puzzling and philosophically trouble-some world of quantum mechanics. As we go further, things only get worse. Light is not the only thing that shares wave and particle properties. In fact, *all* matter shares wave and particle properties. Sometimes a "particle" such as an electron behaves as if it were a wave, while at other times a "wave" of light clearly exhibits the properties of a discrete particle. Early quantum physicists would sometimes joke that on Monday, Wednesday, and Friday, light and matter were particles, while on Tuesday, Thursday, and Saturday, light and matter were waves. (And come Sunday it was best just not to think about them at all!)

Light is what light is, and an electron is what an electron is. The trouble with quantum mechanics lies not with the nature of reality, but with what our brains can easily *think about*. This chapter earlier

provided another example of the limitations of our genetic programming. Our brains deal quite well with objects that are sitting still, or even moving as fast as a hard-hit fly ball. Basically our brains cope best with things moving at the speeds of things in nature that we might want to eat, or that might want to eat us. (Animals whose brains could not deal with such speeds tended not to survive for long enough to pass their genes for those brains on to future generations.) But the brains of our ancestors did not have to deal with things moving at close to the speed of light, so we should not be surprised that special relativity seems to defy our intuition. Likewise, there was no evolutionary pressure for our ancestors to be able to think easily about the wave/particle duality of light and matter.

Quantum mechanics and special relativity are not the only places where our ease in thinking about nature breaks down. Quantum mechanics deals with the very smallest scales in nature. At the other extreme, our brains did not evolve to think about things as large or as massive as stars and galaxies and the Universe. When we move on to these larger scales later in the book, we will find our ideas about the nature of space and time themselves further challenged as we seek ways of visualizing curved space-time or understanding why the question "What came before the beginning of the Universe?" is in some ways much like asking, "What was to the left of last Thursday?"

Our brains exist in a box that is defined by the experiences and circumstances which we and our genetic ancestors had to cope with. One of the exciting things about modern physics and astronomy is that it forces us to break down the walls of that conceptual box and find tools for understanding what lies beyond its boundaries.

in your head. This is much the same picture that scientists in the early 20th century held as well. But it has a fatal problem. In this view, an electron whizzing about in an atom is constantly undergoing an acceleration—the direction of its motion is constantly changing. The wave description of electromagnetic radiation says that

any electrically charged particle that is accelerating must also be giving off electromagnetic radiation. This electromagnetic radiation should be carrying away the orbital energy of the electron. (Imagine that electron

Atoms are not "tiny solar systems."

(a) Parts of an atom

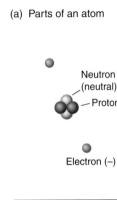

Neutron (neutral)

— Proton (+)

Electron (−)

This is a helium atom (2 protons, 2 neutrons, and 2 electrons).

(b) "Solar system" model

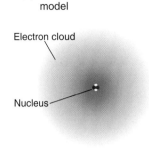

Electrons do not move in orbits like planets...

(c) Quantum mechanical model

Electron cloud

Nucleus

...but rather are waves "smeared out" in a cloud of probability held in place by the electric attraction of the nucleus.

Figure 4.15 (a) *An atom is made up of a nucleus consisting of positively charged protons and electrically neutral neutrons, surrounded by less massive negatively charged electrons.* (b) *Atoms are often drawn as miniature "solar systems", but this model is incorrect.* (c) *Electrons are actually smeared out around the nucleus in quantum mechanical clouds of probability.*

as the wiggling electric charge in Figure 4.3b.) If you calculate how much energy should be carried off by radiation from the electron, you find that only a tiny fraction of a second should be needed for the electrons in an atom to lose all of their energy and fall into the atom's nucleus! Fortunately for us this does not happen. Atoms exist for very long periods of time, and electrons never "fall into" the nuclei of atoms. So something must be wrong with this concept of an atom. A way out of this difficulty came when scientists realized that, just as waves of light have particle-like properties, so too do particles of matter have wavelike properties. With this realization, the *miniature solar system* model of the atom was modified

so that a positively charged nucleus is surrounded *not* by planetlike electrons moving in their orbits, but by electron "clouds" or electron "waves" as illustrated in **Figure 4.15(c).**"

The strings on a guitar can vibrate only at certain discrete frequencies, giving rise to the discrete notes that we hear. In much the same way, the electron waves in an atom can take on only certain specific forms. So, instead of being able to take on *any* arbitrary energy, atoms can

Atoms can have only certain discrete energies, much as guitar strings play only certain notes.

only take on certain specific energies corresponding to the allowed wave forms of their electron clouds. A given atom may have a tremendous number of different energy states available to it, but these states are *discrete*. An atom might have the energy of one of these allowed states, or it might have the energy of the next allowed state, *but it cannot have an energy somewhere in between.* We can imagine the energy states of atoms as being a bookcase with a series of shelves as shown in **Figure 4.16.** The energy of

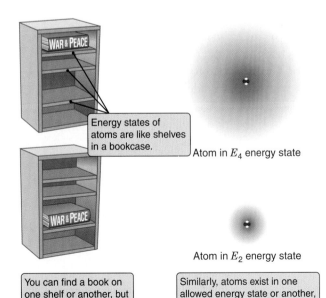

Energy states of atoms are like shelves in a bookcase.

Atom in E_4 energy state

Atom in E_2 energy state

You can find a book on one shelf or another, but not in between.

Similarly, atoms exist in one allowed energy state or another, but never in between.

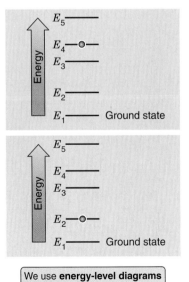

E_5

E_4

E_3

E_2

E_1 —— Ground state

Energy

E_5

E_4

E_3

E_2

E_1 —— Ground state

Energy

We use **energy-level diagrams** to represent the allowed states of an atom.

Figure 4.16 *Atoms can only have certain discrete energies.*

an atom might correspond to the energy of one shelf, or to the energy of the next shelf, but the energy of the atom will *never* be found *between* the two shelves.

The lowest-possible energy state of an atom—the "floor"—is called the **ground state** of the atom. Allowed states with energies lying above the ground state are called **excited states** of the atom. When the atom is in its ground state, it has nowhere to go. An electron cannot "fall" into the nucleus, because there is no allowed state with less energy for it to move down into. It cannot move up to a higher energy state without getting some extra energy from somewhere. For this reason an atom will remain in its ground state forever unless something happens to knock it into an excited state. A book sitting on the floor has nowhere left to fall to, and cannot jump to one of the higher shelves of its own accord.

An atom in an excited state is a very different matter, however. Just as a book on an upper shelf might fall to a lower shelf, an atom in an excited state might **decay** down to a lower-lying state by getting rid of some of its extra energy. An important difference between the atom and the book on the shelf, however, is that while a snapshot might catch the book in between the two shelves, the atom will never be caught between two energy states. When the transition from one state to another occurs, the energy difference between the two states must be carried off all at once. A common way for an atom to do this is to give off a photon. But not just any photon will do. The photon emitted by the

atom must carry away exactly the amount of energy lost by that atom as it goes from the higher energy state to the lower energy state.

THE ENERGY LEVELS OF AN ATOM DETERMINE THE WAVELENGTHS OF LIGHT IT CAN EMIT AND ABSORB

To better understand the relationship between the energy levels of an atom and the radiation it can emit or absorb, imagine a hypothetical atom that has only *two* available energy states. Call the energy of the lower energy state (the ground state) E_1 and the energy of the higher energy state (the excited state) E_2. The energy levels of this atom can be represented in an energy-level diagram like those in Figure 4.16, but with only two levels (see **Figure 4.17a**).

To understand the process of *emission,* imagine that the atom begins in the upper state (E_2) and then spontaneously drops down to the lower energy state (E_1). This is shown in **Figure 4.17(b),** where the downward arrow indicates that the atom went from the upper state to the lower state.

When an atom drops to a lower energy state, the lost energy is carried away as a photon.

Figure 4.17 (a) *The energy levels of a hypothetical two-level atom.* (b) *A photon with energy* $hf = E_2 - E_1$ *is emitted when an atom in the more energetic state decays to the lower energy state.*

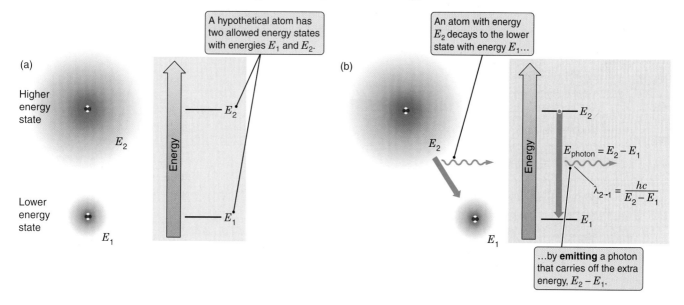

The atom just lost an amount of energy equal to the difference between the two states, or $E_2 - E_1$. However, energy is never truly lost or created, so the energy lost by the atom has to show up somewhere. In this case, the energy shows up in the form of a photon that is emitted by the atom. The energy of the photon emitted must just match the energy lost by the atom, so the energy of the photon must be $E_{photon} = E_2 - E_1$.

We have already seen the relationship between the energy of a photon and the frequency or wavelength of electromagnetic radiation. Using this relationship we can say that the frequency of the photon emitted by a transition from E_2 to E_1, which we will denote as $f_{2\to1}$, is just the energy difference divided by Planck's constant (h):

$$f_{2\to1} = \frac{E_{photon}}{h} = \frac{E_2 - E_1}{h}.$$

Similarly, the wavelength of the photon is just $\lambda = c/f$, or

$$\lambda_{2\to1} = \frac{c}{f_{2\to1}} = \frac{hc}{E_2 - E_1}.$$

This shows that the wavelengths of photons emitted by an atom—the color of the light that the atom gives off—are determined by the energy level structure of the atom. An atom can only emit photons with energies corresponding to the difference between two of its allowed energy states.

Imagine what the light coming from a cloud of gas consisting of our hypothetical two-state atoms would be like. This case is illustrated in **Figure 4.18,** which shows a collection of our two-state atoms. Any atom that finds itself in the upper energy state (E_2) will very quickly decay and emit a photon in some random direction. A cubic meter of the air around you contains about 10^{25} atoms. Even if only a tiny fraction of these atoms emit a photon each second, an enormous number of photons would still come pouring out of the cloud of gas. But instead of containing photons of all different energies (that is, light of all different colors), like sunlight, this light would instead contain only photons with the specific energy $E_2 - E_1$ and wavelength $\lambda_{1\to2}$. In other words, all of the light coming from the cloud would be the same color.

We have all seen what happens to sunlight when it passes through a prism. Sunlight contains photons of all different colors, so when sunlight passes through a prism, it spreads out into all of the colors of a rainbow. But if we were to pass the light from our cloud of gas through a slit and a prism, as in Figure 4.18, the results would be very different. This time there would be no rainbow. Instead, all of the light from the cloud of gas would show up on the screen as a single, bright line. The process we have just described—the production of a photon when an atom decays to a lower energy state—is referred to as **emission.** The bright,

The spectrum of a cloud of glowing gas contains emission lines.

Figure 4.18 *A cloud of gas containing atoms with two energy states, E₁ and E₂, emits photons with an energy* E = hf = E₂ − E₁, *which appear in the spectrogram (right) as an emission line.*

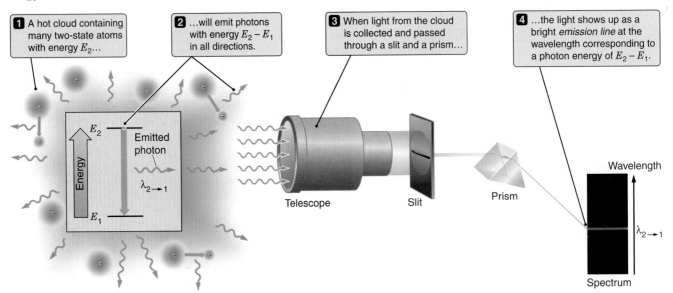

1 A hot cloud containing many two-state atoms with energy E_2...

2 ...will emit photons with energy $E_2 - E_1$ in all directions.

3 When light from the cloud is collected and passed through a slit and a prism...

4 ...the light shows up as a bright *emission line* at the wavelength corresponding to a photon energy of $E_2 - E_1$.

Energy

E_2

Emitted photon

$\lambda_{2\to1}$

E_1

Telescope

Slit

Prism

Wavelength

$\lambda_{2\to1}$

Spectrum

single-colored feature in the spectrum of the cloud of gas is referred to as an **emission line.**

So far in this discussion we have ignored an important question: "How did the atom get to be in the excited state E_2 in the first place?" If an atom is sitting in its ground state, then it will remain in the ground state unless it is somehow given just the right amount of energy to kick it up to an excited state. Most of the time, this extra energy comes in one of two forms: (1) The atom absorbs the energy of a photon (we will talk about this possibility shortly); or (2) the atom collides with another atom, or perhaps an unattached electron, and the collision knocks the atom into an excited state. This is how a neon sign works. When a neon sign is turned on, an alternating electric field is set up inside the glass tube that pushes electrons in the gas back and forth through the neon gas inside the tube. Some of these electrons crash into atoms of the gas, knocking them into excited states. The atoms then drop back down to their ground states by emitting photons, causing the gas inside the tube to glow. (In like fashion, an electron beam in a television picture tube collides with atoms in the screen, knocking them into excited states. When those atoms decay, they emit the photons that we perceive as the picture on the television.)

> High energy states can be excited by photons or by collisions with other atoms or electrons.

So far we have focused on the emission of photons by atoms in an excited state, but what about the opposite process? An atom in a low energy state can absorb the energy of a passing photon and jump up to a higher energy state, as shown in **Figure 4.19,** but not just any photon can be absorbed by the atom. As before, the energy that it takes to get from E_1 to E_2 is the difference in energy between the two states, or $E_2 - E_1$. For a photon to cause an atom to jump from E_1 to E_2, it must provide just this much energy. Using the relationship that $E_{\text{photon}} = hf$ or $f = E_{\text{photon}}/h$, we find that the *only* photons capable of exciting atoms from E_1 to E_2 are photons whose frequency and wavelength are, respectively,

$$f_{1\rightarrow 2} = \frac{E_{\text{photon}}}{h} = \frac{E_2 - E_1}{h}$$

and

$$\lambda_{1\rightarrow 2} = \frac{c}{f_{1\rightarrow 2}} = \frac{hc}{E_2 - E_1}.$$

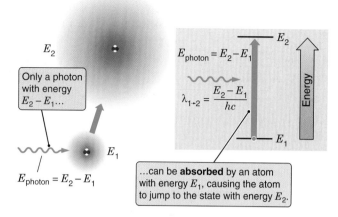

Figure 4.19 *A photon of energy* hf $= E_2 - E_1$ *may be absorbed by an atom in the lower energy state, leaving the atom in the higher energy state.*

This is exactly the same energy photon—the same color of light—that is emitted by the atoms when they decay from E_2 to E_1. This is not a coincidence. The energy difference between the two levels is the same whether the atom is emitting a photon or absorbing one, so the energy of the photon involved will be the same in either case.

> Atoms can absorb photons with the same energies as the photons they emit.

What might the spectrum of a light look like when viewed through a cloud composed of our hypothetical gas of two-state atoms? If we shine photons of all different wavelengths (that is, light of all different colors) through the gas from one side, almost all of these photons will pass through the cloud of gas unscathed. So counting how many photons come *out* the other side of the cloud of gas should give us the number of photons that we shined *into* the gas. There is only one exception. Rather than passing through the gas, some of the photons with just the right energy ($E_2 - E_1$) might instead be absorbed by atoms.

If we shine the light from a lightbulb directly though a glass prism, we will see that a rainbow of colors comes out as shown in **Figure 4.20(a).** If we instead shine the light through a cloud of our two-state atoms before putting the light through a prism, the rainbow of colors will be unchanged, except for one detail. Since some of the photons with energies equal to $E_2 - E_1$ will have been absorbed by the gas, these photons will be missing in the light passing through the prism. If we look at the screen, we will see a

> When viewed through a cloud of gas, the spectrum of a lightbulb contains absorption lines.

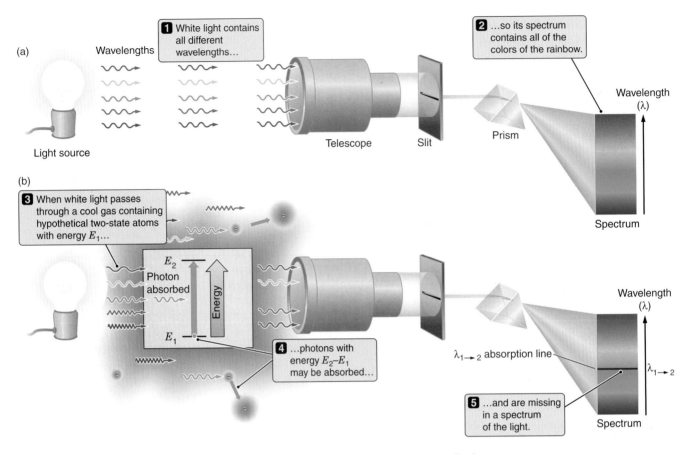

Figure 4.20 (a) *When passed through a prism, white light produces a spectrum containing all colors.*
(b) *When light of all colors passes through a cloud of hypothetical two-state atoms, photons with energy*
$hf = E_2 - E_1$ *may be absorbed, leading to the dark absorption line in the spectrogram.*

sharp, dark line at the color corresponding to these photons **(Figure 4.20b).** The process of atoms capturing the energy of passing photons is referred to as **absorption,** and the dark feature seen in the spectrum is called an **absorption line.**

There is one final point worth making before leaving the subject of emission and absorption of radiation. When an atom absorbs a photon and jumps up to an excited energy state, there is a good chance that the atom will quickly decay back down to the lower energy state by emitting a photon with the same energy as the photon it just absorbed. If the atom re-emits a photon just like the one it absorbed, you might reasonably ask why the absorption matters at all. After all, the photon that was taken out of the passing light was replaced, was it not? The answer is yes and no. The photon was replaced, true enough, but while all of the photons that were absorbed were originally traveling in the *same direction,* the photons that are re-emitted travel off in *random directions.* In other words, some of the photons with energies equal to $E_2 - E_1$ are in effect diverted from their original paths by their interaction with atoms. If you look at a lightbulb *through*

the cloud, you will notice an absorption line at a wavelength of $\lambda_{1\to2}$, but if you look at the cloud from the side (looking perpendicular to the original beam), you will see it as a glowing light with an emission line at this wavelength.

EMISSION AND ABSORPTION LINES ARE THE SPECTRAL FINGERPRINTS OF TYPES OF ATOMS

In the previous section we used a hypothetical atom with only two allowed energy states to help us think about emission and absorption of photons. Real atoms have many more than just two possible energy states

Real atoms have many possible energy states.

that they might occupy, so a given type of atom will be capable of emitting and absorbing photons at many different wavelengths. An atom with three energy states, for example, might jump from state 3 to state 2, or from state 3 to state 1, or from state 2 to state 1. The emission lines from such a gas might have wavelengths of $hc/(E_3 - E_2)$, $hc/(E_3 - E_1)$, and $hc/(E_2 - E_1)$.

The allowed energy states of an atom are determined by the complex quantum mechanical interactions among the electrons and the nucleus that comprise the atom. Every hydrogen atom consists of a nucleus containing one proton, plus a single electron in a cloud surrounding the nucleus. As a result, every hydrogen atom has the *same* energy states available to it. It follows that every hydrogen atom is capable of emitting and absorbing photons with the same wavelengths as any other hydrogen atom. **Figure 4.21(a)** shows the energy-level diagram of hydrogen, along with the spectrum of emission lines for hydrogen in the visible part of the spectrum (**Figures 4.21b** and **c**).

Every hydrogen atom has the *same* energy states available to it, and so every hydrogen atom is in principle capable of producing the same spectral lines as any

other hydrogen atom. But the energy states of a hydrogen atom are *different* from the energy states available to a helium atom, a lithium atom, or a boron atom, just as the energy states of these kinds of atoms are different from each other. Each different *type* of atom has its own unique set of available energy states, and therefore its own unique set of wavelengths at which it can emit or absorb radiation. **Figure 4.21(d)** shows the set of emission lines that are given off by discharge tubes (like those in a neon sign) containing different kinds of atoms. These unique sets of wavelengths serve as unmistakable spectral fingerprints for each type of atom.

Spectral fingerprints are of crucial importance to astronomers. They let us figure out what types of atoms (or molecules) are present in distant objects by doing nothing more than looking at the spectrum of light from those objects. If we see the spectral lines of hydrogen or helium or carbon or oxygen or any other element in the

> The wavelengths at which atoms emit and absorb radiation form unique spectral fingerprints for each type of atom.

Figure 4.21 (a) *The energy states of a hydrogen atom are shown. Decays to level* E_2 *emit photons in the visible part of the spectrum.* (b) *This is what you might see if you looked at the light from a hydrogen lamp projected through a prism onto a screen.* (c) *This graph of the brightness of lines versus their wavelength is an example of how spectra are traditionally plotted.* (d) *Emission lines from several other types of gases: helium, argon, neon, and sodium.*

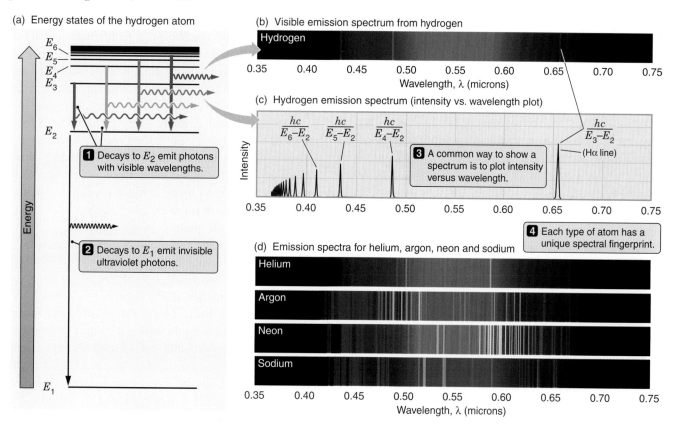

light from a distant object, then we know that some of that element is present in that object. The strength of a line is determined in part by how many atoms of that type are present in the source. By measuring the brightness of the lines from different types of atoms in the spectrum of a distant object, astronomers are often able to infer the relative amounts of different types of atoms of which the object is composed. But it gets even better. The fraction of atoms of a given kind that are in some particular energy state (as opposed to some other energy state) is often determined by factors such as the temperature or the density of the gas. By looking at the relative brightness of different lines from the same kind of atom, it is often possible to determine the temperature, density, and pressure of the material as well.

HOW ARE ATOMS EXCITED, AND WHY DO THEY DECAY?

In the last section, we sidestepped an aspect of the emission process that has troubled physicists and philosophers alike since the earliest days of quantum mechanics. To appreciate this question, return to the analogy between emission of a photon and a book falling off a shelf. If we place a book on a level shelf and do not disturb it, the book will sit there forever. Once the book is resting on the upper shelf, something must *cause* the book to fall off the shelf. So what about the atom? Once an atom is in an excited state, what *causes* it to jump down to a lower energy state and emit a photon? What triggers the event? Sometimes an atom in an upper energy state can be "tickled" into emitting a photon—a process called stimulated emission—but under most circumstances the answer is that *nothing causes the atom to jump to the lower energy state.* There is no trigger. Instead the atom decays *spontaneously.* And while we can say *about* how long the atom is *likely* to sit there in the excited state, the rules of quantum mechanics say (and experiment shows) that we cannot know exactly when a given atom will decay until *after* the decay has happened. The atom decays at some *random* time that is not influenced by anything in the Universe, and cannot be known ahead of time.

Usually nothing causes an atom to decay and emit a photon; it just happens.

You have seen many examples of this rather amazing phenomenon in your life. For example, at some time or other you have probably owned a "glow in the dark" toy or a watch with a "glow in the dark" face. Photons in sunlight or from a lightbulb are absorbed by the atoms in the watch face, knocking those atoms into excited energy states. Unlike many excited energy states of atoms, that tend to decay in a small fraction of a second, the excited states of the atoms in the watch face instead tend to live for many seconds before they decay. Suppose, for example, that, on average, these atoms tend to remain in their excited state for 1 minute before decaying and emitting a photon. In other words, suppose that if we wait for 60 seconds, there is a 50/50 chance that any particular atom will have decayed and a 50/50 chance that the atom will remain in its excited state. There are trillions upon trillions of such atoms in the watch face. While it is impossible to say exactly which atoms in the watch face will decay after a minute, we can say with certainty that *about* half of them *will* decay within 60 seconds. From the standpoint of the glow that we see from the watch, it makes little practical difference which half of the atoms decay and which half do not. All we need to know is that if we wait 1 minute, half of the atoms will have decayed, and the brightness of the glow from the watch will have dropped to half of what it was. If we wait another minute, half of the remaining excited atoms will decay, and the brightness of the glow will be cut in half again. Each 60 seconds, half of the remaining excited atoms decay, and the glow from the watch drops to half of what it was 60 seconds earlier. The glow from the watch slowly fades away.

We have now come upon one of the most philosophically troubling aspects of quantum mechanics. In deep space, where atoms can remain undisturbed for long periods of time, there are certain excited states of atoms that on average live for tens of million years or even longer. Envision an atom in such a state. It may have been in that excited state for a few seconds, or a few hours, or 50 million years when in an instant it decays to the lower level state *without anything causing it to do so.* Newton, and virtually every physicist who lived up until the turn of the 20th century, envisioned a clockwork Universe in which every effect had a cause. They imagined that if one knew the exact properties of every tiny bit of the Universe today, it was just a matter of turning the crank on the laws of physics to predict what the state of the Universe would be tomorrow. Then quantum mechanics came along and turned this view on its head. Instead of dealing with strict cause-and-effect relationships, physicists found themselves calculating the *probabilities* of certain events taking place, and facing fundamental limitations on what can ever be known about the state of the Universe. We have mentioned that while Einstein helped to start the sci-

Quantum mechanics undermines the orderly, causal Universe of Newtonian physics.

entific revolution of quantum mechanics, in the end that revolution left him behind. He could never shake himself of his firm belief in Newton's clockwork, causal Universe. "God does not play dice with the Universe!" he insisted emphatically. As more and more of the predictions of quantum mechanics were borne out by experiment, most physicists came to accept the implications of the strange new theory. Einstein, on the other hand, went to his grave looking unsuccessfully for a way to save his notion of order in the Universe.

It is interesting to note that, while Einstein refused to accept quantum mechanics, our understanding of the quantum mechanical nature of reality owes him a great debt. His was one of the greatest minds of all time, and as he searched tirelessly for flaws in quantum mechanics, he presented challenge after challenge to those who were trying to work out the details of the new theory. Time after time it was in devising answers to Einstein's objections that physicists were forced to confront the full implications of their own work. As an epilog to this story, this struggle continues to this day. At the dawn of the 21st century there are still a few theoretical physicists who are pursuing Einstein's dream, trying to recast quantum mechanics in a way that recaptures the strict causality that seemed irretrievably lost shortly after the beginning of the 20th century. So far they have met with little success, and most physicists doubt that they ever will. However, like Einstein before them, their healthy skepticism has led to ever deeper understanding of the implications and limitations of the theory.

THE DOPPLER EFFECT—IS IT MOVING TOWARD US OR AWAY FROM US?

We have begun to see that, to an astronomer, light is far more than just the stuff that bounces off the page and lets you read these words. Light is a tightly packed bundle of information which, when spread into its component wavelengths, can reveal a wealth of information about the physical state of material located tremendous distances away. The nature of light shapes the way we think about space and time, and has forced physicists to abandon many of their most cherished ideas about the nature of matter and energy. Yet we have only begun to explore what light can tell us. It is time to step back from the precipice of the philosophical implications of quantum mechanics, and look instead at how light can be used to measure one of the most straightforward questions about a distant astronomical object: is it moving away from us or toward us, and at what speed?

Have you ever stood on a street corner and listened as a fire truck sped by with sirens blaring? If so, you might have noticed something funny about the way the siren sounded. As it came toward you, its siren had a high pitch, but as it passed by, the pitch of the siren dropped noticeably. If you were to close your eyes and listen, you would have no trouble knowing when the fire truck passed, just from the change in pitch of its siren. You really do not even need a fire engine to hear this effect. The sound of normal traffic behaves in the same way. As a car drives past, the pitch of the sound that it makes suddenly drops.

The pitch of a sound is like the color of light. It is determined by the wavelength or frequency of the wave. What we perceive as higher pitch corresponds to sound waves with higher frequencies and shorter wavelengths. Sounds that we perceive as lower pitch are waves with lower frequencies and longer wavelengths. When an object is moving toward us, the waves that it gives off "crowd together" in front of the object. You can see how this works by looking at **Figure 4.22,** which shows the locations of successive wave crests given off by a moving object. If you are standing in front of an object moving toward you, the waves that reach you have a shorter wavelength and therefore a **The motion of a source toward or away from us changes the wavelength of the waves reaching us.** higher frequency than the waves given off by the object when it is not moving. In the case of sound waves, the

Figure 4.22 *Motion of a light source relative to an observer may cause waves to be spread out (red-shifted) or squeezed together (blue-shifted). This change in wavelength is called a Doppler shift.*

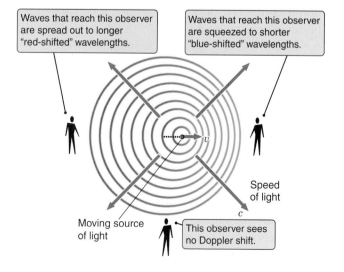

Waves that reach this observer are spread out to longer "red-shifted" wavelengths.

Waves that reach this observer are squeezed to shorter "blue-shifted" wavelengths.

Moving source of light

Speed of light

c

This observer sees no Doppler shift.

sound reaching you from the object has a higher pitch than the sound given off by the object if it were stationary. Conversely, if an object is moving away from you, the waves reaching you from the object are spread out. (Again, refer to Figure 4.22.) In the case of sound, this means that the pitch of the sound drops, in line with our experience with the fire truck.

This same phenomenon, which is referred to as the **Doppler effect,** occurs with light as well as with sound. If an object is moving toward you, the light reaching you from the object has a shorter wavelength than the light emitted by the object. In other words, the light that you see is bluer than if the source were not moving toward you. We say that the light from an object moving toward us is **blue-shifted.** On the other hand, light from a source that is moving away from you is shifted to longer wavelengths. The light that you see is redder in color than if the source were not moving toward you, so we say that the light is **red-shifted.**

4.2 DOPPLER

Light from approaching objects is blue-shifted. Receding objects are red-shifted.

As long as the speed of an object is much less than the speed of light (which means we can ignore relativistic effects), the wavelength of the Doppler-shifted light that you see, λ_1, is given by the equation

$$\lambda_1 = \left(1 + \frac{v_r}{c}\right)\lambda_0.$$

In this expression λ_0 is called the **rest wavelength** of the light. It is the wavelength of the light as measured in the frame of reference of the source of the light. The velocity, v_r, is the velocity at which the object is moving *away* from you. To be more precise, v_r is the rate at which the distance between you and the object is changing. If v_r is positive, it means that the object is getting farther from you. If v_r is negative, it means that the object is getting closer to you.

Take a moment to be sure this expression makes sense to you. Our argument says that, if an object is moving away from you, the wavelength of the light coming to you from that object should be greater than it would be if the object were "at rest"(not moving away from you). If the object is moving away then v_r will be greater than 0. If v_r is greater than 0, then $1 + v_r/c$ is greater than 1, and therefore $\lambda_1 = (1 + v_r/c)\lambda_0$ will be larger than λ_0. This is as we expected. What about the opposite case? If an object is moving toward us, then v_r is less than 0, so $(1 + v_r/c)$ will be less than 1. Now we find that λ_1 is shorter than λ_0. The observed wavelength is shorter than the rest

wavelength, so we see a blue shift. This is again in line with our expectations.

It is important to remember that the Doppler shift provides information only about whether an object is moving toward you or away from you. That is what the subscript r in v_r signifies. This stands for **radial velocity,** which is the rate at which the distance between you and the object is changing. At the moment that the fire truck is passing you, it is neither getting closer to you nor getting farther away from you, and so the pitch that you hear is the same as the pitch heard by the crew riding along on the truck. You can see this directly by looking at Figure 4.22 or by referring to the equation above. If an object is moving perpendicular to your line of sight, then $v_r = 0$, and $\lambda_1 = \lambda_0$. The observed wavelength equals the rest wavelength. The light is neither red-shifted nor blue-shifted.

In the case of light we need to stress again the caveat that this only holds true if the object's velocity is much less than the speed of light. If an object is moving at close to the speed of light, then special relativity says that time will be running slowly for that object. In this case, the frequency of light from the object will be reduced even if the object is moving perpendicular to our line of sight. (The time between ticks of the object's clock will be spread out, which is the same as saying that the frequency of those ticks will be reduced.) The relativistic correction to the Doppler effect can be quite important for some astronomical sources.

The amount by which the wavelength of light is shifted by the Doppler effect is called the **Doppler shift** of the light. Doppler shifts become especially useful if we are looking at an object that has emission or absorption lines in its spectrum. If we can identify a spectral line as being such and such a line from such and such a type of atom, then we know the rest wavelength λ_0 of the line from measurements in terrestrial laboratories. If we

Radial velocity is measured from Doppler shifts of emission or absorption lines.

measure that this line has a wavelength of λ_1 in the spectrum of a distant object, then we know all we need to know to calculate how rapidly the object is moving toward or away from us. Turning the above expression around a bit we can write

$$v_r = \frac{\lambda_1 - \lambda_0}{\lambda_0}c.$$

If you know λ_0, just measure λ_1, plug both values into this expression, and *voila!* you know what v_r is.

It is probably useful to illustrate this with an example. There is a prominent spectral line of hydrogen atoms seen at a rest wavelength of $\lambda_0 = 0.6563$ μm (Figure 4.21). Suppose you go to the telescope and measure the wavelength of this line in the spectrum of a distant object and find that instead of seeing the line at 0.6563 μm, you instead see the line at a wavelength of $\lambda_1 = 0.6590$ μm. You could then infer that the object is moving at a velocity of

$$
\begin{aligned}
v_r &= \frac{\lambda_1 - \lambda_0}{\lambda_0} c \\
&= \frac{0.6590\ \mu\text{m} - 0.6563\ \mu\text{m}}{0.6563\ \mu\text{m}} \times (3 \times 10^8\ \text{m/s}) \\
&= 1.2 \times 10^6\ \text{m/s}.
\end{aligned}
$$

The object is moving away from you with a radial velocity of 1.2×10^6 m/s, or 1,200 km/s.

4.5 WHY MERCURY IS HOT AND PLUTO IS NOT

If you ask a child in elementary school why Mercury is hot and Pluto is cold, the child will probably look at you like you are hopelessly uninformed, then patiently explain to you that Mercury is hot because it is close to the Sun, while Pluto is cold because it is much farther away. This explanation is fine as far as it goes, but we should push on it a bit harder. Why are the planets as hot as they actually are? Why does the surface of Mercury reach temperatures that are hot enough to melt lead, while the surface of Pluto remains so cold that even substances such as methane and ammonia remain permanently frozen? Closer to home, why is the surface of Earth hot enough for water to melt over most of the planet, but cold enough for that water to remain liquid?

The temperature of any object is determined by two things: what is trying to heat up the object, and what is trying to cool it off. If an object's temperature is constant, then these two must be in balance with each other. Your body, for instance, is heated by the release of chemical energy from inside. It is also heated by energy from your surroundings. If you are standing outside in the sunshine on a hot day, the hot air around you and the sunlight falling on you both are working to heat you up. In response to all of this heating, there must be some way that your body manages to cool itself off. This is where perspiration comes in. When you perspire, water seeps from the pores in your skin and evaporates. It takes energy to evaporate

If an object's temperature is constant, then heating must balance cooling.

water, and much of this energy comes from your body. Thus, as the perspiration evaporates, it carries away your body's **thermal energy,** cooling your body down. In order for your body temperature to remain stable, the heating must be balanced by the cooling. Such a balance is referred to as **thermal equilibrium.** If your body is out of thermal equilibrium in one direction—if there is more heating than cooling—then your body temperature climbs. If your body is out of thermal equilibrium in the other direction—if there is more cooling than heating—then your body temperature falls. (When your body is in equilibrium, we can write this down by putting the amount of cooling on one side of an equation, the amount of heating on the other side, and an equals sign in between.)

Planets have a thermal equilibrium as well, and electromagnetic radiation plays a crucial role in maintaining that equilibrium. The energy from light from the Sun heats the surface of Earth, driving its temperature up. This is one side of the equilibrium. The other side of Earth's thermal equilibrium is also controlled by light energy. Earth radiates energy back into space, cooling Earth. Our eyes are not sensitive to the wavelengths of light that Earth radiates, but it is there nonetheless. Overall, Earth must radiate away just as much energy into space as it absorbs from the Sun. Just as we saw with our bodies, if there were more heating than cooling, then the temperature of Earth would climb. (It would be absorbing more energy from the Sun than it was getting rid of, and this imbalance would show up in increasing temperature.) If there were more cooling than heating, then the temperature would fall. For Earth to remain the same average temperature over time—which it has done for many centuries—the energy that Earth radiates away into space must exactly balance the energy absorbed from the Sun. Thermal equilibrium must be maintained.

Earth is heated by sunlight and cooled by energy radiating back into space.

Equilibrium is an important concept in science. There are many kinds of equilibrium besides thermal equilibrium, some of which we will encounter later in the book. See Foundations 4.1 for an explanation of equilibrium's basic properties.

We now have a qualitative understanding that is interesting, but still is not yet useful. We would like to turn this intuitive idea of thermal equilibrium into a real prediction for the temperatures of the planets. To do this we need to find out more about light and temperature and the relationship between the two. Before going too far down this path, however, we should start by better understanding what we mean by "temperature."

FOUNDATIONS 4.1

EQUILIBRIUM MEANS BALANCE

Equilibrium is the term we use to refer to systems that are in balance. Imagine two well-matched teams struggling in a contest of tug-of-war. Each team pulls steadfastly on the rope, but the force of their pull is only enough to match but not overcome the force exerted by their opponent. Muscles flex, but the scene does not change. A picture taken now and another taken five minutes from now would not differ in any significant way. In this book we will frequently encounter this kind of **static equilibrium,** represented by a tug-of-war where opposing forces just balance each other. The equilibrium between the downward force of gravity and the pressure which opposes it will play a central role in the stories of planetary interiors, planetary atmospheres, and the interiors of stars.

However, not all types of equilibrium are static. Equilibrium can also be dynamic, in which the system is constantly changing. However, one source of change is exactly balanced by another source of change, so that the configuration of the system remains the same. **Figure 4.23** shows a simple example of **dynamic equilibrium.** A can with a hole cut in the bottom has been placed underneath an open water faucet. The depth of the water in the can determines how fast water pours out through the hole in the bottom of the can. Once the water reaches just the right depth, as in Figure 4.23(a), then water continues to pour out of the bottom of the can at exactly the same rate it pours from the faucet into the top of the can. The water leaving the can balances the water entering the can, and equilibrium is established. If you were to take a picture now and another picture a few minutes from now, little of the water in the can would be the same. Even so, the pictures would be indistinguishable.

If a system is not in equilibrium, then its configuration will change. Look at Figure 4.23(b). Here the level of the water in the can is too low, so water will not flow out of the bottom of the can fast enough to balance the water flowing into the can, causing the water level to rise. A picture taken now and another a short time later will not look the same. The configuration of the system is changing. Likewise, if the water level in the can is too high, as in Figure 4.23(c), water will flow out of the can faster than it flows into the can, and the water level will fall. Once again, if the system is not in equilibrium, its configuration will change.

Water passing through a can is an example of a **stable equlibrium.** When a stable equilibrium is *perturbed* (forced away from its equilibrium configuration), it will tend to return to its equilibrium state. If the water level is too high or too low, it will move back towards its equilibrium level. The equilibrium that we discuss in this chapter, between sunlight falling on a planet and thermal energy radiated away into space, is a stable equilibrium. If you take Figure 4.23 and replace "water in" with "sunlight absorbed," "water out" with "energy radiated by the planet," and "water level" with "the planet's temperature," then the stable equilibrium that sets the level of the water in the can becomes the stable equilibrium that sets the temperature of a planet.

An equilibrium is not necessarily stable. Consider a book standing on its edge, unsupported on either side by other objects or books. If you give the book a nudge, it will fall over, rather than settling back into its original position. This is an example of an **unstable equilibrium.** When an unstable equilibrium is perturbed, it will move further away from equilibrium rather than back toward it.

TEMPERATURE IS A MEASURE OF HOW ENERGETICALLY PARTICLES ARE MOVING ABOUT

When we say that something is hot or cold, we know exactly what we mean. In everyday life, *hot* and *cold* are defined in terms of our subjective experiences. Something is hot when it "feels" hot, and cold when it "feels" cold. But our perceptions of hot and cold are a layer of subjective experience that comes between us and a definable, quantifiable concept called **temperature.** When we talk of temperature, we talk of "degrees" on a thermometer. The way we define a "degree" is arbitrary—a matter of convention. If you grew up in the United States, for example, you probably think of temperatures in "degrees Fahrenheit," while if you grew up anywhere else in the world,

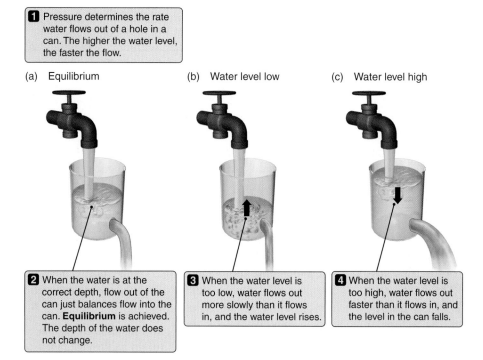

1 Pressure determines the rate water flows out of a hole in a can. The higher the water level, the faster the flow.

(a) Equilibrium (b) Water level low (c) Water level high

2 When the water is at the correct depth, flow out of the can just balances flow into the can. **Equilibrium** is achieved. The depth of the water does not change.

3 When the water level is too low, water flows out more slowly than it flows in, and the water level rises.

4 When the water level is too high, water flows out faster than it flows in, and the level in the can falls.

Figure 4.23 *Water flowing into and out of a can determines the water level in the can. This is an example of dynamic equilibrium.*

you think of temperatures in "degrees Celsius." Both of these are perfectly reasonable scales for measuring temperatures. But what is it that the thermometer is actually measuring?

What we refer to as temperature is actually a measurement of how energetically the atoms that make up an object are moving about. The air around you is composed of vast numbers of atoms and molecules. Those molecules are not sitting still, but are moving about every which way. Some are moving slowly, while some are moving more rapidly. Similarly the atoms that make up the chair you are sitting in or the floor that you are walking on (or the anatomical parts of your body that are involved in those two activities) are constantly wiggling about. We can characterize these motions by talking about the average kinetic energy (energy of motion) that these atoms have. The more energetically the atoms or molecules in something are bouncing about, the higher its temperature. In fact, the random motions of atoms and molecules are often referred to as their **thermal motions** to emphasize the connection between these motions and temperature.

> **Higher temperature means more energetic random motions of atoms and molecules.**

This definition of temperature should make some intuitive sense. If something is hotter than we are,

our experience says that thermal energy flows from that object into us. At the atomic level that means that the atoms in the object are bouncing around more energetically than the atoms in our bodies, so that if we touch the object, its atoms collide with our atoms, causing the atoms in our body to move about faster. Our body gets hotter as thermal energy flows from the object to us. (At the same time, these collisions rob the particles in the object of some of their energy. Their motions slow down. The hotter object becomes cooler.) In general, when we talk about heating, we mean processes that increase the average thermal energy of the particles something is composed of, and when we talk about cooling, we mean a process that decreases the average thermal energy of those particles.

The change in thermal energy associated with a change of one unit, or degree, is arbitrary on any temperature scale. On the Fahrenheit scale there are 180 degrees between the melting point (32°) and the boiling point (212°) of water at sea level. On the Celsius scale there are 100 degrees between those two temperatures. For these two scales, the temperature corresponding to "zero degrees" is also arbitrary. On the Celsius scale, 0°C is chosen to be the temperature at which water freezes, while on the Fahrenheit scale 0°F is placed at a temperature corresponding to −17.78°C. However, there is a lowest-possible physical temperature below which no object can fall. As the motions of the atoms in an object slow down, the temperature drops lower and lower. When the motions of the particles finally stop, things have gotten as cold as they can get. This lowest possible temperature, where thermal motions have come to a standstill, is called **absolute zero.** Absolute zero corresponds to −273.16°C, or −459.7°F.

> **Absolute zero, the temperature at which thermal motions stop, is zero on the Kelvin scale.**

The preferred temperature scale for most scientists is the **Kelvin scale.** For convenience, the size of one unit on the Kelvin scale, called a **kelvin** (abbreviated K) is the same as the Celsius degree. What makes the

Kelvin scale special is that 0 kelvins is set equal to that absolute lowest temperature where thermal motions stop, absolute zero. There are no negative temperatures on the Kelvin scale. The importance of the Kelvin scale is that *when temperatures are measured in kelvins, we know that the average thermal energy of particles is proportional to the measured temperature.* The average thermal energy of the atoms in an object with a temperature of 200 K is twice the average thermal energy of the atoms in an object with a temperature of 100 K.

HOTTER MEANS MORE LUMINOUS AND BLUER

So far we have focused our attention on the way discrete atoms emit and absorb radiation. This led us to a very useful understanding of emission lines and absorption lines and how we might use these lines to learn a great deal about the physical state and motion of distant objects. But not all objects have spectra that are dominated by discrete spectral lines. For instance, if you pass the light from an incandescent lightbulb through a prism, as we saw in Figure 4.20(a), then instead of discrete bright and dark bands you will see light spread out smoothly from the blue end of the spectrum to the red. Similarly, if you look closely at the spectrum of the Sun, you will see absorption lines, but mostly what you will see is light smoothly spread out across all of the colors of the spectrum—"Roy G. Biv." What is the origin of such **continuous radiation,** and what clues might this kind of radiation carry about the objects which emit it?

Objects like lightbulbs emit continuous radiation at all wavelengths.

We can think of a dense material as being composed of a collection of charged particles that are being jostled about as their thermal motions cause them to run into their neighbors. The hotter the material is, the more violently the particles that make it up are being jostled. Any time a charged particle is subjected to an acceleration it radiates, and so the jostling of particles due to their thermal motions causes them to give off a continuous spectrum of electromagnetic radiation. This is why *any* material that is sufficiently dense for its atoms to be jostled by their neighbors will emit light *simply because of its temperature.* Radiation of this sort is called **thermal radiation.**

Thermal motions of charged particles cause dense objects to radiate.

We can guess how the radiation from an object changes as the object heats up or cools off. Start with the question of the **power** (energy per second, measured in Watts) radiated, which is referred to as the object's **luminosity.** The hotter the object, the more energetically the charged particles within it wiggle about. (Again, this is what it *means* to be hot.) The more energetically that the charged particles wiggle about, the more energy they emit in the form of electromagnetic radiation. So, as an object gets hotter and hotter, we expect the light that it emits to get more and more intense. Here is our first intuition about thermal radiation—*hotter means more luminous.*

Making an object hotter makes its thermal radiation more luminous.

Now we move on to the question of what color the light is that an object emits. Again, as the object gets hotter, the thermal motions of the particles within it get more and more energetic. These more energetic motions are capable of producing more energetic photons. So, as an object gets hotter, we might expect the average energy of the photons that it emits to become greater. In other words, we might expect the average wavelength of the emitted photons to get shorter. The light from the object gets bluer in color. Here is our second intuition about thermal radiation—*hotter means bluer.*

Making an object hotter also makes its thermal radiation bluer.

Both of these intuitive predictions are borne out by a simple experiment you can do. An incandescent lightbulb is a good example of an object that emits thermal radiation. The electric current in the filament in the lightbulb heats the filament. (More precisely, electrons being pushed through the filament by electric fields collide with atoms in the filament, increasing the thermal motions of those atoms.) The hot filament then glows. Somewhere in your home or dorm or office, you can probably find a lightbulb with a rheostat, or a "dimmer," on it. When you turn the knob on the dimmer, it changes the amount of electric current in the filament in the lightbulb. Turning the dimmer up increases the current, which increases the number and strength of collisions between electrons and atoms, which in turn increases the temperature of the filament. The hotter filament is more luminous. This confirms the first of our expectations: *hotter means more luminous.*

But what about the color of the emitted light? Look again at the lightbulb as you turn up the dimmer. When the bulb first comes on, it glows a very dull red, but as the current through the bulb is increased, driving up the temperature of the filament, the perceived color of the light changes. When the dimmer is turned all the way up, the light from the bulb has lost its red tint. The hotter the lightbulb gets, the more energetic blue photons there are mixed in with the less energetic red photons, and the light becomes whiter in color. The color of the light shifts from the red toward the blue, confirming our second intuitive expectation—*hotter means bluer.*

These arguments offer some intuitive grasp of the way that the light given off by an object depends on the temperature of the object, but to be really useful we need to get quantitative. We need to know *how much* more luminous and *how much* bluer. The detailed answers to these questions were worked out around 1900 by Max Planck. Planck was thinking about a special situation—a hollow, totally enclosed cavity of material at a specific temperature, T. The crucial point here is that inside the cavity all of the radiation emitted by the walls of the cavity is also absorbed by the walls of the cavity. In this situation a balance is set up, with each small bit of the wall emitting just as much thermal radiation as it is absorbing from its surroundings. Physicists refer to such a special situation as a **blackbody.** Planck used this balance to calculate the spectrum of the light inside such a cavity. The result of his calculation, which beautifully matches the results of experiments, is called a **Planck spectrum** or a **blackbody spectrum.**

A blackbody emits thermal radiation that has a Planck spectrum.

You might reasonably ask what this hypothetical cavity has to do with the light from the filament of a lightbulb. Surely the filament of a lightbulb is not a cavity of this sort! But in a certain sense it is. The light emitted by charged particles within the filament is mostly absorbed by other charged particles within the filament. This is exactly the assumption that Planck made when calculating the shape of the spectrum of a blackbody. As a result, we expect that the radiation existing *inside the filament of the lightbulb* will have a Planck spectrum. It is this radiation which "leaks out" of the filament, just as light might leak out of a small hole in the side of Planck's cavity. As a result, the radiation from the filament of a lightbulb is very close to a Planck spectrum. The light from stars such as the Sun and the thermal radiation from a planet also often come close to having a blackbody spectrum.

STEFAN'S LAW SAYS THAT HOTTER MEANS MUCH MORE LUMINOUS

Figure 4.24 shows plots of the Planck spectra for objects at several different temperatures. We now ask the question that scientists must always ask: Does the theoretical prediction agree with observation and experiment? Do these spectra agree with our intuitive ideas and with our experiment with the lightbulb and the dimmer? Begin with luminosity. As the temperature of an object increases, Planck's theory says that the object gives off more radiation at every wavelength, and so the luminosity of the object should increase. In fact, it increases in a hurry. Adding up all of the energy in a Planck spectrum shows that the increase in luminosity is proportional to the *fourth power* of the temperature: luminosity $\propto T^4$. This result is known as **Stefan's law,** because it was discovered in the laboratory by Josef Stefan before Planck's theory came along to explain it.

The luminosity of a blackbody is proportional to T^4.

What Stefan's law actually says is that the amount of energy radiated *by each square meter* of the surface of an object is given by the equation

$$\mathcal{F} = \sigma T^4.$$

In this equation, $\mathcal{F}$ is called the **flux.** It is a measurement of the total amount of energy coming through each square meter of the surface each second. The constant σ (pronounced "sigma") is

Stefan's law gives the amount of thermal radiation from each m^2 of a blackbody surface.

Figure 4.24 *Planck spectra emitted by sources with temperatures of 2,000 K, 3,000 K, 4,000 K, 5,000 K, and 6,000 K. At higher temperatures the peak of the spectrum shifts toward shorter wavelengths, and the amount of energy radiated per second from each square meter of the source increases.*

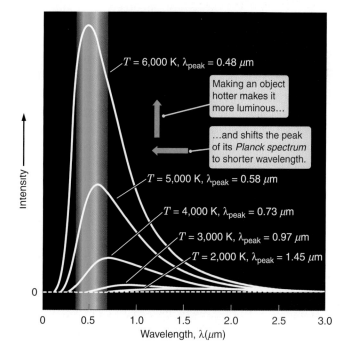

called **Stefan-Boltzmann constant,** named after the discoverer of this relationship. The value of σ is the same for all cases, and is given by 5.67×10^{-8} W/(m² K⁴) (where W stands for watts). To find the total amount of energy emitted by an object in the form of electromagnetic radiation, multiply $\mathcal{F}$ by the surface area of the object.

Returning to our lightbulb example, we can actually use Stefan's law to figure out what the surface area of the filament in a lightbulb must be. Suppose the filament in an incandescent bulb operates at a temperature of about 2,500 K. The amount of energy radiated by the bulb is stamped right on the face of the bulb—a 100 W bulb has a luminosity of 100 W, which means that it radiates away 100 joules each second. (A **joule** is a unit of energy, and is abbreviated J.) If the total amount of light from the filament is equal to the flux ($\mathcal{F}$) times the surface area of the filament (A), then we can write

$$100 \text{ W} = A \times \sigma T^4.$$

Solving this equation for the area, we get

$$A = \frac{100 \text{ W}}{\sigma T^4} = \frac{100 \text{ W}}{\left(5.67 \times 10^{-8} \dfrac{\text{W}}{\text{m}^2 \text{K}^4}\right) \times (2{,}500 \text{ K})^4}$$

$$= 4.5 \times 10^{-5} \text{m}^2.$$

Note that not much filament surface area is needed to provide the light that turns night into day in our homes and cities.

Stefan's law says that not only does an object become more luminous as the temperature increases, but it

Slight changes in temperature mean large changes in brightness.

becomes more luminous *in a hurry* as the temperature increases. If the temperature of an object goes up by a factor of 2, then the amount of energy being radiated each second increases by a factor of 2^4, or 16. If the temperature of an object goes up by a factor of 3, then the energy being radiated by the object each second goes up by a factor of $3^4 = 81$! A lightbulb with a filament temperature of, say, 3,000 K radiates 16 times as much light as it would if the filament temperature were 1,500 K. Even modest changes in temperature can result in large changes in the amount of power radiated by an object.

WIEN'S LAW SAYS THAT HOTTER MEANS BLUER

Look again at Figure 4.24, but this time instead of paying attention to how high each curve is, notice where the peak of each curve falls. As the temperature increases, the *peak* of the Planck spectrum shifts toward

shorter and shorter wavelengths, which means that the average energy of the photons becomes greater and greater. Just as we surmised above, increasing the temperature causes the light from the object to get bluer. The shift in the location of the peak of the Planck spectrum with increasing temperature is given by the equation

$$\lambda_{\text{peak}} = \frac{2{,}900 \ \mu\text{m K}}{T}.$$

This result is referred to as **Wien's law.** In this equation λ_{peak} (pronounced "lambda peak") is the wavelength where the Planck spectrum is at its peak. It is the wavelength where the electromagnetic radiation from an object is greatest. Wien's law

The peak wavelength of a blackbody is inversely proportional to its temperature.

says that the location of the peak in the spectrum is inversely proportional to the temperature of the object. If you increase the temperature by a factor of 2, the peak wavelength becomes half of what it was. If you increase the temperature by a factor of 3, the peak wavelength becomes a third of what it was.

4.3 SW LAWS

It is useful to put a few numbers into Wien's law. The surface of the Sun, for example, has a temperature of about 5,800 K. Wien's law says that the peak in the light from the Sun occurs at a wavelength of

$$\lambda_{\text{peak},\odot} = 0.5 \ \mu\text{m}.$$

The light given off by the Sun is concentrated at a wavelength of about 0.5 μm, which is in the middle of what we refer to as the visible part of the spectrum.

Wien's law will prove to be very handy as we continue our study of the Universe. If we can measure the spectrum of an object emitting thermal radiation and find where the peak in the spectrum is, then we can use Wien's law to calculate the temperature of the object. Turning our previous

The temperature of an object is calculated from its spectrum using Wien's law.

example around, we have no way of dropping a thermometer into the Sun and directly measuring its temperature, but we *can* observe the spectrum of the light coming from the Sun. When we do so, we find that the peak in the spectrum occurs at a wavelength of about 0.5 μm. Wien's law can be rewritten as

$$T = \frac{2{,}900 \ \mu\text{m K}}{\lambda_{\text{peak}}}.$$

If we take the observed peak of the spectrum of the Sun ($\lambda_{\text{peak}} \approx 0.5 \ \mu$m) and plug it into this equation, we get

$$T = \frac{2{,}900 \ \mu\text{m K}}{0.5 \ \mu\text{m}} = 5{,}800 \text{ K.}$$

This is how we know the temperature of the Sun.

4.6 TWICE AS FAR MEANS ONE-FOURTH AS BRIGHT

You might have noticed above that we consistently spoke of the "luminosity" of objects, where in everyday language we probably would have just said that one object is "brighter" than another. This is a case where everyday language is too sloppy for science. To a physicist or astronomer, the **brightness** of electromagnetic radiation refers to the amount of light that is arriving at some location, such as the page of the book you are reading, or the pupil of your eye. Luminosity refers to the amount of light leaving a source. The concept of brightness is certainly related to the concept of luminosity. For example, replacing a lightbulb with a luminosity of 50 W with a 100 W bulb succeeds in making a room twice as bright, since it doubles the light reaching any point in the room. But brightness also depends on the distance from a source of electromagnetic radiation. If you needed more light to read this book by, you *could* replace the bulb in your lamp with a more luminous bulb, but it would probably be easier to just move the book closer to the light. Conversely, if a light were too bright for you, you would move away from it. Our everyday experience says that as we move away from a light, its brightness decreases.

The particle description of light provides a convenient way to think about the brightness of radiation, and how brightness depends on distance. Suppose you had a piece of cardboard that was 1 meter on a side. Intuitively, you might imagine that making the light that falls on the cardboard twice as bright would mean doubling the number of photons that hit the cardboard each second. Tripling the brightness of the light would mean increasing the number of photons hitting the cardboard each second by a factor of 3, and so on. Here is a beginning point for understanding brightness. Brightness depends on the number of photons falling on each square meter of a surface each second.

Brightness measures how much light falls per square meter per second.

Working with this idea of brightness, now imagine a lightbulb sitting at the center of spherical shell, as

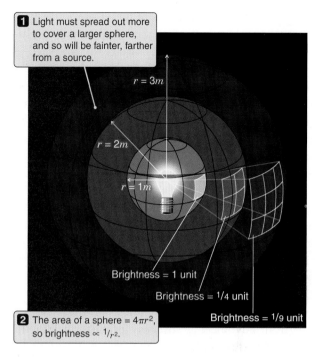

1 Light must spread out more to cover a larger sphere, and so will be fainter, farther from a source.

r = 3m

r = 2m

r = 1m

Brightness = 1 unit

Brightness = 1/4 unit

Brightness = 1/9 unit

2 The area of a sphere = $4\pi r^2$, so brightness ∝ $1/r^2$.

Figure 4.25 *Light obeys an inverse square law as it spreads out away from a source. Twice as far means one fourth as bright.*

shown in **Figure 4.25.** Photons from the bulb travel off in all directions, and land on the inside of the shell. To find the number of photons landing on each square meter of the shell during each second (that is, to find the brightness of the light), take the *total* number of photons given off by the light bulb each second, and divide by the number of square meters those photons have to be spread over. The number of square meters of surface area of a sphere is given by the formula $A = 4\pi r^2$, where r is the distance between the bulb and the surface of the sphere (thus r = the radius of the sphere). So when written as a formula, we find that

$$\begin{aligned} \text{Number of photons} \atop \text{striking one square} \atop \text{meter each second} &= \frac{\text{total number of photons} \atop \text{emitted per second}}{\text{number of square meters} \atop \text{the photons are spread over}} \\[2mm] &= \frac{\text{total number of photons} \atop \text{emitted each second}}{4\pi r^2}. \end{aligned}$$

The next step in building an understanding of brightness is to change the size of the spherical shell, while keeping the total number of photons given off by the lightbulb the same. As the shell becomes larger, the photons from the lightbulb must spread out to cover

Like gravity, light obeys an inverse square law.

a larger surface area. Each square meter of the shell receives fewer photons each second, and so the brightness of the light falls. If the surface of the shell is moved twice as far from the light, then the area over which the light must be spread increases by a factor of $2^2 = 2 \times 2 = 4$. The photons from the bulb are spread out over four times as much area, and so the number of photons falling on each square meter each second becomes $\frac{1}{4}$ of what it was. If the surface of the sphere is three times as far from the light as illustrated in Figure 4.25, then the area over which the light must be spread increases by a factor of $3 \times 3 = 3^2 = 9$, and the number of photons falling on each square meter becomes $\frac{1}{9}$ of what it was originally. We encountered just this kind of relationship earlier when we talked about gravity (Chapter 3). Just like gravity, light obeys an inverse square law. The brightness of the light from an object is inversely proportional to the square of the distance from the object. Twice as far means one-fourth as bright.

4.4 ISL

It is nice to think of brightness in terms of photons streaming onto a surface from a light, because this provides us with a nice mental picture of the physical nature of brightness and why brightness follows an inverse square law. In practice, however, it is usually more convenient to speak of the *energy* coming to a surface each second, rather than the number of photons received.

The luminosity of an object is the total number of photons given off by the object times the energy of each photon. Instead of talking about how the number of photons must spread out to cover the surface of a sphere (brightness), we now talk about how the *energy* carried by the photons must spread out to cover the surface of a sphere. When speaking of brightness in this way, we mean the amount of energy falling on a square meter in a second. If L is the luminosity of the bulb, then the brightness of the light at a distance r from the bulb is given by

$$\text{Brightness} = \frac{\text{energy radiated per second}}{\text{area over which energy is spread}}$$

$$= \frac{L}{4\pi r^2}.$$

Before moving on we offer the following aside. Usually, the only information that astronomers have to work with is the light from a distant object. For this reason, we will use our understanding of radiation over and over again throughout our journey. Time spent now thinking carefully about the electromagnetic spectrum, emission and absorption of photons, Planck radiation, and the inverse square law for brightness will be a *very* good investment for what is to come.

4.7 RADIATION LAWS ALLOW US TO CALCULATE THE EQUILIBRIUM TEMPERATURES OF THE PLANETS

We began our discussion of thermal radiation by asking a straightforward question: "Why does a planet have the temperature that it does?" In a qualitative way we said that the temperature of a planet is determined by a balance between the amount of sunlight being absorbed and the amount of energy being radiated back into space. We now have the tools that we need to turn this qualitative idea into a real prediction of the temperatures of the planets.

Begin with the amount of sunlight being absorbed. The amount of energy absorbed by a planet is just the area of the planet that is absorbing the energy times the brightness of sunlight at the planet's distance from the Sun. When we look at a planet, we see a circular disk with a radius equal to the radius of the planet. The area of this circular disk is πR^2, where R is the radius of the planet. We found in our discussion in Section 4.6 that the brightness of sunlight at a distance d from the Sun is equal to the luminosity of the Sun ($L_\odot$ in watts) divided by $4\pi d^2$. (This d is the same as the r in the previous section. We use d here to avoid confusion with the planet's radius, R.) There is one additional factor which we must consider. Not all of the sunlight falling on a planet is absorbed by the planet. The fraction of the sunlight that is reflected from a planet is called the **albedo** of the planet. The corresponding fraction of the sunlight that is absorbed by the planet is 1 minus the albedo. A planet with an albedo of 1 reflects all of the light falling on it. A planet which absorbs 100% of the sunlight falling on it has an albedo of 0.

We can calculate the energy absorbed by a planet . . .

Writing this as an equation, we say that

$$\begin{pmatrix} \text{Energy absorbed} \\ \text{by the planet} \\ \text{each second} \end{pmatrix} = \begin{pmatrix} \text{absorbing} \\ \text{area of} \\ \text{planet} \end{pmatrix} \times \begin{pmatrix} \text{brightness} \\ \text{of sunlight} \end{pmatrix} \times \begin{pmatrix} \text{fraction} \\ \text{of sunlight} \\ \text{absorbed} \end{pmatrix}$$

$$= \pi R^2 \times \frac{L_\odot}{4\pi d^2} \times (1 - a),$$

where a is the albedo of the planet.

Moving on to the other piece of the equilibrium, the amount of energy that the planet radiates away into space each second is just the number of square meters of surface area that the planet has times the

... **as well as the thermal radiation emitted by a planet.**

power radiated by each square meter. The number of square meters of surface area for the planet is given by $4\pi R^2$. Stefan's law tells us that the power radiated by each square meter is given by σT^4. So we can say that

$$\begin{pmatrix}\text{Energy radiated by}\\\text{planet per second}\end{pmatrix} = \begin{pmatrix}\text{surface area}\\\text{of planet}\end{pmatrix} \times \begin{pmatrix}\text{energy radiated by}\\\text{each m}^2\text{ each second}\end{pmatrix}$$

$$= 4\pi R^2 \times \sigma T^4.$$

In equilibrium, these two must balance each other.

If the planet's temperature is to remain stable—if it is to keep from heating up or cooling off—then it must be radiating away just as much energy into space as it is absorbing in the form of sunlight, as indicated in Figure 4.26. That means that we can equate these two expressions. We can set the quantity "Energy radiated by planet" equal to the quantity "Energy absorbed by planet." When we do this, we arrive at the expression

$$\begin{array}{c}\text{Energy radiated by}\\\text{the planet each second}\end{array} = \begin{array}{c}\text{energy absorbed by}\\\text{the planet each second,}\end{array}$$

or,

$$4\pi R^2 \sigma T^4 = \pi R^2 \frac{L_\odot}{4\pi d^2}(1-a).$$

Look at this equation for a moment. It may seem fairly complex, but when broken into pieces it becomes more digestible. On the left side of the equation, $4\pi R^2$ tells how many square meters of the planet's surface are radiating energy back into space, while σT^4 tells how much energy each one of those square meters radiates each second. Put them together, and you get the total amount of energy radiated away by the planet each second. On the right side of the equation, πR^2 is the area of the planet as seen from the Sun. That amount times the brightness of the sunlight reaching the planet, $L_\odot/4\pi d^2$ tells how much energy is falling on the planet each second. The final $1-a$ tells how much of that energy the planet actually absorbs. Put everything on the right side of the equation together, and you get the amount of energy absorbed by the planet each second. The equals sign says that the energy radiated away needs to balance the sunlight absorbed. There is no magic here. In fact, when broken down, the formidable equation above embodies little more than a few straightforward ideas such as "hotter means more luminous," "twice as far means one-fourth as bright," and "heating and cooling must balance each other." The

math just gives us a convenient way to work with these concepts.

We started down this path hoping to find a way to predict the temperatures of the planets, and a bit of algebra gets us the rest of the way there. Rearranging the previous equation to put T on one side and everything else on the other gives

$$T^4 = \frac{L_\odot(1-a)}{16\sigma\pi d^2}.$$

If we take the fourth root of each side, we wind up with

$$T = \left[\frac{L_\odot(1-a)}{16\sigma\pi}\right]^{1/4} \times \frac{1}{\sqrt{d}}.$$

Figure 4.26 *Planets are heated by sunlight and cooled by emitting thermal radiation into space. If there are no other sources of heating or means of cooling, then the equilibrium between these two processes determines the temperature of the planet.*

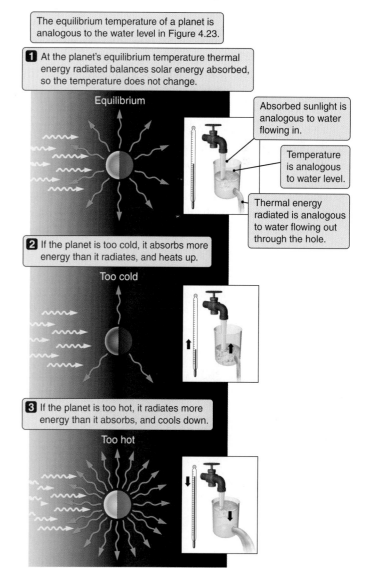

The equilibrium temperature of a planet is analogous to the water level in Figure 4.23.

1 At the planet's equilibrium temperature thermal energy radiated balances solar energy absorbed, so the temperature does not change.

Equilibrium

Absorbed sunlight is analogous to water flowing in.

Temperature is analogous to water level.

Thermal energy radiated is analogous to water flowing out through the hole.

2 If the planet is too cold, it absorbs more energy than it radiates, and heats up.

Too cold

3 If the planet is too hot, it radiates more energy than it absorbs, and cools down.

Too hot

We have now produced a full-fledged physical model for why the temperatures of planets are what they are. Restating the meaning in words, T is the temperature at which the energy radiated by a planet exactly balances the energy absorbed by the planet. If the planet were hotter than this equilibrium temperature, it would radiate energy away faster than the planet absorbed sunlight, and the temperature would fall. If the planet were cooler than this temperature, it would radiate away less energy than was falling on it in the form of sunlight, and the temperature of the planet would rise. Only at this equilibrium temperature do the two balance.

Balancing cooling and heating sets an equilibrium temperature.

The equation tells us that as the distance from the Sun increases—in other words, as d gets bigger—the temperature of the planet decreases. No surprise there. But now we know how much the temperature should decrease. It should be inversely proportional to the square root of the distance. Using this formula can turn our intuition about why planets that are close to the Sun are hot into a prediction of just how hot they should be.

Figure 4.27 shows a graph of the predicted temperatures of the planets. The vertical bars show the range of temperatures found on the surfaces of each planet (or, in the case of the giant planets, at the tops of their clouds). The black dots show our predictions using the equation above. From the figure, you can see that, overall, we are not too far off. That should give us a sense of accomplishment, because it says that our basic understanding of *why* planets have the temperatures that they do is probably not too far off. Mercury, Mars, and Pluto agree particularly well. (The agreement for Mercury would get much better if we took into account the huge difference in temperature between the daytime and nighttime sides of the planet, and recomputed our equilibrium accordingly.)

In other cases, however, our predictions are wrong. For Earth and the giant planets the actual temperatures are a bit higher than the predicted temperatures. In the case of Venus the actual surface temperature is wildly higher than our prediction. Rather than cause for despair, these discrepancies between theory and observation are cause for excitement. As we built our physical model for the equilibrium temperatures of planets, we made a number of assumptions. For example, we assumed that the temperature of the planet was the same everywhere. This is clearly not true, since we might expect planets to be hotter on the day side than on the night side. We also assumed that a planet's only source of energy is the sunlight falling on it. Finally, we assumed that a planet is able to radiate energy into space freely as a blackbody. The discrepancies between our theory and the measured temperatures of some of the planets tell us that for these planets, some or all of these assumptions must be incorrect. In other words, the places where the predictions of our theory are not confirmed by observation point to areas where there is something still to be discovered and understood. The question of *why* these planets are hotter than the prediction will lead us to a number of new and interesting insights into how these planets work. Scientific theories sometimes succeed and sometimes fail, but even when they fail they can teach us a lot about the Universe.

Discrepancies between predictions and actual planetary temperatures lead to new knowledge.

Figure 4.27 *Predicted temperatures for the planets, based on the equilibrium between absorbed sunlight and thermal radiation into space, are compared with ranges of observed surface temperatures. Some predictions are correct. Interestingly, others are not.*

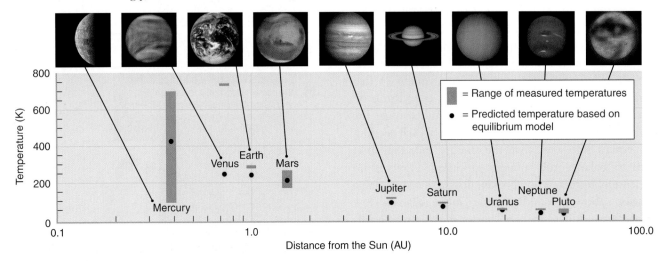

SEEING THE FOREST THROUGH THE TREES

In our daily lives, light is the ideal messenger, faithfully telling us about the world around us. As with any good courier, we usually concentrate on the information that light carries, taking the messenger itself for granted. It is easy to forget that our seemingly immediate visual perception of reality is actually a derived experience—the result of a sophisticated interplay between our eyes, our brains, and the flood of electromagnetic radiation that is emitted, absorbed, transmitted, and reflected by objects in the world around us.

The light that our eyes see is only a tiny portion of the full span of the electromagnetic spectrum. Electromagnetic radiation carries with it a wealth of information about the temperature, density, composition, state of motion, and other physical characteristics of the place of its origin and the material it interacts with *en route*. In place of the carefully controlled laboratory experiments of many other sciences, the astronomer uses a combination of ingenuity and technology to "slice up" the light reaching us and tease out the information it carries about conditions throughout the Universe. The role of electromagnetic radiation in astronomy is far more than that of messenger. Radiation is also a participant in the processes that we study. For example, it is light that carries energy from the Sun outward through the Solar System, heating the planets, and it is light that carries energy away from each planet, allowing it to cool. The balance between these two processes establishes the conditions of our existence.

Light may be both an informative messenger and an important player in the ebb and flow of the Universe, but it is the very nature of light itself that plays havoc with our commonsense ideas about the world. When we explored the motions of Earth, the Moon, and the planets, we relied heavily on our intuition about the world. The pull or shove of one object on another and the force of gravity that holds us tightly to the surface of Earth are well within the realm of our everyday experience. But when we consider the properties of light, the boundaries of our experience and intuition are shattered. We are forced beyond the confines of the "box" within which our brains evolved. Answering a simple question like "How fast does light travel?" demands that we abandon our most cherished ideas about the nature of space, time, matter, and energy. We ask, "What is light?" and confront abstract concepts like electric and magnetic fields, while running headlong into the seemingly impossible question of how something can be both a wave *and* a particle. We delve into the interaction between light and matter, and find that at the scale of atoms and photons, Newton's clockwork Universe crumbles. In its place we discover a world of random chance and uncertainty, governed not by the strict march of cause and effect, but by laws of probability and statistics. We ask about the nature of light—and we collide squarely with the shortcomings of our intuitive ideas about the nature of reality itself.

We are now ready to resume our outward journey, armed with a new appreciation of how important it is to proceed carefully. Common sense is not enough. We must instead rely on our growing understanding of physical law, and on rigorously tested predictions of carefully constructed theories. These are the tools that will allow us to look beyond the surface of spectacular vistas that otherwise would be devoid of sense or meaning or connection. Kepler's laws, Newton's laws, relative motion, Doppler shifts, Wien's law, Stefan's law, energy states, spectral lines, and the rest—here are the keys we will use to build an understanding of planets, stars, galaxies, and ultimately the Universe itself. The first place in which we will bring these tools to bear will be in our immediate neighborhood, as we consider the nature and origin of our Solar System.

STUDENT QUESTIONS

THINKING ABOUT THE CONCEPTS

1. Patterns of emission or absorption lines in spectra can uniquely identify individual atomic elements, just as DNA testing uniquely identifies individual human beings. Explain how positive identification of atomic elements can be used as one way of testing the validity of the cosmological principle.

2. The Sun loses more than 4 million tons of mass each second. Explain what happens to this mass.

3. Imagine a future cosmonaut traveling in a spaceship at 0.866 times the speed of light. Special relativity says that the length of his spaceship along the direction of flight is only half of what it was when it was at rest on Earth. He checks this out with a meter stick that he brought along with him. Would his measurement confirm the contracted length of his spaceship? Explain.

4. During a popular art exhibition, the museum staff finds that, to protect the artwork, they must limit

the total number of viewers in the museum at any one time. Therefore, new viewers are admitted at the same rate that others leave. Is this an example of static or of dynamic equilibrium? Explain.

5. Many physical properties are proportional to the temperature of an object, raised to some power. For example, as we have seen in this chapter, the luminosity of a radiating body is proportional to T^4. Why must the temperature here be expressed in Kelvins rather than degrees Celsius or Fahrenheit?

6. The difference between brightness and luminosity can be very confusing to many people. How would you explain the difference to a family member or a friend who is not taking this class?

7. Think up and describe a practical experiment that would prove the special relativity prediction that time passes more slowly in a moving frame of reference.

8. Consider two hypothetical airless planets. One orbits the Sun at an average distance of 5.0 AU and the other at an average distance of 10.0 AU, yet both have the same average surface temperature. Explain how this could be possible.

APPLYING THE CONCEPTS

9. If a spaceship approching us at 0.9 times the speed of light shines a laser beam at Earth, how fast will the photons in the beam be moving when they arrive at Earth?

10. We are given a bar of tungsten, a metal with a melting point of 3,640 K and a boiling point of 6,170 K. We heat the bar until it starts to melt, and plot a graph of its Planck energy distribution. We then continue to heat the tungsten until it begins to boil and plot a similar graph.
   a. Sketch and label these graphs on the same set of axes. At what wavelength does the spectrum of each peak?
   b. How many times more luminous is the boiling tungsten than the melting tungsten?

11. In our sky, the angular diameter of the Sun is about 30 minutes of arc. Now imagine that you have been transported to Neptune, 30 AU from the Sun.
   a. What would be the angular diameter of the Sun as seen in Neptune's sky?

b. Would the Sun appear to your eye as a small disk or as a point of light?
   c. How bright would the Sun appear compared to its brightness as seen from Earth?

12. Imagine you are riding in a car, tuned in to 790 on the AM radio dial. This station is broadcasting at a frequency of 790 kilohertz (7.9×10^5 Hz). What is the wavelength of the radio signal? You switch to an FM station at 98.3 on your dial. This station is broadcasting at a frequency of 98.3 megahertz (9.83×10^7 Hz). What is the wavelength of this radio signal?

13. An airless planet at 1 AU from the Sun would have an average blackbody surface temperature of 279 K if it absorbed all electromagnetic energy emitted by the Sun that falls on the planet.
   a. What would be the average temperature on this planet if its albedo were 0.1, typical of a rock-covered surface?
   b. What would be the average temperature if its albedo were 0.9, typical of a snow-covered surface?

14. On a dark night, you notice that a distant lightbulb happens to be the same brightness as a firefly that is 5 m away from you. If the lightbulb is a million times more luminous than the firefly, how far away is the lightbulb?

15. Refer to Question 13 in Chapter 3 about the hypothetical planet named Vulcan. If such a planet existed in an orbit $\frac{1}{4}$ the size of Mercury's, what would be the average temperature on Vulcan's surface? Assume that the average temperature on Mercury's surface is 450 K.

16. The average temperature of Earth's surface is approximately 290 K. If the Sun were to become 5% more luminous (1.05 times its present luminosity), how would Earth's temperature increase? (You might consider these to be only modest increases in solar luminosity and Earth's average temperature, but they would nevertheless have a drastic effect on our planet's climate.)

17. An isolated, rarified cloud of gas once contained atoms in an excited energy state. A thousand years after the excitation process stopped, half of the excited atoms decayed to the ground state, radiating photons in the process. What fraction of the original atoms would remain in the excited state after five thousand years have passed?

THE SOLAR SYSTEM

A BRIEF HISTORY OF THE SOLAR SYSTEM

5.1 FORMING STARS AND EVOLVING PLANETS

Until the last part of the 20th century, every discussion of the Solar System would necessarily start with an accounting of its pieces. "There are nine planets with such and such properties. There are moons, there are asteroids, and there are comets. And so forth." Any mention of the origin of the Solar System would wait until the end of the discussion—the part of a work to which speculation is generally relegated. Yet the last few decades have seen this ordering turned on its head. Today astronomers studying the formation of stars and planetary scientists studying clues about the history of the Solar System find themselves arriving at the same picture of our early Solar System but from two very different directions. This unified understanding provides the foundation for the way we now think about the Sun and the myriad objects that orbit about it.

Later in our journey we will extend our attention beyond the confines of our Solar System and learn about the incomprehensibly vast and tenuous clouds of gas that fill the expanse of interstellar space. We will discuss how, when conditions are right, this gas collapses under the force of its own gravity to form stars. In this chapter we will skip the details of this process, except

KEY CONCEPTS

At the opening of the 21st century we have come to see our Solar System as an unmistakable byproduct of the birth of the Sun, rather than a random collection of planets and moons. It is this understanding that brings order to what we see around us today. Here we begin our investigation of the Sun's family by recounting something of the story of the formation of the Solar System—a story in which we will learn that:

* A star forms when a cloud of interstellar gas and dust collapses under its own weight;
* When a star forms it is surrounded by a flat, rotating disk that provides the raw material from which a planetary system might form;
* Dust grains in the disk around a young star stick together to form larger and larger solid objects;
* Differences in temperature from place to place within the disk determine the kinds of materials from which solid objects can form;
* Giant planets form when solid planet-sized bodies capture gas from the surrounding disk;
* The atmospheres of smaller terrestrial planets are gases released by volcanism and volatile materials that arrived on comets; and
* Planetary systems are common around other stars.

(a)

(b)

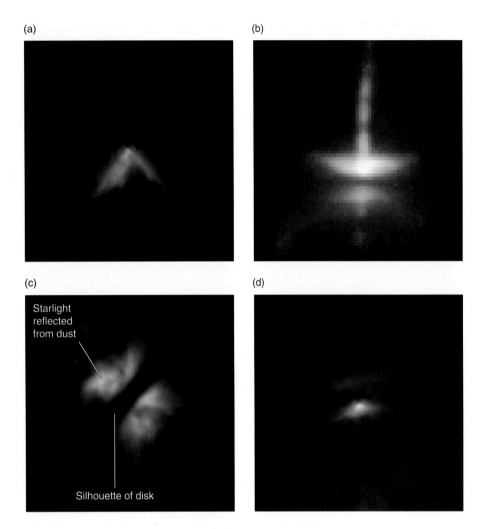

(c)

Starlight
reflected
from dust

Silhouette of disk

(d)

Figure 5.1 Hubble Space Telescope *images showing accretion disks around newly formed stars. The dark bands are the shadows of the disks seen more or less edge on. Bright regions are dust illuminated by starlight.*

The orbits of all of the planets in the Solar System lie very close to a single plane, which says that the early Solar System must have been flat. The fact that all of the planets orbit the Sun in the same direction says that the material from which the planets formed must have been swirling about the Sun in the same direction as well. Other clues about what the early Solar System was like were harder to puzzle out. *Meteorites,* for example, include bits and pieces of material that are left over from the Solar System's youth. These fragments of the early Solar System can be captured by Earth's gravity and fall to the ground, where they can be picked up and studied. Many meteorites, such as the one in **Figure 5.2,** look something like a piece of poorly mixed concrete in which pebbles and sand are mixed in with a much finer filler. This structure is surely telling us *something* about how these pieces of interplanetary debris formed, but *what?*

Beginning in the 1960s a flood of information about Earth and other objects in the Solar System poured in from a host of sources including space probes, ground-based telescopes, laboratory analysis of meteorites, and theoretical calculations. Scientists working with this wealth of information began to see a pattern. What they were learning made sense only if they assumed

The Solar System formed from a rotating disk of gas and dust.

that the larger bodies in the Solar System had grown from the aggregation of smaller bodies. Following this chain of thought back in time, they came to envision an

for one important fact. For reasons that have to do with why a spinning ball of pizza dough spreads out to form a flat crust, the cloud that eventually produced the Sun collapsed first not into a ball but instead into a rotating disk. Most of the material in the disk eventually either traveled inward onto the forming star at its center or was thrown back into interstellar space. However, a small fraction of the material in the disk was left behind. As astronomers' tools got better and better, this scenario of how stars form was confirmed by discovery after discovery of disks of gas and dust surrounding young stellar objects like the ones shown in **Figure 5.1.**

Young stars are surrounded by rotating disks.

During the same years that astronomers were beginning to ferret out the secrets of star formation, another group of scientists with different backgrounds—mainly geologists—was piecing together the history of our Solar System. Some of the characteristics that the early Solar System must have had are fairly obvious.

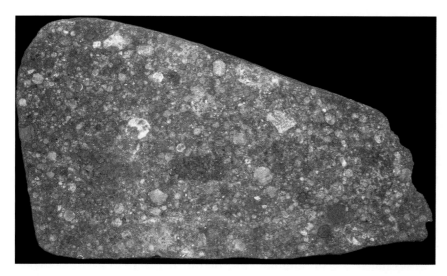

Figure 5.2 *Meteorites are fragments of the young Solar System that have fallen to Earth. It is clear from this cross section that this meteorite formed from many smaller pieces that stuck together.*

early Solar System in which the young Sun was surrounded by a flattened disk of both gaseous and solid material. This swirling disk of gas and dust provided the raw material from which the objects in our Solar System would later form.

Reading over the previous few paragraphs, you may have noticed a remarkable similarity between the disks that astronomers find around young stars, and the disk that planetary scientists hypothesize as the cradle of the Solar System. This similarity is not happenstance. As astronomers and planetary scientists compared notes, they realized they had arrived at the *same* picture of the early Solar System from two completely different directions. The rotating disk from which the planets formed was

none other than the remains of the disk which accompanied the formation of the Sun. The planet we live on, along with all of the other orbiting bodies that make up our **Solar System,** evolved from the remnants of the interstellar cloud that collapsed to form our local star, the Sun.

The connection between the formation of stars and the origin and subsequent evolution of the Solar System has become one of the cornerstones of both astronomy and planetary science—a central theme around which a great deal of our understanding of our Solar System revolves. As we begin the 21st century, it is the story of the Sun's formation and the history of the material in the surrounding disk that brings order to our understanding of our Solar System.

5.2 IN THE BEGINNING WAS A DISK

Later in the book we will turn our attention beyond the boundaries of our Solar System to the process of star formation itself. For now it is enough to jump into this story of star formation, midstream. Begin by holding the picture shown in **Figure 5.3** firmly in mind.

Figure 5.3 *When you think of the young Sun, think of it as being surrounded by a flat but flared rotating disk of gas and dust.*

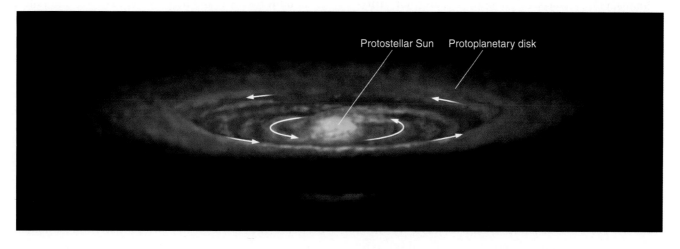

Protostellar Sun   Protoplanetary disk

Roughly 5 billion years ago the newly formed Sun was adrift in interstellar space. The Sun was not yet a star in the true sense of the word, for the nuclear fires that power the Sun today had yet to ignite. It was still a **protostar**—a large hot ball of gas that shone due to gravitational energy being turned into thermal energy as the protostar collapsed from a cloud of interstellar gas. (The often-used prefix *proto-* means "early form" or "in the process of formation.") Surrounding the proto-stellar Sun was a flat, rotating disk of gas and dust called the **protostellar disk.** "Orbiting" is perhaps a better word than "rotating," as the material in this thin disk was orbiting around the Sun according to the same laws of motion and gravitation that govern the orbits of the planets today. The disk around the Sun was much like the disks that astronomers see today surrounding protostars and newly formed stars elsewhere in our Galaxy. This disk is referred to as a **protoplanetary disk,** which is a protostellar disk capable of producing planets. It probably contained less than 1% as much mass as the nascent star at its center, but this amount was more than enough to account for the bodies that make up the Solar System today.

Okay, so we know the Solar System formed from a protoplanetary disk, and that disks are seen around newly formed stars—but *why* is this the case? What is it about the process of star formation that leads not only to a star itself, but to a flat orbiting collection of gas and dust as well? The answer to this question lies with a skater spinning on the ice.

THE COLLAPSING CLOUD ROTATES

We have all seen an Olympic skater spinning on the ice. Like any rotating object or isolated group of objects, the spinning ice skater has some amount of **angular momentum.** The amount of angular momentum that an

Angular momentum measures the rotation of an object.

object possesses depends on three things. First, it depends on how fast the object is rotating: the faster an object is rotating, the more angular momentum it has. A top that is spinning rapidly has more angular momentum than the same top does when it is spinning slowly. Second, an object's angular momentum depends on the mass of the object: Suppose that you compare two spinning tops. Both tops have the same size, shape, and rate of spin. They are the same, except for the fact that one top is made of lead while the other top is made of balsa wood. The lead top has more angular momentum. The third thing

that angular momentum depends on is how the mass of the object is distributed—how "spread out" the object is: for an object of a given mass and rate of rotation, the more spread out the object is, the more angular momentum it has. An object that is rotating slowly but is very spread out might have more angular momentum than a more rapidly rotating but more compact object.

The thing that makes angular momentum such an important and useful idea in physics and astronomy is that *the amount of angular momentum possessed by an object or isolated group of objects does not change unless those objects are acted on in the correct way by something other than themselves.* This statement is referred to as the **law of conservation of angular momentum.** In the parlance of physics, if

Angular momentum is conserved.

something is "conserved," it means that the amount of that quantity does not change by itself. This idea might remind you of Newton's first law of motion, which says that in the absence of some external force, an object continues to move in a straight line at a constant speed. Indeed, both Newton's first law and the conservation of angular momentum are examples of **conservation laws.** There are many other conservation laws in physics, including the law of conservation of momentum, the law of conservation of energy (discussed below), and the law of conservation of electric charge.

Conservation of angular momentum brings us back to the ice skater we opened the section with, and to the collapsing interstellar cloud. You have probably noticed that an ice skater can control how rapidly he spins by doing nothing other than pulling in or extending his arms. A compact object must spin more rapidly to have the same amount of angular momentum as a more extended object with the same mass. As our skater spins, his angular momentum does not change much. (The slow decrease is due to friction, an external force.) When his arms and leg are fully extended, he spins slowly, but as he pulls his arms and leg in he spins faster and faster. With the skater's arms held tightly in front of him and one leg wrapped around the other, his spin becomes a blur. He finishes with a flourish by throwing his arms and leg out, which abruptly slows his spin. Despite the dramatic effect, his angular momentum remains the same throughout. This impressive athletic spectacle comes courtesy of the law of conservation of angular momentum, and from the difference between an extended object and a compact object.

Now we turn our attention to how the conservation of angular momentum affects a forming star. We begin with a cloud of interstellar gas that is collapsing under the force of its own gravity. It might seem most natural for the cloud to collapse directly into a ball—and so it

5.1
BASICS

would, but for the cloud's own angular momentum. *Interstellar clouds* are truly vast objects, light-years in size. (A *light-year* is the distance traveled by light in one year, or about 9.5 trillion kilometers.) As they orbit about the Galaxy center, they are constantly being pushed around by stellar explosions or by collisions with other interstellar clouds. This constant "stirring" guarantees that all interstellar clouds will have *some* amount of rotation. As spread out as an interstellar cloud is, even a tiny amount of rotation corresponds to a huge amount of angular momentum. Imagine our ice skater now with arms that reach from here to the other side of Earth. Even if he were rotating very slowly at first, think how fast he would be spinning by the time he pulled those long arms to his sides! Just as the ice skater speeds up when he pulls in his arms, the cloud rotates faster and faster as it collapses. Suppose, for example, that we start with a cloud that is about a

Interstellar clouds have far more angular momentum than the stars they form.

light-year across—say 10^{16} meters—and is rotating so slowly that it takes a million years to complete one rotation. By the time such a cloud collapsed to the size of our Sun—a mere 1.4×10^9 meters across, or only one 10-millionth the size of the original cloud—it would be spinning 50 trillion times faster, completing a rotation in only 0.6 of a second! This is over 3 million times faster than our Sun is actually spinning. At this rate of rotation, the Sun's self-gravity would have to be almost 200 million times stronger to hold the Sun together!

AN ACCRETION DISK FORMS

There is a puzzle here. Conservation of angular momentum would seem to say that stars cannot form from collapsing interstellar clouds, and yet there is no other way for stars *to* form. When scientists find what appears to be a contradiction—like the apparent contradiction between the principle of conservation of angular momentum and the fact that a star has far less angular momentum than the cloud that formed it—it is cause for great excitement. Such seeming contradictions do not mean that nature is breaking the rules. (Nature *never* breaks its own rules!) Instead, it means that our understanding of what is going on is incomplete, or that we have the wrong rule. It means that we have found a place where there are new things to be learned.

The key to solving the riddle of angular momentum in a collapsing interstellar cloud lies in realizing that the *direction* of the collapse is important. The cloud's rotation may thwart the collapse of the cloud

toward its axis of rotation, but there is nothing to prevent collapse *parallel* to the axis of rotation (see **Figure 5.4**). Instead of collapsing into a ball, the interstellar cloud becomes flattened as it collapses. As the cloud collapses more and more, the strength of gravity causing the collapse gets greater and greater. Eventually the

Figure 5.4 *A rotating interstellar cloud is free to collapse parallel to its axis of rotation, but not perpendicular to that axis. As a result, the cloud collapses into a disk.*

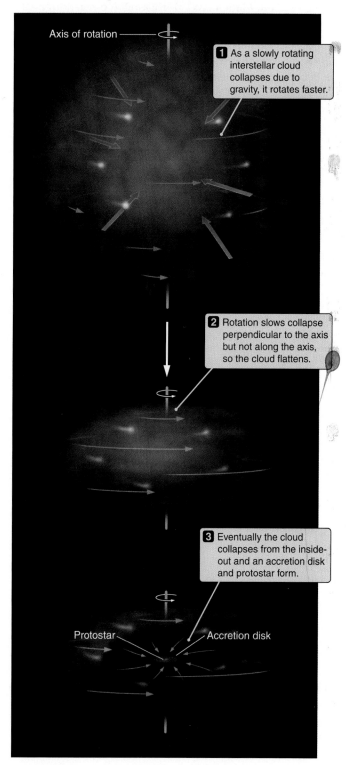

Axis of rotation

1 As a slowly rotating interstellar cloud collapses due to gravity, it rotates faster.

2 Rotation slows collapse perpendicular to the axis but not along the axis, so the cloud flattens.

3 Eventually the cloud collapses from the inside-out and an accretion disk and protostar form.

Protostar   Accretion disk

flattening cloud reaches a point where the inner parts of the cloud begin to fall freely inward, raining down on the growing object at the center. As this happens, the outer portions of the cloud lose the support of the collapsed inner portion of the cloud, and they start falling inward too. The whole cloud collapses inward, much like a house of cards with the bottom layer knocked out. As this material makes its final inward plunge, it lands on a thin, rotating, pizza-dough-like structure called an **accretion disk.** The accretion disk serves as a way station for material on its way to becoming part of the star that is forming at its center.

The cloud collapses into a disk rather than directly into a star.

Formation of accretion disks occurs in many situations in astronomy, so it is worth taking a moment to think carefully about this process. We can use what we learned about orbits in Chapter 3 to better understand what happens during this final stage of the collapse of an interstellar cloud. As the material falls toward the forming star, it travels on curved, almost always elliptical paths, just as Kepler's laws say that it should. These orbits would carry the material around the forming star and back into interstellar space, but for one problem: the path inward toward the forming star is a one-way street. When material nears the center of the cloud, it runs headlong into material that is falling in from the *other* side. Imagine a huge rotary, or traffic circle with lots of entrances but no exits **(Figure 5.5).** As traffic flows into the traffic circle, it has nowhere else to go, resulting in a continuous growing line of traffic driving around and around. Eventually, as more and more cars try to pack in, the traffic piles up. This situation is analogous to an accretion disk. As material falls onto the disk, its motion perpendicular to the disk stops abruptly, but its motion parallel to the surface of the disk adds to the disk's angular momentum. The angular momentum of the infalling material has been transferred to the disk.

Traffic in a traffic circle moves on a flat surface, but the accretion disk around a protostar forms from material coming in from all directions in three-dimensional space. Where the disk forms—the *plane* of the accretion disk—is determined by a balance between the amounts of material falling onto the disk from each side of the disk. There is only one plane that satisfies this requirement: the plane perpendicular to the axis the cloud is rotating about. The gas settles down into a rotating accretion disk that has a radius of hundreds of astronomical units—and thousands of times greater than the radius of the star that will eventually form at its center—making it large enough to accommodate the angular momentum of the infalling material.

The next obvious question concerns how material falling onto the accretion disk finds its way inward onto the growing protostar. We will pick up this question in Chapter 14, when we return to follow the story of the

Figure 5.5 (a) *Traffic piles up in a traffic circle with entrances but no exits.* (b) *Similarly, gas from a rotating cloud falls inward from above and below, piling up onto a rotating disk.*

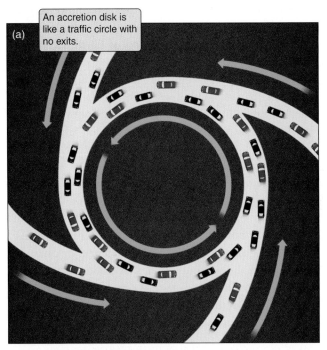

(a) An accretion disk is like a traffic circle with no exits.

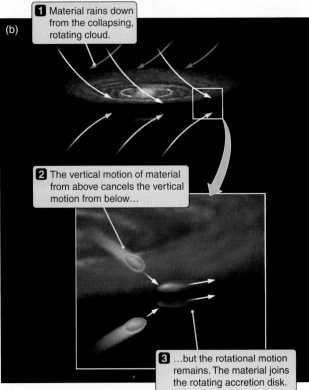

(b)

1 Material rains down from the collapsing, rotating cloud.

2 The vertical motion of material from above cancels the vertical motion from below...

3 ...but the rotational motion remains. The material joins the rotating accretion disk.

growth and evolution of the star at the center of the disk. For now it is enough to know that most of the matter that lands on the accretion disk either ends up as part of the star, or is ejected back into interstellar space. However, a small amount of material is left behind in the disk. It is this leftover disk—the dregs of the process of star formation—to which we next turn our attention.

Before going on with our story, however, we should stress that theoretical calculations by astronomers have long predicted that accretion disks should be found around young stars, and illustrations like Figure 5.3 have been a mainstay in textbooks for years. In the last decade or so of the 20th century, however, these cartoon drawings had to make room for images of the real thing. Figure 5.1 shows *Hubble Space Telescope* images of edge-on accretion disks around young stars. The dark bands are the shadows of the edge-on disks, the top and the bottom of which are illuminated by light from the forming star. It would be nice if we could go back 5 billion years and watch as our own Sun formed from a cloud of interstellar gas, but we do not have to. All we have to do is look at objects like the ones in Figure 5.1 to know what we would have seen.

5.3 SMALL OBJECTS STICK TOGETHER TO BECOME LARGE OBJECTS

The chain of events that connects the accretion disk around a young star to a planetary system such as our

Gas motions push small particles into larger particles.

own begins with the same basic physics that leaves you with dust in your eye on a windy day. A stiff breeze picks up dust and sand and blows them about, but leaves the pebbles and rocks behind. In like fashion, the motions of the gas within the protoplanetary disk push the smaller grains of solid material back and forth past the larger grains, and as this happens, the smaller grains stick to the larger grains—just as wind-borne dust blows into and sticks to your eye.

In matters of social dynamics we sometimes hear that "the rich get richer at the expense of the poor." While we can always debate this social axiom, the principle certainly holds true when it comes to the dynamics within a protoplanetary disk, as seen in **Figure 5.6.** The larger dust grains get larger at the expense of the smaller grains. Starting out at only a few microns (mi-

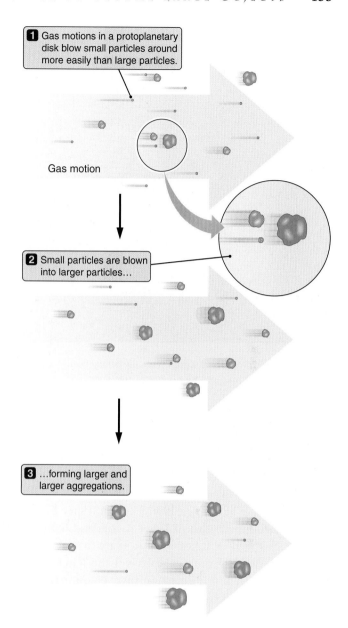

Figure 5.6 *Motions of gas in a protoplanetary disk blow smaller particles of dust into larger particles, making the larger particles larger still. This process continues, eventually making objects many meters in size.*

crometers) across, the slightly larger bits of dust grow to the size of pebbles, then to the size of boulders. The rate at which objects grow in this way is thought to decrease when clumps of boulders are about 100 me-

Aggregates grow to about 100 m across.

ters across. Such objects are so few and far between in the disk that chance collisions become less and less frequent. Even so, the process of growth continues at a slower pace, as 100 meter clumps join together to produce still larger bodies.

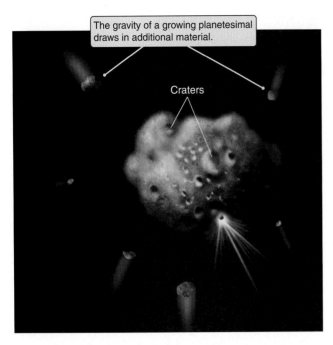

The gravity of a growing planetesimal draws in additional material.

Craters

Figure 5.7 *The gravity of a planetesimal is strong enough to attract surrounding material, causing the planetesimal to grow.*

In order for two such clumps to stick together, they must bump into each other very gently. Otherwise, the energy of collision would cause the two colliding bodies to fragment into many smaller pieces instead of forming a single larger one. Typical collision speeds cannot be much greater than 0.1 m/s if colliding boulders are to stick together. If you were to walk that slowly, it would take you 15 minutes to travel the length of a football field. In a real accretion disk, collisions more violent than this certainly happen on occasion, breaking these clumps back into smaller pieces. Likely there are many reversals in this growth of clumps of boulders.

Up until this point, larger objects have grown mainly by "sweeping up" smaller objects that run into them or that get in their way. As the clumps reach a size of about a kilometer, a different process becomes important (see **Figure 5.7**). These kilometer-sized objects, now called **planetesimals** (literally "tiny planets"), are massive enough that their gravity starts to be important, exerting a significant attraction on nearby bodies. No longer is growth of the planetesimal fed only by chance collisions with other objects, for the planetesimal's gravity can now pull in and capture other smaller planetesimals that lie outside its direct path. The growth of planetesimals speeds up, with the larger planetesimals quickly consuming most of the remaining bodies in the vicinity of their orbits.

Gravity helps planetesimals grow into planets.

The final survivors of this process are now large enough to be called **planets.** As with the major bodies in orbit about the Sun, some of the planets may be small and others quite large.

5.4 THE INNER DISK IS HOT, BUT THE OUTER DISK IS COLD

The accretion disks surrounding young stars form from interstellar material that may have a temperature of only a few kelvins, but the disks themselves reach temperatures of hundreds of kelvins or more. What is it that heats up the disk around a forming star? The answer lies with our old friend gravity. Material from the collapsing interstellar cloud falls inward toward the protostar, but because of its angular momentum it "misses," falling instead onto the surface of the disk. When the material that is raining onto the disk hits the disk, its bulk motion comes to an abrupt halt, and the velocity that the atoms and molecules in the gas had before hitting the disk is suddenly converted into random *thermal* velocities instead. That is to say, the cold gas that was falling toward the disk gets very hot when it lands on the disk.

To help you visualize this, imagine dumping a box of marbles from the top of a tall ladder onto a rough, hard floor below **(Figure 5.8).** The marbles fall, picking up speed as they go. Even though the marbles are speeding up, however, they are all speeding up *together.* As far as one marble is concerned, the other marbles are not moving very fast at all. (If you were riding on one of the marbles, the other marbles would not appear to you to be moving very much; it would be the rest of the room that was whizzing by.) The atoms and molecules in the gas falling toward the protostar are like these marbles. They are picking up speed as they fall as a group toward the protostar, but the gas is still *cold* because the random *thermal* velocities of atoms and molecules with respect to each other are still low. Now imagine what happens when the marbles hit the rough floor. They bounce every which way. They are still moving rapidly, but they are no longer moving together. A change has taken place from the ordered motion of marbles falling together, to the random motions of marbles traveling in all directions. The atoms and mol-

Motions of atoms become random thermal motions when infalling gas hits the disk.

ecules in the gas falling toward the central star behave in the same fashion when they hit the disk. They are no longer moving as a group, but their random "thermal" velocities are now very large. The gas is now *hot.*

Another way to think about why the gas that falls on the disk makes the disk hot is to apply the ever useful concept of conservation of energy. The law of **conservation of energy** means that unless energy is added to or taken away from a system from the outside, the total amount of energy in the system must remain constant. But the *form* the energy takes *can* change.

Imagine lifting a heavy object—say a brick. It is hard to do, because you are working against gravity. It takes energy to lift the brick, and conservation of energy says that energy is never lost. But where does that energy go? It is changed into a form called **gravitational potential energy**. In a sense, this energy is "stored" in the brick in a way that is reminiscent of how energy is stored in a battery. Potential energy is energy that "has potential"—it is waiting to show up in some more obvious form. If you drop the brick, it falls, and as it falls, it speeds up. The gravitational potential energy that was stored in the brick is being converted instead to energy of motion, which is called **kinetic energy.** When the brick hits the floor, it stops suddenly. The brick loses its energy of motion, so what form does this energy take now? If the brick cracks, part of the energy goes into breaking the chemical bonds that hold it together. Some of the energy is converted into the sound the brick makes when it hits the floor. But *most* of the energy is converted into thermal energy. The atoms and molecules that make up the brick are moving about within the brick a bit faster than they were before the brick hit, and so the brick and its surroundings grow a tiny bit warmer. Similarly, as gas falls toward the disk surrounding a protostar, gravitational potential energy is converted first to kinetic energy, so the gas picks up speed. When the gas hits the disk and stops suddenly, that kinetic energy is turned into thermal energy. (Foundations 5.1 discusses why it can be useful to think about the same thing in different ways.)

> The gravitational energy of infalling material turns into thermal energy.

In this way, as material falls onto the accretion disk around a forming star, the disk is made hot. Material hitting the inner part of the disk (which we will call the *inner disk*) has fallen far in the gravitational field of the forming star. Like a rock dropped from a tall building, material hitting the inner part of the disk is moving quite rapidly when it hits the disk, and so heats the inner disk to high temperatures. In contrast, material falling onto the outer part of the disk (which we will call the *outer disk*) is moving much more slowly (envision a rock

> The inner disk is hotter than the outer disk.

Figure 5.8 (a) *Marbles dropped as a group fall together until they hit a rough floor, at which point their motions become randomized.* (b) *Similarly, atoms in a gas fall together until they hit the accretion disk, at which point their motions become randomized, which raises the temperature of the gas.*

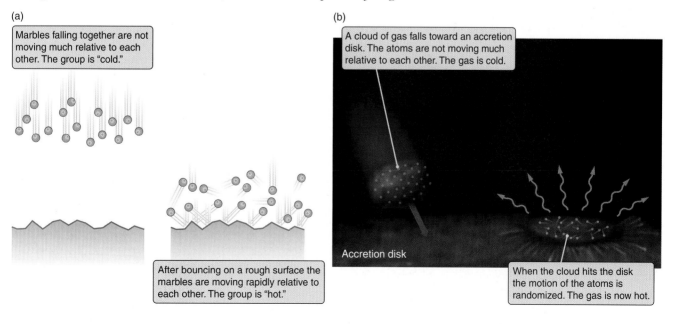

(a)

Marbles falling together are not moving much relative to each other. The group is "cold."

After bouncing on a rough surface the marbles are moving rapidly relative to each other. The group is "hot."

(b)

A cloud of gas falls toward an accretion disk. The atoms are not moving much relative to each other. The gas is cold.

Accretion disk

When the cloud hits the disk the motion of the atoms is randomized. The gas is now hot.

FOUNDATIONS 5.1

THINKING ABOUT ENERGY IN DIFFERENT WAYS

In the text we discuss why the gas falling onto a protostellar disk heats the disk. First, we give an example of marbles falling on a floor, noting how the motion of marbles is jumbled up when they hit, like the atoms in the gas hitting the disk. Then we give a very different explanation, talking about how energy is conserved but changes form from gravitational energy to kinetic energy, and finally to thermal energy. These may be two different explanations of why the disk gets hot, but they are *not* two different *reasons* why the disk gets hot. The protostar gets hot as a result of the same physical process—the "reason" is the same—regardless of the words we use to describe it. Instead, these two explanations offer two different ways to *think about* why the disk gets hot.

Both ways of thinking about the process are correct. Both are included here because sometimes students—or scientists—need to look at the same thing from several different directions before they understand it. In this case, even though both ways of thinking about the process are correct, most scientists would agree that the second way of thinking about the problem is far more *powerful* than the first. The idea of energy changing forms connects how disks are heated with a wide variety of other phenomena. Properly stated, the heating of protostellar disks is an example of one of the most far-reaching patterns in nature: the conservation of energy.

Once you understand the different forms of energy, and how energy is conserved, you begin to see this pattern of nature everywhere. For example, when water falls through the turbines in a hydroelectric generator, it turns the generator and produces electric energy that is carried out over power lines and lights the lights in your home. Where did the energy to heat the water that pours from your tap come from? If you have an electric water heater and get your power from a hydroelectric plant, it comes from the gravitational potential energy of the water in a reservoir near you—just as the energy to heat the newly formed Sun came from the gravitational potential energy of the reservoir of gas from which the Sun formed.

Conservation of energy is also a more powerful way to think about the heating of disks around young stars because it allows astronomers to *calculate* how hot these disks get. If we know the mass of the protostar and the surrounding disk, and how far away from the protostar the gas is falling from, we can calculate how much gravitational energy the gas started out with. According to the conservation of energy, this gravitational energy must eventually be converted to thermal energy, so we can calculate the expected temperature of the protostar. So without calculating any of the details of how the disk formed, we can say right off the bat how much thermal energy will be deposited onto the disk.

Scientists spend much of their time trying to come up with new ways of "thinking about" problems, looking for particularly powerful ways that point to new insights and discoveries. The most powerful means of thinking about a problem usually tie the problem to ever grander patterns in nature. Conservation of energy is one of the grandest and most useful patterns around.

---

dropped from only a foot or so), leaving the outermost parts of the disk at temperatures that may be little higher than the original interstellar cloud. Stated another way, material falling onto the inner disk converts more gravitational energy into thermal energy than material falling onto the outer disk.

The energy released as material falls onto the disk is not the only source of thermal energy in the disk. Even before the nuclear fires that will one day power the new star have ignited, conversion of gravitational energy into thermal energy drives the temperature at the surface of the protostar to several thousand kelvins, and drives the luminosity of the huge ball of glowing gas to many times the luminosity of the present-day Sun. For the same reasons that Mercury is hot while Pluto is not (see Section 4.5), the radiation streaming outward from the protostar at the center of the disk drives the temperature in the inner parts of the disk up even higher, increasing the difference in temperature between the inner and outer disk.

ROCK, METAL, AND ICE

We have all noticed that temperature affects which materials can and cannot exist in a solid form. On a hot summer day, ice melts and water quickly evaporates, while on a cold winter night even the water in our breath freezes into tiny ice crystals before our eyes. Some materials, such as iron, silicates, and carbon—rocky materials and metals—remain solid even at quite

Refractory materials remain solid even at high temperatures.

high temperatures. Such materials, which are capable of withstanding high temperatures without melting or being vaporized, are referred to as **refractory materials.** Other materials, such as water, ammonia, and methane, can only remain in a solid form if their temperature is quite low. These less refractory substances are called **volatile materials.** The solid form of a volatile material is often referred to as an **ice.**

Differences in temperature from place to place within the disk have a significant effect on the makeup of the dust grains in the disk. In the hottest parts of the disk (closest to the protostar), only the most refractory substances can exist in solid form. In the inner disk, dust grains are composed of refractory materials only. Somewhat farther out in the disk, some of the hardier volatiles such as water ice and certain organic substances can survive in solid form, adding to the materials that dust grains are made of. Highly

Volatile ices survive in the outer disk, but only refractory solids survive in the inner disk.

volatile components such as methane, ammonia, and carbon monoxide ices and some organic molecules survive in solid form only in the coldest, outermost parts of the accretion disk, far from the central protostar. The differences in composition of dust grains within the disk are reflected in the composition of the planetesimals formed from that dust. Planets that form closest to the central star tend to be made up mostly of refractory materials such as rock and metals. Those that form farthest from the central star also contain refractory materials, but also contain large quantities of ices and organic materials.

As we turn to a study of our own Solar System, we will find that the trend in composition expected in a protoplanetary disk is closely echoed in the makeup of the solid bodies orbiting the Sun. The inner planets are composed of rocky material surrounding metallic cores of iron and nickel. In contrast, objects in the outer Solar System, including moons, giant planets, and comets, are composed largely of ices of various types. In the years to come, as astronomers learn more about planetary systems around other stars, it seems likely that this trend will prove to be very common. In fact, based on our understanding of the way stars and planetary systems form, this change in composition with distance from the central star would seem to be almost unavoidable. Even so, chaotic encounters like those we will discuss in Chapter 9 can shuffle the deck, adding diversity to the organization of planetary systems.

SOLID PLANETS GATHER ATMOSPHERES

Once a solid planet has formed, it may have a chance to continue growing by capturing gas from the protoplanetary disk. However, if it is going to do so, it must act quickly. Young stars and protostars are known to be sources of strong "winds" and intense radiation that can quickly disperse the gaseous remains of the accretion disk. Gaseous planets such as Jupiter probably have only about 10 million years or so to form and to grab whatever gas they can. Tremendous mass is a great advantage in a planet's ability to accumulate and hold on to the hydrogen and helium gas that makes up the bulk of the disk. Because of their strong gravitational fields, more massive young planets are thought to create their own mini-accretion disks as gas from their surroundings falls toward them. What follows is much like the formation of a star and protoplanetary disk, but on a smaller scale. Just as happened in the accretion disk around the

A mini-accretion disk forms around a new massive planet.

star, gas from a mini-accretion disk moves inward and falls onto the solid planet.

The gas that is captured by a planet at the time of its formation is referred to as the planet's **primary atmosphere.** The primary atmosphere of a large planet can become massive enough to dominate the mass of the planet, as in the case of giant planets such as Jupiter. Some of the solid material in the mini-accretion disk might stay behind to coalesce into larger bodies in much the same way that dust in the protoplanetary disk came together to form planets. The result is a "mini-solar system"—a group of moons that orbit about the planet.

A less massive planet may also capture some gas from the protoplanetary disk, only to lose its prize. Here again, more massive planets have the advantage. As we will learn in Chapter 7, the gravity of small planets may not be strong enough to prevent less massive atoms and molecules such as hydrogen or helium from escaping back into space. Even if a small planet is able to gather an amount of hydrogen and helium from its surroundings, this tempo-

Less massive planets lose their primary atmospheres, then form secondary atmospheres.

rary primary atmosphere will be short-lived. The atmosphere that remains around small planets like our Earth is a **secondary atmosphere.** A secondary atmosphere forms later in the life of a planet. Carbon dioxide and other gases released from the planet's interior by widespread volcanoes is probably one important source of a planet's secondary atmosphere. Also, as we will see later, volatile-rich comets that formed in the outer parts of the disk continue to fall in toward the new star long after the planets have formed, and sometimes collide with planets. Comets possibly provide a significant source of water, organic compounds, and other volatile materials on planets close to the central star.

5.5 A TALE OF NINE PLANETS

We are now at a point in our discussion where we can take our general ideas about the evolution of the material in a protoplanetary disk and apply them to our own Solar System. Only in the closing years of the 20th century did our knowledge progress to the point that this story could be told. It is still sketchy in places, and doubtless wrong in some ways. But error and uncertainty are part of the advance of science. In this section we bring together a tremendous wealth of information taken both from what we know of our local star and planetary neighbors, and from what we have learned about stars forming around us today. It is a synthesis of the painstaking efforts of hundreds of astronomers and planetary scientists over the course of decades. Many more scientists will devote their lives to unraveling this story before its details are fully known. There are good reasons to believe that the explanations we give in this section are basically correct, although we know they are not yet complete. Returning to our early discussion of what science is and how science works, the story we tell here is one that many people have "tried to prove wrong," but which has withstood all the tests . . . so far.

Nearly 5 billion years ago, our Sun was still a protostar surrounded by a protoplanetary disk of gas and dust. Over the course of a few hundred thousand years, much of the dust in the disk had collected into planetesimals—clumps of rock and metal near the emerging Sun and aggregates of rock, metal, ice, and organic materials in the more distant parts of the disk. Within the inner few astronomical units (AU) of the disk, several rock and metal planetesimals, probably less than a half dozen, quickly grew in size to become the dominant masses at their respective distances from the Sun. With their ever-strengthening gravitational fields, they

Our Solar System is a concrete test of our ideas about star and planet formation.

Rocky terrestrial planets formed in the inner Solar System.

either captured most of the remaining planetesimals or ejected them from the inner part of the disk. These dominant planetesimals had now become planet-sized bodies with masses ranging between that of Earth and about one-twentieth of that value. They were to become the *terrestrial planets*. Mercury, Venus, Earth, and Mars are the surviving terrestrial planets. One or two others are thought to have formed in the young Solar System, but were later destroyed. For several hundred million years following the formation of the four surviving terrestrial planets, leftover pieces of debris still in orbit around the Sun continued to rain down on their surfaces. Much of this barrage may have originated in the outer Solar System where the gravitational tug of the massive, newly formed outer planets acted like a slingshot shooting debris inward. Today we can still see the scars of these post–formation impacts on the cratered surfaces of all of the terrestrial planets, such as the surface of Mercury shown in **Figure 5.9.** This rain of debris continues today, albeit at a much lower rate.

Before the proto-Sun emerged as a true star, gas in the inner part of the protoplanetary disk was still plentiful. During this early period the two larger terrestrial planets, Earth and Venus, may have held onto weak primary atmospheres of hydrogen and helium. If so, these thin atmospheres were soon lost to space. For the most part, the terrestrial planets were all born devoid of thick atmospheres and remained so until the formation of the secondary atmospheres that now surround Venus, Earth, and Mars. Mercury's proximity to the Sun and the Moon's small mass must have prevented these bodies

Figure 5.9 *Large impact craters on Mercury (and on solid bodies throughout the Solar System) record the final days of the Solar System's youth, when planets and planetesimals grew as smaller planetesimals rained down on their surfaces.*

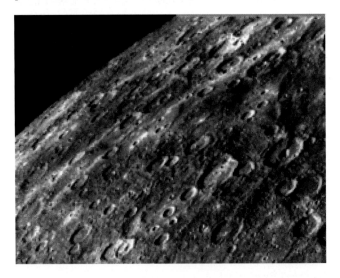

from retaining a significant secondary atmosphere. They remain nearly airless today.

Farther out in the nascent Solar System, 5 AU from the Sun and beyond, planetesimals coalesced to form a number of bodies with masses around 10 to 15 times that of Earth. Why such large bodies formed in the region beyond the terrestrial planets remains an unanswered question.

The cores of the outer giant planets formed from planetesimals, as did the inner planets . . .

Located in a much colder part of the accretion disk, these planet-sized objects formed from planetesimals containing volatile ices and organic compounds in addition to rock and metal. Four such massive bodies, found between 5 AU and 30 AU from the Sun, would later become the cores of the *giant planets,* which we know today as Jupiter, Saturn, Uranus, and Neptune. Some planetologists believe that all the giant planets may have formed closer to where Jupiter is now and that their mutual gravitational interactions caused them to migrate to their present orbits.

. . . but the outer planets were able to capture and hold many times more gas.

Mini-accretion disks formed around these planetary cores, capturing massive amounts of hydrogen and helium and funneling this material onto the planet. Jupiter's solid core was able to capture and retain the most gas—a quantity roughly 300 times the mass of Earth, or 300 $M_\oplus$. (The symbol $\oplus$ signifies "Earth.") The other planetary cores captured much smaller amounts of hydrogen and helium, perhaps because they were located in more remote parts of the protoplanetary disk where the density of gas was lower. Saturn ended up with less than 100 $M_\oplus$ of gas, while Uranus and Neptune were able to grab only a few $M_\oplus$ of gas.

For the same reasons that a forming protostar is hot—namely, conversion of gravitational into thermal energy—the gas surrounding the cores of the giant planets compressed under the force of gravity and became hotter. Proto-Jupiter and proto-Saturn probably became so hot that they actually glowed a deep red color, similar to the heating element on an electric stove. Their internal temperatures may have reached as high as 50,000 K. In a sense, the more massive protoplanets were trying to become stars. But for them this was an unreachable goal. In Chapter 14 we will find that a ball of gas must have a mass of at least 0.08 times that of the Sun if it is to become a star. This minimum mass is some 80 times the mass of Jupiter. Science fiction films notwithstanding, Jupiter never had a chance.

Some of the material remaining in the mini-accretion disks surrounding the giant planets coalesced into small bodies, which became moons. (A **moon** is any natural satellite in orbit about a planet.) The composition of these giant-planet moons followed the same trend as the planets

Moons formed from the mini-accretion disks around the giant planets.

which formed around the Sun: the innermost moons formed under the hottest conditions, and therefore contained the least amounts of volatile material. As very young moons, Io and Europa may have experienced nearby Jupiter glowing so intensely that it rivaled the distant Sun. The high temperatures created by the glowing planet would have evaporated most of the volatile substances in the inner part of its mini-accretion disk. Io today contains no water at all. However, water is relatively plentiful on Europa, Ganymede and Callisto.

Not all of the planetesimals in the protoplanetary disk went on to become planets. Jupiter is a true giant of a planet. Its gravity kept the region of space between it and Mars so "stirred up" that the planetesimals there never coalesced into a single

Asteroids and comets are planetesimals that survive to this day.

planet. This region, now referred to as the **asteroid belt,** contains many planetesimals that remain from this early time. In the outermost part of the Solar System planetesimals also persist to this day. Born in a "deep freeze," these objects retained most of the highly volatile materials found in the grains present at the formation of the protoplanetary disk. Unlike conditions in the crowded inner part of the disk, planetesimals in the outermost parts of the disk were too sparsely distributed for large planets to grow. Icy planetesimals in the outer Solar System remain today as **comets**—relatively pristine samples of the material from which our planetary system formed. Frozen Pluto may be a large example of these denizens of the outer Solar System.

The early Solar System must have been a very violent and chaotic place. Many of the objects in the Solar System show evidence of cataclysmic impacts that reshaped worlds. A dramatic difference in the terrain in the northern and southern hemispheres on Mars, for example, has been interpreted as the result of one or more colossal collisions. Mercury has a crater on its surface from an impact so devastating that it caused the crust to buckle on the opposite side of the planet. In the outer Solar System, one of Saturn's moons, Mimas, sports a crater roughly one-third the diameter of the moon itself. Uranus suffered a collision that was violent enough to literally knock the planet on its side. Today its axis of rotation is tilted at an almost right angle to its orbital plane.

Not even our own Earth escaped devastation by these cataclysmic events. In addition to the four terrestrial planets that remain, there was at least one other terrestrial planet in the early Solar System—a planet about the same size and mass as Mars. As the newly formed planets were settling down into their present-day orbits, this fifth planet suffered a grazing collision with Earth and was completely destroyed. The remains of the planet, together with material knocked from Earth's outer layers, formed a huge cloud of debris encircling Earth. For a brief period, Earth may have displayed a magnificent group of rings like those of Saturn. In time, this debris coalesced into the single body we know as our Moon.

Earth's Moon formed out of the debris from a collision.

Figure 5.10 *Planetary systems have been discovered around scores of stars other than the Sun, confirming what astronomers have long suspected—that planets are a natural and common by-product of star formation. A few of these systems are represented here. (Masses of planets are given in units of Jupiter masses, M_J.)*

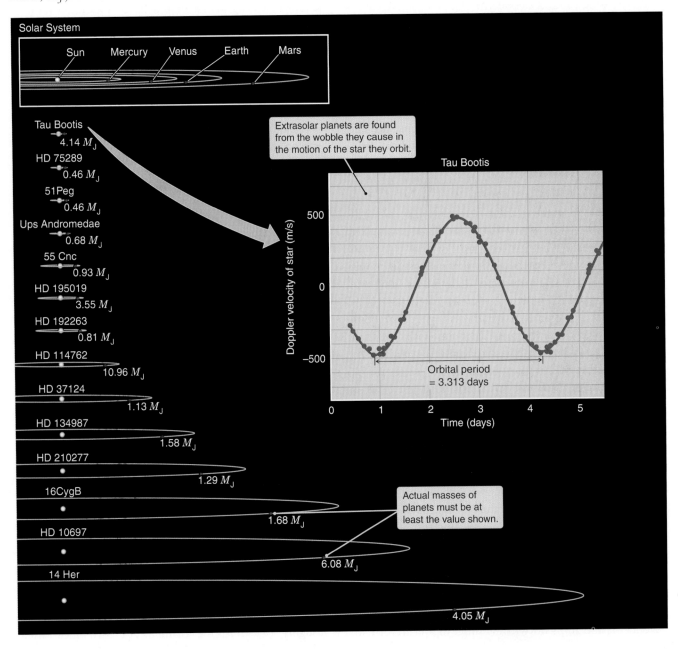

5.6 THERE IS NOTHING SPECIAL ABOUT IT

We began this chapter by discussing the formation of a generic planetary system around a generic star. Only later did we turn to the specifics of our own Solar System. We did this to make a very important point: according to our current understanding, *there is absolutely nothing special about the conditions that led to the formation of our Sun and our Solar System.* When astronomers turn their telescopes on Sun-like stars forming around us today, they see just the sort of disks from which our own Solar System formed. Based on what we know, the physical processes that led to the formation of our Solar System should be commonplace wherever new stars are being born. In fact, as the 20th century came to a close, the astronomical community was abuzz with reports confirming what astronomers have long suspected—that planets exist around stars other than our Sun.

Planets have been discovered around other stars.

As a planet orbits about a star, the planet's gravity causes the position of the star to wiggle ever so slightly. By observing the Doppler shift of the star, this wiggle can sometimes be detected and the properties of the planet—its mass and distance from the star—can be inferred. (We will use an almost identical method in Chapter 12 to measure the masses of pairs of stars that orbit about each other.) A planetary system like our own around a nearby star would not have been detected in the searches for planets that have been carried out to date, but the sensitivity of this technique is improving. We now know of planets around many stars. A few of these planetary systems are shown in **Figure 5.10.** Many astronomers were surprised at these discoveries because it is not yet clear how such massive planets can form so close to stars. It seems likely that these planets actually formed farther out from their parent stars and later moved inward. Yet even though these findings challenge some aspects of our understanding of planet formation, and even though discovery of *Earth-like planets*—planets similar to Earth in size and distance from their star—around other stars is still years or decades away, the message conveyed by these discoveries is clear: the formation of planets frequently, and perhaps always, accompanies the formation of stars.

The implications of this conclusion are profound. Planets are a common by-product of star formation.

So, in a galaxy of a hundred billion stars, and a Universe of hundreds of billions of galaxies, how many Earth-like planets (or Earth-like moons) might exist? And with all of those Earth-like worlds in the Universe, how many might play host to the particular category of chemical reactions that we refer to as "life"?

In learning about the formation of stars and then our Solar System, our journey has come full circle. We began our journey by looking outward in wonder at the lights in the night sky. Now that outward exploration becomes instead a look inward, as we discover the processes and events that set the stage for our own existence. In a sense, the insights we have gained in this chapter "complete" the revolution Copernicus started long ago when he had the audacity to challenge authority and suggest that Earth was not the center of all things. Not that long ago, our ancestors viewed Earth and the heavens as fundamentally different regimes—forever apart, each with its own rules and reality. Today we know there is nothing unusual about the conditions of our own existence. All we have to do to see our own history is to look at the stars that continue to form around us today.

Modern discoveries about the formation of the Sun and Solar System "complete" the Copernican revolution.

It is ironic that at a time when science has turned such a dazzling spotlight on our place in the Universe, many otherwise educated individuals continue to cling to outdated and fanciful notions about the heavens and Earth.[1] For someone interested in truly mind-boggling insights, the speculations and "revelations" of the mystic pale beside the discoveries of the scientist. Through the sometimes stodgy, often painstaking, and always uncompromising standards of science, we have come to appreciate that we are not *apart from* the Universe. Rather we are *a part of* the Universe. And the processes and events that link us to this larger Universe are fascinating and wondrous indeed.

---

[1]We cannot resist the temptation to quote the famous science fiction author Arthur C. Clarke at this juncture. "In one sense, of course, every age renews itself, as indeed it should. But the nitwits currently parroting this slogan [New Age] seem unable to understand that *their* 'New Age' is exactly the opposite, being about a thousand years past its sale date."

SEEING THE FOREST THROUGH THE TREES

In this chapter we have seen two great lines of investigation merge into a single picture that shapes our understanding of the context of our own existence. Working first from the perspective of the stellar astronomer, basic physical principles—in particular the conservation of angular momentum—demand that when a star like our Sun forms, it will be surrounded by a thin, orbiting disk of gas and dust. This conclusion went from hypothesis to fact with the discovery of such disks around numerous young stars. In the meantime, as stellar astronomers were trying to understand star formation, planetary scientists were scrutinizing the worlds that make up our Solar System. The pieces of the planetary scientists' puzzle range from meteorites collected in Antarctica, to samples of the Moon brought back by Apollo astronauts, to data sent back by spacecraft visiting remote worlds. Those pieces fit together to form a clear picture of a flat, swirling cloud of gas and dust from which Earth and its neighbors coalesced 4.6 billion years ago. When the stellar astronomers and the planetary scientists compared notes, they realized that they had arrived at exactly the same description. The disk from which the Solar System formed was none other than the stellar accretion disk that surrounded the young Sun.

The joining of these two great rivers of thought, observation, and exploration—the link between the accretion disks that surround young stars and the local collection of planets that we call home—is the starting point for a modern study of the Solar System. It also ties our journey of understanding within the Solar System to the course we will later chart outward into the larger Universe. Stated more bluntly, all that we know about the Sun, planets, and the world of our birth makes sense only when viewed within the context of the evolving Universe as a whole. We have seen a first glimpse of the power of this insight in the current chapter. As we move on to look at the planets, moons, asteroids, and comets that orbit the Sun, we will time and again see the fingerprints left behind by our birth among the stars.

The journey that we are taking in *21st Century Astronomy* is one of scientific discovery, but the philosophical implications of the connection between star formation and the formation of the Solar System should not be overlooked. Our very existence is set within the context of the larger Universe. We are the legacy of the processes we see at work all around us, even today.

STUDENT QUESTIONS

THINKING ABOUT THE CONCEPTS

1. In a poetic sense, planets are sometimes referred to as "children of the Sun." A more accurate description, though, would be "siblings of the Sun." Explain.

2. Physicists describe certain properties, such as angular momentum and energy, as being "conserved." What does this mean? Do these conservation laws imply that an individual object can never lose or gain angular momentum or energy? Explain.

3. Planetary bodies have two kinds of angular momentum, orbital and spin. Orbital angular momentum is determined by the orbital motion of a planet moving in orbit around the Sun (or a satellite orbiting a planet). Spin angular momentum is determined by the rotation of the planet about its own spin axis. A body's total angular momentum is the sum of the two. Remembering that angular momentum depends on mass, size of orbit/object, and velocity of rotation/revolution, which do you think contributes most to a planet's total angular momentum, orbital or spin? Explain.

4. Nearly all of the total angular momentum in the Earth-Moon system resides in Earth's spin angular momentum and the Moon's orbital angular momentum in its orbital motion around Earth. (Note that Earth's orbital angular momentum is not included here.) Friction caused by ocean tides is very gradually slowing down Earth's rate of rotation (causing our days to become longer).

   a. As Earth slows down, it loses spin angular momentum. Explain.

   b. Earth's lost spin angular momentum *must* be transferred to the Moon's orbital angular momentum. Why?

   c. What effect do you suppose this has on the Moon's orbit around Earth? Explain.

5. More than 99% of the total mass in the Solar System is in the Sun. Yet Jupiter, with only 1/1,000 of the Sun's mass, possesses more of the Solar System's total angular momentum than any other body, including the Sun. Explain.

6. Nearly all of the planets astronomers have found orbiting other stars have been giant planets with masses more like that of Jupiter than Earth, and with orbits located very close to their parent stars. Does this mean that our Solar System is unusual? Explain.

7. Why do we find rocky material everywhere in the Solar System, but large amounts of volatile material only in the outer regions? Would you expect the same to be true of other solar systems? Explain.

8. The most prominent pre-1950 theory for the formation of the Solar System was that the gravitational interaction with a passing star caused matter to be tidally extracted from the Sun, and that this material formed the planets. Describe and explain evidence we now have that tends to reject this theory.

APPLYING THE CONCEPTS

9. Using the information on the planets given in the Appendices to answer the following:
   a. What is the total mass of all the planets in our Solar System, expressed in Earth masses ($M_\oplus$)?
   b. What fraction of this total planetary mass does Jupiter represent?
   c. What fraction does Earth represent?

10. The outer planets are all made up of rocky cores surrounded by volatiles and gas. Assume that Jupiter's rocky core is 15 $M_\oplus$, Saturn's is 8 $M_\oplus$, Uranus's is 3 $M_\oplus$, and Neptune's is 2 $M_\oplus$. What is the fraction of rock to total mass for each of these four planets? What does this tell you about the ability of a rocky core to capture gases from the surrounding protoplanetary disk?

11. Orbital angular momentum is defined as $L_o = mvr$, where m is the mass, v the speed, and r the orbital radius of the orbiting body. Spin angular momentum (for a uniform sphere) is defined as $L_s = (4\pi mR^2)/5P$, where m is the mass of the sphere, R the radius of the sphere, and P is the period of rotation. Compare Earth's orbital angular momentum with its spin angular momentum using the following values: $m = 6 \times 10^{24}$ kg, $v = 2.98 \times 10^4$ m/s, $r = 1.5 \times 10^{11}$ m, $R = 6.4 \times 10^6$ m, and $P = 8.64 \times 10^4$ s. What fraction does each contribute to Earth's total angular momentum?

12. Assume the Sun is a uniform sphere with a radius of 700,00 km and a rotation period of 26 days. Near the end of its life, the remnant of the Sun will be a white dwarf with a radius of only 5,000 km. Assuming the mass of this white dwarf is still one solar mass, in its dying moments, what will be the Sun's rotation period?

13. Jupiter has a radius 11.2 times that of Earth and a mass 318 times that of Earth. Its rotation period is 9.9 hours. What is the ratio of Jupiter's spin angular momentum to that of Earth?

14. Jupiter has an orbital radius of 5.2 AU and an orbital velocity of 13.1 km/s. Earth's orbital velocity is 29.8 km/s. What is the ratio of Jupiter's orbital angular momentum to that of Earth?

We've sent a man to the Moon, and that's 240,000 miles away. The center of the Earth is only 4,000 miles away. You could drive that in a week, but for some reason nobody's ever done it.

ANDY ROONEY (1919–)

THE TERRESTRIAL PLANETS AND EARTH'S MOON

6.1 HOW ARE PLANETS THE SAME, AND HOW ARE THEY DIFFERENT?

Most of what we know of our planetary neighbors we have learned only in the recent few decades. The second half of the 20th century was a wonderfully exciting time of exploration and discovery about Earth and its sibling worlds. It was a time that saw unmanned probes visit every planet except Pluto, and saw men walk on the surface of the Moon. The variety of techniques used to explore the Solar System is discussed in Tools 6.1. The information returned from these missions has revolutionized our understanding of our planetary system, offering us insights into the current state of each of our neighbors and clues about their histories.

The vast quantity of information about the planets returned by space probes can be hard to digest. What information is truly fundamental and exciting, and what information is flashy but less important? The key to sorting through this information has proven to be comparison among the different planets. The ways the planets are alike and how they differ draws our attention to the most fundamental issues, helping us to ask the right questions and to answer them as well. The correct explanation for some aspect of one planet must

KEY CONCEPTS

The objects that formed in the inner part of the protoplanetary disk around the Sun were relatively small rocky worlds, one of which we call home. Comparison of those worlds with one another teaches us lessons about what shapes a planet's fate. Among the lessons we will learn are:

* That each terrestrial planet is shaped by impacts, tectonism, volcanism, and gradation;
* How impacts scarred planets early in the history of the Solar System, and still occur on occasion today;
* Why the concentration of craters on a planetary surface tells us how old the surface is;
* How radioactive dating of lunar rocks is used to calibrate the cratering clock;
* That larger worlds remain geologically active longer because smaller worlds cool off sooner;
* How predictions of models of Earth's interior are tested using seismic waves from earthquakes;
* That tectonism takes different forms on different planets, but plate tectonics is unique to Earth;
* That among the volcanoes found on Earth, Venus, and Mars, the most colossal are on Mars; and
* The many ways that gradation wears down what other processes form.

TOOLS 6.1

Exploring The Solar System with Spacecraft

As we have stressed, we live in a remarkable time of discovery, when our newfound technological prowess has allowed us to begin the process of exploring our local corner of space. The general strategy for exploring our Solar System begins with a reconnaissance phase, utilizing spacecraft that fly by or orbit a planet or other body. At the opening of the 21st century, we have conducted preliminary reconnaissance of most of the Solar System. We have sent spacecraft flying by all of the planets except Pluto, giving humanity its first ever close-up views of these distant worlds and their moons. We have even seen comets and asteroids at close range. As they sped by, instruments aboard these spacecraft briefly probed the physical properties of their targets and their environments.

Flyby missions have several distinct advantages in the reconnaissance phase of exploration. First, they are the easiest missions to design and execute. Second, flyby spacecraft such as the *Voyager* missions may be able to visit several different worlds during their travels. The downside of flyby missions is that, thanks to the physics of orbits, these spacecraft must move very swiftly. They are limited to just a few hours or, at most, a few days in which to conduct close-up studies of their targets. More detailed reconnaissance work utilizes spacecraft that orbit around planets. These are intrinsically more difficult missions than flyby missions, but orbiters are able to linger, looking in detail at more of the surface of the object they are orbiting, and studying things that change with time, like planetary weather. As of this writing, spacecraft have orbited the Moon, Venus, Mars, Jupiter, an asteroid, and will soon orbit Saturn.

Reconnaissance spacecraft utilize **remote sensing** instrumentation, very much like the remote sensing techniques used by Earth-orbiting satellites to study our own planet. These include tools such as cameras capable of taking images in different wavelength ranges, radar for mapping surfaces hidden beneath obscuring layers of clouds, and spectrometers that spread out the light from their target into a spectrum. Remote sensing allows planetary scientists to map other worlds, measure the heights of mountains, identify geological features, learn about types of rocks present, watch weather patterns develop,

measure the composition of atmospheres, and in general get a feeling for the "lay of the land." Still other instruments make in situ measurements of the extended atmospheres and space environment through which they travel.

Reconnaissance spacecraft provide a wealth of information about a planet, but there is no better way of obtaining "ground truth" than to put our instruments where they can get right to the heart of it—within the atmosphere of a planet or on solid ground. We have landed spacecraft on the Moon, Mars, Venus, and the asteroid Eros. These spacecraft have returned pictures of the surfaces, made measurements of surface chemistry, and conducted experiments to determine the physical properties of the surface rocks and soils. We have also sent small probes into the atmospheres of Venus and Jupiter, from which they returned temperature, pressure, and compositional data while descending.

One of the disadvantages of using landed spacecraft is that only a few landings in limited areas can be made because of the expense, and the results may apply only to the small area of the landing site. Imagine, for example, what a different picture of Earth we might get from a spacecraft that landed in Antarctica, as opposed to a spacecraft that landed in the caldera of a volcano or the floor of a dry riverbed. Sites to be explored with landed spacecraft must be very carefully chosen on the basis of reconnaissance data if we are to know what to make of the information they provide. We can mitigate some of the limitations of landers by putting their instruments on wheels and sending them from place to place, exploring the vicinity of the landing site. Such vehicles, called *rovers,* were used by the Soviets on the Moon more than a quarter century ago and by the United States on Mars in 1997 with the *Pathfinder* mission.

If you pick up a rock by the side of the road, there is a lot you might learn from the rock using the tools that you could easily carry in your pocket. On the other hand, the sophistication of the tools you could carry with you would be limited. It would be much better to pick up a few samples and carry them back to a laboratory equipped with a full range of state-of-the-art instruments capable of measuring chemical compositions, mineral types, radiometric ages, and other information needed to reconstruct the story of their origin and evolution. So, too, is the case in Solar System exploration. One of the most powerful methods for investigating remote objects is to collect samples of the object and bring them back to Earth for detailed study. So far, only samples of the Moon have been collected and returned to Earth. (As we will learn in Chapter 11, however, we do have meteorites that are considered to be parts of Mars.) Plans are currently under way for unmanned "sample and return" missions to Mars. Before these samples are received, spacecraft will also have collected small samples of cometary debris and samples of the solar wind. These materials will be scooped up from interplanetary space by individual spacecraft, which will then return to Earth with their precious cargoes.

Of course, one could not collect specimens in a national park without permission and a scientifically valid reason. Similarly, the return of extraterrestrial samples to Earth is governed by international treaties and standards to ensure that contamination of Earth does not occur. For example, before the lunar samples brought back by the Apollo missions could be studied, they (and the astronauts) were placed in quarantine and tested for alien life-forms. The same international standards apply to spacecraft landing on planets. The goal of these standards is to avoid *forward contamination,* or transporting life-forms from Earth to another planet. If there is life on other planets, then not only is there concern about introducing potential harm, but from a scientific perspective we do not want to "discover" life that we, in fact, have introduced.

With numerous missions under way and others on the horizon, unmanned exploration of the Solar System is an ongoing, dynamic activity. In our journey we will frequently refer to space missions and the information they return, but today's hot results may be tomorrow's old news in light of other, even more exciting discoveries. We hope that you will make use of the *21st Century Astronomy* Web site as a gateway to the wealth of exciting results that the future holds.

work together with what we know about the other planets. For example, when we explain why the Moon is covered with craters, our explanation must also allow for an understanding of why preserved craters on Earth are rare. An explanation for why Venus has such a massive atmosphere should point to reasons why Earth and Mars do not. Such comparisons—an approach called **comparative planetology**—have provided the guideposts to planetary scientists.

We learn about planets by comparing them with each other.

When making comparisons we need a place to start. Earth is the planet that we know best, so it is here at home that we begin our appraisal of the worlds of the inner Solar System.

6.2 FOUR MAIN PROCESSES SHAPE OUR PLANET

For most of human history we have looked upon Earth as a vast, almost limitless expanse. This view of our planet changed forever with a single snapshot taken by Apollo astronauts looking back at Earth from space. We no longer have an excuse to view the world as anything

From the vantage of space, Earth is a tiny blue ball.

other than what it is: a small blue ball, a tiny and fragile lifeboat adrift in the vastness of space, our home.

The psychological impact of these images is rivaled only by the change in scientific perspective of which they are a part. Seen from space **(Figure 6.1)**, Earth is awash with color. White clouds drift in our atmosphere, and white snow and ice cover the planet's frozen poles. The blue of oceans, seas, lakes, and rivers of liquid water—Earth's **hydrosphere**—covers most of the planet. Brown shows us the outer rocky shell of Earth, referred to as Earth's **lithosphere.** Finally, green is the telltale sign of vegetation, part of Earth's **biosphere,** the most extraordinary among Earth's many distinctions.

Earth is a place of change, as geological processes constantly work to reshape our planet. Some geological processes originate in the interior of Earth, powered by the energy generated there. Earthquakes are sudden reminders of the ongoing deformation of Earth's lithosphere, referred to as **tectonism. Volcanism** is a form of **igneous activity,** the formation and action of gas and molten rock, or **magma.**

Figure 6.1 *Seen from space, the colors of Earth tell of the diversity of our planet.*

Collisions involving planetary objects—a process referred to as **impact cratering**—are extremely important in our planet's history. Most of us have seen meteors, the bright streaks that flash across the sky when chunks of material from outer space hit our atmosphere, but for a few the experience with celestial debris is more intimate. In 1954 a meteorite crashed through the roof of a woman's home in Alabama, striking her hip. (Fortunately the roof of the house slowed the meteorite, so she was left with only a bad bruise.) You should not lose much sleep worrying about meteorites landing in your lap. Unlike being struck by lightning, being hit by a meteorite is even less likely than winning the state lottery. However, when a very large object hits a planet, as still happens occasionally in the Solar System, the resulting devastation can be global. At times in our past, such catastrophic events have altered Earth's climate and have changed the course of the history of life itself.

These three processes—tectonism, volcanism, and impacts—affect Earth's surface in their own characteristic ways. Tectonism folds and breaks Earth's crust, forming mountain ranges, valleys, and deep ocean trenches. Volcanic eruptions, like the one shown in **Figure 6.2,** can spill sheets of lava and ash over vast areas, forming mountains or plains in the process. Distinctive scars in Earth's crust tell of impacts in our past. But as these processes work to form **topographic relief,** a fourth process called **gradation** works slowly but persistently to level Earth's surface. Erosion by running water, wind, and other agents wears down the hills, mountains, and continents. The eroded debris collects downslope, filling in valleys, lakes, and ocean basins. Left on its own, gradation would eventually leave the surface of our planet smooth and featureless. Coupled with biological processes, including the actions of humans, the surface of our planet is an ever changing battleground between processes that build up topography and those that tear it down.

> Tectonism, volcanism, and impacts rough up planetary surfaces; gradation smooths them.

Each of the four types of geological processes at work shaping the surface of Earth leaves its own distinctive signature. Planetary scientists have learned to read these signatures on the surfaces of other planets as well.

6.3 IMPACTS HELP SHAPE THE EVOLUTION OF THE PLANETS

Objects orbiting the Sun and the planets in the Solar System move at very high speeds. Earth moves at about 30 km/s in its orbit around the Sun, and meteors can enter Earth's atmosphere at relative speeds in excess of 70 km/s. Because the *kinetic energy* of an object is propor-

Figure 6.2 *Volcanism, such as this eruption in Hawaii, spills molten rock and other materials onto planetary surfaces.*

(a)

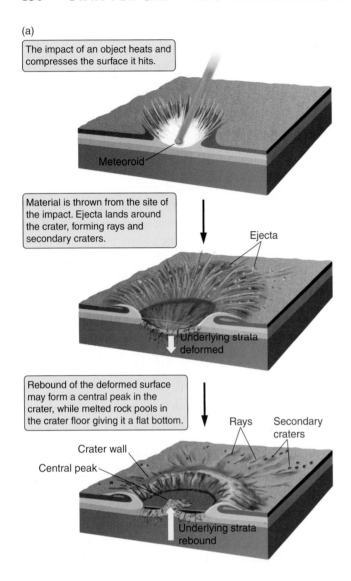

The impact of an object heats and compresses the surface it hits.

Meteoroid

Material is thrown from the site of the impact. Ejecta lands around the crater, forming rays and secondary craters.

Ejecta

Underlying strata deformed

Rebound of the deformed surface may form a central peak in the crater, while melted rock pools in the crater floor giving it a flat bottom.

Rays

Secondary craters

Crater wall

Central peak

Underlying strata rebound

(b)

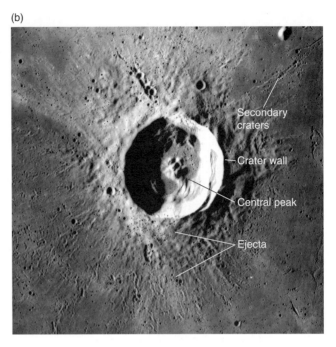

Secondary craters

Crater wall

Central peak

Ejecta

Figure 6.3 (a) *Stages in the formation of an impact crater.* (b) *A lunar crater photographed by Apollo astronauts, showing rays, secondary craters, the crater wall and surrounding material, and a central peak—all typical features associated with impact craters.*

craters are the cooled surfaces of pools of rock melted by the impact. The energy released in an impact can also lead to the formation of new minerals. In fact, some minerals, such as shock-modified quartz, are known to form *only* during the momentary fury of an impact. Geologists look for these distinctive minerals as evidence of ancient impacts on Earth's surface.

Impacts can melt and vaporize rock.

Meteor Crater in Arizona is the best-preserved impact structure on our planet **(Figure 6.5)**. It is thought to be the result of the impact of an iron meteorite about 40 meters in diameter with a mass of about 10^7 kg that hit Earth traveling at 20 km/s about 40,000 years ago. Such a collision would have released about 2,000 times as much energy as the first atomic bomb, detonated in New Mexico in 1945. Yet, at only 1 km in diameter, Meteor Crater is tiny compared with impact craters seen on the Moon or ancient impact scars on Earth. See Connections 6.1 to read about the consequences of an especially violent impact that took place 65 million years ago.

One of the most obvious differences among the terrestrial planets is that, while on some planets the surfaces are covered by impact craters, on others (especially Earth) impact craters seem to be rare. While all terrestrial planets are subject to impacts, tectonism, volcanism, and gradation, the relative intensity of these processes varies among the planets. For example, when we look at

Large impacts produce the greatest concentrations of energy in planetary geology.

tional to the square of its speed (K.E. $= \frac{1}{2}mv^2$), collisions between such objects release very large amounts of energy. In fact, large impacts are by far the most concentrated and sudden release of energy of any geological process, including earthquakes and volcanic eruptions.

When an object hits a planet, its kinetic energy goes into heating and compressing the surface that it strikes, and throwing material far from the resulting **impact crater** (see **Figure 6.3**). Sometimes material thrown from the crater, called **ejecta,** falls back to the surface of the planet with enough energy to cause **secondary craters.** In impacts the rebound of heated and compressed material can also lead to the formation of a central peak or ring of mountains in the floor of the crater, in ways similar to what happens when a drop lands in a glass of milk, as in **Figure 6.4.**

The energy of an impact can be great enough to melt or even vaporize rock. The smooth bottoms of some

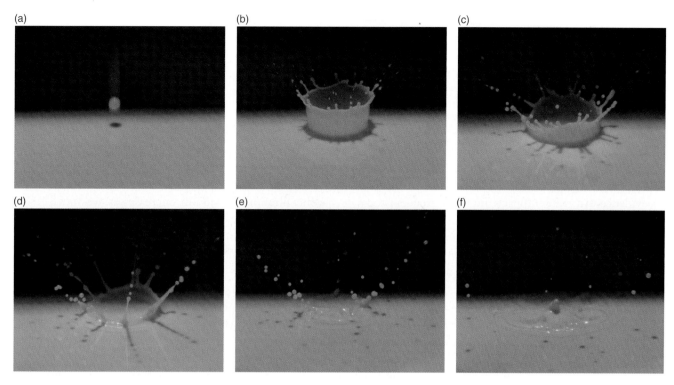

Figure 6.4 *A drop hitting a pool of milk illustrates the formation of features in an impact crater, including crater walls (b and c), secondary craters (d and e), and a central peak (f).*

the Moon, we find millions of craters of all different sizes, one on top of another **(Figure 6.6).** Nearly all of these craters are the result of impacts. By comparison, only about 150 impact craters, or scars from impacts, have been identified on Earth. One reason for the shortage of craters on Earth is that, while the surface of the Moon is directly exposed to this cosmic bombardment, the surface of Earth is partly protected by the blanket of Earth's atmosphere. For example, rock samples from the Moon show craters smaller than a pinhead, formed by micrometeoroids. In contrast, most small meteoroids are burned up by friction in Earth's atmosphere before they reach the surface. The atmospheric pressure on Venus is 92 times that of Earth, offering that planet even more protection—larger objects may be burned up or broken apart by its atmosphere.

The characteristics of a crater also depend on the properties of the planetary surface. An impact in a deep ocean on Earth might create an impressive wave but leave no lasting crater. In contrast, an impact scar formed in granite can be preserved for billions of years. Planetary scientists can tell a lot about the surface of a planet by studying its craters. For example, craters on the Moon are often surrounded by systems of "rays" of material that were thrown out by the impact, like those in Figure 6.3(b). Some craters on Mars have a very different appearance. These craters are surrounded by what appears to be flows of material, much like the pattern

6.1 IMPACTS

The forms of impact scars give clues about the surfaces they are on.

Figure 6.5 *Meteor Crater (also known as Barringer Crater), located in northern Arizona, is a 1.2-km-diameter impact crater formed some 40,000 years ago by the impact of an iron meteoroid.*

CONNECTIONS 6.1

WHERE HAVE ALL THE DINOSAURS GONE?

In 1994 a host of astronomical instruments watched as the pieces of comet Shoemaker-Levy 9 slammed one after another into the clouds of Jupiter, leaving temporary scars that were visible through backyard telescopes. The Jupiter comet crash provided graphic evidence that, while impacts are not as frequent today as they were when the Solar System was young, they *do* still occur. The giant planets are not the only targets for such cosmic bombardment either. In 1996 two students at the University of Arizona discovered a small *asteroid,* a small planetlike object, perhaps 500 m across, as it whizzed past Earth at 93,000 km/hour, missing our planet by only 450,000 km. When large impacts happen on Earth, they can have far-reaching consequences for Earth's climate and for terrestrial life.

Earth's fossil record shows that on occasion large numbers of species vanish from the face of the planet in a geological blink of an eye. The most famous of these extinctions occurred 65 million years ago, when over 50% of all living species, including the dinosaurs, became extinct. This mass extinction is marked in Earth's fossil record by the **Cretaceous-Tertiary boundary,** or simply the K-T boundary. (The **Cretaceous period** lasted from 146 million years ago to 65 million years ago. The **Tertiary period** started when the Cretaceous period ended, and lasted until 1.8 million years ago. We live in the **Quaternary period** of Earth's history, which started at the end of the Tertiary period.) In older layers found below the K-T boundary, fossils of dinosaurs and other now-extinct life forms abound. In the newer rocks above the K-T boundary, more than half of all previous species are absent, and in their place are found fossils of newly evolving species. Big winners in the new order were the mammals—our own distant ancestors—which moved into ecological niches vacated by extinct species.

The K-T boundary is marked in the fossil record in many areas by a layer of clay. Studies at more than 80 locations around the world have found that this layer contains large amounts of the element iridium as well as traces of soot. Iridium is very rare in Earth's crust but is common in meteorites. The soot at the K-T boundary tells of a time when widespread fires burned the world over. The thickness of the layer of clay at the K-T boundary, as well as the concentration of iridium, increases as we move toward what is today the Yucatan Peninsula in Mexico. While the original crater has been erased by gradation, geophysical surveys and rocks from drill holes in this area show a highly deformed subsurface rock structure, similar to that seen at known impact sites. Together, these results provide compelling evidence that 65 million years ago a 10-km-diameter asteroid struck the area, throwing great clouds of red-hot dust and other debris into the atmosphere **(Figure 6.8)** and igniting a worldwide conflagration. The energy of the impact is estimated to have been more than that released by 5 *billion* atomic bombs.

An impact of this magnitude clearly would have had a devastating effect on terrestrial life. Could this cosmic impact have been responsible for the sudden disappearance of forms of life that had ruled Earth for 150 million years? Many scientists believe so. In addition to igniting a near-global fire storm, dust thrown into Earth's upper atmosphere by the impact would have remained there for years, blocking out sunlight and plunging Earth into decades of cold and darkness. The fire storms, temperature changes, and decreased food supplies could have led to mass starvation that would have been especially hard on large animals such as the dinosaurs.

Not all paleontologists believe that this mass extinction was the result of an impact. They point out that the evolution of species is a complex process, and that simple answers are seldom complete. However, the evidence is compelling that a great impact *did occur* at the end of the Cretaceous period. The rock record also shows many other instances of mass extinctions associated with colossal impacts, suggesting that impacts have played a central role in the saga of life on Earth.

---

you might see if you threw a rock into mud **(Figure 6.7).** The apparent flows seem to indicate that, unlike the lunar case, the Martian surface rocks contained water or ice at the time of the impact. Not all Martian crater ejecta deposits look like this, so the water or ice must

have been concentrated in only some areas, and the locations might have changed with time.

It is possible that these craters were formed at a

Mars was once wetter than it is today.

Figure 6.6 *Terrain on the Moon, photographed by the* Apollo 17 *astronauts, has been heavily cratered by impacts.*

Figure 6.7 *Some craters on Mars look like those formed by rocks thrown in mud, suggesting that material ejected from the crater contained large amounts of water. The crater on the right is about 30 km across.*

time in the past when there was liquid water on the surface of Mars. Dry riverbeds on Mars seen today attest to this possibility. But there is another intriguing possibility. Today, the surface of Mars is dry and mostly frozen. This suggests that water which might have once been on the surface of Mars has soaked into the ground, as well as having been trapped in polar ice caps. For example, some planetary scientists speculate that water ice lies below the surface in some areas, much like the water

frozen in the ground in Earth's polar regions. The energy released by the impact of a meteoroid would melt this ice, possibly turning the surface material into a slurry with a consistency much like wet concrete. When thrown from the crater this slurry would have hit the surrounding terrain and slid out across the surface, forming the craters we see today.

CALIBRATING A COSMIC CLOCK

While many factors affect the formation of craters, the biggest difference among planets is in the rate at which craters are destroyed. Earth experienced an impact history similar to that of the Moon and the other terrestrial planets, and yet preserved impact craters are rare on Earth. Obliteration of impact scars on Earth is due to a combination of gradation, tectonic processes, and volcanism acting throughout geological time. The Moon has been nearly

Figure 6.8 *This artist's rendition depicts a comet or asteroid, perhaps 10 km across, striking Earth 65 million years ago in what is now the Yucatan Peninsula in Mexico. The effects of the impact killed off most forms of terrestrial life, including the dinosaurs.*

geologically dead for billions of years, and its surface still preserves the scars of craters dating from the early days of the Solar System. In contrast, geologically active planets such as Earth bear the scars of only more recent impacts. Planetary scientists use this simple relationship—more craters means an older planetary surface—to estimate the ages and geological histories of planetary surfaces. Although we have seen only half of Mercury's surface, its preserved craters suggest that Mercury, like the Moon, has been geologically inactive for a long time. This contrasts to the much less heavily cratered surfaces of Mars, Venus, and Earth.

The number of craters on a surface indicates the age of the surface.

Some impact craters are young, while most are old. But how old is "old," and how young is "young"? We can use the amount of cratering as a clock to measure the ages of surfaces, but first we need to know how fast that clock runs. We need to be able to say that a surface with *this many* craters is *this* old, but a surface with *that many* craters is *that* old. We need a way of "calibrating the cratering clock."

The key to calibrating the cratering clock came mostly from our exploration of the Moon. The surface of the Moon is not uniform. Some parts of the Moon are heavily cratered, with craters overlapping their neighbors. Other parts of the Moon are much smoother, telling of more recent geological activity. Between 1969 and 1976, Apollo astronauts and Soviet machines visited the Moon and brought back samples taken from nine different locations on the lunar surface. By measuring relative amounts of various radioactive elements and the elements they decay into, scientists were able to assign ages to these different lunar regions (see Foundations 6.1). The results of that work were surprising. While smooth areas on the Moon were indeed found to be younger than heavily cratered areas, the differences in age were not very great. The oldest, most heavily cratered regions on the Moon date back to about 4.4 billion years ago, while the smoother parts of the lunar

Moon rocks let us interpret the history of impacts in the Solar System.

Most of the surface of the Moon is older than 3.4 billion years.

FOUNDATIONS 6.1

DETERMINING THE AGES OF ROCKS

Look at a picture of the Grand Canyon. The rock layers tell the story of the canyon's geological history. Most of the layers were laid down by a process called *sedimentation,* in which material carried by water or wind buries what lies below. Volcanism also contributes to the layering, as lavas flow over Earth's surface. At the top of the stack are the latest deposits, such as those found on the rim of the Grand Canyon. Moving down through these layers takes us progressively further and further back into Earth's past.

In order to assign real dates to these different layers—or to rocks from any location, including the Moon—scientists use a technique called **radiometric dating.** Radiometric dating makes use of the steady decay of radioactive **parent elements** into more stable **daughter products.** Some minerals can contain radioactive isotopes as part of their chemical structure. (Different **isotopes** of an **element** have the same number of protons in their nuclei but different numbers of neutrons.) Chemical analysis of such a mineral immediately after its formation would

reveal the presence of the radioactive parents, but the daughter products of the radioactive decay would be absent. Over time, as radioactive atoms decay, however, the amount of the trapped daughter products builds up. Chemical analysis of an old sample of such a mineral would reveal both remaining radioactive parent atoms as well as daughter products trapped within the structure of the mineral.

By comparing the relative amounts of radioactive parent and daughter products, scientists can determine when the mineral was formed, and hence the age of the rock. For example, the most abundant isotope of the element uranium (the parent) decays through a series of intermediate daughters to an isotope of the element lead (its final daughter). The **half-life** of uranium is 4.5 billion years. This means that in 4.5 billion years, a sample that originally contained uranium (the parent) but no lead (its final daughter) would be found instead to contain equal amounts of uranium and lead. If we were to find such a mineral, we would know half the uranium atoms have turned to lead, and that the mineral formed 4.5 billion years ago.

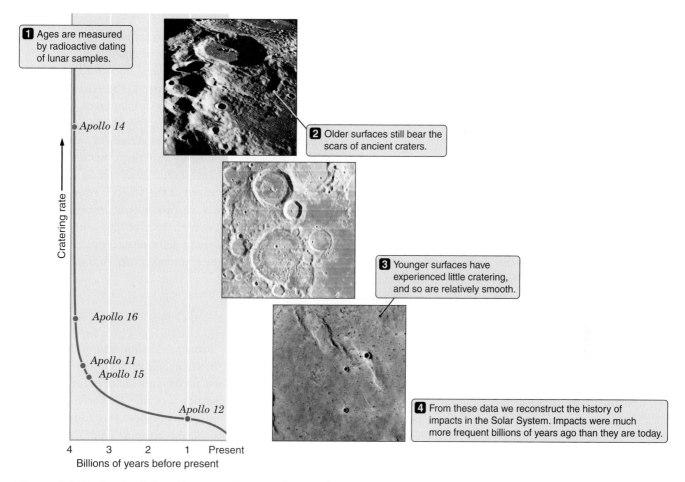

1 Ages are measured by radioactive dating of lunar samples.

2 Older surfaces still bear the scars of ancient craters.

3 Younger surfaces have experienced little cratering, and so are relatively smooth.

4 From these data we reconstruct the history of impacts in the Solar System. Impacts were much more frequent billions of years ago than they are today.

Cratering rate

Apollo 14

Apollo 16

Apollo 11
Apollo 15

Apollo 12

4   3   2   1   Present
Billions of years before present

Figure 6.9 *Radioactive dating of lunar samples returned by Apollo astronauts is used to determine how the cratering rate has changed over time. Using this information, the cratering of a surface tells us its age.*

surface are typically 3.1 billion to 3.9 billion years old. The clear implication is that the vast preponderance of the cratering in the Solar System took place within the first billion years of the Solar System's formation **(Figure 6.9)**. Heavily cratered surfaces such as those of Mercury and the Moon are ancient indeed.

6.4 THE INTERIORS OF THE TERRESTRIAL PLANETS TELL THEIR OWN TALE

To understand the processes responsible for all but wiping away the visible record of impacts in Earth's past, and for continually remaking the surface of our planet, we begin by looking below its surface. What lies hundreds and thousands of kilometers below our feet, and how do we know? These are the questions to which we now turn.

WE CAN PROBE THE INTERIOR OF EARTH IN MANY WAYS

For all the effect that human activity has had on Earth, we have literally only scratched the surface of our planet. The deepest holes ever drilled have gone only about 12 km down. Even so, scientists are confident about what the interior of Earth is like. Information about Earth's interior comes to us in many ways. For example, the size of Earth and the strength of Earth's gravity, together with Newton's universal law of gravitation, tell us the mass and hence the density of Earth. In this way we know that the average density of Earth is 5.5 kg per liter, or five and a half times the density of water. Even though we have no direct samples of material from deep within Earth's interior, this number alone tells us that the composition of the interior of our planet has a density higher than does the surface, which

The density of matter inside terrestrial planets increases with depth.

averages about 2.9 kg/liter. Other clues about Earth's interior come from studies of meteorites. Because these fragments are left over from the time when the Solar System was young and Earth was forming, the overall composition of Earth should resemble the composition of meteorite material, which includes minerals with abundant iron.

By far the most important source of information about Earth's interior comes from monitoring the vibrations from earthquakes. When an earthquake occurs, vibrations spread out through and across the planet as **seismic waves.** There are two different classes of seismic waves. As the name implies, **surface waves** travel across the surface of a planet, much like waves on the ocean. If conditions are right, surface waves from earthquakes can be seen rolling across the countryside like ripples on water. These waves are responsible for much of the heaving of Earth's surface during an earthquake, and cause damage such as the buckling of roadways.

Seismic waves provide information on the interior configuration of Earth.

The other type of wave travels through Earth, rather than along its surface, probing the interior of our planet. These include primary waves and secondary waves. **Primary waves** are longitudinal pressure waves (Figure 4.4b) that result from alternating compression and decompression of material. These are much like sound waves traveling through air or water, or a wave that moves along the length of a spring. **Secondary waves** are more like the motion of a guitar string. They are transverse waves (Figure 4.4a) that result from sideways motion of material. Unlike primary waves, which travel because rock rebounds after being compressed, secondary waves travel because rock springs back after being bent.

The progress of seismic waves through Earth's interior depends upon the characteristics of the material they are moving through **(Figure 6.10).** For example, an important difference between primary and secondary waves in studying Earth's interior is that while primary waves can travel through either solids or liquids, secondary waves cannot travel through liquids because liquids do not "spring back" when they are "bent." In addition, the speed at which seismic waves travel depends on the density and composition of rock. As a result, seismic waves moving through rocks of varying densities or composition are bent in much the same way that waves of light are bent by glass. In fact, when a wave comes to a place where

Seismic waves are affected by their passage through Earth's interior.

the density of rock layers varies abruptly the wave can be refracted (i.e., bent) or even reflected just as light is refracted or reflected by a pane of glass.

For nearly a hundred years, thousands of **seismometers** scattered around the globe have measured the vibrations from countless earthquakes and other seismic events, such as volcanic eruptions and nuclear explosions. When seismic waves arrive at a seismometer station, geologists ask many questions, including: What types of waves were measured? How strong were they? When were they received at the station? Alone, a single seismometer can only record ground motion at one place on Earth. But when we combine our measurements with those of many seismometers placed all over Earth, we get a comprehensive picture of the interior of our planet.

BUILDING A MODEL OF EARTH

How geologists go from raw seismic data to an understanding of Earth's interior is a very good example of the interplay between theory and observation in modern science. To construct a model of Earth's interior, geologists begin with some obvious clues from the seismic data. Are there liquid regions where secondary waves cannot penetrate? Are there jumps in the density of the rock from which waves are reflected? How do waves bend, and what does that say about the density profile of Earth? The model must also be consistent with Earth's average density of 5.5 kg/liter.

To go beyond this basic sketch, geologists turn to the laws of physics, combined with knowledge of the properties of materials and how they behave at different temperatures and pressures. Scientists then construct a model of Earth's interior that follows the rough outline of their basic sketch, but which is also physically consistent. (For example, the pressure at any point in Earth's interior must be just high enough to balance the weight of all of the material above it, as discussed in Foundations 6.2.) They next "set off" earthquakes in their model, calculate how seismic waves would propagate through a model Earth with that structure, and predict what those seismic waves would look like at seismometer stations around the globe. They then test their model by comparing these predictions with actual observations of seismic waves from real earthquakes. The extent to which the predictions agree with observations points out both strengths and weaknesses of the model. The structure of the model is adjusted (always remaining consistent with the known physical properties of materials) until a good match is found between prediction and observation. That is how geologists arrived at our current picture of the interior of Earth. Changing any one part of this picture would so change the way that waves travel through Earth's interior that the model would no longer agree with seismic observations.

The first thing you should notice about the interior structure of Earth is that its composition is far from uniform. Based on physical models and seismograph data, we find that the major subdivisions of Earth's interior, shown in Figure 6.10, include a **core,** surrounded by a thick **mantle,** on top of which is the **lithosphere.** At the center, Earth's core consists primarily of iron, nickel, and other dense metals. In contrast, the outer parts of Earth are made of materials that are of lower density. The **crust,** which is the outermost part of the lithosphere, comes in two forms: low-density continental crust that is rich in silica (SiO_2), and higher-density oceanic crust. Common continental rocks include **granite,** while most oceanic crust is **basalt,** a heavy, dark volcanic rock that is rich in iron and magnesium.

Geologists speak of Earth's interior as being "differentiated" and the process of separating materials by density as **differentiation.** Differentiation of the interiors of Earth, other terrestrial planets, and the Moon is a result of the fact that these planetary interiors were once molten. If rocks of different types are mixed together, they tend to stay mixed. However, once this rock is melted, the denser materials are free to sink to the bottom, and the less dense materials are free to

> **Differentiation of planets shows that they once were molten.**

Figure 6.10 *Different types of seismic waves propagate through the interior of Earth in different ways. Measurements of when and where different types of seismic waves arrive after an earthquake allow us to test predictions of detailed models of Earth's interior.*

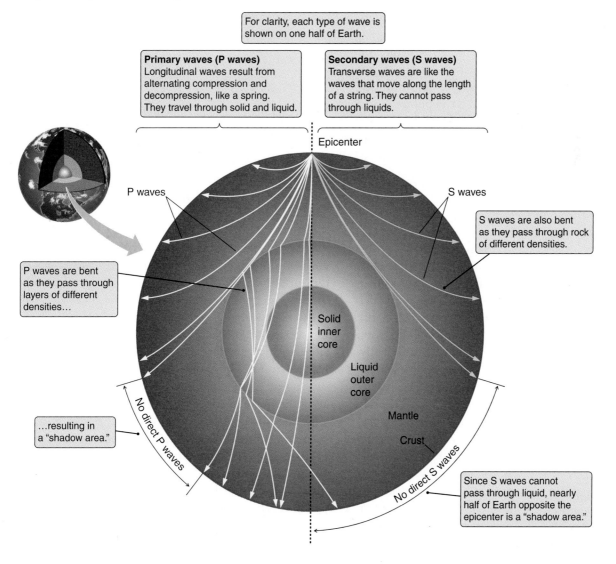

For clarity, each type of wave is shown on one half of Earth.

Primary waves (P waves)
Longitudinal waves result from alternating compression and decompression, like a spring. They travel through solid and liquid.

Secondary waves (S waves)
Transverse waves are like the waves that move along the length of a string. They cannot pass through liquids.

Epicenter

P waves

S waves

S waves are also bent as they pass through rock of different densities.

P waves are bent as they pass through layers of different densities...

Solid inner core

Liquid outer core

Mantle

Crust

No direct P waves

...resulting in a "shadow area."

No direct S waves

Since S waves cannot pass through liquid, nearly half of Earth opposite the epicenter is a "shadow area."

PRESSURE AND WEIGHT

The pressure at any point within a planet's interior is determined by the weight of the material above it. To see why this is so, think about what would happen if it were not. If the outward pressure at some point within a planet were less than the weight per unit area of the overlying material, then that material would fall inward, crushing what was underneath it; if the pressure at some point within a planet were greater than the weight of the overlying material, then the pressure would *not* be confined. The material would be able to expand, lifting the overlying material. The only stable situation is when the weight of matter above is just balanced by the pressure within the whole interior of the planet.

This balance between pressure and weight is a very general and very useful result which scientists refer to as **hydrostatic equilibrium.** We will see this balance again and again as we talk about the structure of planetary interiors, planetary atmospheres, and the structure and evolution of stars.

---

float to the top (just as less dense oil floats on denser vinegar in a bottle of salad dressing). Today, little of Earth's interior is molten, but the differentiated structure of the planet tells of a time when Earth was much hotter, and its interior was liquid throughout.

Figure 6.11 shows the differentiated structure of each of the terrestrial planets and Earth's Moon. As we continue our study of the composition of objects in the Solar System, differentiation will be an important concept. For example, in Chapter 11 we will find that, by analyzing the chemical composition of meteorite material, it is often possible to tell whether it was once a part of a body that was chemically differentiated.

THE MOON WAS BORN FROM EARTH

Models of the interior of the Moon show that it has only a very tiny core, and is composed mostly of material that is similar to that found in Earth's mantle. The best model explaining the Moon's composition is that when Earth was very young a Mars-sized protoplanet collided with Earth, blasting off and vaporizing parts of Earth's partly differentiated crust and mantle. This debris condensed into orbit around Earth, evolving into our Moon. This model explains the similarities in composition between the Moon and Earth's mantle, and can account for the Moon's general lack of **volatile matter** while Earth and its closest neighbors—Mars and Venus—are volatile-rich.

According to this model, during the vaporization stage of the collision, most gases were lost to space, leaving only the nonvolatiles to condense as the Moon.

The Moon was probably formed from a collision between a protoplanet and the young Earth.

Earth, however, was large enough to retain its volatiles, which continued to be released from the interior. Because of the higher gravity of Earth, these gases were retained as part of our atmosphere.

THE EVOLUTION OF PLANETARY INTERIORS DEPENDS ON HEATING AND COOLING

A general feature of planetary interiors is that the deeper we go within the planet, the higher the temperature climbs. A moment's thought about what happens to thermal energy in the interior of a planet shows why this must be so. A planet cools by radiating energy from its surface into space, so we would expect the surface of the planet to be the coolest part. Since thermal energy flows from hotter areas to cooler areas, the temperature must increase as we move inward from the surface if thermal energy is to flow from the interior toward the surface. The pressure in the interior of a planet also increases as we go deeper because the pressure at any given point is determined by the weight of all of the material above it.

Planetary interiors are hotter than their surfaces.

The structure of the core of Earth results from an interesting interplay between the effects of increasing pressure and increasing temperature. We normally think about freezing and melting points—that is, whether a material is solid or liquid—as being a function of temperature. When something gets hot enough, it melts and becomes a liquid. When something gets cold enough, it freezes and becomes a solid. The center of Earth is the hottest location in Earth's interior. With a temperature

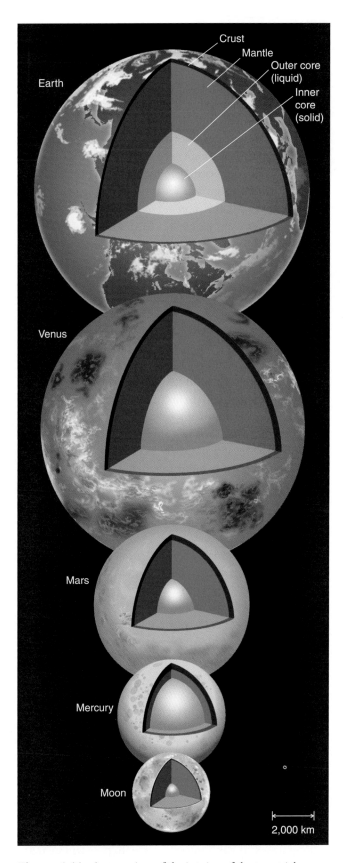

Figure 6.11 *A comparison of the interiors of the terrestrial planets and Earth's Moon. Some fractions of the cores of Mercury, Venus, and Mars are probably liquid.*

of perhaps 6,000 K, it is hotter than the surface of the Sun. Yet the center of Earth is solid! It is the *outer* core of Earth that is molten, despite the fact that it is hundreds of kelvins cooler than the inner core.

Whether a material is solid or liquid depends on pressure as well as on temperature. With most materials (water ice being a rare exception), the solid form of the material is more compact than the liquid form. Putting the material under higher pressure forces atoms and molecules closer together, and makes it more likely for the material to become a solid. Moving toward the center of Earth, the effects of temperature and pressure oppose each other: the higher temperature makes it more likely that material would melt, but the higher pressure favors a solid form. It is only in the outer core of Earth that the high temperature wins, and the material is molten. At the center of Earth, even though the temperature is higher, the pressure is so great that the inner core of Earth is solid.

Part of the thermal energy in the interior of Earth is left over from when Earth formed. We know that the tremendous energy liberated by the collisions responsible for the formation of Earth, together with energy from short-lived radioactive elements, was enough to melt the planet. The differentiated structure of Earth is evidence of this fact. The surface of Earth then cooled rather rapidly by radiating energy away into space, forming a solid crust on top of a molten interior. Because a solid crust does not conduct thermal energy very well, it served as an insulator—a "wool sweater," if you will. (Anyone who has watched a lava flow "crust over" and then walked across its surface has "first-foot" experience of this fact. Molten rock is still there, possibly only centimeters beneath their feet.) But the crust is not a perfect insulator. Over the eons, energy from the interior of the planet continued to leak through the crust and radiate into space. The interior of the planet slowly cooled, and the mantle and the inner core solidified.

Earth was molten when it formed, due to collisions and radioactivity . . .

But there must be more to the story than leftover energy from the time of Earth's formation. If this were the only source of heating in Earth's interior, calculations show that the interior of Earth should be much cooler than it is today, and Earth should have long ago solidified completely. There must be additional sources of energy continuing to heat the interior of Earth if we are to account for the high temperatures that persist there today.

One source of heating Earth's interior is friction generated by tidal effects of the Moon and Sun. When we discuss tides in Chapter 9, we will find that tidal heating is responsible for keeping the interior of the moon Io molten. However, tidal heating fails by a wide margin to explain the elevated temperature of Earth's interior. Instead, *most* of the extra energy in Earth's interior comes from long-lived radioactive elements. As these radioactive elements trapped in the interior of Earth decay, they liberate energy, heating the planet's interior. Today the temperature of Earth's interior is determined by dynamic equilibrium (Foundations 4.1) between heating of the interior and the loss of energy that is radiated away into space. As radioactive "fuel" in Earth's interior is consumed by decay, the amount of thermal energy generated declines, and Earth's interior becomes cooler as it ages.

. . . and has been slowly cooling as it ages.

Because of this equilibrium, the internal temperature of a planet depends on the planet's size. The amount of heating produced in a planet is determined by the planet's *volume,* because it is the volume of the planet that determines how much radioactive material ("fuel") there is. On the other hand, the planet's ability to get rid of the thermal energy in its interior depends on the planet's *surface area,* because thermal energy has to escape through the planet's surface. (If you want to keep warm, you huddle up in a ball, reducing the amount of exposed skin through which thermal energy can escape. But if you want to cool off, you spread out your arms and legs, exposing as much skin as possible and allowing it to get rid of thermal energy.)

The energy-producing volume of a planet increases in proportion to the cube of the planet's radius ($\propto R^3$), while the cooling surface area of the planet increases only as the square of the radius ($\propto R^2$). The ratio of the two—the amount of energy there is to lose divided by the surface area through which thermal energy can escape—is proportional to R^3/R^2, or R. Smaller planets have less energy to lose in relation to their surface areas, so should be cooler. Larger planets have more energy to lose per square meter of surface, and so remain hotter. Indeed, in Section 6.5 we will find that the geological activity of planets is driven by the thermal energy in their interiors. It should not be surprising to learn that the smallest objects—Mercury and the Moon—are geologically inactive in comparison to the largest terrestrial planet, Earth.

Generally, smaller terrestrial planets cool faster than larger terrestrial planets.

MOST PLANETS GENERATE THEIR OWN MAGNETIC FIELDS

As most schoolchildren know, a magnetic compass consists of a small bar magnet that is allowed to swing about freely. The compass needle lines up with Earth's magnetic field and points "north" and "south." But if we were to map the orientation of compass needles at every place on Earth, the north arrows would all converge, not at the North Pole, but at a location in northern Canada. Earth behaves as if it contains a giant bar magnet which is slightly tilted with respect to Earth's rotation axis **(Figure 6.12).** The location in Canada that compasses point to is one of Earth's magnetic poles. An opposite magnetic pole exists near Earth's South Pole, as well.

Earth's magnetic field behaves like a giant bar magnet . . .

Earth's magnetic field actually originates deep within the interior of the planet, and the processes responsible for generating Earth's magnetism are not understood in detail. However, one thing we are certain of is that Earth's magnetic field is *not* due to *permanent magnets* (naturally occurring magnetic materials whose individual atoms are magnetically aligned) buried within the planet. Even though naturally occurring magnetic materials do exist, permanent magnets cannot explain the fact that Earth's magnetic field is constantly changing. Some changes in Earth's magnetic field, such as shifting in the exact location of the magnetic north pole, can be seen over times that are shorter than a human lifetime. In addition, the geological record shows that much more dramatic changes have occurred over the history of our planet.

. . . but the direction of Earth's magnetic poles changes with time.

The study of **paleomagnetism**—the fossil record of Earth's changing magnetic field—is an important part of geology. If a magnetic material such as iron gets hot enough, it loses its "memory" of its previous magnetization. As the material cools, it again becomes magnetized, but along the direction of any externally imposed magnetic field. In this way a memory of that magnetic field becomes "frozen" into the material. (This basic piece of physics is used in many ways, including in some computer data storage technologies. This is also another reason why permanent magnets cannot be responsible for Earth's magnetic field. At the temperatures in the interior of Earth, permanent magnets lose their magnetization.) Lava extruded from a volcano carries a record of Earth's magnetic field at the time that it cooled. Using the same sorts of radiometric

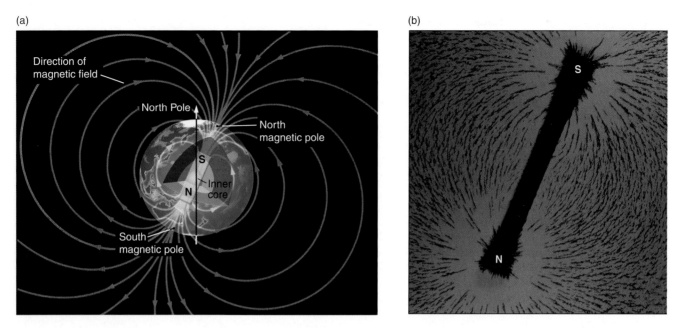

Figure 6.12 (a) *Earth's magnetic field is similar to that of a giant bar magnet tilted relative to Earth's axis of rotation.* (b) *Iron filings sprinkled around a bar magnet help us visualize such a magnetic field.*

techniques discussed in Foundations 6.1 to date these materials gives us a record of how Earth's magnetic field has changed over time. While Earth's magnetic field has probably existed for billions of years, the north/south *polarity* reverses from time to time. On average, these reversals in Earth's magnetic field take place every 500,000 years or so.

Although the details are not known, we do have a general idea of how Earth's magnetic field originates in the motions of material in Earth's interior. Magnetic fields result from electric currents, which are due to moving electric charges. Earth's magnetic field is thought to be a side effect of Earth's rotation about its axis and fluid motions including convection within Earth's liquid outer core. The interior of Earth acts as if it were a giant *dynamo,* converting mechanical energy into magnetic energy. (Other objects in the Solar System, including the Sun and the giant planets, are also thought to act like dynamos.)

The Moon is the most magnetically surveyed object other than Earth. During the Apollo program, astronauts used surface magnetometers (devices for measuring magnetic fields) to make local measurements, and on two missions, small satellites were placed into orbit to search for global magnetism. Results show that the Moon either lacks a magnetic field today or, at the most, has a very weak field. The lack of a lunar mag-

The Moon had a magnetic field long ago, but lacks one today.

netic field can probably be understood because of the small size of the Moon and its correspondingly cooler interior. It also has a very small core. Overall, it would be difficult to make a lunar dynamo work today. However, remnant magnetism is preserved in the rocks from an earlier time when there likely was a magnetic field.

When we study the other terrestrial planets, however, there are a few surprises in store. Other than Earth, Mercury is the only terrestrial planet to have a significant magnetic field today. Mercury's field is understandable. This planet has a large iron core, parts of which may be molten, so the ingredients for a dynamo are present. It is the lack of detected magnetic fields from Venus and Mars that presents a puzzle. Venus should, like Earth, have an iron-rich core and partly molten interior. Its lack of a magnetic field might be attributed to its extremely slow rotation, which could make its dynamo very inefficient. On the other hand, Mercury rotates very slowly as well (once every 58.6 Earth days) but still has a planetary magnetic field.

The lack of magnetic fields on Venus and Mars is a puzzle.

The lack of a magnetic field today on Mars might be the result of its small core, but given that it is expected to have a partly molten interior and rotates rapidly, the lack of a field is still surprising. However, similar to the Moon, Mars has a pronounced remnant magnetic field, as discovered by the *Mars Global Surveyor* orbiter in the late 1990s. The magnetic signature occurs only in the ancient crustal rocks, showing that early in Mars's history some sort of a magnetic field existed. Geologically younger rocks lack this residual magnetism, so the planet's magnetic field has long since disappeared.

6.5 TECTONISM— HOW PLANETARY SURFACES EVOLVE

As you drive through mountainous or hilly terrain, take a look at the places like that shown in **Figure 6.13,** where the roadway has been cut through rock. These cuts show layers of rock that have been bent, broken, or fractured into pieces. Sometimes the force responsible for tectonism—the deformation of Earth's crust—is just gravity. For example, for more than 60 million years, rivers have dumped trillions of metric tons of rock and sediment into the Gulf of Mexico. Layers of sediment over 25 km thick have built up, and some have solidified to rock. This enormous mass has been

Figure 6.13 *Tectonic processes fold and warp Earth's crust, as seen in these rocks along a roadside in Israel.*

EXCURSIONS 6.1

PALEOMAGNETISM: A TICKER-TAPE RECORD OF PLATE TECTONICS

The discovery that plates and continents are moving did not by itself confirm the hypothesis of plate tectonics because it did not demonstrate that these motions continue over geological time spans. An important clue to this puzzle came from studies of Earth's paleomagnetism.

Later in the text we note that ocean-floor rifts such as the mid-Atlantic rift are spreading centers. These are locations where hot material rises toward Earth's surface and fills the gap between tectonic plates to become new ocean floor. In our discussion of paleomagnetism we found that as hot material cools, it becomes magnetized along the direction of the local magnetic field. Ocean floor "remembers" the direction of Earth's magnetic field at the time it cooled as it moves away from the spreading center. In this way, the spreading ocean floor carries with it a record of the changes in Earth's magnetic field over time. The farther away from a spreading center, the older the ocean floor, and the earlier the time that its

magnetization reflects. The ocean floor acts much like a ticker-tape recorder for Earth's magnetic history.

The discovery of this magnetic record was made during surveys of the ocean floor in the 1960s, and became one of the most important pieces of evidence supporting the theory of plate tectonics **(Figure 6.14).** These surveys showed a banded magnetic structure surrounding ocean rifts. Material near the rifts is magnetized in the same sense as Earth's current magnetic field, but the magnetization changes farther away from a rift. This banded magnetic structure is often symmetric about rifts. If a change in the magnetization of the ocean floor is seen 100 km on one side of a rift, then the same change will usually be seen about 100 km on the other side of the rift.

Combined with radiometric dates for the rocks, this magnetic record proved that the spreading of the seafloor and the motions of the plates have been going on over long geological time spans.

pulled downward by gravity, causing the rock layers to bend or to break along **faults.** Faults and **folds** in these rocks form traps for the accumulation of petroleum, creating some of the richest oil fields in the world.

While the weight of the crust is responsible for some of the deformation of Earth's crust, most of the faulting and buckling that we see at Earth's surface originates instead from forces deep within Earth's interior. Early in this century, some scientists recognized that Earth's continents could be fit together like pieces of a giant jigsaw puzzle. The fit was particularly striking between the Americas and Africa-plus-Europe. Other evidence also suggested that this fit was more than coincidence. For example, the layers in the rock on the east coast of South America and the fossil records they hold match those on the west coast of Africa. Based on evidence such as this, in the 1920s the German scientist Alfred Wegener proposed that over millions of years continents had shifted their positions. This theory is popularly referred to as **continental drift.** Wegener pro-

posed that the continents were originally joined in one large landmass that subsequently broke apart and the continents began to "drift" away from each other.

Originally the idea of continental drift met with great skepticism among geologists. However, in subsequent decades, the evidence supporting the once highly controversial theory of continental drift became

> **Earth's lithosphere is slowly but constantly moving.**

impossible to ignore. Paleomagnetism of the ocean floor provided early evidence of continental drift, as discussed in Excursions 6.1. Today, precise surveying techniques and satellite positioning techniques allow locations on Earth to be determined to within a few centimeters. These measurements confirm that Earth's lithosphere is indeed moving. Some areas are being pulled apart by more than 15 centimeters (or about the length of a pencil) each year. This rate is slower than a snail's pace, but over millions of years of geological time such motions add up. To quote from the old proverb

Figure 6.14 (a) *As new seafloor is formed at a spreading center, the cooling rock becomes magnetized. The magnetized rock is then carried away by tectonic motions.* (b) *Maps like this one of banded magnetic structure in the seafloor near Iceland provide support for the theory of plate tectonics.*

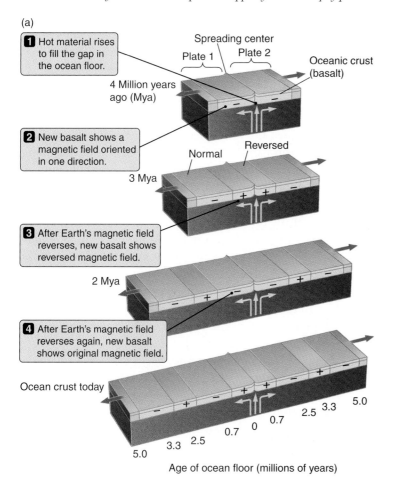

(a)

1 Hot material rises to fill the gap in the ocean floor.

Spreading center

Plate 1   Plate 2

Oceanic crust (basalt)

4 Million years ago (Mya)

2 New basalt shows a magnetic field oriented in one direction.

Normal   Reversed

3 Mya

3 After Earth's magnetic field reverses, new basalt shows reversed magnetic field.

2 Mya

4 After Earth's magnetic field reverses again, new basalt shows original magnetic field.

Ocean crust today

5.0   3.3   2.5   0.7   0   0.7   2.5   3.3   5.0

Age of ocean floor (millions of years)

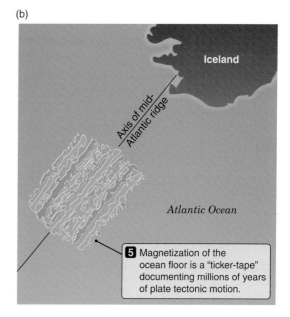

(b)

Iceland

Axis of mid-Atlantic ridge

Atlantic Ocean

5 Magnetization of the ocean floor is a "ticker-tape" documenting millions of years of plate tectonic motion.

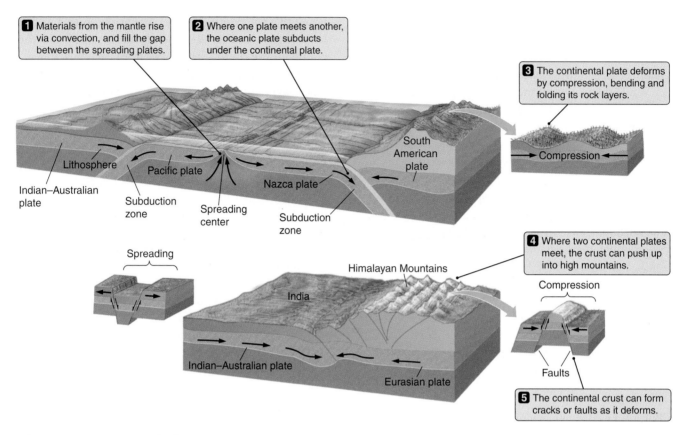

Figure 6.15 *Divergence and collisions of tectonic plates create a wealth of geological features.*

about the hare and the tortoise, "Slow and steady wins the race." Over 10 million years—not a long time by geological standards—15 cm/year becomes 1,500 km, by which time the maps definitely need to be redrawn!

Today geologists recognize that Earth's outer shell is composed of a number of segments, or **lithospheric plates,** and that motion of these plates is constantly changing the surface of Earth. This theory, which is perhaps the greatest advance in 20th-century geology, is referred to as **plate tectonics.** Plate tectonics is ultimately responsible for a wide variety of geological features on our planet.

> The theory explaining motions of Earth's lithosphere is called plate tectonics.

PLATE TECTONICS IS DRIVEN BY CONVECTION

The forces required to set lithospheric plates into motion are immense. We now understand that these forces are the result of thermal energy escaping from the interior of Earth. The motions of lithospheric plates are caused by **convection** in Earth's mantle. If you have ever noticed how water moves about in a pot on a stovetop, then you have seen convection at work. Thermal energy from the stove warms water at the bottom of the pot. The warm water expands slightly, becoming less dense than the water above it, and the warmer water with lower density rises. When the lower-density water reaches the surface, it gives up part of its energy to the air, and in so doing cools, becomes denser, and sinks back toward the bottom of the pot. If you watch the water on the stove carefully, you can learn a lot about this process. Water cannot rise and fall in the same place, and so the convection becomes organized into a circulating pattern in which warm water rises in some locations and cool water sinks in others. As discussed in Connections 6.2, convection is an important process in many contexts.

In the case of Earth, thermal energy generated by radioactive decay in the interior of the planet causes convection to occur in the mantle. Earth's mantle is not molten (if it were, secondary seismic waves could not travel through it), but it is somewhat mobile. You can think of the mantle as having the plastic consistency of hot glass. This allows convection to take place, albeit very slowly.

> Earth's mantle has a plastic consistency like hot glass, allowing slow convection.

Careful mapping shows that Earth's lithosphere is divided into about seven major plates and about a half dozen smaller ones. These plates are driven by convection cells in Earth's mantle, carrying both continents and ocean crust along with them. In some places mantle material rises up and cools to form new crust, which slowly spreads out. The ocean floor around these **spreading centers** is the youngest part of Earth's crust.

Earth's outer rocky shell is divided into seven major plates and about six smaller ones.

Figure 6.15 illustrates the process of plate tectonics and some of its consequences. If you think about convection, you will understand that if material is rising and spreading out in one location, it must be colliding and converging in another. Locations where plates converge and convection currents turn downward are called **subduction zones.** In a subduction zone one plate slides beneath the other, and convection drags the submerged lithospheric material back down into the mantle. The Mariana Trench—the deepest part of Earth's ocean floor—is such a subduction zone. Much

Plates separate, or spread apart, in some regions and collide in other regions.

of the ocean floor lies between spreading centers and subduction zones, and as a result the ocean floor tends to remain the youngest portion of Earth's crust. In fact, the *oldest* seafloor rocks are less than 200 million years old. (This is one reason we do not see evidence for very large impact craters beneath the oceans.) In some places, however, the plates are not subducted, but collide with each other and are shoved upward. For example, the highest mountains on Earth, the Himalayas, result from the collision of the Indian subcontinental plate as it pushes northward into the Asian plate. In still other places, lithospheric plates meet at oblique angles and slide along past each other. The San Andreas Fault in California is one such shear zone.

Locations where plates meet tend to be very active geologically. In fact, one of the best ways to see the outline of Earth's plates is to look at a map of where earthquakes and volcanism occur, like that in **Figure 6.16.** At locations where plates run into each other, enormous stresses build up. Earthquakes result as the friction between the two plates finally gives way, and the plates slip past each

Most volcanoes and earthquakes occur along plate boundaries.

Figure 6.16 *Major earthquakes and volcanic activity are often concentrated along the boundaries of Earth's principal tectonic plates.*

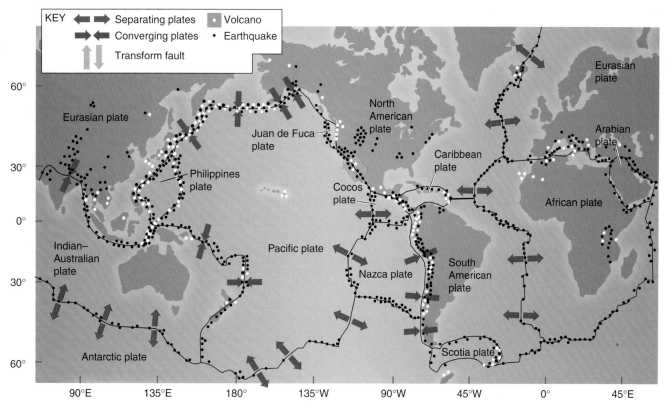

CONNECTIONS 6.2

CONVECTION: FROM THE STOVETOP TO INTERIORS OF STARS

Convection can occur any time energy is introduced at the *bottom* of a gas, fluid, or even a deformable solid. The heated material expands and becomes less dense than the material above it, and so floats upward. At a higher level the material may give up some of its energy to its surroundings, cool and become denser, and so sinks back toward the source of the energy. This sets up a circulating pattern or *cell,* in which some material is rising while other material is sinking **(Figure 6.17).** In this chapter we see how convection in Earth's mantle may explain how Earth's continents move, and how

convection in Earth's core may play a role in the formation of Earth's magnetic field. In Chapters 7 and 8 we will see that convection is also important in the atmospheres of planets, where it helps to explain everything from thunderstorms in the desert and the layer of smog that hangs over Los Angeles to the colorful bands of Jupiter. Convection will appear again when we turn our attention to the structure of the Sun and stars in Part III. All of these phenomena follow from a single physical process—one you can watch in action in a pan of water on the stove.

---

other, relieving the stress. Volcanoes occur when friction between plates melts rock that is then pushed up through cracks to the surface. Lithospheric plates can be thousands of kilometers across and range in thickness from about 5 to 100 km. As they shift, some parts

move more rapidly than others, causing the plates to stretch, buckle, or fracture. These effects are readily seen on the surface as folded and faulted rocks. Mountain chains also are common near converging plate boundaries, where plates buckle and break.

Figure 6.17 (a) *Convection occurs when a fluid is heated from below.* (b) *Convection in Earth's mantle drives plate tectonics.*

TECTONISM ON OTHER PLANETS IS DIFFERENT FROM ON EARTH

Somewhat surprisingly, evidence of plate tectonics is found only on Earth. But while spreading centers and subduction zones are unique to our planet, *all* of the terrestrial planets show evidence of tectonic disruptions. Fractures have cut the crust of the Moon in many areas, leaving fault valleys **(Figure 6.18).** Most of these features are the result of large impacts that crack and distort the lunar crust.

Mercury also has fractures and faults similar to those on the Moon. In addition, there are numerous cliffs on Mercury that are hundreds of kilometers long. These appear to be the result of compression of Mercury's crust. Like the other terrestrial planets, Mercury was once molten. After the surface of the planet cooled and the crust formed, the interior of the planet continued to cool and shrink. As the planet shrank, Mercury's lithosphere cracked and buckled in much the **Mercury's surface shrank after it cooled from a molten state.** same way that a grape skin wrinkles as it shrinks to become a raisin. Planetary scientists estimate that the volume of Mercury must have shrunk by about 5% after the formation of the planet's crust in order to explain the faults seen on the planet's surface.

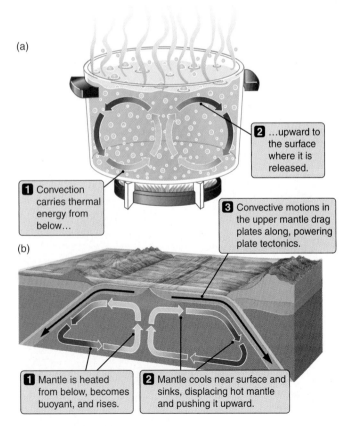

(a)

1 Convection carries thermal energy from below...

2 ...upward to the surface where it is released.

3 Convective motions in the upper mantle drag plates along, powering plate tectonics.

(b)

1 Mantle is heated from below, becomes buoyant, and rises.

2 Mantle cools near surface and sinks, displacing hot mantle and pushing it upward.

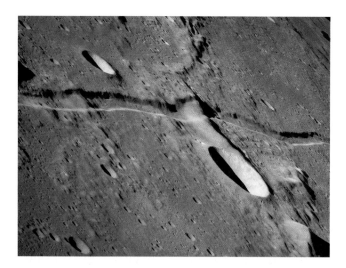

Figure 6.18 *An* Apollo 10 *photograph of Rima Ariadaeus, a 2-km-wide valley between two tectonic faults on the Moon.*

Figure 6.19 *A mosaic of* Viking Orbiter *images showing Valles Marineris, the major tectonic feature on Mars. This canyon system is more than 4,000 km long. The dark spots on the left are shield volcanoes in the Tharsis region.*

Figure 6.20 *Venus's atmosphere blocks our view of the surface in visible light. This false-color image of Venus shows radar measurements made by the* Magellan *spacecraft. Bright yellow and white areas are mostly fractures and ridges in the crust. Some circular features seen in the image may be regions of mantle upwelling, or hot spots. Most of the surface is formed by lava flows, shown in orange.*

On Mars the most impressive tectonic feature, and possibly the most impressive tectonic feature in the Solar System, is Valles Marineris **(Figure 6.19).** Stretching along the equatorial zone for nearly 4,000 km, this system, if occurring on Earth, would link San Francisco with New York. Earth's Grand Canyon would be little more than a minor spur on the side of this chasm. Valles Marineris includes a series of massive cracks in Mars's lithosphere that are thought to have formed as local forces, perhaps related to mantle convection, pushed it upward from below. Once formed, the cracks were eroded by wind, water, and landslides, resulting in the massive chasm that we see today. Other parts of Mars show faults similar to those on the Moon, but the cliffs seen on Mercury are absent.

> **Mars has experienced extensive tectonism, forming large features such as Valles Marineris.**

Venus is similar to Earth in many respects. Venus has a mass about 0.81 times that of Earth, a radius 0.95 times the radius of Earth, and a surface gravity 0.91 times that of Earth. As a result, many scientists speculated that Venus might also show evidence of plate tectonics. However, these speculations were not borne out by the *Magellan* mission, which used radar to peer through the thick clouds which enshroud the planet and map about 98% of the surface of Venus. *Magellan* provided the first high-resolution views of the surface of the planet. *Magellan's* view of one face of Venus is shown in **Figure 6.20.** Although Venus shows a wealth of volcanic features and tectonic fractures, there is no evidence of lithospheric plates or plate motion of the sort seen on Earth. Yet the relative scarcity of

> **Most of the surface of Venus is less than 1 billion years old.**

impact craters on Venus suggests that most of its surface is less than 1 billion years old.

The absence of plate tectonics on Venus is a puzzle. The interior of Venus should be very much like the interior of Earth, and convection should be occurring in its mantle. This presents a problem because it is not clear how the thermal energy from the interior of Venus is able to escape from the planet. On Earth, mantle convection and plate tectonism provide a means for thermal energy to escape from the interior of the planet. Earth also has a few **hot spots,** where upwellings of hot mantle material rise to release thermal energy. The Hawaiian Islands are the result of one such hot spot. On Venus, hot spots may be the principal way that thermal energy escapes from the planet's interior. Circular fractures on the surface of Venus, ranging from a few hundred kilometers to more than 1,000 km across, may be the result of upwelling plumes of hot mantle that have fractured Venus's lithosphere.

Some planetary scientists have suggested that a radically different form of tectonism may be at work on Venus. They believe that hot spots are not adequate to allow the thermal energy generated within the planet to escape, and that as a result energy may continue to build up in the interior until large chunks of the lithosphere melt and overturn. This would suddenly release an enormous amount of energy, following which the surface of the planet would cool and resolidify. This idea remains highly controversial, but it could help explain the relatively young age of the surface. It also serves to drive home the point of how different the geological histories of the various planets appear to be.

Why Venus and Earth should have such different styles of tectonism remains an unsolved puzzle. The segmentation of Earth's lithosphere into moving plates seems to be unique among the planets and moons of the Solar System.

Earth's tectonic plates are unique in the Solar System.

6.6 IGNEOUS ACTIVITY: A SIGN OF A GEOLOGICALLY ACTIVE PLANET

As with earthquakes, some of the most tragic natural disasters result from volcanism. The eruption of Mt. Vesuvius in A.D. 79 that buried Pompeii, the 1883 explosion of Krakatoa in the western Pacific that led to the loss of 36,000 lives, the hot ash flows from Mt. Pelée that demolished the Caribbean port of Saint Pierre in 1902 killing all but one of its 28,000 inhabitants—these are only three of a long list of examples.

Volcanoes are as destructive as earthquakes.

TERRESTRIAL VOLCANISM IS RELATED TO TECTONISM

How do volcanoes form, and why are they found in some regions and not in others? Is there evidence of volcanoes on other planets? To answer these questions we must first understand how and where magma—the main component in volcanism—originates.

Magma does not come to Earth's surface from its molten core, as is sometimes imagined. We know from seismic signals that magma originates in the lower lithosphere and upper mantle, where sources of thermal energy combine. These thermal energy sources include rising convection cells in the mantle, frictional heating generated by movement in the lithosphere and between tectonic plates, and concentrations of radioactive elements that produce energy from radioactive decay.

Because the thermal energy sources are not uniformly distributed inside our planet, volcanoes tend to be located only in specific areas, most notably (but not exclusively) over hot spots and along plate boundaries. Maps of geological activity, such as Figure 6.16, leave little doubt that most terrestrial volcanism is ultimately linked to the same forces responsible for plate motions. For example, there is a tremendous amount of friction as plates slide one under the other at a subduction zone. This friction generates a great deal of thermal energy, raising the temperature and pushing rock toward its melting point.

Thermal energy is generated by friction between moving plates.

When we studied Earth's core, we found the counterintuitive result that even though the inner core of the planet is hotter than its surroundings, it is *solid* while its surroundings are liquid. This is because the inner core is under more pressure than its surroundings, and this higher pressure forces the material to stay solid. A similar effect occurs near Earth's surface. Material at the base of a lithospheric plate is under a great deal of pressure because of the weight of the plate pushing down on it. This pressure drives up the melting point of the material, forcing it to remain solid even though its temperature is above its melting point on Earth's surface. But as this material is forced up through the crust, its pressure drops, and as its pressure drops, so does its melting point. Because of this declining pressure, material that started out solid at the base of a plate may become molten as it nears the surface.

Material forced up through the crust may become molten.

An obvious place to look for volcanic activity is along spreading centers, where convection carries hot mantle material toward the surface. As mentioned in the previous paragraph, the decrease in pressure as the material nears the surface may allow it to become molten. Spreading centers are indeed seen to be frequent sites of eruptions. Iceland, which is one of the most volcanically active regions in the world, sits astride one such spreading center—the Mid-Atlantic Ridge (see Figures 6.14 and 6.16).

Once lava reaches the surface of Earth, it can form many types of structures. Flows from spreading centers often form vast sheets, especially if the eruptions come from long fractures called **fissures.** If very fluid lava flows from a single "point source," it can spread out over the surrounding terrain or ocean floor, forming what is known as a **shield volcano,** so named because it resembles an upside-down warrior's shield. Pasty lava can also form thick masses called **volcanic domes.**

A third setting for terrestrial volcanism is found where convective plumes rise toward the surface in the interiors of lithospheric plates, creating local hot spots. Volcanism over hot spots works much like volcanism at a spreading center, except that the convective upwelling occurs at a single spot, rather than along the length of a spreading center. These hot spots can melt mantle and lithospheric material and force it toward the surface of Earth.

There are numerous hot spots on Earth, including the region around Yellowstone Park as well as the Hawaiian Islands. The Hawaiian Islands are a chain of shield volcanoes that formed as the lithospheric plate on which they ride was dragged across a relatively stationary hot spot. Volcanoes erupt over the hot spot, building an island. The island ceases to grow as the plate motion carries the island away from the hot spot, which is the source for the volcanic activity. The slower process of erosion, going on since the island's inception, continues to wear the island away. In the meantime, a new island grows over the hot spot. Today the Hawaiian hot spot is located off the eastern coast of the big island of Hawaii, where it continues to power the active volcanoes. On top of the hot spot the newest Hawaiian volcano is already forming. This volcano, called Loihi, remains submerged under the surface of the Pacific Ocean. However, viewed another way, Loihi

The Hawaiian Islands result from "hot spot" volcanism.

is already a massive shield volcano, rising more than 5 km above the ocean floor. Geologists expect that it will eventually break the surface of the ocean and merge with the big island of Hawaii.

VOLCANISM ALSO OCCURS ELSEWHERE IN THE INNER SOLAR SYSTEM

While Earth is the only planet on which plate tectonics is an important process, evidence of volcanism is found throughout the Solar System. Even before the Apollo astronauts set foot on the lunar surface, photographs showed it to have what appeared to be flowlike features in the dark regions. Early observers thought these looked like seas, thus the name maria, plural of **mare,** Latin for "seas." Their appearance suggested to planetary scientists that these are vast lava flows similar to basalts on Earth. Because the maria contain relatively few craters, we know that these volcanic flows occurred after the period of heavy bombardment ceased.

The dark areas on the Moon's surface are ancient lava flows.

When the Apollo astronauts returned rock samples from the lunar maria, the rocks were indeed found to be basalts. Many of these Moon rocks contained gas bubbles typical of volcanic materials **(Figure 6.21).** Experiments show that when this lava flowed across the lunar surface, it must have been extremely fluid—something like the consistency of motor oil at room temperature. The fluidity of the lava is due partly to its iron- and titanium-rich chemical composition. This fluidity explains why lunar basalts form vast, thin sheets filling low-lying areas such as impact basins. It also

Figure 6.21 *This sample of the Moon, collected by the* Apollo 15 *astronauts from a lunar lava flow, shows gas bubbles typical of gas-rich volcanic materials. This rock is about 6 by 12 cm in size.*

Figure 6.22 *The lava flowing across the floor of Mare Ibrium on the Moon must have been extremely fluid to spread out for hundreds of kilometers in sheets that are only tens of meters thick.*

areas on Mercury and are thought to be volcanic in origin. Only half of Mercury has been explored, and the other half of the planet is mostly unknown. Moreover, most of the spacecraft images of Mercury have a resolution no better than that of ground-based telescope images of the Moon, so there remains much to be learned about this innermost planet of the Solar System.

More than half the surface of Mars is covered with volcanic rocks. Plain-forming lavas covered huge regions of Mars, flooding the older, cratered terrain. Few of the vents or fissures for these flows are seen, suggesting that most were buried under the lava that poured forth from them **(Figure 6.23).** Among the most impressive features on Mars are the enormous shield volcanoes (Figure 6.19). These volcanoes are the largest mountains in the Solar System. Olympus Mons, standing 25 km high at its peak and nearly 700 km wide at its base, would tower over Mount Everest and dwarf Hawaii's Mauna Loa. Standing a "mere" 9 km above the floor of the Pacific ocean and spreading out to cover an area 120 km across, Mauna Loa is the largest mountain on Earth.

The largest mountains in the Solar System are Martian volcanoes.

Despite the difference in size, the volcanoes of Mars are shield volcanoes, just like their Hawaiian cousins. Olympus Mons and its neighbors grew as the result of hundreds of thousands of individual eruptions that sent

explains the Moon's lack of classic volcanoes like Mt. Rainier: motor oil poured from a can does not pile up; it spreads out **(Figure 6.22).**

The samples also showed that most of the lunar lava flows are older than 3 billion years! Only in a few limited areas of the Moon are younger lavas thought to exist; most of these have not been sampled directly. Samples from the heavily cratered terrain of the Moon also originated from magma, which shows that the young Moon went through a molten stage. These rocks cooled from a "magma ocean" and are more than 4 billion years old, preserving the early history of the Solar System. Thus, most of the sources of heating and volcanic activity on the Moon must have shut down some 3 billion years ago, unlike on Earth, where volcanism continues. This conclusion is certainly consistent with our earlier argument that smaller planets should cool more efficiently and thus be less active than larger planets.

The Moon had an "ocean" of magma early in its history.

There is no conclusive evidence for past or present volcanism on Mercury. There are, however, smooth plains which are similar in appearance to the lunar maria. These sparsely cratered plains are the youngest

Figure 6.23 *This mosaic of* Viking Orbiter *images shows a series of lava flows (shaded red) on Mars extending from the north (left) to the south (right) into cratered terrain. Note the crater rims that are partly "flooded" by the lava flows. The area shown is about 180 by 150 km.*

lava running down their flanks. The difference in size between Olympus Mons and Mauna Loa could be a result of the absence of plate tectonics on Mars. As discussed earlier, the motion of the plate that Hawaii rides on carries the Hawaiian volcanoes away from the hot spot after only a few million years. The Martian volcanoes, on the other hand, have remained over their respective hot spots for billions of years, growing ever taller and broader with each successive eruption.

Although no samples have been returned directly from Mars, analysis by instruments on landed spacecraft, remote sensing data, and the shapes of the lava flows and volcanoes all suggest that the Martian lavas are basalts much like those found on Earth and the Moon. However, chemical analyses at the *Pathfinder* lander site suggest that the rocks contain slightly more silica than typical basalts, which could mean that the magma was partly differentiated chemically before it erupted. In Chapter 11 we will discuss the evidence that certain meteorites found on Earth were probably blasted from the surface of Mars by impacts. Chemical analysis of these meteorites supports the view that volcanism on Mars involved basaltic lavas.

Of all the terrestrial planets, Venus has the largest population of volcanoes. Radar images reveal a wide variety of volcanic landforms. These

Venus has the largest population of volcanoes among the terrestrial planets and the Moon.

include flood lavas covering thousands of square kilometers, shield volcanoes approaching those of Mars in size and complexity, dome volcanoes, and lava channels thousands of kilometers long. These lavas must have been extremely hot and fluid to flow for such long distances. Some of the volcanic eruptions on Venus are thought to have been associated with deformation of Venus's lithosphere above hot spots such as the circular fractures mentioned earlier.

Although we know little about the compositions of the volcanic rocks on Venus, the Soviet *Venera* landers measured some surface compositions and, for the most part, the results suggest that lavas on Venus are basalts, much like the lavas on Earth, the Moon, and Mars. It is presumed that the possible lavas on Mercury are basalts as well.

In summary, what can we say about the volcanic histories of the terrestrial planets and the Moon? After the Moon went though a molten state—a sort of "magma ocean" phase—shortly after its birth, the Moon developed an ever thickening lithosphere overlying a mantle. Depending on locations of radioactive materials, local reservoirs of magma were generated. Some large impacts were able to penetrate these reservoirs or otherwise trigger the release of magma to the surface through fractures. Most of this volcanism ceased about 3 billion years ago, although some minor eruptions could have continued sporadically for another billion years. In all, less than 18% of the lunar surface is covered with volcanic rocks (excluding those cooled from the "magma ocean"). Volcanism on Mercury—if indeed volcanism occurred at all—probably mimicked that of the Moon. Many of the inferred volcanic plains on Mercury are also associated with impact scars. The ages of these plains are not known, but from superposed impact craters we can conclude that the plains are probably billions of years old. Until more and better data are available for Mercury, the formation and eruption of magma on that planet remains speculative.

Lava flows and other volcanic landforms span nearly the entire history of Mars, estimated to extend from the formation of crust some 4.4 billion years ago to geologically recent times, and cover more than half of the surface of the red planet. But remember, "recent" in this sense could still be more than 100 million years ago,

The terrestrial planets have diverse histories of volcanism.

or back to our own age of the dinosaurs. Although some "fresh"-appearing lava flows are identified on Mars, until rock samples are radiometrically dated, we will not know the age of these latest eruptions. Mars could, in principle, experience eruptions today.

Most of Venus is covered with volcanic materials or tectonically disrupted rocks of presumed volcanic origin. A geological time scale for Venus has not yet been devised, but from its relative lack of impact craters, most of the surface is considered to be less than 1 billion years old. When volcanism began on Venus and whether volcanoes are still active today remain unanswered questions.

Earth remains the champion for the diversity of volcanism. Volcanic rocks are found throughout the rock record, while compositions of magma span the spectrum of silica-rich to super-iron-rich materials erupted directly from the mantle.

6.7 GRADATION: WEARING DOWN THE HIGH SPOTS AND FILLING IN THE LOW

Gradation is the "great leveler" of planetary surfaces. The term *gradation* covers a wide variety of processes which together serve to smooth out planetary terrain, wearing down the high spots and filling in the low. The first step in the process of gradation is called *weathering* in which rocks are broken into smaller pieces and may be chemically altered. For example, rocks on Earth are physically weathered along shorelines, where they are broken into beach sand by pounding waves, and along streambeds, where they are slammed together. Other weathering processes include chemical reactions, as when oxygen in the air combines with iron in rocks to form a type of rust. One of the most efficient forms of weathering involves freeze-thaw cycles during which liquid water runs into crevices then freezes, expanding and shattering the rock.

After weathering, the resulting debris can be carried away by flowing water, glacial ice, or blowing wind and deposited in other areas as sediments. Where material is eroded, we can see features such as river valleys, wind-sculpted hills, or mountains carved by glaciers. Where eroded material is deposited, we see features such as river deltas, sand dunes, or piles of rock at the bases of mountains and cliffs. It is not surprising to find that gradation is most efficient on planets where water and wind are present. On Earth, where water and wind are so dominant, most impact craters are worn down and filled in even before they are turned under by tectonic activity. If other processes were not at work to form mountains, valleys, and other topographic relief, gradation would eventually wear planets like Earth as smooth as billiard balls.

The actions of water and wind produce the greatest amount of gradation on Earth.

Yet even on the Moon and Mercury, which have no atmospheres or running water, a type of gradation (albeit *very* slow) is still at work. Radiation from the Sun and from deep space slowly works to decompose some types of minerals, effectively weathering the rock. Such effects are only "skin deep"—generally a few millimeters at most—as a kind of rock "sunburn." Impacts of micrometeoroids also chip away at rocks. In addition, landslides can occur wherever gravity and differences in elevation are present. Although landslide activity is enhanced by the lubricating effects of water, landslides are also seen on dry planets like Mercury and the Moon. Debris from landslides has even been seen on the tiny moons of Mars and on asteroids (see Chapter 11).

Gradation can also occur without wind or water, albeit much more slowly.

Earth, Mars, and Venus, on the other hand, do have atmospheres, and all three planets show evidence of the effects of wind storms. Images of the surfaces of Mars and Venus returned by spacecraft landers show surfaces that have clearly been

The surfaces of Mars and Venus are also modified by winds.

Figure 6.24 *A view of the surface of Mars taken from the* Pathfinder *lander in July 1997 showing the rock-littered surface. Most of the rocks were carried to the site by ancient rivers; other rocks were excavated by impacts onto the surface. Fine-grained material between and partly covering the rocks is wind-transported sand and dust.* Sojourner, *the small rover in this view, is about 50 cm long and was capable of making chemical compositional measurements of the rock and soil.*

subject to the forces of wind (see **Figure 6.24**). Likewise, orbiting spacecraft have returned pictures showing sand dunes, wind-eroded hills, and surface patterns called *wind streaks*. Planet-encompassing dust storms have been seen on Mars. These storms have been known to blot out the visibility of the surface of the planet for months on end.

Sand dunes are common on all three planets wherever moderately strong winds blow and there is a supply of loose grains. The largest field of windblown sand on Mars **(Figure 6.25)** is in the north polar area. Its size, covering more than 700,000 square kilometers, is comparable to the largest fields of sand on Earth. The most common wind-related features on Mars and Venus are wind streaks **(Figure 6.26)**. These are surface patterns that appear, disappear, and change in response to winds blowing sediments around hills, craters, and cliffs. They serve as local "wind vanes," telling planetary scientists about the direction of local prevailing surface winds.

Today, Earth is the only planet where the temperature and atmospheric conditions allow extensive liquid surface water to exist. Water is an extremely powerful

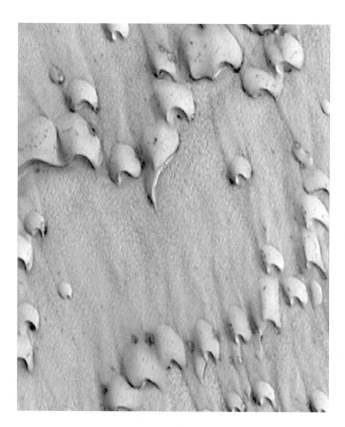

Figure 6.25 *This high-resolution image taken by a camera aboard the* Mars Global Surveyor *shows windblown sand dunes in the north polar region of Mars.*

Figure 6.26 *The bright patterns associated with these craters and hills on Mars resulted from winds that redistribute sand and dust on the surface. The winds forming these streaks blew from right to left. The area shown is about 160 by 185 km.*

agent of gradation and dominates erosion on Earth. Every year, rivers and streams on Earth empty about 10 billion metric tons of sediment into the oceans. Even though today there is no liquid water on the surface of Mars, at one time water flowed across its surface in such vast quantities as to make the Amazon river look like a backyard irrigation ditch. Huge dry riverbeds such as those shown in **Figure 6.27** attest to tremendous floods that poured across the Martian surface. In addition, many regions on Mars show small networks of valleys that are thought

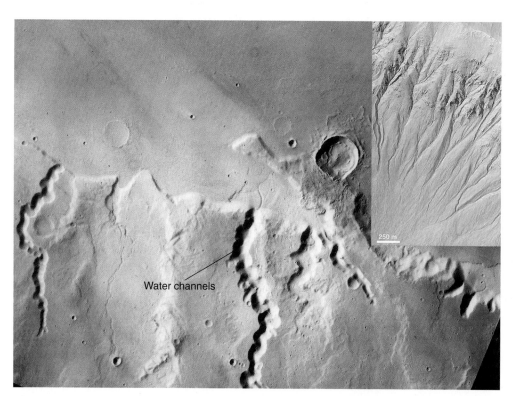

Water channels

Figure 6.27 Viking Orbiter *images showing channels on Mars carved long ago by flowing water. Liquid water cannot exist on the surface of Mars today. The area shown is about 100 km wide. Inset: High resolution* Mars Global Surveyor *image showing geologically-recent "gullies" that may also have been carved by water.*

250 m

to have been carved by flowing water. Some parts of Mars may even have once contained oceans and glaciers.

Where is Mars's water and ice today? At least some appears to be locked in the polar regions, just as the ice caps on Earth hold much of our planet's water. (But most of the material in the Martian polar caps is frozen carbon dioxide rather than frozen water.) However, much if not most of the water on Mars may be trapped below the surface of the planet. It is also intriguing that, while Earth and Mars are the only terrestrial planets that show evidence for liquid water at any time in their histories, water ice could exist on Mercury and the Moon today. There are deep craters in the polar regions of both Mercury and the Moon with floors that are in perpetual shadow. Because these planets lack atmospheres, temperatures in these permanently shadowed areas remain below 180 K. For many years, planetary scientists speculated that ice—perhaps implanted by impacting comets—could be found in these craters. In the early 1990s, Earth-based radar measurements of Mercury's north pole and infrared measurements of the Moon's polar areas by the joint DOD/NASA *Clementine* mission both returned information that seems to support this possibility. Although these observations do not prove that ice exists in these regions, the possibility is exciting and important. The existence of ice on the Moon could make future lunar colonization much more practical than if all water had to be either brought up from Earth or synthesized from hydrogen and oxygen extracted from the lunar soil.

Water ice could exist on the Moon and Mercury, as well as on Mars.

SEEING THE FOREST THROUGH THE TREES

In Chapter 5 we placed the formation of the Solar System and our Earth into the broader context of the ongoing process of star formation that we see occurring around us in the Universe today. Earth is but one of several rocky planets that coalesced from the dust surviving in the hot inner parts of the disk that surrounded the young Sun. Mercury, Venus, Earth, and Mars share this common heritage, and have been further shaped by the same fundamental processes of impacts, tectonism, igneous activity, and gradation over the ensuing 4.5 billion years. Earth is our benchmark for understanding the nature and effects of these processes. The cosmological principle not only suggests that we can apply physical laws discovered in terrestrial laboratories throughout the Universe, it also guides us in our exploration of the Solar System. We live in a remarkable age of exploration and discovery, when our species has launched probes to visit almost all of the major bodies in the Solar System, and many of the minor bodies as well. The resulting flood of information is a testament to 20th-century technology, but it is our understanding of Earth that has guided scientists as they have pieced together the story of what these data all mean.

The impression that we are left with in the face of the results from the last four decades of planetary exploration is one of remarkable diversity. The surfaces of the four terrestrial planets and Earth's moon have all been stressed and fractured over the eons. One of the most startling discoveries of the last century was the fact that the outer shell of Earth itself consists of multiple plates that are in constant motion as they ride slow but inexorable convection currents within Earth's mantle. Exploration of the inner Solar System has provided new perspectives on the drifting motion of our planet's lithosphere by showing us that other possibilities exist. While Mercury's surface is cracked and wrinkled, tectonic activity on Mars fractured the planet, forming a vast canyon system. Even our sister planet, Venus, shows hints of a particularly violent form of tectonism in which the entire surface of the planet may have overturned and released bursts of pent-up thermal energy from the planet's interior. Likewise, igneous activity of several kinds has played an important role in the history of our planet, building such monuments to the power of volcanism as the Hawaiian Islands and the Cascade Range. Yet when looking at our neighbors we find that Earth's volcanoes shrink into relative insignificance beside the towering heights of Mars's Olympus Mons.

Along the way we have again seen the process of science at work. The realm of our species encompasses only the tiniest fraction of our planet. As alluded to in the humorous quote that opened the chapter, we are creatures of the *surface* of Earth, as far from direct exploration of the inner reaches of our own world as we are from direct exploration of the heart of the Sun. Even so, on this leg of our journey we have seen how the methods of science have allowed us to construct a detailed picture of the interior of Earth. We have built up that picture by applying physical laws to construct models of Earth, then tested the predictions of those models by analyzing the echoes from earthquakes. At first glance this might seem an insecure foundation for knowledge, but that is far from the case. We go through our lives building knowledge of the world around us through light waves that reach our eyes and sound waves that reach our ears, both interpreted by brains shaped by evolution to accomplish the task. Is it so different to say that we know of the interior of Earth through the seismic waves that reach our instruments, interpreted through computer models that are shaped for the job by our understanding of physical law? So it will be throughout the rest of our journey, as we use the techniques and tools of science (and a carefully thought-out notion of what it means "to know") to "sense" the nature of the Universe.

We have seen several ways in which the study of terrestrial planets has forced us to change the way we view our planet. Earth is but one of several rocky planets, and our world is only one small corner of the range of the possible. Yet no result of Solar System exploration has more dramatically changed our perspective on our own history—and perhaps our own future—than our understanding of the role of impact cratering in the inner Solar System. The active Earth is very effective at erasing its own memory. Plate tectonics, igneous activity, and gradation are continuously at work, wiping the record of our history clean. It took the ancient surfaces of Mercury and the Moon, and even the less ancient surfaces of Mars and Venus, to show us that impacts of objects from space have played a significant role in shaping Earth. With this realization, the cosmological principle truly comes full circle. Earth may provide many of the clues that we need to understand the Solar System, but it is the Solar System that provides the immediate context for the existence of Earth. Life on our planet has had its course turned time and again by sudden and cataclysmic events when asteroids and comets have slammed into Earth. It seems extremely likely that we owe our existence to the luck of our remote anscestors—small rodentlike mammals—that could live amid the destruction following such an impact 65 million years ago. The time since that event is but a brief moment in the history of our planet. The time until the next such event will almost certainly be no more than another short span of geological time.

STUDENT QUESTIONS

THINKING ABOUT THE CONCEPTS

1. Explain how we know that Earth's interior includes a liquid zone.
2. Can all rocks be dated using radiometric methods? Explain.
3. Explain how we know that rocks at the bottom of Arizona's Grand Canyon are older than those found on the rim.
4. One region on the Moon is covered with craters, while another is a smooth volcanic plane. Which is older? How much older? How do we know?
5. Describe the sources of heating that are responsible for the generation of Earth's magma.
6. Compare and contrast tectonism on Venus, Earth, and Mercury.
7. Describe and explain the evidence for reversals in the polarity of Earth's magnetic field.
8. A current theory suggests that a mass extinction occurred as a consequence of an enormous impact on Earth 65 million years ago. What is the evidence for this theory?
9. Explain the difference between a spreading center and a subduction zone.
10. If Venus is completely enshrouded in clouds, how do we know as much as we do about its surface?
11. Given images with adequate resolution, describe and explain the criteria you would apply in distinguishing between a crater formed by an impact and one formed by a volcanic eruption.

APPLYING THE CONCEPTS

12. Assume that Earth and Mars are perfect spheres with radii of 6,378 km and 3,393 km, respectively.
    a. Calculate the surface area of Earth.
    b. Calculate the surface area of Mars.
    c. If 0.72 (72%) of Earth's surface is covered with water, compare the amount of Earth's land area with the surface area of Mars.
13. Compare the kinetic energy of a 1 g piece of ice (about half the mass of a dime) entering Earth's atmosphere at a speed of 50 km/s with that of a 2 metric ton automobile (mass = 2×10^3 kg) speeding down the highway at 90 km/h.
14. Assume the density of rock is 3.0 kg/liter and iron is 7.9 kg/liter (a liter has a volume of 10^{-3} cubic meters). All other factors being equal, which would produce a larger impact crater, a spherical rocky object 10 m in diameter traveling at 15 km/s, or a spherical iron object 5 m in diameter traveling at 30 km/s? Explain in detail how you arrived at your answer.
15. Earth's mass is 6×10^{24} kg, and the average depth

of its oceans is 3.8 km. Using information from Question 12, and assuming that sea water has a density of 1,000 kg/m^3,

a. Calculate the total mass of Earth's oceans.

b. What fraction of Earth's total mass do our oceans represent?

c. Does this fraction represent Earth's total water inventory? Explain.

16. Say you find a piece of ancient pottery and take it to the laboratory of a physicist friend. He finds that the glaze contains radium, a radioactive element that decays to radon and has a half-life of 1,620 years. He tells you that there could not be any radon in the glaze when the pottery was being fired, but that it now contains 3 atoms of radon for each atom of radium. How old is the pottery? Explain your reasoning.

17. Assume that the east coast of South America and the west coast of Africa are separated by an average distance of 4,500 km. Say that Global Positioning System measurements indicate that these continents are now moving apart at a rate of 3.75 cm/year. If you could assume that this rate has been constant over geological time, how long ago were these two continents joined together as part of a supercontinent?

I come to carry you to yon shore, and lead
Into the eternal darkness, heat, and cold.

DANTE ALIGHIERI (1265–1321)
THE DIVINE COMEDY, INFERNO, CANTO III

ATMOSPHERES OF THE TERRESTRIAL PLANETS

7.1 ATMOSPHERES ARE OCEANS OF AIR

Earth's atmosphere surrounds us like an ocean of air. We see it in the blueness of the sky and feel it in the breezes that tousle our hair. People who live in large cities can sometimes even smell it. It can bring joy into our lives in the form of a spectacular sunset, or apprehension with the approach of a hurricane or tornado. It is responsible for all of our weather, be it pleasant or stormy. Without our atmosphere, there would be neither clouds nor rain; no streams, lakes, or oceans. There would be no living creatures. Without an atmosphere, Earth would look somewhat like the Moon. And we, quite simply, would not exist.

Among the five terrestrial bodies that we discussed in Chapter 6, only Venus and Earth have dense atmospheres **(Figure 7.1).** Mars has a very low-density atmosphere, while the densities of the atmospheres of Mercury and the Moon are so low they can hardly be detected. Why should some of the terrestrial planets have dense atmospheres, while others have little or essentially none? Are atmospheres created right along with the planets they envelop, or do they appear at some later time?

KEY CONCEPTS

Blankets of atmosphere warm and sustain Earth's temperate climates, but also push the surface of Venus beyond Dante's worst nightmares of hell, while leaving Mars frozen. As was the case with their surfaces, comparing these worlds is the key to understanding their atmospheres. Among many interesting insights, on this leg of our journey we will discover:

* That terrestrial planets owe their atmospheres to volcanism and to volatiles captured from comets;
* Why some planets hold on to their atmospheres while others do not;
* That differences among Earth, Venus, and Mars are largely due to the atmospheric greenhouse effect;
* How Earth's atmosphere has been reshaped by life;
* That atmospheres are layered by convection and differences in how they are heated, while pressure steadily falls at higher and higher altitudes;
* How the Coriolis effect redirects convective flows into patterns of winds;
* That Earth's magnetic field interrupts the flow of the solar wind and traps charged particles in a huge magnetosphere responsible for aurorae; and
* The unsettling fact that we are living through an uncontrolled experiment in climate modification.

Figure 7.1 *Global views of the atmospheres of* (left to right) *Venus, Earth, and Mars. Mercury and the Moon have no atmospheres to speak of.*

SOME ATMOSPHERES APPEARED VERY EARLY

For these answers, we need to look back nearly 5 billion years, to the story told in Chapter 5, to a time when the planets were just completing their growth. The phases in the formation of planetary atmospheres are illustrated in

Primary atmospheres are captured gas.

Figure 7.2. The young planets at that time were still enveloped by the remaining hydrogen and helium that filled the protoplanetary disk surrounding the Sun, and they were able to capture some of this surrounding gas. Gas capture must have continued until the gaseous disk ultimately dissipated (soon after the formation of the planets) and the supply of gas ran out. The gaseous envelope collected by a newly formed planet is called its *primary atmosphere*. Although the giant planets still retain most of their original primary atmospheres, the terrestrial planets probably lost theirs soon after the protoplanetary disk was blown away by the emerging Sun. Why is it that only the terrestrial planets lost their primary atmospheres? The answer lies in their relatively small masses.

The terrestrial planets, with their weak gravity, lack the ability to hold light gases such as hydrogen and helium. As soon as the supply of gas in the disk ran out, their primary atmospheres began leaking back into space. How can gas molecules escape from a planet? As discussed in Excursions 7.1, all it takes to escape a planet is a speed greater than the escape velocity (see Chapter 3), pointed in the right direction. Intense radiation from the Sun, especially in the inner parts of the Solar System, raises the kinetic energy of atmospheric atoms and molecules, so they move about more rapidly. The less massive the molecules, the faster they move at any temperature. The very least massive, hydrogen and helium, can have their speeds raised so high that they actually escape from the uppermost levels of a planet's atmosphere and drift off into nearby space. Heated by the Sun and lack-

A planet's atmosphere can sometimes escape into space.

ing a strong gravitational grasp, the terrestrial planets soon lost the hydrogen and helium they had temporarily acquired from the protoplanetary disk. Born naked, they were naked once more.

SOME ATMOSPHERES DEVELOPED LATER

If Earth's primary atmosphere was lost early in its history, what is the source of the air we breathe today? There are probably two principal sources. During the accretion process, minerals containing water, carbon dioxide, and other volatile matter collected in the interiors of the terrestrial planets. Later, as the interiors heated up, the higher temperatures released these gases from the minerals that had held them. Volcanism then brought the various gases to the surface, where they accumulated and created

Secondary atmospheres are a product of volcanism and comet impacts.

what we call a *secondary atmosphere*. Many planetary scientists now believe that there was another important source of gas that formed the secondary atmospheres of the terrestrial planets: impacts by huge numbers of comets, which had formed in the outer parts of the Solar System and were therefore rich in volatiles (see Chapter 5). Why did these icy bodies come into the inner Solar System? Their orbits were disrupted by the growth of the giant planets.

As the giant planets of the outer Solar System grew to maturity, their gravitational perturbations must have stirred up the entire population of icy planetesimals *(comet nuclei)* that orbited within their domain. Many of these icy bodies were flung outward by the giant planets to form the parts of our Solar System we call the Kuiper Belt and Oort Cloud, to be discussed in Chapter 11. Others were scattered into the inner parts of the Solar System, where they could rain down on the surfaces of the terrestrial planets. These comet nuclei brought with them ices such as water, carbon monoxide, methane, and ammonia. Cometary water mixed together with the local water that was released into the atmosphere by volcanism. Most of the water vapor then condensed as rain and flowed into the lower areas to form the earliest oceans. Some of the other cometary gases were not able to survive in their original form. Ultraviolet (UV) light from the Sun easily fragments cometary molecules such as ammonia and methane. Ammonia, for example, is broken down into hydrogen and nitrogen. When this happened, the lighter hydrogen atoms quickly escaped to space, leaving behind the much heavier nitrogen. Nitrogen atoms then combined to form even more massive nitrogen molecules, making it even less likely for these molecules to escape into space. This decomposition of ammonia by sunlight is likely the primary source of molecular nitrogen in the atmospheres of the terrestrial planets and on one of Saturn's moons, Titan. This molecular nitrogen makes up the bulk of Earth's atmosphere.

Today, among the terrestrial planets, only Earth, Venus, and Mars have significant secondary atmospheres. What happened in the case of Mercury and the Moon? Even if these two bodies had experienced less volcanism than the other terrestrial planets, they could hardly have escaped the early bombardment of comet nuclei from the outer Solar System. Some carbon dioxide and water must have accumulated during volcanic eruptions and comet impacts. Where are these gases now?

Venus, Earth, and Mars have significant secondary atmospheres.

It appears that because of Mercury's relatively small mass and its proximity to the Sun, it lost its secondary atmosphere to space, just as it had earlier lost its primary atmosphere. Even more massive molecules, such as carbon dioxide, can escape from a small planet if the temperature is high enough. Furthermore, intense UV radiation from the Sun can break molecules into less massive fragments, which are lost to space even more quickly. The Moon is much farther from the Sun than Mercury and is therefore cooler, but its mass is so small that even at relatively low temperatures molecules

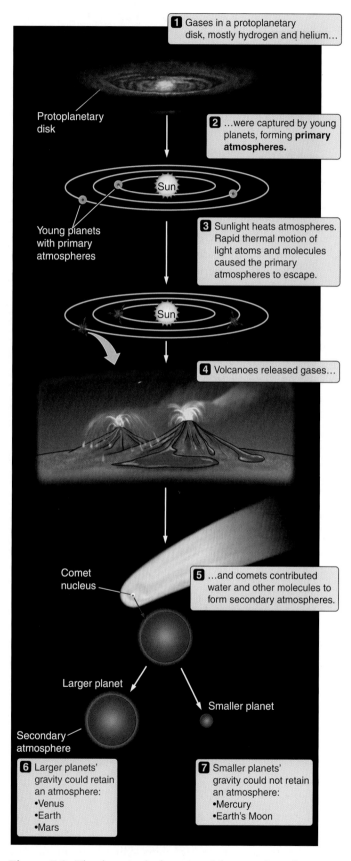

Figure 7.2 *The phases in the formation of the atmosphere of a terrestrial planet.*

EXCURSIONS 7.1

ESCAPE OF PLANETARY ATMOSPHERES

In Chapter 4 we discussed the meaning of temperature and the equilibrium that determines the temperature of a planet. Sunlight is the primary source of thermal or kinetic energy in the atmospheres of the terrestrial planets. Atmospheric temperatures of the terrestrial planets, therefore, depend largely on the distance of the planet from the Sun. What we call the *temperature* of a gas is really just a measure of the average kinetic energy of its molecules. The kinetic energy of any object is determined by its mass and its speed. Hotter molecules have higher kinetic energies than cooler molecules, and therefore move faster.

When a volume of air contains different kinds of molecules having different masses, there is a tendency for the average kinetic energy to be distributed equally among the different types. In other words, all of the types of molecules, from light to massive, will have the same average kinetic energy. But if all types have the same average energy, it means that the less massive molecules must be moving faster than the more massive ones. For example, in a mixture of hydrogen and oxygen at room temperature, hydrogen molecules will be rushing around at about 2,000 m/s on average, while the much more massive oxygen molecules will be poking along at a sluggish 500 m/s. Remember, though, that these are the *average* speeds. Some small fraction of

the molecules will always be moving much slower than the average, while a few will be moving much faster than average. A very few may be moving so fast that they exceed the planet's escape velocity.

Deep within a planet's atmosphere, these high-speed molecules will almost certainly collide with other molecules before they have a chance to escape. During the collision process, there is usually an exchange of energy. The high-speed molecule usually emerges with a lower speed, while the slower one tends to move faster. After a collision, then, both are likely to move with speeds that are closer to the average. Near the top of the atmosphere, where there are fewer surrounding molecules, a high-speed molecule has a very good chance of escaping, if it is heading more or less upward, without first colliding with another molecule. Less massive molecules and atoms, such as hydrogen and helium, move faster and will be more quickly lost to space than more massive molecules, such as nitrogen or carbon dioxide. Thus the primary atmospheres, composed of hydrogen and helium, were lost early in the evolutionary development of the terrestrial planets. The giant planets, on the other hand, are far more massive than their terrestrial cousins, and are situated in the cooler environment of the outer Solar System. Stronger gravity and lower temperatures have allowed them to retain nearly all of their massive primary atmospheres.

Mercury and the Moon are basically airless.

can easily escape. Because of their small mass and their proximity to the Sun, both Mercury and the Moon were doomed from the beginning to remain almost totally airless.

7.2 A TALE OF THREE PLANETS

Two of the terrestrial planets—Venus and Earth—are similar in both mass and composition, and they have adjacent orbits that are less than 0.3 AU apart. Because of these similarities, we might think of them as rather close twins. The third—Mars—is also similar in composition,

but with a mass only about a tenth that of Earth or Venus. We might look at Mars, then, as being related to Venus and Earth but as a somewhat distant cousin. All three of these planets have secondary atmospheres, yet they are all quite different from one another. Would we have expected such differences? To understand this, we need to see how each of them got started.

All three planets are either volcanically active today or have been in their geological past, and all must have shared the intense cometary showers of the distant past. This suggests that their early secondary atmospheres might have been quite similar. In **Table 7.1,** we see that the atmospheres of Venus and Mars today are nearly identical in composition, mostly carbon dioxide with much smaller amounts of nitrogen. This is what we would expect. But Table 7.1 also shows us that their surface pressures are very different. Mars and Venus have vastly different *amounts* of atmosphere. The atmospheric

TABLE 7.1

ATMOSPHERES OF THE TERRESTRIAL PLANETS

Physical properties and composition of the atmospheres of the terrestrial planets.

	Planet		
	Venus	**Earth**	**Mars**
Surface pressure (bars)	92	1.0	0.007
Surface temperature (K)	737	288	210
Carbon dioxide (%)	96.5	0.033	95.3
Nitrogen (%)	3.5	78.1	2.7
Oxygen (%)	0.00	20.9	0.13
Water (%)	0.02	0.1 to 3	0.03
Argon (%)	0.007	0.93	1.6
Sulfur dioxide (%)	0.015	0.02	0.00

pressure on the surface of Venus is nearly a hundred times greater than Earth's, while the average surface pressure on Mars is less than a hundredth of our own. Earth differs in another important respect in that, alone among the planets, its atmosphere is made up primarily of nitrogen and oxygen, containing much less than 1% carbon dioxide. Although all of these planets must have started out with atmospheres of similar composition and comparable quantity, they ended up being very different from one another. Why did they evolve so differently? As with human beings, environment played the major role.

We can most easily see the similarities and differences between the atmospheres of Venus and Mars. Both have experienced widespread volcanism at some time during their history. There is evidence that Venus might still be volcanically active, and Mars has certainly been volcanically active in the recent past. Carbon dioxide and water vapor must have poured out as volcanic gases into the emerging secondary atmospheres of both Venus and Mars. Decomposed cometary ammonia would be the likely source of nitrogen on both planets. As mentioned earlier, Venus and Mars have similar atmospheric compositions—except for water abundance, which we will discuss later. The major difference is that, even after allowing for differences in size and surface gravity, Venus today has nearly 2,500 times more atmospheric mass than Mars. Why such a large difference? We can find the answer by considering the relative masses of the two planets. Venus has nearly eight times as much mass as Mars. This means

Venus retained its atmosphere more effectively than Mars.

that Venus had more bulk material with which to produce an atmosphere in the first place and, equally important, it has the gravitational pull necessary to hang on to its atmosphere. Mars had less material with which to produce an atmosphere, and has less gravitational attraction to keep it. Furthermore, when a planet such as Mars loses much of its atmosphere to space, the process takes on a runaway behavior. With less atmosphere, there are fewer intervening molecules to keep breakaway molecules from escaping (see Excursions 7.1), and the rate of escape increases. This in turn leads to even less atmosphere and increased escape rates, and so on.

In this scenario we can understand the present-day differences between Venus and Mars. But why is the composition of Earth's atmosphere so different from that of the other two? We may find the answer in Earth's special location in the Solar System. Consider the early Earth and Venus, each having about the same mass, but with Venus orbiting just a little closer to the Sun. Volcanism must have poured out large amounts of carbon dioxide and water vapor to form early secondary atmospheres on both planets. Most of Earth's water quickly rained out of the atmosphere to fill vast ocean basins. However, Venus was closer to the Sun, and its surface temperatures were higher than Earth's.

The atmospheres of Earth and Venus have evolved very differently.

Most of the rainwater on Venus immediately re-evaporated, much as it does today in Earth's desert regions. Venus was left with a planetwide surface containing very little liquid water, but with an atmosphere filled with water vapor. The continuing buildup of both water vapor and carbon dioxide in the Venus atmosphere then led to a runaway **atmospheric greenhouse effect** that drove up the surface temperature of the planet, as discussed in Foundations 7.1. The surface of the planet became so hot that no water could survive there.

This early difference between a watery Earth and an arid Venus forever changed the ways their atmospheres and surfaces would evolve. On Earth, water erosion caused by rain and rivers continuously exposed fresh minerals, which then reacted chemically with atmos-

FOUNDATIONS 7.1

THE ATMOSPHERIC GREENHOUSE EFFECT

In Chapter 4 we calculated the expected temperatures of the planets, balancing absorbed solar radiation against emitted thermal radiation, and found that Earth is somewhat warmer than expected, while Venus is wildly hotter than our simple model predicts. When the predictions of a model fail, it means that we are leaving something out of the model. In this case the "something" is the atmospheric greenhouse effect

What we call the atmospheric greenhouse effect in planetary atmospheres is in some ways a misnomer. The **greenhouse effect** traps the Sun's energy in a building somewhat differently from the way planetary atmospheres trap solar energy. A good example of what is commonly called the greenhouse effect happens in a car on a sunny day when you leave the windows closed. The consequences can be severe, especially in midsummer desert environments. Sunlight pours through the windows, heating the interior. With the hot air unable to escape, temperatures of the car's interior can approach 180°F. At this temperature the interior radiates in the infrared (IR). But this radiation cannot escape from the car by the same path through which the sunlight entered. Glass is nearly opaque to IR radiation. Most of the energy ends up heating the air within the car. Heating by solar radiation is most efficient if the enclosure is transparent, which is why the walls and roofs of real greenhouses are made mostly of glass.

In the case of planetary atmospheres, it is not hot air that is trapped, but rather the electromagnetic energy received from the Sun. The atmospheric greenhouse effect is illustrated in **Figure 7.3.** Atmospheric gases such as nitrogen, oxygen, carbon dioxide, and water vapor freely transmit visible solar energy, allowing the Sun to warm the planet's surface. The warmed surface now tries to radiate the

excess energy back into space according to the temperature of that surface, which is much lower than that of the Sun. At the typical temperatures of planet surfaces, this energy is reradiated as infrared energy. But carbon dioxide and water vapor, among other kinds of molecules in our atmosphere, strongly absorb IR radiation, converting it to thermal energy. Some of this thermal energy is subsequently reradiated into space by the same molecules, but much goes back to the ground. The surface is now receiving thermal energy from both the Sun and from the atmosphere. Molecules such as water vapor and carbon dioxide that transmit visible radiation but absorb IR radiation are called **greenhouse molecules.** Methane and chlorofluorocarbons (CFCs) are other greenhouse molecules that are found in our atmosphere. As a result of greenhouse molecules in its atmosphere, the surface temperature of the planet will rise. This rise in temperature continues until the surface is hot enough, and therefore radiating enough energy that even the fraction of infrared radiation leaking out through the atmosphere is enough to balance the absorbed sunlight. Convection will also transport thermal energy to the top of the atmosphere, where radiation to space will help balance absorbed sunlight. In short, the temperature rises until equilibrium between absorbed sunlight and thermal energy radiated away by the planet is reached, just as our earlier discussion in Chapter 4 said it must be.

Even though the mechanisms are different, both the greenhouse effect and the atmospheric greenhouse effect produce the same net result—the local environment is heated by trapped solar radiation.

---

pheric carbon dioxide to form solid carbonates. This removed some of the atmospheric carbon dioxide, burying it within Earth's crust as a rock called limestone. Later, as life developed in Earth's oceans, it accelerated the process of removing atmospheric carbon dioxide. Tiny sea creatures built their protective shells of carbonates and, as they died, they built up massive beds of limestone on the ocean floors. As a result of

Earth's carbon dioxide is locked up in limestone.

water erosion and the chemistry of life, all but a trace of Earth's total inventory of carbon dioxide is now tied up in limestone beds. It seems that Earth's particular location in the Solar System spared it from the runaway atmospheric greenhouse effect. What if Earth had formed a bit closer to the Sun? Look now at **Table 7.2.** If all of the carbon dioxide locked up in limestone beds had remained in Earth's atmosphere, it would have a composition not very different today from that of Venus and Mars.

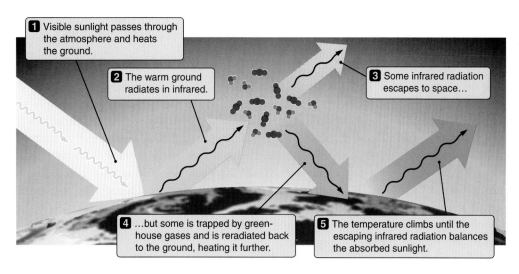

1 Visible sunlight passes through the atmosphere and heats the ground.

2 The warm ground radiates in infrared.

3 Some infrared radiation escapes to space...

4 ...but some is trapped by greenhouse gases and is reradiated back to the ground, heating it further.

5 The temperature climbs until the escaping infrared radiation balances the absorbed sunlight.

Figure 7.3 *Greenhouse gases such as water and carbon dioxide trap infrared radiation, increasing a planet's temperature.*

the planet's surface, where it would be removed from the atmosphere by oxidizing surface minerals.

Differences in the masses of the atmospheres of Venus, Earth, and Mars have a large effect on their present-day surface temperatures. Solar radiation can be trapped by the atmospheric greenhouse effect. What matters here is the actual numbers of greenhouse molecules in the atmosphere, not the percentage that they represent. For example, even though the atmosphere of Mars is composed almost entirely of carbon dioxide, a good greenhouse molecule, its tenuous atmosphere contains few carbon dioxide molecules compared with Venus or Earth. As a result, the atmospheric greenhouse effect on Mars raises the mean surface temperature by only about 5 K. Earth's more massive atmosphere is more efficient. Temperatures on Earth are 35 K warmer than they would be in the

Earth would freeze without the greenhouse effect.

absence of an atmospheric greenhouse effect produced mainly by water vapor and carbon dioxide. Without this greenhouse warming, the mean global temperature of Earth would be well below the freezing point, leaving us with a world of frozen oceans and ice-covered continents!

Yet nowhere in the Solar System is the atmospheric greenhouse effect more dramatic than on Venus. Its massive atmosphere of carbon dioxide and sulfur compounds raises its surface temperature by more than 400 K, to about 735 K. At such high temperatures, any remaining water and most carbon dioxide locked up in surface rocks would long ago have been driven

Differences in the amount of water on Venus, Earth, and Mars are not so well understood. Geological evidence tell us that liquid water was once plentiful on the surface of Mars (see Chapter 6), and we believe that significant amounts may still exist in the form of subsurface ice, far more than the atmospheric abundance indicated in Table 7.1. Earth's liquid and solid water supply is even greater, about 10^{21} kg or 0.02% of its total mass. More than 97% of Earth's water is in the oceans, which have an average depth of about 4 km. If all of that water were to go into the atmosphere, it would produce a crushing surface pressure 300 times greater than what we now experience. Earth today has 100,000 times more water than Venus. What happened to all the water on Venus? One possibility is that water molecules high in the Venus atmosphere were broken apart into hydrogen and oxygen by solar ultraviolet radiation. Hydrogen atoms, being very low mass, were quickly lost to space. Oxygen, however, would eventually have migrated downward to

TABLE 7.2

TERRESTRIAL ATMOSPHERES IF ALL CARBON DIOXIDE WERE INCLUDED

How three terrestrial planets' atmospheres would compare if all of Earth's carbon dioxide were still in its atmosphere. Composition is given in fractional units.

	Planet		
	Venus	**Earth**	**Mars**
Carbon dioxide (%)	96.5	98.5	95.3
Nitrogen (%)	3.5	1.1	2.7
Oxygen (%)	0.00	0.3	0.13
All other constituents (%)	0.5	0.1	2.17

FOUNDATIONS 7.2

WHAT IS A GAS?

As we continue our journey of discovery through the Universe, we will find that it is the gaseous state of matter that is most commonplace. More than those of any other form of matter, it is the properties of gases that we will turn to as we seek to understand the workings of planets, stars, and galaxies.

Matter is composed of atoms and molecules, and different forms of matter result from differences in the way those atoms and molecules interact with each other. In a *solid,* molecules are packed tightly, held in place like bricks in a wall by the presence of their neighbors. In a *liquid,* molecules are able to move about but are constantly being jostled. Molecules in a liquid are like people in a crowded subway station. The crowd is free to flow, but the individuals that make it up are still limited in their movement by the people around them. The molecules in a *gas,* on the other hand, go their own way, traveling relatively long distances without interacting with other molecules. When you think of a gas, picture a swarm of tiny atoms and molecules flying about, each with its own speed and direction.

The temperature of a gas is a measure of the average kinetic energy of the individual molecules as they fly about. Two things determine a molecule's kinetic energy: the mass of the molecule and the speed at which it is moving. A massive, slow-moving molecule might have the same kinetic energy as a less-massive, faster-moving molecule. Because of this, when a gas is composed of different types of molecules (such as the air around you), the average speed of less-massive molecules will be higher than the average speed of more-massive molecules. (They have to be if their average kinetic energies are to be the same.) The average speed of a molecule in a gas is inversely proportional to the square root of its mass. Oxygen molecules, for example, are 16 times more massive than hydrogen molecules. In a gas containing both hydrogen and oxygen, the hydrogen molecules will therefore be moving four ($= \sqrt{16}$) times faster than the oxygen molecules. As we have seen, this difference explains why Earth can hold on to the oxygen in its atmosphere, but loses the hydrogen to space.

As we study gases in different astronomical settings, we will find that one of the most important properties of a gas is how hard it pushes on its sur-

roundings. Imagine a box containing gas, as shown in **Figure 7.4(a).** Molecules are constantly bouncing off the walls of the box, pushing the walls of the box outward. This outward push, measured in units of force per square meter of the surface of the box, is referred to as **pressure.** (Any time that you think about the pressure of a gas, this is the mental picture that you should bring to mind: atoms and molecules slamming against the walls of a box, pushing outward.)

Armed with nothing more than this mental picture, we can draw some interesting conclusions about the pressure of a gas. First, if we increase the number of molecules in the box, then there will be more molecules hitting the walls of the box each second. The pressure will increase. Double the density of the gas but keep the temperature the same **(Figure 7.4b)** and you double its pressure.

Increasing the temperature of the gas increases the average speed at which molecules move, which also increases the pressure of the gas. There are two reasons for this. First, if the molecules in the box are moving faster, they will hit the walls *more frequently* **(Figure 7.4c).** More molecules hitting the walls of the box each second means that the pressure is higher. Second, faster-moving molecules hit the wall *harder,* exerting more force. Together these two effects mean that doubling the temperature of a gas doubles the pressure of the gas.

Pressure is proportional to density and pressure is proportional to temperature. We can combine these two relationships to get

$$\text{Pressure} \propto \text{density} \times \text{temperature.}$$

This relationship between density, temperature, and pressure is called the **ideal gas law.** The ideal gas law has a special place in the history of science. The empirically discovered ideal gas law provided strong support for the atomic theory of matter, in much the same way that Kepler's empirical laws of planetary orbits provided support for Newton's theories of motion and gravitation. The fact that laboratory gases come very close to obeying the ideal gas law provided compelling early evidence that atoms and molecules are real things.

There is one other property of gases that we need to know about as our journey continues. If you compress a gas, you not only increase the density of the gas, but you also increase its temperature. Conversely, when a gas is allowed to expand, it cools off. There are many everyday examples of this behavior. Pump up a bicycle tire, and the pump becomes hot because the air is being compressed. Hold down the nozzle on an aerosol can, and it feels cold because the propellant gas is expanding. An air conditioner works by alternately compressing a gas to make it hot, letting that gas cool then allowing the gas to expand and get really cold.

Over and over again on our journey, our understanding of gases will be our guide, equally as valid when applied to the hellish cores of stars as to a gentle breeze on a summer day. Every time that we talk about an object made of gas, think back to the picture of molecules bouncing around in a box, and remember the commonsense ideas that explain its behavior.

into the atmosphere, further enhancing the atmospheric greenhouse effect.

The conditions existing on Venus today could be created on an Earth-like planet by the runaway atmospheric greenhouse effect. Imagine a hypothetical situation in which large quantities of carbon dioxide or some other greenhouse gas suddenly appeared in Earth's atmosphere. The increased warming would raise surface temperatures, driving more water vapor into the atmosphere, which would in turn increase the strength of the atmospheric greenhouse effect. This would cause even greater warming and still larger amounts of atmospheric water vapor, ultimately creating a surface pressure about 300 times as great as it is now and a surface temperature that might exceed 800 K. Long before reaching this stage, all life on our planet would have ceased to exist.

In reality, the process is more complicated. Increased cloudiness caused by increased water in the atmosphere might decrease the amount of sunlight reaching Earth's surface to a point where the runaway effect would be turned off. The ocean currents are important in transporting energy from one part of Earth to another, and how they would be affected by increased warming is something we really do not know. In fact, the process is so complicated by small changes leading to very large results that we still cannot

We are conducting an uncontrolled experiment on Earth's atmosphere.

Figure 7.4 (a) *The pressure of a gas comes from the motions of atoms and molecules. Making a gas* (b) *denser or* (c) *hotter (i.e., increasing the speed of atoms or molecules) increases the pressure of the gas.*

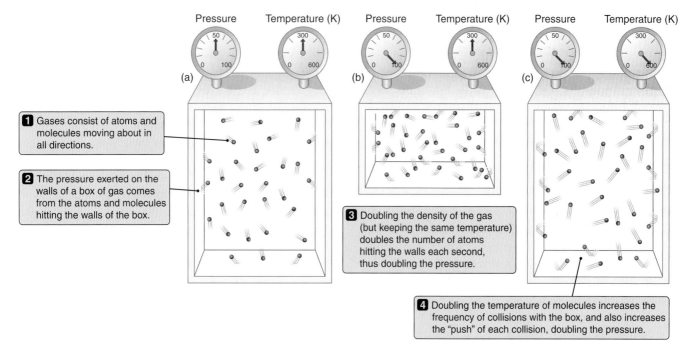

1 Gases consist of atoms and molecules moving about in all directions.

2 The pressure exerted on the walls of a box of gas comes from the atoms and molecules hitting the walls of the box.

3 Doubling the density of the gas (but keeping the same temperature) doubles the number of atoms hitting the walls each second, thus doubling the pressure.

4 Doubling the temperature of molecules increases the frequency of collisions with the box, and also increases the "push" of each collision, doubling the pressure.

predict the long-term outcome of the small changes that humans are now making in the composition of Earth's atmosphere. In a real sense, we are experimenting with the atmosphere of Earth. We are asking the question (whether we know it or not), "What happens to Earth's climate if we steadily increase the number of greenhouse molecules in its atmosphere?" We do not *yet* know the answer to this question, but we will eventually know the answer by seeing the results. Whether we will be happy with the results is another matter. We will explore this issue again toward the end of the chapter.

7.3 EARTH'S ATMOSPHERE— THE ONE WE KNOW BEST

In simplest terms, we can describe Earth's atmosphere as a blanket of gas several hundred kilometers deep, with a total mass of approximately 5×10^{18} kg. As enormous as this may seem, it represents less than one-millionth of Earth's total mass. In Foundations 6.2 we learned about the concept of hydrostatic equilibrium, which tells us that the pressure at any point within a planet must be great enough to balance the weight of the overlying layers. The same thing must hold true in a planetary atmosphere. The atmospheric pressure on a planet is what it needs to be to hold up the weight of the overlying atmosphere. A difference between the two cases is what provides the pressure. In the interior of a solid planet it is the resistance of solid material to being compressed. In our atmosphere, it is the motions of molecules in a gas. The relationship among density, temperature, and pressure of a gas was discussed in Foundations 7.2. The weight of Earth's atmosphere creates a force of approximately 100,000 N acting on each square meter of sur-

Earth's atmosphere is in hydrostatic equilibrium.

face. This amount of pressure is expressed as a unit called a **bar.** Earth's average atmospheric pressure at sea level is approximately 1 bar. (Meteorologists generally quote atmospheric pressures in *millibars,* which are thousandths of a bar.) We can perhaps get a better feeling for how strongly our atmosphere presses on us if we realize that its pressure is equivalent to that of a layer of water 10 meters deep. (Imagine the extra pressure you experience at the bottom of a 33-foot-deep swimming pool.) Yet in air we seem to be completely unaware of atmospheric pressure. How can this be? It is because the very same pressure exists both within and outside of our bodies. The two precisely balance one another—another example of equilibrium.

Two principal gases make up Earth's atmosphere. About four-fifths of our atmosphere is nitrogen and one-fifth is oxygen, although there are many important minor constituents, such as water vapor and carbon dioxide. Atmospheric temperatures near Earth's surface can range from as high as almost 60°C in the deserts to as low as −90°C in the polar regions, with a mean global temperature of about 15°C.

THE COMPOSITION OF EARTH'S ATMOSPHERE IS CONTROLLED BY LIFE

As Table 7.1 shows, the composition of Earth's atmosphere is very different from that of Venus and Mars. We have already discussed the differences in carbon dioxide content, but what sets Earth truly apart from all other known planets is its oxygen. Our atmosphere contains abundant amounts of oxygen and the other planets do not. Why should this be so? Oxygen, it turns out, is a highly reactive gas. It chemically combines with, or *oxidizes,* almost any material it touches. Witness the rust that forms on metals, for example. For a planet to retain significant amounts of this reactive gas in its atmosphere, there would need to be some process to replace what is lost through oxidation. Such a process exists on Earth. We call it plant life.

Only Earth's atmosphere contains abundant oxygen.

Figure 7.5 shows the change in oxygen concentration in Earth's atmosphere over the history of the planet. When Earth's secondary atmosphere first appeared about 4 billion years ago, it was almost totally free of oxygen. About 2.8 billion years ago, an ancestral form of cyanobacteria, single-celled organisms that contain chlorophyll, began releasing oxygen into Earth's atmosphere as a waste product of their metabolism. At first this biologically generated oxygen combined readily with exposed metals and minerals in surface rocks and soils and so was removed from the atmosphere as quickly as it formed. In this way, emerging life dramatically changed the very composition and appearance of Earth's surface, the first of many such planetwide modifications imposed on our planet by living organisms. Ultimately, the explosive growth of plant life accelerated the production

Life is responsible for the oxygen in Earth's atmosphere.

of oxygen, building up atmospheric concentrations that only approached today's levels about 500 million years

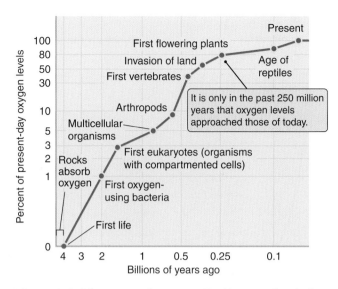

Figure 7.5 *The amount of oxygen in Earth's atmosphere built up over time as a result of plant life on the planet.*

ago. All plants, from tiny green algae to the giant redwoods, use the energy of sunlight to build carbon compounds out of carbon dioxide and produce oxygen as a metabolic waste product. Earth's atmospheric oxygen content is held in a delicate balance primarily by plants. If plant life on our planet were to disappear, so too would nearly all of Earth's atmospheric oxygen.

Several minor gases in Earth's atmosphere also affect our daily lives. Over the current range of terrestrial temperatures, water is a volatile substance, and so its atmospheric abundance varies from time to time and from place to place. In warm, moist climates, water vapor may account for as much as 3% of the total atmospheric composition. In cold, arid climates, it may be less than 0.1%. The continuous process of condensation and evaporation of water involves the exchange of thermal and other forms of energy, making water vapor a major contributor to Earth's weather.

Carbon dioxide is another variable component of Earth's atmosphere, and a complex pattern of *sources* (places where it originates) and *sinks* (places where it accumulates) determines how much of it will be present at any one time. Plants consume carbon dioxide in great quantities as part of their metabolic process. Coral reefs are colonies of tiny ocean organisms that build their protective shells with carbonates produced from carbon dioxide. Fires, decaying vegetation, and human burning of fossil fuels all release carbon dioxide back into the atmosphere. This balance between sources and sinks can and does change with time. Meteorological records show that the amount of carbon dioxide in our atmosphere has been increasing for almost two centuries, since the time of the Industrial Revolution. As

noted earlier, this in turn has had a direct impact on the environment, because carbon dioxide is also a powerful greenhouse gas. It should come as no surprise to learn that Earth's mean global temperature seems to be increasing with the growing abundance of this gas.

Another minor constituent in our atmosphere is *ozone* (O_3). This important molecule is formed when UV light from the Sun breaks molecular oxygen (O_2) into its individual atoms, which can then recombine with other oxygen molecules to form ozone ($O_2 + O \rightarrow O_3$). Most of Earth's natural ozone is concentrated in the upper atmosphere at altitudes between 20 and 50 km. There it acts as a very strong absorber of UV sunlight. Without the ozone layer, this lethal radiation would reach all the way to Earth's surface, making it completely uninhabitable for nearly all forms of life (see Excursions 7.2). Ozone protects life. However, ozone is not always beneficial. Ozone in the lower atmosphere is found primarily in urban environments. It is mostly a human-made pollutant and is a health hazard.

> **High-altitude ozone protects life. Our continuing survival depends on it.**

EARTH'S ATMOSPHERE IS LAYERED LIKE AN ONION

Earth's lowermost atmospheric layer, the one in which we live and breathe, is called the **troposphere (Figure 7.6).** It contains 90% of Earth's atmospheric mass and is the source of all of our weather. At Earth's surface, usually referred to as *sea level,* the troposphere has an average temperature of 15°C (288 K) and an average pressure of 1.013 bars. Within the troposphere, atmospheric pressure, density, and temperature all decrease with increasing altitude. For example, a few thousand feet below the summit of Mt. Denali in Alaska at an altitude of 5.5 km, the atmospheric pressure and density are only 50% of their sea level values and the average temperature has dropped to −20°C. Still higher, at an altitude of 12 km, where commercial jets cruise, the temperature is a frigid −60°C and the den-

> **Pressure and temperature decrease with altitude in the troposphere.**

sity and pressure are less than one-fifth what they are at sea level. Mountain climbers and astronomers are both very much aware of this behavior of Earth's troposphere. At the Mauna Kea Observatory in Hawaii, even the most dedicated astronomers, surrounded by thin air and subfreezing temperatures, have been known to gaze longingly at the sunny beaches some 4 km below, where it is a pleasant 30°C warmer.

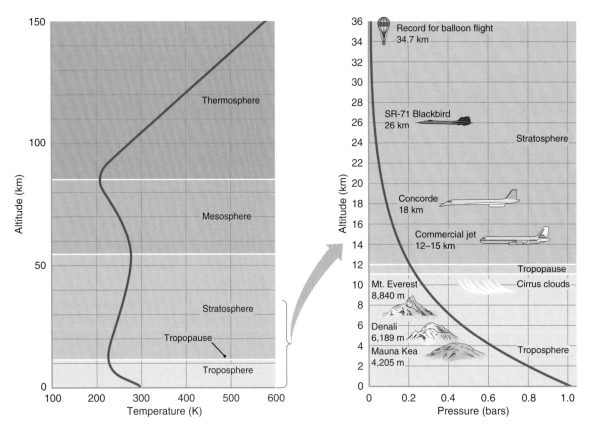

Figure 7.6 *Earth's atmospheric layers, temperature, and pressure are plotted here as a function of height.*

But why does the atmosphere get colder as we climb to higher elevations? From what we have already learned we might guess correctly that it is warmer near Earth's surface because the air is in contact with the sunlight-heated ground, which also warms the air by its infrared radiation. It also makes sense that the atmosphere would be cooler at very high altitudes because there it is able to freely radiate its thermal energy off to space. In fact, it would get colder with increasing altitude even faster if it were not for the process of convection. We encountered convection in our Chapter 6 discussion of the motions of material in Earth's mantle that drive plate tectonics. **Figure 7.7** illustrates how convection carries thermal energy upward through Earth's atmosphere. At a given pressure, cold air is more dense than warm air. So when cold air encounters warm air, the denser cold air slips under the less-dense warm air, pushing the warm air upward. (Rather than "warm air rises" we should perhaps say "cold air sinks.") This convection sets up a circulation of air between the lower and upper levels of the atmosphere, and this tends to diminish the extremes caused by heating at the bottom and cooling at the top.

Convection carries thermal energy upward through the troposphere.

7.1 ENERGY

Convection also affects the vertical distribution of atmospheric water vapor. The ability of air to hold water in the form of vapor depends very strongly on the air temperature. The warmer the air, the more water vapor it can hold. We refer to the amount of water va-

Figure 7.7 *Atmospheric convection carries thermal energy from the Sun-heated surface upward through the atmosphere.*

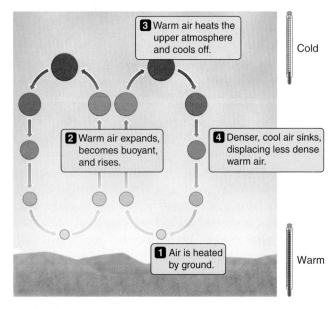

EXCURSIONS 7.2

THE OZONE HOLE

There is a common misconception confusing the greenhouse effect with the so-called ozone hole. Both are caused by the buildup of certain gases in our atmosphere and both are a cause for concern, but the individual causes and effects are very different. Recently, the measured amount of ozone in our upper atmosphere has been decreasing, primarily over the polar latitudes in both the Northern and Southern Hemispheres. The decrease appears to be caused by a buildup of atmospheric *halogens,* mostly chlorine, fluorine, and bromine, such as those found in human-made refrigerants. Halogens readily diffuse upward into the stratosphere, where they destroy ozone without themselves being consumed. We call such agents *catalysts*—materials that

participate in and accelerate chemical reactions but are not themselves modified in the process. Because they are not used up, halogens may remain in Earth's upper atmosphere for decades or even centuries.

Why should we be so concerned about the loss of a minor constituent from so high in our atmosphere? Even in tiny amounts, ozone filters out harmful solar ultraviolet radiation, preventing it from reaching Earth's surface. Measured increases in the levels of UV radiation appear to be related to increases in skin cancer in humans, and we do not yet understand the effects it may have on other life-forms to which we are inexorably linked. There is good reason to be concerned.

---

por in the air relative to what the air could hold at that temperature as the **relative humidity.** Air that is saturated with water vapor has a relative humidity of 100%. As air is convected upward, it cools, limiting its capacity to hold water vapor. If the air temperature decreases to the point where the air can no longer hold all of its water vapor, **saturation** occurs. The water begins to condense out in the form of tiny droplets, which in large numbers become visible to us as clouds. When these droplets coalesce to form large drops, convective updrafts can no longer support them, and they fall as rain. For this reason most of the water vapor in Earth's

Most atmospheric water stays close to Earth's surface.

atmosphere stays within 2 km of the surface. At an altitude of 4 km, for example, the Mauna Kea Observatory is higher than approximately one-third of Earth's atmosphere, but it lies above nine-tenths of the atmospheric water vapor. This is very important for astronomers who observe the heavens in the infrared region of the spectrum, because water vapor strongly absorbs IR light (recall Figure 4.12).

The top of the troposphere, called the **tropopause,** is defined as the height where temperature no longer decreases with increasing altitude (Figure 7.6). This change in atmospheric behavior is caused by heating from absorbed sunlight within the atmospheric layers that lie above the tropopause. And, because temperature no

longer decreases with altitude above the tropopause, atmospheric convection also dies out there. The tropopause varies between 10 and 15 km above sea level, depending on latitude, and is highest at the equator.

Above the tropopause and extending upward to an altitude of 50 km above sea level is the **stratosphere.** This is a region in which very little convection takes place, because the temperature-altitude relationship actually reverses at the tropopause

The stratosphere, mesosphere, and thermosphere lie above the troposphere.

and the temperature begins to *increase* with altitude. The reason for this is the presence of ozone, which warms the stratosphere by absorbing sunlight.

The region above the stratosphere is called the **mesosphere.** It extends from 50 km to an altitude of about 90 km. In the mesosphere there is no ozone to absorb sunlight, so temperatures once again decrease with altitude. The base of the stratosphere and the upper boundary of the mesosphere are two of the coldest levels in Earth's atmosphere (see Figure 7.6).

At altitudes above 90 km, solar ultraviolet radiation and high-energy particles from the solar wind ionize atmospheric atoms and molecules, causing the temperature to once again increase with altitude. We call this region the **thermosphere,** and it is the hottest part of the atmosphere. The temperature can reach 1,000 K near the top of the thermosphere, at an alti-

EXCURSIONS 7.3

BLUE SKIES, WHITE CLOUDS, AND RED SUNSETS

Have you ever noticed the beam from a flashlight or the headlights of a car on a foggy night? As the beam of light shines through the cloud bank, some of the light bounces off of tiny water droplets, and you see the light that happens to bounce in your direction. In this way, you "see the beam of light."

When light bounces off small particles in its path, it is called **scattering.** If the particles that are scattering light are much larger than the wavelength of that light, as is the case for water droplets in a fog bank, then photons of all colors are equally likely to be scattered. As a result, the light scattered by a cloud is the same color as the light shining on the cloud. Sunlight is white, so clouds that are illuminated by direct sunlight are also white.

Things get more interesting if the scattering particles are smaller than the wavelength of the light they are interacting with. In such instances, shorter-wavelength photons are more likely to be scattered than longer-wavelength photons. Molecules in Earth's atmosphere, for example, scatter blue light ($\lambda = 0.4\ \mu m$) about five times more effectively than they scatter red light ($\lambda = 0.6\ \mu m$). When the Sun is high overhead, sunlight follows a short path through the atmosphere, and so is relatively unaffected by scat-

tering. Even so, a small fraction of the blue photons in the sunlight are scattered, and when you look at the sky, your eyes detect the blue photons that were scattered in your direction. As illustrated in **Figure 7.8,** that is why the sky is blue!

This situation changes dramatically as evening approaches. As the Sun drops lower and lower in the sky, the light from the Sun must pass through more and more air before it reaches you. As the Sun nears the horizon, sunlight passes through hundreds of times more air than it did when the Sun was high overhead. So much of the blue light is scattered away that by the time the light reaches you, the Sun looks orange. Tiny particles of dust and other materials in the atmosphere (which are similar in size to the wavelength of light) scatter away additional blue light, leaving only a glorious red.

The next time you are captivated by the beauty of a sunset, remember also to look up at the deepening blue of the sky overhead. The red sunset is white sunlight minus the blue light that was scattered away. The blue sky is scattered blue light only. The red of sunset, the blue of the sky, and the white of a billowing cloud are three facets of the same gem—a gem called scattering.

---

tude of 600 km. The gases within and beyond the thermosphere are ionized by ultraviolet photons and high-energy particles from the Sun to form a **plasma.** (A plasma is any gas that is made up largely of electrically charged particles rather than only neutral atoms and molecules.) This region of ionized atmosphere is called the **ionosphere.** The ionosphere is important to us in part because it reflects certain frequencies of radio waves back to the ground. Amateur radio operators, for example, are able to communicate with each other around the world by bouncing their signals off of the ionosphere.

Earth and its atmosphere are surrounded by a large region filled with electrons, protons, and other charged particles from the Sun that have been captured by Earth's magnetic field. This region, called Earth's **magnetosphere,** has a radius approximately 10 times the radius of Earth, filling a volume over a thousand times the volume of the planet itself. In order to appreciate Earth's magnetosphere, we need to begin by looking more carefully at magnetic fields and the force they ap-

ply to charged particles. Magnetic fields have no effect on charged particles unless the particles are moving. Charged particles are free to move *along* the direction of the magnetic field, but if they try to move *across* the direction of the field, they experience a force that is perpendicular both to their motion and to the magnetic field direction. This force causes them to loop around the direction of the magnetic field, as illustrated in **Figure 7.9(a).** It is almost as if charged particles were beads on a string, free to slide along the direction of the magnetic field but unable to go across it.

The picture of how charged particles move in a magnetic field gets even more interesting if the field is pinched together at some point. As particles move into the pinch, they feel a magnetic force that (if conditions are right) pushes them back along the direction from which they came. If charged particles are located in a region in which the field is pinched on both ends, as shown in **Figure 7.9(b),** then they may bounce back

Earth's magnetic field traps charged particles from the Sun.

and forth many times. This magnetic field configuration is called a *magnetic bottle.* Earth's magnetic field is pinched together at the two poles, and spreads out around the planet. This configuration is like taking many magnetic bottles and bending them over, attaching them to Earth at either end. Earth and its magnetic field sit in a constant stream of charged particles from the Sun, called the **solar wind.** As these particles stream by, they are diverted by Earth's magnetic field, as a river is diverted around a boulder. As they stream past, some of these charged particles are trapped by Earth's magnetic field, where they bounce back and forth between Earth's magnetic poles, as illustrated in **Figure 7.9(c).**

An understanding of Earth's magnetosphere is of great practical importance for space travel. Regions in the magnetosphere that contain especially strong concentrations of energetic charged particles, called **radiation belts,** can be very damaging both to electronic equipment and to astronauts. Yet we need not leave the surface of the planet to witness beautiful and dramatic effects of the magnetosphere. Disturbances in Earth's magnetosphere can lead to changes in Earth's magnetic field that are large enough to trip power grids, causing blackouts, and to wreak havoc with communications. Earth's magnetic field also funnels energetic charged particles down into the ionosphere in two rings located around the magnetic polls. These charged particles (mostly electrons) collide with atoms such as oxygen, nitrogen, and hydrogen in the upper atmosphere, causing them to glow like the gas in a "neon" sign. These glowing rings, called **aurorae,** can be seen from space **(Figure 7.10a).** When viewed from the ground **(Figure 7.10b),** they appear as eerie, shifting curtains of multicolored light. People living far from the equator are often treated to spectacular displays of the *aurora borealis* ("northern lights") in the Northern Hemisphere, or *aurora australis* in the Southern Hemisphere.

Figure 7.8 *The blue of the sky and the red of a sunset are both due to differences in the way Earth's atmosphere scatters light of different wavelengths.*

(a)

1 The blue of the sky is short wavelength light scattered from sunlight by Earth's atmosphere, but at midday most sunlight reaches the ground.

(b)

2 The red of sunset is sunlight after short wavelength blue light has been scattered away.

Although our discussion has concentrated on the atmosphere of our own planet, it is important to know that the structure we have described here is not limited to Earth's atmosphere. The major vertical structural components—troposphere, tropopause, stratosphere, and ionosphere—also exist in the atmospheres of Venus and Mars, as well as in the atmospheres of the giant planets. And as we will see in the following chapter, the magnetospheres of the giant planets are among the largest structures in the Solar System.

> **All planetary atmospheres have layered vertical structure.**

WHY THE WINDS BLOW

Variation in solar heating is the chief reason for differences in the ground-level temperature of Earth's atmosphere from place to place at similar altitudes and throughout the year. It is usually warmer in the daytime than at night, warmer in the summer than in winter, and warmer at the equator than in the polar regions. Large bodies of water, such as oceans, also affect atmospheric surface temperatures. As we discovered in Foundations 7.2, heating a gas increases its pressure, which in turn causes it to push into its surroundings. It is because of these horizontal pressure differences that we have winds, and the strength of the winds is closely related to the magnitude of the difference in temperature from place to place. Think about what happens as air in Earth's equatorial regions, heated by the warm surface, begins to rise due to convection. This displaces the air above it, which then has no place to go but toward the poles. Cooling and becoming denser as it moves pole-

> **Nonuniform solar heating creates our weather.**

Figure 7.9 (a) *The motion of charged particles, in this case electrons, in a uniform magnetic field.* (b) *When the field is pinched, charged particles can be trapped in a magnetic bottle.* (c) *Earth's magnetic field acts like a bundle of magnetic bottles, trapping particles in Earth's magnetosphere.*

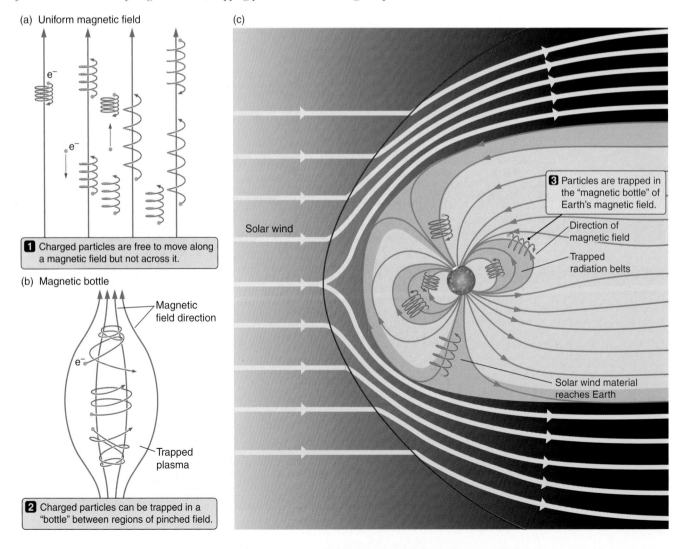

(a) Uniform magnetic field

1 Charged particles are free to move along a magnetic field but not across it.

(b) Magnetic bottle

Magnetic field direction

Trapped plasma

2 Charged particles can be trapped in a "bottle" between regions of pinched field.

(c)

Solar wind

3 Particles are trapped in the "magnetic bottle" of Earth's magnetic field.

Direction of magnetic field

Trapped radiation belts

Solar wind material reaches Earth

(a)

(b)

Figure 7.10 *Aurorae result when particles trapped in Earth's magnetesphere collide with molecules in the upper atmosphere.* (a) *An auroral ring around Earth's north magnetic pole, as seen from space.* (b) *Aurorae borealis viewed from the ground—the "northern lights."*

air between equator and poles is called **Hadley circulation.** Global Hadley circulation, it turns out, seldom occurs in planetary atmospheres, because other factors break up the planetwide flow into a series of smaller *Hadley cells.* Planet rotation is a major factor here. Most planets and their atmospheres are rotating rapidly, and the Coriolis effects produced by this rotation strongly interfere with Hadley circulation by redirecting the horizontal flow **(Figure 7.11b).**

On a rapidly rotating planet, air is not free to flow in just any direction. As we first learned in Chapter 2, when a volume of air starts to move directly toward or away from the poles, the Coriolis effect diverts it into relative motion that is more or less parallel to the planet's equator. This creates winds that blow predominantly in an east-west direction (Figure 7.9b). Meteorologists call these **zonal winds.** In general, the more rapid the planet's rotation,

ward, the displaced air now descends in the polar regions. There it displaces the surface polar air which is forced back toward the equator, completing the circulation **(Figure 7.11a).** This has the effect of keeping the equatorial regions cooler and the polar regions warmer than they otherwise would be. Such planetwide flow of

Figure 7.11 *Schematic diagrams of Hadley circulation.* (a) *The classic Hadley circulation.* (b) *Hadley flow often breaks up into smaller circulation cells. The poleward-equatorward flow is diverted into zonal flow by the Coriolis effect.*

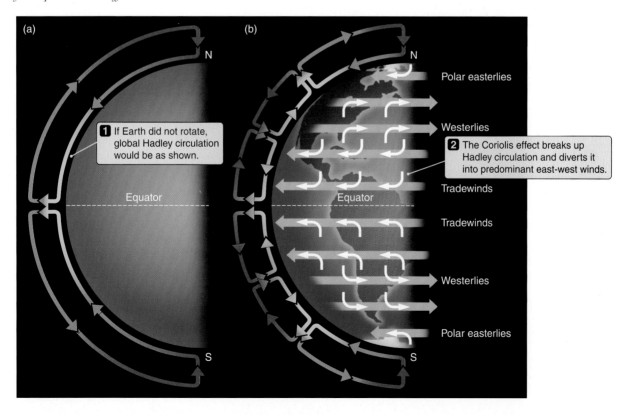

Nonuniform heating together with the Coriolis effect cause east-west zonal winds.

the stronger the Coriolis effect and the stronger its zonal winds will be. Zonal winds are often confined to relatively narrow bands of latitude. Between the equator and the poles in most planetary atmospheres, the zonal winds alternate between "easterly," those blowing toward the *west,* and "westerly," those blowing toward the *east.* Confusing? Very! This unfortunate terminology is a historical carryover from early terrestrial meteorology, in which winds were labeled not by the direction *toward* which they are blowing, but by the direction from which they come!

In Earth's atmosphere, several bands of alternating zonal winds lie between the equator and the poles of both hemispheres. We call this zonal pattern Earth's **global circulation,** because its extent is planetwide. The best-known zonal currents are the subtropical trade winds, more or less easterly winds that once carried sailing ships from Europe westward to the New World, and the midlatitude prevailing westerlies that carried them home again.

Embedded within Earth's global circulation pattern are systems of winds associated with large high- and low-pressure regions. We saw in Chapter 2 that a combination of a low-pressure region and the Coriolis effect produces a circulating pattern called cyclonic motion. Cyclonic motion is associated with stormy weather, including hurricanes. Similarly, high-pressure

7.2
WINDS

systems are localized regions where the air pressure is higher than average. We think of these regions of greater-than-average air concentration as "mountains" of air. Owing to the Coriolis effect, high-pressure regions rotate in a direction opposite to that of low-pressure regions. These high-pressure circulating systems experience anticyclonic motion, and are generally associated with fair weather.

THUNDERSTORMS ARE INTEGRAL TO THE WATER CYCLE

It takes the absorption of thermal energy to turn liquid water into vapor. Water in Earth's oceans, lakes, and rivers is evaporated by thermal energy acquired from the absorption of sunlight. The water vapor then carries this thermal energy along with it as it circulates throughout the atmosphere. When the water vapor recondenses, it gives up its thermal energy to its surroundings. This is the process that powers hurricanes and thunderstorms. We can most easily see how it works with a thunderstorm. A *thunderstorm* begins when Earth's surface, heated by the Sun, warms moist air close to the ground **(Figure 7.12).** The moist air is convected upward, cooling as it gains altitude. Cooling causes the water vapor in the moist air to condense as rain. As it condenses the water vapor gives up its thermal energy to the surrounding air, warming it and thus increasing the strength of the convection. With strong so-

Figure 7.12 *Thunderstorms are powered by convection and by thermal energy released as water vapor condenses to form droplets. The "anvil" top is caused by stratospheric winds shearing the top of the convective system.*

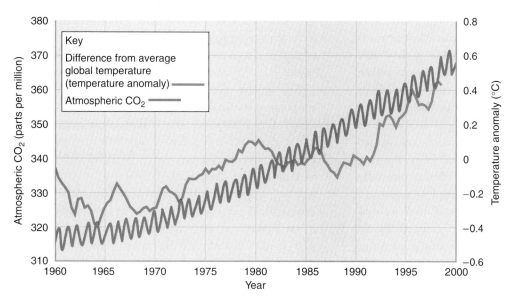

Figure 7.13 *Annual variations in atmospheric CO_2 are due to seasonal variations in plant life and fossil fuel use, while the steady climb is due to human activities. Global temperatures are climbing along with CO_2 concentration. Our climate is delicately balanced, and carbon dioxide plays a vital role in that balance.*

lar heating and an adequate supply of moist air, this self-feeding process can grow within minutes to become a violent thunderstorm. Water, falling back to the surface as rain, eventually returns to the lakes and oceans, wearing down mountains, eroding the soil, and nourishing life as it flows. From the oceans to the air and back again—this is the *water cycle.*

IS EARTH GETTING WARMER?

Climate is the term we use to define the average state of Earth's atmosphere, including temperature, humidity, winds, and so on. Earth's climate appears to go through very lengthy temperature cycles, usually lasting hundreds of thousands of years and occasionally producing shorter cold periods called *ice ages.* These oscillations in the mean global temperature are far smaller than typical geographical or seasonal temperature variations. But Earth's atmosphere is so sensitive to even small temperature changes that it takes a drop of only a few degrees in the mean global temperature to plunge our climate into an ice age. We still do not know the reason for these climate-changing temperature swings. An external influence, such as small changes in the Sun's energy output, may be the cause. Changes in Earth's orbit or the inclination of its rotation axis also have been suggested. Or they may be triggered internally by volcanic eruptions (which can produce global sunlight-blocking clouds or hazes) or long-term interactions between Earth's oceans and its atmosphere.

Many scientists believe that changes in Earth's climate have been accelerating recently and that

Climate is the average state of Earth's atmosphere.

Earth is becoming warmer. Temperature measurements over the past century show a steady increase in the mean global temperature. But whether this represents the beginning of a long-term change—caused, perhaps, by the buildup of human-made greenhouse gases **(Figure 7.13)**—or merely a temporary short-term cycle is a question still being debated. However, we know that our atmosphere is a delicately balanced mechanism. Tiny changes can produce enormous and often unexpected results. Earth's climate is an example of the sort of complex, chaotic system we will discuss in Chapter 9. We see an example of this in *El Niño,* where small shifts in ocean temperature cause much larger global changes in air temperature and rainfall. Are we unknowingly jeopardizing our future by meddling with our atmosphere? In a sense, we are much like children playing with matches.

Are we treating our atmosphere irresponsibly?

7.4 VENUS HAS A HOT, DENSE ATMOSPHERE

Venus and Earth are similar in many ways—so similar that they might be thought of as sister planets. Indeed, when we used the laws of radiation in Chapter 4 to predict temperatures for the two planets, we predicted that they should be very close. But that was before considering the greenhouse effect, and the role of carbon dioxide in blocking the infrared radiation that a planetary surface radiates. The atmospheric pressure at the surface of Venus is a crushing 92 times greater than at the surface of our own planet, equal to the water pressure at an ocean depth of

900 m. The hull of a World War II vintage submarine would be crushed on the surface of Venus. Most (96%) of this massive atmosphere is carbon dioxide, with a small amount (3.5%) of nitrogen. This thick blanket of carbon dioxide effectively traps the IR radiation from Venus, driving the temperature at the surface of the planet to a sizzling 737 K, which is hot enough to melt lead.

> **Venus has a massive carbon dioxide atmosphere, making it a "poster child" for the greenhouse effect.**

While Earth is a lush paradise, a true Eden, the runaway greenhouse effect has turned Venus into a convincing likeness of Hell, an analogy made complete by the presence of choking amounts of sulfurous gases. Venus may be our "sister" planet in many respects, but it will be a *very* long time before humans visit its surface, if ever.

At altitudes between 50 and 80 km, the atmosphere is cool enough for sulfuric acid vapor to react with water vapor to form dense clouds of concentrated sulfuric acid droplets (H_2SO_4). Except for the low-resolution views provided by cloud-penetrating radar, these dense clouds had prevented us from seeing the surface of Venus until the Soviet Union succeeded in landing cameras there in 1975. But it is only with radar that we can globally map the surface of Venus, as so ably done by the spacecraft *Magellan* in the early 1990s.

Sunlight also has difficulty penetrating Venus's dense clouds. Noontime on the surface of Venus is no brighter than a very cloudy day on Earth. High temperatures keep the lower atmosphere of Venus free of clouds and hazes. But though the local

> **High clouds hide Venus' surface from view, but the actual surface of Venus is free of clouds.**

horizon could be seen clearly, the distant mountains would not be so clear. Molecules in even a pure gas will scatter light, and the scattering efficiency increases sharply with decreasing wavelength. Strong scattering by molecules in the very dense atmosphere of Venus would greatly soften any view we might have of distant scenes. We see this same effect, but to a lesser extent, in our own atmosphere. Molecular scattering, always stronger at the shorter wavelengths as discussed in Excursions 7.3, causes a loss of contrast and adds a bluish cast to distant terrain. The high atmospheric temperatures also mean that neither liquid water nor liquid sulfurous compounds can exist on the surface of Venus, leaving a very dry lower atmosphere with only 0.01% of water and sulfur dioxide vapor.

Unlike most planets, Venus rotates on its axis in the opposite sense of its motion around the Sun. Relative to the stars, Venus spins once every 243 Earth days, but a solar day on Venus—the time that it takes for the Sun to return to the same place in the sky—is only 117 Earth

days. Regardless, this slow rotation means that Coriolis effects on the atmosphere are small, resulting in a global circulation that is very close to a classical Hadley pattern (Figure 7.11a). Venus is the only planet known to behave in this way. Its massive atmosphere is very efficient in transporting thermal energy around the planet, so the polar regions are only a few degrees cooler than equatorial regions, and there is almost no temperature difference between day and night.

> **Surface temperatures on Venus vary little from pole to equator or from day to night.**

Because the Venus equator is nearly in the plane of its orbit, seasonal effects are very small, producing only negligible changes in surface temperature. (Recall the discussion in Chapter 2 about how the seasons change due to the tilt of Earth's equator relative to the plane of its orbit around the Sun.) Such small temperature variations also mean that wind speeds near the surface of Venus are quite low, typically about a meter per second. High in the rarefied atmosphere, though, where temperature differences can be larger, winds reach speeds of 110 m/s, circling the planet in only four days. The variation of this high-altitude wind speed with latitude can be seen in the chevron shape of the cloud patterns in **Figure 7.14.**

Large variations in the observed amounts of sulfurous compounds in the high atmosphere of Venus suggest to us that the source of sulfur may be sporadic episodes of volcanic activity. This strengthens the possibility that Venus remains a volcanically active planet.

Figure 7.14 *This* Pioneer *image, taken in ultraviolet light, shows clouds of sulfur compounds in the atmosphere of Venus. Variations in wind speed with latitude cause the clouds to streak out into a "chevron" pattern.*

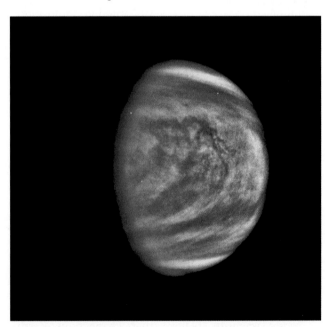

7.5 MARS HAS A COLD, THIN ATMOSPHERE

Compared to Venus, the surface of Mars is almost hospitable. Because of this, we can confidently expect that humans will eventually set foot on the red planet, quite likely before the end of the 21st century and possibly much sooner. What they will see is a stark, waterless landscape, colored reddish by the oxidation of iron-bearing surface minerals. The sky will sometimes be a dark blue, but more often a pinkish color caused by windblown dust **(Figure 7.15).** The lower density of the Mars atmosphere makes it more responsive than Earth's to heating and cooling, so temperature extremes are greater. Near the equator at noontime, future astronauts may experience a comfortable 20°C, about the same as a cool room temperature. Nighttime temperatures typically drop to a frigid −100°C and, during the polar night, the air temperature can reach −150°C, cold enough to freeze carbon dioxide out of the air in the form of a dry-ice frost. For human visitors, the low surface pressure will certainly be uncomfortable. With less than 1% of our own sea-level pressure, the surface of Mars is equivalent to an altitude of 35 km on Earth, far higher than our highest mountain. Like Earth, Mars does have some water vapor in its atmosphere, but the low temperatures condense much of it out as clouds of ice crystals. Early morning ice fog in the lowlands **(Figure 7.16a)** and clouds hanging over the mountains will give terrestrial visitors some reminders of their home planet. While the cold, thin air might be endured, what is seriously lacking and what future astronauts must carry with them is oxygen. Without plants, Mars has only a tiny trace of this life-sustaining gas. Like Venus, the atmosphere of Mars is composed almost entirely of carbon dioxide (95%), with a lesser amount of nitrogen (2.7%). The near absence of oxygen also means that Mars can have very little ozone, and this lack of ozone allows solar ultraviolet radiation to reach all the way to the surface. To survive on Mars, any surface life forms would have to develop protective layers that could shield against the lethal ultraviolet rays.

The surface of Mars is cold and the air is thin.

The inclination of the Mars equator to its orbital plane is similar to Earth's, so it has similar seasons (see Chapter 2). But the effects are larger, for two reasons: Mars varies more in its annual orbital distance from the Sun than does Earth, and the low density of the

Seasonal effects on Mars's climate are larger than Earth's.

Figure 7.15 *A view of the dust-filled pink sky of Mars as seen from the* Viking *lander. In the absence of dust, the sky's thin atmosphere would appear deep blue.*

Mars atmosphere makes it more responsive to seasonal change. The large diurnal, seasonal and latitudinal temperature differences on Mars often create locally strong winds, some estimated to be as high as 100 m/s. For more than a century, astronomers have watched the seasonal development of springtime dust storms on Mars. The stronger ones spread quickly, and within a few weeks can envelope the entire planet in a shroud of dust **(Figure 7.16b).** Such large amounts of windblown dust can take many months to settle out. Seasonal movement of dust from one area to another alternately exposes and covers large areas of dark, rocky surface. This phenomenon led some astronomers of the late 19th and early 20th centuries to believe that they were witnessing the seasonal growth and decay of vegetation. Public imagination carried the astronomers' interpretations a step further, to stories about advanced civilizations on Mars and invasions of Earth by warlike Martians (a theme found in some movies of today).

Mars likely had a more massive atmosphere in the distant past, and there is geological evidence that liquid water once flowed across its surface, as we saw in Chapter 6, but its low gravity was responsible for the loss of much of this earlier atmosphere. This is a good example of a runaway atmosphere, as discussed earlier in Section 7.2.

June 26, 2001   September 4, 2001

Figure 7.16 (a) *Patches of early morning, water vapor fog forming in canyons on Mars.*
(b) Hubble Space Telescope *images of the development of a global dust storm that enshrouded Mars in September 2001.*

7.6 MERCURY AND THE MOON HAVE HARDLY ANY ATMOSPHERE

Our story would not be complete if we did not include the kinds of "atmosphere" that exist around the remaining two terrestrial objects, Mercury and the Moon. There is actually very little to say, as the atmospheres of Mercury and the Moon would hardly be noticed by a visitor from Earth. They are less than a million-billionth (10^{-15}) as dense as our own, and probably vary somewhat with the strength of the solar wind. Such ultrathin atmospheres can have no effect whatsoever on local surface temperatures. Unless you are one of the few astronomers interested in the interaction between solar wind and solid bodies, you may comfortably ignore the atmospheres of Mercury and the Moon.

SEEING THE FOREST THROUGH THE TREES

As these words are being written, your author is sitting in his backyard on a pleasant summer morning. As birds fly by overhead, he is warmed by the rays of the Sun and cooled by a gentle morning breeze. Such an idyllic setting might seem far from the path of our journey through astronomy, but nothing could be further from the truth. Some of the atoms drifting past arrived at Earth as it formed, later to be expelled into Earth's forming secondary atmosphere by volcanism. The rest, especially volatile materials, arrived in the rain of ice-laden comets that fell upon the young Earth. Over the ensuing billions of years, a delicate chemical and physical dance has taken place. The temperature of the young Earth was such that water could condense and fall on the surface as rain. The presence of liquid water served to scrub our atmosphere, absorbing carbon dioxide and locking it up in carbon-bearing rock such as limestone.

Liquid water also provided the bath in which organic chemistry could take place, leading first to molecules with the useful property of being able to reproduce themselves at the expense of other molecules, and later to life itself. Early life on Earth consumed molecules such as carbon dioxide. The waste products formed by this life include what, at the time, was a deadly poison—oxygen. Over the eons the amount of oxygen in our planet's atmosphere increased, and new forms of life evolved to take advantage of its presence. We are among these later life-forms. To us the reactive, corrosive gas called oxygen is the breath of life.

In contrast to Earth, Mercury and the Moon were simply too small to hold on to either their primary or their secondary atmospheres, leaving them as airless rocks. Looking inward from Earth toward the Sun, we find a planet that would seem more like our own home. Venus has much the same size and mass as Earth and is a similar distance from the Sun. We might have imagined Venus to be the ideal world for future human colonization. Yet its history has been different from that of Earth. Slightly warmer than the young Earth, the young Venus was too hot for liquid water to pool on its surface, allowing the same greenhouse effect responsible for maintaining Earth's balmy climate to run away with itself. The result is an environment that is hellish beyond even the fevered imagination of Dante Alighieri.

If we step outward from the Sun, we find Mars. Like Venus, Mars is a planet that is not so different from Earth. While smaller and less massive, it still has strong enough gravity to hold on to its atmosphere. But while Venus was too hot, Mars when it was young was too cold. It had a thick atmosphere, and rain fell in torrents. Images of the surface of the planet show flood basins into which the Amazon would have been but a minor tributary. Liquid water on the surface of Mars was too effective at scrubbing the planet's atmosphere of carbon dioxide. The process that prevented Earth from becoming a Venus-like hothouse got out of hand on Mars, and the temperature fell until the planet's atmosphere nearly froze out.

We owe our lives to the thin blanket of atmosphere that covers our planet. At one time we may have felt justified in taking this atmosphere for granted, but after looking around us in the Solar System we have come to appreciate that our world is maintained by the most delicate of balances. Over billions of years, life has shaped the atmosphere of our planet, and today through the activities of humans, life is reshaping our planet's atmosphere once again.

Great political debate surrounds issues such as the release of ozone-destroying chlorofluorocarbons in Earth's atmosphere, as well as the destruction of oxygen-producing plant life, and the release of huge amounts of carbon dioxide through the burning of fossil fuels. In the rush for profit and convenience, there are those who say that it is too early to worry about the effects that humans are having on the planet. "Earth is too complex," they say, "and our models too primitive to accurately predict the effect that human activities will have on our atmosphere and climate."

But this burden of proof is incorrectly placed. We know that effects such as the greenhouse effect make our planet what it is today. In the absence of the greenhouse effect, Earth would have an average temperature below the freezing point of water. We also know that human activities are measurably changing the chemical balance of our atmosphere. We are undeniably playing with the knobs that regulate our climate. The only intellectually honest way to look at the situation is to start with the hypothesis that unchecked human activity such as the destruction of rainforests and the burning of fossil fuels will have a significant, adverse impact on the future of our planet. This hypothesis has not been disproven.

STUDENT QUESTIONS

THINKING ABOUT THE CONCEPTS

1. Describe the origins of primary and secondary atmospheres.
2. Primary atmospheres of the terrestrial planets were composed almost entirely of hydrogen and helium. Explain why they contained only these gases and not others.
3. Consider a molecule of hydrogen located close to the ground in young Earth's primary atmosphere. Some time later it is moving through interplanetary space, free of Earth's gravitational grip. Trace the journey of this molecule, explaining the important events that led to its escape.
4. Explain how the size and mass of the terrestrial planets, coupled with their distance from the Sun, have determined the kind of atmosphere that each (including the Moon) has today.
5. Contrast the atmospheres and the histories of the atmospheres of Venus, Earth, and Mars.
6. Nitrogen, the principal gas in Earth's atmosphere, was not a significant component of the protostellar disk from which the Sun and planets formed. Where did Earth's nitrogen come from?
7. You check the barometric pressure and find that it

is reading only 920 millibars. There are two possible effects that could be responsible for this low reading. What are they?
8. Why is Venus very hot and Mars very cold if both of their atmospheres are dominated by carbon dioxide, a good "greenhouse" molecule?
9. Astronomers who observe in the infrared region of the spectrum prefer to work at observatories that are located high above sea level. Identify the two important physical reasons for their preference.
10. In what ways does plant life affect the composition of Earth's atmosphere?
11. Global warming appears to be responsible for increased melting of ice in Earth's polar regions.
    a. Why does the melting of Arctic ice, which floats on the surface of the Arctic Ocean, not affect the level of the oceans?
    b. What effect is the melting of glaciers in Greenland and Antarctica having on the level of the oceans?
12. Describe how the atmospheric greenhouse effect works, and how it can both help and harm terrestrial life.
13. What is the principle cause of winds in the atmospheres of the terrestrial planets?

14. If Mercury had an atmosphere at least as dense as Earth's, would you expect its atmospheric circulation to more closely resemble that of Venus or Earth? Explain.

15. Eventually astronomers will have the technical capability to detect Earth-sized planets around other stars. When we find such objects, what observations would indicate with near certainty that they harbor some form of life as we know it?

16. Draw a diagram of a glass greenhouse.
    a. Explain how its interior (or that of a locked car parked in a sunny area) is heated.
    b. In what ways is this similar to the greenhouse effect in Earth's atmosphere?
    c. In what ways is it different?

17. In 1975 the Soviet Union landed two camera-equipped spacecraft on Venus, providing planetary scientists with their first (and only) close up views of the planet's surface. Both cameras ceased to function after only an hour. What environmental condition most likely led to their demise? Explain.

18. Imagine yourself on Mars on a clear, dustless, sunny day. Describe the appearance of the sky above you. How would it appear different from Earth's sky seen under the same conditions?

APPLYING THE CONCEPTS

19. The average mass of the atmosphere above a square meter of Earth's surface is about 10^4 kg. Assume carbon dioxide (CO_2) molecules represent 6×10^{-4} of Earth's total atmospheric mass.
    a. What is the average mass (kg) of CO_2 above each square meter of Earth's surface?
    b. What average pressure (N/m^2) does this CO_2 contribute to Earth's total atmospheric pressure?

20. The average mass of the atmosphere above a square meter of the surface of Mars is close to 190 kg and the Martian atmosphere is about 95% CO_2.
    a. What is the average mass (kg) of CO_2 above each square meter of Mars?
    b. How does this compare with the mass of atmospheric CO_2 above each square meter of Earth's surface?

21. The average mass of the atmosphere above a square meter of the surface of Venus is about 10^6 kg and the atmosphere of Venus is about 96% CO_2.
    a. What is the average mass (kg) of CO_2 above each square meter of Venus?
    b. How does this compare with the mass of CO_2 above each square meter of Earth's surface?
    c. How does this compare with the mass of CO_2 above each square meter of Mars' surface?

22. The ability of wind to erode the surface of a planet is related in part to the wind's kinetic energy.
    a. Compare the kinetic energy of a cubic meter of air at sea level on Earth (mass = 1.3 kg) moving at a speed of 10 m/s with a cubic meter of air at the surface of Venus moving at 1 m/s.
    b. Compare the terrestrial case with a cubic meter of air at the surface of Mars moving at a speed of 100 m/s.

23. Say you seal a rigid container that has been open to air at sea level when the temperature is 0°C (273 K). The pressure inside the sealed container is now exactly equal to the outside air pressure, 10^5 N/m^2.
    a. What would be the pressure inside the container if it were left sitting in the desert shade where the surrounding air temperature is 50°C (323 K)?
    b. What would be the pressure inside the container if were left sitting in an Antarctic night where the surrounding air temperature is −70°C (203 K)?
    c. What would you observe in each case if the walls of the container were not rigid?

24. Using the average density of air at sea level (1.293 kg/m^3) and the average mass of Earth's atmosphere above sea level per square meter (1.035×10^4 kg/m^2), what would be the total depth of Earth's atmosphere (in km) if its density were the same at all altitudes? This is called a scale height (H), a useful quantity for comparing Earth's atmosphere with the atmospheres of other planets.

Before the starry threshold of Jove's Court
My mansion is.
Above the smoke and stir of this dim spot
Which men call earth.

JOHN MILTON (1608–1674)

WORLDS OF GAS— THE GIANT PLANETS

8.1 THE GIANT PLANETS— DISTANT WORLDS, DIFFERENT WORLDS

The four largest planets in our Solar System are Jupiter, Saturn, Uranus, and Neptune, and all have characteristics that clearly distinguish them from the terrestrial, or Earth-like, planets. The most obvious difference is that they are all enormous compared with their terrestrial cousins. Even the smaller of them, Uranus and Neptune, are nearly four times the size of Earth. Another distinguishing feature is their very low density. All are composed almost entirely of light materials such as hydrogen, helium, and water, rather than the rock and metal that make up the terrestrial planets. Collectively, we call Jupiter, Saturn, Uranus, and Neptune the **giant planets,** although you may sometimes hear them referred to as the "jovian planets" after Jupiter, the largest: Jove is another name for Jupiter, the highest-ranking Roman deity.

Giant planets are large, massive, and of lower density than terrestrial planets.

Even though the giant planets share many characteristics, there are significant differences. Jupiter and Saturn are similar to one another in size and are composed primarily of hydrogen and helium. Uranus and

KEY CONCEPTS

Unlike the solid planets of the inner Solar System, four worlds in the outer Solar System were able to capture and retain gases and volatile materials from the Sun's protoplanetary disk and swell to enormous size and mass. Examining these giant planets, we will discover:

* Atmospheres and oceans, but no solid surfaces;
* Two "gas giants," primarily composed of hydrogen and helium, and two "ice giants," primarily composed of water and other volatile materials;
* How changes in temperature and pressure with increasing depth lead to changes in the chemical composition of clouds in giant planet atmospheres;
* Gravitational energy being converted into thermal energy in the interiors of three giant planets, driving strong convection in their atmospheres;
* How the Coriolis effect on these rapidly rotating worlds turns convective motions into powerful winds, huge storms, and planet-spanning bands of multihued clouds;
* Extreme conditions deep within the interiors of the giant planets, and;
* Brilliant aurorae and glowing donuts of gas associated with the huge magnetospheres and strong magnetic fields of the giant planets.

TABLE 8.1

PHYSICAL PROPERTIES OF THE GIANT PLANETS

A comparison of the properties of the four giant planets.

	Jupiter	Saturn	Uranus	Neptune
Orbital radius (AU)	5.2	9.6	19.2	30.0
Orbital period (years)	11.9	29.5	84.0	164.8
Orbital velocity (km/s)	13.1	9.7	6.8	5.4
Mass ($M_\oplus = 1$)	318	95	14.5	17.1
Equatorial diameter (1,000 km)	143	120.5	51.1	49.5
Equatorial diameter ($D_\oplus = 1$)	11.2	9.45	4.0	3.9
Density (water = 1)	1.3	0.7	1.3	1.6
Rotation period (hours)	9.9	10.7	17.2	16.1
Obliquity (degrees)	3.1	26.7	97.8	28.3

Neptune are also similar in size to one another, and both contain much larger amounts of water and other ices than Jupiter and Saturn. These differences are sufficiently large that some astronomers feel it makes sense to divide the giant planets into two classes: Jupiter and Saturn as *gas giants* and Uranus and Neptune as *ice giants*.[1] Other differences between the gas giants and the ice giants will become evident as we more closely examine their physical and chemical characteristics.

Jupiter and Saturn are gas giants. Uranus and Neptune are ice giants.

The giant planets orbit the Sun far beyond the orbits of Earth and Mars (see **Table 8.1**). The closest to the Sun, and to us as well, is Jupiter, and even it is more than five times as far from the Sun as Earth is. Neptune, the most distant, is $4\frac{1}{2}$ billion kilometers away, or some 30 times farther from the Sun than we are. To put this distance into perspective: if you were traveling at the speed of a

Giant planet orbits are larger than those of the terrestrial planets.

commercial jetliner, it would take you more than 500 years to reach Neptune.

The Sun shines only dimly and with very little warmth in these remote parts of the Solar System. If we were to journey to Jupiter, we would see the Sun as but a tiny disk, $\frac{1}{27}$ as bright as it appears from Earth. At Neptune, the Sun would no longer show a disk, but would appear as a brilliant starlike point of light about 500 times brighter than the full Moon in our own sky. How can a point of light be so much brighter than a large disk? It is because the **surface brightness** of the solar disk is 400,000 times greater than the surface brightness of the Moon. Daytime on Neptune is equivalent to a perpetual twilight here on our own planet. With so little sunlight available for warmth, daytime temperatures hover around 123 K at the cloud tops on Jupiter, and they can dip to just 37 K on Neptune's moon Triton.

TWO WERE KNOWN TO ANTIQUITY AND TWO WERE DISCOVERED

Although very far away, the giant planets are so large that all but Neptune can be seen with the unaided eye. Jupiter and Saturn are a familiar sight in the nighttime sky, comparable to the brightest stars. Along with Mercury, Venus, and Mars, they were among the five planets (a Greek word meaning "wandering stars") known to ancient cultures. In contrast, Uranus appears only slightly brighter than the faintest stars visible on a dark night, and Neptune cannot be seen without the aid of binoculars.

Uranus was the first planet to be "discovered," but it was not found until late in the 18th century, more than 170 years after Galileo made the first astronomical observations with a telescope. In 1781, William

Uranus was the first planet discovered in modern times.

---

[1]Strictly speaking, *ice* refers to the solid, frozen form of a volatile substance such as water or methane. However, planetary scientists often refer to these volatile substances as *ices,* even when they are in their liquid form. Uranus and Neptune may be *ice giants,* but that does not mean that they are solid, frozen worlds.

Herschel, a German-English professional musician and amateur astronomer, came upon it quite by accident. Herschel was producing his own catalog of the sky at his home in Bath, England, when he noticed a tiny disk in the eyepiece of his six-inch telescope. At first he thought he had found a comet, but the object's slow nightly motion soon convinced him that it was a new planet beyond the orbit of Saturn. Politically astute, Herschel proposed calling his new planet Georgium Sidus ("George's Star") after King George III of England. Obviously pleased, the monarch rewarded Herschel with a handsome lifetime pension. The astronomical community, however, later rejected Herschel's suggestion, preferring the name Uranus instead. For 65 years Uranus would remain the most distant known planet.

Over the decades that followed Herschel's discovery, astronomers found to their dismay that Uranus was straying from its predicted path in the sky. This aberrant behavior was viewed by mathematicians with grave concern. By then, Newton's laws of motion had been firmly established and were the basis for predicting the motion of the newly discovered planet. Could something be wrong with the theory? As a reasonable explanation, a few astronomers suggested that the gravitational pull of some unknown planet might be responsible for this "unacceptable" behavior of Uranus. Using measured positions of Uranus provided by astronomers, two young mathematicians, Urbain-Jean-Joseph Le Verrier in France and John Couch Adams in England, independently predicted where the hypothetical planet should be. Although Adams was first to compute his predictions, he was unable to gain the attention of England's Astronomer Royal, so the opportunity for England to gain credit for a second planetary discovery was lost. Meanwhile, Le Verrier was having similar problems convincing French astronomers. It was a German who would finally triumph. Armed with Le Verrier's predictions, Johann Galle began a search at the Berlin Observatory. He found the planet on his first observing night, just where Le Verrier and Adams had predicted it would be. The discovery of Neptune in 1846 became a triumph for mathematical prediction based on physical law, and for the subsequent confirmation of theory by observation.

Neptune was discovered 65 years after Uranus.

Neptune became the eighth known planet, and it would remain the outermost for another 84 years until the discovery of Pluto in 1930. Although officially designated a planet, Pluto's characteristics are much closer to those of an icy moon or a comet nucleus than to either a giant or a terrestrial planet. For this reason, we will discuss Pluto in Chapter 10, when we talk about the moons of the Solar System.

8.2 HOW GIANT PLANETS DIFFER FROM TERRESTRIAL PLANETS

Historically, the vast distances to the giant planets have made them difficult objects for scientific study. All of this changed with the arrival of the space age, and with the simultaneous development of more powerful optical and electronic ground-based instruments during the latter decades of the 20th century. Modern instruments on both ground-based telescopes and the *Hubble Space Telescope (HST)* have provided new insight into the composition and physical structure of the giant planets, but our greatest leaps of knowledge have come from close-up observations made possible by the planetary probes: *Pioneer, Voyager, Galileo* and *Cassini*. Over the past several decades, one or another of these spacecraft has visited all four giant planets, sending back a wealth of scientific data with a level of detail that cannot be obtained from the vicinity of Earth. **Figure 8.1** compares the true relative sizes of the giant planets with their relative sizes as seen from Earth. Although ground-based telescopes and the *HST* have made significant contributions to our knowledge of the giant planets, much of what follows is based on what we have learned from the probes.

THE GIANT PLANETS ARE LARGE AND MASSIVE

The giant planets represent an overwhelming 99.5% of all the nonsolar mass in our Solar System. The vast multitude of other Solar System objects—terrestrial planets, moons, asteroids and comets—are all included in the remaining 0.5%. Jupiter alone is 318 times as massive as Earth and some $3\frac{1}{2}$ times as massive as Saturn, its closest rival. Jupiter, in fact, weighs in at more than twice the mass of all the other planets combined. Uranus and Neptune, the ice giants, are the lightweights among the giant planets, but each still has the equivalent of approximately 15 Earth masses.

How do we measure the mass of the planets? In Chapter 3 we learned how a planet's gravitational attraction can affect the motion of a nearby small body, say one of its moons or a passing spacecraft. We found that the motion of the small body can be accurately predicted if we know the planet's mass and apply Newton's law of gravitation and Kepler's third law. Seeing how this works, we should now be able to invert the procedure and calculate the planet's mass by observing the motion of a small body. Prior to the space age, we measured a planet's mass by observing the motions of its moons. This, of course, worked only with planets that have moons. We had to estimate the masses of Mercury and Venus from their size. The accuracy of those early calculations was always limited by how precisely we could measure the positions of the moons with our ground-based telescopes. Planetary spacecraft have now made it possible to measure the masses of planets with unparalleled accuracy. As a spacecraft flies by, the planet's gravity tugs on it and deflects its trajectory. By tracking and comparing the spacecraft's radio signals using several antennae here on Earth, we can detect minute changes in the spacecraft's trajectory, thereby providing a highly accurate measure of the planet's mass.

The motions of a planet's moons reveal its mass . . .

Jupiter is not only the largest of the nine planets, it is more than a tenth the size of the Sun itself. Saturn is only slightly smaller than Jupiter, with a diameter of 9.5 Earths. Uranus and Neptune are each about 4 Earth diameters across, with Neptune being slightly the smaller of the two. The most accurate measurements of planet sizes have come from observing the length of time it takes a planet to eclipse, or *occult,* a star **(Figure 8.2)**. We call these events **stellar occultations.** For example, Newton's laws may tell us that a particular planet is moving along in its orbit at precisely 25 km/s relative to Earth's own motion, and we observe that this planet takes exactly 2,000 seconds to pass directly in front of a star. The planet then must have a diameter equal to the distance it traveled during the 2,000 seconds, or 50,000 km. The center of the planet rarely passes directly in front of a star, of course, but observations of occultations from several widely separated observatories provide the geometry necessary to calculate both the planet's size and its shape. Occultations of the radio signals transmitted from orbiting spacecraft and images taken by the spacecraft's cameras have also provided accurate measures of the sizes and shapes of planets. You may be surprised to learn that the giant planets are not perfectly round. We will see the reason for this later in the chapter.

. . . and stellar occultations reveal a planet's diameter.

Figure 8.1 Top Row: *Images of the giant planets, shown to the same physical scale.*
Bottom Row: *The same images, scaled according to how the planets would appear as seen from Earth.*

Jupiter   Saturn   Uranus   Neptune

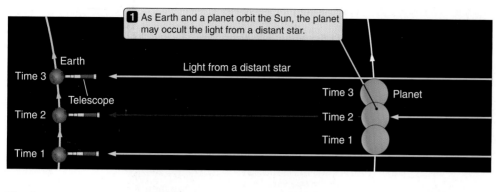

1 As Earth and a planet orbit the Sun, the planet may occult the light from a distant star.

Earth

Time 3

Light from a distant star

Telescope

Time 2

Time 1

Time 3   Planet

Time 2

Time 1

Figure 8.2 *Occultations occur when a planet, moon, or ring passes in front of a star. Careful measurements of changes in the starlight's brightness and the duration of these changes give information about the size and properties of the occulting object.*

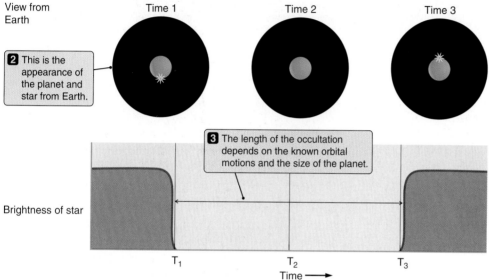

View from Earth

Time 1   Time 2   Time 3

2 This is the appearance of the planet and star from Earth.

3 The length of the occultation depends on the known orbital motions and the size of the planet.

Brightness of star

T_1 T_2 T_3

Time ⟶

of the atmosphere prevents us from seeing them. Neptune displays a few high clouds with a deep, clear atmosphere showing between them.

JUPITER'S CHEMISTRY IS MORE LIKE THE SUN'S THAN EARTH'S

WE SEE ONLY ATMOSPHERES, NOT SURFACES

The giant planets are made up entirely of gases and liquids. Their characteristic structure is that of a relatively

The giant planets are all gas and liquid.

shallow atmosphere merging seamlessly into a deep liquid "ocean," which in turn merges smoothly into a more dense liquid core. Although shallow compared with the depth of the liquid layers below, the atmospheres of the giant planets are still much thicker than those of the terrestrial planets—thousands of kilometers as compared to hundreds. As with Venus, only the very upper levels of their atmospheres are visible to us. In the case of Jupiter or Saturn, we see the tops of a layer of thick clouds, the highest of many others that lie below. Although a few thin clouds are visible on Uranus, we find ourselves looking mostly into a clear, seemingly bottomless atmosphere. Atmospheric models tell us that thick cloud layers must lie below, but molecular scattering (see Chapter 7) in the clear part

In Chapter 6 we learned that the terrestrial planets are composed mostly of rocky minerals, such as silicates, along with various amounts of iron and other metals. It is true that the atmospheres of the terrestrial planets contain lighter materials, but the masses of these atmospheres—and even of Earth's oceans—are insignificant compared with the total planetary mass. The terrestrial planets are thus very compact, the densest objects in the Solar System.

In contrast, the giant planets are composed almost entirely of lighter materials such as hydrogen, helium, and water. This makes their densities much lower than the rock-and-metal terrestrial planets. Neptune is the most compact among the giant planets, having a density about 1.6 times that of water. Saturn is the least compact—only 0.7 times the density of water. This means that, placed in a large enough (and deep enough) body of water, Saturn would actually float with 70% of its volume submerged. Jupiter and Uranus have densities intermediate between those of Neptune and Saturn.

Jupiter's chemical composition is quite similar to that of the Sun and the rest of the cosmos. Astronomers take the relative abundance of the elements

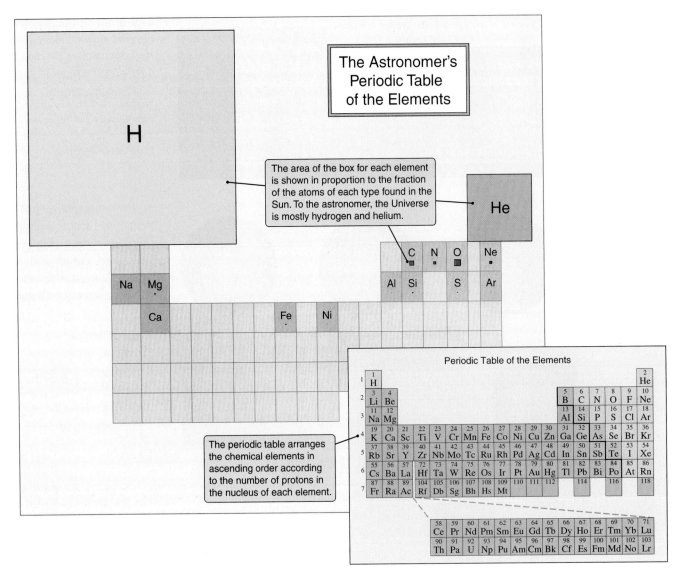

Figure 8.3 *The periodic table of the elements shows the chemical elements laid out in order of the number of protons in the nucleus of each. But, in this "astronomer's periodic table," the Universe is made up of mostly hydrogen and helium.*

Jupiter's chemistry is similar to the Sun's.

in the Sun as a standard reference, termed **solar abundance**. As illustrated in **Figure 8.3,** hydrogen (H) is the most abundant element, followed by helium (He). Jupiter has about a dozen hydrogen atoms for every atom of helium, which is typical of the Universe as a whole. Only 2% of its mass is made up of **massive elements** (atoms more massive than helium)—mostly oxygen (O), carbon (C), neon (Ne), nitrogen (N), magnesium (Mg), silicon (Si), iron (Fe), and sulfur (S). Many of

Massive elements are all the elements other than hydrogen and helium.

these elements have combined chemically with one another and, because of its great abundance, most are linked up with hydrogen. Thus, atoms of oxygen, carbon, nitrogen, and sulfur have combined with hydrogen to form molecules of water (H_2O), methane (CH_4), ammonia (NH_3), and hydrogen sulfide (H_2S), respectively. More complex combinations produce materials such as ammonium hydrosulfide (NH_4HS). Helium and neon are what we call **inert gases,** meaning they do not normally combine with other elements or with themselves. Most of the iron, the remains of the original rocky planet around which the gas giant grew, and even much of the water has ended up in Jupiter's liquid core.

The chemical compositions of the giant planets are not all the same. Proportionally, Saturn has a somewhat larger inventory of massive elements than Jupiter. In Uranus and Neptune, massive elements are so abundant that they are major compositional components of these two planets. What is especially interesting is that the total mass of massive elements in each of the four giant planets is nearly the same—approximately 10 to 15 $M_⊕$ (Earth masses). The principal compositional differences among the four of them lie in the amounts of hydrogen and helium they contain. This turns out to be an important clue to the process by which the giant planets formed, a subject we will return to later in the chapter.

SHORT DAYS AND NIGHTS ON THE GIANT PLANETS

Still another distinguishing characteristic of the giant planets is their rapid rotation, meaning that the lengths of their days are short. A day on Jupiter is just under 10 hours long, and Saturn's is only a little longer. Neptune and Uranus have rotation periods of 16 and 17 hours, respectively, giving them days that are intermediate in length between those of Jupiter and Earth.

The rapid rotation of the giant planets distorts their shapes. If they did not rotate at all, these fluid bodies would be perfectly spherical. In Chapter 5 we used the analogy of a spinning ball of pizza dough to see why a collapsing, rotating cloud must settle into a disk. The same principle is at work in the rapidly rotating giant planets as well, causing the planets to bulge at their equators, and giving them an overall flattened appearance. We call this flattening **oblateness.** Saturn's oblateness is especially noticeable (see Figure 8.1), with an equatorial diameter that is 10% greater than its polar diameter. In comparison, the oblateness of Earth is only 0.3%.

Rapid rotation distorts the shapes of the giant planets.

A planet's **obliquity**—the inclination of its equator to its orbital plane—is a major factor in determining the prominence of its seasons. The obliquities of each of the giant planets are given in Table 8.1. With an obliquity of only 3°, Jupiter has almost no seasons at all. The obliquities of Saturn and Neptune are slightly larger than those of Earth or Mars, giving these planets moderate but well-defined seasons. Curiously, Uranus spins on an axis that is nearly in the plane of its orbit. This creates the appearance from Earth of a planet that is either spinning face on to us or rolling along on its side (or something in between), depending on where Uranus happens to be in its orbit. The obliquity of Uranus is 98°, where a value greater than 90° indicates that the planet rotates in a clockwise direction when seen from above its orbital plane (see Chapter 2). (Venus (Chapter 7) and Pluto (Chapter 10) are the other two planets that behave this way.) Seasons on Uranus are thus extreme, with each polar region alternately experiencing 42 years of continuous sunshine followed by 42 years of total darkness. If we calculate the average amount of sunlight absorbed over an entire orbit, it turns out that the poles are warmed more than the equator, a situation quite different from that of any other planet. Why does Uranus have an obliquity so different from the other planets? As we learned in Chapter 5, many astronomers believe the planet was "knocked over" by the impact of a huge planetesimal near the end of its accretion phase.

Seasons differ greatly among the giant planets.

8.3 A VIEW OF THE CLOUD TOPS

Even when viewed through small telescopes, Jupiter is perhaps the most striking and colorful of all the planets (see Figure 8.1). A dozen or more parallel bands, ranging in hue from bluish gray to various shades of orange, reddish brown, and pink, stretch out across its large, pale yellow disk. Traditionally, astronomers call the darker bands *belts* and the lighter ones *zones*. Many small clouds—some dark and some white, some circular and others more oval in shape—appear along the edges of, or within, the belts. The most prominent of these is a large, often brick-red feature in Jupiter's southern hemisphere known as the **Great Red Spot** (GRS). Oval in shape, with a length of 25,000 km and a width of 12,000 km, two Earths could fit comfortably side by side within its boundaries. No one really knows how long this huge feature has been circulating in Jupiter's atmosphere, but it was first spotted more than three centuries ago, shortly after the invention of the telescope. Since then, the GRS has varied unpredictably in size, shape, color, and motion as it drifts among Jupiter's clouds. Small clouds seen moving around the periphery of the GRS show that it circulates like a giant **vortex.** Its cloud pattern looks a lot like that of a terrestrial hurricane, but since it rotates in

The Great Red Spot (GRS) is a giant anticyclone.

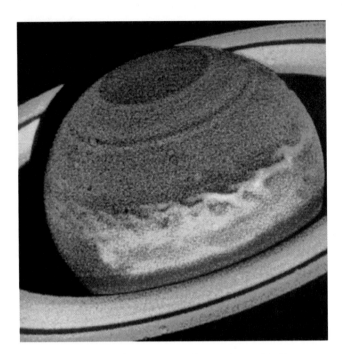

Figure 8.4 *Tremendous storms are known to occasionally erupt on Saturn. This Hubble Space Telescope image shows the storm that began shortly after the launch of the telescope in 1990.*

matched by other celestial objects. But it is the rings that make it so spectacular; farther away and somewhat smaller in size, the planet itself appears less than half as large as Jupiter. Saturn, like Jupiter, displays atmospheric bands, but their colors and contrasts are much more subdued. Individual

Saturn's rings are magnificent, but the planet itself is rather bland.

clouds on Saturn are seen only rarely from Earth. On these infrequent occasions, large, white cloudlike features may suddenly erupt in the tropics, spread out in longitude, and then fade away over a period of a few months **(Figure 8.4)**.

In even the largest telescopes, Uranus and Neptune appear only as tiny, featureless disks with a pale bluish green color. Yet in the near-infrared, beyond the spectral range of our eyes, Uranus and Neptune take on very different appearances, showing limb hazes and small clouds **(Figure 8.5)**. This illustrates how observations made in different spectral regions can add significantly to our understanding of astronomical objects.

the opposite direction, it exhibits *anticyclonic* rather than *cyclonic* flow (see Chapter 2). Because of its colorful appearance and unpredictable changes, the Great Red Spot has long been a favorite among amateur astronomers.

Our first look at Saturn through a telescope, in some ways even more impressive than Jupiter, is a moment not quickly forgotten (see Figure 8.1). Adorned by its magnificent system of rings, Saturn provides a sight un-

Observed from close up, the giant planets suddenly appear as real and tangible worlds (see Figure 8.1). The clouds of Jupiter are a landscape of variegated color and intricate formations. Time-lapse imaging shows a roiling, swirling giant with atmospheric currents and vortices so complex in nature that we still do not understand the details of how they interact with one another, even after decades of analysis. The GRS alone displays more structure than was visible over all of Jupiter prior to the space age. Dynamically, it also reveals some

8.1
CLOUDS

Figure 8.5 Hubble Space Telescope *images of Uranus* (left) *and Neptune* (right), *taken at a wavelength of light that is strongly absorbed by methane. The visible clouds are high in the atmosphere. Uranus's rings show prominently in this image, which subdues the brightness of the planet.*

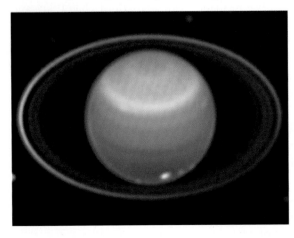
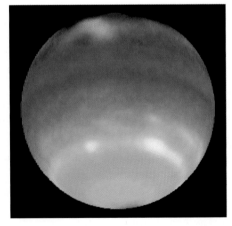

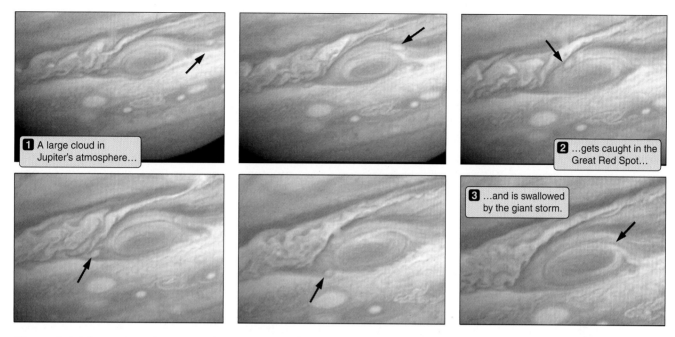

1 A large cloud in Jupiter's atmosphere...

2 ...gets caught in the Great Red Spot...

3 ...and is swallowed by the giant storm.

Figure 8.6 *This sequence of images, obtained by the* Voyager *spacecraft during its encounter with Jupiter, shows the swirling, anticyclonic motion of Jupiter's Great Red Spot.*

rather bizarre behavior, such as cloud cannibalism. In a series of time-lapse images, *Voyager* observed a number of Alaska-sized clouds being swept into the GRS vortex. Some of these clouds were carried around the vortex a few times and then ejected, while others were swallowed up and never seen again **(Figure 8.6)**. Other, smaller clouds with structure and behavior similar to that of the GRS are seen in Jupiter's middle latitudes.

The GRS cannibalizes smaller atmospheric features.

Broad bands and individual clouds show prominently in *Voyager* images of Saturn's atmosphere. A relatively narrow, meandering band in the midnorthern latitudes encircles the planet in a manner similar to our own terrestrial jet stream **(Figure 8.7)**. The largest features are about the size of the continental United States, but many that we see are smaller than terrestrial hurricanes.

Saturn has jet streams similar to Earth's.

A small oval feature, red in color and resembling a miniature version of Jupiter's Great Red Spot, appeared in images taken by *Voyager* in 1980–1981. We do not know whether Saturn's "mini–Red Spot" still survives, because it is too small to see from Earth. The inability of occasional planetary probes to provide continuous monitoring of the giant planets is one of their more serious shortcomings.

A few clouds and a weak, orange-colored band surrounding the south pole are recognizable in images of Uranus (see Figure 8.1). Several muted bands appear at lower latitudes. Even from close up, Uranus presents a rather bland appearance at visible wavelengths.

A number of small bright clouds appear in *Voyager 2* images of Neptune's atmosphere (see Figure 8.1). In some instances, we can see their shadows cast down through the clear upper atmosphere onto a dense cloud layer 75 km below. Some of Neptune's clouds are

Figure 8.7 (a) Voyager *image of a jet stream in Saturn's northern hemisphere, similar to jet streams in the terrestrial atmosphere.* (b) *The dynamical relationship between the jet stream and vortices around regions of low and high pressure nested within its peaks and troughs.*

(a)

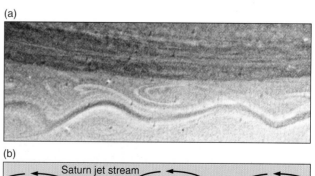

(b)

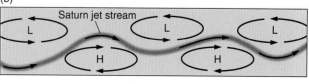

Saturn jet stream

Neptune has both bright and dark clouds.

dark. A large, oval-shaped dark feature in the southern hemisphere **(Figure 8.8)** is reminiscent of Jupiter's Great Red Spot. Predictably, astronomers called it the Great Dark Spot (GDS). However, the Neptune feature was gray rather than red in color, changed its length and shape more rapidly than the GRS does, and lacked the permanence of the GRS. When the *HST* trained its corrected optics on Neptune in 1994, the Great Dark Spot had disappeared. It has not been seen since.

8.4 A JOURNEY INTO THE CLOUDS

Our visual impression of the giant planets is based on our "two-dimensional" view of their cloud tops. Atmospheres, though, have depth. They are three-dimensional structures whose temperature, density, pressure, and even chemical composition vary with height and over horizontal distances. As a rule, atmospheric temperature, density, and pressure all decrease with increasing altitude, although temperature will sometimes reverse itself at very high altitudes (see Chapter 7). The *stratospheres* above the cloud tops appear relatively clear, but closer inspection shows that they contain layers of thin haze that show up best when seen in profile above the limbs of the planets. The composition of the haze particles remains unknown, but we believe that they are photochemical,

smog-like products created when ultraviolet sunlight acts on hydrocarbon gases, such as methane.

What lies beneath the cloud tops of the giant planets? Imagine riding along in the *Galileo* probe as it descends through Jupiter's atmosphere. At first, the only change we note is the expected increase in the outside pressure and temperature as our altitude rapidly decreases. Suddenly, we find ourselves passing through dense layers of cloud, separated by regions of relatively clear atmosphere. Each of these cloud layers is composed of a different chemical substance. In Earth's atmosphere, water is the only substance that can condense into clouds, but the atmospheres of Jupiter and the other giant planets contain a variety of volatiles that can condense at different temperatures and atmospheric pressures. A descent through Jupiter's cloud layers thus becomes a journey that explores the many minor constituents of its atmosphere. Each kind of volatile, such as water, condenses at a particular temperature and pressure, and each therefore forms clouds at a different altitude, as illustrated in **Figure 8.9.** Below the condensation cloud layer, each volatile is freely mixed as a gaseous atmospheric constituent. Above, it is highly depleted. We can see the reason for this: As the planet's atmosphere convects (a process explained in Connections 6.2), volatile materials are carried upward along with all other atmospheric gases. When a particular volatile reaches an altitude where the temperature is low enough, the condensation process removes most of it from the other gases, leaving it depleted in the air above.

Different volatiles produce different clouds at different heights.

Figure 8.8 (a) *The Great Red Spot (GRS) on Jupiter and* (b) *Great Dark Spot (GDS) on Neptune reproduced approximately to scale. Earth is shown for comparison. The Great Red Spot has persisted for centuries, but the Great Dark Spot disappeared between the time* Voyager *flew by Neptune and HST images were obtained.*

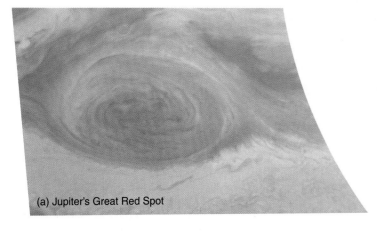

(a) Jupiter's Great Red Spot

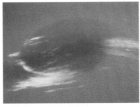

(b) Neptune's Great Dark Spot

Earth

During our descent we find that ammonia has condensed near the top of Jupiter's *troposphere* (see Chapter 7) at a temperature around −140°C (133 K). Next we pass through a layer of ammonium hydrosulfide clouds at a temperature of about −80°C (193 K). Not long after this, information from the probe ends. (While descending slowly via parachute near an atmospheric pressure of 22 bars and a temperature of about 100°C (373 K), the *Galileo* probe failed, presumably because its transmitter got too hot.) What lies below this level in Jupiter's atmosphere must for now be left to our theories and atmospheric models.

The distance of each planet from the Sun partly determines its tropospheric temperatures. The farther the planet is from the Sun, the colder its troposphere will be. This now determines the altitude at which a particular volatile, such as ammonia or water, will condense to form a cloud layer on each of the planets (see Figure 8.9). If temperatures are too high, some volatiles may not condense at all. The highest clouds in the frigid atmospheres of Uranus and Neptune are crystals of methane ice. The highest clouds on Jupiter and Saturn are made up of ammonia ice. Methane exists only in gaseous form throughout the warmer atmospheres of Jupiter and Saturn.

Volatiles form clouds at different heights on different planets.

In their purest form, the ices that make up the clouds of the giant planets are all white, similar to snow on our own planet. Why, then, are some clouds so colorful, especially Jupiter's? These tints and hues must come from impurities in the ice crystals, similar to the way syrups color "snow cones." While the identities of these impurities remain unknown, our prime suspects include elemental sulfur and phosphorus and various organic materials produced by the photochemical action of sunlight on atmospheric hydrocarbons. Ultraviolet photons from the Sun, absorbed by molecules of hydrocarbons such as methane, acetylene, and ethane, among others, have enough energy to break these molecules apart. The molecular fragments can then recombine to form complex, organic compounds that condense into solid particles, many of which are quite colorful. Photochemical reactions also occur in our own terrestrial atmosphere. Some of the photochemical products produced close to the ground are quite obvious. We call them "smog."

Clouds on Jupiter and Saturn are colored by impurities.

The upper tropospheres of Uranus and Neptune, unlike those of Jupiter and Saturn, are relatively clear, with only a few white clouds—probably methane ice—appearing here and there. Uranus and Neptune are not bluish green because of clouds. Instead, they are blue for much the same reason Earth's oceans are blue. Methane gas is much more abundant in the atmospheres of Uranus and Neptune than in Jupiter and Saturn. Like water, methane gas tends to selectively absorb the longer wavelengths of light—yellow, orange, and red. This leaves only the shorter wavelengths—green and blue—to be scattered from the relatively cloud-free atmospheres of Uranus and Neptune, giving them a characteristic greenish blue color. We earlier described the atmosphere of Uranus as appearing nearly "bottomless." Molecular scattering also contributes to the bluish color and is so strong in the clear Uranus atmosphere that it completely hides the thick ammonia and water cloud layers that lie far below.

8.5 WINDS AND STORMS— VIOLENT WEATHER ON THE GIANT PLANETS

The rapid rotation and resulting strong Coriolis effects (see Chapter 2) in the atmospheres of the giant planets create much stronger zonal winds (see Chapter 7) than we see in the atmospheres of the terrestrial planets, even though there is less thermal energy available. **Figure 8.10(a)** shows the zonal wind pattern on Jupiter. On Jupiter the strongest winds are equatorial westerlies, with speeds of up to 550 km/h (150 m/s). (Remember from Chapter 7 that westerly winds are those that blow *from*, not *toward*, the west.) At higher latitudes, the winds alternate between easterly and westerly in a pattern that seems to be related to Jupiter's banded structure, but scientists are not sure. Near 20° south latitude, the Great Red Spot appears to be caught between a pair of easterly and westerly currents with opposing speeds of more than 200 km/h. If you think this might imply something about the relationship between zonal flow and vortices, you are right.

The equatorial winds on Saturn are also westerly, but here the wind speed rises to an impressive 1,650 km/h (460 m/s) **(Figure 8.10b)**. Alternating easterly and westerly winds also occur at higher latitudes, but, unlike Jupiter's case, this alternation seems to bear no clear association with Saturn's atmospheric bands. This is but one example of unexplained dif-

Equatorial winds are fast on Jupiter and even faster on Saturn.

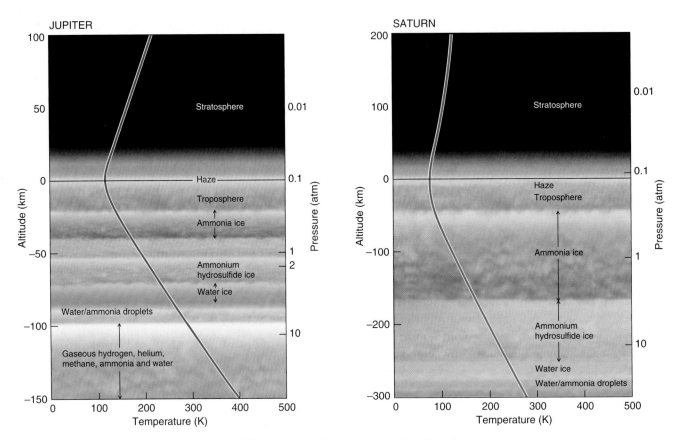

Figure 8.9 *Volatile materials condense out at different levels in the atmospheres of the giant planets, leading to chemically different types of clouds at different depths in the atmosphere. The 1.0-atm level corresponds to atmospheric pressure at sea level on Earth.*

ferences between the giant planets. As mentioned earlier, there is a narrow meandering river of air with alternating crests and troughs (see Figure 8.7a). This feature, located near 45° north latitude, is similar to our own terrestrial jet streams, where high-speed winds blow generally from west to east but with alternate wanderings toward and away from the pole. Nested within the crests and troughs of Saturn's jet stream are anticyclonic and cyclonic vortices. They appear remarkably similar in both form and size to terrestrial high- and low-pressure systems, which bring us alternating periods of fair and stormy weather as they are carried along by our terrestrial jet streams. This similarity in jet streams on Saturn and Earth is a good illustration of the many reasons we study other planets— for the ability to compare them to each other and to similar phenomena on Earth. From observing and analyzing similar atmospheric systems on other planets, we often learn more about the way our own weather works.

Our knowledge of global winds on Uranus is much poorer than that of the other giant planets because of the relatively few clouds we have been able to see and track. In Excursions 8.1 we see that winds are measured by observing the motion of clouds. When *Voyager 2* flew by Uranus in 1986, even those few clouds

that we did see were all in the southern hemisphere. The northern hemisphere was in complete darkness at the time. The strongest winds observed were 650 km/h westerlies in the middle to high southern latitudes. No easterly winds were detected on the part of the planet that could be seen. The most important finding was that the winds are zonal. In 1986 the Sun was shining almost straight down on the south pole of Uranus. Some astronomers had thought earlier that Uranus might have a global wind system very different from that of the other giant planets: because of its peculiar orientation, Uranus has a "reversed" temperature pattern wherein the poles are warmer than the equator. That the dominant winds on Uranus also turned out to be zonal, as they are on the other giant planets, indicates how dominant the Coriolis effect is, and how completely it determines the fundamental structure of the global winds on all the giant planets.

As expected, the strongest winds on Neptune occur in the tropics. The surprise was that they are easterly rather than westerly, and have speeds of up to 1,600 km/h. Westerly winds with speeds of 900 km/h and

Neptune has wind speeds of up to 1,600 km/h.

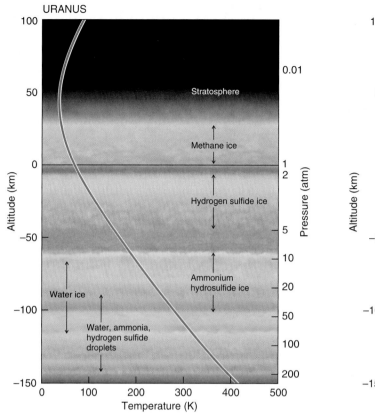

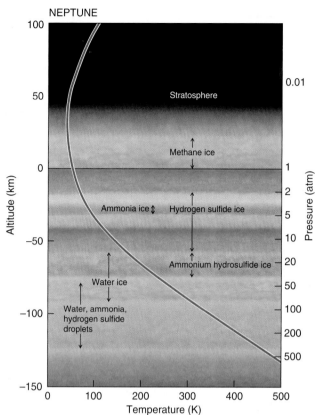

higher were seen in the south polar regions. With wind speeds five times greater than those of the fiercest hurricanes on Earth, Neptune and Saturn are the windiest planets we know of.

As we learned in Chapter 7, vertical temperature differences can create the localized type of atmospheric motion called *convection*. On the giant planets, the ther-

mal energy that drives convection comes from the Sun and also from the hot interiors of the planets themselves. As heating drives air up and down, the Coriolis effect shapes that convection into atmospheric vortices, examples of which are familiar to us on Earth as high-pressure systems, hurricanes, and thunderstorms. On the giant planets, convective vortices are visible as iso-

Figure 8.10 *Strong winds blow on Jupiter and Saturn, driven by powerful convection and the Coriolis effect on these rapidly rotating worlds.*

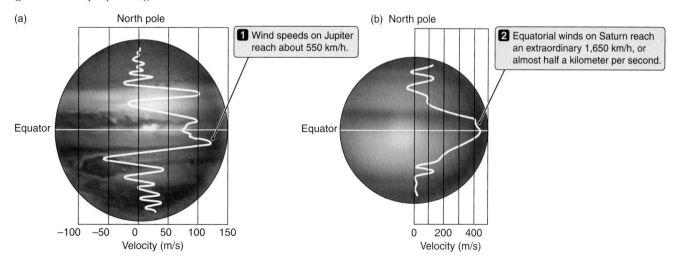

EXCURSIONS 8.1

HOW WIND SPEEDS ON DISTANT PLANETS ARE MEASURED

How are we able to measure wind speeds on planets that are so far away? It turns out to be surprisingly easy. If we can see individual clouds in their atmospheres, we can measure their winds. As on Earth, clouds are typically carried right along with the local winds. By measuring the positions of individual clouds and noting how much they move during an interval of a day or so, we are able to calculate the local wind speed. We need to know one important additional piece of information though: how fast the planet itself is rotating. This is because we want to measure the speed of the winds with respect to the planet's rotating surface. In the case of the giant planets, of course, there is no solid surface against which to measure the winds. We must instead assume a hypothetical surface, one that rotates as though it were somehow "connected" to the planet's deep interior. How can we know how rapidly the invisible interior of a planet is rotating? Periodic bursts of radio energy caused by the rotation of a planet's magnetic field tell us how fast the interior of the planet is rotating.

Thermal energy drives powerful convection on the giant planets.

lated circular or oval cloud structures. The Great Red Spot (GRS) on Jupiter and the Great Dark Spot (GDS) on Neptune are classic examples. Observations of small clouds distributed within the GRS show that it is an enormous atmospheric whirlpool or eddy, swirling around in a counterclockwise direction with a period of about one week. On a rapidly rotating planet, winds generated by Coriolis effects can be very strong. Clouds circulating around the circumference of the GRS have been clocked at speeds of up to 1,000 km/h. Such weather systems dwarf our terrestrial storms in both size and intensity. Comparable behavior is observed in many of the smaller oval-shaped clouds found elsewhere in Jupiter's atmosphere, as well as in similar clouds observed in the atmospheres of Saturn and Neptune.

As the atmosphere ascends near the centers of the vortices, it expands and cools. Cooling condenses certain volatile materials into liquid droplets, which then fall as rain. As they fall, the raindrops collide with surrounding air molecules, stripping electrons from the molecules and thereby developing tiny electric charges in the air. The cumulative effect of countless falling raindrops is often to generate an electric charge so large that it breaks down a conductive path through the air, creating a surge of current and a flash of lightning. A single observation of Jupiter's night side by *Voyager 1* revealed several dozen lightning bolts within an inter-

On the giant planets, as on Earth, falling raindrops generate lightning.

val of 3 minutes. We estimate the strength of these bolts to be equal to or greater than the "superbolts" that occur in the tops of high convective clouds in the terrestrial tropics. Although various constraints prevented an imaging search for lightning on the other giant planets, radio receivers on *Voyager* picked up lightning static on all of them.

Are atmospheric vortices somehow connected to the global circulation, that is, to zonal flow? We have noted that the narrow zonal jet on Saturn moves with wavelike motion (Figure 8.7a). Nestled in each of its crests and troughs are clockwise (anticyclonic) and counterclockwise (cyclonic) features, strongly suggesting a dynamical relationship between these systems and the zonal jet. Jupiter's Great Red Spot is situated between pairs of strong zonal winds flowing in opposite directions, as was Neptune's Great Dark Spot. Is this by chance? Similar relationships observed between other iso-

Vortices created by convection appear to drive the strong zonal winds of the giant planets.

lated vortical clouds and the zonal wind flow suggest that it is not. In looking at Figure 8.7(b), we might ask, then, whether the zonal winds are dragging the edges of vortices around or, conversely, the vortices are driving the zonal wind currents. From a careful study of the interaction between vortices and zonal winds, we believe the latter to be true. As shown in **Figure 8.11,** the enormous wind energy developed within numerous convective vortices seems to drive the alternating easterly and westerly zonal winds that characterize the global circulation of the giant planets.

8.6 SOME THERMAL ENERGY COMES FROM WITHIN

We have seen how uneven heating of planets, together with their rotation, drives global atmospheric circulation, and how temperature-related differences in pressure from place to place drive winds. In fact, virtually all weather on every planet is driven by the interplays of thermal energy within the planet's atmosphere. On Earth and the other terrestrial planets, the source of this thermal energy is as clear as the Sun shining in a summer sky. Sunlight powers our climate.

On the terrestrial planets, nearly all thermal energy comes from the Sun.

This is not a new insight. In Chapter 4 we learned about the equilibrium between the absorption of sunlight and the radiation of infrared light into space, while in Chapter 7 we saw how the resulting equilibrium temperature is modified by the greenhouse effect on Earth and Venus. Yet when we calculate this equilibrium for the giant planets, as we did in Chapter 4, we find something amiss. According to these calculations, the equilibrium temperature for Jupiter, for example, should be 105 K, but when it is measured we find instead an average temperature of about 119 K. A difference of 14 K might not seem like much, but remember that according to Stefan's law (see Chapter 4), the energy radiated by an object depends on its temperature raised to the fourth power. Applying this to Jupiter, we get $(T_{\text{true}}/T_{\text{expected}})^4 = (119 \text{ K}/105 \text{ K})^4 = 1.65$. The implications of this are somewhat startling: *Jupiter is radiating roughly two-thirds more energy into space than it absorbs in the form of sunlight.* Almost half of the thermal energy powering Jupiter's weather comes from somewhere other than the Sun. Where else could this extra energy arise, but from within the planet itself?

Jupiter has a large internal heat source, as do Saturn and Neptune.

Similarly, the internal energy escaping from both Saturn and Neptune is observed to be almost twice as great as the sunlight that each of them absorbs. Strangely, whatever internal energy may be escaping from Uranus, it is very small compared with the absorbed solar energy.

We have already learned that winds on the giant planets are considerably stronger than on Earth. It is interesting to note that this is true even though lesser amounts of thermal energy are available to drive these winds. Available solar energy per unit area at Jupiter is less than 4% of that received by Earth, and at Neptune only 0.1% of Earth's. Even with the additional internal energy, the total energy per unit area falls far short of that available to the terrestrial planets. We do not fully understand why these winds should be so strong.

Thermal energy from the hot interiors of the giant planets diffuses slowly outward to warm their upper atmospheres. The rate at which this energy is delivered to the atmosphere depends both on the temperature and the thermal properties of the interior. The most direct

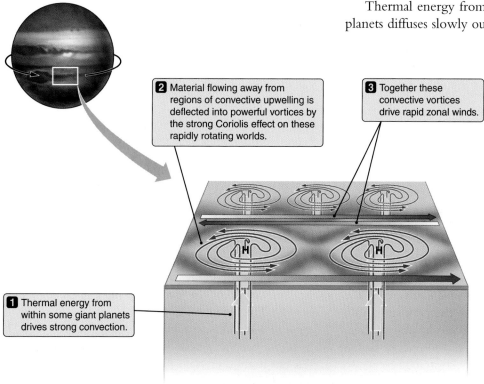

2 Material flowing away from regions of convective upwelling is deflected into powerful vortices by the strong Coriolis effect on these rapidly rotating worlds.

3 Together these convective vortices drive rapid zonal winds.

1 Thermal energy from within some giant planets drives strong convection.

Figure 8.11 *Material rises within a giant planet due to convection. Anticyclonic flow around the resulting region of high pressure is caused by the Coriolis effect. Such convective vortices drive zonal winds on the giant planets.*

EXCURSIONS 8.2

PRIMORDIAL ENERGY

In Chapter 5 we learned about the collapse of a protostar. A mass of gas collapses under the force of its own self-gravity and, as it does so, converts its gravitational potential energy to thermal energy. If the mass of the protostar is large enough to form a star, the core warms up to a temperature so high that self-sustaining thermonuclear reactions take place. Once thermonuclear energy becomes available, internal thermal energy can be generated and maintain enough outward pressure to prevent further collapse, thereby stabilizing the newly formed star.

A giant planet, with its smaller mass, does not generate central temperatures that are high enough to generate thermonuclear energy, but is otherwise much like a collapsing protostar. The gaseous planet continues to contract indefinitely, releasing its gravitational potential energy as it shrinks. This is the primary energy source for replacing the internal energy that leaks out of the interior of Jupiter and is probably an important source for the other giant planets as well.

In Saturn's case, and perhaps Jupiter's as well there is an additional source of internal energy. In Excursions 8.3, we will see that under the right conditions, liquid helium separates from a hydrogen-helium mixture and "rains" downward toward the core. As the droplets of liquid helium sink, they release their gravitational potential energy as thermal energy. Planetary physicists believe that most of Saturn's internal energy and perhaps some of Jupiter's comes from this separation of liquid helium.

We can think of this internal energy that lies deep within the giant planets as being *primordial*. In other words, it is left over from their creation. The giant planets are still contracting and converting their gravitational potential energy into thermal energy today as they did when they first formed, only they are doing it more slowly (see Connections 8.1). The annual amount of contraction necessary to sustain their internal temperature is only a tiny fraction of their radius. For Jupiter it means only 1 mm or so—a hundred-billionth of its radius—per year. If this same rate were to continue for the next billion years, Jupiter would shrink by only a thousand kilometers, a little more than 1% of its radius.

In the popular literature, Jupiter is frequently and quite erroneously labeled as "a star that failed," probably because it is gaseous, has a high internal temperature, and releases a large amount of primordial energy. While these statements are true, to imply that it is "almost a star" is most certainly an exaggeration. Jupiter's central temperature is probably about 17,000 K, while the temperatures necessary to initiate the self-sustaining thermonuclear reactions that occur in stars are in the tens of millions of degrees. To achieve such temperatures, Jupiter would require a total mass 80 times greater than it actually has. In other words, the least-massive stars must still have masses that are approximately 80 times greater than Jupiter's. Jupiter is merely a large planet and never came close to being a star.

---

path for thermal energy to escape is by convection outward through the liquid and gaseous layers. The final step in this process is seen in the convective vortices that we discussed in the previous section.

With energy continually escaping from the interiors of the giant planets, it is easy to wonder how they have maintained their high internal temperatures over the past $4\frac{1}{2}$ billion years. In Excursions 8.2, we find that the giant planets are still converting the gravitational potential energy of their creation into thermal energy. This continuous production of thermal energy is sufficient to replace the energy that is escaping from their interiors. In contracting ever so slightly each year, they are continuing the process of their creation.

8.7 THE INTERIORS OF THE GIANT PLANETS ARE HOT AND DENSE

At depths of a few thousand kilometers, the atmospheric gases of Jupiter and Saturn are so compressed by the weight of the overlying atmosphere that they liquefy. This transition from a gas to a liquid is so subtle as to be hardly noticeable. To put it another way, the *physical* difference between a liquid and a highly com-

Hydrogen-helium oceans lie at depths of a few thousand kilometers on Jupiter and Saturn.

pressed, very dense gas is something that could be appreciated only by—well—a physicist. Thus, unlike the well-defined surface between Earth's atmosphere and its oceans, on Jupiter and Saturn there is no clear boundary between the atmosphere and the "ocean" of liquid hydrogen and helium that lies below. The depths of these hydrogen-helium oceans are measured in tens of thousands of kilometers, making them the largest structures within the interiors of any of the giant planets. Uranus and Neptune, as we will see, do not have these oceans of liquid hydrogen and helium.

Figure 8.12 shows the interior structure of the giant planets. At the center of each of the giant planets is a dense, liquid core consisting of a very hot mixture of heavier materials such as water, rock, and metals. Here temperatures may be in the tens of thousands of degrees, with pressures of tens of megabars. As we learned in Chapter 7, a pressure of 1 bar corresponds closely to Earth's atmospheric pressure at sea level. A megabar is a million times as great as 1 bar. For comparison, when the submersible research vessel *Alvin* cruises Earth's ocean bottoms 10,000 meters below the surface, it is experiencing a surrounding pressure of about 1,000 bars, or 1/1,000 of a megabar. It may seem strange to be talking about water that is still liquid at temperatures of tens of thousands of degrees, but there is really nothing peculiar about this. Like a super pressure cooker, the extremely high pressures at the centers of the giant planets prevent the water from turning to

Jupiter's core is liquid water and rock at a temperature of about 17,000 K.

steam. The temperature at Jupiter's center is thought to be as high as 17,000 K, and the pressure may reach 70 megabars. Central temperatures and pressures of the other, less massive giant planets are correspondingly lower than those of Jupiter.

URANUS AND NEPTUNE ARE DIFFERENT FROM JUPITER AND SATURN

You might suppose that the average densities of the giant planets would tell us how much heavy material they contain. In practice, it is not so simple. In each planet, the core mass is roughly the same, perhaps 10 $M_\oplus$ for Jupiter and Saturn and less for Uranus and Neptune. As Table 8.1 shows, Jupiter and Saturn have total masses of 318 and 95 $M_\oplus$, respectively. The heavy materials in their 10 $M_\oplus$ cores, then, contribute very little to their average chemical composition. This means we can think of both Jupiter and Saturn as having approximately the same composition as the Sun and the rest of the Universe: about 98% hydrogen and helium and only 2% of everything else. Why then, with nearly identical compositions, should Jupiter's density be nearly twice as great as Saturn's? In Excursions 8.3 we find that the internal pressure created by Jupiter's much greater mass compresses its hydrogen and helium and its core to a higher average density than in the core of Saturn. What does this imply about Uranus and Neptune?

Figure 8.12 *The central cores and outer liquid shells of the interiors of the giant planets. Even though each core has roughly the same mass (10 Earth masses), only Jupiter and Saturn have significant amounts of the molecular and metallic forms of liquid hydrogen surrounding their cores.*

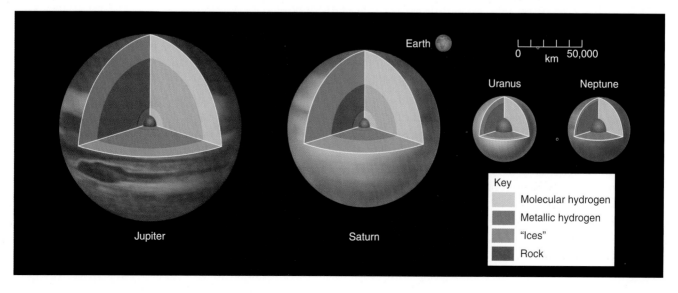

CONNECTIONS 8.1

DIFFERENTIATION IN THE GAS GIANTS

In Chapter 6 we learned that denser materials, such as iron and other metals, sank to the centers of the terrestrial planets when they were in an earlier, more molten state. This process, which we call *differentiation,* deposited most of the metals in what became the cores of the terrestrial planets, leaving their mantle and crust relatively depleted in metals.

The cores of the giant planets did not form in the same manner, however. Most of the material now in their cores was already there in the original bodies that captured hydrogen and helium from the Sun's protoplanetary disk to form what ultimately became giant planets. Differentiation, however, has occurred and is still occurring in Saturn, and perhaps Jupiter as well. In Excursions 8.2 and 8.3 we find that helium tends to condense out of the hydrogen-helium oceans on Saturn. Because these droplets of helium are denser than the hydrogen-helium liquid in which they condense, they sink toward the center of the planet. This tends to enrich helium concentration in the core while depleting it in the upper layers. At the same time, the process heats the planet by converting gravitational potential energy to thermal energy.

If Uranus and Neptune were also of solar composition—that is, if they were made primarily of hydrogen and helium—their average densities would be less than half the density of water, even less than that of Saturn. This is because their lower mass would not be as effective as Saturn's in compressing their hydrogen and helium. But such low average densities are not what we observe. Instead, we find Uranus and Neptune to be about twice as dense as Saturn (see Table 8.1). We now have a clear indication that, unlike Jupiter and Saturn, denser materials must dominate the chemical composition of Uranus and Neptune. Is this denser material water or rock? Our observations should be able to tell us. Neptune, the densest of the giant planets, is about $1\frac{1}{2}$ times as dense as uncompressed water and only about half as dense as uncompressed rock. This alone tells us there must be more water than rock. But we must also keep in mind that the high pressures within the interiors of Uranus and Neptune cause both water and rock to be more dense than in their uncompressed form. Thus water and other low-density ices, such as ammonia and methane, must be the major compositional components of Uranus and Neptune, with lesser amounts of silicates and metals. The total amount of hydrogen and helium in these planets is probably limited to no more than 1 or 2 $M_\oplus$, with most of it residing in their relatively shallow atmospheres. Based on density alone, as we suggested at the beginning of this chapter, neither Uranus nor Neptune very well fit the description of a "gas giant." It is more appropriate to refer to them as "ice giants."

> **Unlike Jupiter and Saturn, the composition of Uranus and Neptune is far from solar.**

The water that makes up so much of Uranus and Neptune is probably in the form of very deep oceans. Dissolved gases and salts would make these oceans electrically conducting. All of the giant planets have deep layers of a conducting fluid—metallic hydrogen in the case of Jupiter and Saturn, and a saltwater brine in the case of Uranus and Neptune. Currents in these conducting layers are very likely the source of the giant planets' intense magnetic fields.

> **Uranus and Neptune are "ice giants," with deep, salty oceans.**

DIFFERENCES ARE CLUES TO ORIGINS

That each of the giant planets formed around cores of similar mass but turned out very differently is an important clue to their origins. Why do Jupiter and Saturn have so much hydrogen and helium compared with Uranus and Neptune? Why is hydrogen-rich Jupiter so much more massive than hydrogen-rich Saturn? The answers may lie both in the time that it took for these planets to form and in the distribution of material from which they formed. We think that all of the hydrogen and helium in the giant planets was captured from the protoplanetary disk by the strong gravitational attraction of their massive cores. The much lower hydrogen-helium content of Uranus and Neptune suggests that these cores formed much later than those of Jupiter and Saturn, at a time when most of the gas in the protoplanetary disk had been blown away by the emerging Sun. Why should the cores of Uranus and Neptune have formed so late? Probably because the icy planetesimals from which they formed were more widely dispersed at their greater distances from the Sun. With more space between planetesimals, it would take longer to build up their cores. Saturn may have captured less gas than Jupiter, both because its core formed somewhat later and because less gas was available at its greater distance from the Sun.

8.8 THE GIANT PLANETS ARE MAGNETIC POWERHOUSES

All of the giant planets have magnetic fields that are much stronger than Earth's. Planetary magnetic fields are produced by the motions of electrically conducting liquids deep within planetary interiors. In Jupiter and

EXCURSIONS 8.3

STRANGE BEHAVIOR IN THE REALM OF ULTRAHIGH PRESSURE

In the interiors of the giant planets, the transition from a gas to a liquid state for hydrogen and helium is so gradual that there is no well-defined boundary between the two states—as there is, say, between Earth's atmosphere and oceans. When the interior pressure climbs to 4 megabars and the temperature reaches 10,000 K, hydrogen molecules are battered so violently that their electrons are stripped free and the hydrogen becomes electrically conducting like a liquid metal. This happens at a depth of about 20,000 km in Jupiter's atmosphere and 30,000 km in Saturn's. Uranus and Neptune are less massive, have lower interior pressures, and contain a smaller fraction of hydrogen—conditions that do not favor the formation of metallic hydrogen. Thus their interiors probably contain only a small amount of liquid hydrogen, with little or none of it in the metallic state.

Helium can also be compressed to a liquid, but it does not reach a metallic state under the physical conditions existing in the interiors of the giant planets. At the temperatures found in Jupiter's interior, the liquid helium is mostly dissolved together with the liquid hydrogen. Within Saturn's interior, temperatures are lower, making the helium less soluble. Those who cook know that you can easily dissolve large amounts of sugar in hot water, but relatively little when the water is cold. So it is with helium and hydrogen. Physicists believe that some of the helium in Saturn's interior is separating out into tiny droplets, like water from oil, and sinking slowly toward the planet's center. This could explain an observed depletion of helium relative to hydrogen in Saturn's upper atmosphere. And through the release of gravitational potential energy as the helium sinks,

the separation process could provide an additional source of internal energy that would not be available in the interiors of other giant planets.

Jupiter's average density is nearly twice as great as Saturn's, even though both have approximately the same composition and are nearly the same size. This is because Jupiter's self-gravity due to its greater mass compresses its material more than Saturn's can. For planets the size of Jupiter and Saturn, adding additional hydrogen does not make them much bigger; it makes them denser instead. We now find ourselves with a curious situation that goes against our intuition: If we were to add much more hydrogen to Jupiter, its diameter would actually get *smaller* rather than larger. A planet with a mass, say, 10 times greater than Jupiter would be about 10% smaller. The additional overlying mass creates higher internal pressures, which in turn compresses the interior still further. The decreased volume caused by increased pressure more than makes up for the increased volume of the additional hydrogen. It turns out that, by chance, Jupiter is almost the largest that any planet can be in this or any other solar system. Planets that are either less massive or more massive will be smaller! Does this mean that stars should also be smaller than Jupiter? They would be, except for one important difference. Stars have nuclear reactions going on within their cores, and thus have very much higher internal temperatures than any planet. These ultrahigh temperatures create enormous internal pressures that better resist gravitational compression. This keeps even the smallest "normal" stars much larger than the largest planets. We say "normal" stars, because later, in Chapters 15 and 16, we will introduce you to dying stars that are even smaller than Earth.

Saturn, magnetic fields are generated in deeply buried layers of metallic hydrogen. In Uranus and Neptune, magnetic fields arise in salt-brine oceans. While their origins are complex, we can schematically illustrate the geometry of the magnetic fields of the giant planets as if they came from bar magnets in the interiors of the planets, as shown in **Figure 8.13.**

Jupiter's magnetic field is inclined 10° to its rotation axis, an orientation similar to Earth's but displaced about a tenth of a radius from the planet's center. Note that the direction of Jupiter's magnetic field is opposite to that of Earth, as defined by where the north end of a compass would point. The total strength of Jupiter's

magnetic field is nearly 20,000 times that of Earth's. On the other hand, Jupiter is huge compared to Earth. By the time Jupiter's magnetic field emerges from the cloud tops, the field has dropped to about 4.3 gauss, only 15 times Earth's surface field.

> **Jupiter's magnetic field is 20,000 times as strong as Earth's.**

The bar magnet used to approximate Saturn's magnetic field is located almost precisely at the center of Saturn and is almost perfectly aligned with the planet's rotation axis. Saturn's magnetic field is much weaker than Jupiter's, but overall it is still more than 500 times stronger than Earth's. Because Saturn's diameter is

Figure 8.13 *The magnetic fields of the giant planets can be approximated by the fields from bar magnets offset and tilted with respect to the planets' axes. Compare these with Earth's magnetic field, shown in Figure 6.12(a).*

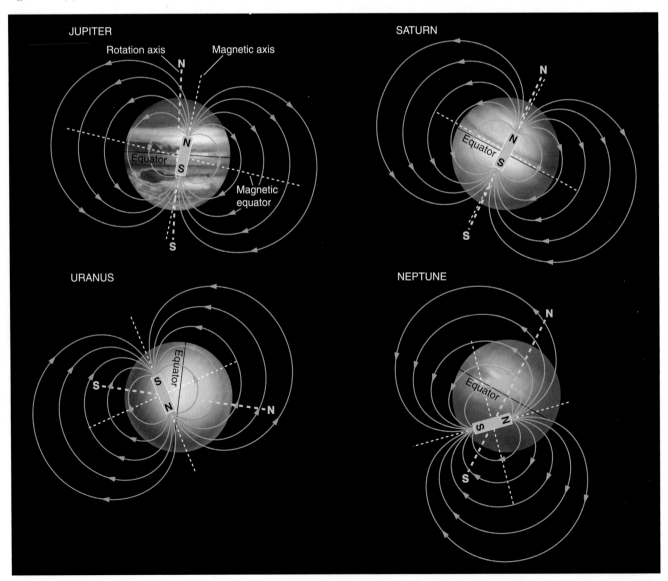

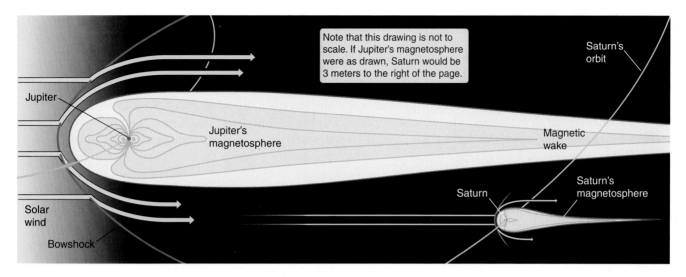

Figure 8.14 *The solar wind compresses Jupiter's (or any) magnetosphere in the sunward direction and draws it out into a magnetic wake in the direction away from the Sun. Jupiter's wake stretches beyond the orbit of Saturn.*

much greater than Earth's, the magnetic field strength at its cloud tops is about 0.5 gauss, very similar to the strength at the Earth's surface. As on Jupiter, a compass would point south on Saturn.

Voyager 2 found that the magnetic field of Uranus is inclined nearly 60° to its rotation axis and its center is displaced by a third of a radius from its center (see Figure 8.13). Considering the strange spin orientation of Uranus, this did not come as any great surprise to the Voyager scientists. The total strength of the field averages 50 times Earth's field, but the large displacement of Uranus's field from the planet's center causes the field at the cloud tops to vary between 0.1 and 1.1 gauss.

Uranus and Neptune have highly inclined magnetic fields.

The really big surprise came when *Voyager 2* reached Neptune. The orientation of Neptune's rotation axis is similar to that of Earth, Mars, and Saturn. But Neptune's magnetic field is inclined 47° to its rotation axis, and the center of *this* magnetic field is displaced from the planet's center by more than half the radius, an offset even greater than that of Uranus (Figure 8.13). The overall strength of Neptune's magnetic field is only half as great as that of Uranus.

The reason for the unusual geometry of the magnetic fields of Uranus and Neptune remains unknown, but it is clearly not related to the orientations of their rotation axes. The displacement of the field is primarily toward Neptune's southern hemisphere, thereby creating a large asymmetry in the field strength at the cloud tops, with 1.2 gauss in the southern hemisphere and only 0.06 gauss in the north.

GIANT PLANETS HAVE GIANT MAGNETOSPHERES

Just as Earth's magnetic field traps energetic charged particles to form Earth's magnetosphere, the magnetic fields of the giant planets also trap energetic particles to form magnetospheres of their own. While Earth's magnetosphere may have a radius over 10 times that of the planet itself, our magnetosphere is tiny in comparison with the vast clouds of plasma held together by the much more powerful magnetic fields of the giant planets. By far the most colossal of these is Jupiter's magnetosphere. Its radius is as much as 100 times that of the planet itself, or around 7 million kilometers. That is roughly 10 times the radius of the Sun! While the magnetospheres of the other giant planets are much

The magnetospheres of the giant planets are enormous.

smaller, even the relatively weak magnetic fields of Uranus and Neptune form magnetospheres that are comparable in size to the Sun.

The solar wind does more than supply some of the particles for a magnetosphere. The pressure of the solar wind also pushes on and compresses a magnetosphere. The size and shape of a planet's magnetosphere can change a great deal depending on how the solar wind is blowing at a particular time. Planetary magnetic fields also divert the solar wind, which flows around magnetospheres the way a stream flows around boulders. Just as a rock in a river creates a wake that extends downstream, the magnetosphere of a planet produces a wake that can extend downstream for great distances. The wake of Jupiter's magnetosphere **(Figure 8.14)** ex-

FOUNDATIONS 8.1

SYNCHROTRON EMISSION—FROM PLANETS TO QUASARS

In Chapter 4 we found that any time a charged particle experiences an acceleration, the particle emits electromagnetic radiation. The accelerations resulting from forces exerted by magnetic fields are no exception. As we have seen, charged particles moving in a magnetic field experience a force that is perpendicular both to the motion of the particle and to the direction of the magnetic field. This force produces an acceleration, causing the particles to follow helical paths around the direction of the magnetic field. The accelerations also cause the particles to radiate.

If the particles are traveling at a significant fraction of the speed of light, then relativistic effects cause the radiation they emit to be beamed in the direction in which they are traveling. The situation is illustrated in **Figure 8.15.** The resulting radiation is called **synchrotron radiation,** so named because it was first discovered in a type of particle accelerator called a synchrotron.

The amount of radiation from a particle depends on the amount of acceleration the particle experiences. For a given amount of force, a less massive particle experiences more acceleration. Since electrons are much less massive than protons or any ion, it is

the electrons in a magnetized plasma that experience the greatest accelerations. Combining these ideas, we see that it must be the electrons in a magnetized plasma that produce the overwhelming majority of its synchrotron radiation.

The magnetospheres of the giant planets contain energetic electrons moving in strong magnetic fields, and so satisfy the requirements for synchrotron radiation. Synchrotron radiation is unlike thermal (that is, Planck) radiation because instead of being strongly peaked in one part of the electromagnetic spectrum, synchrotron radiation from a single source can range from radio waves to X-rays. The spectrum of synchrotron radiation is determined by the strength of the magnetic field and how energetic the radiating particles are. Synchrotron emission from planetary magnetospheres is concentrated in the low-energy radio part of the spectrum. This is our first encounter with synchrotron radiation, but it will not be our last. As we move outward into the Universe we will find many objects, from quasars to the remnants of supernovae, that emit synchrotron radiation throughout the electromagnetic spectrum.

---

tends for over 6 AU outward from the planet, well past the orbit of Saturn. Jupiter's magnetosphere is the largest permanent "object" in the Solar System, surpassed in size only by the tail of an occasional comet. The magnetic wakes of Uranus and Neptune have a curious structure. Because of the tilt and the large displacements of their magnetic fields from the centers of these planets, their magnetospheres wobble as the planets rotate. This causes the wakes of their magnetospheres to twist like corkscrews as they stretch away from the planets.

MAGNETOSPHERES PRODUCE SYNCHROTRON RADIATION

We need not send spacecraft to the outer Solar System, or even call on telescopes orbiting Earth, to see evidence of the giant planets' magnetospheres. Rapidly moving electrons in planetary magnetospheres spiral around the direction of the magnetic field, and as they do so they emit synchrotron radiation, as discussed in Foundations 8.1. If our eyes were sensitive to radio

waves, then the second brightest object in the sky would be Jupiter's magnetosphere. The Sun would still be brighter, but it would not appear larger: even at a distance from Earth of 4.2 to 6.2 AU, Jupiter's magnetosphere would still appear roughly twice the size of the Sun in the sky. Saturn's magnetosphere would also be large enough to see, but would be much fainter than Jupiter's. Even though Saturn has a strong magnetic field, pieces of rock, ice, and dust in Saturn's spectacular rings act like sponges, soaking up magnetospheric particles. Magnetospheric particles typically collide with ring material soon after those particles enter the magnetosphere. With far fewer magnetospheric electrons, there is much less radio emission from Saturn.

We can learn a great deal about planets by studying the synchrotron emission from their magnetospheres. For example, precise measurement of periodic variations in the radio signals "broadcast" by the giant

Radio signals broadcast a planet's true rotation period.

planets tells us the planets' true rotation periods. This is because the magnetic field of each planet is locked to the conducting liquid layers deep within the planet's interior, and so the magnetic field rotates with exactly the same period as the deep interior of the planet. Given the fast and highly variable winds that push around clouds in the atmospheres of the giant planets, measurement of radio emission is the *only* way we have to determine the true rotation periods of the giant planets.

RADIATION BELTS AND AURORAE

As a planet rotates on its axis, it drags its magnetosphere around with it. In the enormous magnetosphere of a rapidly rotating planet like Jupiter, charged particles are swept around at very high speeds. These fast-moving charged particles slam into neutral atoms (which do not share the motion of material in the magnetosphere), and the energy released in the resulting high-speed collisions heats the plasma to very high temperatures. In 1979, while passing through Jupiter's magnetosphere, *Voyager 1* encountered a region of tenuous plasma with a temperature of over 300 million kelvins, 20 times the temperature at the center of the Sun! The density of the plasma (around 10,000 atoms/m^3) was much lower than that of the best vacuum we can produce on Earth, however, so the spacecraft was in no danger. The high temperatures are impressive nonetheless.

Charged particles trapped in planetary magnetospheres are concentrated in certain regions, called **radiation belts.** While Earth's radiation belts are enough for astronauts to worry about, the radiation belts that surround Jupiter are searing in comparison. In 1973 the *Pioneer 11* spacecraft passed through the radiation belts of Jupiter. During its brief encounter, *Pioneer 11* picked up a radiation dose of 400,000 rads, or around 1,000 times the lethal dose for humans. Several of the instruments on board were permanently damaged as a result, and the spacecraft itself barely survived to continue on its journey to Saturn.

Jupiter has intense radiation belts.

In addition to protons and electrons from the solar wind, the magnetospheres of the giant planets also contain large amounts of other elements including sodium, sulfur, oxygen, nitrogen, and carbon. These elements come from several sources, including the planets' atmospheres and the moons that orbit within them. The most intense radiation belt in the Solar System is a toroidal (that is, torus- or doughnut-shaped) ring of plasma associated with Io, the innermost of Jupiter's four Galilean moons. Io is the most volcanically active body in the Solar System. Because of its low surface gravity and the violence of its volcanism, some of the gases erupting from its interior are able to escape the moon and become part of Jupiter's radiation belt. As charged particles are slammed into the moon by the rotation of Jupiter's magnetosphere, even more material is sputtered into space. If the dislodged atoms are electrically neutral, they will continue to orbit the planet in close to the same orbit as the moon from which they escaped. Images of the region around Jupiter, taken in the light of emission lines from atoms of sulfur or sodium, show a faintly glowing torus of plasma supplied by the moon **(Figure 8.16)**.

Io is a source of magnetospheric particles.

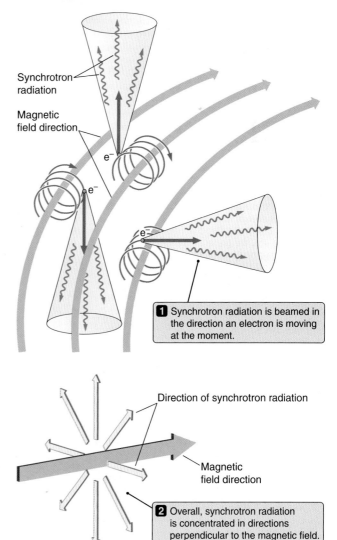

Figure 8.15 *A rapidly moving charged particle loops around the direction of a magnetic field, and gives off electromagnetic radiation as a result of the acceleration it experiences. The radiation, called synchrotron radiation, is beamed by relativistic effects in the direction of particle motion.*

Synchrotron radiation

Magnetic field direction

e$^-$

e$^-$

e$^-$

1 Synchrotron radiation is beamed in the direction an electron is moving at the moment.

Direction of synchrotron radiation

Magnetic field direction

2 Overall, synchrotron radiation is concentrated in directions perpendicular to the magnetic field.

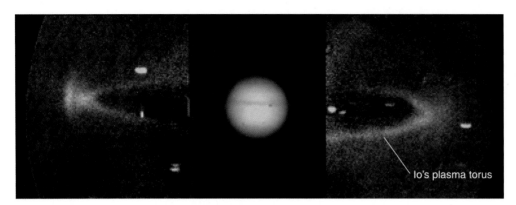

Figure 8.16 *Atoms from Io, the innermost of Jupiter's Galilean moons, fill a faintly glowing torus of plasma that surrounds the planet.*

Io's plasma torus

Other moons also influence the magnetospheres of the planets they orbit. The atmosphere of Saturn's largest moon, Titan, is rich in nitrogen. Leakage of this gas into space is the major source of a plasma torus that forms in Titan's wake. The density of this rather remote radiation belt is highly variable because the orbit of Titan is sometimes within and sometimes outside Saturn's magnetosphere, depending on the strength of the solar wind. When Titan is outside Saturn's magnetosphere, any nitrogen molecules lost from the moon's atmosphere are carried away by the solar wind.

Charged particles spiral along the magnetic field lines of the giant planets, bouncing back and forth be-

tween the two magnetic poles, just as they do around Earth. As with Earth, these energetic particles collide with atoms and molecules in a planet's atmosphere, knocking them into excited energy states that decay and emit radiation. The results are bright auroral rings **(Figure 8.17)** that surround the poles of the giant

> **Bright aurorae ring the magnetic poles of the giant planets.**

planets, just as the *aurora borealis* and *aurora australis* ring the north and south magnetic poles of Earth.

Jupiter's aurorae have an added twist that we do not see on Earth, however. As Jupiter's magnetic field sweeps past Io, it behaves like a dynamo, generating an electric potential of 400,000 volts. Electrons, accelerated by this

Figure 8.17 Hubble Space Telescope *images of auroral rings around the poles of* (a) *Jupiter and* (b) *Saturn. These images were taken in ultraviolet light, at which wavelength haze in the atmospheres of the planets obscures the view of structure in the cloud layers. The bright spot and trail outside the main ring of Jupiter's aurorae are the footprint and wake of Io's flux tube.*

1 Auroral rings appear where the planet's magnetic field channels energetic particles into the planet's atmosphere near the poles.

2 Flux tubes associated with moons can form auroral footprints that move through a planet's atmosphere.

(a) JUPITER

Footprint of Io's flux tube

(b) SATURN

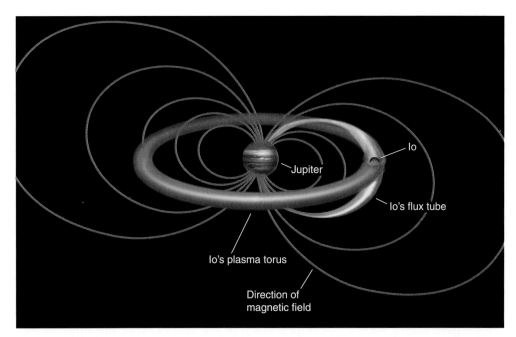

Figure 8.18 *A diagram showing the geometry of Io's plasma torus and flux tube.*

enormous electric field, spiral along the direction of Jupiter's magnetic field. The result is a magnetic channel, called a *flux tube,* that connects Io with Jupiter's atmosphere near the planet's magnetic poles **(Figure 8.18).** Io's flux tube carries an electric current of 5 million amperes, amounting to 2 trillion watts of power, or about $\frac{1}{6}$ of the total power produced by generating stations on Earth. Much of the power generated within the flux tube is radiated away as radio energy. These radio signals are received at Earth as intense bursts of synchrotron radiation. However, a substantial fraction of the energy of particles in the flux tube is deposited into Jupiter's atmosphere. At the location where Io's flux tube intercepts Jupiter's atmosphere, there is a spot of intense auroral activity. As Jupiter rotates, this spot leaves an auroral trail behind in Jupiter's atmosphere. The footprint of Io's plasma torus, along with its wake, can be seen outside the main auroral ring in Figure 8.17(a).

SEEING THE FOREST THROUGH THE TREES

In our discussion of the terrestrial planets a major theme was diversity. We used Earth as the basis for understanding our planetary neighbors, and at the same time learned from those neighbors what it is about Earth that sets it apart. On this leg of our journey, we have come to understand that even the range of conditions from the frozen poles of Mars to the hellish surface of Venus are modest in comparison with what is possible. As the Solar System formed, planets in the cold outer reaches grew more massive than the planets closer to the Sun. They even became massive enough to capture and hold on to hydrogen and helium gas that was so abundant in the disk surrounding the young Sun.

As we explored the giant planets, we discovered worlds with scales beyond human experience. In place of Earth's gentle trade winds, we found vast atmospheric bands streaming around the giant planets at up to 1,600 km/h. The most powerful terrestrial hurricanes would be but inconsequential eddies within Jupiter's Great Red Spot, a storm so enormous that it could swallow two Earths whole. On Earth we marvel at the sight of a terrestrial thunderstorm towering over-

head, lit up by powerful flashes of lightening. Can we begin to imagine the appearance of multihued clouds of water, ammonia, methane, and other compounds, tinted by sulfur, phosphorus, and photochemical smog, as they roil up from within the depths of a giant planet? Can we envision the bolts of lightning that sear through those clouds?

Of course, these analogs of our terrestrial experience only scratch the surface of the differences between the giant planets and Earth. In the case of the terrestrial planets the distinction between planet and atmosphere is clear-cut. On our Earth, clouds billow in a blue sky, adrift over the palpable surface of land and ocean. As we descend into the giant planets, there is no sudden transition to solid or liquid. Instead, there is only the steady and inexorable increase in pressure, a smooth transition from the tops of the clouds to a core that in the case of Jupiter may have an absolute temperature almost three times that at the surface of the Sun, and a crushing pressure 70 million times that at the surface of Earth. The higher-than-expected temperatures of the giant planets that we found in Chapter 4 are signposts of these internal differences. While Earth's climate is defined by the interplay between energy from the Sun and the physics and chemistry of our atmosphere and oceans, the fierce weather systems on the giant planets are powered largely by conversion of gravitational energy to thermal energy within their interiors.

While the bulk of the giant planets dwarf that of their terrestrial cousins, their influence does not end here. Powerful magnetic fields trap energetic charged particles streaming outward from the Sun, leading to the formation of giant magnetospheres—the largest permanent structures in the Solar System.

The time has come for a brief digression. The systems of moons and rings that surround the giant planets are gravitational playgrounds, rich with phenomena that go far beyond Kepler's laws. Before we take the next step outward, we now pause and return our attention to our old friend, gravity.

STUDENT QUESTIONS

THINKING ABOUT THE CONCEPTS

1. In what manner was Uranus observed to "stray from its path," and how was this used as a clue by Adams and Le Verrier to predict the location of Neptune?
2. Describe the ways in which the giant planets differ from the terrestrial planets.
3. Identify and describe the two subclasses of giant planets, and indicate which planets fall into which subclass.
4. Describe what we are seeing when we look at the visible "surfaces" of Jupiter and Saturn, and explain why this is what we see.
5. Uranus and Neptune, when viewed through a telescope, are distinctly bluish-green in color. There are two reasons for their striking appearance. What are they?
6. Astronomers take the unusual position of lumping together all atomic elements other than hydrogen and helium into a single category, which they call "massive elements." Why would this be a reasonable thing to do?
7. Is Jupiter's chemical composition more like that of the Sun or that of Earth? Why?
8. Three of the giant planets radiate into space approximately twice as much energy as they absorb from the Sun. What is the source of this excess energy?
9. Discuss evidence demonstrating that the Coriolis effect is more important than solar heating in determining the global atmospheric circulation of the giant planets.
10. The Great Red Spot (GRS) is a long-lasting atmospheric vortex in Jupiter's southern hemisphere. Winds rotate around its center in a counterclockwise direction. Is the GRS cyclonic or anticyclonic? Is it a region of high or low pressure? Explain.
11. In which portions of the electromagnetic spectrum can we detect electromagnetic radiation from Jupiter? What are the sources of this radiation?

APPLYING THE CONCEPTS (Information from the Appendices might be needed to answer some of these questions.)

12. The Sun appears 400,000 times brighter than the full Moon in our sky. How far out from the Sun (in AU) would you have to go for the Sun to appear only as bright as the full Moon appears in our nighttime sky? Compare your answer with the radius of Neptune's orbit.
13. As Uranus occults a star the relative motion between Uranus and Earth is 23.0 km/s. An observer on Earth sees the star disappear for 37 minutes and 2 seconds, and notes that the center of Uranus passed directly in front of the star. Based on these observations, what value would he get for the diameter of Uranus?
14. Jupiter is an oblate planet with an average radius of 69,800 km, compared to Earth's average radius of 6,370 km. Remembering that volume is proportional to the cube of the radius, how many Earths could fit inside Jupiter?
15. Jupiter is 318 times as massive as Earth. Show that Jupiter's average density is about $\frac{1}{4}$ that of Earth's.
16. A small cloud in Jupiter's equatorial region is observed to be at 122.0° west longitude in a coordinate system that rotates at the same rate as the deep interior of the planet. (West longitude is measured along a planet's equator toward the west.) Another observation made exactly 10 Earth-hours later finds the cloud at 118.0° west longitude. What is the wind speed in km/h? Is this an easterly or westerly wind?
17. The average temperature at the cloud tops of Saturn is 95 K. The equilibrium temperature for Saturn based solely on absorbed solar radiation should be 82 K. How much energy is Saturn radiating into space compared to the amount it absorbs from the Sun?

Then there are the Tides, so useful to man. . .
We must be grateful for the Moon's existence on that account alone.

JAMES NASMYTH (1808–1890)

GRAVITY IS MORE THAN KEPLER'S LAWS

9.1 GRAVITY ONCE AGAIN

Long before Kepler wrote down the laws describing the motions of the planets about the Sun, or Newton explained and interpreted these motions in terms of the effects of gravity, our ancestors knew that the Moon and the Sun have a direct influence on Earth. Those living near the ocean were especially attuned to these effects. Twice each day—in harmony with the changing position of the Moon in the sky—they saw the seawater rise and then recede once again. They would have noted that this effect, Earth's **tides,** is greatest when the Sun and the Moon are either together in the sky or at opposite extremes (that is, during a full Moon or a new Moon), and is more subdued when the Sun and Moon are separated in the sky by 90° (first-quarter Moon or third-quarter Moon). Doubtless the unarguable association between the Moon and tides had a great deal to do with the development of superstitions about the power of the Moon over our lives.

Today we understand that tides are the result of the gravitational pull of the Moon and the Sun on Earth. More to the point, tides are the result of *differences* between how hard the Moon or Sun pulls on one part of our planet in comparison with the pull that they exert on other parts of the planet. So far in our discussion of

KEY CONCEPTS

Newton's law of universal gravitation is simplicity itself. Yet when applied to extended rather than point-like masses, or when acting among three or more objects, this simple rule gives rise to a surprising diversity of phenomena. In this chapter we look beyond Kepler's elegant laws of planetary motion and discover:

✳ Symmetries that allow us to say a great deal about the gravity of an object without actually calculating anything;

✳ That the gravity within a spherical object is determined only by the mass within a given radius;

✳ Tides on Earth resulting from the fact that gravity from the Moon and Sun pulls harder on one side of Earth than the other;

✳ Tidal interactions between planets and moons (including Earth's Moon) that lock a moon's rotation to its orbit;

✳ Comets that have been shattered by tides, and tortured moons alive with tide-powered volcanism;

✳ Orbital resonances that nudge asteroids from their orbits and sweep out gaps in planetary rings; and

✳ Chaotic, unpredictable orbits in which the tiniest difference at one point in time leads to huge differences later on.

gravity we have focused on the gravitational interaction between two whole objects. Such interactions explain the motion of the planets about the Sun, the Moon about Earth, and a space shuttle about our globe. Yet gravity is far more than Newton's derivation of Kepler's laws. While concentrating on the elliptical, parabolic, and hyperbolic orbits of one mass about another, we have overlooked many other fascinating and important manifestations of gravity. We now turn our attention to these effects.

9.2 GRAVITY DIFFERS FROM PLACE TO PLACE WITHIN AN OBJECT

A good place to begin a discussion of the additional effects of gravity is with the way an object interacts gravitationally with *itself*. You can think of Earth, for example, as a collection of small masses, each of which feels a gravitational attraction toward every other small part of Earth. This gravitational attraction between different parts of Earth is what holds our planet together.

Self-gravity holds Earth together.

In a certain sense, you are one of these small pieces of Earth. You are perhaps slightly more mobile than the average lump of clay, but as far as the gravitational makeup of our planet goes, you serve more or less the same purpose. As you sit in your chair reading this book, you are exerting a gravitational attraction on every other fragment of Earth, and every other fragment of Earth is exerting a gravitational attraction on you (see **Figure 9.1a**). Your gravitational interaction is strongest with the parts of Earth closest to you. The parts of Earth that are on the other side of our planet are much farther away from you, so their pull on you is correspondingly less.

As you know from experience, the net effect of all of these forces is to pull you toward the center of Earth. If you drop a hammer, it falls directly down toward the ground. We can understand why this is so just by thinking about the shape of Earth. Since Earth is nearly spherical, for every piece of Earth pulling you toward your right, there is a corresponding piece of Earth pulling you toward your left with just as much force. For every piece of Earth pulling you forward, there is a corresponding piece of Earth pulling you backward. Because Earth is **spherically symmetric,** all of these "sideways" forces cancel out, leaving behind an overall force that points toward Earth's center.

Understanding the size of the force is a bit trickier. Some parts of Earth are closer to you while others are farther away, but there must be some "average" or "characteristic" distance between you and each of the small fragments of Earth that is pulling on you. Not surprisingly, this average distance turns out to be the distance between you and the center of Earth. So, as illustrated in **Figure 9.1(b),** *the overall pull that you feel is the same as if all the mass of Earth were concentrated at a single point located at the very center of the planet.* This is true for any spherically symmetric object. As far as the rest of the Universe is concerned, the gravity from such an object behaves as if all of the mass of that object were concentrated at a point at its center. We have already made extensive use of this result. It was an implicit assumption in our discussion in Chapter 3 of orbits and Kepler's laws, for example, where we said that the force acting on a space shuttle in

Gravity from a sphere is like gravity from mass concentrated at the sphere's center.

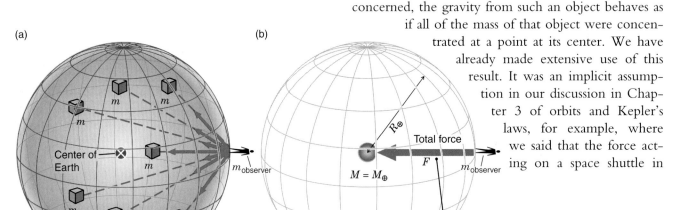

(a)

Center of Earth

m m m m m m m

m_{observer}

Total mass = $M_\oplus$

An object on the surface of a spherical mass (such as Earth) feels a gravitational attraction toward each small part of the sphere.

(b)

$R_\oplus$

Total force F

$M = M_\oplus$

m_{observer}

$$F = G\frac{M_\oplus\, m_{\text{observer}}}{R_\oplus^2}$$

The net force is the same as if we scooped up the mass of the entire sphere and concentrated it at a point at the center.

Figure 9.1 *The net gravitational force due to a spherical mass is the same as the gravitational force from the same mass concentrated at a point at the center of the sphere.*

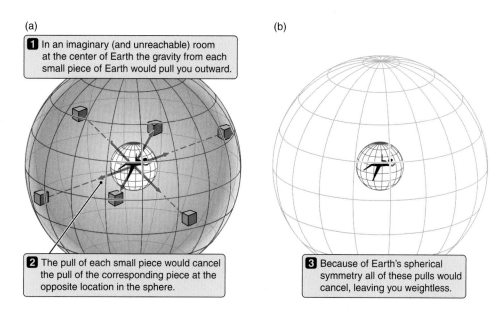

(a)

1 In an imaginary (and unreachable) room at the center of Earth the gravity from each small piece of Earth would pull you outward.

2 The pull of each small piece would cancel the pull of the corresponding piece at the opposite location in the sphere.

(b)

3 Because of Earth's spherical symmetry all of these pulls would cancel, leaving you weightless.

Figure 9.2 *In an imaginary room at the center of Earth, the gravitational pull from each small piece of Earth would cancel, leaving you weightless.*

orbit about Earth is equal to $GM_{\oplus}m_{\text{shuttle}}/r^2$, where $M_{\oplus}$ is the mass of Earth and r is the distance between the center of Earth and the space shuttle.

GRAVITY INCREASES OUTWARD FROM AN OBJECT'S CENTER

We now have a way of accounting for the net gravitational attraction of Earth on an object on or above the surface of Earth, but what about the effect of gravity *within* Earth? What is the overall gravitational attraction felt by an object buried deep inside our planet? A good place to begin is by thinking about the gravitational attraction felt by an object at the very center of Earth. Imagine that you were in a room at the center of Earth, as shown in **Figure 9.2**. (Ignore for a moment the crushing pressure or immense temperatures that would surround you!) Every small part of Earth would be exerting a slight gravitational force on you, each of which would be *outward* away from the center of Earth. Again, Earth's spherical symmetry tells us that overall these gravitational forces must sum to zero. For each part of Earth pulling you outward in one direction, there would be a corresponding piece of Earth on the other side pulling you outward in exactly the opposite direction, as shown in Figure 9.2(a). These two forces would cancel out so that the net force that you feel from these two small parts of the planet would be zero (Figure 9.2b). The same holds true for *each and every* small part of Earth. The net effect of the sum

Net gravity is zero at the center of Earth.

of all of the gravitational forces acting on an object at the center of Earth is zero! If you were in a room at the center of Earth, you would hang there, surrounded by the enormous mass of a planet, but truly weightless. (See Foundations 9.1 for more discussion about symmetry.)

So much for a room at the center of Earth. Now imagine that your room is located part way out from the center of the planet. To tackle this question we need to think of Earth as being made up of two different pieces, as shown in **Figure 9.3.** The first piece is just a spherical ball containing all of the parts of Earth that are closer to the center of Earth than your imaginary room is. You could think of this inner ball as a "planet within a planet." Your room is sitting on the surface of this imaginary sphere. From the discussion in the preceding paragraphs, we know the net effect of the gravity of this inner sphere. It is equivalent to taking all of the mass contained within that imaginary ball and placing it at the center of Earth. The force that you would feel from this inner planet is

$$F = \frac{G \times M_{\text{inner}} \times m}{r^2},$$

where M_{inner} is the mass of this inner ball (which depends on where within the planet your room is located), m is your mass, and r is the distance from the center of Earth to your room.

So far so good, but what about the gravitational attraction that you would feel from the *rest* of Earth—the shell of material that surrounds this "inner planet"? The parts of that shell that are closest to you, and so individually pull on you most strongly, are above you. The force from this part of the planet would pull you away from the center of Earth. However, *most* of the mass in the shell is on the side where its gravitational attraction pulls you *toward* the center of Earth. While this material is farther away from you—which means that each small

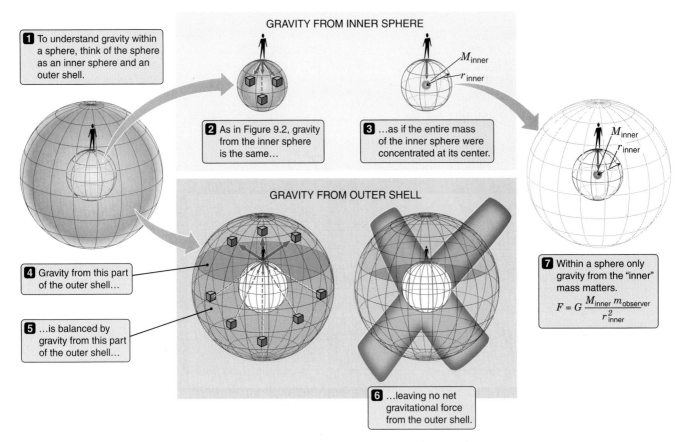

Figure 9.3 *The gravitational force felt in a room in the interior of a spherical body such as Earth is equal to the force due to the mass closer to the center of the body than the room is.*

part of it pulls on you less strongly—there is a good bit more of it than the material immediately over your head. If we were to work it out in detail, we would find that the gravitational attraction from the part of the shell pulling you away from the center of Earth is exactly canceled by the gravitational attraction of the part of the shell that is pulling you toward the center of Earth. The net effect—the overall force—is zero. If you were inside a huge spherical shell, then the overall gravitational attraction from the material in that shell is zero. (This is just a more general case of the "room at the center of Earth" problem.)

Let's put this all together. Regardless of where we are within Earth, *we only feel a net gravitational force from the part of the planet that lies deeper within the planet than we are, and that mass acts as if it were all concentrated at the center of Earth.* If we want to know what the net gravitational attraction is at any point within a spherical object, all we need to do is determine how much mass is

Only mass closer to the center of a sphere exerts a net gravitational force.

closer to the center of the sphere than our point of interest. This mass then acts as if it were concentrated at the center of the sphere.

This is a general and very important result. We have already seen something of this effect in our discussion of the interior of planets. This result will become even more important as we consider the forces responsible for the structure of stars.

TIDES ARE DUE TO DIFFERENCES IN GRAVITY FROM EXTERNAL OBJECTS

We have seen that each small part of an object feels a gravitational attraction toward every other small part of the object, and that this **self-gravity** differs from place to place. It is also the case that each small part of an object feels a gravitational attraction toward every other mass in the Universe, and that these *external* forces differ from place to place within the object as well.

The most notable local example of this involves the effect of the Moon's gravity on Earth. Overall, the Moon's gravity pulls on Earth as if the mass of Earth were concentrated at the planet's center, just as we saw in the previous section. When astronomers calculate the orbits of Earth and the Moon assuming that the mass of each is concentrated at the center point of each body, they get the right answer. Yet there is more going on than this. The side of Earth that faces the Moon is *closer* to the Moon than the rest of Earth, and so feels a stronger-than-average gravitational attraction toward the Moon. In contrast, the side of Earth facing away from the Moon is *farther* than average from the Moon, and so feels a weaker-than-average attraction toward the Moon. Putting in numbers, we find that the pull on the near side of Earth is about 7% larger than the pull on the far side of Earth.

The Moon's gravity pulls harder on the side of Earth facing the Moon.

Imagine holding three rocks at different altitudes high above the surface of the Moon, as shown in **Figure 9.4(a).** If you let them go at the same time, they will all fall toward the Moon. However, rock 1, located closest to the Moon, will fall faster than rocks 2 and 3. Rock 3, located farthest from the Moon, will fall more slowly than either of the other two rocks. As the rocks fall toward the Moon, the separation between them gets larger and larger. A person falling along with rock 2 would see both rocks 1 and 3 moving away from him. Now suppose the three rocks are connected by springs, as in **Figure 9.4(b).** As the rocks fall toward the Moon, the differences in the gravitational forces that they feel cause the springs to be stretched. To someone falling along with rock 2 it would seem as if there were forces pulling rocks 1 and 3 in opposite directions. Exactly the same thing happens with Earth. If we replace the three rocks with different parts of Earth, as shown in **Figure 9.4(c),** we see that differences in the Moon's gravitational attraction on different parts of Earth try to stretch Earth out along a line pointing toward the Moon.

This difference in the Moon's gravity stretches Earth.

We can also think about this situation by applying the idea of relative motion from our discussions of Newton's laws (Chapter 3) and special relativity (Chapter 4). Earth as a whole is constantly falling toward the Moon.

FOUNDATIONS 9.1

THE POWER OF SYMMETRY

In our discussion of gravity we have managed to arrive at some very interesting results based on nothing but the way one part of an object matches up with another—a property called **symmetry.** We were able to say that the overall gravitational attraction of all parts of Earth must point along a line connecting us with the center of Earth, simply by noting that each sideways pull from a different part of Earth is balanced by a corresponding pull in the other direction. We could make this claim because the distribution of mass to our right as we stand on Earth is just the same as the distribution of mass to our left, which is just the same as the distribution of mass to our front or back. That is to say, Earth is *symmetric.* Similarly, we were able to argue that the net effect of gravity on an object at the center of Earth must be zero, because every small force from one part of Earth is balanced by a corresponding force from the corresponding part of Earth on the opposite side of us. Again, no calculation was needed—only an appeal to the symmetry of the distribution of mass within Earth.

If you look around, you will find symmetry everywhere, and when you find it, you will often find a clue to why something works the way it does.

The tire on a bicycle rolls because it has *circular symmetry,* meaning that its shape is the same regardless of how it rotates about the axis running through its center. Your image in a mirror is flipped right for left and left for right from your actual appearance—an example of *reflection symmetry.* You hardly notice the difference, however, because your body itself is nearly symmetrical between its right and left sides. A child's cubical blocks can be stacked any of six ways because when you rotate them by 90° they still look the same—a special type of *rotational symmetry.* A soccer ball can be kicked in any direction at any time, and will roll any which way—both results of its *spherical symmetry.* An American football makes for a lousy game of soccer but a fantastic spiraling forward pass because of its rotational symmetry about a single axis.

Science progresses by finding and making use of patterns that exist in the Universe. Symmetry turns out to be an especially elegant and powerful type of pattern. It is often true that physicists and astronomers are able to learn a great deal about the properties and behavior of an object or system without making a single calculation, simply by understanding the symmetry of the object or system that they are studying.

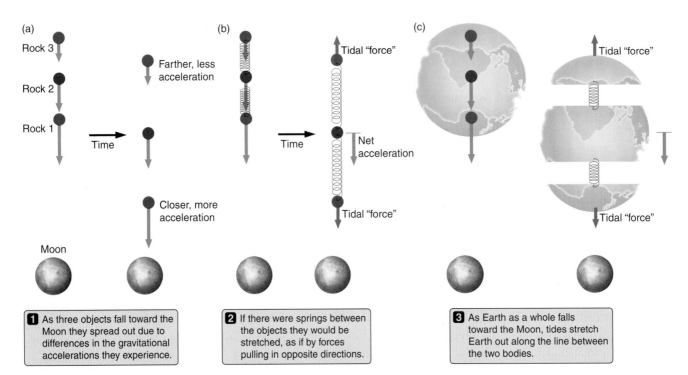

Figure 9.4 (a) *As three objects fall toward the Moon, they experience different gravitational accelerations.* (b) *If the three objects were connected by springs, the springs would be stretched as if there were forces pulling outward on each end of the chain.* (c) *This is the cause of Earth's tides.*

The acceleration that Earth experiences is not very great—only about 3.3×10^{-5} m/s^2. The gravitational acceleration that we feel toward the center of Earth is 300,000 times greater. Yet the gravitational pull of the Moon on Earth is a substantial influence on our planet, causing it to wobble more than 9,300 km back and forth over the course of a month. Despite this fact, we do not personally perceive any sensation from the gravitational attraction of the Moon on Earth. Just as every-

thing in a traveling car shares the motion of the car, everything on Earth falls *together* toward the Moon.

What is *not* exactly the same everywhere on Earth are the "leftover" accelerations—the differences between the gravitational acceleration due to the Moon at any given location and the average gravitational acceleration acting on Earth as a whole. **Figure 9.5** shows the effect of these leftover accelerations. On the side of Earth closer to the Moon, the actual accelera-

Figure 9.5 *Tides stretch Earth along the line between the Earth and the Moon, but compress the Earth perpendicular to this line.*

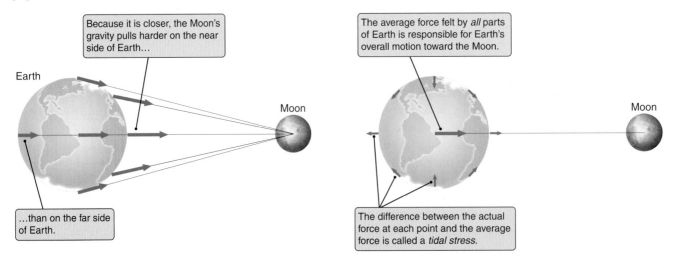

tion is greater than the average acceleration. The result is that 1 kilogram of material on the side of Earth closer to the Moon is pulled *toward* the Moon with a force of 1.1×10^{-6} newtons *relative to Earth as a whole.* On the other hand, on the side of Earth away from the Moon, the actual force is *less than* the average force. The difference between the actual force and the average force points *away from the Moon!* Figure 9.5 shows that there is also a net force squeezing inward on Earth in the direction perpendicular to the line between Earth and the Moon. Together the stretching of tides along the line between Earth and the Moon and the squeezing of the tidal forces perpendicular to this line distort the shape of Earth like a rubber ball caught in the middle of a tug-of-war.

Tides stretch out Earth in one direction and squeeze it in the other.

Out of convenience we speak of **tidal stresses** as if there is actually a force pulling the back side of Earth away from Earth's center. However, it is important to remember that the Moon is not pushing the back side of Earth away. Rather, it simply is not pulling on the back side of Earth as hard as it is on the planet as a whole. The back side of Earth is "left behind" as the rest of the planet falls more rapidly toward the Moon.

EARTH'S OCEANS FLOW IN RESPONSE TO TIDAL FORCES

The tidal acceleration of a mass on either the near side or the far side of Earth is equal to about a ten-millionth of the acceleration due to Earth's gravity. This may not seem like much, until we consider the enormous masses involved. Since $F = ma$, a really huge m times a small a results in a large force. In addition, the liquid water covering the majority of Earth's surface is free to move in response to tidal forces. In the idealized case—where the surface of Earth is perfectly smooth and covered with a uniform ocean, and where Earth does not rotate—the **lunar tides** (tides on Earth due to the gravitational pull of the Moon) would pull our oceans into an elongated

Earth's oceans have a tidal bulge.

tidal bulge like that in **Figure 9.6(a).** The water would be at its deepest on the side toward the Moon and on the side away from the Moon, and at its shallowest at the points midway in between. Of course, our Earth is *not* a perfectly smooth, nonrotating body covered with perfectly uniform oceans, and there are many effects which complicate this simple picture. One complicating effect is Earth's rotation. As a point on Earth rotates through the ocean's tidal bulges, that point experiences the ebb and flow of the tides. In addition,

Rotation drags Earth's tidal bulge around.

friction between the spinning Earth and its tidal bulge drags the oceanic tidal bulge around in the direction of Earth's rotation, as shown in **Figure 9.6(b).**

Follow along in **Figure 9.6(c)** as we take a ride through the course of a day. We begin as the rotating Earth carries us through the tidal bulge on the Moon-ward side of the planet. Because the tidal bulge is not

Figure 9.6 *Tidal stresses pull Earth and its oceans into a tidal bulge. Earth's rotation pulls its tidal bulge slightly out of alignment with the Moon. As the Earth's rotation carries us through these bulges, we experience the well-known ocean tides.*

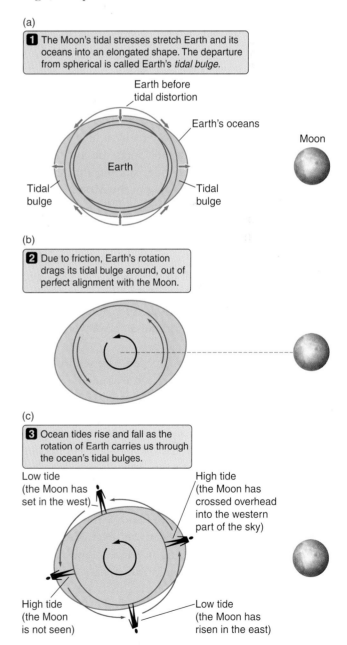

(a)

1 The Moon's tidal stresses stretch Earth and its oceans into an elongated shape. The departure from spherical is called Earth's *tidal bulge.*

Earth before tidal distortion

Earth's oceans

Moon

Earth

Tidal bulge

Tidal bulge

(b)

2 Due to friction, Earth's rotation drags its tidal bulge around, out of perfect alignment with the Moon.

(c)

3 Ocean tides rise and fall as the rotation of Earth carries us through the ocean's tidal bulges.

Low tide (the Moon has set in the west)

High tide (the Moon has crossed overhead into the western part of the sky)

High tide (the Moon is not seen)

Low tide (the Moon has risen in the east)

exactly aligned with the Moon, the Moon is not exactly overhead, but is instead high in the western sky. When we are at the high point in the tidal bulge, the ocean around us is deeper than average—a condition referred to as *high tide*. About six and a quarter hours later, probably somewhat after the Moon has settled beneath the western horizon, the rotation of Earth carries us through a point where the ocean is shallowest. It is *low tide*. If

Tides rise and fall twice each day.

you wait another six and a quarter hours, it is again high tide. You are now passing through the region where ocean water is "pulled" (relative to Earth as a whole) into the tidal bulge on the side of Earth that is away from the Moon. The Moon, which is responsible for the tides you see, is itself hidden from your view on the far side of Earth. Six and a quarter hours later, probably sometime after the Moon has risen above the eastern horizon, it is low tide. About 25 hours after you started your journey—the amount of time the Moon takes to return to the same point in the sky from which it started—you again pass through the tidal bulge on the near side of the planet. This is the age-old pattern by which mariners have lived their lives for millennia: the twice-daily coming and going of high tide, shifting through the day in lock-step with the passing of the Moon.

The shape of Earth's landmasses and ocean basins is another complicating factor in the tides seen at any particular point. In the open ocean, at a point passing directly "beneath" the Moon as Earth rotates, the difference between the depth of the ocean at low tide and the depth of the ocean at high tide is about a meter. However, the actual tides that are seen along coastlines can be less than this, or in some cases very much greater than this.

In order to respond to the tidal stresses from the Moon, Earth's oceans must flow around the various landmasses which break up the water covering our planet. Some places, like the Mediterranean Sea and the Baltic Sea, are protected from tides by their relatively small sizes and the narrow passages connecting these bodies of water with the larger ocean. In other places the shape of the land funnels the tidal surge from a large region of ocean into a relatively small area, concentrating its effect. The Bay of Fundy lies between the Canadian provinces of Nova Scotia and New Brunswick. This bay, along with the Gulf of Maine, forms a great basin in which water naturally rocks back and forth with a period of about 13 hours. This is very close to the $12\frac{1}{2}$-hour period of the rising and falling of the tides. The characteristics of the basin amplify the tides, sending the water sloshing back and forth like the

water in a huge bathtub. **Figure 9.7** shows a picture of one location on the Bay of Fundy at both high and low tide. The average difference between low and high tide on the bay is around 14.5 meters, and under the right conditions can exceed 16.6 meters!

The Sun is another complicating factor in Earth's tides. As in the case of the Moon, the side of Earth closer to the Sun is pulled toward the Sun more strongly than the side of Earth away from the Sun. The absolute strength of the Sun's pull on Earth is nearly 200 times greater than the strength of the Moon's pull on Earth. Even so, the Sun is much farther away than

Solar tides may reinforce or diminish lunar tides.

the Moon, so the Sun's gravitational attraction does not change much from one side of Earth to the other. As a result, **solar tides** (tides on Earth due to differences in the gravitational pull of the Sun) are only about half as strong as lunar tides. **Figure 9.8** illustrates

Figure 9.7 *The world's most extreme tides are found in the Bay of Fundy, located between Nova Scotia and New Brunswick, Canada. The difference in water depth between low tide (top) and high tide (bottom) is typically about 14.5 meters, and may reach as much as 16.6 meters.*

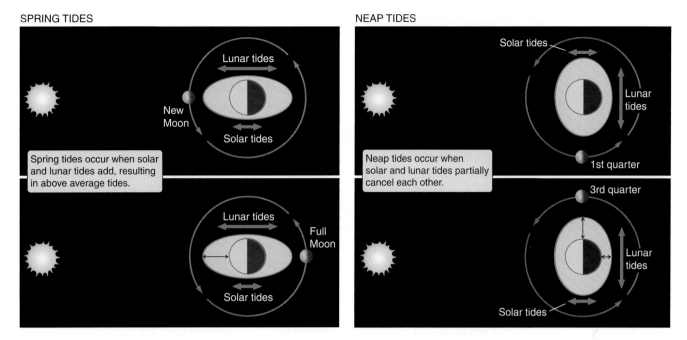

Figure 9.8 *Earth experiences solar tides that are about half as strong as tides due to the Moon.*

the interaction between solar and lunar tides. If the Moon and the Sun are lined up with Earth, as occurs at either new Moon or full Moon, then the tides on Earth due to the Sun are in the same direction as the tides on Earth due to the Moon. In this case the solar tides reinforce the lunar tides. At these times, tides are about 50% stronger than average. The especially strong tides near the time of new or full Moon are referred to as **spring tides.** Conversely, around the first- and third-quarter Moon, the stretching due to the solar tide is at right angles to the stretching due to the lunar tide. The solar tides pull water into the dip in the lunar tides, and pull water away from the tidal bulge due to the Moon. This has the effect of making low tides higher and high tides lower. Such times, when lunar tides are diminished by solar tides, are called **neap tides.** Neap tides are only about half as strong as tides are on average, and only a third as strong as spring tides.

9.3 TIDES TIE AN OBJECT'S ROTATION TO ITS ORBIT

So far in our discussion we have implicitly assumed that the solid body of Earth itself is rigid, and that the liquid of Earth's oceans is the only thing that actually moves in response to the tidal stresses from the Moon and Sun. In reality this is not the case. Earth is some-

what elastic, and like the tortured rubber ball alluded to earlier, the solid body of Earth itself is deformed by tidal stresses. The tidal deformation of the solid Earth amounts to a vertical displacement of about 30 cm between high tide and low tide, or roughly a third of the displacement of the oceans.

As Earth rotates through its tidal bulge, the solid body of our planet is constantly being deformed. It takes energy to deform the shape of a solid object. (If you want a practical demonstration of this fact, **Tides on Earth slow its rotation.** hold a rubber ball in your hand, and squeeze and release it a few dozen times. While you are shaking out your sore hand, imagine the energy that it must take to compress Earth by a third of a meter twice a day!) This energy is converted into thermal energy by *friction* in Earth's interior. This friction opposes the rotation of Earth, causing it to gradually slow. This internal friction within Earth adds to the even greater slowing caused by friction between Earth and its oceans as the planet rotates through the oceans' tidal bulge. Right now the length of Earth's day is getting longer by about 0.0015 second every century.

THE MOON IS TIDALLY LOCKED TO EARTH

The Moon has no bodies of liquid to make tides obvious, but it is distorted in the same manner as Earth.

In fact, because of Earth's much greater mass, the tidal effects of Earth on the Moon are about 20 times as great as the tidal effects of the Moon on Earth. While the average tidal deformation of Earth is about 30 cm, the average tidal deformation of the Moon is closer to 20 meters! This extreme tidal deformation is the reason why the Moon always keeps the same face toward Earth.

Very early on in its history, the period of the Moon's rotation was almost certainly different from its orbital period. However, as the Moon rotated through its extreme tidal bulge, friction was tremendous, rapidly slowing the Moon's rotation. After a fairly short period of time, the period of the Moon's rotation equaled the period of its orbit. When this happened, the Moon no longer rotated *with respect to its tidal bulge.* Instead, the Moon and its tidal bulge rotated *together* in lockstep with the Moon's orbit about Earth. With the frictional forces within the Moon gone, the Moon's rotation no longer slowed, but instead remained equal to the period of the Moon's orbit about Earth. This is how things remain today, as the tidally distorted Moon orbits about Earth, always keeping the same face and the long axis of its tidal bulge toward Earth **(Figure 9.9).** The synchronous rotation of the Moon discussed in Chapter 2 is a result of the **tidal locking** of the Moon to Earth.

Tides lock the Moon's rotation to its orbit around Earth.

Figure 9.9 *Tides due to Earth's gravity lock the Moon's rotation to its orbital period.*

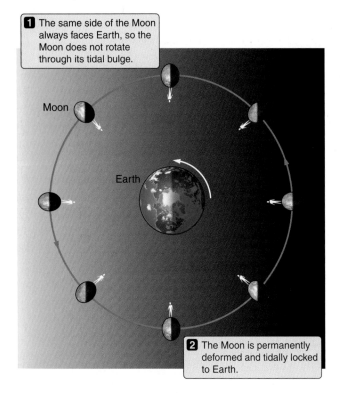

1 The same side of the Moon always faces Earth, so the Moon does not rotate through its tidal bulge.

Moon

Earth

2 The Moon is permanently deformed and tidally locked to Earth.

In addition to their effects on the rotations of the Moon and Earth, tides also influence the orbits of these two objects. Because of its tidal bulge, Earth is not really a spherical object. The material in Earth's tidal bulge on the side nearer the Moon pulls on the Moon more strongly than material in the tidal bulge on the back side of Earth. Since the tidal bulge on the Moon-ward side of Earth "leads" the Moon somewhat, as shown in **Figure 9.10,** the gravitational attraction of the bulge causes the Moon to accelerate slightly along the direction of its orbit about Earth. It is as if the rotation of Earth were dragging the Moon along with it, and in a sense this is exactly what is happening. As discussed in Connections 9.1, the angular momentum lost by Earth as its rotation slows is exactly equal to the angular momentum gained by the Moon as it accelerates along in its orbit.

The acceleration of the Moon in the direction of its orbit causes the orbit of the Moon to grow larger. At the present day, the Moon is drifting away from Earth at a rate of 3.83 cm/year; as the Moon grows more distant, the length of the lunar **The Moon's orbit is growing larger and Earth's rotation is slowing.** month increases by about 0.014 second each century. At this rate, within slightly over a billion years the Moon will be far enough away from Earth that a total solar eclipse will no longer be a possibility. If this were to continue for long enough (about 50 billion years), Earth would become tidally locked to the Moon, just as the Moon is now tidally locked to Earth. At that point the period of rotation of Earth, the period of rotation of the Moon, and the orbital period of the Moon would all be exactly the same—about 47 of our present days—and the Moon would be about 43% farther from Earth than it is today. However, this situation will never come to be, or at least not before the Sun itself has burned out. The effects of tides can be seen throughout the Solar System. Most of the moons in the Solar System are tidally locked to their parent planet, and in the case of Pluto and its moon, Charon, each is tidally locked to the other.

TIDES AND NONCIRCULAR ORBITS: AN IMPERFECT MATCH

The tidal locking between a moon and a planet (or between any two objects) can only really be perfect if both objects are in circular orbits. Even though *on average* an object in an elliptical orbit may rotate at exactly the same rate at which it moves on its orbit,

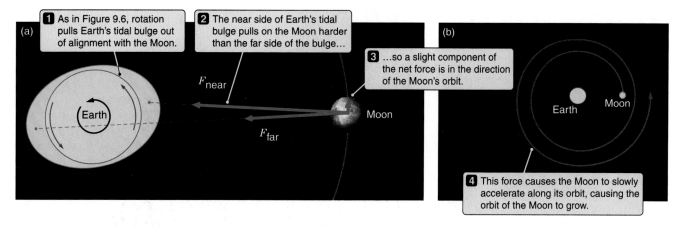

Figure 9.10 *Interaction between Earth's tidal bulge and the Moon causes the Moon to accelerate in its orbit, and the Moon's orbit to grow.*

as the object follows its orbit, it is constantly speeding up and slowing down. Like a spinning top, the Moon rotates on its axis at a steady rate that is equal to the *average* rate of the Moon's progress around Earth. However, when the Moon is closest to Earth and moving fastest on its orbit, the Moon's orbital motion "gains" a little on its rotation. Conversely, when the Moon is farthest from Earth and moving most slowly on its orbit, the Moon's orbital motion lags behind its steady rotation. From our perspective on Earth, the face of the Moon appears to be rocking back and forth. This effect is referred to as lunar **libration.** The motions responsible for lunar libration are illustrated in **Figure 9.11.** One consequence

CONNECTIONS 9.1

EARTH, THE MOON, AND CONSERVATION OF ANGULAR MOMENTUM

In this chapter we learned that the tidal interaction between Earth and the Moon is causing Earth's rotation to slow while it causes the Moon's orbit to grow larger. Given enough time, Earth and the Moon will become tidally locked to each other, with each keeping the same face toward the other. Looking at Figure 9.10 you might imagine that it would be very difficult to calculate the exact size of the net force causing the Moon's orbit to grow, or the net frictional forces within Earth that cause its rotation to slow. You would be correct. However, we can understand the relationship between these two effects, as well as their end result, without having to worry about any of those difficult details.

As Earth's rotation slows, Earth loses angular momentum. How can this be? When we discussed the formation of stars and planetary systems in Chapter 5, we found that angular momentum is conserved. Where does the angular momentum that

Earth loses as tides slow it down go? The answer is that angular momentum is transferred to the Moon. The angular momentum that the Moon picks up as its orbit grows balances exactly the angular momentum that Earth loses as its rotation about its axis slows down. This may seem like some grand conspiracy, but it is not. The frictional forces in Earth cause its rotation to slow, but they are also what causes its tidal bulge to be pulled around out of alignment with the Moon. This misalignment in turn causes the Moon's orbit to grow. In the end it *must be* the case that these effects work together to balance Earth's angular momentum loss and the Moon's angular momentum gain. Conservation of angular momentum is a grand pattern that *always* holds true, whether we are talking about the formation of stars, rotations and orbits of planets, a spinning top, or the pinwheel motion of the Galaxy itself.

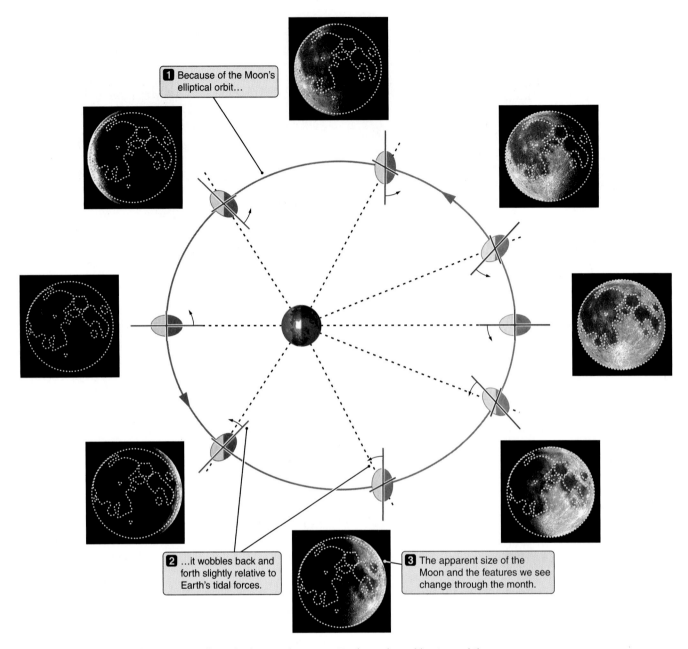

Figure 9.11 *Because of the Moon's elliptical orbit, an observer on Earth sees lunar libration, while an observer on the Moon sees Earth moving slightly back and forth in the sky.*

Tides work to circularize orbits.

of these motions is that the Moon is constantly moving back and forth through its tidal bulge. Friction within the Moon, working together with the gravitational interaction between the Moon's tidal bulge and Earth's gravity, is slowly causing the Moon's orbit to become more circular.

The masses of the giant planets are far greater than the mass of Earth, so the tidal stresses they exert on their satellites are correspondingly stronger. The tidal effect of Jupiter on its innermost moon, Io, is 250 times greater than the effect of Earth's tides on the Moon. As

with most of the moons of the giant planets, Io is tidally locked to Jupiter, rotating once on its axis in exactly the same amount of time that it takes Io to complete one orbit about the planet. Normally we would have expected Io long ago to have settled into a perfectly circular orbit. However, its orbit is constantly being perturbed by the gravity of Jupiter's other satellites, which has prevented it from achieving a steady circular orbit. As a result, Io experiences libration, just as the Moon does, but with much more dramatic consequences. As Io's libration forces it to

Tidal heating of Io associated with libration powers its active volcanism.

wobble back and forth through the satellite's tremendous tidal bulge, there is enough frictional heating to keep the interior of Io molten, powering the perpetually active volcanism **(Figure 9.12)** seen on this small world. In Chapter 10 we will learn that Io is the most volcanically active object in the Solar System. The powerhouse behind that activity is tides.

Tidal locking, in which an object's rotation period is exactly equal to its orbital period, is only one example of **spin–orbit resonance.** Other types of spin–orbit resonance are also possible. Mercury is a case in point. Mercury is in a very elliptical orbit about the Sun. As with the Moon, tidal stresses have coupled Mercury's rotation to its orbit. Yet unlike the Moon's synchronous rotation, Mercury is in a 3:2 spin–orbit resonance, spinning on its axis three times

Mercury has a 3:2 spin-orbit resonance.

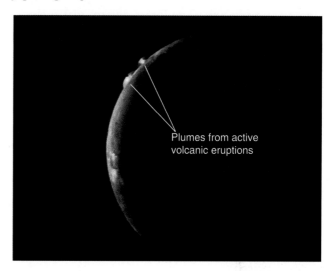

Figure 9.12 *Volcanoes are constantly erupting on Jupiter's moon Io. The energy to power these volcanoes comes from tides due to Jupiter's gravity.*

Plumes from active volcanic eruptions

CONNECTIONS 9.2

TIDES ON MANY SCALES

This section of the book concerns itself with the objects that make up our Solar System and the processes which influence them. As such, it is fitting that we have concentrated on the role tides play in the Solar System. On the other hand, we could have included a discussion of tides and their effects at almost any point in the book. Tidal stresses result from differences in gravitational forces from place to place, and such differences are a general consequence of the inverse square law of gravity. *Anytime* two objects of significant size or two collections of objects interact gravitationally, the gravitational forces will differ from one place to another within the objects, giving rise to tidal effects.

We might easily have included a discussion of tides in Chapter 12, where we will find that stars are often members of binary pairs in which the two stars orbit about each other. Strong tidal interactions between such stars can tidally lock the rotation of each star to its orbital period about its companion, and can even pull material from one star onto the other. When we study this process in Chapter 15, we will learn of the *Roche lobes* of a binary star. These are named after the same **Edouard A. Roche** (1820–1883) who did the pioneering work on the disruption of satellites, and for whom the Roche limit is named. Under some circumstances in the Universe, tides become unimaginably powerful. In the vicinity of very massive, very dense objects the pull of gravity can change by huge

amounts over very small distances. In Chapter 16 we will find that the tidal effects near the surface of a black hole can be billions of times stronger than the force of gravity holding us to the surface of Earth. Such tides are enough to rip normal matter apart atom by atom.

Tides continue to play a role at even larger scales. Tidal effects can strip stars from clusters consisting of thousands of stars. Beginning in Chapter 17 we will also learn about *galaxies*—vast collections of hundreds of billions of stars which all orbit about each other under the influence of gravity. It often happens that two galaxies pass close enough together to strongly interact gravitationally. When this happens, as in **Figure 9.13,** both galaxies taking part in the interaction can be grossly distorted by tidal effects. Tides even play a role in shaping huge collections of galaxies—the largest known structures in the Universe.

The next time you are at the beach "watching the tide roll in," you might take a moment to think about tides and the role they play throughout the Universe. Vast collections of billions of stars strewn about millions of light-years of space, matter shredded on its way into a black hole, gas stripped from a burgeoning star, comets and moons pulled to pieces, planets "flipped" like pancakes—all are the result of the same gravitational effect that is responsible for the gentle twice-daily rise and fall of the waterline on an oceanside dock.

for every two trips around the Sun. The period of Mercury's orbit—87.97 Earth days—is exactly one and a half times the 58.64 days that it takes Mercury to spin once on its axis. The forces due to solar tides on Mercury flip Mercury about, like a juggler throwing a stick into the air and catching it first by one end, then by the other.

TIDES CAN BE DESTRUCTIVE

We normally think of the effects of tides as small compared with the force of gravity holding an object together, yet this is not always the case. Tidal effects act to pull an object apart, which means that when tides become large, they can be extremely destructive. Consider for a moment the fate of a small moon, asteroid, or comet that wanders too close to a massive planet such as Jupiter or Saturn. All of the objects in the Solar System larger than about a kilometer in size are held together by their self-gravity. However, the self-gravity of a small object such as an asteroid, comet, or small moon is quite feeble. In contrast, the tidal stresses close to a massive object such as Jupiter can be fierce. If the tidal stresses trying to tear an object apart become greater than the self-gravity trying to hold the object together, then the object will begin to break into pieces.

Astronomers have a special name for the distance from the center of a planet at which tidal forces on a second, smaller body just equal that body's self-gravity. It is called the planet's **Roche limit.** An object bound together solely by its own gravity can remain

Tidal forces exceed self-gravity inside the Roche limit.

intact when it is outside a planet's Roche limit, but not when it is inside the limit. Tidal disruption of small bodies is thought to be the source of the particles that make up the rings of the giant planets. Comparison of the major ring systems around the giant planets shows that most rings lie within their respective planets' Roche limits. In the mid-1990s, astronomers were treated to a rare look at tidal disruption in action. Comet Shoemaker-Levy 9 was discovered in March 1993. As astronomers studied the comet, they came to understand that it had been captured by Jupiter and that in July of 1992, the comet had passed within 1.4

Comet Shoemaker-Levy 9 was shattered by Jupiter's tides.

Jupiter radii of the planet's center, well within the planet's Roche limit. The comet was broken into pieces by this encounter, so pictures taken by the *Hubble Space Telescope* **(Figure 9.14a)** showed not a single comet, but instead a chain of orbiting fragments. In the summer of 1994, Comet Shoemaker-Levy 9 treated the world to an even more unusual spectacle, as one after another the pieces of the comet slammed into Jupiter's atmosphere creating enormous fireballs and scars larger than Earth. **Figure 9.14(b)** shows the shattered remains of another comet, comet West, which was broken apart by the Sun's tides as it passed close to the Sun in 1976.

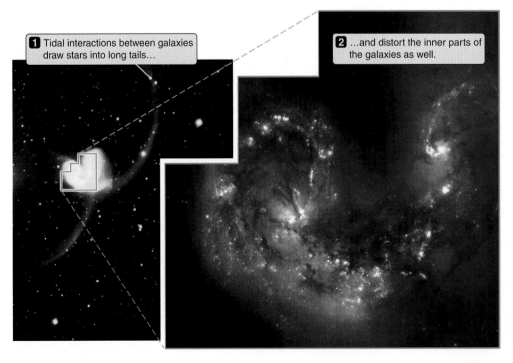

1 Tidal interactions between galaxies draw stars into long tails...

2 ...and distort the inner parts of the galaxies as well.

Figure 9.13 *Tidal interactions distort the appearance of two galaxies. The tidal "tails" seen in this image are characteristic of tidal interactions between galaxies.*

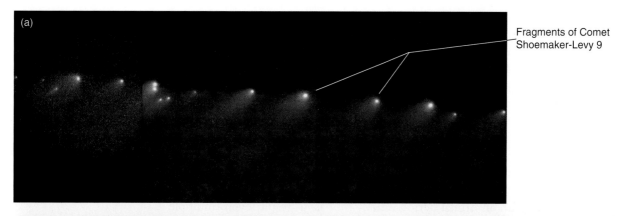

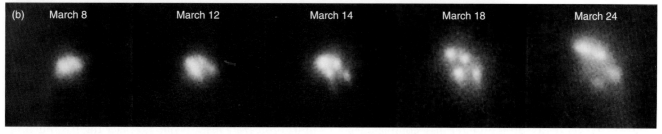

Figure 9.14 (a) *Comet Shoemaker-Levy 9 was shattered into pieces as it passed within Jupiter's Roche limit in July 1992.* (b) *Comet West was shattered by solar tides during its 1976 pass through the inner Solar System.*

We have seen that tides are at work throughout the Solar System. In Connections 9.2 we find they are at work throughout the Universe, as well.

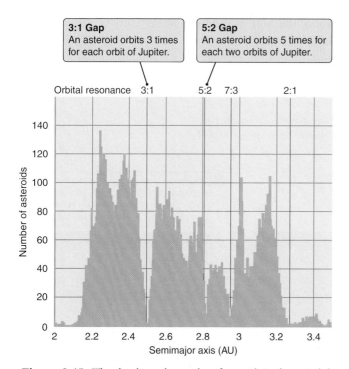

Figure 9.15 *The plot shows the number of asteroids in the main belt with a given orbital period. The gaps in the distribution of asteroids, called Kirkwood gaps, are caused by orbital resonances with Jupiter.*

9.4 MORE THAN TWO OBJECTS CAN JOIN THE DANCE

We have seen that tidal interactions can lead to simple relationships or resonances between the orbital period of an object and the period of the object's rotation. Resonances can also exist between the *orbital* periods of two or more bodies orbiting about the Sun or a planet. In Chapter 11 we will study the smaller objects that orbit about the Sun, including the small rocky or metallic worlds called *asteroids*. Most of the asteroids in the Solar System are located in a "belt" between the orbits of Mars and Jupiter. You might imagine that asteroids would be distributed randomly throughout this region. However, if you look at the distribution of the sizes of the orbits of the asteroids in the main belt, shown in **Figure 9.15,** you will see an amazing thing. Rather than asteroids with orbits of all sizes in this region, there are certain-sized orbits that are seldom occupied. These gaps in the asteroid belt are referred to as the **Kirkwood gaps.** Astronomers studying the Kirkwood gaps came to realize that there is a simple relationship between the "missing" orbits and the orbit of Jupiter. For example, none of the asteroids orbiting the Sun have an orbital period equal to half the orbital period of that planet. In fact, all of the Kirkwood gaps in the asteroid belt correspond to orbits which are related to the orbital period of Jupiter by the ratio of two small integers. Such a simple relationship between the periods of the orbits of two or more objects is referred to as an **orbital resonance**. (Note that this is different from spin–orbit resonance, which we discussed in Section 9.3.)

Orbital resonances with Jupiter cause the Kirkwood gaps.

The easiest way to understand the origin of the Kirkwood gaps is to imagine what would happen to an asteroid that started out with an orbital period exactly half that of Jupiter. Follow along in **Figure 9.16** as we watch the orbit of such an asteroid, beginning when the asteroid and Jupiter are at their closest point to each other. Because the asteroid is closer to the Sun than Jupiter is, its orbital velocity is greater than that of Jupiter, so it leaves Jupiter behind as they continue on their respective orbits. By the time the asteroid has completed half of its orbit, Jupiter has completed a fourth of its orbit. By the time the asteroid has gone around its orbit once, Jupiter is halfway. When the asteroid has completed one and one-half orbits, Jupiter has been through three-quarters of its orbit. It is only after the asteroid has completed two complete orbits and Jupiter has been around the Sun once that the two objects again line up. Courtesy of the relationship between the periods of their two orbits, when Jupiter and the asteroid line up, they are in the same place they started out in. As Jupiter and the asteroid continue along in their orbits, they line up again and again at this same location every 11.86 years (the orbital period of Jupiter).

The gravitational force that Jupiter exerts on an asteroid at its closest approach is tiny compared with the gravitational force of the Sun on the asteroid, which is over 360 times stronger. A single close pass between Jupiter and the asteroid does very little to the asteroid's orbit. For an asteroid that is *not* in orbital resonance with Jupiter, the tiny nudges from Jupiter come at a different place in its orbit each time. The effects of these "here and there" nudges average out, and as a result even multiple passes close to Jupiter have little overall effect on the asteroid's orbit. However, as we just saw, for an asteroid with a 2:1 orbital resonance with Jupiter (in other words, with an orbital period equal to half that of Jupiter), the nudge from Jupiter comes at the *same place* in its orbit *every time*. While no *single* tug has much influence on the asteroid, the *repeated* tugs from Jupiter at the same place add up, changing the asteroid's orbit. This is why there are no asteroids with orbital periods equal to half the orbital period of Jupiter. If we were to put an asteroid in such an orbit, it would not stay there for long. The same phenomenon works for other combinations of orbital periods as well. With a piece of paper and a pencil, you ought to be able to convince yourself that other orbital resonances, such as a 3:1 resonance (in which an asteroid completes three orbits about the Sun in the time that it takes Jupiter to com-

If Jupiter tugs on an asteroid at the same place over and over, it changes the asteroid's orbit.

Figure 9.16 *An asteroid with an orbital period exactly half that of Jupiter would line up with Jupiter at exactly the same place in every other orbit. The repeated gravitational pull from Jupiter at the same point in the asteroid's orbit would nudge the asteroid enough to change its orbital period. There are no asteroids with a 2:1 orbital resonance with Jupiter.*

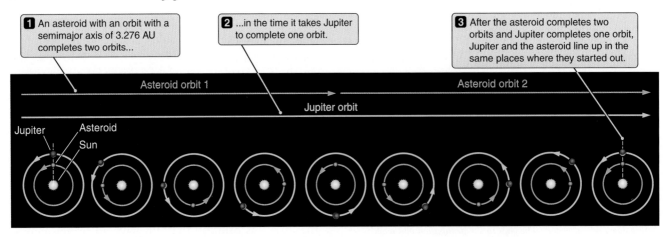

1 An asteroid with an orbit with a semimajor axis of 3.276 AU completes two orbits...

2 ...in the time it takes Jupiter to complete one orbit.

3 After the asteroid completes two orbits and Jupiter completes one orbit, Jupiter and the asteroid line up in the same places where they started out.

Asteroid orbit 1

Asteroid orbit 2

Jupiter orbit

Jupiter   Asteroid
Sun

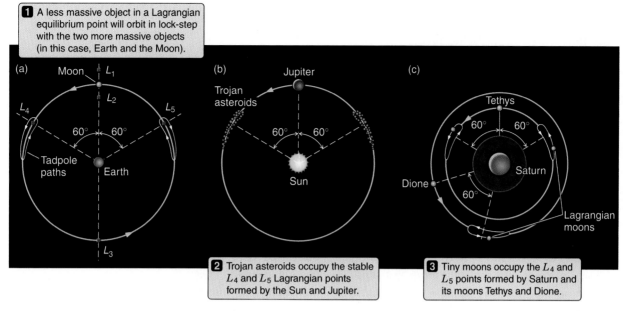

1 A less massive object in a Lagrangian equilibrium point will orbit in lock-step with the two more massive objects (in this case, Earth and the Moon).

2 Trojan asteroids occupy the stable L_4 and L_5 Lagrangian points formed by the Sun and Jupiter.

3 Tiny moons occupy the L_4 and L_5 points formed by Saturn and its moons Tethys and Dione.

Figure 9.17 (a) *The pattern of Lagrangian equilibrium points of a two-body system, in this case comprised of Earth and the Moon. This entire pattern rotates like a turntable.* (b) *Trojan asteroids.* (c) *Small moons occupy Lagrangian points formed by Saturn and Tethys, and by Saturn and Dione.*

plete one orbit) will also have a similar effect. Asteroids are not found in the Kirkwood gaps because their gravitational interaction with Jupiter prevents them from staying there.

Asteroids are not the only objects in the Solar System that are affected by orbital resonances. Exactly the same phenomenon occurs in the ring systems around the giant planets. In Chapter 10 we will learn of a very famous gap in the rings around Saturn called the **Cassini Division.** This gap became famous because of its visibility—it is so large that it could be seen from Earth even with crude, small telescopes. The Cassini Division corresponds to a 2:1 orbital resonance with Saturn's moon Mimas. That is to say, a ring particle located in the Cassini Division would have an orbital period about the planet equal to half the orbital period of Mimas. Just as Jupiter's gravity pulls asteroids out of resonant orbits, leading to the Kirkwood gaps, Mimas pulls ring particles out of the resonant orbits found within the Cassini Division. Orbital resonances and similar effects are responsible for a great deal of the structure in ring systems.

One of the most interesting examples of orbital resonance occurs when two massive bodies (such as a planet and the Sun, or a moon and a planet) move about their common center of mass in circular or

> **Orbital resonances shape Saturn's rings.**

nearly circular orbits. In this situation, there are five locations where the combined gravity of the two bodies adds up in such a way that a third, lower-mass object will orbit in lock-step with the other two. These five locations are called **Lagrangian equilibrium points,** or simply Lagrangian points.

> **Lagrangian points are orbital resonances formed by the gravity of two objects.**

The exact locations of the Lagrangian points depend on the ratio of the masses of the two primary bodies. The Lagrangian points for the Earth–Moon system are shown in **Figure 9.17(a).** (Remember that the pattern shown in the figure rotates like a turntable as the two massive objects orbit about each other.) Three of the Lagrangian points lie along the line between the two principle objects. These points are designated L_1, L_2, and L_3. These three points are equilibrium points, but they are unstable in the way that the top of a hill is unstable. A ball placed on the top of a hill will sit there if it is perfectly perched, but give it the slightest bump and it goes rolling down one side. Similarly, an object displaced slightly from L_1, L_2, or L_3 will move away from that point. Thus, while L_1, L_2, and L_3 are equilibrium points, they do not capture and hold on to objects in their vicinity. Even so, they can be useful. For example, the *SOHO* spacecraft sits at the L_2 point of the Sun–Earth system, where tiny nudges from its onboard

A WORLD OF CHAOS

Imagine that you head for class but are one second later leaving than you might have been. That single second causes you to get caught at a traffic light, which delays you enough to wind up first in line waiting for a passing train. While you await the train, there is an accident up ahead, and as you sit in the resulting traffic jam, your car's engine overheats. Eventually, while your classmates are listening to an inspiring lecture about orbital motions, you find yourself sitting in a mechanic's waiting room learning the hard way about how much expensive damage you can do to an engine by letting it get too hot. You do not realize it, but the only thing that separated you from a normal day in class was the extra sip of coffee that put you out the door a second later than you might have been. You could never have predicted such a thing would happen. Welcome to the wonderful world of **chaos.**

Chaos is the technical term to describe interrelated systems in which tiny differences at one point in time lead to large differences at later times. Any system capable of exhibiting chaotic behavior is referred to as a **complex system.** Complex systems need not be extremely complicated, like the example in the previous paragraph, but can be seemingly very simple. Newton's law of gravity is simplicity itself, and as long as only two objects are involved, the resulting motions are simple as well. When we think of a planet orbiting about the Sun, we think of the regular, repeating elliptical orbits described by Kepler's laws. However, when more than two objects are involved, the resulting motions can be anything but simple and regular.

Figure 9.18(a) shows a calculation of the orbit of Jupiter's moon Pasiphae over the course of approximately 38 years. Pasiphae orbits far enough from Jupiter that its orbit is strongly perturbed by the Sun's gravity. Rather than a simple ellipse, Pasiphae's orbit looks more like what a five-year-old might produce if handed a crayon and asked to quickly trace out a circle again and again. Pasiphae's orbit is an example of chaos. Chaotic orbits are complex, irregular motions in which extremely small differences in an object's position or speed grow into large differences in the object's subsequent motion.

Capture of satellites by planets is another example of chaos at work. **Figure 9.18(b)** shows the path of a small planetesimal, such as a comet, as it approaches a planet such as Jupiter. Very tiny differences in the position of the comet lead to very different outcomes. In the illustration, there is a gnat's eyelash of difference between the starting point of a path in which the comet impacts the planet, a path in which the comet swings around Jupiter and continues to orbit the Sun, and a path that winds up with the planetesimal orbiting the planet. Each of the giant planets has a handful of moons that revolve about the planet in the wrong direction, indicating they must have been captured in such chaotic events.

Chaos also affects the orbits of planets themselves. Each of the planets in our Solar System moves under the combined gravitational influence not only of the Sun, but of all the other planets as well. While these extra influences are small, they are not negligible. Over the course of millions of years, they lead to significant differences in the locations of planets in their orbits. While analysis of our Solar System indicates that the orbits of at least eight of the planets (excluding Pluto) are fairly stable, at least over times of a few billion years, that is not always the case. In many possible planetary systems, chaotic interactions among planets might cause planets to dramatically change their orbits, or even be ejected from the system entirely.

In Chapter 5 we discussed the fact that planetary systems have been found around many other stars. Some of these systems contain giant planets that seem much too close to their parent star to have formed there. Models suggest that these planets might have formed farther out in these systems, and then moved inward toward the star as a result of chaotic interactions. Indeed, events such as the collision between Earth and a Mars-sized body that formed the Moon tell of a more chaotic period early in the history of our own Solar System.

One of the key characteristics of chaotic systems is that while they are *deterministic,* they are not *predictable.* The law of gravity is well known, and there is in principle no problem with calculating the path that Earth will follow as a result of its ongoing interactions with other planets. It is a perfectly well determined situation. However, even Earth's orbit is subject to chaos. An uncertainty of only 1 *centimeter* in our knowledge of the position of Earth along its orbit today is enough to make it completely impossible to predict where Earth will be in its orbit 200 million years from now.

Scientists studying the orbits of celestial bodies were among the first to recognize the existence of chaotic behavior, and in the last decades of the 20th century they began to develop the mathematical tools needed to describe and even manipulate chaos. These tools are finding more and more common applications in our lives. The beating human heart, weather patterns, and traffic control are all examples of complex systems to which these tools have been applied. There are even pacemakers that use these techniques to recognize when the regular beating of a heart is about to give way to the deadly, uncoordinated fluttering called fibrillation. A tiny nudge on the part of the pacemaker, slightly changing the timing of a handful of heartbeats before fibrillation begins, is enough to prevent disaster. Chaos—the physics of complexity—is but one of many examples in which studies of the Universe have spilled over into better understanding of the immediate world in which we live.

jets are enough to keep it orbiting directly between Earth and the Sun, giving a constant, unobscured view of the side of the Sun facing Earth.

The situation is different for the other two Lagrangian points, L_4 and L_5. These two points are located 60° in front and 60° behind the less massive of the two main bodies in its orbit about the center of mass of the system. These are *stable* equilibrium points, like the stable equilibrium that exists at the bottom of a bowl. If you bump a marble sitting at the bottom of a bowl, it will roll around a bit, but it will stay in the bowl. In like fashion, objects near L_4 or L_5 follow elongated, tadpole-shaped paths relative to these gravitational "bowls."

Under the correct circumstances, L_4 and L_5 are able to capture and hold on to passing objects. Objects are often found orbiting about the stable L_4 and L_5 Lagrangian points of two-body systems. For example, the stable Langrangian points of the Sun–Jupiter system are home to a collection of planetesimals called the **Trojan asteroids (Figure 9.17b)**. There are also small moons in the L_4 and L_5 Lagrangian points of Tethys and Dione, two moons of Saturn **(Figure 9.17c)**. A well-known group that advocates human colonization of space calls itself the L_5 *Society* in recognition of the fact that the L_4 and L_5 Lagrangian points of the Earth–Moon system would make excellent locations to park a number of large orbital colonies constructed of material mined on the Moon.

Figure 9.18 (a) *The chaotic orbit of Jupiter's moon Pasiphae.* (b) *Tiny differences in the motion of a comet lead to dramatic differences in the comet's fate.*

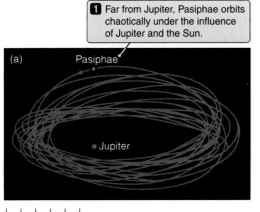

1 Far from Jupiter, Pasiphae orbits chaotically under the influence of Jupiter and the Sun.

(a) Pasiphae
Jupiter

0 5 10 15 20 25
Millions of km

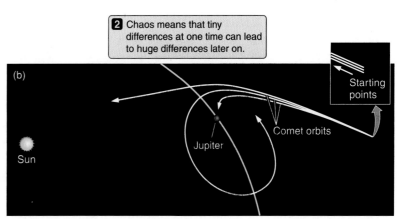

2 Chaos means that tiny differences at one time can lead to huge differences later on.

(b)
Sun
Jupiter
Comet orbits
Starting points

9.5 SUCH WONDROUS COMPLEXITY COMES FROM SUCH A SIMPLE FORCE

The tides and resonances that we have discussed in this chapter are but a small sample of the tremendous wealth and variety of phenomena that result from the gravitational interactions between extended objects or collections of more than two objects. There are a great many other such phenomena that we might have mentioned. The 26,000-year precession of Earth's axis, for example, is a result of the Sun's gravity causing Earth's axis to wobble like a top's. How a gravitational tug in one direction can cause the axis of a planet (or a top) to move in another direction is a truly fascinating story.

Gravity causes other effects, including the precession of Earth's axis.

It is with regret that we leave the gravitational ballet behind, but in the interest of space (no pun intended) we must. Before leaving, however, it is worth pointing out just how broad and deep the general problem of gravitational interactions is. While the motion of *two* masses orbiting about each other was solved centuries ago by Newton, the general problem of the motion of *three* bodies remains unsolved today! More than a few Ph.D. theses, scholarly journal articles, and monographs in mathematics, astronomy, and physics have been filled with years of work on various aspects of this seemingly simple problem. Even in this day of powerful computers which can crank out brute-force solutions to problems that seemed unapproachable a few decades ago, we remain humble in the face of gravity. As discussed in Excursions 9.1, even the gravitational interactions between a few objects such as the planets in our Solar System are *chaotic*. Unbelievably tiny differences in the initial positions and velocities of the planets—differences that are far too tiny to keep track of with the best computers—can, over the course of time, lead to dramatically different end results.

As we leave this interlude about gravity behind, we can only marvel at the extraordinary complexity and diversity that arise in nature from this deceptively simple force. What an amazing array of implications and insights and puzzles and surprises there are within Newton's elegant statement that the force of gravity is proportional to the product of the masses of two objects and inversely proportional to the square of the distance between them!

SEEING THE FOREST THROUGH THE TREES

In Chapter 3 we learned about the birth of modern science. We saw how a few straightforward ideas about forces and the inverse square law of gravity opened a new window on the Universe. When Newton used his laws of motion and gravitation to explain the motions of the planets about the Sun, he changed the face of science and the course of history. On this leg of our journey, we have pursued the implications of Newton's insight even further. We have seen how that same inverse square law of gravitation explains not only the simple elliptical orbits of planets about the Sun, but also an amazing wealth of different phenomena. The coming and going of the tides, the synchronous rotation of the Moon, the delicate structure of Saturn's rings—all of these and more are the logical consequences of the fact that gravity is one-fourth as strong when two objects are twice as far apart. The insights that we have arrived at in this chapter will have meaning again and again as we continue our journey. For example, when we follow the birth, evolution, and death of stars like our Sun, much of what we will see will be determined by the sizes and masses of the cores of those stars. Because of what we have learned in this chapter, we will be able to say that the force of gravity at the surface of a stellar core is determined by the core alone, regardless of what the rest of the star is doing.

Even more important, this leg of our journey has shown us again the aesthetics of science. If you ask a physicist today about the ultimate goal of physics, you will hear about the desire to explain the way the Universe behaves in terms of the fewest and simplest possible physical laws. In Chapter 3 we discovered Newton's laws of motion. In a sense, once we know those laws, everything—from the flight of a bumble-bee, to the swirl of a hurricane, to the motions of an Olympic gymnast—is just arithmetic. We began this chapter with the single statement about the inverse square law of gravitation, and went on to tell stories ranging from the ebb and flow of Earth's tides to the shredding of distant galaxies. In Chapter 4 we encountered the principles of quantum mechanics. Earth, wind, fire, water—even thought itself—are but practical applications of these principles.

At the turn of the 21st century, physicists have come to believe that even the basic physical laws that we are using on our journey—Newton's laws, relativity, gravitation, and quantum mechanics—are themselves embodiments of patterns in nature which are more fundamental still. Toward the end of our journey we will learn that today the frontiers of science are located at the two extremes. Theoretical physicists and cosmologists push ever backward in time toward the beginning of the Universe in search of the single theory—the Theory Of Everything—from which all else follows. Some scientists believe we may be just a few decades (or perhaps only a few years) from knowing the shape of such an all-encompassing theory. We

may even come to know whether it was possible for the Universe to become other than it is. At the same time, physicists, astronomers, planetary scientists, biologists, and others are pushing forward with their understanding of the complex world around us. They are developing new tools and approaches, such as the tools describing chaos, to allow them to find the hand of simple physical law in all that we see. So the next time you find yourself looking up at the Moon, gazing on the same face of our sister world that our ancestors saw from time immemorial, think about gravity and tides and the interplay between sister worlds. But at the same time, think about what it represents—the glorious complexity of this fascinating Universe in which we live, and the beautiful simplicity of physical law from which it all derives.

STUDENT QUESTIONS

THINKING ABOUT THE CONCEPTS

1. The acceleration due to gravity (g) at Earth's surface is 9.832 m/s^2 at the poles and 9.781 m/s^2 at the equator (note that both values are close to the value of 9.8 m/s^2 we use in this text). Give two reasons why the value of g is less at Earth's equator than at the poles.

2. The best time to dig for clams along the seashore is when the ocean tide is at its lowest. What phases of the Moon and times of day would be best for clam digging?

3. A pendulum clock, adjusted to run accurately at one geographical location, may not maintain that same accuracy at some other location. Why not?

4. Lunar tides raise Earth's land surfaces about 15 cm above their average level and raise the ocean surface about 50 cm above mean sea level. What does this tell you about the density of Earth's crust compared to that of water?

5. If Lunar tides can raise the ocean surface only about 1 m above mean sea level, how do tides as large as 5 to 10 meters occur?

6. If the Moon spins on its axis once every $29\frac{1}{2}$ days relative to Earth, why do we not find Earth tides on the Moon rising and falling over this same interval?

7. In our daily lives we encounter many types of symmetry, including rotational, reflection, and bilateral (which is when left and right sides are identical). List some daily-life examples of each type of symmetry.

8. A circus seal balances a ball on his nose, and the ball remains on the tip of his nose for as long as the seal wants to keep it there. Is this an example of stable or unstable equilibrium? Explain your answer.

9. In 1959, a spacecraft launched by the USSR photographed the far side of the Moon, showing us a large part of the Moon's surface that had never before been seen. Yet, more than half (59%) of the Moon's surface had already been mapped prior to this Soviet achievement. If the Moon always keeps the same face toward Earth, how was this possible?

10. Tidal friction slows Earth's rotation and causes the Moon to orbit ever farther from Earth. Is this an example of conservation of angular momentum or conservation of energy? Explain your answer.

11. Most commercial satellites orbit Earth well inside the Roche limit. Why are they not torn apart?

APPLYING THE CONCEPTS

12. Assume for the moment that Earth is homogeneous, that is, its density is constant throughout, and it is spherical in shape. (We know, of course, from Chapter 6 that both of these assumptions are not true.) If you were in a deep well half way to Earth's center, how would your weight there compare with your weight at Earth's surface?

13. Saturn's small moon, Hyperion, is in a 4:3 orbital resonance with its largest moon, Titan, which orbits closer to Saturn than does Hyperion. The

orbital period of Titan is 15.945 days. What is the orbital period of Hyperion?

14. Tidal influence is proportional to the mass of the disturbing body and is inversely proportional to the cube of its distance. (The proof of this relationship is beyond the scope of this text.) Some astrologers claim that your destiny is determined by the "influence" of the planets that are rising above the horizon at the moment of your birth. Compare the tidal influence of Jupiter (mass = 1.9×10^{27} kg; distance = 7.8×10^{11} m) with that of the doctor in attendance (mass = 80 kg, distance = 1 m).

15. The mass of the Sun is 27 million times greater than the mass of the Moon, but the Sun is about 390 times farther away. Using the relationship for tidal influence given in Question 14, show why solar tides on Earth are only about half as strong as lunar tides.

Although we are mere sojourners on the surface of the planet, chained to a mere point in space, enduring but for a moment in time, the human mind is not only enabled to number worlds beyond the unassisted ken of mortal eye, but to trace the events of indefinite ages before the creation of our race. . . .

SIR CHARLES LYELL (1797–1875)

PLANETARY MOONS AND RINGS, AND PLUTO

10.1 MOONS AND RINGS— GALILEO'S LEGACY

In 1610, the Italian astronomer Galileo Galilei observed that Jupiter was accompanied by four "stars" that changed their positions nightly. He quickly realized that, like Earth, Jupiter has moons of its own. We honor his discovery by calling them the *Galilean* moons of Jupiter. Galileo showed that Jupiter and its moons resemble a miniature Copernican planetary system: just as planets revolve around the Sun, so too do the many moons of Jupiter revolve around it. By the end of the 17th century, astronomers had found five moons orbiting around Saturn as well, and this further strengthened their belief in the Copernican system (see Chapter 3). Today, we realize that the Solar System abounds with moons. There are 91 that we know of as of February 2002, and probably many others—especially in the outer Solar System—that we have not yet found.

If Galileo's discovery of Jupiter's moons was personally satisfying, his other important discovery was decidedly less so. When he first pointed his telescope at Saturn in 1610, the tiny disk seemed to be accompanied by smaller companions on both sides. Unlike the moons of Jupiter that he had found a few months earlier, these features did not move. Galileo was troubled

KEY CONCEPTS

For centuries, celestial wonders such as Saturn's rings and the Galilean moons of Jupiter delighted those who looked through telescopes. But with the dawn of the space age, robotic explorers traveling through the Solar System have shown us much more than points of light, revealing wondrous, diverse families of worlds orbiting other planets. Among them we will find:

* Scores of worlds composed of rock and solid ice, some of which formed with their planets, and others that were captured later on;
* Geologically active moons, freckled with volcanoes and geysers, and geologically dead moons covered with impact craters;
* Shattered remains of moons and comets forming the rings that surround each giant planet;
* Exquisite, delicate structure in ring systems resulting from subtle gravitational interactions among planets, moons, and ring particles;
* Moons that may harbor deep liquid oceans beneath their ice-covered surfaces, and may conceivably provide a home for extraterrestrial life; and
* Two frozen, unexplored worlds, Pluto and Charon, orbiting at the boundary between the inner and outer Solar System.

by this, for Saturn was like nothing else that he had observed. Two years later, the "companions" had vanished. Their unexpected disappearance upset the Italian astronomer greatly, because he feared his earlier observations had been in error. A few years later, the mysterious features reappeared. For more than four decades, astronomers puzzled over Galileo's discovery.

In 1655, a 26-year-old Dutch instrument maker, **Christiaan Huygens** (1629–1695), pointed a superior telescope of his own design at Saturn and saw what the astronomers of his day had failed to see. Saturn is surrounded by an apparently continuous, flat **ring,** and, as Huygens correctly deduced, the variations in its visibility are caused by changes in the apparent tilt of the ring as Saturn orbits the Sun. What was the nature of this strange ring encircling Saturn? Astronomers assumed it

was a solid disk spinning around the planet, an interpretation that lasted for more than a century after Huygens's discovery. That notion began to weaken when, in 1675, the great Italian-French astronomer, **Jean-Dominique Cassini** (1625–1712), found a gap in the planet's seemingly solid ring. Saturn now appeared to have two rings rather than one, and the gap that separated them became known as the *Cassini Division* that we learned of in the previous chapter.

In 1850, a fainter ring, located just inside the two bright rings, was found independently by English and American observers, giving Saturn a known total of three rings. For illustrators and cartoonists, Saturn had become an icon for depicting all the planets. Yet for more than three and a half centuries, it was the only planet known to have rings.

Figure 10.1 *Images of the major moons in the Solar System obtained by various spacecraft. The images are shown to scale. The planets Mercury and Pluto are shown for comparison. Mars's moons, Phobos and Deimos, are too small to be shown. (Pluto and Charon are artists' conceptions based on ground- and space-based images.)*

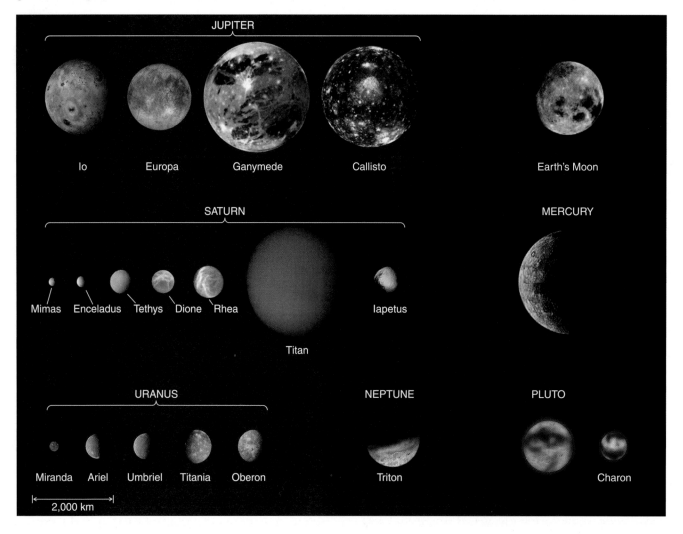

Most Planets Are Adorned with Moons

The moons of our Solar System are not distributed equally among the nine planets. Mercury and Venus have none, Earth has one, and Mars has two; thus there are only three moons in the inner part of the Solar System. Pluto possesses a single known moon, and all of the remaining moons belong to the giant planets.

Figure 10.1 shows images of many of the major moons in the Solar System, shown to scale. In many ways, moons resemble smaller versions of the terrestrial planets. Some, such as our own, are made of rock. Others, especially in the outer Solar System, are mixtures of rock and water ice, with densities intermediate between the two. A few seem to be made almost entirely of ice. Several moons are comparable in size to, or even larger than, Mercury or Pluto, while the smallest known moons would fit within the expanse of a large metropolitan airport. Although most moons are airless, one has an atmosphere denser than Earth's and several have very low-density atmospheres. Some of the larger moons appear to have differentiated chemically, and two are known to have active volcanoes or geysers.

Moons are made of rock, ice, or mixtures of both.

Our own Moon is not covered in this chapter; we discussed it in Chapter 6 because of its close similarity to the four terrestrial planets of the inner Solar System. On the other hand, Pluto—nominally a planet—*is* included in this chapter because of its similarity to the icy moons of the outer Solar System.

As discussed in Chapter 5, we believe that many moons were formed at the same time as the planets they orbit, and that they were created in much the same way that the planets themselves grew—from the accumulation of planetesimals orbiting the Sun. In Excursions 10.1, we follow the formation and evolution of one such moon as it coalesced from grains in the disk that orbited the young Jupiter. We call those moons that formed together with their parent planet **regular moons.** Regular moons revolve around their planet in the same direction as the planet rotates and in orbits that lie nearly in the planet's equatorial plane. This is because the debris from which the regular moons formed was orbiting in the planet's equatorial plane and in same direction as the evolving planet was rotating. With very few exceptions, regular moons are *tidally locked* to their parent planet. Recall from Chapter 9 that tidal locking causes a body to rotate synchronously with respect to its orbit, as does Earth's Moon. A moon in synchronous rotation around its planet has fixed leading and trailing hemispheres, whose surfaces can appear very different from one another. An especially strange moon is Saturn's Hyperion. Its rotation is *chaotic* (see Chapter 9), tumbling in its orbit with a rotation period and a spin-axis orientation that are constantly changing.

Moons that formed together with their planet are called regular moons.

Some moons revolve in a direction that is opposite to the rotation of their planet, and some are situated in distant, unstable orbits. These are almost certainly bodies that had formed elsewhere and were later captured by the planet. We call them **irregular moons.** The largest irregular moon is Neptune's Triton. It orbits Neptune in a **retrograde,** or "backward," direction. Other moons, such as Saturn's Phoebe and Pluto's Charon, also have retrograde orbits. Many of the irregular moons of the outer planets are only a few kilometers across.

Some captured or "irregular" moons have retrograde orbits.

We might ask why Mercury and Venus failed to form or capture any moons of their own. But is this really surprising? As we noted earlier, the smaller terrestrial planets and Pluto tend to have far fewer moons than the giant planets. Mercury and Venus, after all, have only one less than Earth or Pluto. Mars apparently captured a couple of asteroids, but Mars is situated adjacent to the main asteroid belt (see Chapter 11). Finally, Earth would be without a moon if not for a cataclysmic collision when the planet was young. What seems important is that the larger planets had greater amounts of debris around them while they were forming, which is what gave rise to their greater number of moons.

10.2 Rings Surround the Giant Planets

Over the centuries following the discovery of Saturn's rings, exhaustive searches failed to detect rings around any other planet. Papers were published by more than one distinguished astronomer explaining why, theoretically, only Saturn could have rings. Then, in the latter part of the 20th century, a new search technique became available: observation of stellar occultations (see Chapter 8). In 1977, a

The rings of Jupiter, Uranus, and Neptune are recent discoveries.

EXCURSIONS 10.1

FORMATION OF A LARGE MOON

Ganymede, Jupiter's largest moon, is a good example of what we mean by a regular moon. Scientists think that a moon like Ganymede formed around young Jupiter in much the same way that the planets formed around the young Sun. Whereas the planets formed from a protoplanetary accretion disk surrounding the Sun, Ganymede grew from the accretion of ice and dust grains that were orbiting in a similar disk surrounding the young, hot proto-Jupiter

At the distance of Ganymede's orbit from the glowing proto-Jupiter core, temperatures were low enough for grains of water ice to survive and, along with silicate materials, they coalesced to form planetesimals. It took less than a half million years for these planetesimals to accrete and create the moon Ganymede. Heat generated from accretion melted parts of the moon to form an outer water layer, an inner silicate zone, and an ice-silicate core. These layers, however, were not stable. As cooling took place, much of the outer water layer froze to form a dirty ice crust. Most of the denser silicate materials sank to the center to form a core, leaving an intermediate ice-silicate zone **(Figure 10.2).**

team of American astronomers were using the occultation technique to study the atmosphere of Uranus when they saw brief, minute changes in the brightness of the star as it first approached and then receded from the planet. The interpretation was immediately obvious: Uranus has rings! Over the next several years, stellar occultations revealed a total of nine rings surrounding the planet. Later, in 1986, *Voyager 2* imaged two additional Uranus rings, bringing the total to 11.

Stellar occultations not only show the existence of rings; they also tell us something about the rings themselves. The duration of an occultation event is a measure of the width of the ring. The observed decrease in the brightness of the star is an indication of the ring's transparency and therefore of the amount of material it contains. The very brief interruption of starlight as the Uranus rings passed in front of the star showed that they are very narrow, far too narrow to have been seen in the earlier, unsuccessful searches by more conventional methods.

More ring discoveries were to follow from still another technology: close-up studies by planetary probes. In 1979, cameras on *Voyager 1* recorded a faint ring around Jupiter, and *Pioneer 11* found a very narrow ring just outside the bright rings of Saturn.

For a while, Neptune seemed to be the only giant planet deprived of rings. Then, in the early to mid-1980s, occultation searches by teams of American and French astronomers began yielding positive but confusing results. Several occultation events that appeared to be due to rings were seen on only one side of the planet. The astronomers concluded that Neptune was surrounded not by complete rings but rather by several arclike ring segments. Only when *Voyager 2* reached Neptune in 1989 did we learn that its rings are indeed complete, and that the **ring arcs** are merely high-density segments within one of its narrow rings. All of Neptune's rings are faint and, with the exception of the ring arcs, contain too little material to be detected by the stellar occultation technique.

One of Neptune's rings contains higher-density segments called ring arcs.

All of the giant planets are now known to have ring systems, although each is quite different from the others. The most complex system of all belongs to what was once thought to be the only ringed planet—spectacular Saturn. Moreover, it turns out that the giant planets are the only planets in our Solar System that have rings; none of the terrestrial ones do.

Now that we have considered this brief history of ring discoveries, what do we know about the rings themselves? We start our discussion of ring structure using Saturn's densely packed, bright rings as an example. Huygens, with his mid–17th-century understanding of physics, believed Saturn's ring to be a solid disk surrounding the planet (as we saw in Section 10.1). This view was challenged in later years, but it was not

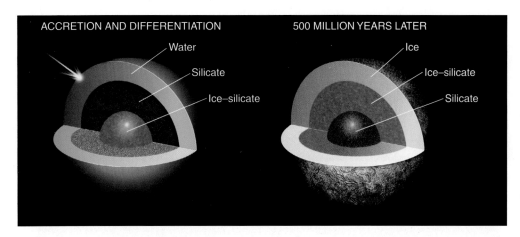

Figure 10.2 *Diagram showing the evolution of Ganymede's interior. In the final stages of accretion, Ganymede was covered with a global ocean of liquid water, underlain by a zone of silicates and a mixed core of ice and silicates. Five hundred million years later, the dense silicates gravitationally had sunk to form a core and much of the ocean had frozen.*

until the middle of the 19th Century that the brilliant Scottish mathematician, James Clerk Maxwell, showed that solid rings would be unstable and would quickly break apart. They must instead consist of countless numbers of small particles, like so many tiny moons in individual orbits around Saturn. What keeps all these small ring particles together? We saw in Chapter 9 how gaps in rings are caused by orbital resonances with satellites. Later in the chapter, we will see that there is much more to the gravitational dance performed by the complex system involving the planet, its moons, and the countless individual ring particles.

Rings are swarms of tiny moons orbiting according to Kepler's laws.

Kepler's laws dictate that the speed and orbital periods of all ring particles must vary with their distance from the planet, with the closest moving the fastest and having the shortest orbital periods. The orbital periods of Saturn's bright rings range from 5^h45^m at their inner edge to 14^h20^m at the outer one. Ring particles can vary in size from tiny grains to house-sized boulders, and in these densely packed rings of Saturn, the orbits of all of them must be perfectly circular and in precisely the same plane.

To understand why the orbits of particles in dense rings must be so orderly, imagine yourself riding along on a large particle right in the middle of Saturn's bright rings. Around you is a swarm of particles of all shapes and sizes, many of them close enough for you to reach out and touch. They are all moving along in almost prefect step with your own, like members of a well-disciplined marching band. We say "almost" because the particles slightly closer to the planet are moving just a bit faster than you are, and those a little farther out are moving just a bit slower, as Kepler's laws demand. The differences in speed are very small. The particle orbiting 1 meter inward from you is moving only a tenth of a millimeter per second faster than you are—a relative speed difference slower than a snail's pace.

Now consider what would happen if that particle were moving in a slightly noncircular orbit, perhaps because it was bumped by one of its other neighbors. Its eccentricity means that it will drift alternately a bit outward and then inward as it completed each orbit. But as it moved outward, the particle would eventually nudge up against you and be unable to go any farther. With no room for maneuvering, the orbits of errant particles would (and do) quickly become circular. The same would be true for any particle in an orbit with even the smallest inclination: During each orbit, the deviant ring fragment would be carried alternately above and below the ring plane. As it approached your particle from, say, above, the mild encounter would quickly remove its out-of-plane motion and bring it into a coplanar (in the same plane) orbit. Saturn's densely packed rings have no place for nonconformists. Like the well-disciplined band, all particles must march along together in unison.

Dense rings must be circular.

SATURN'S MAGNIFICENT RINGS—A CLOSER LOOK

We have chosen the rings of Saturn as our introduction to planetary rings, because they alone exhibit such phenomenal complexity. **Figure 10.3** shows the ring systems of the giant planets, including the individual

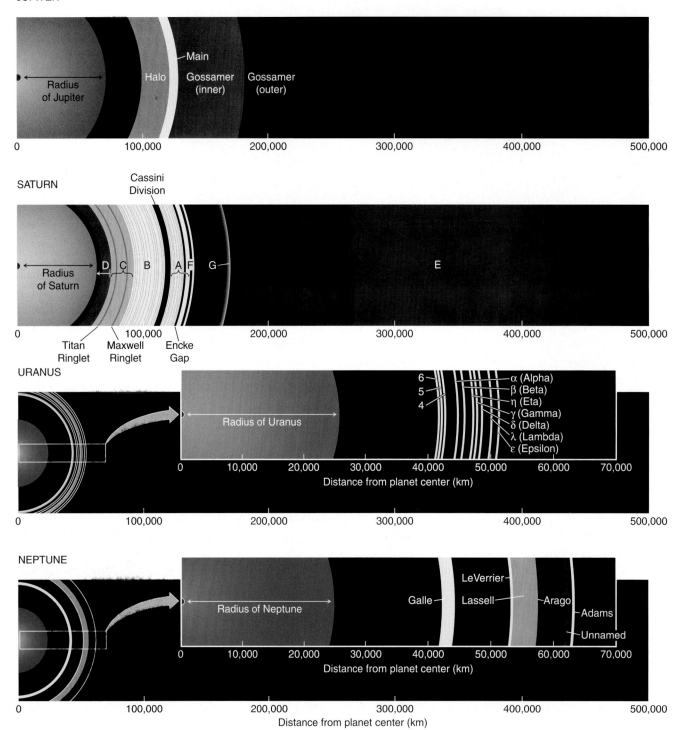

Figure 10.3 *A comparison of the ring systems of the four giant planets.*

components of Saturn's bright-ring system, and its major divisions and gaps. The most conspicuous are the expansive bright rings, which dominate all photographs of Saturn **(Figure 10.4).** Among the four giant planets, only Saturn has rings so wide and so bright.

Photographs usually show only the two outer and brighter rings, separated by the Cassini Division. The

outermost, or A Ring, is the narrowest of the three bright rings. It has a very sharp outer edge and contains several narrow gaps. The conspicuous Cassini Division is so wide (4,700 km) that the planet Mercury would almost fit within it. Astronomers once thought that it was completely empty. In fact, space scientists planning

10.1 RINGS

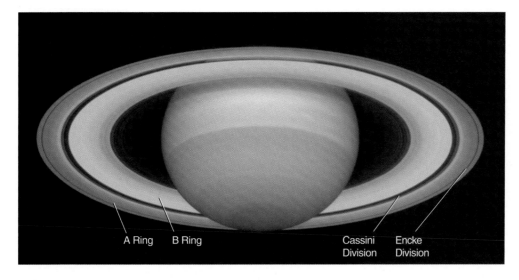

Figure 10.4 *A* Hubble Space Telescope *image showing Saturn, the A Ring (the outermost bright ring), the B Ring (the middle bright ring), the Cassini Division (the wide, relatively dark division between the A and B Rings), and the Encke Division (the narrow division near the outer edge of the A Ring).*

the 1979 encounter of *Pioneer 11* with Saturn gave serious consideration to flying the spacecraft through the Cassini Division to get as close to the planet as possible. Had they carried out this plan, the *Pioneer 11* mission would certainly have ended right there. Images taken the following year by *Voyager 1* show the Cassini Division to be filled with material. Why, then, did astronomers think it was empty? The Cassini Division and many other smaller "gaps" in the bright rings are simply regions with less ring material. In contrast with their denser surroundings, they appear darker and thus empty.

"Divisions" and "gaps" in Saturn's rings are not truly empty.

The B Ring is the brightest of Saturn's rings. With a width of 25,500 km, two Earths could fit side by side between its inner and outer edges. Strangely, the B Ring seems to have no gaps at all, at least on the scale of those seen in the other bright rings. The C Ring often fails to show up in normally exposed photographs, because of the limited ability of film to record a wide range of brightnesses. At the eyepiece of the telescope, though, this beautifully translucent ring appears like delicate gossamer, and hence is often called the Crepe Ring. There is no known gap between the C Ring and either of the adjacent rings. Only an abrupt change in brightness marks the boundary between them. What could cause such a sharp change in the amount of ring material remains an unanswered question. Too dim to be seen next to Saturn's bright disk, the D Ring is a fourth wide ring that was unknown until imaged by *Voyager 1.* It shows more coarse structure than any of the bright rings, and appears to have no inner edge. The D Ring may extend all the way down to the top of Saturn's atmosphere, where its ring particles would enter and burn up as meteors.

Saturn's bright rings are far from homogeneous. The A and C Rings contain hundreds, and the B Ring thousands, of individual **ringlets,** some only a few kilometers wide **(Figure 10.5).** Each of these ringlets is a narrowly confined concentration of ring particles bounded on both sides by regions of relatively little material. Spacecraft images of Saturn's rings turned up many other surprises, as we will see later.

Saturn's rings contain thousands of individual narrow rings called ringlets.

Each time the plane of Saturn's rings lines up with Earth, as it does about every 15 years, the rings all but

Figure 10.5 *This* Voyager 2 *image of the outer B Ring shows so many ringlets and minigaps that it looks like a close-up of an old-fashioned phonograph record. The narrowest features are only 10 km across, at the limit of resolution. Even finer structure was noted during stellar occultations by the rings, as observed by the* Voyager *photometer.*

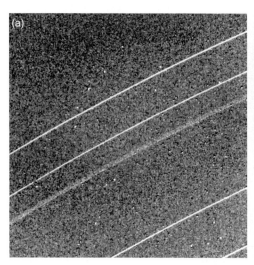

Figure 10.6 *The appearance of rings depends dramatically on lighting conditions. Here, the rings of Uranus (a) appear as narrow, faint bands when viewed with the Sun at our back but (b) burst into dazzling brilliance when illuminated from behind. (Bright lines in the images are stars "streaked out" by the spacecraft motion during exposures.)*

vanish for a day or so in even the largest telescopes. With the glare of the rings temporarily gone, we can search for undiscovered moons or other faint objects close to Saturn. In 1966, an astronomer looking for moons found weak but compelling evidence for a faint ring near the orbit of Saturn's moon Enceladus. In 1980, *Voyager 1* confirmed the existence of this faint ring, now called the E Ring, and found another closer one, known as the G Ring. Both the E and G Rings are diffuse rings, with no distinct boundaries. Unlike the bright and the narrow rings, diffuse rings do not have well-defined inner and outer edges, nor are they confined to a thin plane.

Although Saturn's bright rings are very wide—more than 62,000 km from the inner edge of the C Ring to the outer edge of the A Ring—they are extremely thin. From our previous discussion, you might guess that they could be no thicker than the diameter of the larger ring particles. But in these densely packed rings, there is simply not enough room to jam all of the particles into the same plane, so they settle down as close as they can get to the ring plane. Saturn's bright rings are thus no more than a hundred meters and probably only a few tens of meters from their lower to upper surface. The extremes between their width and their thickness can be difficult to picture, but let's try.

> **Saturn's bright rings are exquisitely thin.**

Say you'd like to make a scale model of Saturn and its rings using a basketball to represent Saturn. The basketball is about 20 cm in diameter. You could make the three bright rings out of paper by cutting a circle 45 cm in diameter, with a 25 cm hole cut from the center. To represent the Cassini Division, you could paint a dark stripe 1.5 cm wide and about 12 cm in from the outer edge. After mounting the paper ring around the basketball, you have a splendid model of Saturn and its rings. Unfortunately, your model is not completely to scale: the paper rings are more than a thousand times too thick! If you wanted to make your rings from paper similar to that used in this book, the planet would have to be a ball 250 meters in diameter and the paper rings would have to extend over the length of six football fields! The diameter of Saturn's bright-ring system is 10 million times the thickness of the rings.

In diffuse rings, where the separation between particles may be very large, an occasional collision between two particles can cause their orbits to become eccentric, inclined, or both. Since these disturbed orbits get restored rather infrequently, the particles are likely to remain in their noncircular, noncoplanar orbits. For this reason, diffuse rings become spread out and thick, sometimes without any bounds.

OTHER PLANETS, OTHER RINGS

Ring structure among the other giant planets is not as diverse as Saturn's. Most rings other than Saturn's are quite narrow, although a few are very diffuse. What we see when looking at a ring system depends dramatically on the lighting conditions under which we view the rings. Pebbles and boulders are easiest to see if the light is coming from behind us and reflecting off these rather large pieces of material. On the other hand, if you have ever tried to drive into sunlight on a dusty day, you may have noticed that particles of dust stand out most strongly when looking *into* the light. As discussed in Excursions 10.2, photographers call this effect **back-lighting.** When *Voyager* scientists looked at Jupiter's

> **Most rings of the other giant planets are relatively narrow.**

EXCURSIONS 10.2

THE BACKLIGHTING PHENOMENON

Photographers often place their models in front of a bright light to highlight the model's hair, a technique called *backlighting*. Under backlighted conditions, individual strands of hair shine brightly, creating a halo effect around the model's face. This happens when light falls on very small objects—those with dimensions a few times to several tens of times the wavelength of light. Human hair is near the upper end of this range. Light falling on the strands of hair is not scattered uniformly in all directions, but tends to continue in the direction away from the source of illumination. Very little of the light is scattered off to the side, and almost none back toward the source.

Some of the dustier planetary rings are filled with particles whose size is just a few times the wavelength of visible light. To a spacecraft approaching from the direction of the Sun, such rings may be very difficult or even impossible to see as in **Figure 10.6(a)**. This is because the tiny ring particles scatter very little sunlight back toward the Sun and the approaching spacecraft. When the spacecraft passes by the planet and looks backward in the general direction of the Sun **(Figure 10.6b)**, these dusty rings will suddenly appear as a circular blaze of light, much like a halo surrounding the nighttime hemisphere of the planet. As illustrated by the images of the rings of Uranus in Figure 10.6, many planetary rings are best seen with backlighting, and some have been observed only under these conditions.

---

ring with the Sun behind the camera, all they saw was a narrow, faint strand. However, when they looked back toward the Sun while in the shadow of the planet, Jupiter's rings suddenly blazed into prominence. Jupiter's rings are mostly made up not of rocks but of fine dust.

Ten of the 11 rings of Uranus are very narrow and widely spaced relative to their widths. Most have widths of only a few kilometers but are many hundreds of kilometers apart (see Figures 10.3 and 10.6a). The outermost, called the Epsilon Ring, is the widest of the narrow rings. Its width varies between 20 and 100 km. The 11th ring of Uranus is wide and diffuse, with an undefined inner edge. As with Saturn's D Ring, material in the 11th ring may be spiraling into the top of the Uranus atmosphere. When viewed under backlighted conditions by *Voyager 2* (Figure 10.6b), the space between Uranus's rings turned out to be filled with dust, much like Jupiter's rings.

Three of Neptune's six rings are narrow, with widths of a few tens of kilometers (see Figure 10.3). They are named after 19th-century astronomers who made major contributions to Neptune's discovery. Much of the material in the Adams Ring is clumped together into several arclike segments, with lengths of 4,000 to 10,000 km and a width of about 15 km **(Figure 10.7)**. Another ring, as yet unnamed, is 5,800 km wide and lies between the Adams and Le Verrier Rings.

MOONS ARE THE ORIGIN OF RING MATERIAL

Planetary rings are a wonderland of the sorts of gravitational interactions that we discussed in Chapter 9. Tides, for example, are thought to be largely responsible for much of the material found in planetary rings. If a moon or other planetesimal held together by gravity alone happens within the Roche limit of a giant planet, it will be pulled apart by tides. The location of Saturn's bright rings inside the planet's Roche limit, for example, sug-

Figure 10.7 *A* Voyager 2 *image of the three brightest arcs in Neptune's Adams Ring. Neptune itself is very overexposed in this image.*

Saturn's rings are the remains of icy moons disrupted by tides.

gests that the rings formed from the breakup of one or more moons or captured planetesimals that were perturbed by larger moons into an orbit that took them too close to the planet.

If rings are made of material that originated on moons, we might expect to find the composition of a planet's rings mimicking the makeup of the moons that surround the planet. Such is indeed the case. Saturn's bright rings appear bright because they reflect about 60% of the sunlight falling on them. We might suspect from their brightness alone that they are made of water ice, and spectral observations confirm this suspicion, clearly showing the distinct signature of water. A slight reddish tint to the rings tells us that they are not made of pure ice but must be contaminated with other materials such as silicates. The icy moons around Saturn or the frozen comets that prowl the outer Solar System could easily provide this material.

Saturn's bright rings are the brightest in the Solar System, and are the only ones that we know are composed of water ice. In stark contrast, the rings of Uranus and Neptune are among the darkest objects known in our Solar System. Only 3% of the sunlight falling on them is reflected back into space, making the

The rings of Uranus and Neptune are darker than coal.

ring particles blacker than coal or soot. No silicates or similar rocky materials are this dark. However, a number of carbon-rich meteorites are this dark, as is the nucleus of comet Halley. This suggests that the rings of Uranus and Neptune may be formed largely of similar material, rich in organic compounds. Jupiter's rings are neither as bright as Saturn's nor as dark as those of Uranus and Neptune, suggesting they may be rich in dark silicate materials, like the innermost of Jupiter's small moons.

The jumble of fragments that make up Saturn's rings is easily understood as a product of tidal disruption of a moon or planetesimal, but moons can contribute material to rings in other ways as well. The last looks that *Voyager* had at Jupiter's rings carried hints of things unseen, and so matters were to remain for 20 years. It is ironic that until recently we knew less about Jupiter's rings than any other ring system in the Solar System, yet as a result of *Galileo*'s arrival at the planet, Jupiter's ring system is now the best observed of all. **Figure 10.8** shows a *Galileo* image of Jupiter's rings. Jupiter's ring system turns out largely to be the product of the strong gravity of the planet itself, combined with the presence of a handful of small, rocky moons close to the planet. As interplanetary meteoroids are pulled toward

Jupiter by its strong gravity, a few of them strike the surface of one of Jupiter's four innermost moons. These moons are so tiny that some of the dust from these impacts is kicked off with speeds in excess of the escape velocities of the moons. This dust provides a steady supply of material for the rings.

The dust in Jupiter's rings is blasted from moons.

The brightest of Jupiter's rings is a narrow strand only 6,500 km across, consisting of material from Metis and Adrastea (Figure 10.8). These two moons orbit in Jupiter's equatorial plane, and the ring they form is very thin. Beyond the main ring, however, are the very different Gossamer rings, so called because they are extremely tenuous. The Gossamer rings are supplied by dust from the moons Amalthea and Thebe. Unlike the main ring, the Gossamer rings are very thick; the inner Gossamer ring, associated with Amalthea, is actually located within the outer Gossamer ring formed of material from Thebe. These rings are so thick because the orbits of the moons that supply the ring material are slightly tilted with respect to Jupiter's equatorial plane. The orbital planes of these moons wobble, as does the orbital plane of Earth's Moon, but instead of taking almost 19 years to complete one wobble (as our Moon does), these moons complete a wobble in only a few months. The Gossamer rings, made up of material from these wandering moons, are spread as far below and above Jupiter's equatorial plane as the orbits of the satellites that form them.

The innermost ring in Jupiter's system is called the Halo Ring, which consists mostly of material from the main ring. As the dust particles in the main ring drift slowly inward toward the planet, they pick up electric charges and are pulled into this rather thick torus, or doughnut-shaped ring, by electromagnetic forces associated with Jupiter's powerful magnetic field.

Finally, moons may contribute ring material through volcanism. Volcanoes on Jupiter's moon Io continuously eject sulfur particles into space, many of which drift inward under the influence of pressure from sunlight and find their way into the Jupiter ring. The particles in Saturn's E Ring appear to be ice crystals that have been ejected from the moon Enceladus, which is located in the very densest part of the E Ring.

MOONS MAINTAIN ORDER AND CREATE GAPS

Planetary rings are ephemeral; they do not have the long-term stability of most Solar System objects. Ring particles are constantly colliding with one another in

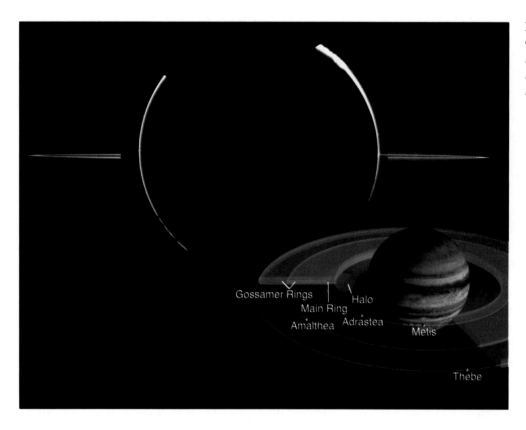

Figure 10.8 *A backlit* Galileo *view of Jupiter's system of rings and a diagram of the small moons that form them.*

Janus passes the particle, once out of every six Janus orbits and every seven particle orbits, it does so at the same position in the particle's orbit. And each time, the particle suffers a small gravitational nudge from the moon in the same direction. The effects build up and eventually the particle loses enough energy to fall back into the ring. In short, the sharp edge of the ring itself is just like the sharp edge of a resonant gap like the Cassini Division.

Each time a moon nudges a ring particle inward the minute amount of energy lost by the particle is absorbed by the more massive moon, causing it to move imperceptibly farther from the ring and the planet. Over long periods of time, a moon may move so far from the ring edge that it can no longer provide stability to the ring. The ring is then free to dissipate. In the case of Saturn's bright rings, the situation is even more complicated, because Janus itself is also in resonance with other moons of Saturn, meaning that all these moons must be pushed away before the rings can dissipate. Thus, the bright rings of Saturn may be much more stable than most rings.

Moons can also influence particles deep within a ring. All that is necessary is for the moon to be massive enough to create a significant gravitational tug on the particles, and be in orbital resonance with them. In Chapter 9 we discussed the 2:1 orbital resonance with Saturn's moon, Mimas, that creates the Cassini Division. This is the same mechanism by which Jupiter creates the Kirkwood gaps in the asteroid belt. Such resonances are known to produce some of the gaps that appear in Saturn's bright rings. For example, we know that one of the gaps in the C Ring is caused by a 4:1 resonance between the ring particles and Mimas. Unfortunately, the cause of many—perhaps we should say most—other gaps in Saturn's rings remains unexplained. If they also are produced by resonances, we

Rings don't last forever.

their tightly packed environment, either gaining or losing orbital energy as they do so. This redistribution of energy can cause particles at the ring edges to leave the rings and drift away, aided by nongravitational influences such as the pressure of sunlight. Under these conditions, all rings should quickly dissipate. Yet we see that many rings have very sharp edges with no visible material appearing in the space beyond. For these rings at least, something is holding most of the particles in place, preventing the rings from quickly dissipating.

The key to the sharp edges and stability of the rings lies with the same sort of orbital resonances discussed in Chapter 9. As moons orbit a planet, their gravity can nudge ring particles that are in resonant orbits, keeping them in line. Consider, for example, the abrupt outer edge of Saturn's A Ring. There is no visible material beyond this boundary. The ring particles at the edge of the A Ring are in a 7:6 orbital resonance with the co-orbital moons Janus and Epimetheus, meaning that the ring particles make *precisely* seven orbits for every six orbits of Janus and Epimetheus.

Moons maintain sharp ring edges and create gaps in the rings.

Now picture a particle suffering a collision and being forced a little outside the edge of the ring. Each time

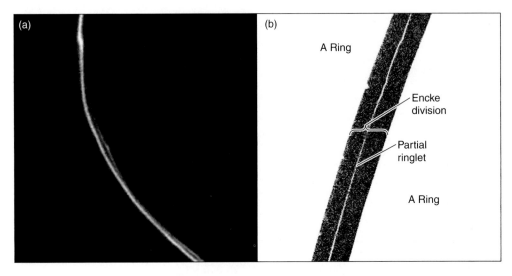

Figure 10.9 (a) *A Voyager 1 image of Saturn's "twisted" F Ring.* (b) *Two partial rings reside within the Encke Division.*

have yet to identify the source. One possibility is that these gaps are the result of collisions between ring particles. Any collision between two ring particles will cause one of the particles to move to an orbit farther out, and the other to an orbit farther in. Over time, this process could sweep some areas clean of ring particles, forming gaps, while piling those ring particles up in narrow ringlets between the gaps.

There are also important gravitational interactions between ring particles themselves. These interactions determine the ring shapes at the edges of gaps. In fact, analysis of the shapes of ring edges allows the masses of the rings to be estimated. Even though planetary rings can be large and prominent, they account for only the tiniest fraction of the mass of the material around a giant planet. Saturn's bright rings are by far the most massive rings in the Solar System. In fact, they contain more material than all other planetary rings combined. Even so, their total mass is estimated to be less than that of Mimas, a

The mass of all Saturn's rings combined is about the same as a small icy moon.

small icy moon of Saturn about 390 km in diameter. The amount of material in the narrow rings is, of course, very much less. All of the particles in the largest ring of Uranus, the Epsilon Ring, could be compressed into single body no more than 20 km across. All of the material in both the Neptune rings and the Jupiter ring could fit into single objects only a few kilometers in diameter.

Other kinds of orbital resonances are possible in ring systems. For example, most narrow rings are caught up in a periodic gravitational tug-of-war with nearby moons. These are called **shepherd moons** in recognition of the way they shepherd a flock of ring particles. Shepherd moons are usually very small, are located close to the narrow

Moons that keep narrow rings from spreading are called shepherd moons.

rings, and often come in pairs, one orbiting just inside and the other just outside the narrow ring. The shepherding mechanism is much like the resonances discussed earlier. A shepherd moon just outside a ring will rob orbital energy from any particles that drift outward beyond the edge of the ring, causing them to move back inward. A shepherd moon just inside a ring will give up orbital energy to a ring particle that has drifted too far in, nudging it back in line with the rest of the ring. In some cases, very narrow rings are trapped in between two shepherd moons in slightly different orbits.

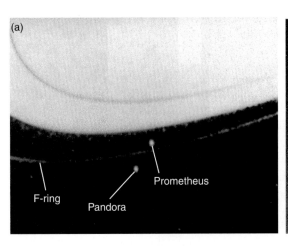

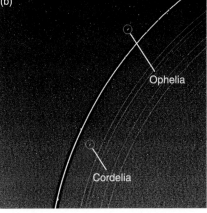

Figure 10.10 Voyager 2 images of (a) *Saturn's F Ring and its shepherd moons, Prometheus and Pandora;* and (b) *Uranus' Epsilon Ring with its shepherds, Cordelia and Ophelia.*

STRANGE THINGS AMONG THE RINGS

Among the many strange rings imaged by *Voyager,* the archetype is clearly Saturn's F Ring, shown in **Figure 10.9(a).** Images of the ring taken by *Pioneer 11* a year earlier had shown nothing out of the ordinary, but the first high-resolution images of the F Ring made by *Voyager* had the startled spacecraft scientists staring in disbelief. The ring was separated into several strands that *appeared* to be intertwined. Some media reporters quickly claimed that the F Ring was "disobeying the laws of physics." This, of course, was not the case, but ready explanations for this seemingly non-Keplerian behavior of the ring particles were not immediately forthcoming. And, as if the multiple strands were not perplexing enough, the ring also displayed what appeared to be a number of knots and kinks.

Saturn's F Ring is now understood to be a dramatic example of the action of a pair of shepherd moons. The F Ring is flanked by Prometheus, a moon that orbits 860 km inside the ring, and Pandora, the second moon, which orbits 1,490 km on the outside **(Figure 10.10).** Both moons are irregular in shape, with average diameters of 110 and 90 km, respectively. Because of their relatively large size and proximity, they exert significant gravitational

Rings can be distorted by the gravitational influence of nearby moons.

forces on nearby ring particles. The resulting tug-of-war between Prometheus pulling ring particles in its vicinity to larger orbits and Pandora drawing its neighbors into smaller orbits—leads to the bizarre structure that originally baffled scientists and reporters alike.

The F Ring is not an isolated case. The 360-km-wide Encke Division in the outer part of Saturn's A Ring contains two narrow rings, both of which also seem to show knots and kinks **(Figure 10.9b).** If shepherd moons are in eccentric or inclined orbits, they cause the confined ring to also be eccentric or inclined, as is the case for Uranus's Epsilon Ring. Because ring shepherds can be so small, they often escape detection. According to current theories of ring dynamics, there must be a number of still-unknown shepherd moons interspersed among the ring systems of the outer Solar System.

The rings in the Encke Division are unusual for another reason. Each is a discontinuous, partial ring, like the arcs in Neptune's Adams Ring. Arclike segments appear in Uranus's Lambda Ring as well. As yet, there is no generally satisfactory explanation for the origin of ring arcs. Recent infrared imaging by the *Hubble Space Telescope* shows that the arcs in Neptune's Adams Ring have changed very little over the decade since they were first seen by *Voyager 2,* but we know nothing about their long-term stability.

Various orbital resonances with moons may help to guide the orbits of ring particles and delay the dissipation of the rings themselves, but at best this can only be a temporary holding action. All rings eventually face their inevitable fate. It is very unlikely that the planetary rings we see today have existed in their current form since the Solar System's beginning. Indeed, it is far more likely that many ring systems may have come and gone over the history of the Solar System. Even our own planet has probably had several short-lived rings at various times during its long history. Any number of comets or asteroids must

Earth doesn't have a ring because it lacks shepherding moons.

have passed within Earth's Roche limit[1] and disintegrated catastrophically into a swarm of small fragments, thereby creating a temporary ring. Yet Earth lacks shepherding moons to provide orbital stability to rings. Interactions between ring particles would have caused such a ring to spread out and dissipate, while the inner parts of a ring around Earth would feel the drag of Earth's extended atmosphere and spiral inward, creating spectacular meteor displays as they fell. A similar absence of small, inner moons also prevents Venus and Mercury from keeping rings over geological time scales. We leave Mars off the list for now. While we know of no ring around Mars, its two tiny moons, Phobos and Deimos, might easily serve to shepherd a collection of orbiting debris. As we continue to explore our Solar System, Mars may surprise us yet.

It is a wild understatement to say that many of the physical properties of planetary rings were unexpected by planetary scientists before spacecraft began surveying the giant planets at close range. We now have come to appreciate that all of this structure—from the narrowest of strands in the F Ring, to the countless ringlets of the A, B, and C Rings—is the result of the same gravitational dance between moons and rings that we encountered in Chapter 9. Here and in that earlier discussion, we have only scratched the surface of what is possible. Indeed, some of the most extraordinary structure seen in planetary rings is still not well understood, decades after its discovery. In the end we must tip our hats to the seemingly boundless diversity of unpredictable, chaotic behavior that is possible in such complex systems.

Gravitational interactions do not hold a monopoly on strange behavior among ring systems. One of the more puzzling discoveries made by *Voyager 1* at Saturn was the

[1]Earth's Roche limit is about 25,000 km for rocky bodies and over twice that for icy bodies.

(a)

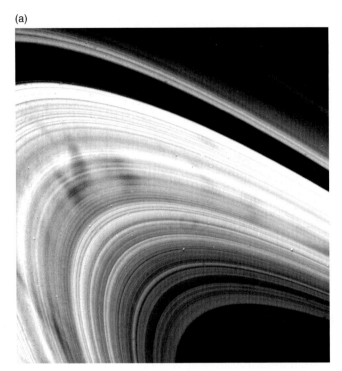

(b)

Figure 10.11 *Spokes in Saturn's B Ring appear* (a) *dark in normal viewing, but* (b) *bright with backlighting, indicating that the spoke particles are very small.*

10.3 MOONS AS SMALL WORLDS

There are several ways that we could group the moons of our Solar System. Some schemes are based on the sequence of the moons in their orbits around the parent planet, and others on the sizes or compositions of the moons. For example, earlier in the chapter, we saw that a few moons are predominantly rocky objects, some are mostly ices, while most appear to be mixtures of ice and rock.

The scheme we will use here is based on the amount and timing of a moon's geological activity, as expressed by the features we see on their surfaces. As we learned when looking at the terrestrial planets and Earth's Moon, surface features observed on planets and moons provide critical clues to their geological history. For example, water ice is a common surface material among the moons of the outer Solar System and the freshness of that ice tells us something about the ages of those surfaces. Meteorite dust darkens the icy surfaces of moons, just as dirt darkens snow late in the season in our own urban areas. In other words, a bright surface often means a fresh surface.

> **Brightness, structure, and crater density of moon terrains give clues to geological activity.**

In addition, as we discussed in Chapter 6, the size and number of impact craters gives us the relative timing of events such as volcanism, and this timing allows us to gauge if and when a moon may have been active in the past. Terrains having a large number of craters are older than those having only a few. Observations of erupting volcanoes, as on Io, are direct evidence that some moons are geologically active today.

What happens when we apply these age-dating techniques to the moons in the Solar System? We find an immense diversity—some moons have been frozen in time

> **Saturn's B Ring has transient radial features called spokes.**

appearance of dozens of dark spokelike features in the outer part of the B Ring **(Figure 10.11).** The spokes rotated at the same Keplerian rate as the ring particles, and grew in a radial direction as they circled Saturn. Yet no individual spoke was seen to last for more than half an orbit. This half-orbit survival tells us that the particles in the spokes must be suspended above the ring plane. Why is this obvious? Any particle that is not in the ring plane must be in an inclined orbit, and it thus has to pass through the plane twice during each orbit of the planet. As the spoke particles try to pass through the densely packed B Ring, they run into the B Ring particles and are absorbed. Such a model can nicely explain what causes the spokes to disappear, but it does not tell us why they appear.

One plausible suggestion links the origin of the spokes with meteoroid impacts on large ring particles. Meteoroids, as they strike these particles, can collide with so much energy that they create an ionized cloud of tiny charged particles—a plasma—that becomes briefly suspended above the ring plane. Before the cloud can descend back onto the ring plane, Saturn's magnetic field causes the charged particles to drift outward, creating the radial, spokelike features. Still, questions remain. For example, our model has no explanation for the fact that *Voyager* imaged spokes only in the outer part of the B Ring, and not in the inner part of the ring or in either of the other two wide rings.

10.2 MOONS

There is a great diversity of activity among moons.

since their formation during the early development of the Solar System, while others are even more geologically active than Earth. In our classification scheme of moons, we include four categories of geological activity: (1) definitely active today, (2) possibly active today, (3) active in the past but not today, and (4) apparently not active at any time since their formation.

GEOLOGICALLY ACTIVE MOONS: IO AND TRITON

One of the more spectacular surprises in Solar System exploration was the discovery of active volcanoes on Io, the innermost of the four large Galilean moons of Jupiter.

Io is the Solar System's most volcanically active body.

Yet, in one of those rare events that happen in science, Io's volcanism was predicted by planetologists just two weeks before its discovery, based on the same tidal stresses as those discussed in Chapter 9. Did you ever take a piece of metal and bend it back and forth, eventually breaking it in half? Touch the crease line and you can burn your fingers! Just like the metal in your hands, the continual flexing of Io's crust caused by the changing strength and direction of the tides generates enough energy to melt parts of the crust. In this way Jupiter's gravitational energy is converted into thermal energy powering the most active volcanism in the Solar System.

As *Voyager* approached Jupiter, images showed that Io's surface is literally covered by volcanic features, including vast lava flows, volcanoes, and volcanic craters. Amazingly, however, pictures from *Voyager* and *Galileo* failed to show a single impact crater, making Io unique among all the solid planetary bodies seen so far in planetary exploration. With a surface so young, Io must be volcanically active indeed, with lava flows and volcanic ash burying impact craters as quickly as they form. Scientists working on the *Voyager* mission discovered just how volcanically active Io is in spectacular and undeniable fashion when post-encounter images looking back toward this moon showed explosive volcanic eruptions sending debris hundreds of kilometers above Io's surface! As of this writing, Io has more than 300 known volcanic vents, and more than 60 active volcanoes were observed during the *Galileo* mission between 1996 and 2000. The most vigorous eruptions, with vent velocities of up to 1 km/s, spray sulfurous gases and solids as high as 300

Io's volcanic plumes reach heights of several hundred kilometers.

km above the surface. Ash and other particles rain onto the surface as far as 600 km from the vents. The moon is so active that frequently several huge eruptions are occurring at the same time. One look at an image such as **Figure 10.12** leaves little doubt about the source of the

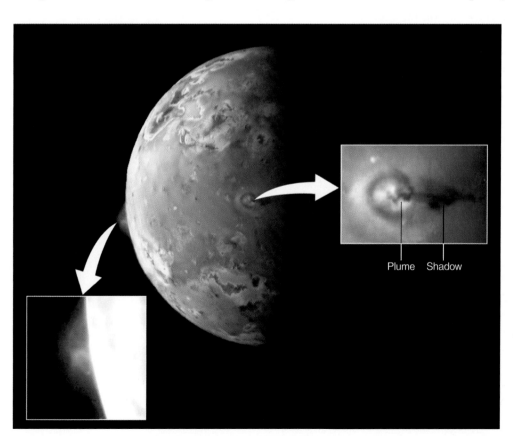

Plume   Shadow

Figure 10.12 *When this image of Io was obtained by Galileo, two volcanic eruptions could be seen at once. The plume of Pillan Patera rises 140 km above the limb of the moon on the left, while the shadow of a 75-km-high plume can be seen to the right of the vent of Prometheus, near the moon's terminator.*

Figure 10.13 *An image of Jupiter's volcanically active moon Io, constructed of images obtained by* Galileo.

material supplying the plasma torus and Io flux tube we discussed in Chapter 8.

Io's surface, shown in **Figure 10.13**, displays a wide variety of colors—pale shades of red, yellow, orange, and brown. Mixtures of sulfur, sulfur dioxide frost, and sulfurous salts of sodium and potassium on the moon's surface are the likely cause of the color. Bright patches may be fields of sulfur dioxide snow. You may well wonder how snow can fall from Io's nearly nonexistent atmosphere. According to current understanding, liquid sulfur dioxide must flow beneath Io's surface, held at high pressure by the weight of overlying material. Like water from an artesian spring, this pressurized sulfur dioxide can be pushed out though fractures in the crust, producing sprays of sulfur dioxide snow crystals that travel for up to hundreds of kilometers before settling back to the moon's surface. (To see a similar process in action, just put a carbon dioxide fire extinguisher to use. These fire extinguishers contain liquid carbon dioxide at high pressure that immediately turns to "dry ice" snow as it leaves the nozzle.)

Voyager and *Galileo* images show parts of the surface of Io at high resolution. They reveal a variety of plains, irregular craters, and flows, all related to eruption of mostly silicate magmas onto the surface of the moon. They also show high-standing mountains, some nearly twice the height of Mt. Everest,

Silicate magmas dominate Io's volcanism.

Earth's highest mountain. Huge structures, some 65 km across, show multiple calderas and other complex structures telling of a long history of repeated eruptions followed by collapse of the partially emptied magma chambers. Many of the floors are very hot **(Figure 10.14)**, and might still contain molten material similar to magnesium-rich lavas that erupted on Earth more than 1.5 billion years ago. It is important to note that volcanoes on Io are spread around the moon in a much more random fashion than on Earth, where tectonic patterns influence the location of volcanism.

Io probably formed at the same time as the other giant planets' moons. Based on its current volcanic activity, Io's entire mass may have recycled, or turned inside-out, more than once over the past, leading to chemical differentiation. Volatiles, such as water and carbon dioxide, probably escaped into space long ago, with most heavier materials sinking to the interior to form a core. Sulfur and various sulfur compounds, aided by silicate magmas, are constantly being recycled, forming the complex surface we see today.

Volcanism may have turned Io inside-out several times.

Triton, the largest moon in the Neptune system, is an irregular moon that must have been captured by its planet **(Figure 10.15)**. To achieve its present circular, synchronous orbit, it must have experienced extreme tidal stresses from Neptune following its capture. Such stresses would be similar to those on Io, and would have generated large amounts of thermal energy. The interior might even have melted, allowing Triton to become chemically differentiated.

Triton's surface shows the torment of being a captured body.

Figure 10.14 *A* Galileo *image showing flows of molten lava on Io.*

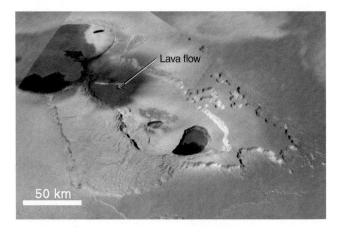

Like all moons in the outer reaches of the Solar System, Triton is a cold place. Its surface temperature is only about 38 K. Triton has a thin atmosphere, with a surface composed mostly of ices and frosts of methane and nitrogen. From the relative lack of craters, we know that the surface is geologically young. Part of Triton is covered with *cantaloupe terrain,* so named because it looks like the skin of a cantaloupe. Irregular pits and hills may represent surface deformation and extrusion of slushy ice onto the surface from the interior. Veinlike features include grooves and ridges that could result from the extrusion of ice along fractures. The rest of Triton is covered with smooth plains of volcanic origin. Irregularly shaped depressions, as wide as 200 km **(Figure 10.16),** formed when mixtures of water, methane, and nitrogen ice melted in the interior of Triton and erupted onto the surface, much as rocky magmas erupted onto the lunar surface and filled impact basins on the Moon.

Voyager 2 found four active geyserlike volcanoes on Triton. Each consisted of a plume of gas and dust as much as a kilometer wide rising 8 km above the surface, where the plume was caught by upper atmospheric winds and carried for hundreds of kilometers downwind. How do the eruptions on Triton work? Their association with Triton's southern hemispheric ice cap may provide some clues to the process. We know that nitrogen ice in its pure form is transparent and allows easy passage of sunlight. Thus, clear nitrogen ice could create a localized greenhouse effect (see Chapter 7), in which solar energy trapped beneath the

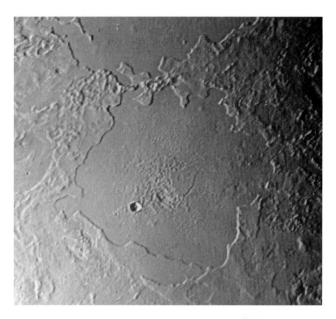

Figure 10.16 *This irregular basin on Triton has been partly filled with frozen water, forming a relatively smooth ice surface. The state of New Jersey could just about fit within the basin's boundary.*

ice raises the temperature at the base of the ice layer. A temperature rise of only 4°C would be enough to vaporize the nitrogen ice. As gas is formed, the expanding vapor would exert very high pressures beneath the ice cap. Eventually the ice would rupture and vent the gas explosively into the low-density atmosphere. Dark material, perhaps silicate dust or radiation-darkened, methane ice grains, is carried along with the expanding vapor into the atmosphere, where it then settles to the surface forming dark patches streaked out by local winds (Figure 10.15).

Nitrogen propels Triton's geyserlike volcanism.

POSSIBLY ACTIVE MOONS

One of the most fascinating objects in the outer Solar System is Jupiter's Europa, a rock world about the size of our Moon, but with an outer shell of water. We know that the surface of the water is frozen, but we do not know what lies beneath the ice. Like Io, Europa experiences a continuously changing tidal stress from Jupiter that generates internal energy and possibly volcanism. However, some calculations show that the tidal heating may be too small to melt the interior and produce volcanism. On the other hand, sufficient thermal energy might be present to melt the ice (or to keep the water in a liquid state) and there is the possibility of a global ocean below the ice.

Figure 10.15 *This* Voyager *mosaic shows various terrains on the Neptune-facing hemisphere of Triton. "Cantaloupe" terrain is visible at the top; its lack of impact craters indicates a geologically younger age than the bright, cratered terrain at the bottom.*

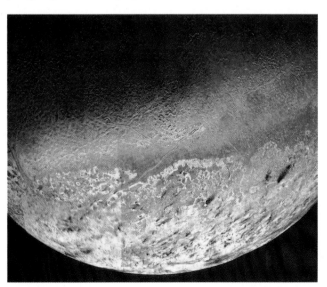

Figure 10.17 *A high-resolution Galileo image of Jupiter's moon, Europa, showing where the icy crust has been broken into slabs that, in turn, have been rafted into new positions. These areas of chaotic terrain are characteristic of a thin, brittle crust of ice floating atop a liquid or slushy ocean.*

Europa's surface is young, with few impact craters, but is deformed by tectonic activity driven by heating from its interior. Regions of chaotic terrain, as shown in **Figure 10.17,** are places where the icy crust has been broken into slabs that have been rafted into new positions. These slabs can be fit back together like pieces of a jigsaw puzzle. In other areas, the crust has split apart and been filled in with new material from the interior, in some ways similar to seafloor spreading on Earth (see Chapter 6). Of the handful of large impact structures preserved on Europa, all are shallow features resembling the patterns formed when a rock is dropped into stiff mud. The chaotic terrain, the spreadinglike features, the impact structures, and other surface features all suggest that the icy crust of Europa consisted of a thin brittle shell overlying either liquid water or warm, slushy ice at the time the features formed. Although we do not know whether these are the conditions on Europa today, the geologically young surface holds open that possibility.

Europa is covered with broken slabs of ice.

Like Mars, Europa, is a high-priority target in the search for extraterrestrial life. The necessary ingredients for life include liquid water, an energy source, and the presence of organic compounds. A search of the Solar System shows that there are relatively few places where these conditions might be present; Europa is one such place. We know that water is present (at least as ice) and that tidal energy generates heat. That leaves only the presence of organic materials as an unknown. However, comets are known to contain organic compounds and, as we will see in Chapter 11, comet nuclei have likely crashed into all of the planets and moons, implanting organic materials throughout the Solar System.

Saturn's moon Titan is larger than Mercury and has a density that suggests a composition of about 45% water and 55% rocky material. What makes Titan especially remarkable is a thick atmosphere (some 30% denser than Earth's), which obscures our view of its surface. Titan's atmosphere is reminiscent of Los Angeles on a bad day. Like Earth, its atmosphere is mostly nitrogen, but views of Titan's limb show layers in the atmosphere that are probably photochemical hazes, much like smog.

A dense, hazy atmosphere obscures Titan's surface.

As Titan heated up and differentiated chemically, various ices emerged from the interior to form an early atmosphere. Ultraviolet photons from the Sun have enough energy to cause the **photodissociation** of ammonia molecules into nitrogen and hydrogen. As we learned in Chapter 7, the resulting hydrogen molecules, being much lighter than nitrogen, would then have rapidly escaped into surrounding space, leaving Titan with an atmosphere that, like our own, was mostly nitrogen. Planetary scientists suggest that Titan is covered with a liquid ethane-methane ocean to a depth of about a kilometer. Ground-based radar observations indicate that both liquids and solids exist on the surface. This possibility is bolstered by *HST* images taken in the near-infrared showing dark patches and bright, continent-sized areas that could be solid surfaces **(Figure 10.18).** We can guess that much of the solid surface material must be water ice. In this model, we visualize islands of water ice rising from the ocean floor, with a mixture of liquid ethane and methane playing the same role on Titan as liquid water does on Earth.

Whether geological processes, such as volcanism, tectonics, or impact cratering, have left landforms is open to question. As with cloud-covered Venus, an orbiting radar-imaging system that can pierce the thick cloud cover is needed to view the surface of Titan with the resolution required for geological interpretation. Such a system has been developed for *Cassini,* a joint NASA–European Space Agency mission that includes the probe *Huygens* for descending to Titan's surface. In many respects, Titan resembles a primordial Earth, and the anticipated presence of organic compounds that could be biological precursors in the right environment makes Titan another high-priority target for exploration.

One of Saturn's moons, Enceladus, is only 500 km in diameter. Even so, it shows a wide variety of ridges, faults, and smooth plains, evidence of tectonic processes unexpected for such a small object. Volcanism on Enceladus may involve eruptions of liquid water, perhaps mixed with ammonia or water-ice slush, onto the surface from melt zones beneath. Is Enceladus active today? It could be. The orbit of Enceladus is located in the densest part of Saturn's E Ring and appears to be providing the ring with a continuous supply of tiny particles, most probably ice crystals. Active volcanic plumes on Enceladus, energetic enough to overcome the moon's low gravity, could be sending eruptive products into space to repopulate particles that are being continuously lost from the E Ring.

Enceladus may still be volcanically active.

FORMERLY ACTIVE MOONS: GANYMEDE AND SOME MOONS OF SATURN AND URANUS

Some moons show clear evidence of past ice volcanism and tectonic deformation, but no current geological activity. Ganymede is the largest moon in the Solar System, even larger than the planets Mercury and Pluto.

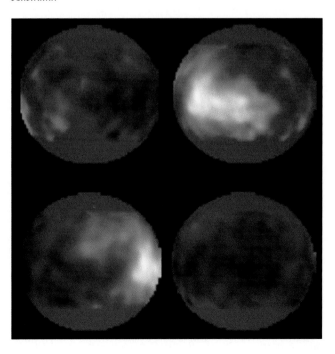

Figure 10.18 Hubble Space Telescope *views of Saturn's largest moon, Titan, taken in the near-infrared. At this wavelength, the atmosphere is nearly transparent, allowing the surface to be seen. The bright area is a surface feature about the size of Australia.*

The surface is composed of two prominent terrains: a dark, heavily cratered (and therefore ancient) terrain, and a bright terrain characterized by ridges and grooves. Ganymede's low density (1.9 times that of water) indicates that its bulk composition is about half water and half rocky materials. Overall, the moon's surface is very bright. Even the so-called dark terrains are brighter than the highlands on Earth's Moon. The high number of impact craters superposed on the dark terrain reflects the period of intense bombardment during the early history of the Solar System. The most extensive region of ancient, dark terrain includes a semicircular area more than 3,200 km across (about the size of Europe) on the leading hemisphere. Furrowlike depressions occurring in many dark areas are among Ganymede's oldest surface features. They may represent surface deformation from internal processes, or they may be relics of impact cratering processes.

Ganymede is the Solar System's largest moon.

Impact craters range up to hundreds of kilometers in diameter, with the larger craters being proportionately more shallow. With time, the icy crater rims deform by viscous (very slow) flow, as might a lump of soft clay, and can ultimately lose nearly all of their topography. Such features are seen as flat, circular patches of bright terrain, called **palimpsests,** which are characteristic of Ganymede's icy lithosphere **(Figure 10.19a)**. Palimpsests are found principally in the dark terrain of Ganymede and are believed to be scars left by early impacts onto a thin icy crust overlying water or slush, as illustrated in **Figure 10.19(b)**.

Palimpsests are scars of ancient impacts.

When astronomers first viewed the bright terrain in *Voyager* images, it was thought to represent regions that had been flooded by water or slush erupted from the interior of Ganymede. *Galileo* images, however, failed to show any indications of such flooding, other than in a few local places. Then how did the bright terrain form? The answer seems to be related to a style of surface formation by tectonic processes not previously considered. In Chapter 6 we discussed how planetary surfaces can be fractured by faults or folded by compression resulting from movements such as those initiated in the mantle. On Ganymede, the tectonic processes have been so intense that the fracturing, faulting, and folding have completely deformed the icy crust, destroying all signs of older features such as impact craters.

Many other moons show evidence of an early period of geological activity that has resulted in a dazzling array of terrains. A 400 km impact crater scars Saturn's

(a)

(b)

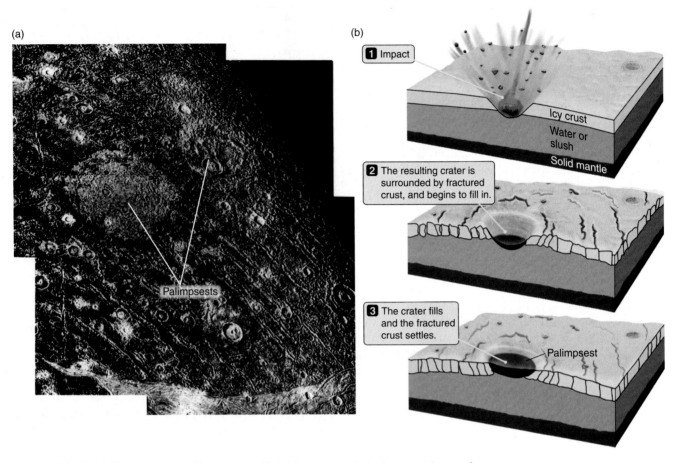

Figure 10.19 (a) Voyager *image of impact scars called palimpsests on Jupiter's moon, Ganymede.* (b) *Palimpsests form as viscous flow smooths out structure left by impacts on icy surfaces.*

moon Tethys, covering 40% of the diameter of the moon itself, while an enormous canyonland wraps at least three-quarters of the way around the moon's equator. Dione's surface shows bright rays that could be frost deposited from the explosive release of interior gases through fractures in the crust. Some craters on Rhea show very bright patches on the walls, which may be ice deposits exposed relatively recently by landslides. The trailing hemisphere of Iapetus is bright, reflecting half the light that falls on it, while much of the leading hemisphere is as black as tar.

Saturn's Mimas, no larger than Ohio, is heavily cratered with deep, bowl-shaped depressions. The most striking feature on Mimas is a huge impact crater on the leading hemisphere. Named Herschel, after the German-English astronomer Sir William Herschel, who discovered many of Saturn's moons, the crater is 130 km across, or a third the size of Mimas itself. (**Figure 10.20** illustrates the striking similarity between Mimas and the "Death Star" from *Star Wars,*

Mimas may have been broken apart and reassembled many times.

which came out three years before the image of Mimas was taken.) It is doubtful that Mimas could have survived the impact of a body much larger than the one that created Herschel. Some astronomers believe that Mimas, and perhaps other small, icy moons as well, was hit many times in the past by objects so large as to fragment the moon into many small pieces. Each time this happened, those individual pieces still in Mimas's orbit would coalesce to reform the moon.

Areas on Uranus's Miranda have been resurfaced by eruptions of icy slush or glacierlike flows (**Figure 10.21**). Other Uranus moons—Oberon, Titania, and Ariel—are covered with faults and other signs of early tectonism. On Ariel, in particular, very old, large craters appear to be missing, perhaps obliterated by earlier volcanism.

Geologically Dead Moons

Geologically dead moons, including Jupiter's Callisto, Uranus's Umbriel, and a large assortment of irregular

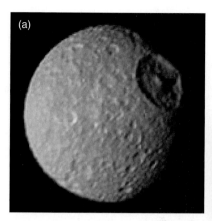

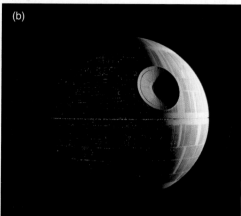

Figure 10.20 (a) Voyager *image of Saturn's moon, Mimas, and the crater Herschel.* (b) *The Death Star from* Star Wars. *(The movie came out three years before the image of Mimas was taken.)*

materials. Except for terrains related to large impact events, the surface is essentially uniform, consisting of relatively dark, heavily cratered terrain. High-resolution images reveal that Callisto's surface has been modified by local landslides and places where the small craters have been erased by some unknown process. Its most prominent feature is a 2,000-km-sized, multiringed structure of impact origin named Valhalla (the largest bright feature visible on Callisto's face in Figure 10.1). The impact may have occurred in a rather thin, rigid crust overlying a fluidlike interior. The fluid mass then rapidly filled the initial crater bowl, leaving only a trace of the impact scar. Geophysical measurements obtained from the *Galileo* spacecraft suggest that Callisto is not differentiated, implying that it never went through a molten phase. On the other hand, the magnetometer on board *Galileo* returned results suggesting that a liquid ocean could exist beneath the heavily cratered surface, implying that some sort of differentiation has occurred. These observations simply point out that many times in science we are faced with conflicting ideas that usually can be resolved only with additional measurements or observations.

Umbriel, the darkest of the large Uranus moons, appears uniform in color, reflectivity, and general surface features, indicative of an ancient surface. The real puzzle posed by Umbriel is why it has been dead for so long while the surrounding large moons of Uranus have been so active.

moons, are those for which there is little or no evidence of internal activity having occurred at any time since their formation. Their surfaces are heavily cratered and show no modification, other than the cumulative degradation caused by a long history of impacts.

Callisto is about the size of Mercury. It is also the darkest of the Galilean moons, yet it is still twice as reflective as Earth's Moon. This indicates that it is rich in water ice, but with a mixture of dark, rocky

Figure 10.21 Voyager *image of fault zones and cratered terrain on Miranda, a 472-km-diameter moon of Uranus.*

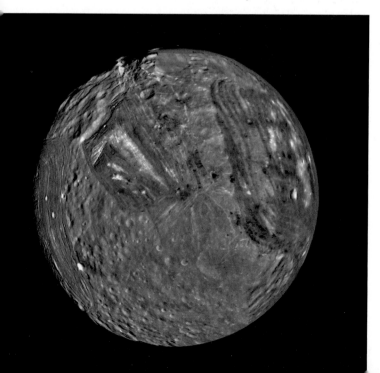

10.4 PLUTO: TINY PLANET OR GIGANTIC COMET?

The last planet to be discovered and the smallest in our Solar System, Pluto has been an enigma since its discovery. The story begins with what appeared to be unacceptable differences between the observed and predicted orbital positions of Uranus and Neptune throughout the 19th century. Inspired by these apparent discrepancies, astronomers early in the 20th century began a search for the unseen body they believed was perturbing the orbits of Uranus and Neptune. They called it Planet X and estimated that it had six times Earth's mass and was located somewhere beyond Neptune's orbit. Planet X was finally found by Clyde Tombaugh in 1930, not far from its predicted position. It was named Pluto for the Roman god of the underworld.

Over the years that followed, observational evidence began to indicate that the mass of Pluto was

Although Pluto's existence was predicted, finding it turned out to be a coincidence.

far too small to have produced the presumed perturbations in the orbits of Uranus and Neptune. When astronomers reanalyzed the 19th-century observations, they found that the orbital discrepancies were erroneous. Pluto's discovery thus turned out to be a strange and improbable coincidence based on faulty data.

Pluto's moon, Charon, was found in 1978. By applying Kepler's and Newton's laws to observations of the moon's motion around Pluto, astronomers were finally able to accurately "weigh" the Pluto-Charon system. Its total mass is only $\frac{1}{418}$ that of Earth **(Figure 10.22).**

Figure 10.22 (a) *An HST image and* (b) *an artist's rendition of Pluto and Charon.*

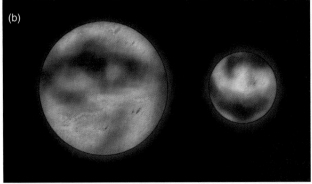

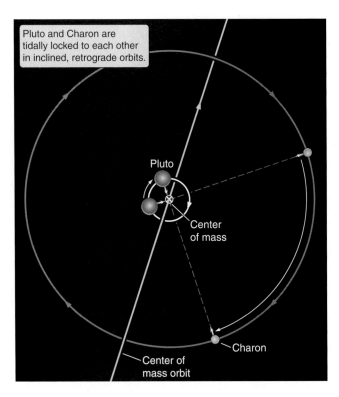

Figure 10.23 *A diagram showing the doubly synchronous rotation and revolution in the Pluto–Charon system.*

Although it is usually the outermost of the Sun's nine known planets, Pluto's eccentric orbit periodically brings it inside the orbit of Neptune, which is nearly circular in shape. From 1989 to 1999, Pluto was closer to the Sun than Neptune, making Neptune temporarily the most distant planet. More than 248 Earth years are required for Pluto to complete one orbit, meaning that Neptune must wait until the year 2237 before it can once again claim title to being the most distant planet. Since its discovery, Pluto has traveled only about a quarter of its way around the Sun.

Pluto is not a typical planet. It is only two-thirds as large as the Moon and only twice as big as its own moon, Charon. Because of the similarity in size between Pluto and Charon, we might think of the Pluto–Charon system

Pluto has a moon half as big as it is.

as a "double planet." Moreover, Pluto and Charon are a dynamically locked pair, the only known example in the Solar System. Both are in synchronous rotation with one another, so that each has one hemisphere that always faces the other body and another that never sees the other body **(Figure 10.23).** If immobile inhabitants were living on opposite sides of Pluto and Charon, they would never know of each others' existence.

From observations made with ground-based telescopes and *HST,* we know that Pluto's surface contains methane and carbon monoxide ice, while Charon's is mostly water ice and lacks methane. Pluto also has a low-density atmosphere of methane and possibly nitrogen, which may contain argon or other heavy gases as well. With densities greater than twice that of water, both Pluto and Charon resemble Jupiter's Europa more than most icy moons. They probably consist of a rocky core that makes up 75% of their mass, surrounded by a water-ice mantle. Because of their great distance from Earth, Pluto and Charon were not included in the great program of spacecraft exploration that took place late in the 20th century. We thus know little about the properties of their surfaces and nothing at all about their geological history. However, a mission to Pluto and Charon will probably take place in the near future.

Pluto is mixture of rock and ice and has a thin methane atmosphere.

Pluto's inclusion in the list of objects we call planets may be little more than an accident of history. In the chapter that follows we will learn that in the outermost parts of the disk of the Solar System, beyond the realm of the planets, countless frozen icy planetesimals orbit. These objects, called *Kuiper Belt objects*, or KBOs, are comet nuclei. We now know of at least two KBOs similar in size to Charon, including one larger than Charon, as well as a pair of KBOs orbiting each other. We have included Pluto and Charon in our discussion of planets and their moons, but they might more properly belong in the following chapter. If discovered today, they would certainly be considered especially large KBOs rather than a planet and its moon.

SEEING THE FOREST THROUGH THE TREES

We normally think of the Solar System as consisting of the planets we learned about in grade school. Yet there are far more moons in our Solar System than there are planets, and each moon is a unique world in its own right. There was no way in 1610 for Galileo to know that the four points of light he discovered circling giant Jupiter were the flagships of a vast armada of strange and amazing worlds. The diversity of the solid worlds of the inner Solar System, while remarkable, is nothing compared with the diversity of the worlds that surround Jupiter, Saturn, Uranus, and Neptune. Some of these worlds are frozen remnants of the time long ago when the giant planets formed at the centers of their own swirling disks. Others are testament to chaotic events in which passing objects were captured in gravitational interactions between planets and other moons. At the other extreme are moons such as Io, a world so geologically active that its surface is remaking itself through volcanism as we watch.

These moons are not the uniform worlds, with iron-and-nickel cores and silicate mantles, found close to the Sun. Instead they, along with Pluto, are assembled from the diverse mix of metals, rocks, and ices that existed in solid form in the outer reaches of the disk that surrounded the newly formed Sun 4.5 billion years ago. As such, these bodies offer important clues about the history of our Solar System, allowing us to test our ideas of what that young Solar System must have been like. These worlds may even give us clues about the history of life in the Universe. The fractured ice flows that make up the surface of Europa may well cover a vast, deep ocean, warmed and enriched by geothermal vents not unlike those that dot the floors of Earth's oceans. When scientists came to realize that Mars and Venus were not as Earth-like as once imagined, prospects for finding life elsewhere in the Solar System seemed dim. Now, at the opening of the 21st century, we are turning our attention with fresh excitement and hope to these small worlds that circle far from the Sun.

Systems of moons may offer a fascinating future for geologists, but they also represent gravitational playgrounds in which the simple two-body interactions described by Kepler's laws give way to complex, chaotic interactions among many different objects. The tidal stresses we learned about in Chapter 9 also rip and pull at these worlds, locking them into synchronous orbits and in some cases churning their interiors into seething cauldrons of geological activity. Sometimes these tidal stresses are so great that a moon or passing comet pays the ultimate price and is ripped into tiny fragments. Borne of these violent events are the majestic rings that circle Saturn and the less spectacular but equally fascinating systems of rings surrounding the other giant planets. The extraordinary complexity of these rings is also a testament to Newton's remarkable inverse square law of gravitation as planets, moons, and rings do their gravitational dance. These interactions go so far as to distort rings into structures so remarkable that they seemed all but impossible until revealed by the eye of the *Voyager* spacecraft.

It is with sadness that we leave this realm of moons and rings so quickly, for these worlds deserve much more attention than we can give them on our journey. Like a package tour promising to show us Europe in three days, we look at the marvels through the window, glance at our guidebooks, and promise to return some day when we have more time. Planetary scientists themselves share our wistful regret. Our glances from passing spacecraft at the moons and rings surrounding the giant planets are really only enough to whet our appetites. No doubt lifetimes of wonder and surprise remain as we continue to explore the smaller worlds of our Solar System.

There may be more moons than there are planets in the Solar System, but planets, moons, and rings do not complete the accounting of objects adrift within the gravitational realm of the Sun. Still remaining are the uncounted swarms of asteroids and comets that orbit within the domain of the planets, and also stretch halfway to the nearest star. In the next stage of our journey we will discover that, far from being insignificant, this flotsam and jetsam carries with it the most direct record of the history of our Solar System, and provides essential pieces to the puzzle of the origin and history of life on Earth.

STUDENT QUESTIONS

THINKING ABOUT THE CONCEPTS

1. Why does Earth not have a ring?
2. Many planetary rings are best seen by backlighting. Give three examples of ordinary objects you might encounter that appear more conspicuous under backlighting conditions.
3. If Triton has ice geysers, is it likely that such geysers are also present on Pluto and its moon, Charon? Explain your answer.
4. In what ways do planetary rings differ from moons? In what ways are they similar?
5. Astronomers once thought that Saturn's rings were a solid disk. Explain why this would not be physically possible.
6. Using data given in Appendix 4, describe and discuss the differences in physical and orbital properties typical of regular and irregular moons.
7. Will the particles in Saturn's bright rings eventually stick together to form one solid moon orbiting at the mean distance of all the ring particles? Explain your answer.
8. Explain the process that drives volcanism on Jupiter's moon, Io.
9. Could there be tidal heating effects on a moon located in a perfectly circular orbit around a planet? Think carefully about this and explain your answer.
10. Discuss evidence that supports the idea that Europa might have a subsurface ocean of liquid water.
11. Explain the differences and similarities in the orbits of Ganymede, Pluto-Charon and Triton.
12. Describe and explain a mechanism that helps keep planetary rings from dissipating.
13. Describe some of the ways Pluto differs greatly from all of the other Solar System planets.
14. Why does Earth pass through the plane of Saturn's rings only every 15 years? Why don't we ever see Saturn's rings pole-on?

APPLYING THE CONCEPTS

15. The inner and outer diameters of Saturn's B Ring are 184,000 km and 235,000 km, respectively. If the average thickness of the ring is 10 m and the average density is 150 kg/m^3, what is the mass of Saturn's B ring?
16. The mass of Saturn's small icy moon, Mimas is 3.8×10^{19} kg. How does this mass compare with the mass of Saturn's B Ring? Why is this comparison meaningful?
17. Particles in the very outer edge of Saturn's A Ring are in a 7:6 orbital resonance with the moon, Janus. If the orbital period of Janus is 16^h41^m, what is the orbital period of the outer edge of Ring A?
18. Planetary scientists have estimated that Io's extensive volcanism could be covering the moon's surface with lava and ash to an average depth of up to 3 mm per year.
    a. Io's radius is 1,815 km. If we model Io as a sphere, what are its surface area and volume?
    b. What is the volume of volcanic material deposited on Io's surface each year?
    c. How many years would it take for volcanism to do the equivalent of depositing Io's entire volume on its surface?
    d. How many times might Io have "turned inside out" over the age of the Solar System?
19. Pluto and its moon, Charon, are a dynamically locked pair, that is, both are in synchronous rotation with each other. Draw a diagram showing how each rotates with respect to the other. Show how some immobile residents of Pluto and Charon would always be looking at one another, while others would never know of one another's existence.

Almost in the center of it, above the Prechistenka Boulevard, surrounded and sprinkled on all sides by stars . . . shone the enormous and brilliant comet of 1812—the comet which was said to portend all kinds of woes and the end of the world.

LEO TOLSTOY (1828–1910)
WAR AND PEACE

ASTEROIDS, METEORITES, COMETS, AND OTHER DEBRIS

11.1 GHOSTLY APPARITIONS AND ROCKS FALLING FROM THE SKY

The Sun, Moon, and planets are usually the brightest, most obvious members of our Solar System, but they are not all there is. Like an ocean afloat with flotsam and jetsam, our Solar System swarms with smaller members that might totally escape our attention until one of them washes ashore. It would be a mistake to assume this planetary debris is unimportant, however. Comets, asteroids, and meteorites provide astronomers with some of their most important clues about how our Solar System formed and evolved.

Throughout human history, the sudden and unexpected appearance of a bright *comet* has elicited fear, wonder and superstition. Early cultures viewed these spectacular visitors as omens. Comets were often seen as dire warnings of disease, destruction, and death, but sometimes as portents of victory in battle or as heavenly messengers announcing the impending birth of a great leader. The earliest records of comets date from as long ago as the 23rd century B.C. Mediterranean and Far Eastern literatures are especially full of references and popular superstitions about comets.

KEY CONCEPTS

Asteroids and comets are tiny in comparison to planets, yet these objects, and their fragments that fall to Earth as meteorites, have told us much of what we know about the early history of the Solar System. As we explore these bits of interplanetary flotsam and jetsam, we will discover:

* Small, irregular worlds called asteroids, which are made of rock and metal;
* That some asteroids are primitive, while others are differentiated or are pieces of larger, differentiated bodies;
* Different types of meteorites that are fragments of these varieties of asteroids;
* That most asteroids orbit between Mars and Jupiter, but some have orbits that cross Earth's;
* Comet nuclei—pristine, icy planetesimals—adrift in the frozen outer reaches of the Solar System;
* Spectacular active comets that are warmed by the Sun as they dive through the inner Solar System;
* Meteor showers that occur when Earth passes through a comet's trail of debris; and
* World-jarring meteorite impacts that still occur today, and have played a vital role in shaping the history of life on Earth.

Meteorites have also engendered fascination and awe for as long as humankind has watched the heavens. Nearly all ancient cultures venerated these rocks from the sky. Early Egyptians preserved meteorites along with the remains of their pharaohs, Japanese placed them in Shinto shrines, and ancient Greeks worshipped them. In 1492—the same year Columbus voyaged to the New World—the townsfolk of Ensisheim (now in France) watched as a 172 kg stone fell to Earth. The event was recorded in a famous woodcut **(Figure 11.1),** and for a while the meteorite was enshrined in a local church. Despite numerous eyewitness accounts of meteorite falls, however, many were slow to accept that these peculiar rocks actually came from far beyond Earth. In 1807 two Yale professors investigating the meteorite that landed in Weston, Connecticut concluded that the object truly did drop from the sky. Thomas Jefferson, third president of the United States and an enlightened scientist, is rumored to have responded, "I would rather believe that two yankee professors would lie, than that stones fall from heaven." But as the evidence continued to mount, it became impossible to ignore. By the early 1800s, scientists had documented so many meteorite falls that their true origin was indisputable. Today, hardly a year passes without a recorded meteorite fall, including some that have smashed into cars, houses, pets, and (occasionally) people.

A third type of interplanetary debris was not discovered until the age of telescopes. On New Year's Day 1801, a Sicilian astronomer named Giuseppe Piazzi found a bright object between the orbits of Mars and Jupiter. Piazzi named the new object Ceres. When Piazzi discovered Ceres, he thought he might have found a hypothetical "missing planet." With a diameter of 940 km, Ceres is larger than most moons and smaller than any planet. But as more objects were discovered orbiting between Mars and Jupiter, astronomers soon realized that Ceres was the flagship of a new kind of

Solar System object. Because Ceres and its cousins appeared in astronomers' eyepieces as nothing more than faint points of light, William and Carolyn Herschel (the brother-sister pair of English amateur astronomers who discovered Uranus) named them *asteroids*, a Greek word meaning "starlike." As we begin the 21st century, we have found more than 30,000 asteroids, of which only about 10,000 have well-determined orbits.

Apart from their role as fascinating curiosities and reminders that there is more to heaven and Earth than meets the eye, this interplanetary debris plays a very significant role in our quest to discover humankind's cosmic origins. We now know that meteorites, asteroids, and comet nuclei are fragments of our Solar System's ancient past. Like archeologists who assemble the history of civilizations from fossil bones and shards **Comets and asteroids provide scientific clues to our past.** of pottery, planetary scientists work to piece together the story of our Solar System from these relics of a time when the Sun, Earth, and planets were young. Fortunately, while scientists have had to reconstruct the puzzle of the early Solar System from the pieces they find, on this journey we have the luxury of starting with the results of their work. We begin by describing what the completed puzzle looks like, and then talk about the pieces and how they fit together.

11.2 ASTEROIDS AND COMETS: PIECES OF THE PAST

We began our discussion of the Solar System in Chapter 5 with the story of its history. Our Solar System was born 4.6 billion years ago inside an enormous interstellar cloud of gas and dust. At the same time that our Sun was becoming a star, tiny dust grains of primitive material were sticking together to form swarms of planetesimals. Some kinds of materials,

Figure 11.1 *A woodcut of the meteorite fall at Ensisheim in 1492.*

such as iron-nickel alloys and rock, were able to stick together even in the hot, inner parts of the protoplanetary disk that surrounded the Sun. Those planetesimals that formed near the young Sun were composed mostly of these heat-resistant materials. Other materials—such as ices of water, methane, and ammonia, and organic compounds[1]—could coalesce only in the outer parts of the disk where temperatures were much lower. In time, many of these planetesimals stuck together to build still larger bodies, the moons and planets. When the planet-building process finally ended, some of the original planetesimals remained in orbit around the Sun.

Most **asteroids** are relics of rocky or metallic planetesimals that originated in the region between the orbits of Mars and Jupiter. Although early collisions between these planetesimals created several bodies large enough to differentiate, Jupiter's tidal disruption prevented them from forming a single moon-sized planet. As they orbit the Sun, asteroids continue to collide with one another, producing small fragments of rock and metal. Most meteorites are pieces of these asteroidal fragments that have found their way to Earth and crashed to its surface.

Asteroids are relics of rocky planetesimals that formed between Mars' and Jupiter's orbits.

Icy planetesimals surviving today form the solid cores, or *nuclei,* of comets. While asteroids can survive in the inner Solar System, comet nuclei cannot. Today, most comet nuclei are preserved in the frigid regions of space beyond the planets. Sometimes, however, something happens to cause one of these icy bodies to change direction and dive inward toward the domain of the inner planets. When a comet nucleus approaches too close to the Sun, solar heating unglues its surface structure, gradually breaking it up into tiny clumps of dusty ice. When Earth passes through swarms of dusty ice from a disintegrating comet nucleus or pebbles from the breakup of an asteroid, the particles can burn up in the atmosphere, creating an atmospheric phenomenon we refer to as a **meteor** (see Excursions 11.1).

Comet nuclei are remnants of icy planetesimals that formed far from the Sun.

Planetesimals that became part of the planets or their moons were so severely modified by planetary processes that nearly all information about their original physical condition and chemical composition has been hopelessly lost. On the other hand, asteroids and comet nuclei constitute an ancient and far more pristine record of what the early Solar System was like. If you visit your local planetarium or science museum, you may find a meteorite on display that you can walk right up to and touch. Do so with respect. The meteorite under your hand may be older than Earth itself. Some meteorites contain tiny grains of material that predate the formation of our Solar System. These grains include diamonds and carbon compounds that originated as material ejected from dying stars.

Some meteorites contain presolar grains.

11.3 METEORITES: A CHIP OFF THE OLD ASTEROID BLOCK

Most asteroids are so small compared with planets that they appear as nothing but unresolved points of light through telescopes on Earth. Historically, this has made it difficult to learn much about them directly. Even determining gross properties such as size, shape, and rate of spin has required sophisticated analysis of telescopic, spacecraft, and radar observations. Despite this fact, we probably know more about the structure and composition of asteroids than any Solar System object other than Earth! As asteroids orbit about the Sun, they occasionally collide with each other, chipping off smaller rocks and bits of dust. Occasionally, one of these fragments is captured by Earth's gravity and survives its fiery descent through Earth's atmosphere as a meteor, to be picked up and added to someone's collection of meteorites.

Meteorites are pieces of asteroids. They tell us about the body they came from.

Thousands of meteorites reach the surface of Earth every day, but only a tiny fraction of these are ever found and identified as such. The best meteorite hunting in the world is in Antarctica. Meteorites are no more likely to fall in Antarctica than anywhere else, but in Antarctica they are far easier to distinguish from their surroundings. Antarctic snowfields serve as a huge net for collecting meteorites. Glaciers then carry the meteorites along, concentrating them in regions where the glacial ice is eroded by wind. The meteorites are left lying on the surface for collectors to pick up. The big advantage of hunting on Antarctic ice is that in

---

[1]The term *organic* does not mean "life," but refers to a large class of chemical compounds containing the element carbon. All terrestrial life is organic, but not all organic compounds come from living organisms.

EXCURSIONS 11.1

METEORS, METEOROIDS, METEORITES, AND COMETS

Meteors are familiar to all of us. Just stand outside for a few minutes on a starry night, away from bright city lights, and you will almost certainly be rewarded by a glimpse of a "falling star," a meteor. Most of us realize we are not witnessing the demise of a real star, but are instead seeing a piece of Solar System debris entering our atmosphere. Few are aware that it is seldom the actual debris we see, but rather a trail of heated atmospheric gas glowing from the friction caused by a small unseen object entering at speeds of up to about 70 km/s. A meteor, then, is an *atmospheric phenomenon*.

The small[2] solid body that creates the meteor is called a **meteoroid.** These small objects are of either cometary or asteroidal origin. Cometary fragments are typically less than a centimeter across and have about the same density as cigarette ash. What they lack in size and mass, they make up for in speed. A 1 g meteoroid (about half the mass of a dime) entering Earth's atmosphere at 50 km/s has a kinetic energy comparable to that of an automobile cruising along at the fastest highway speeds. Before plunging into our atmosphere, the meteoroid may have been orbiting the Sun for millions of years after being chipped from an asteroid or left behind by a disintegrating comet nucleus. Most meteoroids are so small and fragile that they burn up completely before reaching Earth's surface.

A meteoroid large enough to survive all the way to the ground is called a **meteorite.** All meteoroids are pieces of Solar System debris. The larger pieces that survive the plunge through Earth's atmosphere are usually fragments of asteroids. Most of the smaller pieces that burn up before reaching the ground are cometary fragments.

Comets appear as fuzzy patches of light in the nighttime sky. They are bodies of ice, dust, and gas that shine by reflected sunlight. Comets frequently remain visible in our skies for weeks at a time, with tails that can be 100 million km long and stretch halfway across our sky. In contrast, a meteor may streak across 100 km of our atmosphere and last a few seconds at the most.

When meteorites are found on the snowfields of Antarctica, their true identity is often obvious.

many places the *only* stones to be found on the ice are meteorites. Antarctic meteorites have also spent their time on Earth literally in the deep freeze; many are thus relatively well preserved, showing little contamination from terrestrial dust or organic compounds.

Meteorites are extremely valuable because they are samples of the same relatively pristine material from which asteroids are made. Also, unlike planets or asteroids themselves, we can take meteorites into the laboratory and study them using sophisticated equipment and techniques. Scientists compare them to rocks found on Earth and the Moon, and contrast their structure and chemical makeup with those of rocks studied by spacecraft that have landed on Mars and Venus. Meteorites are also compared with asteroids and other objects on the basis of what colors of sunlight they reflect and absorb.

Meteorites are normally grouped into three categories based on the kinds of materials they are made of and the degree of differentiation they experienced within their parent body. **Figure 11.2** shows cross sections of these several meteorite types.

There are three classes of meteorites: stony (including chondrites and achondrites), iron, and stony-iron.

Over 90% of meteorites are **stony meteorites,** which are similar to terrestrial silicate rocks. A stony meteorite can be recognized by the thin coating of melted rock that forms as it passes through the atmosphere. Many stony meteorites contain small round spherules called chondrules, which range in size from that of sand grains up to marble-sized objects. Chondrules are all held together by a matrix of finer-grained material, much like the gravel in concrete held together by a matrix of cement. Stony meteorites containing chondrules are called **chondrites,** while those that do not contain chondrules are called **achondrites.** Unlike chondrites, achondrites have

---

[2]By definition, meteoroids are those objects that fall within a size range of 100 μm to 100 m. Particles smaller than this are considered *interplanetary dust*, while larger objects are called *planetesimals*.

crystals that appear to have formed in the same place and at the same time.

The second major category of meteorites, **iron meteorites,** are the easiest to recognize. Their surfaces have a melted and pitted appearance generated by frictional heating as they streaked through the atmosphere. (The meteorite in your local museum is probably an iron meteorite.) Even so, many are never found, either because they land in water or simply because no one happens to notice and recognize them for what they are.

The final category is the **stony-iron meteorites,** which consist of a mixture of rocky material and iron-nickel alloys. Stony-iron meteorites are relatively rare.

One of the most interesting stories that meteorites have to tell about the Solar System is its age. In Chapter 6 we learned that radioactive materials locked up in solids provide a kind of clock that begins ticking the moment the material solidifies. By comparing how much of a radioactive substance remains in a rock with how much of the radioactive material's decay products have accumulated, we can tell how long this clock has been ticking and therefore how long the rock has been solid. When we apply this technique to meteorites, nearly all turn out to be about 4.54 billion years old. One type of meteorite stands out with particular interest. **Carbonaceous chondrites,** chondrites that are rich in carbon, are thought to be the very building blocks of the Solar System. Although these meteorites

Meteorites tell us the age of the Solar System.

cannot be dated directly because they are composed of aggregates of many individual grains, indirect measurements suggest they are about 4.56 billion years old. This is our best measurement of the time that has passed since the formation of the Solar System.

The diversity among meteorites and asteroids reflects the range of physical conditions under which they were formed. For example, laboratory experiments simulating formation show that chondrules formed at higher temperatures than the finer-grained minerals that surround them. Chondrules must once have been molten droplets that rapidly cooled to form the crystallized spheres we see today. Meteorites thus hold clues to the physical conditions existing in the Solar System at the time they were formed.

Not surprisingly, the different classes of meteorites are related to different classes of asteroids. As we learned in Chapter 5, the objects that we find in the Solar System today are the result of small objects sticking together to form larger objects. This process is called *accretion.* Refer to **Figure 11.3** as we follow the possible fates of a planetesimal. As was the case in the interior of the terrestrial planets, large amounts of energy were released in collisions as planetesimals accreted smaller objects. Additional thermal energy was released as radioactive elements inside the planetesimals decayed. Some planetesimals never reached the temperatures needed to melt their interiors. These planetesimals—aggregations

C-type asteroids never differentiated.

of the primordial material from which they formed—simply cooled off, and have since remained pretty much as they were when they formed. These are the **C-type asteroids,** which are composed of primitive material that we believe has largely been unmodified since the origin of the Solar System, almost 4.6 billion years ago.

(a) Chondrules

(c) Stony material / Once-molten iron and nickel

(b)

(d)

Figure 11.2 *Cross sections of several kinds of meteorites: (a) a chondrite, (b) an achondrite, (c) a stony-iron meteorite, and (d) an iron meteorite. Chondrites and achondrites are stony meteorites.*

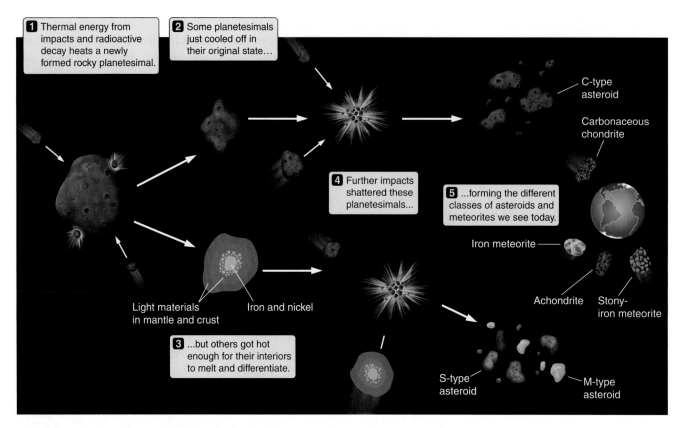

Figure 11.3 *The fate of a rocky planetesimal in the young Solar System depends on whether it gets large and hot enough to melt and differentiate, and the impacts it experiences. Different histories lead to the different types of asteroids and meteorites found today.*

In contrast to the C-type asteroids, some planetesimals were heated enough by impacts and radioactive decay to cause them to melt. Once the interior of a planetesimal melted, it differentiated, just as all the planets and large moons did. Denser matter such as iron sank to the center of the planetesimal, while lower density material such as compounds of calcium, silicon, and oxygen floated toward the surface and combined to form a mantle and crust of silicate rocks. **S-type asteroids** (and the stony achondritic meteorites that come from them) may be pieces of the mantles and crusts of such differentiated planetesimals. They are chemically more similar to igneous rocks found on Earth than to primitive chondrites and C-type asteroids. This means that they were hot enough at some point to lose their carbon compounds and other volatile materials to space. The large interlocking crystals that make up achondrites are also like those seen in rocks that slowly cooled from molten material. Similarly, **M-type asteroids** (from which iron meteorites come) are easy to understand as fragments of the iron- and nickel-rich

> **S- and M-type asteroids may be pieces of larger, differentiated bodies.**

cores of one or more differentiated planetesimals that shattered into pieces in collisions with other planetesimals. Indeed, slabs cut from iron meteorites show large interlocking crystals characteristic of iron that cooled very slowly from molten metal. The rare stony–iron meteorites may come from the transition zone between the stony mantle and the metallic core of such a planetesimal.

A few planetesimals in the region between the orbits of Mars and Jupiter seem to have evolved toward becoming tiny planets before being shattered by collisions. They even became volcanically active, complete with eruption of lava flows onto their surfaces. Vesta (the third largest known asteroid, with a diameter of 525 km) seems to be a piece of such a highly evolved planetesimal. Vesta does not fit well into any of the main asteroid groups, but its spectrum is strikingly similar to the reflection spectrum of a peculiar group of meteorites. These meteorites, which are probably pieces of Vesta, look like rocks taken from iron-rich lava flows on Earth and the Moon. The main difference is that they are only 4.5 billion years old, a bit younger than most meteorites. It is hard to escape the conclusion that early in the history of the Solar System, some planetesimals were large enough to become differentiated and to develop volcanism. But rather than going on to form planets, these planetesimals broke into pieces in collisions with other planetesimals.

The story of asteroids and meteorites is a great success of planetary science. A vast wealth of information about this diverse collection of objects has come together into a single consistent story of planetesimals growing and possibly becoming differentiated, then being shattered by subsequent collisions. The story is even more satisfying because it, in turn, fits so well with the even grander story of how most planetesimals were accreted into the planets and their moons.

But what of the "missing planet" that Piazzi thought he found in 1801? Was there once a single planet between present-day Mars and Jupiter that shattered in a tremendous collision? While this may be a wonderfully romantic notion, the evidence says that events did not happen that way. A more detailed look at the wide variety of chemical properties and orbits of today's asteroids seems to indicate that they are fragments of at least a dozen or more parent bodies, rather than of a single planet. As discussed in Chapter 9, we now understand that it was the "stirring" of this region by Jupiter's gravity that prevented the asteroids from coalescing into a planet.

As we stress throughout this book, patterns are extremely important in science, not only when they are followed, but also when they are broken. Some types of meteorites fail to follow the patterns discussed above. While most achondrites have ages of 4.5 billion to 4.6 billion years, some members of one group are less than 1.3 billion years old. Some are chemically and physically similar to the rocks, soil, and atmospheric gases that were measured on Mars by the *Viking* landers. The similarities are so strong that most planetary scientists believe these meteorites are pieces of Mars that were knocked into space by asteroidal impacts.[3] This means we have pieces of another planet we can study in laboratories here on Earth. The martian meteorites support the general belief that much of the surface of Mars is covered with iron-rich volcanic materials.

To hold certain meteorites is to hold a piece of Mars.

If pieces of Mars have reached Earth from its orbit almost 80 million kilometers beyond our own, we might expect that pieces of our companion Moon would find their way to Earth as well. Indeed, another group of meteorites bears striking similarities to samples returned from the Moon. Like the meteorites from Mars, these meteorites are chunks of the Moon that were blasted into space by impacts and later fell to Earth.

11.4 ASTEROIDS ARE FRACTURED ROCK

With a few exceptions, such as Ceres, asteroids are too small for their self-gravity to have pulled them into a spherical shape. Some asteroids imaged by spacecraft or by Earth-based radar have highly elongated shapes, suggesting objects that are either fragments of larger bodies or objects created haphazardly from collisions between several smaller bodies. The masses of a number of asteroids have been measured by noting the effect of their gravity on spacecraft passing nearby. Knowing the mass and the shape of an asteroid allows us to determine its density. The densities of these asteroids range between 1.3 and 3.5 times the density of water. Those at the lower end of this range are considerably less dense than the meteorite fragments they create. How can asteroids be less dense than the rock that likely came from them? Planetologists think that some of them are shattered heaps of rubble, with large voids between the fragments. Once again, this is as we would expect of objects that were assembled from smaller objects and then suffered a history of violent collisions.

Asteroids can have very irregular shapes.

Asteroids rotate just as planets and moons do, although the rotation of irregularly shaped asteroids is more of a tumble than a spin. A day on a typical asteroid is about 9 hours long, although the rotation periods of some asteroids are as short as 2 hours, and others are longer than 40 Earth days. How can we measure the rotation period of an object that appears starlike in a telescope? Unless the asteroid is perfectly round, and we know that very few are, we measure rotation periods by watching changes in their brightness as they alternately present their broad and narrow faces to Earth.

A day on a typical asteroid is only 9 hours long.

Asteroids are found throughout the Solar System. Most orbit the Sun in several distinct zones, with the majority residing between the orbits of Mars and Jupiter in the **main asteroid belt.** The main-belt asteroids are not distributed uniformly within the belt.

---

[3]In 1996, a NASA research team announced that a particular Martian meteorite (ALH84001) showed physical and chemical evidence of early life on Mars. These conclusions have since been challenged by many scientists and remain controversial.

As discussed in Chapter 9, several empty regions, the Kirkwood gaps, have been depleted by perturbations from Jupiter.

The main belt contains at least 1,000 objects larger than 30 km (about the size of Washington, D.C.), of which about 200 are larger than 100 km. Small asteroids are more difficult to find than large ones, but there are many more of them. Taking this into account, it is estimated that there are about 30 times as many asteroids larger than 30 km as there are asteroids larger than 100 km. Similarly, there seem to be about 30 times as many asteroids over 10 km in size as there are asteroids larger than 30 km. By assuming that this pattern continues for smaller and smaller bodies, we estimate that there may be as many as 10 million asteroids larger than 1 km in the main asteroid belt. However, while asteroids are great in number, they account for only a tiny fraction of the matter in the Solar System. If all of the asteroids were combined into a single body, it would be about a third the size of Earth's Moon.

There are many asteroids, but their total mass is small.

Figure 11.4 shows orbits of several asteroids that are representative of their classes. In Chapter 9 we discussed the Trojan asteroids that occupy the L_4 and L_5 Lagrangian points of the Sun–Jupiter system. Asteroids whose orbits bring them close to Earth's orbit are called **near-Earth asteroids.** Members of one group, called the *Amors,* cross the orbit of Mars but do not cross Earth's orbit. Members of two other groups, the **Atens** and the **Apollos,** have orbits that cross Earth's orbit (see Figure 11.3). The difference between these two populations is that Apollos also cross the orbit of Mars, while the Atens remain inside Mars's orbit. Atens and Apollos occasionally collide with Earth or the Moon. Astronomers estimate that there are more than 2,000 Atens and Apollos with diameters larger than a kilometer. As we discovered in Chapter 6 (see Connections 6.1) such collisions are geologically important, and have dramatically altered the history of life on Earth.

Apollo and Aten asteroids cross Earth's orbit and could collide with us one day.

11.5 ASTEROIDS VIEWED UP CLOSE

Until the space age, scientists had no good idea of what asteroids looked like. As the 1980s drew to a close planetary scientists were about to have their theories put to the test. Would the first close-up views of aster-oids pull the rug out from under our carefully constructed story of asteroids, meteorites, and the early Solar System? (If so, it would not be the first time a closer examination of some phenomenon revealed that people's ideas about it were wrong!) Fortunately, as spacecraft began to return detailed information about the asteroids they visited, those data served to place our ideas about the history of the Solar System on even more solid footing.

In 1991, the *Galileo* spacecraft was following a circuitous path through the inner Solar System, gaining momentum from encounters with Earth and Venus to help send it on its journey to Jupiter. While flying through the main asteroid belt, it got close enough to the asteroid Gaspra to get dozens of good images. Gaspra is an S-type asteroid, about 12 by 20 by 11 km. The *Galileo* images show it to be cratered and irregular in shape. Faint, groovelike patterns on its surface may be fractures resulting from the same impact that chipped Gaspra from a larger planetesimal. Images taken in several different wavelengths of light show distinctive colors on Gaspra's surface, implying Gaspra is covered with a variety of different types of rock.

11.1 GALILEO

Later in its mission, *Galileo* made another pass through the main belt and this time sent back images of the asteroid Ida **(Figure 11.5).** Ida orbits in the outer part of the main asteroid belt, and it too is an S-type asteroid. *Galileo* flew so close to Ida that it was able to see details as small as 10 m—about the size of a small house or cottage. Ida is shaped like a croissant, 56 km long and 24 by 21 km across. Like Gaspra, Ida shows the scars of a long history of impacts with smaller bodies. Just as planetary scientists use the number and sizes of craters to estimate the ages of the surfaces of planets and moons (as we saw in Chapter 6), the cratering on Ida indicates that its surface is a billion years old, twice the age estimated for Gaspra. Also like Gaspra, the *Galileo* images of Ida revealed the presence of fractures in the asteroid. The fractures seen in the two asteroids indicate that they must be made of relatively solid rock. (You can't "crack" a loose pile of rubble!) This is an important confirmation of the idea—crucial to our story of asteroids and meteorites—that some asteroids must be pieces chipped from larger, solid objects.

Ida seems to be a piece of a still larger asteroidal body.

Several features on Ida appear to be landslides, indicating that the surface of Ida must be covered with a "soil" of sorts. Other features are clearly rocks, some larger than football fields, exposed on the surface of the asteroid. Together these data show that Ida's gravity was able to hold on to some of the material thrown

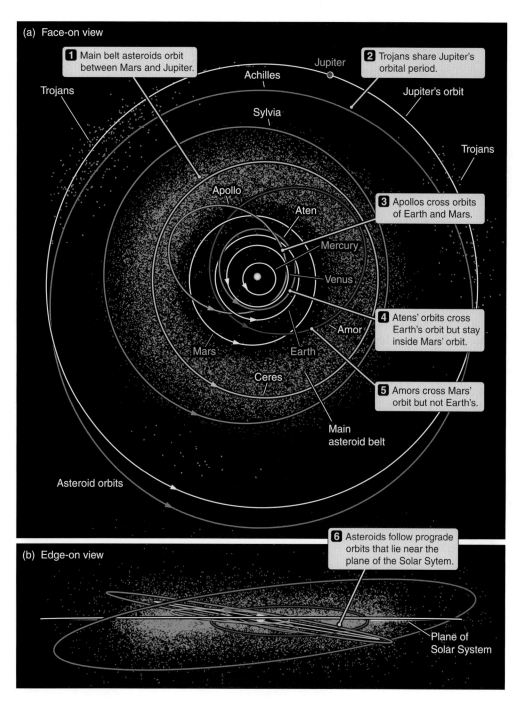

(a) Face-on view

1 Main belt asteroids orbit between Mars and Jupiter.

Trojans

Jupiter

Achilles

Sylvia

Jupiter's orbit

2 Trojans share Jupiter's orbital period.

Trojans

Apollo

Aten

Mercury

Venus

3 Apollos cross orbits of Earth and Mars.

Amor

Mars

Earth

Ceres

Main asteroid belt

4 Atens' orbits cross Earth's orbit but stay inside Mars' orbit.

5 Amors cross Mars' orbit but not Earth's.

Asteroid orbits

(b) Edge-on view

6 Asteroids follow prograde orbits that lie near the plane of the Solar Sytem.

Plane of Solar System

Figure 11.4 *Blue dots show the locations of known asteroids at a single point in time. Most families of asteroids take their names from prototype members of the family. For example, Apollo family asteroids cross the orbits of Earth and Mars, like the asteroid Apollo. Most asteroids, such as Ceres and Sylvia, are main belt asteroids.*

tem objects with solid surfaces, its surface is cratered from impacts. While planetary scientists had long speculated that asteroids might themselves have moons, Dactyl was the first such object to be discovered. But is Dactyl really a moon? To planetary scientists, any significant Solar System body orbiting a larger body other than the Sun is called a *moon*. Dactyl may not be the stuff poetry is written about, but by the planetary scientist's definition of moon, it counts.

Since *Galileo*'s visits to Gaspra and Ida, spacecraft have visited three other asteroids: Mathilde, Eros, and Braille. This last object was seen only briefly in 1999 during the flyby of *Deep Space 1*. It was found to have a highly reflective surface, similar to Vesta, leading to speculation that Braille could be a chunk knocked from Vesta by a collision sometime in the past.

The *NEAR Shoemaker*[4] spacecraft was designed to do more than make a brief encounter with an

about when smaller objects hit its surface. This hardly seems worth mentioning until you consider that Ida's surface gravity is feeble—only a thousandth as strong as that of Earth. The escape velocity from Ida is only about 20 meters per second—about half the speed of a good fastball pitch.

Ida's gravity has managed to hold on to more than a bit of soil—*Galileo* images show a tiny moon orbiting about the asteroid. Ida's moon, called Dactyl, is only 1.4 km across, and like nearly all Solar Sys-

Ida's feeble gravity holds a tiny moon called Dactyl.

[4]*NEAR* stands for *Near Earth Asteroid Rendezvous*. The spacecraft was renamed *NEAR Shoemaker* in March 2000 in honor of Gene Shoemaker, a pioneer of the study of comets and asteroids who died in an automobile accident in 1997 while studying impact craters in the Australian outback.

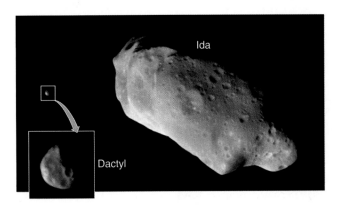

Figure 11.5 *A* Galileo *spacecraft image of the asteroid Ida with its tiny moon, Dactyl (also shown enlarged in inset).*

NEAR Shoemaker **was the first spacecraft to orbit and then land on an asteroid.**

asteroid—in early 2000 it was placed into orbit around the asteroid Eros to begin long-term observations of this S-type object. **Figure 11.6** shows a number of images of Eros obtained by *NEAR Shoemaker*. As a near-Earth asteroid, Eros is one of the 800 known objects whose orbits bring them within 1.3 AU of the Sun. Eros measures about 33 by 13 by 13 km and, like its counterparts Gaspra and Ida, shows a surface with grooves, rubble, and impact craters, including a crater 8.5 km across. However, the scarcity of smaller craters suggests that its surface could be younger than Ida's. One of the most important discoveries made by *NEAR Shoemaker* was analysis of X-rays emitted by the asteroid as it was being irradiated by X-rays from an eruption on the Sun. The X-ray/Gamma-ray spectrometer aboard *NEAR Shoemaker* detected emission lines from a number of elements, allowing the chemical composition of the asteroid to be measured. These observations confirmed the similarity in composition between Eros and primitive meteorites. At the end of its mission, *NEAR Shoemaker* became the first spacecraft to land on an asteroid, when on February 12, 2001, the orbiter was gently crashed into the surface of Eros. On its way in, *NEAR Shoemaker* obtained images showing features as small as a few centimeters in size.

On its way to Eros, *NEAR Shoemaker* flew past the asteroid Mathilde and provided our first information about a C-type body. Mathilde was found to measure 66 by 48 by 44 km and to have a surface only half as reflective as charcoal. Its color properties suggest a composition of materials such as carbonaceous chondrules. However, estimates of the overall density of Mathilde are about 1,300 kg/m³ (1.3 times the density of water),

only half that of carbonaceous chondrite meteorites measured in the laboratory. This implies that Mathilde is "porous" inside, as expected if it is composed of chunks of rocky material stuck together by the accretion process with open spaces between the chunks. The surface of Mathilde is dominated by craters, the largest of which is more than 33 km across. The great number of craters means that Mathilde probably dates back to the very early history of the Solar System.

Some asteroids are porous inside.

The spectrum of sunlight reflected by Mars's tiny moons, Phobos and Deimos **(Figure 11.7),** is similar to that of C-type asteroids, and many scientists believe these moons must be asteroid-like objects that were somehow captured by Mars. Images of Phobos and Deimos returned by *Mariner 9* in 1971 showed them to be irregular shaped objects with cratered surfaces, but planetary scientists got an even better look at the two moons a few years later. NASA controllers used the *Viking* orbiters to take close-up views of both moons, commanding *Viking 2*'s orbiter into an orbit that caused it to whiz by within 28 km of Deimos, and raising a few concerns about a possible unplanned "impact experiment" in Mars orbit! Fortu-

Phobos and Deimos may be asteroids captured by Mars.

Figure 11.6 *Images of the asteroid Eros obtained by the* NEAR Shoemaker *spacecraft.*

Figure 11.7 Viking *orbiter images of Mars's two tiny moons:* Deimos *(left), and* Phobos *(right).*

nately, the delicate maneuver came off without a hitch, and the spacecraft sent back images showing features as small as a meter in size. While the *Viking* images of Phobos and Deimos were spectacular, it is still not clear that they answered the question at hand. Are Phobos and Deimos really asteroids? Controversy about this question has never been put to rest. Some scientists argue that it is unlikely that Mars could have "captured" two asteroids, and that Phobos and Deimos must somehow have evolved together with Mars. A third possibility is that the bodies must have been parts of a much larger body that was fragmented by a collision early in the history of Mars.

Once called the "vermin of the skies," asteroids have now achieved considerable respectability in scientific circles. Not only do they hold fascinating and unique clues about how the Solar System formed, but they also play a part in the continuing (albeit slow) evolution of the Solar System. And if an asteroid were to get *too* close to Earth, as happens from time to time, it would quickly become a *very* respectable topic for conversation!

11.6 THE COMETS: CLUMPS OF ICE

As with many astronomical objects, our concept of comets changed markedly during the 16th and 17th centuries, when we looked upon nature with an increasingly refined and rational eye. In much earlier times, comets were regarded simply as ghostly apparitions, pale luminous patches or streaks of light in the nighttime sky that would mysteriously appear, remain for a few days or weeks, and then vanish. Until the end

of the Middle Ages, the Aristotelian view of comets prevailed, regarding them as atmospheric phenomena rather than astronomical objects. The popular beliefs that comets were very nearby and that their tails contained poisonous vapors contributed to the fear and panic that comets frequently generated.

A major change in our view came in the 16th century at the hands of Tycho Brahe. Tycho reasoned that if comets were atmospheric phenomena like clouds, as Aristotle had supposed, then their appearance and location in the sky should be very different to observers located many miles away from each other. However, when Tycho compared sightings of comets made by observers at several different sites, he found no evidence of such differences, and concluded that comets must be at least as far away as the Moon. Lest we become arrogant in our faith in humankind's rationality, it is worth noting that the proof that comets are far outside Earth's atmosphere did not completely dispel the fear they evoke. As recently as 1910, the news that Earth would pass through the tail of Halley's Comet was accompanied by widespread apprehension throughout Europe and the United States. And during the approach of comets Kohoutek in 1973 and Hale-Bopp[5] in 1997, many otherwise rational people experienced similar anxieties, regrettably fueled by certain opportunistic writers who exploit human fear and superstition for power or profit.

> **Comets are astronomical objects, *not* atmospheric phenomena.**

THE ABODE OF THE COMETS

When we think of a comet, we think of a spectacular display of the sort produced by comet Hyakutake in 1996, but for most of their lives, comets are merely icy planetesimals—**comet nuclei**—adrift in the frigid outer reaches of the Solar System. Comet nuclei put on a show only when they come deep enough into the inner Solar System to suffer destructive heating from the Sun. When they are close enough to show the effects of solar heating, we call them **active comets.** Comet nuclei spend most of their lives as nothing more than icy planetesimals—small clumps of primordial material that managed to escape the planet-building

---

[5]Believing that comet Hale-Bopp was accompanied by a UFO that would carry their souls away to a "better place," 39 members of the cult Heaven's Gate committed mass suicide in March 1997.

Swarms of comets spend most of their lives far beyond the planets.

process—that follow orbits that keep them far beyond the planets in the outermost reaches of the Solar System. Drifting in these distant regions, most are much too small and far away to be seen and counted by telescopes on Earth, and so no one really knows their total number. Estimates range as high as a trillion (10^{12}) comet nuclei—more than all the stars in the Milky Way Galaxy. Despite this, we have seen hardly more than a thousand, perhaps only a billionth of the presumed total.

We know where comets reside by observing their orbits as they pass through the inner Solar System. Based on these studies, they seem to fall into two distinct groups: *Kuiper Belt* comets and the *Oort Cloud* comets. These two populations of comet nuclei are named for scientists Gerard Kuiper and Jan Oort, who first proposed their existence in the mid–20th century.

The Kuiper Belt and the Oort Cloud are reservoirs of comets.

The extent of the realm of comets is illustrated in **Figure 11.8.** Comet nuclei from the **Kuiper Belt** orbit about the Sun within a disk-shaped region that begins just beyond the orbits of Neptune and Pluto and extends out to several thousand astronomical units from the Sun. The innermost part of the Kuiper Belt appears to contain tens of thousands of icy planetesimals, which we call **Kuiper Belt objects** (KBOs). The largest KBOs are over a thousand kilometers across, and the closest of these are just within the range of ground-based telescopes. Astronomers have already found several hundred, but many smaller KBOs must also be there, just beyond the reach of our telescopes. While they are too far from the Sun to be "active," it seems very likely that these planetesimals are comet nuclei.

Unlike the flat disk of the Kuiper Belt, the **Oort Cloud** is a spherical distribution of comet nuclei that are much too remote to be seen by even the most powerful telescopes. We know the size and shape of the Oort Cloud because comet nuclei from the Oort Cloud approach the inner Solar System from seemingly random directions in the sky, and follow orbits that bring them in from as far as 50,000 AU from the Sun, or about a fifth of the way to the nearest stars!

We can think of the Kuiper Belt and Oort Cloud as enormous reservoirs of icy planetesimals that now and then fall into the inner Solar System. But why do they sometimes leave their frigid haven and take what will probably be a fatal plunge inward toward the Sun? One idea is that comet nuclei run into each other from time to time, knocking each other from their orbits. But even with the large number of nuclei that the Oort Cloud

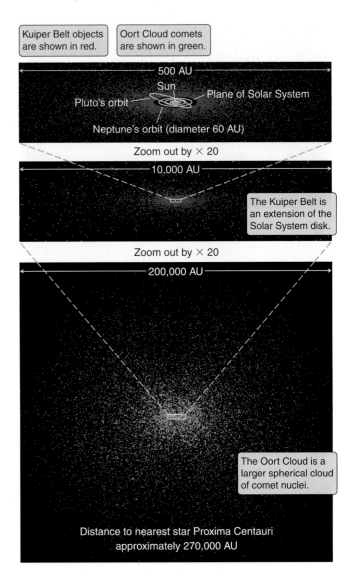

Figure 11.8 *Most comet nuclei near the innner Solar System form an extension to the disk of the Solar System called the Kuiper Belt. The spherical Oort Cloud is far larger and contains many more comets.*

must contain, the immense volume of this region means that the distances between comet nuclei are great. The average separation between an Oort Cloud object and its nearest neighbor must be at least 10 AU—as great as the distance between the Sun and Saturn—making collisions between comet nuclei *extremely* uncommon events. It seems more likely that gravitational perturbations by objects beyond the Solar System—such as stars and interstellar clouds—disturb the orbits of comet nuclei, kicking the nuclei in toward the Sun.

By looking at the distribution of stars around us, astronomers estimate that every 5 million to 10 million years, one star or another, traveling in nearly the same orbit around the Milky Way as does the Sun, passes within about 100,000 AU of the Sun, perturbing the orbits of comet nuclei. The gravitational attraction from huge clouds of dense interstellar gas concentrated in the plane of the Milky Way Galaxy[6] might also stir up the Oort Cloud as the Sun passes back and forth through the galactic plane during its 220-million-year orbital journey around the center of our Galaxy. These disturbances from beyond the Solar System have little effect on the orbits of the planets and other bodies in the inner Solar System. Inner Solar System objects are close enough to the Sun that external disturbances never exert more than a tiny fraction of the gravitational force of the Sun. In the distant Oort Cloud, however, things are very different. Oort Cloud comet nuclei are so far from the Sun, and the Sun's gravitational force on them is so feeble, that they are just barely bound to the Sun at all. The tug of a slowly passing star or interstellar cloud can compete with the Sun's gravity, significantly changing the orbits of Oort Cloud objects. If the interaction adds to the orbital energy of a comet nucleus, it may move outward to an even more distant orbit, or perhaps escape from the Sun completely to begin an eons-long odyssey through interstellar space. On the other hand, a comet nucleus that loses orbital energy falls inward. Some of these come all the way into the inner Solar System, where they may appear briefly in our skies as active comets before returning once again to the Oort Cloud.

The gravity of other stars may send comets into the inner Solar System.

Unlike comet nuclei in the Oort Cloud, those in the Kuiper Belt are packed close enough together to interact gravitationally from time to time. In such events, one nucleus gains energy while the other loses it. The "winner" may gain enough energy to be sent into an orbit that reaches far beyond the boundary of the Kuiper Belt. It seems likely that the Oort Cloud was populated in just this way (by comet nuclei ejected from the very inner edge of the Kuiper Belt in what is now the Uranus-Neptune region). The experience could prove fatal for the "loser" if enough orbital energy is lost for it to fall inward toward the Sun.

ANATOMY OF AN ACTIVE COMET

Figure 11.9 shows a diagram of a fully active comet. The small object at the center of the comet—the icy planetesimal itself—is the comet nucleus. The nucleus is by far the smallest component of a comet, but it is the

Figure 11.9 (a) *The principal components of a fully developed active comet are the nucleus, the coma, and two types of tails called the dust tail and the ion tail. Together, the nucleus and coma are called the head.* (b) *Comet Hale-Bopp.*

---

[6]Interstellar clouds and the structure of the Milky Way Galaxy will be discussed in Chapters 14 and 18, respectively.

An active comet has a tiny nucleus at its heart.

source of all the mass that we see stretched across the skies as the comet nears the Sun. Comet nuclei are anywhere from a few tens of meters to several hundred kilometers across. We might formally describe comet nuclei as "planetesimals of modest size composed of a mixture of various ices of volatile compounds, organics, and dust grains loosely packed together to form a porous conglomerate," but mercifully a shorter description is possible. In the middle of the 20th century, astronomer Fred Whipple offered an elegant way to sum all of this up in just two words: comets are "dirty snowballs."

Unlike asteroids, which have been through a host of chemical and physical changes as a result of collisions, heating, and differentiation, most comet nuclei have been preserved over the past 4.6 billion years by the "deep freeze" of the outer Solar System. Comet nuclei are made of the most nearly pristine material remaining from the formation of the Solar System.

As a comet nucleus nears the Sun, sunlight heats its surface, turning volatile ices into gases, which then stream away from the nucleus, carrying embedded dust particles along with them. This process of conversion from solid to gas is called **sublimation.** An example much closer at hand illustrates this process. Dry ice (frozen carbon dioxide), does not melt like water ice, but instead turns directly into carbon dioxide gas. Water ice melts. Dry ice sublimates. That is why we call it "dry." Set a piece of dry ice out in the sun on a summer day, and you will get a pretty good idea of what happens to a comet.

The gases and dust driven from the nucleus of an active comet form a nearly spherical atmospheric cloud around the nucleus called the **coma** (plural comae). The nucleus and the inner part of the coma are sometimes referred to collectively as the comet's **head.** Pointing from the head of the comet in a direction more or less away from the Sun are long streamers of dust, gas, and ions called the **tail.**

The coma is an extended cometary atmosphere surrounding the nucleus.

A VISIT TO HALLEY

Typical comet nuclei at typical distances appear so small that even the most powerful telescopes see them only as points of light. Even when a rare comet comes very close to Earth, we are still out of luck because the nucleus remains hidden within the dusty coma that surrounds it. While astronomers have studied the gases that make up the comae and tails of comets with great interest for centuries, direct observations of comet nuclei have remained an elusive prize.

This all changed in 1986, when Halley's Comet made its "once in a lifetime" trip into the inner Solar System, providing a rare opportunity to send space probes to observe a nucleus at close range. This opportunity was so rare because, despite the fact that even relatively pristine comet nuclei are frequent visitors to the inner Solar System, we seldom have enough advance warning of a comet's visit or sufficiently detailed knowledge of its orbit to mount a successful mission to intercept it. But while planetary scientists knew the orbit of Halley extremely well, sending probes to the comet still turned out to be a difficult task as space missions go. The problem was that the orbit of comet Halley is *retrograde*—it goes around the Sun in a direction opposite to that of Earth and the other planets. As a result, the closing speed between an Earth-launched spacecraft and the comet would be 68 km/s, or about 75 times the speed of a bullet from a high-powered rifle. Observations close to the nucleus would have to be made very quickly, and there would be the ever present danger of high-speed collisions with debris from the nucleus.

Halley's Comet comes around once in a lifetime.

Despite the difficulties, working in cooperation with each other, three different space agencies rose to the challenge of sending a five-spacecraft armada to meet comet Halley as it passed through the inner Solar System. The whole operation was rather like driving down the highway at 70 miles per hour and trying to thread a needle held out of the window by someone in the oncoming lane, but in the end, it was a glorious success. Almost all of what we know about comet nuclei and the innermost parts of the coma were learned from data sent back by this armada of Soviet, European, and Japanese (but not U.S.) spacecraft.

An international armada of spacecraft visited comet Halley.

Images taken by these spacecraft showed the nucleus of Halley to be an irregular peanut-shaped object measuring 8 by 8 by 14 km **(Figure 11.10a).** As the Soviet *VEGA* and the European *Giotto* spacecraft passed close by the nucleus, they observed jets of gas and dust moving at speeds of up to 1,000 m/s from its surface, far exceeding the ability of its feeble gravity to contain the material. Here was dramatic evidence that the gas and dust that make up a comet's coma and tail come directly from the nucleus. But instead of streaming out from the whole surface of the nucleus, the spacecraft images showed that material was coming instead from several hot spots, or jets, that covered only about a tenth of the surface. Ninety percent of the surface was "quiet" at the time of the observations. This is most easily understood if the surface of the nucleus is covered by a thick crust of rubble, with the escaping gas coming from beneath this crust through a number of small cracks or fissures.

By observing the jets of material streaming away from the nucleus of Halley, planetary scientists estimated that 20,000 kg of gas and 10,000 kg of dust were being lost by the nucleus each second. During its 1986 passage around the Sun, comet Halley must have lost at least 100 billion (10^{11}) kilograms of material. But these jets also indirectly allowed planetary scientists to measure another fundamental property of this comet. Just like the engines on a jet airplane, the jets of material streaming away from the nucleus push on it, subtly altering its orbit. By observing the jets flowing away from the nucleus, scientists could tell how hard the jets were pushing on it. And by observing the comet's orbit, they learned how the nucleus was responding to these shoves. Using these observations, scientists were then able to apply Newton's laws of motion, which require the change in the momentum of the nucleus to balance the change in momentum of the material streaming away. They thus estimated the mass of the nucleus to be about 3×10^{14} kg. This mass divided by the volume as measured from the images of the nucleus, tells us that the nucleus is only about a fourth as dense as water. This is more like a loosely packed powdery snow than normal ice. As suggested by Whipple years earlier, the nucleus of comet Halley must be a very fragile object that is porous throughout. Even with a mass loss of 100 billion kg of material during each encounter, this is less than 0.1% of its total mass, so Halley's Comet should still be entertaining terrestrial onlookers hundreds of generations from now.

> **Comet Halley loses 100 billion kg each time it comes around.**

The surface of the nucleus is also much darker than expected. It reflects only 3% of the sunlight falling on it, making it even blacker than coal or black velvet. The "dirty snowball" at the heart of comet Halley is not only dirty, it is among the *darkest known objects* in the Solar System. This low reflectivity means that comet nuclei, or at least the nucleus of Halley, must be rich in organic matter that is far more complex than simple hydrocarbons such as methane, propane, butane, or octane. Simple organic compounds such as these form clear liquids and ices at low temperatures. In contrast, more complex hydrocarbons such as tar are usually very dark.

> **The nucleus of comet Halley is blacker than coal.**

The implication is that very complex organic matter was present as dust in the disk around the young Sun, perhaps even in the interstellar cloud from which the Solar System formed.

The coma of a comet is a tenuous and temporary atmosphere surrounding the nucleus, consisting of gas and dust that are driven from the nucleus by solar heating. **Figure 11.10(b)** shows an image of the coma of comet Hyakutake obtained with the *Hubble Space Telescope*. As a comet nucleus falls from the outer Solar System inward toward the Sun, the coma begins to form at about the time the nucleus passes within the orbit of Jupiter. (While from our vantage point so close to the heart of the Solar System Jupiter may seem far removed from the Sun's warmth, from the comet's perspective the solar warming at 5 AU is hundreds to many millions of times greater than it is in the Kuiper Belt or the Oort Cloud.) As the nucleus approaches still closer to the Sun, rapid sublimation of ices near its surface can cause the coma to swell to a diameter of a million kilometers, three times the distance between Earth and the Moon. But while the coma is huge compared with the nucleus, it is still more tenuous than the very best vacuums we can produce on Earth.

VEGA and *Giotto* entered the coma of comet Halley when they were still nearly 300,000 km from the

Figure 11.10 (a) *The nucleus of comet Halley as imaged by the* Giotto *spacecraft in 1986.* (b) *A* Hubble Space Telescope *image of the coma of comet Hyakutake. (Images are false color.)*

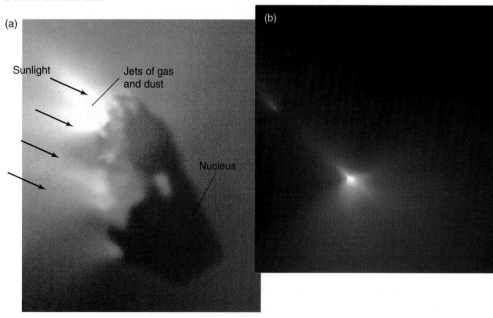

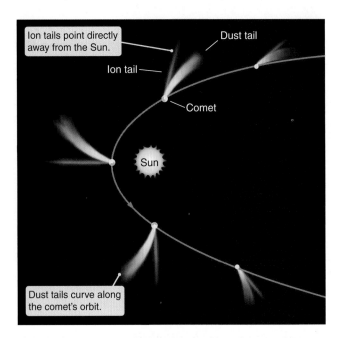

Figure 11.11 *The orientation of the dust and ion tails at several points in a comet's orbit. The ion tail points directly away from the Sun, while the dust tail curves along the comet's orbit.*

nucleus. In addition to cameras, these spacecraft also carried instruments capable of measuring the size and chemical composition of the dust particles in Halley's coma. These experiments found that the dust grains escaping from the nucleus were generally between 0.01 and 10 μm in size. This dust is very fine, finer even than the dust responsible for that annoying thin film that collects on your car and ruins its shine. Chemically, the dust from comet Halley was found to consist of a mixture of light organic substances (combinations of hydrogen, carbon, nitrogen, and oxygen) and heavier rocky material (combinations of magnesium, silicon, iron, and oxygen). Instruments on the spacecraft designed to study the gases surrounding the nucleus found it consisted of about 80% water (H_2O) and 10% carbon monoxide (CO), with smaller amounts of carbon dioxide (CO_2), methane (CH_4), ammonia (NH_3), and hydrogen cyanide (HCN).

COMETS HAVE TWO TYPES OF TAILS

The tail, which is the largest and most spectacular part of a comet, is also the "hair" for which comets are named. (*Comet* comes from the Greek word *kometes,* which means "hairy one.") Active comets have two different types of tails, as shown in Figure 11.9. One type of tail is called the **ion tail.** Many of the atoms and molecules that make up a comet's coma are ions. Because they are electrically charged, ions in the coma feel the effect of the solar wind, the magnetized stream of charged particles that blows continuously away from the Sun. The solar wind pushes on these ions, rapidly accelerating them to speeds of more than 100 km/s—far greater than the orbital velocity of the comet itself—and sweeps them out into a long wispy structure. Because the particles that make up the ion tail are so quickly picked up by the solar wind, ion tails are usually very straight and point from the head of the comet directly away from the Sun.

Dust particles in the coma can also have a net electric charge and feel the force of the solar wind. In addition, sunlight itself exerts a force on cometary dust. But dust particles are much more massive than individual ions, and so they are accelerated more gently and do not reach such high relative speeds as the ions. As a result, the **dust tail** of a comet often appears to gently curve away from the head of the comet as the dust particles are gradually pushed from the comet's orbit in the direction away from the Sun.

Figure 11.11 shows the tails of a comet at various points in its orbit. Remember that both types of tails always point *away from the Sun,* regardless of the direction in which the comet is moving. As it approaches the Sun, the comet's two tails trail behind its nucleus, but, like the flag on a sailboat sailing with the wind, the comet's tails extend *ahead* of the nucleus as it moves outward from the Sun.

> **Comet tails always point away from the Sun.**

Comet tails are fascinating structures, and vary greatly from one comet to another. Some comets, such as comets West and Halley, display both types of tails simultaneously, while for reasons we do not understand, some comets produce no tails at all! A tail often forms as the comet crosses the orbit of Mars, where the increase in solar heating drives gas and dust away from the nucleus at a prodigious rate. In October 1965, early morning risers were held spellbound by the sight of comet Ikeya-Seki's glowing tail **(Figure 11.12)** extending 150 million kilometers across the sky, a length as great as the distance between Earth and the Sun. For a brief few weeks, comet Ikeya-Seki achieved temporary splendor as the longest object in the entire Solar System.

The gas in a comet's tail is even more tenuous than the gas in its coma, with densities of no more than a few hundred molecules per cubic centimeter. Compare this with Earth's atmosphere, which at sea level

> **Comet tails are extremely tenuous.**

contains more than 10^{19} molecules per cubic centimeter. Dust particles in the tail are typically about a micrometer in diameter, about the size of smoke particles. One astronomer's often-repeated quip that "all of the material in a long comet tail could be packed into a suitcase" is less an exaggeration than you might imagine. When Earth passed through the tail of comet Halley in 1910, there were no observable effects. In contrast to the widespread predictions of disaster, it was all "much ado about nothing."

THE ORBITS OF COMETS

How long a dying nucleus will survive depends on its orbital period and **perihelion** distance—in other words, how frequently it passes by the Sun and how close it comes. Each passage takes its toll of ice and dust. By convention, comet orbits are generally referred to as short-period or long-period orbits. The

Comet orbits are divided into short-period and long-period.

division between the two is rather arbitrarily set at a period of 200 Earth years. Comets with periods of less than 200 years are called **short-period comets.** The total number of short-period comets known today is about 200. Comets with orbital periods longer than 200 years are termed **long-period comets.** The total number of long-period comets observed to date is about 770, with an average of six new ones being discovered each year.

The orbits of nearly all comets are highly elliptical, in sharp contrast to the relatively circular orbits of the planets. Only a few comets have orbital eccentricities that are less than Pluto's, whose orbit is the least circular of all the planets.

The orbits of most comets are very elliptical.

With a little thought, we might have anticipated this result. When a comet nucleus first enters the inner Solar System, it *must* be on a very elongated orbit since one end of the orbit is close to the Sun while the other end is in the distant parts of the Solar System. Because of this, we might expect all comets seen in the inner Solar System to have *extremely* elliptical orbits and extremely long orbital periods that carry them again and again back to the Oort Cloud or the Kuiper Belt. The surprise is that more than a hundred short-period comets remain relatively close to the Sun. How do these denizens of the frigid hinterlands become resident aliens in the realm of the planets?

The key to solving this mystery lies in an interesting difference between the orbits of long- and short-period comets. **Figure 11.13** shows the orbits of a number of comets. While long-period comets come diving into the inner Solar System from all directions, short-period comets have orbits that are strongly concentrated in the ecliptic plane. Most have orbital inclinations less than Pluto's, which has the highest inclination among the planets. Because their orbits are nearly in the plane of the ecliptic, and cross the orbits of many of the planets, short-period comets frequently get close enough to a planet for the planet's gravity to change the comet's orbit about the Sun in the sorts of chaotic encounters discussed in Chapter 9. Presumably, short-period comets originated in the Kuiper Belt, but, as they fell in toward the Sun, were forced into their present short-period orbits by gravitational encounters with planets.

11.2 COMETS

SHORT-PERIOD COMETS   Figure 11.13(a) shows the orbits of a number of short-period comets in the inner Solar System. Nearly all short-period comets have *prograde* orbits. That is, they orbit about the Sun in the same direction as the planets. Comet Halley, with its retrograde orbit, is an exception. This along

Short-period comets tend to have prograde orbits.

Figure 11.12 *Comet Ikeya-Seki as it appeared in the predawn hours in October 1965. Soon afterward, just before it passed behind the Sun, the head of the comet became so bright that it was visible in broad daylight. When it reappeared a few days later, it had split into two separate components.*

with the concentration of short-period comets in the plane of the Solar System is an important piece of evidence supporting our theory of the origin of short-period comets and how they were captured. The disk-like Kuiper Belt appears to share the prograde rotation of the rest of the Solar System (excepting the Oort Cloud). Since Kuiper Belt comet nuclei start out on prograde orbits, they are more likely to be captured into short-period prograde orbits as well. Also, the encounter between a planet and a comet nucleus on a prograde orbit will last longer than the encounter between a planet and a comet nucleus on a retrograde orbit. This means that a prograde comet's orbit will be more strongly influenced by such an encounter than the orbit of a retrograde comet will be. Jupiter, the most massive of the planets, is the planet most likely to affect the orbit of a Kuiper Belt comet. There are about 60 known short-period comets, referred to as the *Jupiter family,* that all have their **aphelion** (farthest point from the Sun) near Jupiter's orbit and travel in prograde orbits around the Sun.

Halley's Comet is the brightest and most famous of the short-period comets. It also has a very special place in the history of science because it was the first comet whose return was predicted! In 1705, Edmund Halley was studying the orbits of comets employing the gravitational laws that had recently been discovered by his friend and colleague, Isaac Newton. (Halley was instrumental in encouraging Newton to publish his work on orbits and gravitation— work that became the foundation for the science of physics.) Halley noted that a bright comet that had appeared in 1682 had an orbit remarkably similar to those of comets seen in 1531 and 1607. He concluded that all three were actually one and the same, and made the daring prediction that this comet would return in 1758. Unfortunately, Halley did not live to see it. His comet reappeared on Christmas Eve in the very year he had predicted. Astronomers quickly named it

> **The return of Halley's Comet was an important early prediction of Newton's physics.**

Figure 11.13 *Orbits of a number of comets in face-on and edge-on views of the Solar System. Populations of short- and long-period comets have very different orbital properties.*

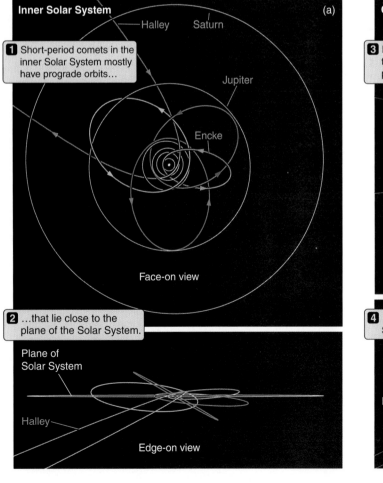

Inner Solar System   (a)

1 Short-period comets in the inner Solar System mostly have prograde orbits…

Halley   Saturn

Jupiter

Encke

Face-on view

2 …that lie close to the plane of the Solar System.

Plane of Solar System

Halley

Edge-on view

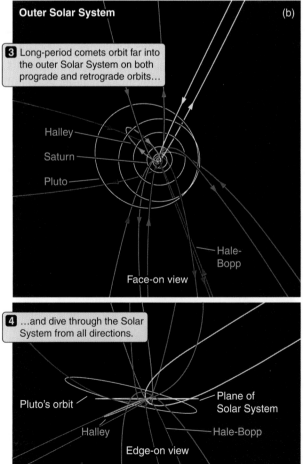

Outer Solar System   (b)

3 Long-period comets orbit far into the outer Solar System on both prograde and retrograde orbits…

Halley

Saturn

Pluto

Hale-Bopp

Face-on view

4 …and dive through the Solar System from all directions.

Pluto's orbit

Plane of Solar System

Halley   Hale-Bopp

Edge-on view

Halley's Comet and heralded it as a triumph for the genius of both Newton and Halley.

Comet Halley's highly elongated orbit takes it from perihelion, about halfway between the orbits of Mercury and Venus, out to aphelion beyond the orbit of Neptune. Astronomers and historians have now identified sightings of the comet that go back to 467 B.C. Halley's Comet has an average period of 76 years, and many of us mark the "once in our lifetime" when we will see an apparition of comet Halley. Mark Twain is famous for saying that he came in with the comet in 1834 and he would go out with the comet in 1910—a promise that he kept. For the authors of this book, our opportunity to see Halley's Comet came in 1986. While that appearance was not very spectacular compared to the one in 1910—in 1986, Halley's Comet and Earth were on opposite sides of the Sun when the comet put on its display—seeing the ghostly sight of Halley is an experience we will all remember. Halley will reach aphelion in 2024 and then begin its long journey back to the inner Solar System, becoming visible to the naked eye once again in 2061.

LONG-PERIOD COMETS   Of the known long-period comets, more than 600 have well-determined orbits. Some have orbital periods of hundreds of thousands or even millions of years, and their nuclei spend almost all of their time in the frigid, outermost regions of the Solar System. Orbits of a few long-period comets are shown in Figure 11.13(b). Unlike the mostly prograde, ecliptic-plane orbits of the short-period comets, long-period comets split about evenly into prograde and retrograde, and fall into the inner Solar System from all directions. These are the comets that tell us of the existence of the Oort Cloud. Because of their very long orbital periods, these comets cannot make more than a single appearance throughout the entire course of recorded history and, in most cases, in all of human history.

Some long-period comets come by only once in a million years.

Long-period comets differ from short-period comets in another way as well. With a few exceptions, Halley's Comet among them, the nuclei of short-period comets have been badly "worn out" by their repeated exposure to heating by the Sun. As the volatile ices are driven from a nucleus, some of the dust and organics are left behind on the surface. As this covering builds up, it slows down cometary activity. (Envision how, as a

With the exception of comet Halley, all "great" comets are long-period comets.

pile of urban—and therefore dirty—snow melts, the dirt left behind is concentrated on the surface of the snow.) That is why most short-period comets create little excitement. In contrast, long-period comets are usually relatively pristine. More of their supply of volatile ices still remains close to the surface of the nucleus, and they can produce a truly magnificent show. A half dozen or so long-period comets arrive each year. Most pass through the inner Solar System at relatively large distance from Earth or Sun and never become bright enough to attract much public attention. However, once or twice per decade, we are treated to the unexpected, spectacular appearance of a truly "great" comet. The next one could appear at any time.

Comet Ikeya-Seki (Figure 11.12) is a member of a family of comets called **sungrazers,** comets whose perihelia are located very close to the surface of the Sun. Many sungrazers fail to survive even a single pass by our local star. Ikeya-Seki became so bright as it neared perihelion in 1965 that it was visible in broad daylight, only two solar diameters away from the noontime Sun, and when it reappeared from behind the Sun, it had split into two pieces. Sungrazers sometimes come in groups, with successive comets following in very similar orbits. It seems likely that each member of such a group started as part of a single larger nucleus that broke into pieces during an earlier perihelion passage.

Sungrazing comet Ikeya-Seki was visible in broad daylight near the noontime Sun.

The closing years of the 20th century witnessed two spectacular long-period comets sporting long, beautiful tails: Hyakutake in 1996 (Figure 11.10b) and Hale-Bopp in 1997 (Figure 11.9b). Both were widely seen by the viewing public, comet Hale-Bopp being perhaps the most observed comet ever. It was bright enough to spot even under urban nighttime skies and dazzled those who were fortunate enough to see it far from city lights.

Comet Hyakutake is not very large as comets go, with a nucleus only a few kilometers across. The comet's awesome appearance was primarily a consequence of its unusually close trajectory past Earth. It approached within a mere a 10 million kilometers less than two months after its discovery by Japanese amateur astronomer Yuji Hyakutake.

Comet Hale-Bopp, on the other hand, is a huge comet, with a nucleus perhaps as large as 40 km in diameter. It was an especially important scientific object for professional astronomers because it was discovered far from the Sun near Jupiter's orbit two years before its perihelion passage. This extended the total time available to study its development as it approached the Sun and provided am-

EXCURSIONS 11.2

COMET SHOEMAKER-LEVY 9

From a wealth of ground- and space-based observations, planetary scientists have pieced together the events leading to the collision of comet Shoemaker-Levy 9 with Jupiter. We have already encountered comet Shoemaker-Levy 9 twice on our journey, first in Chapter 5 in our discussion of the devastation caused by impacts, and later in Chapter 9 when we learned of the destructive nature of tides. Early in the 20th century a comet nucleus from the Kuiper Belt was perturbed from its path and sent on an orbital journey that carried it close to Jupiter. In 1992, it passed so close to the planet that tidal stresses broke it into two dozen major fragments, which subsequently spread out along more than 7 million km of its orbit. The trajectory carried the fragments around for one more two-year orbit about the planet.

Then in July 1994, the entire string of fragments crashed into Jupiter. Over the course of a week, one after another of the fragments, each traveling at 60 km/s, plunged through Jupiter's stratosphere. Even though the impacts occurred just behind the limb of the planet, where they could not be observed from Earth, the *Galileo* spacecraft was on its way to Jupiter and was able to image some of the impacts. In addition, astronomers using ground-based telescopes and the *HST* could see immense plumes rising from the impacts to heights of more than 3,000 km above the cloud tops at the limb. The debris in these plumes then rained back onto Jupiter's stratosphere, causing ripple effects like pebbles thrown into a pond. **Figure 11.14** shows the sequence of events that took place as each fragment of the comet slammed into the giant planet. Sulfur and carbon compounds released by the impacts formed giant, Earth-sized scars in the atmosphere that persisted for months **(Figure 11.15)** and were visible even through small amateur telescopes.

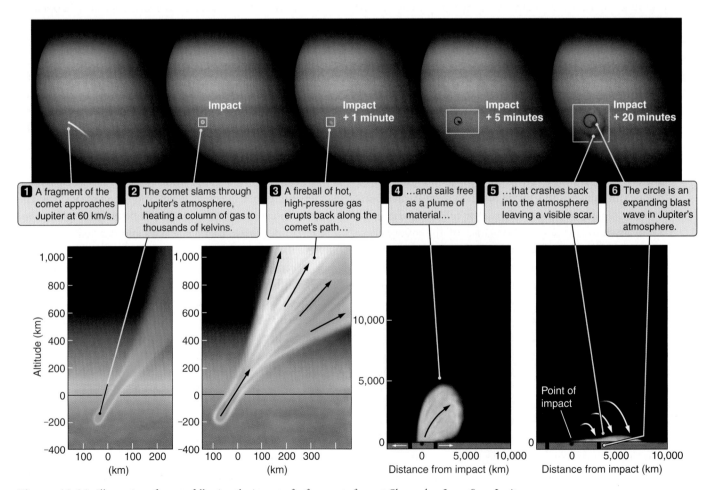

Figure 11.14 *Illustration of events following the impact of a fragment of comet Shoemaker-Levy 9 on Jupiter.*

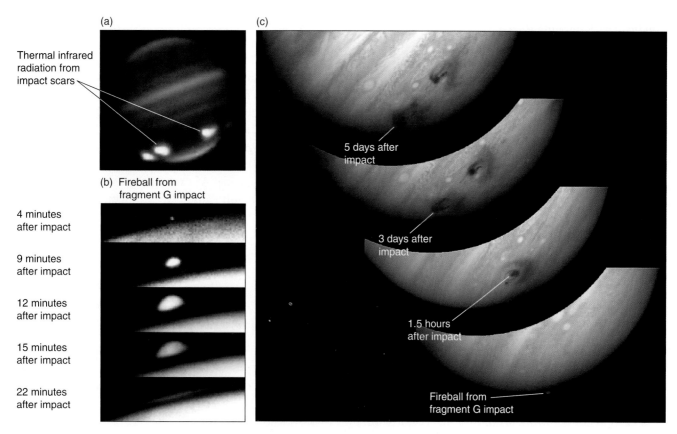

Figure 11.15 (a) *Ground-based infrared images of Jupiter showing the hot glowing scars left by fragments of comet Shoemaker-Levy 9.* (b) *While the fragments struck Jupiter on the back side, these* HST *images show the fireball from one fragment rising above the limb of the planet.* (c) HST *images of the evolution of the scar left by one fragment of the comet.*

ple opportunity to plan the important observations as it neared perihelion. Warmed by the Sun, the nucleus produced large quantities of gas and dust, as much as 300 tons of water per second, with lesser amounts of carbon monoxide, sulfur dioxide, cyanogen, and other gases. As this book goes to press, comet Hale-Bopp is still being observed with large telescopes and remains active even though it is now far beyond the orbit of Saturn. Comet Hale-Bopp will continue its outward journey for well over a thousand years. It will not return to the inner Solar System until sometime around the year 4377.

11.7 COLLISIONS STILL HAPPEN TODAY

As we studied the inner planets and the moons of the outer planets, we learned that almost all hard-surfaced objects in the Solar System still bear the scars of a time when the Sun and planets were young and tremendous impact events were common occurrences. The collision of comet Shoe-

In 1994, much of the world watched as a comet smashed into Jupiter.

maker-Levy 9 with Jupiter in 1994 (see Excursions 11.2) focused attention on the fact that, while such impacts are far less frequent today than they once were, they still happen. Nearly every major observatory in the world observed this spectacular and dramatic event, as did the *Hubble Space Telescope* and *Galileo* spacecraft. The "Jupiter comet crash" was also a landmark event in the history of the public's access to fast-breaking scientific events. Occurring at a time when the World Wide Web was growing in popularity, the latest images of the impacts were downloaded daily by millions of people around the world, scientists and laypersons alike. (While such access may be commonplace to you now, it was heady and exciting stuff to the authors back in 1994.)

In addition to providing a convincing demonstration that major collisions still occur in the Solar System today, the collision of Shoemaker-Levy 9 with Jupiter also let planetary scientists study the role that planetary atmospheres play

Collisions with asteroids and comets still occur in the Solar System.

in such events. In Chapter 6, we found that the dense atmosphere of Venus caused some objects that collided with the planet to break up before reaching the surface, creating clusters of craters in a pattern similar to a shotgun blast, or sometimes leaving only a dark "splotch" on the surface with no crater at all. The Shoemaker-Levy 9 collision also gave direct evidence that the tremendous amounts of energy released in such impacts can create huge atmospheric fireballs. Having seen the effects of Shoemaker-Levy 9's collision with Jupiter, it is easy to imagine the global firestorms on Earth that accompanied the impact that occurred at the Cretaceous–Tertiary boundary 65 million years ago (see Chapter 6). Connections 11.1 discusses several ways that such impacts affected the course of life on Earth.

What we learned from the Jupiter comet crash also helps explain an event that has puzzled scientists for the past century. The Tunguska River flows through

In 1908 a small asteroid or comet smashed into Siberia.

a remote region of western Siberia, inhabited mostly by reindeer and a few reindeer herders. In the summer of 1908, the region was blasted with the energy equivalent of 2,000 Hiroshima-type atomic bombs. **Figure 11.16** shows a map of the region, along with a photo of the destruction caused by the blast. Eyewitness ac-

CONNECTIONS 11.1

COMETS, ASTEROIDS, AND LIFE

One of the great questions of modern astronomy involves understanding how life on Earth is connected with processes at work in the greater cosmos. In a certain sense, this is not so different a quest from that pursued by many of the world's great religions and philosophies. At most of the stops along our journey through the Universe, we find threads of the cosmic tapestry of which terrestrial life is a part, and this stop is no exception.

For example, it is possible that without the existence of comet nuclei, water on Earth might have been less plentiful than it is today. We think that some of Earth's water was contributed during impacts of icy planetesimals (in other words, comet nuclei) early in the history of the Solar System. How could this have come about? The icy planetesimals likely condensed from the protoplanetary disk surrounding the young Sun and grew to their present size near the orbits of the giant planets. These planetesimals subsequently suffered strong orbital disturbances from those same planets. In such interactions, about half of the planetesimals would be flung outward to form the Kuiper Belt and Oort Cloud and half inward toward the Sun. Some of the objects flung toward the Sun would likely have hit Earth, the largest planet in the inner Solar System. Since most of the mass in comet nuclei appears to be in the form of water ice, it is likely that some of Earth's current water supply came from this early bombardment. Water, as we know, is essential to life.

Yet, the existence of comets can threaten life on Earth as well. It seems certain that occasional collisions of comet nuclei and asteroids with Earth have resulted in widespread devastation of Earth's ecosystem, and the extinction of many species. Passing stars or galactic plane crossings may send showers of comet nuclei into the inner Solar System every 10 million to 100 million years—intervals similar to those between mass extinctions found in fossil records of life on our planet. While such events certainly qualify as global disasters for the plants and animals alive at the time, they also represent global opportunities for new life-forms to evolve and fill the niches left by species that did not survive. Such a collision probably played a central role in ending the 180-million-year reign of dinosaurs and provided our ancient mammalian ancestors an opportunity for world dominance.

In our study of comets we may also have found a key to the chemical origins of life on Earth. Comets turn out to be rich in complex organic material—the chemical basis for terrestrial life—and cometary impacts on the young Earth may have played a role in chemically seeding the planet. Also, since comets are believed to be pristine samples of the material from which the Sun and planets formed, it means that organic material must be widely distributed throughout interstellar space. This conclusion, which is supported by radio telescope observations of vast interstellar clouds throughout our Galaxy, might have significant implications as we consider the possibility of life elsewhere in the Universe.

counts included the destruction of dwellings, the incineration of reindeer (including one herd of 700), and the deaths of at least five people. Although trees were burned or flattened for over 2,150 square kilometers—an area greater than metropolitan New York City—no crater was left behind!

For many years, scientists and nonscientists alike speculated about the cause of the Tunguska event. Ideas included such fanciful notions as a collision with a mini–black hole that passed through Earth, to a collision with an object made of antimatter, to an act of aggression by UFO aliens (a favorite of the supermarket tabloids). It now seems fairly clear that the Tunguska event was the result of a tremendous high-altitude explosion that occurred when a small asteroid or comet nucleus hit Earth's atmosphere, ripped apart, and formed a fireball before reaching Earth's surface. This is just the sort of event seen on Jupiter and hypothesized to account for the "splotches" seen on Venus. Recent expeditions to the Tunguska area have recovered resin from the trees blasted by the event. Chemical traces in the resin suggest that the impacting object may have been a stony asteroid.

The Tunguska event and the collision of Shoemaker-Levy 9 with Jupiter are sobering events when you really think about them. We know of several impacts occurring in the Solar System within a human lifetime, and more than one of them involved our planet![7] We know enough about the distribution of relatively large asteroids in the inner Solar System to say that it is highly improbable a populated area on Earth will experience a major collision with an asteroid within our lifetime. Comets and smaller asteroids, however, are less predictable. There may be as many as 10 million asteroids larger than a kilometer, but only about 10,000 have well-known orbits, and most of the unknowns are too small to see until they come very close to Earth. Several previously unknown long-period comets enter the inner Solar System each year.

---

[7]In 1947 yet another planetesimal struck Earth, this time in the Sikhote-Alin region of eastern Siberia. It had an estimated diameter of about 100 m and broke into a number of fragments before hitting the ground, leaving a cluster of craters and widespread devastation. Witnesses reported a fireball brighter than the Sun.

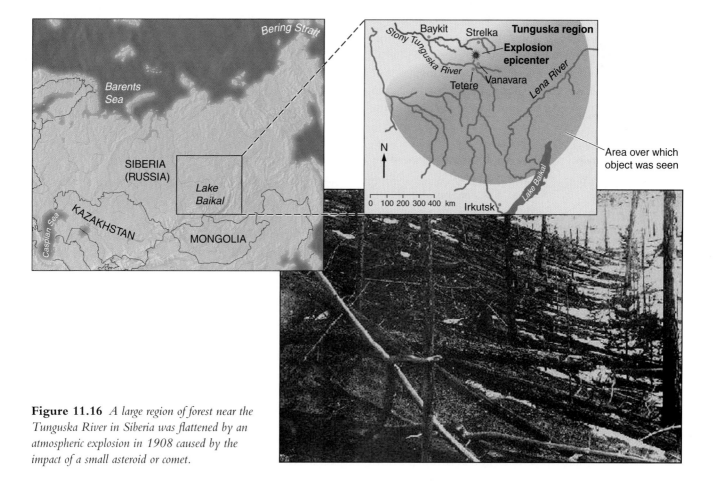

Figure 11.16 *A large region of forest near the Tunguska River in Siberia was flattened by an atmospheric explosion in 1908 caused by the impact of a small asteroid or comet.*

If one happens to be on a collision course with Earth, we might not notice it until just a few weeks or months before impact. For example, comet Hyakutake was discovered only two months before passing near Earth, and a small asteroid that just missed Earth in 1996 was discovered only a few days before its closest approach. While this has become a favorite theme of science fiction disaster stories (*Lucifer's Hammer* by Larry Niven and Jerry Pournelle being a favorite of several of the authors of this text), Earth's geological and historical record suggests that actual impacts by large bodies are infrequent events. Does this mean we need not lose any sleep over a possible collision with a large comet or asteroid? Probably, but remember that even though the probability may be small, the consequences of such an event are enormous. Just ask the dinosaurs.

Comet or asteroid impacts are infrequent but devastating.

11.8 SOLAR SYSTEM DEBRIS

While some comets and asteroids meet the spectacular fates of comet Shoemaker-Levy 9 or the Tunguska planetesimal, most go out with more of a fizzle than a bang. Comet nuclei that enter the inner Solar System generally disintegrate within a few hundred thousand years as a result of their repeated passages close to the Sun. Asteroids have much longer lives, but still are slowly broken into pieces from occasional collisions with each other. Disintegration of comet nuclei and asteroid collisions are the source of most of the debris that fills the inner part of the Solar System. As Earth and other planets move along in their orbits, they continuously sweep up this fine debris. This is the source of most of the meteoroids that Earth encounters. When they burn up in our atmosphere, these meteoroids become the meteors that you can see streaking across the sky on any clear dark night. As discussed in Excursions 11.1, it is only when they are large enough to reach Earth's surface that we call them meteorites.

Comet nuclei and asteroids are the source of Solar System debris.

METEOROIDS AND METEORS

Some 100,000 kg of meteoritic debris is swept up by Earth every day, and what does not burn up (mostly particles smaller than 100 μm) eventually settles to the ground as fine dust. We can demonstrate that meteors are atmospheric phenomena using the same method Tycho Brahe did to figure out that comets are not. If the same meteor is captured on photographs taken from two different locations, it appears to be in different places as seen against the background of stars. We can also measure meteor heights by radar because radio waves bounce off the ionized gas in meteor trails just as they bounce off the metal in an airplane or an automobile. Using these techniques, the altitudes of meteors are found to be between 50 and 150 km. For comparison, commercial jet aircraft typically fly at heights of about 10 km.

Meteors are atmospheric phenomena.

Fragments of asteroids are much denser than cometary meteoroids. If an asteroid fragment is large enough—about the size of your fist—it can survive all the way to the ground to become one of the meteorites discussed in Section 11.3. The fall of a 10 kg meteoroid can produce a fireball so bright that it lights up the night sky more brilliantly than the full Moon. Such a large meteoroid, traveling many times faster than the speed of sound, may create a sonic boom heard hundreds of kilometers away. It may even explode into multiple fragments as it nears the end of its flight, becoming a **bolide.** Some fireballs glow with a brilliant green color, caused by metals in the meteoroid that created them.

METEOR SHOWERS AND COMETS

Standing under a dark nighttime sky with a clear horizon, you can expect to see about a dozen meteors per hour on any night of the year. These are called **sporadic meteors,** and they occur as Earth "sweeps up" random bits of cometary and asteroidal debris in its annual path around the Sun. However, if you happen to be meteor watching on the night of August 11, for instance, you may see four to five times this number. If you pay close attention, you might also notice that nearly all of the meteors seem to be coming from the same region of the sky. This phenomenon is called a **meteor shower.** We call the particular meteor shower that peaks in mid-August the **Perseids,** because the trails they leave all point back to the constellation Perseus.

Meteor showers happen when Earth's orbit crosses the orbit of a comet. Bits of dust and other debris released by a comet nucleus as it rounds the

Sun remain in orbits of their own which are similar to the orbit of the nucleus itself. When Earth crosses through this concentration of cometary debris, the result is a meteor shower.

Meteor showers occur when Earth passes through cometary debris.

Since the meteoroids that are being swept up are all in similar orbits, they all enter our atmosphere moving in the same direction. As a result, the paths of all shower meteors are parallel to one another. But, just as the parallel rails of a railroad track appear to vanish to a single point in the distance, as in **Figure 11.17,** our perspective makes all the meteors appear to originate from the same point in the sky. This point is called the shower's **radiant.**

The Perseids are the result of Earth crossing the orbit of comet Swift-Tuttle. Although spread out along the comet's orbit, the debris is more concentrated in the vicinity of the comet itself. In 1992, comet Swift-Tuttle returned to the inner Solar System for the first time since its discovery in 1862. An exceptional Perseid meteor shower resulted, with counts of up to 500 meteors per hour.

There are more than a dozen comets whose orbits come close enough to Earth's to produce annual meteor showers. Around November 16 each year, Earth passes almost directly through the orbit of comet Temple-Tuttle, a short-period comet with an orbital period of 33.2 years. We call the meteor shower that Temple-Tuttle produces the **Leonids.** In most years, the Leonids fail to produce much of a show because this comet distributes very little of its debris around its orbit. However, in 1833 and 1866, comet Temple-Tuttle was not far away when Earth passed through its orbit. The Leonid showers in those two years were so intense that meteors filled the sky, with as many as 100,000 per hour **(Figure 11.18).** In 1900, one comet orbit later, nothing out of the ordinary happened. Again, in 1933, the Leonids were disappointing. Perturbations of comet Temple-Tuttle by Jupiter had moved the orbit of the comet's nucleus slightly away from Earth's, causing a sharp decrease in shower strength. What Jupiter took away though, it gave back. Further perturbations of the comet's orbit caused a spectacular Leonid shower in 1966 that may have produced as many as a half million meteors per hour! The Leonid shower put on less spectacular but still impressive shows in 1999 and 2001, when several thousand meteors per hour were seen.

ZODIACAL DUST

In the same way that we can "see" sunlight streaming through an open window by observing its reflection from motes of dust drifting in the air, we can see the

Figure 11.17 (a) *Meteors appear to stream away from the radiant of the Perseid meteor shower. These streaks are actually parallel paths that appear to emerge from a vanishing point, like the railroad tracks in* (b).

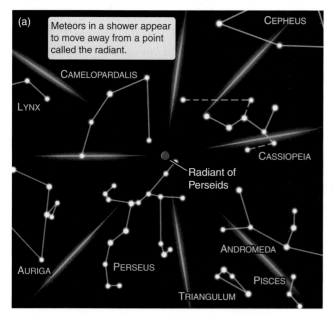

(a) Meteors in a shower appear to move away from a point called the radiant.

CEPHEUS
CAMELOPARDALIS
LYNX
CASSIOPEIA
Radiant of Perseids
AURIGA   PERSEUS
ANDROMEDA
PISCES
TRIANGULUM

(b) Actually the meteor paths are parallel. The radiant is a vanishing point.

Figure 11.18 *An engraving of the 1833 Leonid shower seen in France. One of the authors of this text recalls witnessing the 1966 shower, in which the sky was covered with meteor trails—perhaps as many as a half million per hour!*

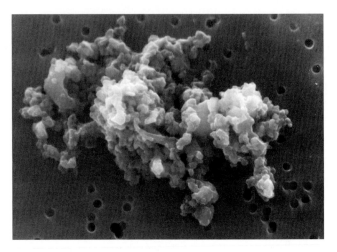

Figure 11.19 *A 10-μm-diameter cluster of interplanetary dust grains collected in Earth's stratosphere by a NASA U-2 airplane.*

sunlight reflected off tiny **zodiacal dust** particles that fill the inner parts of the Solar System close to the plane of the ecliptic. On a clear, moonless night, not long after the western sky has gotten dark, this dust is visible as a faint column of light slanting upward from the western horizon along the path of the ecliptic. This band, called the **zodiacal light,** can also be seen in the eastern sky before dawn. With good eyes and an especially dark night, you may be able to follow the zodiacal dust band all the way across the sky. In its brightest parts, the zodiacal light can be several times brighter than the Milky Way, for which it is sometimes mistaken.

Like meteoroids, zodiacal dust is a mixture of cometary debris and ground-up asteroidal material. The dust grains are roughly a millionth of a meter in diameter, the size of smoke particles. In the vicinity of Earth, there are only a few particles of zodiacal dust in each cubic kilometer of space. The total amount of zodiacal dust in the entire Solar System is estimated to be 10^{16} kg, equivalent to a solid body about 25 km across, or roughly the size of a large comet nucleus. Grains of zodiacal dust are constantly being lost as they are swept up by planets or pushed out of the Solar System by the pressure of sunlight. Such interplanetary dust grains have been recovered from Earth's upper atmosphere by very high flying aircraft **(Figure 11.19)**. If not replaced by new dust from comets, all of the zodiacal dust would be gone within a brief span of 50,000 years.

> **Zodiacal dust is cometary debris and ground-up asteroidal material.**

In the infrared region of the spectrum, thermal emission from the band of warm zodiacal dust makes it one of the brightest features in the sky. It is so bright that astronomers wanting to observe faint infrared sources are frequently hindered by the foreground glow it creates.

SEEING THE FOREST THROUGH THE TREES

Long before their true nature was understood, comets were granted great significance by humankind. These spectacular celestial displays were taken as omens of great events, and more often than not, as harbingers of the end of the world. How ironic that they have turned out instead to be messengers from the time when the world was born.

Anyone who has read a mystery novel knows that sometimes things that seem least significant at first glance turn out to hold the crucial clues to the biggest questions. In our study of the Solar System, two questions rise above all others: How did the Solar System form, and what is its history? Planets may dominate the environment around the Sun, but planets are imperfect in the records they keep of the earliest days of the Solar System. The violence of the planet formation process and aeons of geological activity together effectively erase most clues to their origin. The best samples of the early Solar System that have survived to the present day are instead the smallest bodies—asteroids and especially comets, frozen denizens of the outer reaches of the Sun's influence. Buried within these dirty snowballs are grains that predate even the Solar System itself. These tiny bits of solid material formed in the atmospheres of stars and in material blasted into space in tremendous stellar explosions, survived for a time in the vast reaches between the stars, participated in the collapse of the interstellar cloud that would become the Solar System, and wound up embedded within the small bodies that are the most numerous citizens of our planetary system. The discovery of such grains provides a direct link between our existence and our origins in the stars. How remarkable and fortuitous it is that these very pieces of our Solar System's past are delivered to our doorstep, falling to Earth as meteorites to be picked up from a cornfield in the Midwest or a glacier in the Antarctic, and then deciphered using the most advanced tools modern science has to offer.

Grains that predate the Sun make up only the tiniest fraction of the material found in meteorites. Most of this material was instead formed along with the Solar System itself. This is material that was vaporized in the violence that accompanied the formation of the disk around the young Sun, and then condensed again into solid form. Some asteroids and all comets are formed directly from this pristine material. Meteorites broken off from such bodies provide a window on the conditions that existed at that time. Other asteroids instead carry clues about how planets formed. Iron and stony-iron meteorites, for example, are pieces broken off

from what were once larger, differentiated bodies. Pick up a stony-iron meteorite, and you hold in your hand a snapshot, frozen in time, of the processes that shaped the world.

But comets and asteroids are far more than scientific curiosities; they have played a major role in shaping the history of life on our planet. Your body is largely water, and it is very likely that much of that water (along with the water that covers the surface of our world) arrived on Earth billions of years ago as volatile-rich comets slammed into the surface of our young world. Across the ages, occasional impacts of comets and asteroids on Earth have dramatically altered the planet's climate for a time and redirected the course of life's flow. Intelligent life on Earth descended from mammals rather than dinosaurs only because of a cosmic fluke—the impact of a comet, 65 million years ago, in the region that is now the Yucatan Peninsula. The awe-inspiring fireballs that accompanied the impact of comet Shoemaker-Levy 9 on Jupiter and the devastation of a remote corner of Siberia by a small piece of a comet or asteroid that hit Earth's atmosphere in 1908 are reminders that there is a grain of truth in humankind's deep-seated superstitions about these objects. Sometimes the appearance of a comet *does* mean the end of the world as we know it. It has happened in the past, and unless we develop the technologies to prevent such events (not an easy task), it *will* happen again in the future.

However, as we look to the future of human exploration and utilization of the Solar System, asteroids and comets may play a vital and positive role as ready-made way stations and caches of raw materials that we will need if we are to expand beyond our planet. The history and destiny of our kind, as well as the course of our intellectual journey through the Universe, are inextricably tied to this flotsam and jetsam adrift in interplanetary space.

So far on our journey, we have spent our time digging in our own backyard. The Solar System may dwarf the scales of our everyday lives, but it is vanishingly small compared with the Universe. Just as we found that we could understand our own planet and its history only by putting it within the context of the Solar System, we are unable to understand the Sun and the worlds of our Solar System without placing them within the context of the broader Universe. The starting point on the next leg of our journey begins as we gaze at the myriad of stars that fill the night sky and wonder about what we see there.

STUDENT QUESTIONS

THINKING ABOUT THE CONCEPTS

1. Describe the differences among meteoroids, meteors, and meteorites.

2. How does the composition of an asteroid differ from that of a comet nucleus?

3. Describe the differences between a comet and a meteor in terms of size, distance, and how long they are visible.

4. Comets contain substances closely associated with the development of life, such as water (H_2O), ammonia (NH_3), methane (CH_4), carbon monoxide (CO), and hydrogen cyanide (HCN). Which of these are organic compounds?

5. In 1910, Earth passed through the tail of Halley's Comet. Among the various gases in the tail was hydrogen cyanide, deadly to humans. Yet, nobody became ill from this event. Why not?

6. Why are most asteroids very irregular in shape? What does this tell you about their history?

7. How could you and a friend, armed only with wireless phones and a knowledge of the night sky, prove conclusively that meteors are an atmospheric phenomenon?

8. Most meteorites (pieces of S-type and M-type asteroids) are 4.54 billion years old. Carbonaceous chondrites (pieces of C-type asteroids), however, are 20 million years older. What determines the time of "birth" of these pieces of rock? What does this tell you about the history of the parent bodies?

9. Name the three parts of a comet. Which part is the smallest? Which is the most massive?

10. Suppose you found a rock that has all of the characteristics of a meteorite. You take it to a physicist friend who confirms that it is a meteorite, but that radioisotope dating indicates an age of only a billion years. What might be the origin of this meteorite?

11. Picture a comet leaving the vicinity of the Sun and racing toward the outer Solar System. Does its tail point backward toward the Sun or forward in the direction it is moving? Explain your answer.

APPLYING THE CONCEPTS

12. One recent estimate concludes that nearly 800 meteorites with mass greater than 100 g (massive enough to cause personal injury) strike the surface of Earth each day. Assuming that you present a target of 0.25 m^2 to a falling meteorite, what is the probability that you will be struck by a meteorite during your lifetime?

13. The orbital periods of comet Encke, Halley's Comet and comet Hale-Bopp are 3.3 years, 76 years and 2,380 years, respectively.
    a. What are the semimajor axes (in astronomical units, or AU) of the orbits of these comets?
    b. What are the maximum distances from the Sun (in AU) reached by Halley's Comet and comet Hale-Bopp in their respective orbits?
    c. Which would you guess is the most pristine comet among the three of them? Which is the least? Explain your reasoning.

14. The estimated amount of zodiacal dust in the Solar System remains constant at approximately 10^{16} kg. Yet, zodiacal dust is constantly being swept up by planets or removed by the pressure of sunlight.
    a. If all of the dust would disappear (at a constant rate) over a span of 30,000 years, what must be the average production rate in kg/s?
    b. Is this an example of static or dynamic equilibrium? Explain your answer.

15. The total number of asteroids larger than 1,000 m that cross Earth's orbit (Atens and Apollos) is estimated to be about 2,100, with 5 times as many having diameters larger than 500 m. Assuming this progression remains constant for still smaller asteroids, how many would there be with diameters larger than 125 m. (Note that the impact on Earth of any asteroid larger than 100 m in diameter would cause major, widespread damage.)

16. Zodiacal dust is concentrated in the plane of the ecliptic, and like back-lit dust in planetary rings (Excursions 10.2), is best seen along parts of the ecliptic nearest the Sun. Draw a diagram illustrating why the zodiacal light is brightest shortly after sunset in the western sky or shortly before sunrise in the eastern sky. For an observer at mid-northern latitudes, why is the zodiacal light best seen after evening twilight in February, but just before dawn in October? (*Hint:* Refer to a star chart and look at the location of the ecliptic in the sky at these times.)

STARS AND
STELLAR EVOLUTION

To man, that was in th' evening made,
Stars gave the first delight;
Admiring, in the gloomy shade,
Those little drops of light.

EDMUND WALLER (1606–1687)

Taking the Measure of Stars

12.1 Twinkle, Twinkle, Little Star, How I Wonder What You Are

As children we look at the sky, see the stars, and wonder. What are those points of light? How far away are they? How bright are they? How hot is a star? How big is a star? How much does a star "weigh"? What are stars made of? How long does a star last? What makes a star shine? These questions have always been a part of the human experience.

For most of human history, **stars,** like so much of nature, have seemed mysterious and unknowable. Even so, their passage follows the daily and annual rhythms that shape our lives. It is no wonder that in the absence of any real understanding of the stars, ancient humans turned their imaginations loose and viewed the stars as the province of gods and magic, with power over our lives. Today we understand that the patterns of the constellations have no more mystical influence over the course of events than the random toss of a coin (even if we do persist in paying the salaries of astrologers and reading their ramblings!).

The story of how we know what we know about the stars is a wonderful tale. It starts with simple but in-

KEY CONCEPTS

To even the most powerful of telescopes, a star is just a point of light in the night sky. However, by applying our understanding of light, matter, and motion to what we see, we are able to build a remarkably detailed picture of the physical properties of stars. As we take this first step beyond our local Solar System, we will:

* Apply a form of "stereoscopic vision," extended by Earth's orbit, to measure distances to nearby stars;
* Use the brightness of stars and their distances from Earth to discover how luminous they are;
* Use our knowledge of thermal radiation to infer the temperatures and sizes of stars from their colors;
* Classify stars, and organize this information on a plot of luminosity versus temperature, called the Hertzsprung-Russell or H-R diagram;
* Measure the composition of stars from spectra;
* Discover that 90% of the stars we see lie along a well-defined "main sequence" in the H-R diagram;
* Study the orbits of binary stars and use Kepler's laws to calculate their masses;
* Discover that the mass of a main sequence star determines its luminosity, temperature, and size; and
* Explore the range of stellar properties, learning how our Sun compares to other stars.

genious observations of the sky, borrows from lessons learned about the behavior of matter on Earth and in the Solar System, and ends with clear and straightforward answers to the very questions that a child might ask as he looks at the sky. The path to knowledge of the stars is also wonderful because, unlike many stories of modern science, it is a path that can be traveled and appreciated by anyone willing to take the time to follow it. A bit of geometry, a bit about radiation, a bit about orbits—all things that we have studied in earlier chapters—and we begin to find ourselves with solid answers to age-old questions.

12.2 THE FIRST STEP IS MEASURING THE DISTANCE, BRIGHTNESS, AND LUMINOSITY OF STARS

In a sense, one of the most amazing feats that a human can perform is to catch a fly ball. Here comes the ball, traveling at speeds of 150 km/h or more. To put a glove on the ball the fielder must judge not only the direction to the ball, but also its *distance*. But how does the fielder do it? Like most predators, we have two foward-looking eyes, one on each side of our face. Each of our eyes has a somewhat different view of the world. Exactly how different these perspectives are depends on the distance to the object you are looking at. If you are looking at a house down the street and you blink back and forth between your two eyes, the view that you get changes very little. But if you are looking at a nearby object—say, a finger held up at arm's length—the perspectives of your two eyes differ quite a lot. This difference in our eyes' perspectives on objects at different distances is the basis of our **stereoscopic vision.** Put another way, stereoscopic vision is the major key to the way we perceive distances. (**Figure 12.1** shows that the brain can even be fooled into perceiving distance where none exists by showing each eye a different view.)

> **Stereoscopic vision allows us to perceive the distances of objects.**

Our stereoscopic vision allows us to judge the distances of objects as far away as a few hundred meters, but beyond that it is of little use. Our eyes are only separated by a distance of about 6 cm, so the view that

your right eye gets of a mountain several kilometers away is indistinguishable from the view seen by your left eye. Comparing the two views tells your brain only that the mountain is too far away for you to judge its distance. (Evolution only provided us with enough depth perception to judge distances to things that we might be trying to eat, or that might be trying to eat us!) The distance over which our stereoscopic vision works is limited by the separation between our two eyes. If you wanted to increase the differences between the views your two eyes see, the obvious thing to do would be to somehow move them farther apart. If you could take your eyes and separate them by several meters instead of only a few centimeters, then their perspectives would be different enough to allow you to judge the distances to objects that are kilometers away.

> **The small separation between our eyes limits the range over which we can judge distance.**

Of course, we cannot literally take our eyes out of our heads and hold them apart at arm's length, but we *can* compare pictures taken from two widely separated locations. The greatest separation we can get without leaving Earth is to let Earth's orbital motion carry us from one side of the Sun to the other. If we take a picture of the sky tonight, then wait for six months and take another picture, our point of view between the two pictures will have changed by the diameter of Earth's orbit, or two astronomical units (AU). With 2 AU separating our two "eyes," we should have very powerful stereoscopic vision indeed. **Figure 12.2** shows how our view of a field of stars changes as our perspective changes during the year. This change in perspective is what allows us to measure the distances to nearby stars.

> **Distances to nearby stars are measured by comparing the view from opposite sides of Earth's orbit.**

DISTANCES TO NEARBY STARS ARE MEASURED USING PARALLAX

The eye cannot detect the changes in position of a nearby star throughout the year, but telescopes can. **Figure 12.3** shows Earth, the Sun, and three stars. Look first at star 1. When Earth, the Sun, and the star are in this position, they form a long, skinny right triangle. The short leg of the triangle is the distance from Earth to the Sun, or 1 AU.

> **Stellar parallax measures the apparent change in a star's position due to Earth's orbital motion.**

(a) View from the direction of the north celestial pole

(b) View from the direction of the vernal equinox

(c) View from the direction of the summer solstice

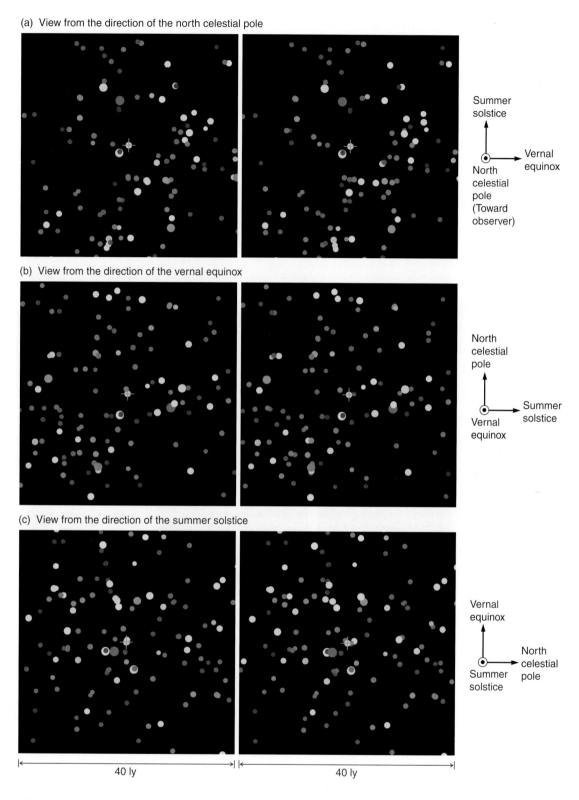

Summer solstice

Vernal equinox

North celestial pole (Toward observer)

North celestial pole

Summer solstice

Vernal equinox

Vernal equinox

North celestial pole

Summer solstice

40 ly   40 ly

Figure 12.1 *Your brain uses the slightly different views offered by your two eyes to "see" the distances and three-dimensional character of the world around you. These stereoscopic pairs show the stars in the neighborhood of the Sun as viewed* (a) *from the direction of the north celestial pole,* (b) *from the direction of the vernal equinox, and* (c) *from the direction of the summer solstice. The field shown is 40 light-years on a side. The Sun is at the center, marked with a green cross. The observer is 400 light-years away, and has "eyes" separated by about 30 light-years. To view these stereoscopic images, hold a card between the two images and look at them from about a foot and half away. Relax, and look straight at the page until the images merge into one.*

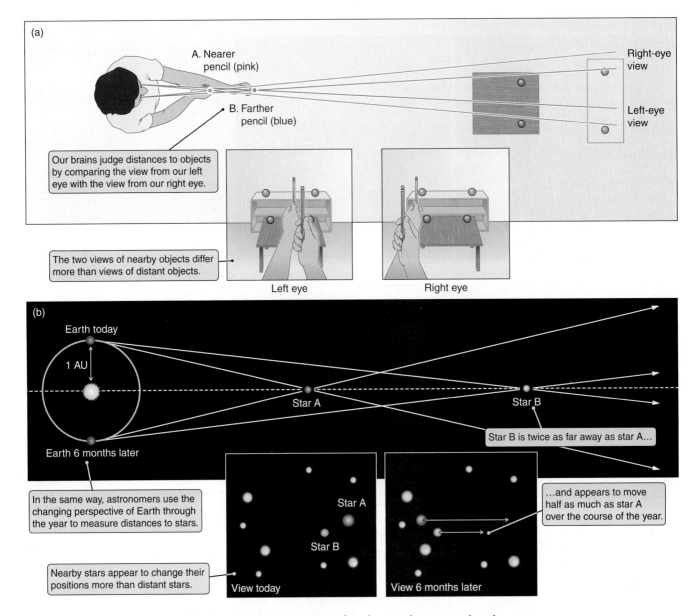

Figure 12.2 *As we move around the Sun, the apparent positions of nearby stars change more than the apparent positions of more distant stars. This is the starting point for measuring the distances to stars.*

The long side of the triangle is the distance from the Sun to the star. The small angle at the end of the triangle is called the *parallactic angle* of the star, or simply the **parallax** of the star. Over the course of a year, the star's position in the sky appears to shift back and forth, returning to its original position one year later. The amount of this shift—the angle between one extreme in the star's apparent motion and the other—is equal to twice the parallax.

The more distant the star, the longer and skinnier the triangle that it forms, and the smaller the star's parallax. Look again at Figure 12.3. Star 2 is twice as far away as star 1, and its parallax is only $\frac{1}{2}$ as great. If you were to draw a number of such triangles for different stars, you would find that increasing the distance to the star always reduces the star's parallax. Move a star three times farther away, as with star 3 in Figure 12.3, and you reduce its parallax to $\frac{1}{3}$ of its original value. Move a star 10 times farther away, and you reduce its parallax to $\frac{1}{10}$ of its original value. The parallax of a star (p) is inversely proportional to its distance (d):[1]

The farther away a star, the smaller its parallax.

$$p \propto \frac{1}{d} \quad \text{or} \quad d \propto \frac{1}{p}.$$

---

[1]Note that parallax is only inversely proportional to distance if the parallax is tiny, as it is for stars. For nearby objects, we must say that the tangent of the parallax is inversely proportional to the distance.

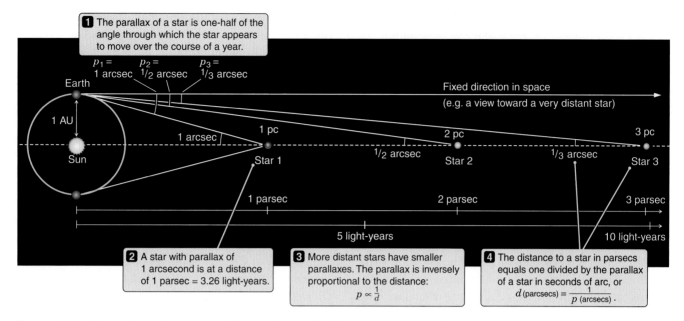

Figure 12.3 *The parallax of three stars at different distances. Parallax is inversely proportional to distance.*

The parallaxes of real stars are tiny. Rather than talking about parallaxes of 0.0000028° or 4.8×10^{-8} radians (1 radian = 57.3°), astronomers normally measure parallaxes in units of seconds of arc. Just as an hour on the clock is divided into minutes and seconds of time, a degree can be divided into minutes and seconds of arc. A **minute of arc** (or **arcminute**) is 1/60 of a degree, and a **second of arc** (or **arcsecond**) is 1/60 of a minute of arc. That makes a second of arc 1/3,600 of a degree, or 1/1,296,000 of a complete circle. An arcsecond is about equal to the angle formed by the diameter of a Ping-Pong ball at the distance of 5 miles. (Arcseconds are often denoted by the symbol ″; one second of arc is written 1″, and one arcminute is written 1′. However, we shall spell out the unit in this text.)

If the angle at the apex of a triangle is 1 arcsecond, and the base of the triangle is 1 AU, then the length of the triangle is 206,264.81 AU. This distance, which corresponds to 3.09×10^{16} m, or 3.26 light-years, is referred to as a **parsec** (abbreviated pc). The relationship between distance measured in parsecs and parallax measured in arcseconds is illustrated in Figure 12.3. Using the fact that a star with a parallax of 1 arcsecond is at a distance of 1 parsec, we can turn the inverse proportionality between distance and parallax into an equation:

> A parsec is defined as the distance at which the parallax equals 1 arcsecond.

$$\begin{pmatrix} \text{Distance measured} \\ \text{in parsecs} \end{pmatrix} = \frac{1}{\begin{pmatrix} \text{parallax measured} \\ \text{in arcseconds} \end{pmatrix}}$$

If you measure the parallax of a star to be 0.5 arcsecond, then you know that the star is located at a distance of 1/0.5 = 2 parsecs. A star with a parallax of 0.01 arcsecond is located at a distance of 1/0.01 = 100 parsecs.

In this book we will usually use units of **light-years** to measure distances to stars and galaxies. One light-year is the distance that light travels in 1 year—about 9 trillion kilometers. We use this unit because it is the unit you are most likely to see in a newspaper article or a popular book about astronomy. But astronomers very seldom use light-years among themselves. When astronomers discuss distances to stars and galaxies, the unit they use is the parsec. (You can always convert between the two units. One parsec is 3.26 light-years.)

The star closest to us (other than the Sun) is Proxima Centauri. Proxima Centauri is located at a distance of 4.3 light-years, or 1.3 parsecs, and is a faint member of a system of three stars called Alpha Centauri. This star has a parallax of only about $\frac{3}{4}$ arcsecond. It is no wonder that ancient astronomers were unable to detect the apparent motions of the stars over the course of a year!

> The closest star beyond the Sun has a parallax of only $\frac{3}{4}$ arcsecond.

When astronomers began to apply this technique, they discovered that stars are very distant objects indeed. The first successful measurement of the parallax of a star was made by F. W. Bessel, who in 1838 reported a parallax of 0.314 arcsecond for the star 61 Cygni. This implied that 61 Cygni was 3.2 parsecs away, or 660,000 times as far away as the Sun. With this one measurement, Bessel increased the known size of the Universe 10,000-fold! Today we know of 54 stars in 37 single-, double-, or triple-star systems within 15 light-years of Earth. A sphere with a radius of 15 light-years has a volume of about 14,000 cubic light-years, so this corresponds to a local density of 37 systems per 14,000 cubic light-years. That is about 0.0026 star systems per cubic light-year. Stated another way, in the neighborhood of the Sun, each system of stars has on average about 380 cubic light-years of space (a volume about 4.5 light-years in radius) all to itself.

Stars are few and far between in our neighborhood.

Knowledge of our stellar neighborhood took a tremendous step forward during the 1990s with the completion of the *Hipparcos* mission. The *Hipparcos* satellite measured the positions and parallaxes of 120,000 stars. These measurements, taken from a satellite well above Earth's obscuring atmosphere, are better than the measurements that can typically be made from telescopes located on the surface of Earth. But even this catalog has its limits. The accuracy of any given *Hipparcos* parallax measurement is about 0.002 arcsecond. Because of this **observational uncertainty,** our measurements of the distances to stars are not perfect. For example, a star with a parallax measured by *Hipparcos* of 0.004 arcsecond might really have a parallax of anywhere between 0.002 and 0.006 arcsecond. So, instead of knowing that the distance to the star is exactly 250 parsecs (1 divided by 0.004 arcsecond), we know only that the star is probably between about 170 parsecs (1/0.006 arcsecond) and 500 parsecs (1/0.002 arcsecond). Even with modern technology, parallax becomes useless as a way of measuring stellar distances for stars more than a few hundred parsecs away. If you are measuring stellar distances using parallax, and need to know the distance to an accuracy of 10% or better, you are restricted to stars that are less than about 50 parsecs (160 light-years) away.

Parallax is useful for measuring distances only within a few hundred light-years.

ONCE DISTANCE AND BRIGHTNESS ARE KNOWN, LUMINOSITY CAN BE CALCULATED

When we talk about the brightness of an object, we are making a statement about how that object appears *to us.* In Chapter 4 we saw that brightness corresponds to the amount of energy falling on a square meter of area each second in the form of electromagnetic radiation. (When astronomers talk about the brightness of stars, they usually use a system called **magnitudes,** which is discussed in Appendix 6.) But, while the brightness of a star is directly measurable, that does not immediately tell us much about the star itself. As illustrated in **Figure 12.4,** a bright star in the night sky may in fact be a dim bulb, appearing bright only because it is close by. Conversely, a faint star may be a powerful beacon, still visible despite its tremendous distance.

Brightness depends on the observer's perspective while luminosity does not.

To learn about the stars themselves, we need to know the total energy radiated by a star each second—the star's luminosity. In Chapter 4 we studied the relationship between the brightness, luminosity, and distance of objects. Borrowing from that earlier work, we recall that the brightness of an object that has a known luminosity and is located at a distance d is given by

$$\text{Brightness} = \frac{\text{total light per second}}{\text{area of a sphere of radius } d}$$
$$= \frac{\text{luminosity}}{4\pi d^2}.$$

We can rearrange this equation, moving the quantities we know how to measure (distance and brightness) to the right side and the quantity we would like to know (luminosity) to the left, giving us

$$\text{Luminosity} = 4\pi d^2 \times \text{brightness}.$$

This equation answers the question: how much total light must a star be giving off to be as distant as it is while still looking as bright as it does? In the process, we take two measurable quantities that depend on our particular perspective (distance and brightness) and from them obtain a quantity that is a property of the star itself, namely luminosity.

When measuring the luminosity of stars as just described, we find that stars vary tremendously in the amount of light they give off. The Sun provides a convenient yardstick for measuring the properties of stars, including their luminosity. The luminosity of the Sun is written as $L_\odot$. The most luminous stars can exceed 1,000,000 $L_\odot$, or 10^6 $L_\odot$ (a million times the luminosity of the Sun). The least luminous stars have luminosities less than 0.0001 $L_\odot$, or 10^{-4} $L_\odot$ (1/10,000 that of the Sun). The most luminous stars

Some stars are 10 billion times more luminous than other stars.

There are many more low-luminosity stars than high-luminosity stars.

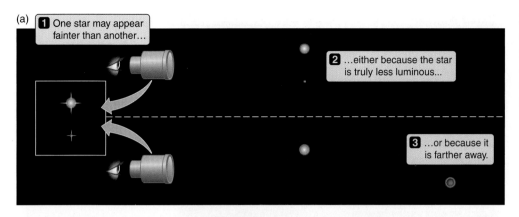

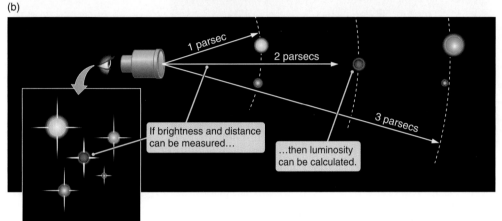

Figure 12.4 *The brightness of a visible star depends on both its luminosity and its distance. If brightness and distance are measured, luminosity can be calculated.*

are over 10 billion (10^{10}) times more luminous than the least luminous stars. Only a very small fraction of stars are near the upper end of this range of luminosities. The vast majority of stars are at the faint end of this distribution, less luminous even than our Sun. **Figure 12.5** shows the distribution of luminosities for the known stars within 1,000 parsecs (3,260 light-years) of Earth.

12.3 RADIATION TELLS US THE TEMPERATURE, SIZE, AND COMPOSITION OF STARS

Two everyday concepts—stereoscopic vision, and the fact that the closer an object is the brighter it appears—have given us the tools that we need to measure the distance and luminosity of stars. In these two steps, stars have gone from being mere faint points of light in the night sky, to being extraordinarily powerful beacons located at distances almost impossible for the mind to comprehend. To go further along this journey of discovery, we again turn to the laws of radiation that we studied in Chapter 4.

Stars are gaseous, but they are fairly dense—dense enough that the radiation from a star comes close to obeying the same laws as the radiation from objects like

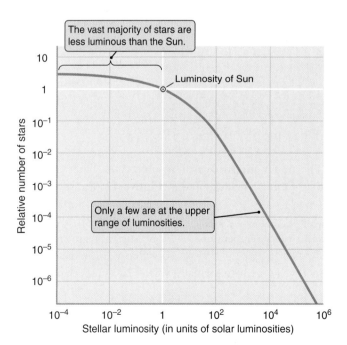

Figure 12.5 *The distribution of luminosities of known stars within 1,000 pc (3,260 ly) of Earth.*

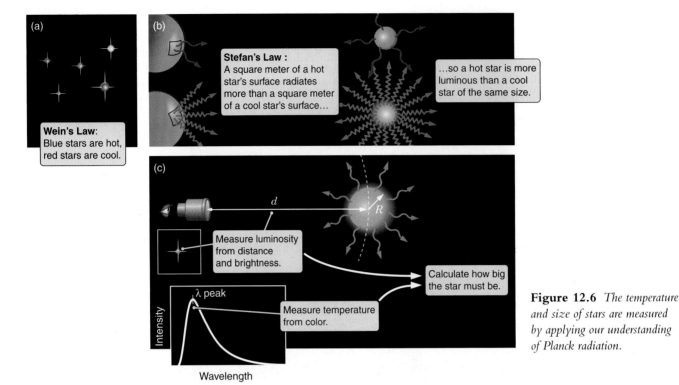

Figure 12.6 *The temperature and size of stars are measured by applying our understanding of Planck radiation.*

the heating element on an electric stove or the filament in a lightbulb. That means we can use our understanding of Planck radiation—results such as Stefan's law (hotter at same size means more luminous), and Wien's law (hotter means bluer)—to understand the radiation from stars. In particular, Wien's law and Stefan's law will allow us to measure the temperatures and sizes of our stellar neighbors.

Wien's Law Revisited: The Color and Surface Temperature of Stars

Measuring the color of a star tells us its surface temperature.

The hotter the surface of an object, the bluer the light that it emits. As illustrated in **Figure 12.6(a),** stars with especially hot surfaces are blue, stars with especially cool surfaces are red, and our Sun is a middle-of-the-road yellow. More formally, Wien's law states that

$$T = \frac{2{,}900 \ \mu\text{m K}}{\lambda_{\text{peak}}},$$

where λ_{peak} is the wavelength at which the electromagnetic radiation from a star is most intense. Obtain a

spectrum of a star, measure the wavelength at which the spectrum peaks, and Wien's law will tell you the temperature of the star's surface.[2]

In practice, it is usually not necessary to obtain a complete spectrum of a star to determine its temperature. Instead, stellar temperatures are often measured in a way that is very similar to the way that your eye sees color. Your eye and brain distinguish color by comparing how bright an object is at one wavelength with how bright it is at another wavelength. Some of the light-sensitive cells in your eye are primarily sensitive to red light, others are mostly sensitive to green light, and still others are sensitive to blue light. If you look at a very red lightbulb, the red-sensitive cells in your eye report a bright light, while the blue- and green-sensitive cells see less light. If you look at a yellow light, the green- and red-sensitive cells are strongly stimulated, but the blue-sensitive cells are not. White light stimulates all three kinds of cells. By combining the signals from cells that are sensitive to different wavelengths of light, your brain is able to distinguish between fine shades of color.

---

[2]It is important to remember that the color of a star tells us only about the temperature at the star's *surface,* since the surface is the part of the star giving off the radiation that we see. Star surfaces may be hot, but later in our journey we will discover that star interiors are far hotter still.

Astronomers often measure the colors of stars in much the same way. The brightness of a star is usually measured through a **filter**—sometimes just a piece of colored glass—that lets through only a certain range of wavelengths. Two of the most common filters used by astronomers are a blue filter that transmits light with wavelengths around 0.44 μm, and a yellow-green filter that passes light with wavelengths around 0.55 μm. The first of these filters is, sensibly enough, called a "blue" filter. However, by convention the second of these filters is referred to as a "visual" filter rather than a "yellow-green" filter, because it transmits roughly the range of wavelengths to which our eyes are most sensitive.

Figure 12.7 shows Planck spectra with temperatures of 2,500 K through 10,000 K, adjusted so that they are all the same brightness at 0.55 microns (the wavelength of the center of the range transmitted by the visual filter). A hot star, with a spectrum like the 20,000 K Planck spectrum shown, gives off more light in the blue part of the spectrum than in the visual part of the spectrum. If we were to divide the brightness of the star as seen through the blue filter (written b_B) by the brightness of the star as seen through the visual filter (written b_V), the ratio b_B/b_V would be greater than 1. On the other hand, a cool star with a spectrum more like the 2,500 K Planck spectrum shown is much fainter in the blue part of the spectrum than in the visual part of the spectrum. The ratio of blue light to visual light b_B/b_V is less than 1 for the cool star. This ratio

of brightness between the blue and visual filters is referred to as the $\boldsymbol{b_B/b_V}$ **color** of the star. By convention astronomers usually adjust the way they measure the blue and visual brightnesses of stars so that the star Vega (which has a surface temperature of around 10,000 K) has a b_B/b_V color of 1. On this scale, the Sun has a b_B/b_V color of 0.56.

The fact that the b_B/b_V color of a star depends on the star's temperature is an extremely handy tool for astronomers. It means that a single pair of snapshots of a group of stars, each taken through a different filter, is enough to allow us to measure the surface temperature of every star in our camera's field of view! In this way it is often possible to measure the temperatures of hundreds or even thousands of stars at once. When we do, we find that just as there are many more low-luminosity stars than high-luminosity stars, there are also many more cool stars than hot stars. Most stars have surface temperatures lower than that of the Sun.

> **There are many more cool stars than hot stars.**

STEFAN'S LAW REVISITED: FINDING THE SIZES OF STARS

Once we know the temperature of a star, we also know how much radiation each square meter of the star is giving off each second. As shown in **Figure 12.6(b)**, each square meter of the surface of a hot, blue star gives off more radiation than a square meter of the surface of a cool, red star. A hot star will be more luminous than a cool star of the same size. A small hot star might even be more luminous than a larger cool star. As noted in **Figure 12.6(c)**, we can use the relationship between temperature and luminosity of each square meter of a surface to infer the sizes of stars.

According to Stefan's law (see Chapter 4), the amount of energy radiated each second by each square meter of the surface of a star is equal to the constant σ times the surface temperature of the star raised to the fourth power. Written as an equation, this says:

$$\begin{array}{c}\text{Energy radiated each}\\ \text{second by 1 m}^2 \text{ of surface}\end{array} = \sigma T^4.$$

To find the total amount of light radiated each second by the star, we need to multiply the radiation per second from each square meter times the number of square meters of the star's surface:

$$\begin{array}{c}\text{Total energy}\\ \text{radiated by the}\\ \text{star each second}\end{array} = \begin{array}{c}\text{energy radiated}\\ \text{by a square meter}\\ \text{each second}\end{array} \times \begin{array}{c}\text{number of square}\\ \text{meters of surface}\\ \text{of the star}\end{array}$$

Figure 12.7 *A star's* b_B/b_V *color depends on its temperature. The Plank spectra shown here are adjusted so they have the same brightness at 0.55 microns.*

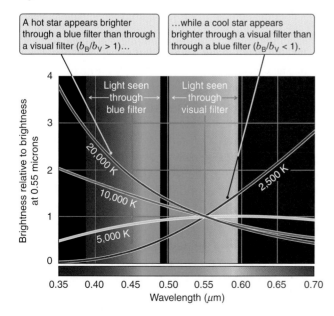

The left item in this equation—the total energy emitted by the star per second (in units of joules per second, abbreviated J/s)—is the star's luminosity L. The middle item in this equation—the energy radiated by each square meter of the star in a second (in units of joules per square meter per second, or $J/[m^2/s]$)—can be replaced with the σT^4 from Stefan's law. The remaining item in the equation—the number of square meters covering the surface of the star—is the surface area of a sphere, $A_{sphere} = 4\pi R^2$ (in units of m^2), where R is the radius of the star. If we replace the words in the equation with the appropriate mathematical expressions for Stefan's law and the area of a sphere, our equation for the luminosity of a star looks like this:

$$\text{Luminosity} = \left(\frac{J}{m^2 s}\right) \times m^2$$
$$= (\sigma T^4) \times (4\pi R^2)$$

Combining the right sides of these two equations, we get

$$L = 4\pi R^2 \sigma T^4 \qquad \text{in units of J/s.}$$

It is worth staring at this equation for a minute to see the sense behind it. Because the constants (4, π, and σ) do not change, the luminosity of a star is proportional only to $R^2 T^4$. The R^2 makes sense: The surface area of a star is proportional to the square of the radius of the star. Make a star three times as large, and its surface area becomes $3^2 = 9$ times as large. There is nine times as much area to radiate, so there is nine times as much radiation. The T^4 is understandable as well: Make a star twice as hot, and each square meter of the star's surface radiates $2^4 = 16$ times as much energy. (As we have seen before, an apparently complex mathematical expression is really a shorthand way of writing our understanding of what is going on physically—in this case the common sense result that larger, hotter stars are more luminous than smaller, cooler stars.)

The larger and hotter a star is, the more luminous it will be.

The equation above answers the question, "How luminous will a star of a given size and a given temperature be?" Now turn this question around and ask, "How large does a star of a given temperature need to be to have a total luminosity of L?" Rearrange the equation above, moving the things that we know how to measure (temperature and luminosity) to the right side of the equation, and the thing that we would like to know (the radius of the

If the temperature and luminosity of a star are known, its size can be calculated.

star) to the left side. After a couple of steps of algebra we find

$$R = \sqrt{\frac{L}{4\pi\sigma}} \times \frac{1}{T^2}.$$

Again, the right side of the equation contains only things that we know or can measure. The constants 4, π, and σ are always the same. L is the luminosity of the star, and can be found by combining measurements for the star's brightness and parallax. T is the surface temperature of the star, which can be measured from its color. From these we now know something new: the size of the star. We will refer to this last equation as the **luminosity-temperature-radius relationship** for stars.

The luminosity-temperature-radius relationship has been used to measure the sizes of many thousands of stars. When we talk about the sizes of stars, we again use our Sun as a yardstick. The radius of the Sun, written $R_\odot$, is 696,000 km, or about 700,000 km. When we look at stars around us, we find that the smallest stars that we see, called *white dwarfs,* have radii that are only a percent or so of the Sun's radius ($R = 0.01\ R_\odot$). The largest stars that we see, called *red supergiants,* can have sizes thousands of times the radius of the Sun. There are many more stars toward the small end of this range, smaller than our Sun, than there are giant stars.

There are many more small stars than large stars.

STARS ARE CLASSIFIED ACCORDING TO THEIR SURFACE TEMPERATURE

So far we have concentrated on what we can learn about stars by applying our understanding of thermal radiation, an approach that seems from the previous section not to have led us too far astray. However, the spectra of stars are far from perfect Planck spectra. Instead, when we look at the spectra of stars, we see a wealth of absorption lines and occasionally emission lines. In Chapter 4 we found that absorption lines occur when light passes through a cloud of gas, and the atoms and molecules of the gas absorb light at certain specific wavelengths, characteristic of the kind of atom or molecule that is doing the absorbing. Simi-

Absorption lines form as light escapes through the atmosphere of the star.

larly, the atoms and molecules in diffuse hot gas will emit light at specific wavelengths as well. Both of these processes are at work in stars.

While the hot surface of a star emits radiation with a spectrum very close to a smooth Planck curve, this light must then escape through the outer layers of the star's atmosphere. The atoms and molecules in the atmosphere of the star leave their absorption line fingerprints in this light, as shown in **Figure 12.8.** The atoms and molecules in the star's atmosphere and any gas that might be found in the vicinity of the star can also produce emission lines in stellar spectra. Although absorption and emission lines complicate how the laws of Planck radiation are used to interpret light from stars, spectral lines more than make up for this trouble by providing a wealth of information about the state of the gas in a star's atmosphere.

The spectra of stars were first classified during the late 1800s, well before stars or atoms or radiation were well understood. Stars were classified, not on the basis of their physical properties, but on the appearance of the mysterious dark bands (now known as absorption

lines) seen in their spectra. The original ordering of this classification was arbitrarily based on the prominence of particular absorption lines known to be associated with the element hydrogen. Stars with the strongest hydrogen lines were labeled *A stars,* stars with somewhat weaker hydrogen lines were labeled *B stars,* and so on.

> **Stars are classified by the appearance of their spectra.**

This classification scheme was refined and turned into a real sequence of stellar properties in the early 20th century. The new system, originally proposed in 1901, was the work of **Annie Jump Cannon,** who led an effort at the Harvard Observatory to systematically examine and classify the spectra of hundreds of thousands of stars. She dropped many of the earlier spectral types, keeping only seven, which she reordered into a sequence that was no longer arbitrary, but instead was based on surface temperatures. Spectra of stars of different types are shown in **Figure 12.9.** The hottest stars, with surface temperatures over 30,000 K, are labeled *O stars. O* stars show relatively featureless spectra, with

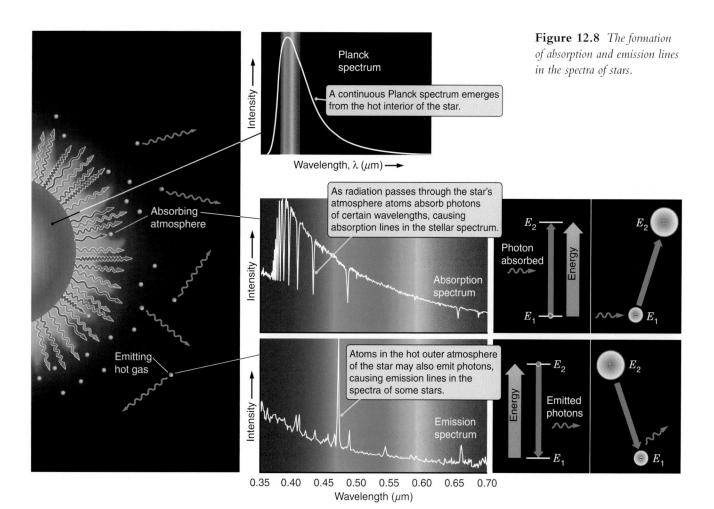

Figure 12.8 *The formation of absorption and emission lines in the spectra of stars.*

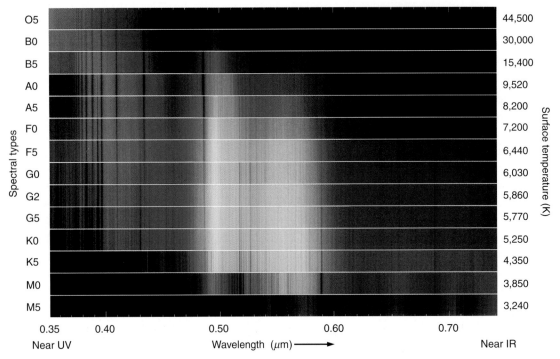

Figure 12.9 *Spectra of stars with different spectral types, ranging from hot blue O stars to cool red M stars. Hotter stars are brighter at shorter wavelengths. The dark lines are absorption lines.*

only weak absorption lines from hydrogen and helium. The coolest stars—*M stars*—have temperatures as low as about 2,800 K. M stars show myriad lines from many different types of atoms and molecules. The complete sequence of **spectral types** of stars, from hottest to coolest, is O, B, A, F, G, K, M; a mnemonic offered by Henry Norris Russell for remembering this sequence—*Oh Be A Fine Girl, Kiss Me*—has been a mantra for students of astronomy for over 70 years. (About half of a typical class may want to substitute the word *Guy* in the appropriate location.) The spectral sequence of OBAFGKM has undergone several modifications over time. As this book goes to press two new spectral types, L and T, are coming to be used to classify stars that are even cooler than M stars.

> The series of spectral types, **OBAFGKM, is a sequence of surface temperatures.**

The scheme of spectral classification is discussed in more depth in Foundations 12.1. Since different spectral lines are formed at different temperatures, the absorption lines in a star's spectrum can be used to measure the star's temperature directly. It is important to note that the surface temperatures of stars measured in this way agree very closely with the surface temperatures of stars measured using Wien's law, again confirming a prediction of the cosmological principle: the same physical laws that apply on Earth apply to stars as well.

STARS ARE MOSTLY COMPOSED OF HYDROGEN AND HELIUM

The most obvious differences in the lines seen in stellar spectra are due to temperature, but the details of the absorption and emission lines found in starlight carry a wealth of other information as well. By applying our knowledge of the physics of atoms and molecules to the study of stellar absorption lines, we are

> Spectral lines are used to **measure many properties of stars including chemical composition.**

able to determine accurately not only surface temperatures of stars, but also pressures, chemical compositions, magnetic field strengths, and other physical properties of stars. Also, by making use of the Doppler shift of emission and absorption lines, we are able to measure rotation rates, motions of the atmosphere, expansion and contraction, "winds" driven away from stars, and other dynamical properties of stars.

As we continue to learn about stars, we will come to appreciate that one of the most interesting and important things about a star is its chemical composition. We already saw in Chapter 4 that if we look at a source of thermal radiation through a cloud of gas, the strength of various absorption lines tells us what kinds of atoms are present in the gas and in what abundance.

While great care must be taken in interpreting spectra to properly account for the temperature and density of the gas in the atmosphere of the star, in most respects stars are ready-made laboratories for carrying out just such an experiment.

When analyzed in this way, most stars are found to have atmospheres that consist predominately of the least massive of elements. Hydrogen typically makes up over 90% of the atoms in the atmosphere of a star, with helium accounting for most of what remains. All of the other chemical elements, which are collectively referred to as *heavy elements* or *massive elements,* are present in only trace amounts. (This would be a good time to go back and review Figure 8.3, the "astronomer's periodic table.") **Table 12.1** shows the chemical composition of the atmosphere of the Sun, which is fairly typical for stars in our vicinity. On the other hand, the chemical composition can vary tremendously from star to star. In particular, some stars show even smaller amounts of elements

Most stellar atmospheres are composed primarily of hydrogen and helium.

TABLE 12.1

THE CHEMICAL COMPOSITION OF THE SUN

The relative amounts of different chemical elements in the atmosphere of the Sun. The second column tells what fraction of the Sun's atoms are the given element. The third column tells what fraction of the Sun's mass consists of a given chemical element.

Element	Percent by Number	Percent by Mass
Hydrogen	92.5%	74.5%
Helium	7.4%	23.7%
Oxygen	0.064%	0.82%
Carbon	0.039%	0.37%
Neon	0.012%	0.19%
Nitrogen	0.008%	0.09%
Silicon	0.004%	0.09%
Magnesium	0.003%	0.06%
Iron	0.003%	0.13%
Sulfur	0.001%	0.04%
Total of others	0.001%	0.03%

FOUNDATIONS 12.1

SPECTRAL TYPE

Figure 12.9 shows representative spectra of stars of different spectral types. You can see in this figure that not only are hot stars bluer than cool stars, but the absorption lines in their spectra are quite different as well. This is because differences in the temperature of the gas in the atmosphere of a star affect the state of the atoms in that gas, which in turn affects the energy level transitions available to absorb radiation.

At the hot end of the spectral classification scheme, the temperature in the atmospheres of O stars is so high that most atoms have had one or more electrons stripped from them by energetic collisions within the gas. There are few transitions available in these ionized atoms that cause absorption lines in the visible part of the electromagnetic spectrum, so the visible spectrum of an O star is relatively featureless. At lower temperatures there are more atoms in energy states that can absorb light in the visible part of the spectrum, so the visible spectra of cooler stars are far more complex than the spectra of O stars. Most absorption lines have an optimum temperature at which they are formed most strongly. For example,

absorption lines from hydrogen are most prominent at surface temperatures of about 10,000 K, which is the surface temperature of an A star. (This should be no surprise. A stars were so named because they are the stars with the strongest lines of hydrogen in their spectra.) At the very lowest stellar temperatures, atoms in the atmosphere of the star begin to react with each other, forming molecules. Molecules such as titanium oxide (TiO) are responsible for much of the absorption in the atmospheres of cool M stars.

It should be stressed that the sequence of spectral classification is really a sequence of surface temperatures, and that the boundaries between types are completely artificial. A hotter-than-average G star is very similar to a cooler-than-average F star. Astronomers break the main spectral types down into a finer sequence of subclasses by adding numbers to the letter designation. For example, the hottest B stars are called B0 stars, slightly cooler B stars are called B1 stars, and so on. The coolest B stars are B9 stars, which are only slightly hotter than A0 stars. The Sun is a G2 star.

other than hydrogen and helium in their spectra. The existence of such stars, all but devoid of more massive elements, provides important clues about the origin of chemical elements and the chemical evolution of the Universe.

STELLAR MASSES ARE MEASURED BY ANALYZING BINARY STAR ORBITS

The most important property of stars that we have yet to investigate is their mass. Determining the mass of an object can be a tricky business. We certainly cannot rely on the amount of light from an object or the object's size as a measure of its mass. Massive objects can be large or small, faint or luminous. The only thing that *always* goes with mass is gravity. Mass is responsible for gravity, and gravity in turn affects the way masses move. When astronomers are trying to determine the masses of astronomical objects, they almost always wind up looking for the effects of gravity.

To measure mass, astronomers look for the effects of gravity.

When discussing the orbits of the planets in Chapter 3, we found that Kepler's laws of planetary motion are the result of gravity. We even went so far as to show that the properties of the orbit of a planet about the Sun can be used to measure the mass of the Sun. If we could find a planet orbiting about a distant star, then we could apply the same technique to determine the mass of that star. Unfortunately, today's telescopes are not yet powerful enough to see directly planets orbiting around other stars, but we can watch as two *stars* orbit about each other. About half of the stars in the sky are actually multiple systems consisting of several stars moving about under the influence of their mutual gravity. Most of these are **binary stars** in which two stars orbit about each other on elliptical orbits predicted by Kepler's laws.

12.3 BINARY

BINARY STARS ORBIT A COMMON CENTER OF MASS

When discussing the motions of the planets, we usually say that the planets orbit around the Sun, without worrying about the effect of their gravity on the Sun's motion. On the other hand if two objects—say, two stars—are closer to the same mass, we have to abandon this simple mental picture. Each object's motion will be noticeably affected by the gravitational force from the other, and we must worry about what happens when the two objects *orbit about each other*. (In fact, as we learned in Chapter 5, it is only from the wobbles that they cause in the motions of stars that we know about planets beyond the Solar System at all.)

Think about what would happen if we were to set two masses—m_1 and m_2—adrift in space with no motion of one mass relative to the other. As soon as we release the two masses, gravity will begin to pull them together. The force of mass 1 on mass 2 equals the force of mass 2 on mass 1, and each mass will begin to fall toward the other. But even though the forces on each mass are equal, the accelerations experienced by the two masses are not. Acceleration equals force divided by mass, so the *less massive object will experience the greater acceleration*.

The forces on two stars in a binary system are equal, but the accelerations of the stars are not.

Suppose, for the moment, that m_1 is three times as massive as m_2. That means that the acceleration of m_2 will be three times as great as the acceleration of m_1. At any given point in time, m_2 will be moving toward m_1 three times as fast as m_1 is moving toward m_2. When the two objects collide, m_2 will have fallen three times as far as m_1. The point where the two objects meet, called the **center of mass** of the two objects, will be three times as far from the original position of the less massive m_2 than from the original position of the more massive m_1. (If the two objects were sitting on a seesaw in a gravitational field, the support of the seesaw would have to be directly under the center of mass for the objects to balance, as shown in **Figure 12.10**.)

Two falling objects meet at their common center of mass.

If we take our two masses and give them some motion perpendicular to the line between them, then instead of simply falling into each other, they will instead orbit about each other. When Newton applied his laws of motion to the problem of orbits, he found that two objects move on elliptical orbits with their center of mass at one focus of both of the ellipses, as shown in **Figure 12.11**. The center of mass, which lies along the line between the two objects, remains stationary. The two objects will always be found on exactly opposite sides of the center of mass, and the elliptical orbit of the more massive ob-

Each star in a binary system follows an elliptical orbit around the system's center of mass.

ject is just a smaller version of the elliptical orbit of the less massive object.

Since the perimeter of the orbit of the less massive star is larger than that of the more massive star's orbit, the less massive star must be moving *faster* than the more massive star. The less massive star has farther to go than the more massive star,

> **The less massive star has a larger orbit and so must move faster.**

but must cover that distance in the same amount of time. As a result, the velocity v of a star in a binary system is inversely proportional to its mass:

$$\frac{v_1}{v_2} = \frac{m_2}{m_1}.$$

This relationship between the velocities and the masses of the stars in a binary system is a crucial part of how binary stars are used to measure stellar masses. Imagine that you are watching a binary star edge-on, as shown in **Figure 12.12.** As one star is moving toward

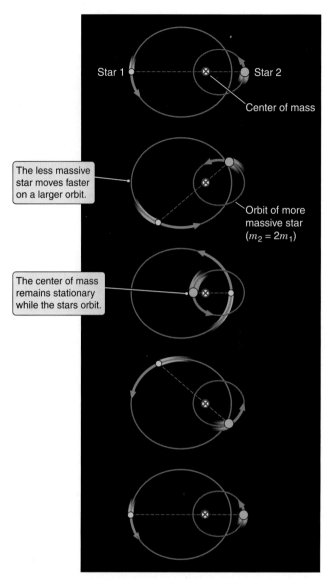

Figure 12.11 *In a binary star system, the two stars orbit on elliptical paths about their common center of mass. In this case star 2 has twice the mass of star 1. The eccentricity of the orbits is 0.5. There are equal time steps between the frames.*

Figure 12.10 *The center of mass of two objects is the "balance" point on a line joining the centers of two masses.*

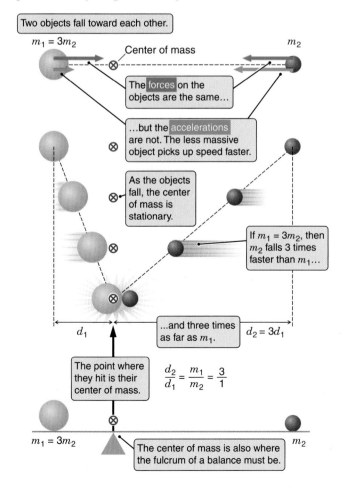

you, the other star will be moving away from you, and vice versa. If you were to look at the spectra of the two stars, you might see that the absorption lines from star 1 are red-shifted relative to the overall center-of-mass velocity, while the absorption lines from star 2 are blue-shifted. If you were to look again half an orbital period later, the situation would be reversed. Lines from star 2 would be red-shifted and lines from star 1 would be Doppler shifted to the blue. Because the two stars are exactly on opposite sides of their orbits, the two stars are always moving in opposite directions. The ratio of the masses of the two stars can be measured by comparing the size of the Doppler shift in the spectrum of star

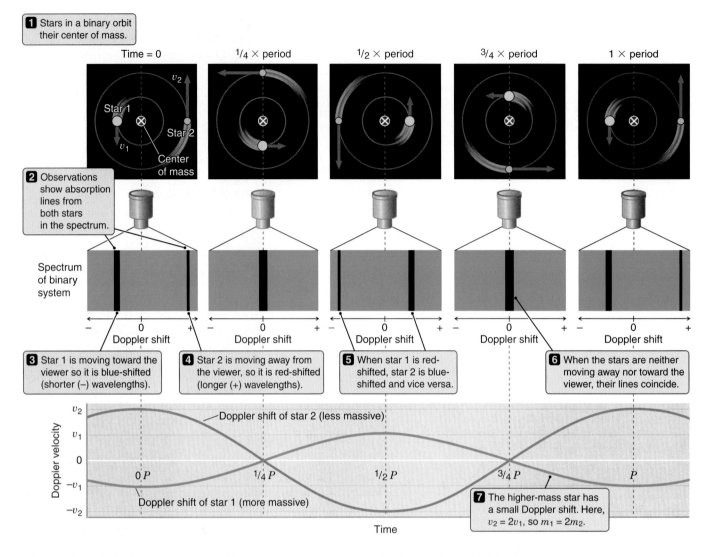

Figure 12.12 *The orbits of two stars in a binary system are shown, along with the Doppler shifts of the absorption lines in the spectrum of each star. Star 1 is twice as massive as star 2 and so has half the Doppler shift.*

1 with the size of the Doppler shift in the spectrum of star 2.

KEPLER'S THIRD LAW GIVES THE TOTAL MASS OF A BINARY SYSTEM

In Chapter 3 we conveniently ignored all of this complexity having to do with the motion of two objects about their common center of mass. Now, however, it is this very complexity that allows us to measure the masses of the two stars in a binary system. In his derivation of Kepler's laws, Newton showed that if two objects with masses m_1 and m_2 are in orbit about each other, then the period of the orbit, P, is related to the average distance between the two masses, A, by the equation

$$P^2 = \frac{4\pi^2 A^3}{G(m_1 + m_2)}.$$

Rearranging this a bit turns it into an expression for the sum of the masses of the two objects:

$$m_1 + m_2 = \frac{4\pi^2}{G} \times \frac{A^3}{P^2}$$

We can cleverly ignore the value of $4\pi^2/G$ by using what we know about Earth's orbit around the Sun. If $A = 1$ AU and $P = 1$ year, then we know that $m_1 + m_2 = M_\odot + M_\oplus \approx M_\odot$ (since $M_\odot$ is much larger than $M_\oplus$). So, if we express the masses, A, and P in that equation in terms of Solar System units (such as

$m_1(M_\odot) = m_1/(1\ M_\odot)$, $A_{AU} = A/(1\ AU)$, and $P_{years} = P/(1\ year)$), then this equation simplifies to

$$m_1(M_\odot) + m_2(M_\odot) = \frac{(A_{AU})^3}{(P_{years})^2}.$$

We now know all that we need to know to measure the masses of two stars in a binary system. If we can measure the period of the binary and average separation between the two stars, then Kepler's third law gives us the total mass. Refering back to the previous subsection, if we can measure the sizes of the orbits of the two stars independently, or independently measure their velocities, then we can determine the ratios of the masses of the two stars. If we know the total mass of the system, as well as the ratio of the two masses, we have all we need to determine the mass of each star separately. Or, to express this in symbols, if we know both m_1/m_2 and $m_1 + m_2$, a bit of algebra yields separate values for m_1 and m_2.

Kepler's third law gives the total mass of a binary system.

There are two ways to go about measuring the orbital properties that we need to determine the masses of stars. If a binary system is a **visual binary**—that is, if we can take pictures that show the two stars separately—then we can watch over time as the stars orbit each other. From these observations we can measure the shapes and period of the orbits of the two stars. When combined with Doppler measurements of the line-of-sight velocities of the stars, this is all the information we need. But in many binary systems, the two stars are so close together and so far away from us that we cannot actually see the stars separately. We know these stars belong to binary systems only because we see the spectral lines of the two stars as they are Doppler shifted away from each other first in one direction, and then in the other direction. Yet even those binary stars that are known only by the Doppler shifts in their spectra can sometimes be used to measure the masses of the stars. If a binary star is an **eclipsing binary,** in which we see a dip in brightness as one star passes in front of the other, then we know we are looking at the system edge on. In this case, the peak Doppler shift of each star gives its total orbital velocity. The period of the orbit is given by the time it takes for a set of spectral lines to go from approaching to receding and back again. If we know the orbital velocities of the stars, and we know the period of the orbit, we also know the size of the orbit, since distance equals velocity times time.

Masses can be measured for visual binaries and eclipsing binaries.

Historically, most stellar masses were measured for stars in eclipsing binary systems, because all that is needed to do the measurement is a series of spectra of the system. Recently, however, new observational capabilities have greatly improved our ability to see the stars in a binary directly. As of this writing, accurate measurements of masses have been obtained for about 250 binary stars, about half of which are eclipsing binaries. (Foundations 12.2 gives a concrete example of how the masses of two stars in an eclipsing binary are determined.) These measurements show that some stars are less massive than the Sun, while others are more massive. However, the range of stellar masses is not nearly as great as the range of stellar luminosities. The least massive stars have masses of about 0.08 $M_\odot$, while the most massive stars have masses of around 100 $M_\odot$. Thus, while stellar luminosities change by a factor of 10^{10}, or 10 billion, from most luminous to least luminous, the most massive stars are only about 10^3, or 1,000 times more massive than the least massive stars.

The H-R Diagram Is Key to Understanding Stars

We have come a long way in our effort to measure the physical properties of stars. **Table 12.2** summarizes the techniques we have used. On the other hand, just knowing some of the bulk properties of stars does not mean that we understand stars. The next step in our journey involves looking for patterns in the properties we have determined. The first astronomers to take this step were a Dane by the name of **Einar Hertzsprung,** and the American astronomer mentioned earlier, **Henry Norris Russell.** In the early part of the 20th century (from 1906 to 1913, to be precise), Hertzsprung and Russell were independently studying the properties of stars. Each scientist plotted the luminosities of stars versus some measure of their surface temperatures (such as color or spectral type). The resulting plot is referred to as the **Hertzsprung-Russell diagram,** or simply the **H-R diagram.** The H-R diagram proved to be the Rosetta stone for understanding stars.

12.4 H-R

Take Time to Understand the H-R Diagram

Before going further in the text, it would be *very much* worth your while to study the H-R diagram carefully

FOUNDATIONS 12.2

USING A CELESTIAL BATHROOM SCALE TO MEASURE THE MASSES OF AN ECLIPSING BINARY PAIR

It is worth looking at an example to see how the relationships discussed in the text are actually put to work. Suppose you are an astronomer studying a binary star system. After observing the star for several years, you build up the following information about the binary system:

1. The star is an eclipsing binary.
2. The stars are in circular orbits.
3. The period of the orbit is 2.63 years.
4. Star 1 has a Doppler velocity that varies between $+20.4$ km/s and -20.4 km/s.
5. Star 2 has a Doppler velocity that varies between $+6.8$ km/s and -6.8 km/s.

These data are summarized in **Figure 12.13.** You begin your analysis by noting that the star is an eclipsing binary, which tells you that the orbit of the star is edge-on to your line of sight. The Doppler velocities tell you the total orbital velocity of each star, and you determine the size of the orbits using the relationship distance = speed × time. In one orbital period, star 1 travels a distance of

$$(20.4 \tfrac{\text{km}}{\text{s}}) \times (2.63 \text{ years}) \times (3.156 \times 10^7 \tfrac{\text{s}}{\text{year}})$$
$$= 1.69 \times 10^9 \text{ km}.$$

This distance is 2π times the radius of the star's orbit, so star 1 is following an orbit with a radius of 2.7×10^8 km, or 1.8 AU. A similar analysis of star 2 shows that its orbit has a radius of 0.6 AU.

Next, you apply Kepler's third law, which says that

$$m_1(M_\odot) + m_2(M_\odot) = \frac{(A_{\text{AU}})^3}{(P_{\text{years}})^2},$$

where A is the average distance between the two stars. Here you have $A = 1.8$ AU $+ 0.6$ AU $= 2.4$ AU. Since you know A and the period P, you can calculate the total mass of the two stars:

$$m_1(M_\odot) + m_2(M_\odot) = \frac{(A_{\text{AU}})^3}{(P_{\text{years}})^2} = \frac{2.4^3}{2.63^2} = 2.0 \; M_\odot$$

To sort out the individual masses of the stars, you use the fact that the more massive the star, the less it moves in response to the gravitational attraction of its companion:

$$\frac{m_2}{m_1} = \frac{v_1}{v_2} = \frac{20.4 \; \frac{\text{km}}{\text{s}}}{6.8 \; \frac{\text{km}}{\text{s}}} = 3.0,$$

so star 2 is three times as massive than star 1. Algebra gets you the rest of the way. Substituting $m_2 = 3 \times m_1$ into the equation $m_1 + m_2 = 2.0 \; M_\odot$ gives you

$$m_1 + (3 \times m_1) = 4 \times m_1 = 2.0 \; M_\odot,$$

or $m_1 = 0.5 \; M_\odot$. Substituting $m_1 = 0.5 \; M_\odot$ into the equation $m_2 = 3 \times m_1$ gives you $m_2 = 1.5 \; M_\odot$.

Star 1 has a mass of 0.5 $M_\odot$, and star 2 a mass of 1.5 $M_\odot$. You can tell your stars to step off the scale now.

Figure 12.13 *Doppler velocities of the stars in an eclipsing binary are used to measure the masses of the stars.*

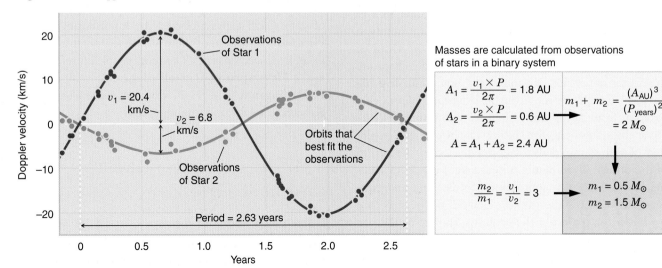

TABLE 12.2

TAKING THE MEASURE OF STARS
A brief summary of the methods used to measure basic properties of stars.

Property	Method
Distance	Measure the **parallax** of the star over the course of the year.
Luminosity	For a star with a known distance, measure the brightness, then apply the **inverse square law of radiation:** $$\text{Luminosity} = 4\pi \times \text{distance}^2 \times \text{brightness}$$
Temperature	Measure the **color** of the star using b_B/b_V or from the star's spectrum. Use **Wien's law** to relate the color to a temperature.
Size	For a star with **known luminosity and temperature,** use **Stefan's law** to compute how large the star must be (the **luminosity-temperature-radius relationship**).
Mass	Measure the motions of the stars in a **binary system,** use these to determine the orbits of the stars, then apply Newton's form of **Kepler's laws.**
Composition	Knowing the temperature of a star, analyze the lines in its **spectrum** to measure chemical composition.

The H-R diagram is the single most used and useful diagram in astronomy.

until you understand what it can tell you. *The H-R diagram is the single most used and useful diagram in astronomy.* Everything about stars, from their formation to their old age and eventual death, is discussed using the H-R diagram.

Begin with the layout of the H-R diagram itself, shown in **Figure 12.14.** Along the horizontal axis (the *x*-axis, if you will) we plot the surface temperature of stars, but it is plotted backward: temperature starts out hot on the left side of the diagram and *decreases* going to the right. Remember that. Hot blue stars are on the left side of the H-R diagram, while cool red stars are on the right side of the H-R diagram. This sequence is plotted *logarithmically,* which is to say that the step along the axis from a point representing a star with a surface temperature of 30,000 K to one with a surface temperature of 10,000 K (that is, a temperature change by a factor of 3) is the same as the step between points representing a star with a temperature of 9,000 K and a star with a temperature of 3,000 K (also a factor-of-3 change). The temperature axis can also be labeled with the spectral types or colors of stars. Hot blue O stars

Hot stars are on the left, cool stars are on the right.

with large b_B/b_V colors are on the left side of the diagram, while cool red M stars with small b_B/b_V colors are on the right side of the diagram.

Along the vertical axis (the *y*-axis) we plot the luminosity of stars—the total amount of energy a star radiates each second. This time the plot is the "right way" around: More-luminous stars are toward the top of the diagram, while less luminous stars are toward the bottom. As with the temperature axis, luminosities are plotted logarithmically, with a step along the axis correspond-

High-luminosity stars are at the top of the H-R diagram.

ing to a multiplicative factor in the luminosity. To understand why the plotting is done this way, recall that the most luminous stars are 10 billion times more luminous than the least luminous stars, yet all of these stars must fit on the same plot.

It turns out that the H-R diagram tells you about more than just the surface temperatures and luminosities of stars. Earlier in the chapter we discovered that if the temperature and the luminosity of a star are known, then we can calculate its radius. Since each point in the H-R diagram is specified by a surface temperature and a luminosity, we can use the luminosity-temperature-radius relationship to find the size of a star at that point as well. We can think through how this works using what we know of radiation. A star in

Size increases in the H-R diagram from lower left to the upper right corner.

the upper right corner of the H-R diagram is very cool, which according to Stefan's law means that each square meter of its surface is radiating only a small amount of energy. But at the same time, this star is extremely luminous. Such a star must be huge to account for its high overall luminosity despite the feeble radiation coming from each square meter of its surface. Conversely, a star in the lower left corner of the H-R diagram is very hot, which means that a large amount of energy is coming from each square meter of its surface. However, this star has a very low

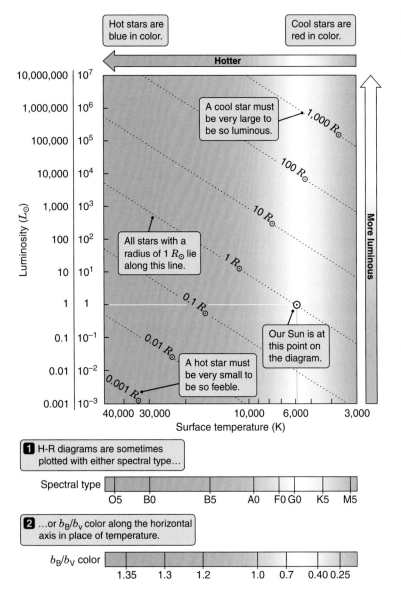

Figure 12.14 *This is the layout of the Hertzsprung-Russell, or H-R, diagram used to plot the properties of stars. More luminous stars are at the top of the diagram. Hotter stars are on the left of the diagram. Stars of the same radius (R) lie along the dotted lines moving from upper left to lower right.*

THE MAIN SEQUENCE IS A GRAND PATTERN IN STELLAR PROPERTIES

Figure 12.15(a) shows the H-R diagram for 16,600 nearby stars based on observations obtained by the *Hipparcos* satellite. A quick look at this diagram immediately reveals a remarkable fact. Instead of being a scatter plot, with stars strewn all about the diagram, we find instead that about 90% of the stars in the sky lie along a well-defined sequence running across the H-R diagram from lower right to upper left. This sequence of stars is called the **main sequence**. On the left end of

> **Most stars lie along the main sequence of the H-R diagram.**

the main sequence are the O stars: hotter, larger, and more luminous than the Sun. On the right end of the main sequence are the M stars: cooler, smaller, and fainter than the Sun. If you know where a star lies on the main sequence, then you know its luminosity, surface temperature, and size.

From a practical standpoint, the main sequence is extremely handy. For example, you may have wondered in our discussion of stellar parallax how we measure the distances to stars that are farther away than a few hundred light years. The main sequence comes to the rescue. It is possible to determine whether a star is a main sequence star by looking at the absorption lines in its spectrum. If a star is on the main sequence, then by virtue of its location on the H-R diagram you know its luminosity. If you know its luminosity and you measure its brightness, you can use the inverse square law of radiation to find its distance. (How far away must a star of that lumi-

overall luminosity. The conclusion is that its surface area cannot be very large. Stars in the lower left corner of the H-R diagram are small.

This result persists everywhere in the H-R diagram. Moving up and to the right takes you to larger and larger stars. Moving down and to the left takes you to smaller and smaller stars. All of the stars of the same radius lie along lines running across the H-R diagram from the upper left to the lower right, as shown in Figure 12.14.

Like an experienced surveyor who "sees" the lay of the land when looking at a topographical map, or a classical musician who "hears" the rhythms and harmonies when looking at a piece of sheet music, you should get to the point where you automatically "see" the properties of a star—its temperature, color, size, and luminosity—from a glance at its position on the H-R diagram.

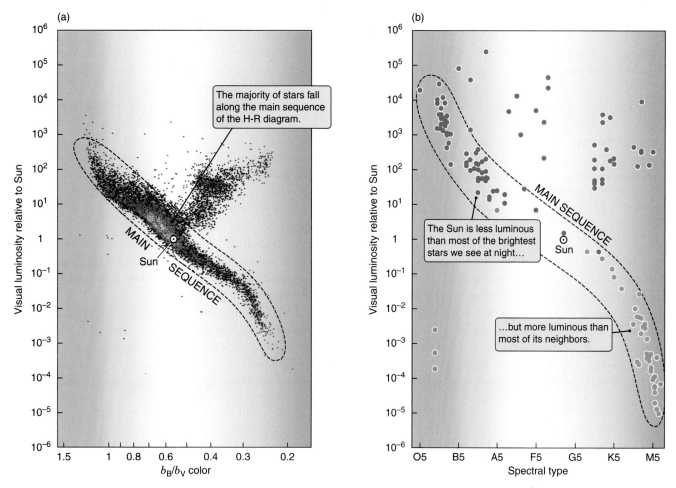

Figure 12.15 (a) *An H-R diagram for 16,600 stars obtained by the* Hipparcos *satellite. Most of the stars lie in a band running from the upper left of the diagram toward the lower right called the main sequence. (Dot color represents number of stars.) (b) An H-R diagram for two different samples of stars. The red symbols show the H-R diagram for 46 stars that are especially close to the Sun. The blue symbols show the 97 brightest stars in the sky. Note that since these are observational H-R diagrams they are plotted against observed quantities,* b_B/b_V *color in (a) and spectral type in (b).*

nosity be to have the brightness we measure?) This method of determining distances to stars is called **spectroscopic parallax.** Parallax and spectroscopic parallax are the first two steps along a chain of reasoning that will let us build our knowledge of distances all the way to the edge of the observable Universe.

Distances to main sequence stars are measured using spectroscopic parallax.

Figure 12.15(b) makes another important point for astronomy, as well as for all sciences. The red symbols are 46 of the nearest stars to Earth, while the blue symbols show the 97 brightest stars as seen in our sky. There are many more cool, low-luminosity stars in the sky than there are hot, high-luminosity stars. If we restrict our attention to stars that are near the Sun, these are all that we

see. Yet if we look at the brightest stars in the sky, we get a very different picture. Even though these stars are farther away, they are so luminous that they dominate what we see. Think about how different your impression of the properties of stars would be if all you knew about were the nearest stars, or if all you knew about were the brightest stars. Neither of these groups alone accurately represents what stars as a whole are like. Whether you are an astronomer studying the properties of stars or a political pollster measuring the sense of the people, the validity of your results depends on the care with which you choose the sample (of stars or people) that you study.

When we add mass to the picture, the main sequence of the H-R diagram gets even more interesting. Stellar mass increases smoothly as we go from the lower right to the upper left along the main sequence. The

faint, cool stars on the right side of the main sequence

The mass of a star determines its position along the main sequence.

have low masses, while the luminous, hot stars on the left side of the main sequence are high-mass stars. Stated another way, if we know the mass of a main sequence star, then we know *where* on the main sequence the star is, which means we know the star's tempera-

Low-mass main sequence stars are faint and cool. High-mass main sequence stars are hot and luminous.

ture, size, and luminosity. If a main sequence star is less massive than the Sun, it will be smaller, cooler, redder, and less luminous than the Sun. It will be located to the right of the Sun on the main sequence. On the other hand, if a star is more massive than the Sun, it will be larger, hotter, bluer, and more luminous than the Sun. It will be located to the left of the Sun on the main sequence. *The mass of a star determines where on the main sequence the star will lie.*

To see this directly, look at **Table 12.3** and **Figure 12.16**, which show the properties of stars along the

main sequence. If, for example, a main sequence star has a mass of 18 $M_{\odot}$, it will be a B0 star. It will have a surface temperature of 30,000 K, a radius of over 9 $R_{\odot}$, and a luminosity around 50,000 times that of the Sun. If a main sequence star instead has a mass of 0.21 $M_{\odot}$, it will be an M5 star. It will have a surface temperature of 3,240 K, a radius of about 0.32 $R_{\odot}$, and a luminosity of about 0.01 $L_{\odot}$. A main sequence star with a mass of 1 $M_{\odot}$ will be a G2 star like the Sun, and will have the same surface temperature, size, and luminosity as the Sun.

It is difficult to overemphasize the importance of this result, and so we state it again for clarity. For stars of similar chemical composition, *the mass of a main sequence star alone determines all of its other characteristics.* **Figure 12.17** illustrates this result. Knowing the mass (and

The mass of a star determines what its fate will be.

chemical composition) of a main sequence star tells us how large it is, what its surface temperature is, how bright it is, what its internal structure is, how long it will live, how it will evolve, and what its final fate will be!

TABLE 12.3

THE MAIN SEQUENCE OF STARS

Approximately 90% of the stars in the sky fall on the main sequence of the H-R diagram. This table gives the properties of stars along the main sequence.

Spectral Type	b_B/b_V Color[1]	Temperature (K)	Mass ($M_{\odot}$)	Radius ($R_{\odot}$)	Luminosity ($L_{\odot}$)
O5	1.36	44,500	60	17.8	794,000
B0	1.32	30,000	18	9.3	52,500
B5	1.17	15,400	5.9	3.8	832
A0	1.02	9,520	2.9	2.5	54
A5	0.87	8,200	2.0	1.74	14
F0	0.76	7,200	1.6	1.35	6.5
F5	0.67	6,440	1.3	1.2	3.2
G0	0.59	6,030	1.05	1.05	1.5
G2 (Sun)	0.56	5,860	1.00	1.00	1.0
G5	0.53	5,770	0.92	0.93	0.8
K0	0.47	5,250	0.79	0.85	0.4
K5	0.35	4,350	0.67	0.74	0.15
M0	0.27	3,850	0.51	0.63	0.08
M5	0.22	3,240	0.21	0.32	0.011
M8	0.19	2,640	0.06	0.13	0.0012

[1]Normally astronomers refer to the "B-V color" of a star as the difference between the B and V magnitudes of the star. To avoid the complication of magnitudes, in this text we instead refer to the brightness ratio b_B/b_V as the "b_B/b_V color" of the star.

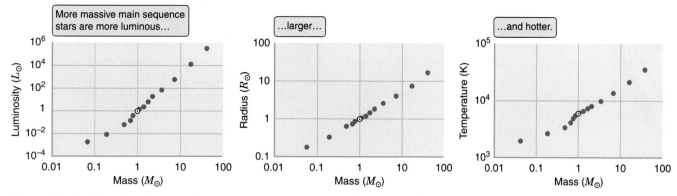

Figure 12.16 *Plots of luminosity, radius, and temperature versus mass for stars along the main sequence. The mass (and chemical composition) of a main sequence star determines all of its other properties.*

Upon reflection, this result is both sensible and possibly the most important and fundamental result in all of astrophysics. If you have a certain amount and type of material to make a star, there is only one kind of star you can make. As we go on to discuss what physical processes give a star its structure, this will make even more sense. We will find that a star is a "battle" between gravity (which is trying to pull the star together) and the energy released by nuclear reactions in the interior of the star (which are trying to blow it apart). The mass of the star determines the strength of its gravity, which in turn determines how much energy must be generated in its interior to prevent it from collapsing under its own weight. The mass of a star determines where the balance is struck.

NOT ALL STARS ARE MAIN SEQUENCE STARS

Before leaving our discussion of the observed properties of stars, we need to point out that while most stars in the sky are main sequence stars, some are not. Some stars are found in the upper right portion of the H-R diagram, well above the main sequence. From their position we know that they must be bloated, luminous, cool giants, with radii hundreds or thousands of times the radius of the Sun. If the Sun were such a star, its atmosphere would swallow the orbits of the inner planets, including Earth. At the other extreme are stars found in the extreme lower left corner of the H-R diagram. These stars must be tiny, with sizes comparable to the size of Earth. Their small surface areas explain why they have such low luminosities, despite having temperatures that can rival or even exceed the surface temperature of the hottest main sequence O stars.

The existence of the main sequence, together with the fact that the mass of a main sequence star controls where on the main sequence it will lie, are grand patterns that point to the possibility of some deep understanding of what stars are and what makes them tick. By the same token, the existence of stars that do *not* follow these grand patterns raises yet more questions. What is it about a star that determines whether it is

Figure 12.17 *The main sequence of the H-R diagram is a sequence of masses.*

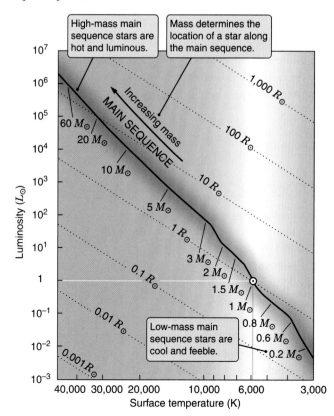

part of the main sequence or not? In the decades that followed the discovery of the main sequence, few problems in astronomy attracted more attention than these questions. Their answers turned out to be as fundamental as stellar pioneers Annie Jump Cannon, Einar Hertzsprung, and Henry Norris Russell could ever have hoped. The existence of the main sequence holds the essential clue to what stars are and how they work. The properties of stars that are not on the main sequence point to an understanding of how stars form, how they evolve, and how they die. Much of the rest of the text discussing stars will be spent on the lessons learned from patterns in the H-R diagram.

SEEING THE FOREST THROUGH THE TREES

When you walk down a path, each step that you take covers only a short distance. However, by persistently putting one foot in front of the other, you find after a time that you have gotten somewhere. The same is true for the road we have traveled in this chapter. Each step along the way has been relatively small and understandable—involving the application of a physical principle that we see at work in the world around us, or the use of a tool from algebra or geometry. However, when we compare our understanding of stars reached by the end of the chapter with the understanding we started out with, it is amazing how far we have come.

Our path in this chapter followed the course of the triumph of our physical understanding of the Universe. The properties of electromagnetic radiation, the structure of atoms and molecules, the gravitational attraction between masses—all of these and more came into play as we built up our understanding of the physical nature of stars one piece at a time. We saw many occasions in this chapter when the cosmological principle might have collapsed. We might have seen emission and absorption lines in the spectra in stars that were different from those measured in terrestrial laboratories. We might have found that the temperatures of stars inferred from their absorption spectra disagreed with the temperatures of stars inferred from their peaks in blackbody emission. We might have found that measurements of the sizes of stars based on Stefan's law disagreed with measurements of the sizes of stars in eclipsing binaries. We might have found that the motions of stars in binary systems failed to follow the predictions of Newton's physics. We might have found any or all of these things, *but we did not!* The successes of this chapter give us strong reason to believe that the same physical laws at work right here on Earth and in our Solar System also describe the fundamental character and behavior of matter and energy throughout the rest of the Universe.

With an understanding of the basic physical properties of stars in place, we are now ready to ask much more fundamental questions about stars. We are ready, figuratively speaking, to "open the hood" and see what lies within. How do stars work? How do they form? How do they evolve? How do they die? We will begin to address these questions by investigating the star that serves as the standard by which we measure other stars: the star we know best, our Sun.

STUDENT QUESTIONS

THINKING ABOUT THE CONCEPTS

1. To know certain properties of a star, you first must determine the star's distance. For other properties, knowledge of distance is not necessary. Into which category would you place each of the following properties: size, mass, temperature, color, spectral type, and chemical composition? In each case, state your reason(s).

2. The distances of nearby stars are determined by their parallaxes. Why is the uncertainty in our knowledge of a star's distance greater for stars that are farther from Earth?

3. Distance to stars can be measured in miles, kilometers (km), astronomical units (AU), light-years (ly) and parsecs (pc). Why do astronomers prefer to use parsecs?

4. Although we tend to think of our Sun as an "average" main sequence star, it is actually hotter and more luminous than average. Explain.

5. Albiero, a star in the constellation of Cygnus, is a binary system whose components are easily separated in a small amateur telescope. Viewers describe the brighter star as "golden" and the fainter one as "sapphire blue."

a. What does this tell you about the relative temperatures of the two stars?

b. What does it tell you about their respective sizes?

6. Explain why the stellar spectral types (O, B, A, F, G, K, M) are not in alphabetical order.

7. Other than the Sun, the only stars whose mass we can measure directly are those in eclipsing or visual binary systems. Explain.

8. How do we estimate the mass of stars that are not in eclipsing or visual binary systems?

9. Star masses range from 0.08 $M_\odot$ to about 100 $M_\odot$.

a. Why are there no stars with masses less than 0.08 $M_\odot$?

b. Why are there no stars with masses much greater than 100 $M_\odot$?

10. Logarithmic (log) plots show major steps along an axis scaled to represent equal factors, most often factors of 10. Why, in astronomy, do we sometimes use a log plot instead of the more conventional linear plot?

APPLYING THE CONCEPTS

11. Sketch a circle and mark an arc along the circumference equal to the radius of the circle. The size of the angle subtended by the arc, measured in *radians,* is given by the length of the arc divided by the radius (r) of the circle. Since the circumference of a circle is $2\pi r$, there must be $2\pi r/r$ or 2π radians in a circle.

a. How many degrees are there in one radian?

b. How many arcseconds are there in one radian? How does this compare to the number of AU in one parsec (206,264.81) that we cite in the text?

c. What angle would a round object that has a diameter of 1 parsec make in our sky if we

see it a distance of 1 parsec? a distance of 10 parsecs?

d. What do your answers to parts (b) and (c) of this question tell you about how the actual size of an object is related to its angular size measured in units of radians?

12. Our eyes are typically 6 cm apart. Suppose your eye separation is average, and that you see an object jump from side to side by one-half degree as you blink back and forth between your eyes. How far away is that object?

13. Sirius, the brightest star in the sky, has a parallax of 0.379 arcseconds. What is its distance in parsecs? In light-years?

14. Sirius is 22 times more luminous than the Sun, and Polaris (the "Pole Star") is 2,350 times more luminous than the Sun. Sirius appears 23 times brighter than Polaris. What is the distance of Polaris in light-years?

15. Proxima Centauri, the star nearest to Earth other than the Sun, has a parallax of 0.772 arcseconds. How long does it take light to reach us from Proxima Centauri?

16. The Sun is about 16 trillion (1.6×10^{13}) times brighter than the faintest stars visible to the naked eye.

a. How far away (in AU) would an identical solar-type star be if it were just barely visible to the naked eye?

b. What would be its distance in light-years?

17. Sirius and its companion orbit around a common center of mass with a period of 50 years. The mass of Sirius is 2.35 times the mass of the Sun.

a. If the orbital velocity of the companion is 2.35 times greater than that of Sirius, what is the mass of the companion?

b. What is the semimajor axis of the orbit?

It is stern work, it is perilous work to thrust your hand in the sun
And pull out a spark of immortal flame to warm the hearts of men.

JOYCE KILMER (1886–1918)

A Run-of-the-Mill G Dwarf: Our Sun

13.1 The Sun Is More Than Just a Light in the Sky

How wonderful, after a long cold night, to see the light and feel the warmth of the rays of the Sun. Energy from the Sun is responsible for daylight, for our weather and seasons, and for terrestrial life itself. No object in nature has been more revered or more worshiped than the Sun. In fact, the modern symbol for the Sun, $\odot$, is nothing other than the ancient Egyptian hieroglyph for the Sun god Ra. Yet, while the Sun may have dominated human consciousness since the dawn of our species, the discussion in Chapter 12 offers a very different perspective. To an astronomer at the opening of the 21st century, the Sun is the prototype for main sequence stars. With a middle-of-the-road spectral type of G2, the Sun is all but indistinguishable from billions of other stars in our Galaxy. It is also the star against which all other stars are measured. The mass of the Sun, the size of the Sun, the luminosity of the Sun—these basic properties of the Sun provide the yardsticks of modern astronomy.

The Sun may be run-of-the-mill as far as stars go, but that makes it no less awesome an object on a human scale. The mass of the Sun, 1.99×10^{30} kilo-

KEY CONCEPTS

To most humans, the Sun is the most important object in the heavens. It lights our days, warms our planet, and provides the energy for life. But to astronomers, the Sun is a typical main sequence star, located conveniently nearby for detailed study. In this chapter, as we take a closer look at our local star we will learn about:

* The balances between pressure and gravity and between energy generation and loss that determine the structure of the Sun;
* Fusion of hydrogen to helium, and how mass is efficiently converted into energy in the Sun's core;
* The different ways that energy moves outward from the Sun's core toward its surface;
* Physical models of the Sun's interior and how they are tested using observations of solar neutrinos and seismic vibrations on the surface of the Sun;
* The structure of the Sun's atmosphere, from its 5,860 K photosphere to its 1 million kelvin corona;
* Sunspots, flares, coronal mass ejections, and other consequences of magnetic activity on the Sun;
* Eleven- and 22-year cycles in solar activity;
* The solar wind streaming away from the Sun; and
* How solar activity affects Earth.

grams, is over 300,000 times that of Earth. The Sun's radius of 696,000 km is over 100 times that of Earth. At a luminosity of 3.85×10^{26} watts, the Sun produces more energy in a second than all of the electrical power plants on Earth could generate in 10 million years. The Sun is also the only star we can study at close range. Much of the detailed information that we know about stars has come only by studying our local star. Even though the Sun has been intensively studied for quite some time, the wealth of phenomena displayed by the Sun will keep solar astronomers busy for many years to come.

In the previous leg of our journey, we looked at the gross physical properties of distant stars, including their mass, luminosity, size, temperature, and chemical composition. Now, as we turn our attention toward our own local star, we face more fundamental questions. How does the Sun work? Where does it get its energy? Why does it have the size, temperature, and luminosity that it does? How has it been able to remain so constant over the billions of years since the Solar System formed? In short, we now confront the question "What is a star?"

13.2 THE STRUCTURE OF THE SUN IS A MATTER OF BALANCE

In Chapter 6, we tackled the question of how we know about the interior of Earth, despite the fact that no machine has ever done more than scratch the surface of our planet. The answer was a combination of physical understanding, detailed computer models, and clever experiments that test the predictions of those models. The task of exploring the interior of the Sun is much the same. As with Earth, the structure of the Sun is governed by a number of physical processes and relationships. Using our understanding of physics, chemistry, and the properties of matter and radiation, we express these processes and relationships as mathematical equations. High-speed computers are then used to simultaneously solve these equations and arrive at a model of the Sun. One of the great successes of 20th-century astronomy was the successful construction of a physical model of the Sun that agrees with our observations of the mass, composition, size, temperature, and luminosity of the real thing.

Our current model of the interior of the Sun is the culmination of decades of work by thousands of physicists and astronomers. Understanding the details that lie

within this model is the work of lifetimes. Even so, the essential ideas underlying our understanding of the structure of the Sun are found in a few key insights. In turn, these insights can be summed up in a single statement: *the structure of the Sun is a matter of balance.*

The first key balance within the Sun is the balance between pressure and gravity illustrated in **Figure 13.1.** The Sun is a huge ball of hot gas. If gravity were stronger than pressure within the Sun, then the Sun would collapse. Likewise, if pressure were stronger than gravity, then the Sun would blow itself apart. We have seen this balance before and have given it a name— hydrostatic equilibrium. *Hydrostatic equilibrium* sets the pressure at each point within a planet, and determines the atmospheric pressure at Earth's surface. Hydrostatic equilibrium says that the pressure at any point within the Sun's interior must be just enough to hold up the weight of all of the layers above that point. If the Sun were not in hydrostatic equilibrium, then forces within the Sun would not be in balance, so the surface of the Sun would *move.* The Sun today is the same as it was yesterday and the day before. That is all the observation that is needed to know that the interior of the Sun is in hydrostatic equilibrium.

The Sun is in hydrostatic equilibrium.

Hydrostatic equilibrium becomes an even more powerful concept when combined with what we know about the way gases behave. As we move deeper into the interior of the Sun, the weight of the material above us becomes greater, and hence the pressure must increase. As we have learned before (see Foundations

Figure 13.1 *The structure of the Sun is determined by balances between forces and in the outward flow of energy.*

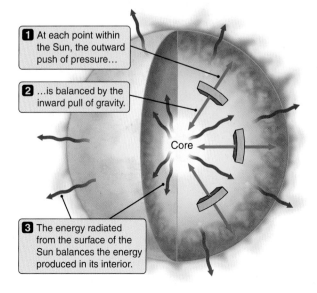

1 At each point within the Sun, the outward push of pressure…

2 …is balanced by the inward pull of gravity.

Core

3 The energy radiated from the surface of the Sun balances the energy produced in its interior.

7.2), in a gas, higher pressure means higher density and/or higher temperature. **Figure 13.2(a)** shows how conditions vary as distance from the center of the Sun changes. As we go deeper into the Sun, the pressure climbs, and as it does, the density and temperature of the gas climb as well.

A second fundamental balance within the Sun is a balance of energy. Stars like the Sun are remarkably stable objects. Geological records show that the luminosity of the Sun has remained nearly constant for billions of years. In fact, the very existence of the main sequence says that stars do not change much over the main part of their lives. To remain in balance, the Sun must produce just enough energy in its interior each second to replace the energy that is radiated away by its surface. This is a new type of balance, one we have not dealt with before. Understanding the balance of energy within the Sun requires thinking about how energy is

Energy production must balance what is radiated away.

generated in the interior of the Sun, and how that energy finds its way from the interior to the Sun's surface, where it is radiated away.

THE SUN IS POWERED BY NUCLEAR FUSION

One of the most basic questions facing the pioneers of stellar astrophysics was where the Sun and stars get their energy. The answer to this question came not from astronomers' telescopes, but from theoretical work and the laboratories of nuclear physicists. At the heart of the Sun lies a nuclear furnace capable of powering the star for billions of years.

The nucleus of most hydrogen atoms consists of a single proton. Nuclei of all other atoms are built from a mixture of protons and neutrons. Most helium nuclei, for example, consist of two protons and two neutrons. Most carbon nuclei consist of six protons and

Figure 13.2 *A cutaway figure showing the interior structure of the Sun.* (a) *Temperature, density, and pressure increase toward the center of the Sun.* (b) *Energy is generated in the Sun's core.*

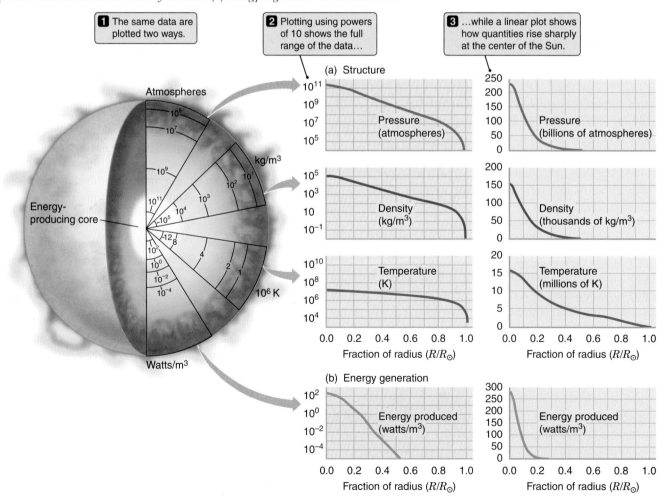

six neutrons. Protons have a positive electric charge, and neutrons have no net electric charge. Like charges repel, so all of the protons in an atomic nucleus must be pushing away from each other with a tremendous force. If electric forces were all there was to it, the nuclei of atoms would rapidly fly apart—yet atomic nuclei *do* hold together. We conclude that there must be some other force in nature, even stronger than the electric force, that "glues" the protons and neutrons in a nucleus together. That force, which acts only over short distances, is called the **strong nuclear force.**

Atomic nuclei are held together by the strong nuclear force.

The strong nuclear force is indeed a very powerful force. It would take an enormous amount of energy to pull apart the nucleus of an atom such as helium into its constituent parts. The reverse of this process says that when you assemble an atomic nucleus from its component parts, this same enormous amount of energy is released. The process of combining two less-massive atomic nuclei into a single more-massive atomic nucleus is referred to as **nuclear fusion.** Many nuclear processes are possible and, as we continue our study of stars, we will find that a wide range of nuclear reactions can occur in stars. In the Sun, as in all main sequence stars, the only significant process going on is the fusion of hydrogen to form helium—a process often referred to as **hydrogen burning.**

13.2
FUSION

Main sequence stars get their energy by fusing hydrogen atoms together to make helium. To judge the efficiency of this reaction, we can make use of one of the key results of special relativity (Chapter 4), the equivalence between mass and energy. Mass can be converted to energy and energy can be converted to mass, with Einstein's famous equation $E = mc^2$ providing the exchange rate between the two. Comparing the mass of the *products* of a reaction with the mass of the *reactants* tells what fraction of the original mass was turned into energy in the process. The mass of four separate hydrogen atoms is 1.007 times greater than the mass of a single helium atom, so when hydrogen fuses to make helium, 0.7% of the mass of the hydrogen is converted to energy.

Main sequence stars burn hydrogen to form helium.

Conversion of 0.7% of the mass of the hydrogen into energy might not seem very efficient—until we compare it with other sources of energy and discover that it is millions of times more efficient than even the most efficient chemical reactions! Fusing a *single gram* of hydrogen into helium releases about 6×10^{11} joules of energy, which is enough to boil all of the

water in about 10 average backyard swimming pools. Scale this up to converting roughly 600 million metric tons of hydrogen into helium every second (with 4 million metric *tons* of matter converted to energy in the process), and you have our Sun. The sunlight falling on Earth may be responsible for powering almost everything that happens on our planet, but it amounts to only about a hundred-billionth of the energy radiated by our local star. The Sun has been burning hydrogen at this prodigious rate for 4.6 billion years, during which time it has converted about half of the hydrogen at its center into helium. A favorite theme of science fiction is the fate that awaits Earth when the Sun dies, but we need not worry anytime soon. The Sun is only about halfway through its 10-billion-year lifetime as a main sequence star.

Nuclear fusion is a very efficient source of energy.

Whether a ball rolling down hill, a battery discharging itself through a lightbulb, or an atom falling to a lower state by emitting a photon, most systems in nature tend to "seek out" the lowest energy state available to them. Going from hydrogen to helium is a big ride downhill in energy, so we might imagine that hydrogen nuclei would naturally tend to fuse together to make helium. Fortunately for us, however, a major roadblock stands in the way of nuclear fusion. The strong nuclear force responsible for binding atomic nuclei together can only act over very short distances—10^{-15} meter or so, or about a hundred-thousandth the size of an atom. To get atomic nuclei to fuse, they must be brought close enough to each other for the strong nuclear force to assert itself, but this is hard to do. All atomic nuclei have positive electric charges, which means that any two nuclei will repel each other. This electric repulsion illustrated in **Figure 13.3,** serves as a barrier against nuclear fusion. Fusion cannot take place unless this electric barrier is somehow overcome.

Electric repulsion between protons is a barrier to nuclear fusion.

As shown in **Figure 13.2(b),** energy in the Sun is produced in its innermost region, called the Sun's **core.** Conditions in the Sun's core are extreme. Matter at the center of the Sun has a density about 150 times the density of water (which is 1,000 kg/m³), and the temperature at the center of the Sun is about 15 million K. The thermal motions of atomic nuclei in the Sun's core are tens of thousands times more energetic than the thermal motions of atoms at room temperature. As illustrated in Figure 13.3(c), under

Hydrogen fuses to helium in the core of the Sun.

(a)

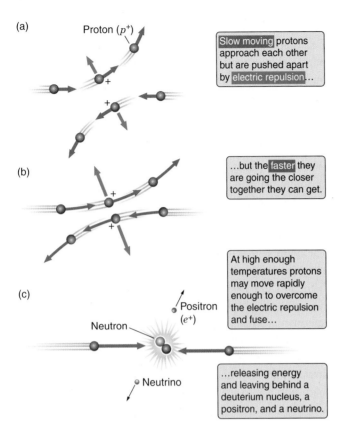

Proton (p^+)

> Slow moving protons approach each other but are pushed apart by electric repulsion...

(b)

> ...but the faster they are going the closer together they can get.

(c)

Positron (e^+)

Neutron

> At high enough temperatures protons may move rapidly enough to overcome the electric repulsion and fuse...

Neutrino

> ...releasing energy and leaving behind a deuterium nucleus, a positron, and a neutrino.

Figure 13.3 *Atomic nuclei are positively charged, and so repel each other. If two nuclei are moving toward each other, the faster they are going, the closer they will get before veering away, as shown in (a) and (b). (c) At the temperatures found in the centers of stars, thermal motions of nuclei are so energetic that nuclei can overcome this electric repulsion, so fusion takes place.*

these conditions atomic nuclei slam into each other hard enough to overcome the electric repulsion between them and allow short-range nuclear forces to act. The hotter and denser a gas, the more of these energetic collisions will take place each second. For

this reason, the rate at which nuclear fusion reactions occur is extremely sensitive to the temperature and the density of the gas. Half of the energy produced by the Sun is generated within the inner 9% of the Sun's radius, or less than 0.1% of the volume of the Sun (Figure 13.2b).

There are several reasons that hydrogen burning is the most important source of energy in main sequence stars. Hydrogen is the most abundant element in the Universe, so it offers the most abundant source of nuclear fuel at the beginning of a star's lifetime. Hydrogen burning is also the most efficient form of nuclear fusion, converting a larger fraction of mass into energy than any other type of reaction. But the most important reason why hydrogen burning is the dominant process in main sequence stars is that hydrogen is also the easiest type of atom to fuse. Hydrogen nuclei—protons—have an electric charge of +1. The electric barrier that must be overcome to fuse protons is the repulsion of a single proton against another. Compare this to the force required, for example, to get two carbon nuclei close enough to fuse. To fuse carbon, we must overcome the repulsion of six protons in one carbon nucleus pushing against the six protons in another carbon nucleus. The resulting force is proportional to the product of the charges of the two atomic nuclei, making the repulsion between two carbon nuclei 36 times stronger than that between two protons. For this reason, hydrogen fusion occurs at a much lower temperature than any other type of nuclear fusion. In the cores of low-mass stars such as the Sun, hydrogen burns primarily through a

Hydrogen burns mostly via the proton–proton chain.

process called the **proton–proton chain.** The proton–proton chain is illustrated in **Figure 13.4,** and discussed in Foundations 13.1.

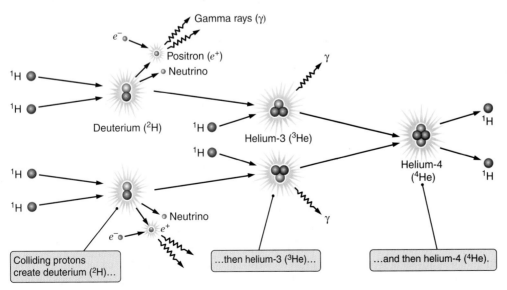

Gamma rays (γ)

e^-

Positron (e^+)

Neutrino

^{1}H

^{1}H

Deuterium (^{2}H)

^{1}H

Helium-3 (^{3}He)

^{1}H

γ

^{1}H

Helium-4 (^{4}He)

^{1}H

^{1}H

^{1}H

Neutrino

e^- e^+

γ

> Colliding protons create deuterium (^{2}H)...

> ...then helium-3 (^{3}He)...

> ...and then helium-4 (^{4}He).

Figure 13.4 *The Sun and all main sequence stars get their energy by fusing the nuclei of four hydrogen atoms together to make a single helium atom. In the Sun, about 85% of the energy produced comes from the branch of the proton–proton chain shown here.*

FOUNDATIONS 13.1

THE PROTON–PROTON CHAIN

Hydrogen burning in the Sun and other low-mass stars takes place through a series of nuclear reactions called the proton–proton chain. There are three different "branches" to the proton–proton chain. The most important of these (Figure 13.4), responsible for about 85% of the energy generated in the Sun, consists of three steps. In the first step, two hydrogen nuclei fuse. In the process, one of the protons is transformed into a neutron by emitting a positively charged particle called a **positron** and another type of elementary particle called a **neutrino.** The conversion of a proton into a neutron by the emission of a positron and a neutrino is one variety of a process referred to as **beta decay.**

The positron is expelled at a great velocity, carrying away some of the energy released in the reaction. Electrons and positrons have opposite electric charges, and so attract each other. As a result, our expelled positron will soon collide with one of the many electrons moving freely about in the center of the Sun. Because the positron is the **antiparticle** of the electron, when the two particles collide they annihilate each other, forming energetic gamma ray photons in the process. These photons carry part of the energy released when the two protons fused, and help to heat the surrounding gas. The neutrino, on the other hand, is a very elusive particle. Its interactions with matter are so feeble that its most likely fate is to escape from the Sun without further interactions with matter.

The new atomic nucleus formed by the first step in the proton–proton chain consists of a proton and a neutron. This is the nucleus of a heavy isotope of hydrogen called *deuterium,* or ^{2}H. The second step of the proton–proton chain occurs when another proton slams into the deuterium nucleus, fusing with it to form the nucleus of a light isotope of helium, ^{3}He, consisting of two protons and a neutron. The energy released in this step is carried away as a gamma ray photon. The third and final step in the proton–proton chain occurs when two ^{3}He nuclei collide and fuse together, producing an ordinary ^{4}He nucleus and ejecting two protons in the process. The energy released in this step shows up as kinetic energy of the helium nucleus and two ejected protons.

This dominant branch of the proton–proton chain can be written symbolically as

$$^1\text{H} + {}^1\text{H} \rightarrow {}^2\text{H} + e^+ + \nu,$$
$$\text{followed by } e^+ + e^- \rightarrow \gamma + \gamma$$
$$^2\text{H} + {}^1\text{H} \rightarrow {}^3\text{He} + \gamma$$
$$^3\text{He} + {}^3\text{He} \rightarrow {}^4\text{He} + {}^1\text{H} + {}^1\text{H}$$

Here the symbols are e^- for an electron, e^+ for a positron, ν (the Greek letter "nu") for a neutrino, and γ ("gamma") for a gamma-ray photon.

ENERGY PRODUCED IN THE SUN'S CORE MUST FIND ITS WAY TO THE SURFACE

Some of the energy released by hydrogen burning in the core of the Sun escapes directly into space in the form of neutrinos (see Foundations 13.1), but most of the energy goes instead into heating the solar interior. To understand the structure of the Sun we must understand how thermal energy is able to move outward through the star. The nature of **energy transport** within a star is one of the key factors determining the star's structure.

Thermal energy can be transported from one place to another by a number of ways. Pick up a bucket of hot coals and carry it from one side of the room to the other, and you have transported thermal energy. A common way in which energy is transported in our everyday lives is by **thermal conduction.** For example, hold one end of a metal rod while you stick the other end into a fire. Soon the end of the rod that is in your hand becomes too hot to hold. Thermal conduction occurs as the energetic thermal vibrations of atoms and molecules in the hot end of the rod cause their cooler neighbors to vibrate more rapidly as well. However, while thermal conduction is the most important way energy is transported in *solid* matter, it is typically inefficient in a *gas.* Thermal conduction is unimportant in the transport of energy from the core of the Sun to its surface. Thermal energy is instead carried outward from the center of the Sun by two other mechanisms: radiation and convection.

The transport of energy from one place to another by **radiative transfer** involves photons moving from hotter regions to cooler regions, carrying energy with them. Imagine a hotter region of a star sitting next to a cooler region, as shown in **Figure 13.5.** Recall from our study of radiation in Chapter 4 that the hotter region will contain more (and more energetic) photons than the cooler region. (There will be a Planck spectrum of photons in both regions, and so the total energy carried by photons will be proportional to the fourth power of the temperature in each region.) More photons will move by chance from the hotter ("more crowded") region to the cooler ("less crowded") region than in the reverse direction. There is a net transfer of photons and photon energy from the hotter region to the cooler region. In this way, radiative transfer carries energy from hotter regions to cooler regions.

Radiation carries energy from hotter regions to cooler regions.

If the temperature differs by a large amount over a short distance, then the concentration of photons will differ sharply as well, which favors rapid radiative energy transfer. The transfer of energy from one point to another by radiation also depends on how freely radiation can move from one point to another within a star. The degree to which matter impedes the flow of photons through it is referred to as **opacity.** The opacity of material depends on many things including the density of material, its composition, its temperature, and the wavelength of the photons moving through it.

Opacity impedes the outward flow of radiation.

Radiative transfer is most efficient in regions where the opacity is low. In the inner part of the Sun, where temperatures are high and atoms are ionized, opacity comes mostly from the interaction between photons and free electrons. Here opacity is relatively low, and radiative transfer is capable of carrying the energy produced in the core outward through the star. The region in which radiative transfer is responsible for energy transport extends 71% of the way out toward the surface of the Sun. This region (see **Figure 13.6**) is referred to as the Sun's **radiative zone.** Even though the opacity of the radiative zone is relatively low, photons are still able to travel only a short distance before interacting with matter. The path that a photon follows is so convoluted that on average it takes the energy of a gamma ray photon produced in the interior of the Sun about 100,000 years to find its way to the outer layers of the Sun. Opacity serves as a blanket, holding energy in the interior of the Sun, and letting it seep away only slowly.

From a peak of 15 million K at the center of the Sun, the temperature falls to about 100,000 K at the outer margin of the radiative zone. At this temperature, atoms are no longer completely ionized, which increases the opacity. As the opacity becomes greater, radiation becomes less efficient in carrying energy from one place to another. The energy that is flowing outward through the Sun "piles up." The physical sign that energy is piling up is that the *temperature gradient*—that is,

13.3 RADIATE

Figure 13.5 *Higher-temperature regions deep within the Sun produce more radiation than lower-temperature regions farther out. While radiation flows in both directions, more radiation flows from the hot regions to the cooler regions than from the cooler regions to the hot regions. In this way, radiation carries energy outward from the inner parts of the Sun.*

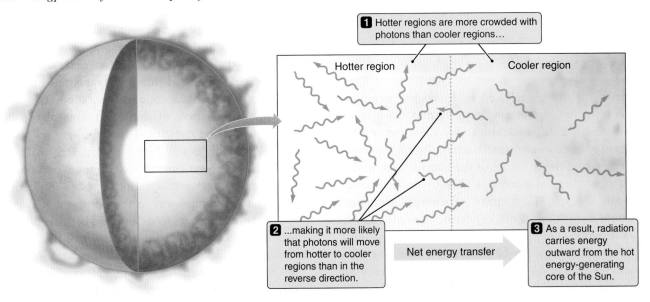

1 Hotter regions are more crowded with photons than cooler regions...

Hotter region   Cooler region

2 ...making it more likely that photons will move from hotter to cooler regions than in the reverse direction.

Net energy transfer

3 As a result, radiation carries energy outward from the hot energy-generating core of the Sun.

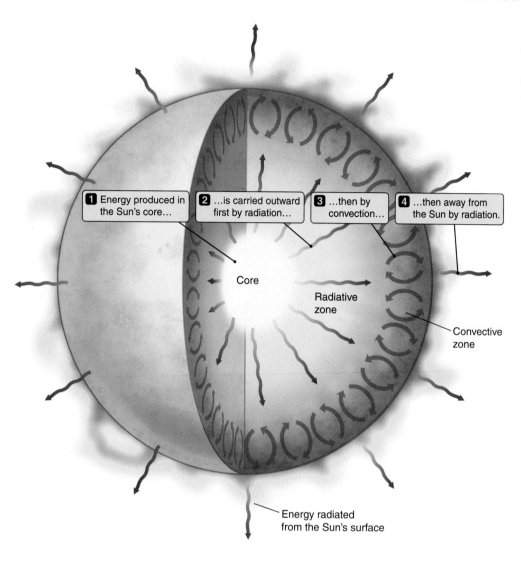

Figure 13.6 *The interior structure of the Sun is divided into zones on the basis of where energy is produced and how it is transported outward.*

1 Energy produced in the Sun's core...

2 ...is carried outward first by radiation...

3 ...then by convection...

4 ...then away from the Sun by radiation.

Core

Radiative zone

Convective zone

Energy radiated from the Sun's surface

how rapidly temperature drops with increasing distance from the center of the Sun—becomes very steep. (Radiative transfer carries energy from hotter regions to cooler regions, smoothing out temperature differences between them. As the opacity increases, radiation becomes less effective in smoothing out temperature differences, so temperature differences between one region and another become greater.)

As we move farther toward the surface of the Sun, radiative transfer becomes so inefficient (and the temperature gradient so steep) that a different way of transporting energy takes over. Like a hot-air balloon, cells (or packets) of hot gas become buoyant and rise up through the lower-temperature gas above them, carrying energy with them. Thus convection begins. Just as convection carries energy from the interior of planets to their surfaces, or from the Sun-heated surface of Earth upward through Earth's atmosphere, convection also

In the outer part of the Sun, energy is carried by convection.

plays an important role in the transport of energy outward from the interiors of many stars, including the Sun. The solar **convective zone** extends from the outer boundary of the radiative zone out to just below the visible surface of the Sun.

In the outermost layers of stars, radiation again takes over as the primary way that energy is transported. (This must be the case. After all, it is radiation that transports energy from the outermost layers of a star off into space.) Even so, the effects of convection can be seen as a perpetual roiling of the visible surface of the Sun.

WHAT IF THE SUN WERE DIFFERENT?

At this point we have seen the pieces that go into calculating a model of the interior of the Sun. To put these pieces together, we again stress that the key is *balance*. The temperature and density in the core of the model Sun must be just right to produce the same amount of energy as would be radiated away by a star of the same size and surface temperature as the Sun. The temperature and density at each point within the model Sun must be just right so that transport of energy away from the core by radiation and convection just balances out the amount of energy produced by fusion in the core. The density, temperature, and pressure of the model Sun must vary from point to point in such a way that the outward push of pressure is everywhere balanced by the inward pull of gravity. Finally, the whole model must depend on only two things: the total mass of gas from which the star is made, and the chemical composition of that gas.

All processes in the Sun must be in balance.

Let us restate that last idea to drive the point home. *In order to be successful, our model of the Sun must start with no more information than the known mass and chemical composition of the real Sun, and from these correctly predict all the other observed properties of the real Sun.* The amazing thing is that it does. Computer models show that a ball of gas with a mass of 1 solar mass and the composition of our Sun can have only one structure and still satisfy all the different balances listed above *at the same time.* That model predicts what the size, temperature, and luminosity of the resulting star should be—predictions that agree remarkably well with the observed properties of the real Sun.

To better understand *why* there is only one possible structure that a $1 M_\odot$ star can have, we can "what-if" the Sun. For example, we might ask, What if a star had the same mass and composition as the Sun, but was somehow larger than the Sun? What properties would such a hypothetical star have? Specifically, what would happen to the balance between the amount of energy generated within such a hypothetical star and the amount of energy that it radiates away into space? Follow along in **Figure 13.7** as we consider what would happen if the Sun were "too large."

Understand the Sun by imagining it were different.

We begin with the second part of this balance. Because our hypothetical star would have more surface area than the Sun, it would be able to more effectively radiate its energy into space. To keep a one-solar-mass star inflated to a size larger than the Sun would require that the star be more luminous than the Sun.

But now let us consider what is going on in the interior of our hypothetical star. Because the star is larger than the Sun but contains the same amount of mass as the Sun, the force of gravity at any point within our hypothetical star would be less than the force of gravity

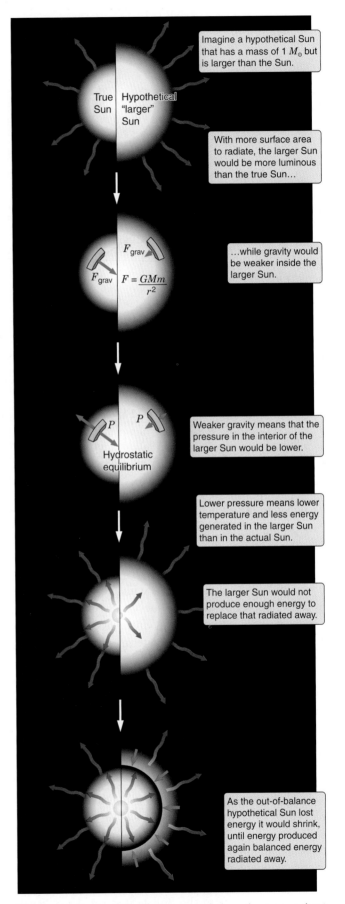

Figure 13.7 *A star like the Sun can only have the structure that it has. Here we imagine the fate of a Sun that had too large a radius.*

at the corresponding location within the Sun. (This is a result of the inverse square law of gravitation. If the radius R is larger in our hypothetical star, then $1/R^2$ will be smaller.) Weaker gravity would mean that the weight of matter pushing down on the interior of our hypothetical star would be less than in the Sun. Since according to hydrostatic equilibrium the pressure at any point within a star is equal to the weight of overlying matter, the pressure at any point in the interior of our hypothetical star would be less than the pressure at the corresponding point in the Sun. This has an effect on the amount of energy that our star produces. The proton–proton chain runs faster at higher temperature and density, and so the lower pressure in the interior of hypothetical star means that less energy is generated there than in the core of the Sun.

But wait a minute—there is a contradiction here. We found that our hypothetical star would have to be more luminous than the Sun, but that at the same time it would be producing less energy in its interior than the Sun does. This violates the balance that must exist in any stable star between the amount of energy generated within the star and the amount of energy radiated into space. Our hypothetical star cannot exist! Stated another way, even if we could somehow magically pump up the Sun to a size larger than it actually is, it would not remain that way. Less energy would be generated in its core, while more energy would be radiated away at its surface. The Sun would be out of balance. As a result, the Sun would lose energy, the pressure in the interior of the Sun would decline, and the Sun would shrink back toward its original (true) size.

If the Sun were any different, it would not be in balance.

We could do the same thought experiment the other way around, asking what would happen if the Sun were smaller than it actually is. With less surface area it would radiate less energy. At the same time the Sun's mass would be compacted into a smaller volume, driving up the strength of gravity and therefore the pressure in its interior. Higher pressure implies higher density and temperature, which in turn would cause the proton–proton chain to run faster, increasing the rate of energy generation. Again there is a contradiction—an imbalance. This time, with more energy being generated in the interior than is radiated away from the surface, pressure in the Sun would build up, causing it to expand toward its original (true) size.

In the end, there is only one possible Sun. If a star contains 1 solar mass of material of solar composition, then there is only one solution for the structure of that star that maintains all of the balances that must be maintained. Our stable, reliable Sun is the result.

13.3 THE STANDARD MODEL OF THE SUN IS WELL TESTED

The standard model of the Sun correctly predicts such bulk properties of the Sun as its size, temperature, and luminosity. This is a remarkable feat, but the model predicts much more than these properties. In particular, the standard model of the Sun predicts exactly what nuclear reactions should be occurring in the core of the Sun, and at what rate. The nuclear reactions that make up the proton–proton chain produce copious quantities of neutrinos. As we noted above, neutrinos are extremely elusive beasts—so elusive that almost all of the neutrinos produced in the heart of the Sun travel freely through the outer parts of the Sun and on into space as if the Sun were not there. The core of the Sun lies buried beneath 700,000 km of dense, hot matter, seemingly buried forever away from our view. Yet the Sun is *transparent* to neutrinos.

Neutrinos escape freely from the core of the Sun.

It may take the thermal energy produced in the heart of the Sun 100,000 years to find its way to the Sun's surface, but the solar neutrinos streaming through you as you read these words were produced by nuclear reactions in the very heart of the Sun only $8\frac{1}{3}$ minutes ago. If we could find a way to capture and analyze these neutrinos, think of what we might learn! In principle, neutrinos offer a direct window into the very heart of the Sun's nuclear furnace.

USING NEUTRINOS TO OBSERVE THE HEART OF THE SUN

Turning the promise of neutrino astronomy into reality turns out to be a formidable technical challenge. The same property of neutrinos that makes them so exciting to astronomers—the fact that their interaction with matter is so feeble that they are able to escape unscathed from the interior of the Sun—also makes them notoriously difficult to observe. Suppose we wanted to build a neutrino detector capable of stopping half of the neutrinos falling on it. Our hypothetical detector would need the stopping power of a piece of lead a light-year thick! Yet despite the difficulties, neutrinos offer such a unique and powerful window into the Sun that they are worth going to great lengths to try to detect.

Neutrino detectors measure neutrinos from the Sun.

Fortunately, the Sun produces a truly enormous number of neutrinos. As you lie in bed at night, about 400 trillion solar neutrinos pass through your body each second, having already passed through Earth. With this many neutrinos about, a neutrino detector does not have to be very efficient to be useful. Several methods have been devised to measure neutrinos from the Sun and from other astronomical sources such as supernova explosions, and a number of such experiments are under way. These experiments have successfully detected neutrinos from the Sun, and in so doing have provided crucial confirmation that nuclear fusion reactions indeed are responsible for powering the Sun.

However, as with many good experiments, measurements of solar neutrinos raise new questions while answering others. After their initial joy at confirming that the Sun really is a nuclear furnace, astronomers became troubled that

Fewer neutrinos are seen from the Sun than originally had been expected.

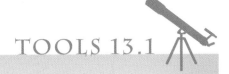

TOOLS 13.1

Neutrino Astronomy

A neutrino telescope hardly fits one's expectation of what a telescope should look like. The first experiment designed to detect solar neutrinos consists of a cylindrical tank filled with 100,000 gallons of dry cleaning fluid—C_2Cl_4, or perchloroethylene—buried 1,500 meters deep within the Homestake gold mine in Lead, South Dakota. A tiny fraction of neutrinos passing through this fluid interact with chlorine atoms, causing the reaction

$$^{37}Cl + \nu \rightarrow {}^{37}Ar + e^-$$

to take place. The ^{37}Ar formed in the reaction is a radioactive isotope of argon. The tank must be buried deep within Earth to shield the detector from the many other types of radiation capable of producing argon atoms. The argon is flushed out of the tank every few weeks and measured.

The Homestake detector **(Figure 13.8a)** began operations in 1965. Over the course of two days, roughly 10^{22} (10 billion trillion) solar neutrinos pass through the Homestake detector. Of these, on average only *one* neutrino interacts with a chlorine atom to form an atom of argon. Even so, this is enough of a signal to measure, and the effects of solar neutrinos are seen. Since then, over a dozen and a half neutrino experiments have been built, each using different reactions to detect neutrinos of different energies. For example, the Soviet-American Gallium Experiment (SAGE) and the European GALLEX experiment make use of reactions involving conversion of gallium atoms into germanium atoms ($^{71}Ga + \nu \rightarrow {}^{71}Ge + e^-$) to detect solar neutrinos. (SAGE uses 60 tons of metallic gallium. GALLEX uses 30 tons of gallium in the form of gallium chloride. These two experiments account for most of the world's current supply of gallium.)

Among the most ambitious neutrino detectors is the Super-Kamiokande detector located in an old zinc mine 2,000 feet under Mount Ikena, near Kamioka, Japan.[1] Super-Kamiokande is a 50,000-ton tank of ultrapure water, surrounded by 11,146 photomultiplier tubes capable of registering extremely faint flashes of light. When a neutrino interacts with an atom in the tank, a faint conical flash of blue light is produced. This flash is seen by some of the photomultipliers. **Figure 13.8(b)** shows a picture of Super-Kamiokande while **Figure 13.8(c)** shows a map of the flash of light from a single neutrino.

Neutrino telescopes observe neutrinos produced in the heart of the Sun, allowing us to directly observe the results of the nuclear reactions going on there. While these observations have provided crucial confirmation that stars are powered by nuclear reactions, they have also challenged our models of the solar interior, and led us to change our ideas about the nature of the neutrino itself (as is evident from the discussion of the solar neutrino problem in the text). In addition to solar neutrinos, a number of experiments detected neutrinos from supernova 1987A. As we will see in Chapter 16, this was an explosion marking the end of the life of a massive star located 160,000 light-years away in a small galaxy called the Large Magellanic Cloud. Neutrino astronomy is one of the great innovations of 20th-century astronomy, and is certain to have many applications in the 21st century.

[1]As this book was going to press, Super-Kamiokande experienced an accident that destroyed most of its photomultiplier tubes. It is expected to be out of commission until late 2002.

(a)   (b)

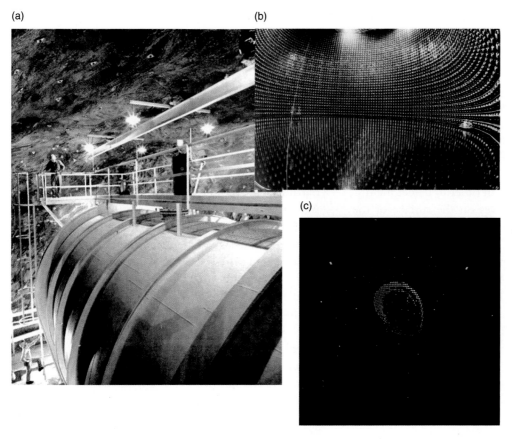

(c)

Figure 13.8 *Neutrino "telescopes" do not look much like visible-light telescopes. (a) The Homestake neutrino detector is a 100,000-gallon tank of dry cleaning fluid located deep in a mine in South Dakota. (b) The Super Kamiokande detector (shown before it was completely filled) is a tank containing 50,000 tons of pure water surrounded by over 11,000 photomultiplier tubes that record flashes of light from reactions within the tank. (c) A map of the flash of light from a single neutrino detected by the Super Kamiokande detector.*

there seemed to be only about a third to a half as many solar neutrinos as predicted by solar models. The difference between the predicted and measured flux of solar neutrinos is referred to as the **solar neutrino problem.**

Understanding the solar neutrino problem is an area of very active research. One possible explanation is that our understanding of the structure of the Sun is somehow very wrong. This seems unlikely, however, because of the many other successes of solar models. A second possibility, considered more likely by most researchers today, is that our understanding of the neutrino itself is incomplete. The neutrino was long thought to have zero mass (like photons) and to travel at the speed of light. However, if neutrinos actually have some mass, then theories from particle physics predict that solar neutrinos should *oscillate* (alternate back and forth) among three different forms. Since only one of these forms of neutrinos can interact with the atoms in neutrino detectors (described in Tools 13.1), neutrino oscillations would provide a convenient explanation for why we see only about a third of the expected number of neutrinos. After three decades of work on the solar neutrino problem, this last idea seems to be winning out. Work currently under way at high-energy physics labs, nuclear reactors, and neu-

Theory and experiment now agree.

trino telescopes around the world is showing that neutrinos *do* have a nonzero mass, and has uncovered evidence of neutrino oscillations.

HELIOSEISMOLOGY OBSERVES ECHOES FROM THE SUN'S INTERIOR

In Chapter 6, we found that models of Earth's interior predict how density and temperature change from place to place within our planet. These differences affect the way pressure waves travel through Earth, bending the paths of these waves. Models of Earth's interior are tested by comparing measurements of seismic waves from earthquakes with model predictions of how seismic waves should travel through the planet.

The same basic idea has now been applied to the Sun. Detailed observations of motions of material from place to place across the surface of the Sun show that the Sun vibrates or "rings," something like a struck bell. Compared to a well-tuned bell, however, the vibrations of the Sun are very complex, with many different frequencies of vibrations occurring simultaneously.

Seismic waves travel through the Sun.

These motions are echoes of what lies below. Just as geologists use seismic waves from earthquakes to probe the interior of Earth, solar physicists use the surface oscillations of the Sun to test our understanding of the solar interior. This science, new to the later years of the 20th century, is called **helioseismology.** Like neutrino astronomy, helioseismology has created quite a stir among astronomers by letting us "see" into the unseeable heart of the Sun **(Figure 13.9).**

Observing the "ringing" motion of the surface of the Sun is a difficult business. In order to detect the disturbances of helioseismic waves on the surface of the Sun, astronomers must use instruments capable of measuring Doppler shifts of less than 0.1 m/s, while detecting changes in brightness of only a few parts per million at any given location on the Sun. In addition, there are tens of millions of different wave motions possible within the Sun. Some waves travel around the circumference of the Sun, providing information about the density of the upper convection zone. Other waves travel through the interior of the Sun, revealing the density structure of the Sun down close to the core. Still others travel inward toward the center of the Sun until they are bent by the changing solar density and return to the surface. All of these wave motions are going on at the same time. Sorting out this jumble requires computer analysis of long, unbroken strings of solar observations.

Helioseismology studies of the Sun got a huge boost in the closing years of the 20th century with two very successful projects. The Global Oscillation Network Group, or *GONG,* is a network of six solar observation stations spread around the world. Between these stations, they are able to observe the surface of the Sun approximately 90% of the time. The other project is the Solar and Heliospheric Observatory, or *SOHO* spacecraft, which is a joint mission between NASA and the European Space Agency. By orbiting at the L_1 Lagrangian point of the Sun–Earth system (see Chapter 9), *SOHO* is able to orbit in lock step with Earth at a location approximately 1,500,000 km (0.01 AU) from Earth on a line directly between Earth and the Sun. *SOHO* carries a complement of 12 scientific instruments designed to monitor the Sun and make measurements of the solar wind upstream of Earth. *SOHO* observations have dramatically improved our detailed knowledge of the Sun.

To interpret helioseismology data, scientists compare the strength, frequency, and wavelengths of observed vibrations with predicted vibrations calculated from models of the solar interior. This technique provides a powerful test of our understanding of the solar interior, and has led both to some surprises and to improvements in our models. For example, some scientists had proposed that the solar neutrino problem might be solved if there were less helium in the Sun than generally imagined—an explanation that was ruled out by analyzing the waves that penetrate to the core of the Sun. Helioseismology also showed that the value for opacity used in early solar models was too low. This led astronomers to recalculate the location of the bottom of the convective zone. Both theory and observation now put the base of the convective zone at 71.3% of the way out from the center of the Sun, with an uncertainty in this number of less than half a percent! The amazing agreement between the predictions of physical models of the Sun and the temperature and density structure within the Sun measured by helioseismology is truly remarkable.

Helioseismology confirms the predictions of solar models.

13.4 THE SUN CAN BE STUDIED UP CLOSE AND PERSONAL

The Sun is a large ball of gas, and so it has no solid surface in the sense that Earth does. Instead, it has the kind of surface that a cloud on Earth does, or the "surface" of one of the Jovian planets in our Solar System. To help

Figure 13.9 *The interior of the Sun rings like a bell as helioseismic waves move through it. This figure shows one particular "mode" of the Sun's vibration. Red shows regions where gas is traveling inward; blue where gas is traveling outward. These motions can be observed by using Doppler shifts.*

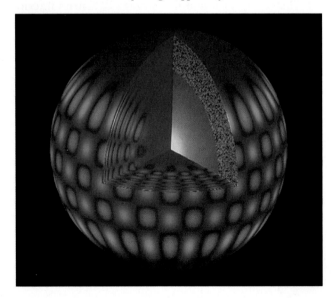

you understand such a surface, imagine a fog bank. The surface of the fog bank is a gradual thing—an illusion really. Imagine watching some people walking into a fog bank. When they disappear from view, you would say that they were definitely inside the fog bank, even though they had never passed through a noticeable boundary. The apparent surface of the Sun is defined by the same effect. Light from the surface of the Sun is able to escape into space, and so we can see it. Light from below the surface of the Sun cannot escape directly into space, and so we cannot see it. The Sun's surface is referred to as the solar **photosphere.** (*Photo* means "light": the photosphere is the place light comes from.) There is no instant when you can say that you have suddenly crossed the surface of a fog bank, and by the same token there is no instant when we suddenly cross the photosphere of the Sun. The surface of the Sun— the photosphere—is a zone about 500 km thick, across which the density and opacity of the Sun increases sharply. The reason the Sun appears to have a well-defined surface and a sharp outline when viewed from Earth (*never* look at the Sun directly!) is because this zone is relatively so shallow; 500 km does not look very thick when viewed from a distance of 150 million km.

> **The apparent surface of the Sun is called the photosphere.**

Look at a photograph of the Sun such as **Figure 13.10(a),** and notice that the Sun appears to be fainter near its edges than near its center. This effect, called **limb darkening,** is an artifact of the structure of the Sun's photosphere. (The *limb* of a celestial body is the outer border of its visible disk.) The cause of limb darkening is illustrated in **Figure 13.10(b).** Near the edge of the Sun you are looking through the photosphere at a steep angle. As a result, you do not see as deeply into the interior of the Sun as when looking directly down through the photosphere near the Sun's center. The light you see coming from near the limb of the Sun is from a layer in the Sun that is shallower, and hence cooler and fainter.

THE SOLAR SPECTRUM IS COMPLEX

In one sense, the surface of the Sun may be an illusion, but in another sense it is not. The transition between "inside" the Sun and "outside" the Sun is quite abrupt. In the outermost part of the Sun, the density of the gas drops very rapidly. This is the region we refer to as the Sun's **atmosphere. Figure 13.11** shows how the pressure and temperature change across the atmosphere of the Sun. The Sun's atmosphere is where all visible solar phenomena take place.

Very nearly all the radiation from below the Sun's photosphere is absorbed by matter and cannot escape—it is trapped. This is exactly our definition from Chapter 4 for the conditions under which blackbody radiation is formed. We should not then be surprised that the radiation able to leak out of the Sun's interior has a spectrum very close to being a Planck (blackbody) spectrum. This is why in Chapter 12 we were able to understand much about the physical properties of stars by applying our understanding of blackbody radiation.

As we look in more detail at the structure of the Sun, however, this simple description of the spectra of stars

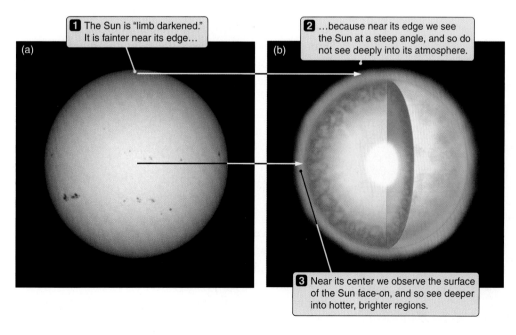

1 The Sun is "limb darkened." It is fainter near its edge…

2 …because near its edge we see the Sun at a steep angle, and so do not see deeply into its atmosphere.

3 Near its center we observe the surface of the Sun face-on, and so see deeper into hotter, brighter regions.

Figure 13.10 (a) *When viewed in visible light, the Sun appears to have a sharp outline, even though it has no true surface. The center of the Sun appears brighter, while the limb of the Sun is darker, an effect known as limb darkening.* (b) *Looking at the middle of the Sun allows us to see deeper into the Sun's interior than looking at the edge of the Sun does. Since higher temperature means more-luminous radiation, the middle of the Sun appears brighter than its limb.*

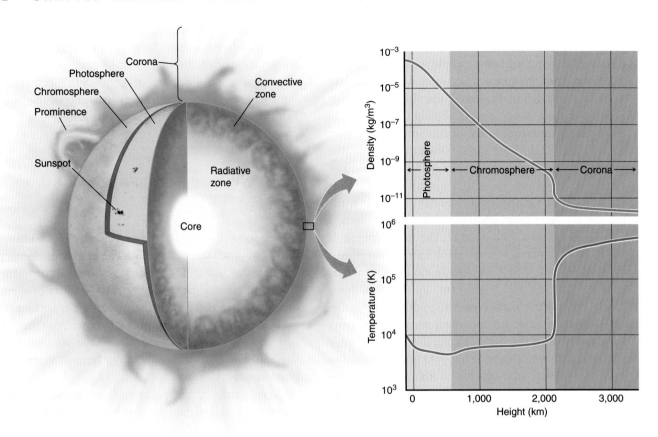

Figure 13.11 *The components of the Sun's atmosphere, along with plots of how the temperature and density change with height at the base of the Sun's atmosphere.*

begins to go wrong. The fact that light from the solar photosphere must escape through the upper layers of the Sun's atmosphere affects the spectrum that we see. In Chapter 12 we discussed the presence of absorption lines in the spectra of stars. Now we can take a closer look at how these absorption lines form. As photospheric light travels upward through the solar atmosphere, atoms in the solar atmosphere absorb light at discrete wavelengths.

Figure 13.12 *A high resolution spectrum of the Sun, stretching from 0.4 μm (lower left corner) to 0.7 μm (upper right corner), showing a wealth of absorption lines.*

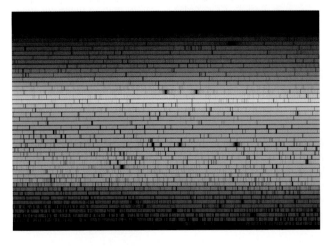

Because from our perspective the Sun is so much brighter than any other star, its spectrum can be studied in far more detail. Today specially designed telescopes and high-resolution spectrographs have been built specifically to study the light from the Sun. The amazing structure present in the solar spectrum can be seen in **Figure 13.12.**

Absorption lines from over 70 elements have been identified. Analysis of these lines forms the basis for much of our knowledge of the solar atmosphere, including the composition of the Sun, and is the starting point for our understanding of the atmospheres and spectra of other stars.

THE SUN'S OUTER ATMOSPHERE: CHROMOSPHERE, CORONA, AND SOLAR WIND

As we move upward through the Sun's photosphere, the temperature continues to fall, reaching a minimum of about 4,500 K at the top of the photosphere. At this point the temperature trend reverses and slowly begins to climb, rising to about 6,000 K at a height of 1,500 km above the top of the photosphere. This re-

(a)

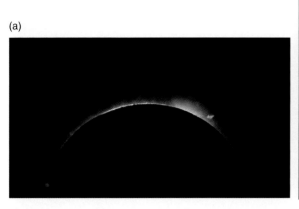

(b)

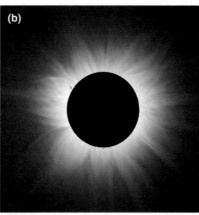

Figure 13.13 (a) *This image of the Sun, taken during a total solar eclipse, shows structure in the Sun's chromosphere.* (b) *This eclipse image shows the Sun's corona, consisting of million-kelvin gas that extends for millions of kilometers beyond the surface of the Sun.*

gion above the photosphere is called the **chromosphere.** The chromosphere was discovered in the 19th century during observations of total solar eclipses (see **Figure 13.13a**). The chromo-

The Sun's chromosphere lies above the photosphere.

sphere is seen most strongly as a source of emission lines, especially the Hα line from hydrogen. In fact, it is from the deep red color of the Hα line that the chromosphere (the place where color comes from) gets its name. It was also from a spectrum of the chromosphere that the element helium was discovered in 1868. Helium is named after *helios,* the Greek word for "Sun."[2]

At the top of the chromosphere, across a transition region that is only about 100 km thick, the temperature suddenly soars (Figure 13.11). Above this transition lies the outermost region of the Sun's atmosphere, called the **corona,** in which tempera-

The corona has a temperature of millions of kelvins.

tures reach 1 million to 2 million K. The corona is probably heated by waves and magnetic fields in much the same way the chromosphere is, but why the temperature changes so abruptly at the transition between the chromosphere and the corona is not at all clear. Since ancient times the Sun's corona has been known—visible during total solar eclipses as an eerie outer glow stretching a distance of several solar radii beyond the Sun's surface (see **Figure 13.13b**). Because it is so hot, the solar corona is a strong source of X-rays. Atoms in the corona are also highly ionized. The spectrum of the corona shows emission lines from ionic species such as Fe^{13+} (iron atoms from which 13 electrons have been stripped) and Ca^{14+} (calcium atoms from which 14 electrons have been stripped).

---

[2]It is rather remarkable that helium, the second most common element in the Universe, was discovered in the Sun before it was identified on Earth!

SOLAR ACTIVITY IS MAGNETIC IN NATURE

Virtually all of the structure seen in the atmosphere of the Sun is imposed on the gas by the Sun's magnetic field. In high-resolution images of the Sun, such as **Figure 13.14,** the Sun looks almost as though it is covered with matted, tangled hair. This fibrous or ropelike texture in the chromosphere is the result of magnetic structures called *flux tubes,* much like the flux tube that connects Io with Jupiter (see Chapter 8). Magnetic fields are responsible for much of the structure of the corona as well.

The corona is far too hot to be held in by the Sun's gravity, but over most of

Magnetic fields cause visible structure.

the surface of the Sun coronal gas is confined instead by magnetic loops, with both ends firmly anchored deep within the Sun. The magnetic field in the corona acts almost like a network of rubber bands which coronal gas is free to slide along but cannot cross. In contrast, about 20% of the surface of the Sun is covered by

Figure 13.14 (a) *Close-up image of the Sun showing tangled structure due to magnetic fields.* (b) *A computer model of the magnetic flux tubes seen in* (a).

(a)   (b)

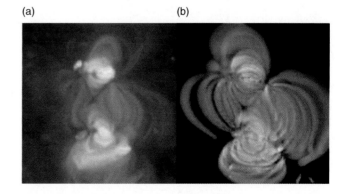

an ever shifting pattern of **coronal holes.** These are apparent in X-ray images of the Sun **(Figure 13.15)** as dark regions, indicating that they are cooler and lower in density than their surroundings. Coronal holes are large regions where the magnetic field points outward, away from the Sun, and where coronal material is free to stream away into interplanetary space.

We have encountered this flow of coronal material away from the Sun before. This is the same **solar wind** responsible for shaping the magnetospheres of planets (Chapters 7 and 8) and for blowing the tails of comets away from the Sun (Chapter 11). The relatively steady part of solar wind consists of lower-speed flows, with velocities of around 350 km/s, and higher-speed flows, with velocities up to about 700 km/s. The higher-speed flows originate in coronal holes. Depending on their speed, particles in the solar wind take about two to five days to reach Earth. Frequently, two to five days after a coronal hole passes across the center of the face of the Sun, there is an increase in the speed and density of the solar wind reaching Earth. The solar wind drags the Sun's magnetic field along with it. The magnetic field in the solar wind gets "wound up" by the Sun's rotation, as shown in **Figure 13.16.** This gives the solar

> A "wind" blows away from the Sun.

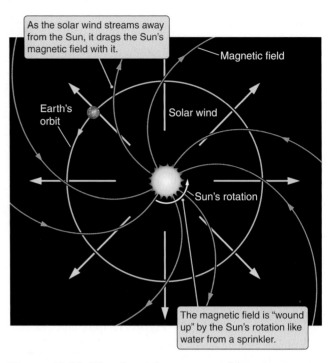

Figure 13.16 *The solar wind streams away from active areas and corona holes on the Sun. As the Sun rotates, the solar wind picks up a spiral structure, much like the spiral of water that streams away from a rotating lawn sprinkler.*

wind a spiral structure something like a stream of water from a rotating lawn sprinkler.

The effects of the solar wind are felt throughout the Solar System. As we have seen, the solar wind causes the tails of comets, shapes the magnetospheres of the planets, and provides the energetic particles that power Earth's spectacular auroral displays. Using space probes, we have been able to observe the solar wind extending out to over 50 AU from the Sun. But the solar wind does not go on forever. The farther it gets from the Sun, the more it has to spread out. Just like radiation, the density of the solar wind follows an inverse square law. At a distance of around 100 AU from the Sun, the solar wind is assumed no longer to be powerful enough to push the **interstellar medium** (the gas and dust that lie between stars in a galaxy, and which surrounds the Sun) out of the way. There the solar wind will stop abruptly, "piling up" against the pressure of the interstellar medium. **Figure 13.17** shows this region of space over which the wind from the Sun holds sway. Sometime in the early part of the 21st century the *Voyager 1* spacecraft is expected to cross this boundary and begin sending back mankind's first "on the scene" measurements of true interstellar space.

The best-known features on the surface of the Sun are relatively dark blemishes in the solar photosphere, called **sunspots,** which come and go over time. Early

Figure 13.15 *X-ray images of the Sun show a very different picture of our star than images taken in visible light. The brightest X-ray emission comes from the base of the Sun's corona, where gas is heated to temperatures of millions of kelvins. This heating is most powerful above magnetically active regions of the Sun. Also visible are dark coronal holes.*

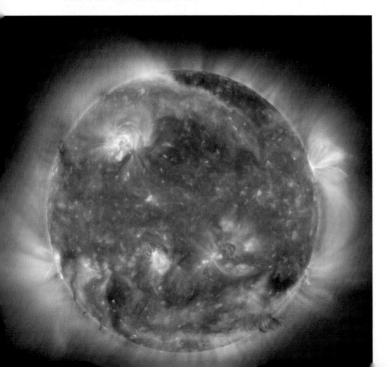

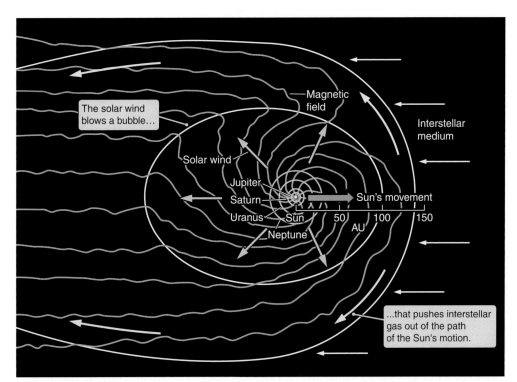

The solar wind blows a bubble...

Magnetic field

Solar wind

Interstellar medium

Jupiter
Saturn
Uranus — Sun
Neptune

Sun's movement

50   100   150
AU

...that pushes interstellar gas out of the path of the Sun's motion.

Figure 13.17 *The solar wind streams away from the Sun for about 100 AU, until it finally piles up against the pressure of the interstellar medium through which the Sun is traveling. At the beginning of the 21st century, the* Voyager 1 *spacecraft is approaching the expected location of this boundary.*

telescopic observations of sunspots made during the 17th century led to the discovery of the Sun's rotation, which has an average period of about 27 days as seen from Earth, and 25 days relative to the stars. Observations of sunspots also show that the Sun rotates more rapidly at its equator than it does at higher latitudes. This effect, referred to as *differential rotation,* is possible only because the Sun is a large ball of gas rather than a solid object. Sunspots appear dark, but only in contrast to the brighter surface of the Sun. Sunspots are about 1,500 K cooler than their surroundings.

Sunspots are cooler than their surroundings.

Figure 13.18 is a photograph of a sunspot group, showing the remarkable structure of these blemishes on the surface of the Sun. Each sunspot consists of an inner, dark core called the **umbra,** which is surrounded by a less dark region called the **penumbra.** The penumbra shows an intricate radial pattern, reminiscent of the petals of a flower. Observations of sunspots show that they are magnetic in origin, with magnetic fields thousands of times greater than the magnetic field at Earth's surface. Sunspots occur in pairs which are connected by loops in the magnetic field, much like the shape of a horseshoe. Sunspots range in size from about 1,500 km across up to complex groups, which may contain several tens of individual spots, and span 50,000 km or more.

Magnetic fields cause sunspots.

The largest sunspot groups are so large that they can be seen with the naked eye, and have been noted since antiquity. (*But do not look directly at the Sun!* More than one solar observer has paid the price of blindness for a

Figure 13.18 *These false-color SOHO images show a group of sunspots. The darkest part of sunspots are called umbrae. The surrounding areas, which look like petals on a flower, are the penumbrae. Sunspots are magnetically active regions that are cooler than the surrounding surface of the Sun.*

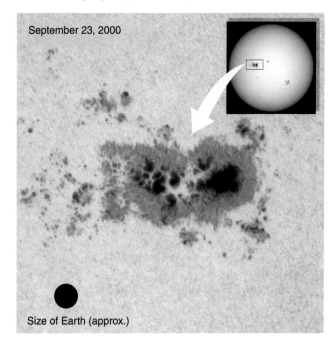

September 23, 2000

Size of Earth (approx.)

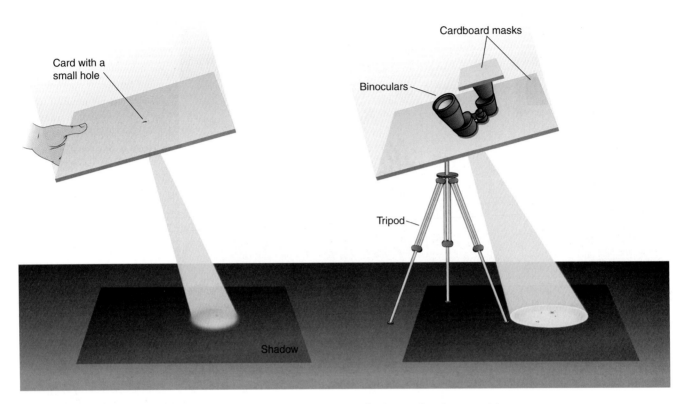

Figure 13.19 *There are a number of ways of safely viewing sunspots on the Sun or observing a partial solar eclipse. Never* look at the Sun directly *(except during totality of a total eclipse)!*

glimpse of a naked eye sunspot. **Figure 13.19** shows a number of ways to look at the Sun in safety.)

Individual sunspots are ephemeral. While a few sunspots last for 100 days or longer, half of all sunspots come and go in about 2 days, and 90% are gone within 11 days. The amount of sunspot activity and the location on the Sun where sunspots are seen changes over time in a pronounced 11-year pattern called the **sunspot cycle (Figure 13.20a).** At the beginning of a cycle, sunspots begin to appear at solar latitudes of around 30° to the north and south of the solar equator. Over the following years the regions where most sunspots are seen move toward the equator, as the number of sunspots increases, and then declines. As the last few sunspots near the equator are seen, sunspots again begin appearing at middle latitudes, and the next cycle begins. A plot showing the number of sunspots at a given latitude plotted against time **(Figure 13.20b)** has the appearance of a series of opposing diagonal bands, and is often referred to as the sunspot "butterfly diagram."

Telescopic observations of sunspots date back for almost 400 years, and there were naked-eye reports of sunspots even before that. **Figure 13.21** shows the historical record of sunspot activity. While astronomers often speak of the 11-year sunspot cycle,

> **Sunspots come and go in an 11-year cycle.**

the cycle is neither perfectly periodic nor especially reliable. The time between peaks in the number of sunspots actually varies between about 9.7 and 11.8 years. The number of spots seen during a given cycle fluctuates as well. Some cycles can be real monsters. There have also been times when sunspot activity has disappeared almost entirely for extended periods. The most recent extended lull in solar activity, called the **Maunder minimum,** lasted from 1645 to 1715. Normally there would be six peaks in solar activity in 70 years, but virtually no sunspots were seen during the Maunder minimum.

In the early 20th century, George Ellery Hale was the first to show that the 11-year sunspot cycle was actually half of a 22-year magnetic cycle during which the direction of the Sun's magnetic field reverses from one 11-year sunspot cycle to the next. In one sunspot cycle the leading sunspot in each pair tends to be a north magnetic pole, while the trailing sunspot tends to be a south magnetic pole. In the next sunspot cycle, this polarity is reversed: the leading spot in each pair is a south magnetic pole. The transition between these two magnetic polarities occurs near the peak of each sunspot cycle. The predominant theory for what causes this magnetic cycle involves a **dynamo** in the interior of the Sun, much like the dynamo that generates the magnetic fields of the planets.

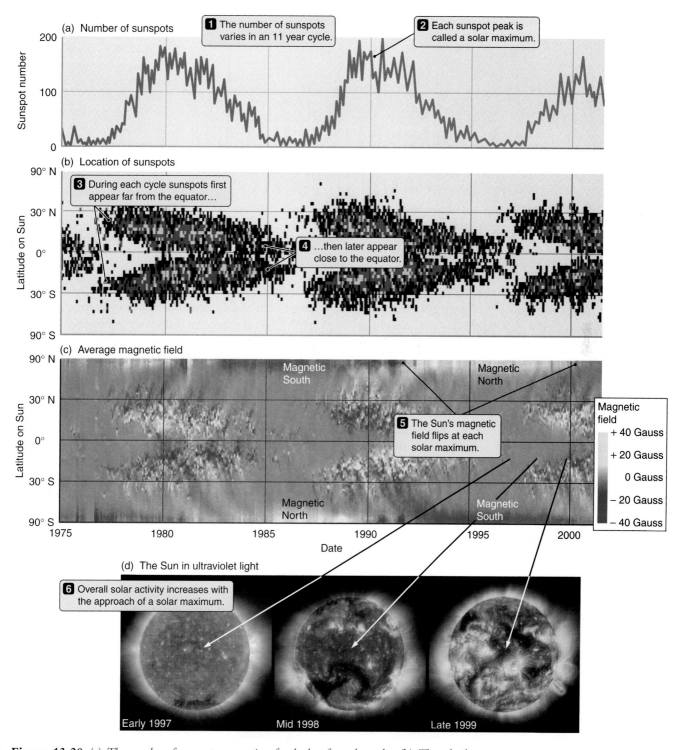

(a) Number of sunspots

1 The number of sunspots varies in an 11 year cycle.

2 Each sunspot peak is called a solar maximum.

(b) Location of sunspots

3 During each cycle sunspots first appear far from the equator…

4 …then later appear close to the equator.

(c) Average magnetic field

Magnetic South

Magnetic North

5 The Sun's magnetic field flips at each solar maximum.

Magnetic North

Magnetic South

Magnetic field
+ 40 Gauss
+ 20 Gauss
0 Gauss
− 20 Gauss
− 40 Gauss

Date

(d) The Sun in ultraviolet light

6 Overall solar activity increases with the approach of a solar maximum.

Early 1997   Mid 1998   Late 1999

Figure 13.20 (a) *The number of sunspots versus time for the last few solar cycles.* (b) *The solar butter-fly diagram, showing the fraction of the Sun covered at each latitude.* (c) *The Sun's magnetic field flips every 11 years.* (d) *The approach of a solar maximum is apparent in the SOHO images taken in ultraviolet light.*

The effects of magnetic activity on the Sun are felt throughout the Sun's photosphere, chromosphere, and corona. Sunspots are only one of a host of phenomenon that follow the Sun's 22-year cycle of magnetic activity. The peaks of the cycle, referred to as **solar maximum,** are times of intense activity, as can be seen in the ultravi-olet images of the Sun in **Figure 13.20(d).** While sunspots are darker-than-average features, they are often accompanied by a brightening of the solar chromosphere that is seen most clearly in the light of emission lines such as Hα. The magnificent loops seen arching through the solar corona are also anchored in solar active regions.

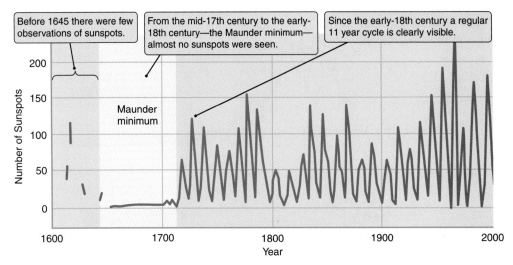

Before 1645 there were few observations of sunspots.

From the mid-17th century to the early-18th century—the Maunder minimum—almost no sunspots were seen.

Since the early-18th century a regular 11 year cycle is clearly visible.

Maunder minimum

Figure 13.21 *Sunspots have been observed for hundreds of years. In this figure, the 11-year cycle in the number of sunspots (half of the 22-year solar magnetic cycle) is clearly visible. Sunspot activity varies greatly over time. The period from the middle of the 17th century to the early 18th century, when almost no sunspots were seen, is called the Maunder minimum.*

These include solar **prominences,** such as those shown in **Figure 13.22.** Prominences are magnetic flux tubes of relatively cool (5,000 to 10,000 K) gas extending through the million-kelvin gas of the corona. While most prominences are relatively quiescent, others can erupt out through the corona, towering a million kilometers or more over the surface of the Sun, and ejecting material into the corona at velocities of 1,000 km/s.

Figure 13.23(a) shows a picture of a **solar flare** erupting from a sunspot group. Solar flares are the most energetic form of solar activity. These are violent eruptions in which tremendous amounts of magnetic energy are released over the course of a few minutes to a few hours. Flares can heat gas to temperatures of 20 million K, and are the source of intense X-ray and gamma ray radiation. Hot plasma is seen to move out-

ward from flares at speeds that can reach 1,500 km/s. These events, called **coronal mass ejections (Figure 13.23b),** send powerful bursts of energetic particles outward through the Solar System. Magnetic effects in flares can accelerate subatomic particles to almost the speed of light.

SOLAR ACTIVITY AFFECTS EARTH

The amount of solar radiation at Earth's distance from the Sun is, on average, 1.35 kilowatts per square meter. Satellite measurements of the amount of radiation coming from the Sun **(Figure 13.24)** show that this value can vary by as much as 0.2% over periods of a few weeks as dark sunspots and bright spots in the chro-

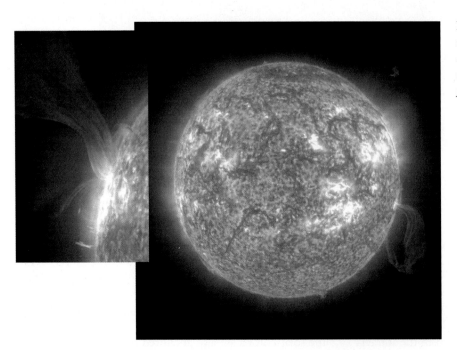

Figure 13.22 *Solar prominences are magnetically supported arches of hot gas that rise high above active regions on the Sun. The inset shows detail at the base of a large prominence.*

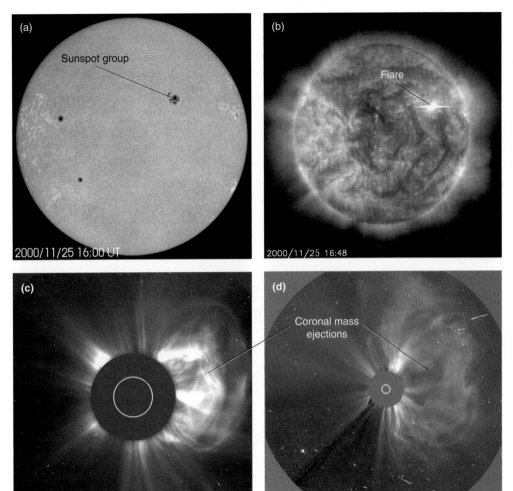

Figure 13.23 *In this sequence of false-color images taken by the SOHO spacecraft a large sunspot group (a) emits a powerful flare, shown in (b) in ultraviolet light. The resulting coronal mass ejection is seen streaming away from the Sun in (c) and (d). (The dark circles in the last two frames obscure the Sun's disk, allowing SOHO to observe the corona. The white circle shows the size of the Sun.)*

mosphere move across the disk. Overall, however, the increased radiation from active regions on the Sun more than makes up for the reduction in radiation from sunspots. On average, the Sun seems to be about 0.1% brighter during the peak of a solar cycle than it is at its minimum.

Solar activity has many effects on Earth. Solar active regions are the source of most of the Sun's extreme ultraviolet and X-ray emission. This energetic radiation heats Earth's upper atmosphere, and during periods of solar activity causes Earth's upper atmosphere to swell. When this happens, it can significantly increase the atmospheric drag on spacecraft orbiting near Earth, causing their orbits to decay. Prior to the launch of the *Hubble Space Telescope,* many people working on the project were concerned that, because it was to be launched into the teeth of a solar maximum, the telescope might not survive for long. One reason for repeated shuttle visits to the *HST* is to "reboost" it up to its original or-

Changes in solar radiation affect Earth's atmosphere.

bit to make up for the slow decay in its orbit caused by drag in the rarefied outer parts of Earth's atmosphere.

Earth's magnetosphere is the result of the interaction between Earth's magnetic field and the solar wind. Increases in the solar wind accompanying solar activity, especially coronal mass ejections directed at Earth, can cause Earth's magnetosphere to become unstable. Spectacular aurorae can accompany such events, as can magnetic storms that have been known to disrupt electrical power grids, causing blackouts across large regions. In fact, the first clear evidence that Earth's environment was affected by solar magnetic activity was the observed relationship between sunspot activity and terrestrial events such as aurorae and perturbations in Earth's magnetic field. Solar flares can have an especially dramatic effect on the environment within the Solar System. Energetic particles accelerated in solar flares pose one of the greatest dangers to human exploration of space.

Solar storms cause aurorae and disrupt electric power grids on Earth.

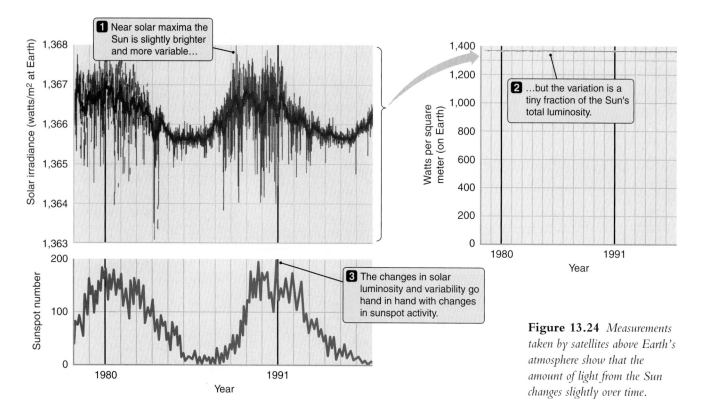

Figure 13.24 *Measurements taken by satellites above Earth's atmosphere show that the amount of light from the Sun changes slightly over time.*

Over the years people have attempted to relate sunspot activity to phenomena ranging from the performance of the stock market to the mating habits of Indian elephants. Few of these supposed relationships have withstood serious scrutiny, however. An especially interesting idea is that variations in Earth's climate might be related to solar activity. We have certainly seen that solar activity affects Earth's upper atmosphere, and it might not be too much of a reach to imagine that it could affect weather patterns as well. It has also been suggested that variations in the amount of radiation from the Sun might be responsible for variations in Earth's climate in the past. While the causes of global climatic change remain controversial, models suggest that observed variations in the Sun could account for only about 0.1 K differences in Earth's average temperature—less than the effects due to the ongoing buildup of carbon dioxide in Earth's atmosphere. On the other hand, triggering the onset of an ice age may only require a drop in global temperatures of around 0.2 to 0.5 K, so the quest for a definite link between solar variability and changes in Earth's climate persists.

SEEING THE FOREST THROUGH THE TREES

The Universe contains more stars than there are grains of sand on all of Earth's beaches. Even "nearby" stars are so distant that at best we see them as faint points of light in the night sky. Only one star is close enough at hand for its complex structure to be studied in great detail. That star is a maelstrom of fierce activity, fueled by nuclear fires burning deep within its interior. Storms rage across its surface. Giant flares burst forth, arching millions of kilometers into space before collapsing back again. That star heats a huge and rarefied corona to millions of kelvins, and blows a bubble in interstellar space large enough to swallow our entire Solar System. That star—a main sequence G2 star, different in no way from main sequence G2 stars throughout our Galaxy and the Universe—is the star we call our Sun.

No machine of human manufacture has ever visited the interior of the Sun, just as no machine has more than scratched the surface of Earth. Even so, our understanding of the interior of the Sun is remarkably complete. Like everything else in the Universe, the Sun is governed by the laws of physics. Computer models based on those physical laws show that there is only one possible structure that a 1 $M_\odot$ ball of gas with the chemical composition of the Sun can have, and remain in balance. According to those models, the Sun draws its energy from the fusion of hydrogen to make helium in its core. That energy is carried from the Sun's core to its surface by radiation and convection, and arrives at just the rate needed to replace the energy radiated away into space. All of this takes place while maintaining a delicate standoff in the battle between the inward pull of gravity, and the outward push of pressure fueled by the nuclear fires in the Sun's interior. Models of the Sun make a host of detailed predictions, ranging from the density and temperature at every point within the Sun's interior to the nuclear reactions that take place in its core. Over the past decades, the predictions of models of the Sun have been tested in ever more sophisticated and detailed ways, and those predictions have been confirmed. At the opening of the 21st century, it is fair to say that we *know* what the interior of the Sun is like.

Our knowledge of the Sun also forms the cornerstone of our knowledge of all stars everywhere. Starting with a successful model of the Sun, we can now go on to ask a host of other questions. What happens if we increase the mass of the model star? What happens if we change its chemical abundances? What happens if we run the model forward in time until all of the hydrogen at the star's center is gone? How do we assemble something that looks like the Sun in the first place? Our study of the Sun has set the stage for the next leg of our journey, as we turn our attention to how stars form, how they differ from one another, how they change with time, and how they die.

STUDENT QUESTIONS

THINKING ABOUT THE CONCEPTS

1. Explain the proton–proton chain process in which the Sun generates energy by converting hydrogen to helium.
2. In the proton–proton chain process the mass of four protons is slightly greater than the mass of a helium nucleus. What happens to the missing mass?
3. Engineers and physicists dream of solving the world's energy supply problem by constructing power plants that would convert globally plentiful hydrogen to helium. What is a major obstacle to this environmentally clean solution?
4. What experiences have you had with energy transmitted to you by radiation alone? by convection alone? by conduction alone? (You are exposed to all three every day.)
5. If neutrinos have mass, as now seems to be the case, they cannot travel at the speed of light. Explain why not.
6. Why are neutrinos so difficult to detect?
7. Discuss the "solar neutrino problem" and how this problem was solved.
8. What techniques do you find in common between how we probe the internal structure of the Sun and the internal structure of Earth?
9. Explain the relationship between the 11-year sunspot cycle and the 22-year magnetic cycle.
10. Solar flares and magnetic storms can adversely affect Earth's power grids, radio communication and the health of humans in orbit. How might observations of the Sun help us forecast such events?

APPLYING THE CONCEPTS

11. The Sun shines by converting mass into energy according to Einstein's well-known relationship, $E = mc^2$. Show that if the Sun produces 3.78×10^{26} J of energy per second it must expend 4.2 million metric tons (4.2×10^9 kg) of mass per second.
12. Assume the Sun has been producing energy at a constant rate over its lifetime of 4.5 billion years (1.4×10^{17} s).
    a. How much mass has it lost creating energy over its lifetime?
    b. The present mass of the Sun is 2×10^{30} kg. What fraction of its present mass has been converted into energy over the lifetime of the Sun?
13. Imagine the source of energy in the interior of the Sun were to abruptly change.
    a. How long would it take before a neutrino telescope could detect the event?
    b. When would a visible-light telescope see evidence of the change?
14. On average, how long does it take particles in the solar wind to reach Earth from the Sun if they are traveling at 300 km/s?
15. Sunspots appear relatively dark because their surface temperature typically is 1,500 K cooler than the surrounding photosphere (5,780 K). Compare the surface brightness (power radiated per square meter) of a dark sunspot to that of the surrounding photosphere.
16. The hydrogen bomb represents humankind's efforts to duplicate processes going on in the core of the Sun. The energy released by a 5 megaton hydrogen bomb is 3×10^{14} J. How much mass did Earth lose each time a 5 megaton hydrogen bomb was exploded?

14

There are no whole truths; all truths are half truths.
It is trying to treat them as whole truths that plays the devil.

ALFRED NORTH WHITEHEAD
(1861–1947)

STAR FORMATION AND THE INTERSTELLAR MEDIUM

14.1 WHENCE STARS?

To mortal humans, the stars seem fixed and unchanging. Indeed, our local star, the Sun, has remained much the same for long enough to allow life on Earth to arise, evolve, and begin to contemplate its own origin and fate. But our perspective is distorted by our brief existence. All of recorded human history amounts to less than a millionth of the age of the Sun and Earth. In Chapter 5 we told the story of how our Solar System formed some 4.6 billion years ago out of a swirling disk of gas and dust that surrounded the protostellar Sun. We are now ready to return to that story, but with a different focus. First, we will step back and look at the interstellar environment which gave birth to the Sun and Solar System. Then we will focus our attention on the protostar at the center of the circumstellar disk, and watch as it becomes a star.

The root of the word *astronomy* may mean "star," but if we ask what fraction of space is filled with stars, we get a very different picture of their importance. The volume of our Sun is about 1.3 million times the volume of Earth. Though this volume might seem enormous to us, it is almost incomprehensibly tiny compared with the volume of space. Our Sun is the only star in a region of space with a volume of over 300 cubic light-years. If our Sun were the size of a Ping-Pong ball sitting on

KEY CONCEPTS

When you look at the night sky it is stars that you notice, yet stars fill only the tiniest fraction of the volume of space. On this leg of our journey we turn our attention to the tenuous medium between the stars, and discover that:

✳ There are various phases of the interstellar medium, ranging from cold, relatively dense molecular clouds to hot, tenuous, intercloud gas;

✳ Different phases of the interstellar medium emit various types of radiation, and can be observed at different wavelengths, ranging from radio waves to X-rays;

✳ Dust in the interstellar medium blocks visible light, but glows with infrared radiation;

✳ Gas in the interstellar medium is heated and ionized by energy from stars and stellar explosions;

✳ Stars form in clusters from dense cores buried within giant molecular clouds;

✳ Forming stars are seen by their infrared emission, and by the effect they have on their surroundings;

✳ We can follow the progress of an evolving protostar by its track across the H-R diagram; and

✳ Protostars collapse, radiating away their gravitational energy until fusion starts in their cores, and they settle onto the main sequence.

Interstellar space is vast.

an author's office desk in Tempe, Arizona, its nearest stellar neighbor would be a similar Ping-Pong ball sitting on another author's desk in Santa Cruz, California, 1,000 km away. The rest of the volume of our Galaxy is filled with the *interstellar medium*. It is here that the story of stars begins and ends. Stars are formed from the gas in the interstellar medium, they live their lives there, and when they die, they bequeath some of their chemical and energetic legacy back to the interstellar medium.

Before taking a look at the interstellar medium, however, we need to put it into another context as well. Remember that galaxies are *crowded* in comparison with the vast expanses of *intergalactic* space. Returning to the Ping-Pong ball analogy, if we ask the whereabouts of a star in the spiral galaxy nearest to our own, that star would be equivalent to a Ping-Pong ball somewhere on Jupiter. But that is a story for later in our journey. For now, we turn our attention to our "local neighborhood" and the interstellar medium that fills the volume of our own Galaxy.

14.2 THE INTERSTELLAR MEDIUM

The Sun formed from the interstellar medium, so it is not too surprising that the chemical composition of the interstellar medium in our region of the Milky Way is similar to the chemical composition of the Sun (see the "Astronomer's Periodic Table" in Chapter 8). In the interstellar medium, about 90% of the atomic nuclei are hydrogen and the remaining 10% are almost all helium. More-massive elements together account for only 0.1% of the atomic nuclei, or about 2% of the mass in the in-

Interstellar gas is extremely tenuous.

terstellar medium. Roughly 99% of that interstellar matter is gaseous, consisting of individual atoms or molecules moving freely about, as do the molecules in the air around us. But interstellar gas is far more tenuous than the thick soup that we breathe. Each cubic centimeter of the air around you contains about 2.5×10^{19} molecules. A good terrestrial vacuum chamber can reduce this density down to about 10^{10} molecules/cm^3—about a billionth as dense. By comparison, the interstellar medium has an average density of less than 1 atom/cm^3—one ten-billionth as dense still! To make this more understandable, imagine a square tube 1 cm on a side and a meter long, filled with air, and sealed at both ends. Also

imagine you can stretch the tube without changing its cross section. As you do so, the air in the tube has to spread out to fill the larger volume, and so is less dense. When the tube is stretched to a length of 10 meters, the air inside will be a tenth as dense as the air around you. When stretched to 100 meters, the air inside will be a hundredth as dense. For the air in the tube to match the average density of interstellar gas—that is, for it to have the same amount of mass per unit volume as the interstellar medium—you would have to stretch it into a tube 25,000 light-years long! Stated another way, there is about as much material in a column of air between your eye and the floor beside you as there is interstellar gas in a column with the same cross section stretching from the Solar System to the center of the Galaxy.

THE INTERSTELLAR MEDIUM IS DUSTY

About 1% of the material in the interstellar medium is in the form of solid grains, referred to as **interstellar dust.** "Interstellar soot" might be a better name. Ranging from little more than large molecules up to particles about 300 nm across, these solid grains more

About 1% of the mass of interstellar medium is in grains.

closely resemble the particles of soot from a candle flame than they do the dust that collects on the window sill. (It would take several hundred "large" interstellar grains to span the thickness of a single human hair.) Interstellar dust gets its start when refractory materials such as iron, silicon, and carbon stick together to form grains in dense, relatively cool environments like the outer atmospheres and "winds" of cool, red giant stars, or in dense material thrown into space by stellar explosions. (More about these topics later.) Once in the interstellar medium, other atoms and molecules may stick to these grains. This process is remarkably efficient: about half of all interstellar matter more massive than helium (1% of the total mass of the interstellar medium) is found in interstellar grains.

If there is only as much material between us and the center of the Galaxy as there is between your eye and the floor, then you might expect our view of distant objects to be as clear as your view of your own big toe. But nothing could be further from the truth. It

Interstellar dust obscures our view.

turns out that interstellar dust is extremely effective at blocking light. If the air around you contained as much dust as a comparable mass of interstellar material, it

Figure 14.1 *The top image shows an all-sky picture of the Milky Way taken in visible light. The dark splotches blocking our view are dusty interstellar clouds. The bottom image shows the same field of view as seen in the near infrared. Infrared radiation penetrates interstellar dust giving us a clearer view of the stars in our Galaxy.*

would be so dirty that you would be hard pressed to see your hand held up 10 cm in front of your face. Go out on a dark summer night in the Northern Hemisphere (or a dark winter night in the Southern Hemisphere) and look closely at the Milky Way, visible as a faint band of diffuse light running through the constellation Sagittarius. You will see a dark "lane" running roughly down the middle of this bright band, splitting it in two. This dark band is the result of vast expanses of interstellar dust blocking our view of distant stars.

When interstellar dust gets in the way of radiation from distant objects, the effect is called **interstellar extinction**. Not all electromagnetic radiation suffers equally from interstellar extinction, however. **Figure 14.1** shows two images of our Galaxy: one taken in visible light, the other taken in the infrared (IR). The dark clouds that block the shorter-wavelength visible light seem to have vanished in the longer-wavelength IR image, allowing us to see through the clouds to the center of our Galaxy and beyond.

As a starting point for understanding why short-wavelength radiation is blocked by dust while long-wavelength radiation is not, think about a different kind of wave—waves on the surface of the ocean. Imagine

you are on the ocean in a strong swell. If you are floating in a small boat, which is much smaller than the wavelength of ocean waves, the swell causes you to bob gently up and down. But that is about all: the waves interact very little with the small boat. The story is quite different if you are on a larger boat that is about half as long as the wavelength of the ocean waves. Now the bow of the boat may be on a wave crest while the stern of the boat is in a trough, or vice versa. The boat tips wildly back and forth as the waves go by. If the size of the boat and the wavelength of the waves are a good match, even fairly modest waves will rock the boat. (You might have noticed this if you have ever been in a canoe or rowboat when the wake from a speedboat came by.) Now imagine viewing these two situations from the perspective of the wave. The wave is hardly affected by the small boat, but it is strongly affected by the large boat. The energy to drive the wild motions of the boat comes from the wave, and so the motion of the wave is affected by the interaction.

Waves interact with things that are about their size.

The interaction of electromagnetic waves with matter is more involved than a boat rocking on the ocean, but the same basic idea often applies. Tiny interstellar dust grains interact most strongly with ultraviolet and blue light, which have wavelengths comparable to the typical size of dust grains. For this reason, ultraviolet and blue light are effectively blocked by interstellar dust. Short wavelengths suffer heavily from interstellar extinction. Infrared and radio radiation, on the other hand, have wavelengths that are too long to interact strongly with the tiny interstellar dust grains, and so can travel largely unimpeded across great interstellar distances. To sum up, at visible and ultraviolet wavelengths, most of the Galaxy is hidden from our view by dust. In the infrared and radio portions of the spectrum, however, we get a far more complete view.

Long-wavelength radiation penetrates interstellar dust.

In the visible part of the spectrum, short-wavelength blue light suffers more from extinction than longer-wavelength red light. As a result, as shown in **Figure 14.2,** an object viewed through dust looks redder (actually, less blue) than it really is. This effect is called **reddening.** Correcting for the fact that stars and other objects appear both fainter and redder than they would in the absence of dust can be one of the most difficult parts of interpreting astronomical observations, and often adds to the uncertainty in the measurement of an object's properties.

Reddening accompanies interstellar extinction.

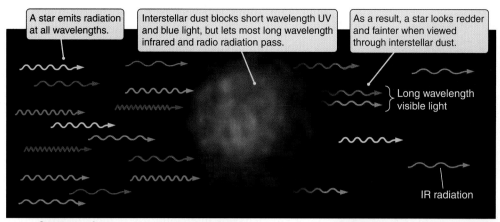

A star emits radiation at all wavelengths.

Interstellar dust blocks short wavelength UV and blue light, but lets most long wavelength infrared and radio radiation pass.

As a result, a star looks redder and fainter when viewed through interstellar dust.

Long wavelength visible light

IR radiation

Figure 14.2 *The wavelengths of ultraviolet and blue light are close to the size of interstellar grains, and so the grains effectively block this light. Grains are less effective at blocking longer wavelength light. As a result, the spectrum of a star (lower left) when seen through an interstellar cloud (lower right) appears fainter and redder.*

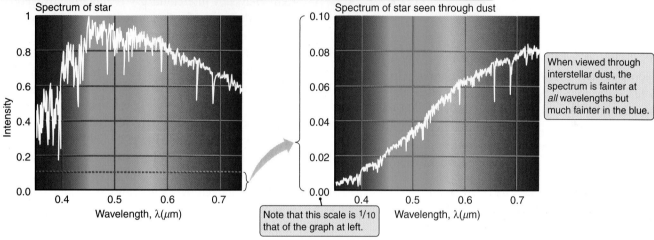

Spectrum of star

Spectrum of star seen through dust

When viewed through interstellar dust, the spectrum is fainter at *all* wavelengths but much fainter in the blue.

Note that this scale is ¹⁄₁₀ that of the graph at left.

Intensity

Wavelength, λ(μm)

Wavelength, λ(μm)

Dust glows in the infrared.

Interstellar extinction may not be much of a concern at infrared wavelengths, but dust still plays an important role in what we see when we look at the sky in the infrared. Grains of dust, like any solid object, glow at wavelengths determined by their temperature. In Chapter 4 we learned how the equilibrium between absorbed sunlight and emitted thermal radiation determines the temperatures of the terrestrial planets (see Figure 4.26). A similar equilibrium is at work in interstellar space, where dust is often heated by starlight and by the gas in which it is immersed to temperatures of tens to hundreds of kelvins **(Figure 14.3).** At a temperature of 100 K, Wien's law says that dust will glow most strongly at a wavelength of about 30 μm:

$$\lambda_{max} = \frac{2{,}900 \ \mu m \ K}{T} = \frac{2{,}900 \ \mu m \ K}{100 \ K} = 29 \ \mu m$$

Cooler dust—say, at a temperature of 10 K—glows most strongly at a wavelength of 290 μm. Both of these wavelengths are in the far infrared part of the electromagnetic spectrum. Much of what we see when looking at the sky with infrared "eyes" is not really starlight

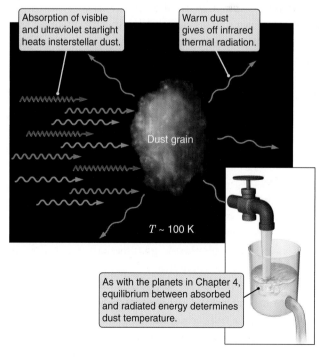

Absorption of visible and ultraviolet starlight heats insterstellar dust.

Warm dust gives off infrared thermal radiation.

Dust grain

$T \sim 100$ K

As with the planets in Chapter 4, equilibrium between absorbed and radiated energy determines dust temperature.

Figure 14.3 *The temperature of interstellar grains is determined by the same type of equilibrium between absorbed and emitted radiation that determines the temperatures of planets.*

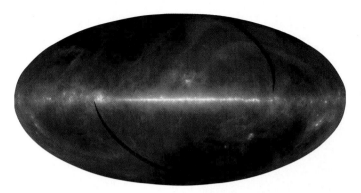

Figure 14.4 *Far infrared images of the sky show clouds of warm, glowing interstellar dust. (Black bands are due to missing data.)*

at all, but thermal radiation from dust. When NASA launched the *Infrared Astronomical Satellite* (*IRAS*) in 1983, astronomers got their first look at the sky at wavelengths out to 100 μm, and the views (like **Figure 14.4**) were unlike any seen before. *IRAS* showed the dark clouds that you see when you look at the Milky Way glowing brilliantly in infrared radiation from dust.

INTERSTELLAR GAS HAS DIFFERENT TEMPERATURES AND DENSITIES

The gas and dust that fill interstellar space within our Galaxy are not spread out evenly. About half of all interstellar gas is concentrated into only about 2% of the volume of interstellar space. These relatively dense regions are referred to as **interstellar clouds.** (Interstellar clouds are *not* like terrestrial clouds, which are locations where water vapor has condensed to form liquid droplets. Rather, interstellar clouds are places where the interstellar gas is more concentrated than in surrounding regions.) The other half of the interstellar gas is spread out through the remaining 98% of the volume of interstellar space. This is called **intercloud gas,** meaning gas that is found "between the clouds."

> Most interstellar material is concentrated in relatively dense clouds.

The properties of intercloud gas vary from place to place. Some intercloud gas is extremely hot, with temperatures in the millions of kelvins, close to those found in the centers of stars. Even so, were you to find yourself adrift in an expanse of hot intercloud gas, your first concern would be freezing to death! The gas is hot, which means that the atoms that make it up are moving about very rapidly, so when one of these atoms runs

into you, it hits you very hard. But the gas is also extremely tenuous. Typically, you would have to search a liter (1,000 cm^3) or more of hot intercloud gas to find a *single* atom. To gather up 1 gram of hot intercloud gas—about the mass found in a liter of air— you would have to collect all of the gas in a cube over 8,000 km on an edge! There are so few atoms in a given volume of hot intercloud gas that atoms would not run into you very frequently, so this million-kelvin gas would do little to keep you warm. You would radiate energy away and cool off much faster than the gas around you could replace the lost energy.

Extremely hot intercloud gas is thought to occupy about half the volume of interstellar space. This gas is mostly heated by the energy of tremendous stellar explosions called **supernovae.** (We will return to the story of supernovae in Chapters 15 and 16. For now, it is enough to know that these are *very* energetic events!) Our Solar System is located inside a bubble of hot intercloud gas that is about 650 light-years across. Some astronomers have suggested that this is the remnant of the hot bubble produced by a supernova explosion 100,000 years ago. Like the million-kelvin gas in the corona of the Sun, hot intercloud gas glows faintly in the energetic X-ray portion of the electromagnetic spectrum. X-rays cannot penetrate Earth's atmosphere, so it is necessary to get above the atmosphere to observe the radiation coming from hot intercloud gas. For rocket- and satellite-borne X-ray telescopes, however, the entire sky is aglow with faint X-rays coming from the bubble of hot gas in which our Solar System is immersed (see **Figure 14.5**).

> We live in a bubble of million-kelvin intercloud gas.

Not all intercloud gas is as hot as our local bubble. Most of the remaining intercloud gas has a temperature of about 8,000 K, and a density ranging from about

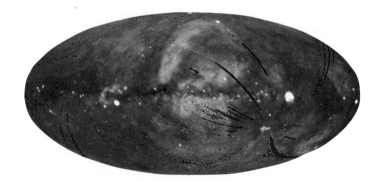

Figure 14.5 *The faint X-ray glow that fills the sky is due largely to emission coming from the bubble of million-kelvin gas which surrounds the Sun. Bright spots are more distant sources, including objects such as bubbles of very hot, high-pressure gas surrounding the sites of recent supernova explosions.*

0.01 to 1 atom per cubic centimeter. To gather up a gram of this warm intercloud gas, we would "only" need to collect all of the gas in a cube of interstellar space 800 km on an edge.

Interstellar space is awash with the light from stars. Some of this light is ultraviolet light with wavelengths shorter than 91.2 nm. Each photon of this ultraviolet light has enough energy that, if absorbed by a hydrogen atom, it will ionize the atom, kicking its electron away from the nucleus. If there are enough of these energetic photons around, then atoms in the interstellar gas cannot remain neutral (since any neutral hydrogen atom will soon be ionized by

Some warm intercloud gas is ionized by starlight.

starlight). In this way, about half of the volume of warm intercloud gas is kept ionized. But if there is a large enough expanse of warm intercloud gas, then the ionizing photons in the starlight get "used up." About half of the volume of warm intercloud gas is mostly neutral. This gas is usually surrounded by regions of ionized intercloud gas that shield the neutral gas from starlight, much as Earth's ozone layer "shields" the surface of Earth from harmful ultraviolet radiation from the Sun.

One way to look for interstellar gas, including both warm and hot interstellar gas, is to study the spectra of distant stars. As starlight passes through interstellar gas, absorption lines form. Just as absorption lines formed in the atmosphere of a star tell us much about the star's temperature, density, and chemical composition, interstellar absorption lines provide similar information about the gas the light has

Absorption lines due to interstellar gas are seen in spectra of stars.

passed through. Interstellar gas can also be studied by looking for the radiation that it emits. In regions of warm *ionized* gas, protons and electrons are constantly "finding each other," and recombining to make hydrogen atoms. Since it takes energy to tear a hydrogen atom apart, conservation of energy says that when a proton and electron combine to form a hydrogen atom, this same amount of energy must be freed up. This freed-up energy must "go" somewhere, and that somewhere is into the form of electromagnetic radiation. Typically when a proton and electron combine, the resulting hydrogen atom is left in an excited state (see our discussion of energy states of atoms in Chapter 4). The atom then drops down to lower and lower energy states, emitting a photon at each step. Together these photons carry away an amount of energy equal to the ionization energy of hydrogen plus whatever energy of motion the electron and protron had before recombining. The photons emitted in this process have wave-

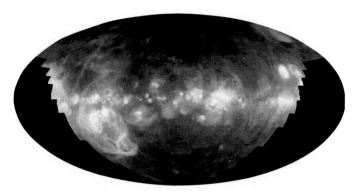

Figure 14.6 *Warm (about 8,000 K) interstellar gas glows in the Hα line of hydrogen. This image, showing the Hα emission from much of the northern sky, shows how complex the structure of the interstellar medium is.*

lengths corresponding to the energy differences between the allowed energy states of hydrogen. In other words, warm ionized interstellar gas glows in emission lines

Warm ionized gas gives off emission lines.

characteristic of hydrogen (as well as other elements). Usually the strongest emission line given off by warm interstellar gas in the visible part of the spectrum is the Hα (hydrogen alpha) line, which is seen in the red part of the spectrum at a wavelength of 656.3 nm.

Figure 14.6 shows an image of a portion of the sky taken in the light of Hα emission. The faint diffuse emission in the figure comes mostly from warm ionized intercloud gas. The especially bright spots, on the other hand, are quite different. These are regions where intense ultraviolet radiation from massive, hot, luminous O and B stars is able to ionize even relatively dense interstellar clouds. These are called **H II regions** (pronounced "H two"), signifying that they are made up

H II regions are ionized by hot, luminous O and B stars.

of the second, or ionized, form of hydrogen. In the chapters to come we will address the question of how long stars live, and discover that O stars live for only a few million years. As a result, they usually do not move too far away from where they formed. The glowing clouds seen as H II regions are the very clouds from which the stars were born. Thus H II regions are signposts to regions where active star formation is taking place.

One of the closest H II regions to the Sun is the "Great Nebula" in the constellation Orion (see **Figure 14.7**). In fact, the Orion Nebula is tiny as H II regions go. It is "great" only because it happens to be very close to us. Almost all of the ultraviolet light that powers the nebula comes from a single hot star, and all

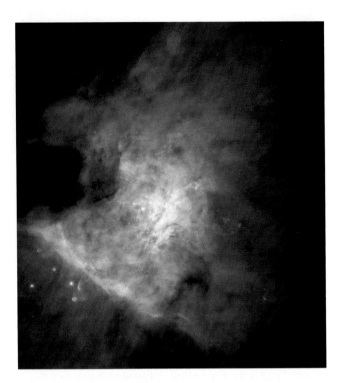

Figure 14.7 *This composite of* HST *images shows the Orion Nebula, a glowing region of interstellar gas surrounding a cluster of young, hot stars. New stars are still forming in the dense clouds surrounding the nebula.*

told there are only a few hundred stars forming in the immediate vicinity. In contrast, **Figure 14.8** shows a giant H II region called 30 Doradus. The 30 Doradus H II region is located in the Large Magellanic Cloud, a small companion galaxy to the Milky Way, 150,000 light-years distant. This enormous cloud of ionized gas is powered by a dense star cluster containing thousands of hot, luminous stars.

Warm *neutral* hydrogen gas gives off radiation in a different way than does warm ionized hydrogen. Many subatomic particles, including protons and electrons, are magnetized. It is as if each of these particles has a bar magnet, with a north and a south pole, built into it. According to the rules of quantum mechanics, a hydrogen atom can exist in only two different configurations. Either the magnetic "poles" of the proton and electron point in the *same* direction, or they point in *opposite* directions. The "electron's magnet" and the "proton's magnet" push on each, so the way they line up affects the energy that the atom has. When the two "magnets" point in opposite directions the atom has slightly more energy than when they point in the same direction. It takes energy to change a hydrogen atom from the lower-energy, magnetically aligned state to the higher-energy, magnetically unaligned state. This

Neutral hydrogen emits 21-cm radiation.

is no problem, since in warm neutral interstellar gas, interactions between atoms easily supply enough energy to do the trick. But once a hydrogen atom is in its higher (magnetically anti-aligned) energy state, getting back into its lower (magnetically aligned) energy state requires nothing but time. If left alone long enough, a hydrogen atom in the higher energy state will spontaneously jump to the lower energy state, emitting a photon in the process. The energy difference between the two magnetic states of a hydrogen atom is extremely small: less than a two-millionth of the energy needed to tear the electron away from the proton completely. The radiation given off by these low-energy transitions has a wavelength of 21 cm, which lies in the radio portion of the spectrum.

The tendency for hydrogen atoms to emit 21-cm radiation is extremely weak. On average, you would have to wait about 11 million years for a hydrogen atom in the higher energy state to spontaneously jump to the lower energy state and give off a photon. Even so, there is a *lot* of hydrogen in the Universe. As seen in **Figure 14.9,** the sky is aglow with 21-cm radiation from neutral hydrogen. Because of its long wavelength, 21-cm radiation

Astronomers use 21-cm radiation to probe the structure of our Galaxy.

Figure 14.8 *This* HST *image shows the 30 Doradus nebula, located in a small companion galaxy to the Milky Way, 150,000 light-years from Earth. Interstellar gas in this region of intense star formation is caused to glow by the ultraviolet radiation coming from the dense cluster of thousands of young, massive stars.*

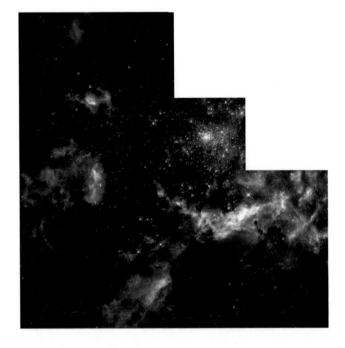

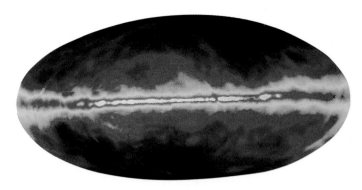

Figure 14.9 *This radio image of the sky taken at a wavelength of 21 cm shows clouds of neutral hydrogen gas throughout our Galaxy. Because radio waves penetrate interstellar dust, 21-cm observations are a crucial probe of the structure of our Galaxy.*

freely penetrates dust in the interstellar medium, allowing us to see neutral hydrogen throughout our Galaxy, while measurements of the Doppler shift of the line tell how fast the source of the radiation is moving. These two attributes make the 21-cm line of neutral hydrogen perhaps the most important kind of radiation there is for understanding the structure of our Galaxy.

REGIONS OF COOL, DENSE GAS ARE CALLED CLOUDS

As mentioned earlier, intercloud gas fills 98% of the volume of interstellar space but accounts for only half of the mass of interstellar gas. The remaining 50% of all interstellar gas is concentrated into much denser interstellar clouds that occupy only 2% of the volume of the Galaxy. Most interstellar clouds are composed primarily of neutral atomic hydrogen (that is, isolated neutral hydrogen atoms). These clouds are much cooler and denser than the warm intercloud gas. They have temperatures of around 100 K, and densities in the range of about 1 to about 100 atoms per cubic centimeter. Even so, at a density of 100 atoms/cm^3, you would still have to gather up all of the gas in a box 170 km on an edge to collect a single gram of this material.

> Most interstellar clouds are neutral hydrogen.

On Earth it is uncommon to find atoms in isolation—most atoms are tied up in molecules. (For example, in Earth's atmosphere, only nonreactive gases such as argon are typically found in their atomic form.) Yet so far, the interstellar gas that we have encountered consists of isolated atoms rather than molecules. The reason is that throughout most of interstellar space, including most interstellar clouds, molecules do not survive for long. If interstellar gas is too hot, then any molecules that do exist will soon collide with other molecules or atoms with enough energy to break the molecules back into their constituent atoms. The temperature in a neutral hydrogen cloud may be low enough for some molecules to survive, but even here a different process will soon destroy any molecules that might form. Photons of starlight with enough energy to break molecules apart are able to penetrate into neutral hydrogen clouds. Only in the hearts of the densest of interstellar clouds, where dust effectively blocks even the relatively low-energy photons required to shatter molecules, does interstellar chemistry get its chance. For obvious reasons, these dark clouds are referred to as **molecular clouds.**

> Molecules survive only in cold, dark interstellar clouds.

In images such as **Figure 14.10**, molecular clouds are evident from their silhouettes against a background of stars. Inside such clouds it is dark and usually very cold, with a typical temperature of only about 10 K— 10 kelvins above absolute zero. Most of these clouds have densities of around 100 to 1,000 molecules/cm^3, but densities as high as 10^{10} molecules/cm^3 have been observed. (Even at 10^{10} molecules/cm^3 this gas is still less than a billionth as dense than the air around you, making it an extremely good vacuum by terrestrial standards.) In this cold, relatively dense environment, atoms combine to form a wide variety of molecules.

> Molecular clouds are dense and cold.

By far the most common component of molecular clouds is molecular hydrogen. Molecular hydrogen consists of two hydrogen atoms, and is the smallest possible molecule. Molecules radiate mostly in the radio and infrared portions of the electromagnetic spectrum. Molecular emission lines are useful to astronomers in much the same way as the spectral lines from atoms: each type of molecule is unique in its properties, and the energies of its allowed states are unique as well. The wavelengths of emission lines from molecules are an unmistakable fingerprint of the kind of molecule responsible.

In addition to molecular hydrogen, over 80 other molecules have been discovered in interstellar space. These molecules range from very simple structures such as carbon monoxide (CO), one of the most common molecules, to complex organic compounds such as methanol (CH$_3$OH) and acetone [(CH$_3$)$_2$CO], and carbon chains

> Many interstellar molecules have been observed.

as large as $HC_{11}N$. Very large carbon molecules, made of hundreds of individual atoms, bridge the gap between large interstellar molecules and small interstellar grains. As mentioned earlier, visible light cannot escape from the molecular clouds where interstellar molecules are concentrated. Fortunately, however, the radio waves emitted by molecules are unaffected by interstellar dust, and so escape easily from dark molecular clouds. Observations of radio waves from molecules give us a remarkable look at the innermost workings of the densest and most opaque interstellar clouds.

Molecular clouds have masses ranging from a few times the mass of our Sun, up to 10^7 (10 million) times the mass of our Sun. The smallest molecular clouds may be less than half a light-year across, while the largest molecular clouds may be over a thousand light-years in size.

Stars form in giant molecular clouds.

The largest molecular clouds qualify for the title **giant molecular cloud.** These behemoths typically have masses of a few hundred thousand times that of our Sun, and on average are about 120 light-years across. There are about 4,000 giant molecular clouds in our Galaxy, and many more smaller ones. Molecular clouds fill only a tiny fraction of the volume of our Galaxy—probably only about 0.1% of interstellar space. These clouds may be rare, but they are extremely important to our story, for they are the cradles of star formation.

Having looked at the many different phases of the interstellar medium, ranging from vast expanses of tenuous million-kelvin intercloud gas to the cold dense interiors of molecular clouds, an obvious question to ask is, "How does it all fit together?" Discussing the structure of the interstellar medium is a lot like discussing the nature of weather on Earth: it is extremely complex, often difficult to understand, and all but impossible to predict precisely.

The structure of the interstellar medium is complex.

In the interstellar medium, the energy to power the "weather" comes from stars. Warm gas heated by ultraviolet radiation from massive, hot stars pushes outward into its surroundings. Blast waves from supernova explosions sweep out vast, hot "cavities," piling up everything in their path like snow in front of a snowplow. Interstellar clouds are destroyed under the onslaught of these violent events, but they are also formed by these same forces. Swept-up intercloud gas becomes the next generation of clouds. Hot bubbles of high-pressure interstellar gas crush molecular clouds, driving up their densities and perhaps triggering the formation of new generations of stars. The interstellar medium is so well stirred that *on average,* interstellar clouds are moving around with random velocities of approximately 20 km/s. The fastest motions of interstellar material are measured in thousands of kilometers per second.

The energy to power interstellar weather comes from stars.

The story of the wide range of phenomena that shape the interstellar medium is fascinating. We could easily stay here and look around for

Figure 14.10 *In visible light, interstellar molecular clouds are seen in silhouette against a background of stars and glowing gas. Interstellar dust in the clouds blocks short-wavelength visible and ultraviolet light.*

quite some time, but instead we must press on in our journey. The time has come to watch as a molecular cloud begins the slow process of collapsing under its own weight, beginning a chain of events that will culminate with the ignition of nuclear fires within a new generation of stars.

14.3 MOLECULAR CLOUDS ARE THE CRADLE OF STAR FORMATION

As we have seen, objects such as planets and stars are held together by gravity. The mutual gravitational attraction between each part of a planet or star and every other part of the object results in a net inward force that pulls all parts of the object toward its center. If an object is stable, then this inward force of the objects' self-gravity must be balanced by something, often the outward push of pressure within the object. In our discussion of planets and stars we referred to the balance between gravity and pressure in a stable object as hydrostatic equilibrium.

All of these same concepts can be applied to clouds in the interstellar medium. As shown in **Figure 14.11**, interstellar clouds also have self-gravity, since each part

Self-gravity is unimportant in most interstellar clouds.

Figure 14.11 *Parcels of gas within a molecular cloud feel the gravitational attraction of all other parts of the molecular cloud, leading to a net gravitational force toward the cloud's center. This is referred to as the cloud's self-gravity.*

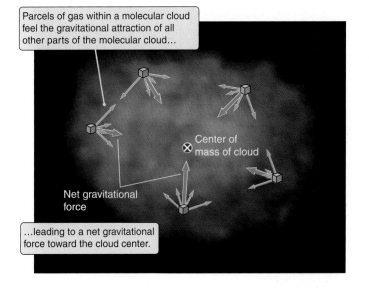

Parcels of gas within a molecular cloud feel the gravitational attraction of all other parts of the molecular cloud...

Center of mass of cloud

Net gravitational force

...leading to a net gravitational force toward the cloud center.

of the cloud feels a gravitational attraction for every other part of the cloud. However, interstellar clouds are very rarely in hydrostatic equilibrium. In most interstellar clouds, pressure is much stronger than self-gravity. (Because gravity follows an inverse square law, the more spread-out an object's mass is, the weaker its self-gravity.) Pressure in these clouds would cause them to expand if not for the pressure of the surrounding intercloud medium holding them in. The intercloud medium is much less dense than interstellar clouds, but it is also much hotter, so its pressure is high enough to confine the clouds. Some molecular clouds, on the other hand, violate hydrostatic equilibrium in the other direction. These clouds are massive enough and dense enough—that is, there is enough mass packed together in a small enough volume—that their self-gravity matters. (More mass along with smaller distances between different parts of the cloud means that gravity is stronger.) Furthermore, these clouds are cool enough that pressure in the clouds is relatively low, despite their high density. In such clouds, self-gravity is much greater than pressure, so the cloud collapses under its own weight.

14.1 COLLAPSE

Some molecular clouds collapse under their own weight.

If self-gravity in a molecular cloud is much more significant than pressure, we might expect that gravity should win outright, and that the cloud should rapidly collapse toward its center. In practice, the process goes very slowly, because there are a number of other effects that stand in the way of the collapse. One of the effects that slows the collapse of a cloud is conservation of angular momentum. We discussed the effects of angular momentum and the flattening of a collapsing cloud in Chapter 5. Other properties of molecular clouds that prevent them from collapsing rapidly are turbulence and the effects of magnetic fields. Even though these effects may slow the collapse of a molecular cloud, in the end gravity will win. Magnetic fields in the cloud can slowly die away. One part of the cloud can lose angular momentum to another part of the cloud, allowing the part of the cloud with less angular momentum to collapse further. Turbulence dies away. The details of these processes are complex, and the subject of much current research. But the crucial point is this: the effects preventing the collapse of a molecular cloud are temporary, but gravity is persistent. As the forces that oppose the cloud's self-gravity gradually fade away, the cloud slowly becomes smaller.

Angular momentum, turbulence, and magnetic fields slow the collapse . . .

. . . but gravity wins in the end.

MOLECULAR CLOUDS FRAGMENT AS THEY COLLAPSE

Crucial to the process of star formation is the fact that, as a cloud becomes smaller, the gravitational forces trying to pull it together grow stronger. This is a result of the inverse square law of gravitation discussed in Chapter 3. Suppose a cloud starts out being 4 light-years across. By the time the cloud has collapsed to 2 light-years across, the different parts of the cloud are, on average, only half as far away from each other as they were when they started. As a result, the gravitational attraction they feel toward each other will be four times stronger than when the cloud was 4 light-years across. When the cloud is $\frac{1}{4}$ as large as it started out to be (or 1 light-year across), the force of gravity will be 16 times as strong. As a collapsing cloud becomes smaller, gravity becomes stronger. As gravity becomes stronger, the collapse picks up speed. And as the collapse picks up speed, gravity increases even more rapidly. The collapse "snowballs."

Denser parts of the clouds collapse more rapidly.

Molecular clouds are never uniform. Some regions within the cloud are invariably denser than others, and so collapse within themselves more rapidly. As these regions collapse, their self-gravity becomes stronger because they are more compact, so they collapse even faster. **Figure 14.12** shows the result of this process. What started out as slight variations in the density of the cloud grow to become very dense concentrations of gas. Instead of collapsing into a single object, the molecular cloud fragments into a number of very dense **molecular cloud cores.** A single molecular cloud may form hundreds or thousands of molecular cloud cores. It is from these dense cloud cores, typically a few light-months in size, that stars may form.

Stars form in molecular cloud cores.

As a molecular cloud core becomes smaller, the gravitational forces that are trying to crush the cloud grow stronger. Eventually, gravity is able to overwhelm the forces of pressure, magnetic fields, and turbulence that have been resisting it. This happens first near the center of the cloud, because it is there that the material in the cloud is most strongly concentrated. The inner parts of the cloud core start to rapidly fall inward. This starts a sequence of events much like a chain of dominos in which one domino topples the next, which topples the next. The pressure from the central part of the cloud core had been supporting the weight of the layers above it. When the center of the cloud collapses, this support is suddenly removed. Without this "pressure support," the next outer layer begins to fall freely inward toward the center as well. But what about the layers farther out? Now there is nothing to hold them up, either, so the process continues: each layer of the cloud core falls inward in turn, thereby removing support from the layers still farther out. As shown in **Figure 14.13,** the cloud core collapses like a house of cards when the bottom layer is knocked out. The whole structure comes crashing down.

The core collapses from the inside out.

Figure 14.12 *As a molecular cloud collapses, denser regions within the cloud collapse more rapidly than less dense regions. As this process continues, the cloud fragments into a number of very dense molecular cloud cores that are embedded within the large cloud. It is these cloud cores which may go on to form stars.*

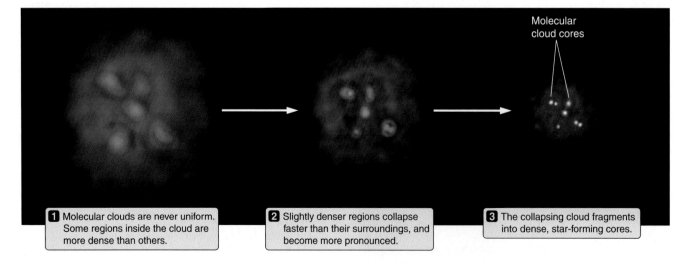

1 Molecular clouds are never uniform. Some regions inside the cloud are more dense than others.

2 Slightly denser regions collapse faster than their surroundings, and become more pronounced.

3 The collapsing cloud fragments into dense, star-forming cores.

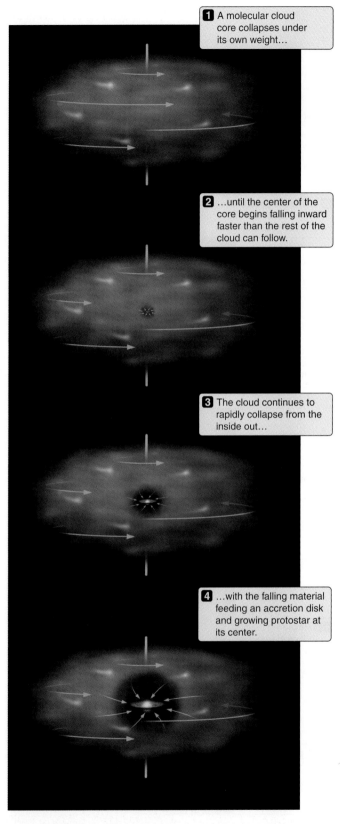

1 A molecular cloud core collapses under its own weight...

2 ...until the center of the core begins falling inward faster than the rest of the cloud can follow.

3 The cloud continues to rapidly collapse from the inside out...

4 ...with the falling material feeding an accretion disk and growing protostar at its center.

Figure 14.13 *When a molecular cloud core gets very dense, it collapses from the inside out. Conservation of angular momentum causes the infalling material to form an accretion disk that feeds the growing protostar.*

We have been here before. It was at this point in the story of star formation that our discussion of the formation of the Solar System in Chapter 5 began. An overview of the process is shown in **Figure 14.14.**

Our Solar System began in a molecular cloud.

Material from the collapsing molecular cloud core falls inward. Because of its angular momentum, this material lands on a flat, rotating accretion disk. Figure 5.1 shows images of several such disks. The dark bands in these images are the disks seen edge on, while the bright regions in the figure are starlight reflected from the surfaces of the disks. Most of this material eventually finds its way inward onto the growing protostar at the center of the disk, but a small fraction of it remains to become the stuff planets are made of. The last time we passed this way, we followed the evolution of the gas and dust left behind in the disk, and saw how it leads naturally to a planetary system with the properties of our Solar System. This time through we will instead follow the story of the protostar as it becomes a star.

14.4 THE PROTOSTAR BECOMES A STAR

We pick up the story of our protostar at a time when the cloud is still collapsing, and more and more material is falling on the disk. At this point, the surface of the protostar has also been heated to a temperature of thousands of kelvins as the gravitational energy of the original molecular cloud is converted into thermal energy. The protostar is also huge—hundreds of times larger than our Sun—which means that the surface of the protostar is tens of thousands of times larger than the surface of the Sun. Each square meter of that enormous surface is radiating energy away in accordance with Stefan's law. As a result, the protostar is thousands of times more luminous than our Sun, even though the nuclear reactions that will power it through its life have yet to begin.

Protostars are large and luminous.

At this stage in its life, a protostar is extremely luminous, and yet it can probably not be seen by astronomers in visible light. There are two reasons for this. First, the protostar is relatively cool as stars go, so most of its radiation is in the infrared part of the spectrum. Even more important, the protostar begins its life buried deep in the heart of a dense and dusty molecular cloud. The dust

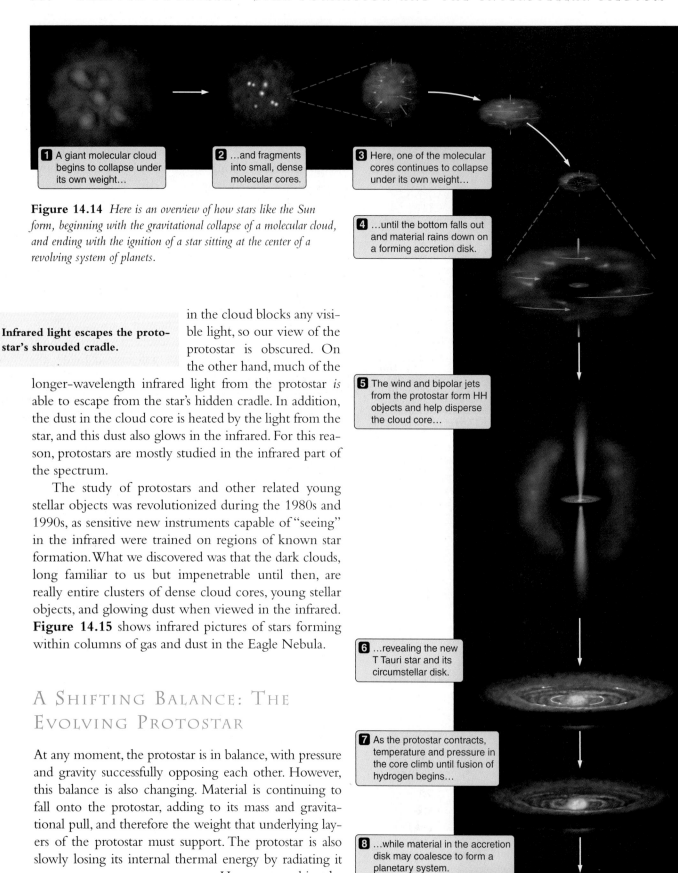

1 A giant molecular cloud begins to collapse under its own weight...

2 ...and fragments into small, dense molecular cores.

3 Here, one of the molecular cores continues to collapse under its own weight...

Figure 14.14 *Here is an overview of how stars like the Sun form, beginning with the gravitational collapse of a molecular cloud, and ending with the ignition of a star sitting at the center of a revolving system of planets.*

4 ...until the bottom falls out and material rains down on a forming accretion disk.

5 The wind and bipolar jets from the protostar form HH objects and help disperse the cloud core...

6 ...revealing the new T Tauri star and its circumstellar disk.

7 As the protostar contracts, temperature and pressure in the core climb until fusion of hydrogen begins...

8 ...while material in the accretion disk may coalesce to form a planetary system.

Infrared light escapes the protostar's shrouded cradle.

in the cloud blocks any visible light, so our view of the protostar is obscured. On the other hand, much of the longer-wavelength infrared light from the protostar *is* able to escape from the star's hidden cradle. In addition, the dust in the cloud core is heated by the light from the star, and this dust also glows in the infrared. For this reason, protostars are mostly studied in the infrared part of the spectrum.

The study of protostars and other related young stellar objects was revolutionized during the 1980s and 1990s, as sensitive new instruments capable of "seeing" in the infrared were trained on regions of known star formation. What we discovered was that the dark clouds, long familiar to us but impenetrable until then, are really entire clusters of dense cloud cores, young stellar objects, and glowing dust when viewed in the infrared. **Figure 14.15** shows infrared pictures of stars forming within columns of gas and dust in the Eagle Nebula.

A SHIFTING BALANCE: THE EVOLVING PROTOSTAR

At any moment, the protostar is in balance, with pressure and gravity successfully opposing each other. However, this balance is also changing. Material is continuing to fall onto the protostar, adding to its mass and gravitational pull, and therefore the weight that underlying layers of the protostar must support. The protostar is also slowly losing its internal thermal energy by radiating it away. How can an object be in perfect balance and yet be changing at the same time? To answer this question,

Gravity and pressure balance in the protostar . . .

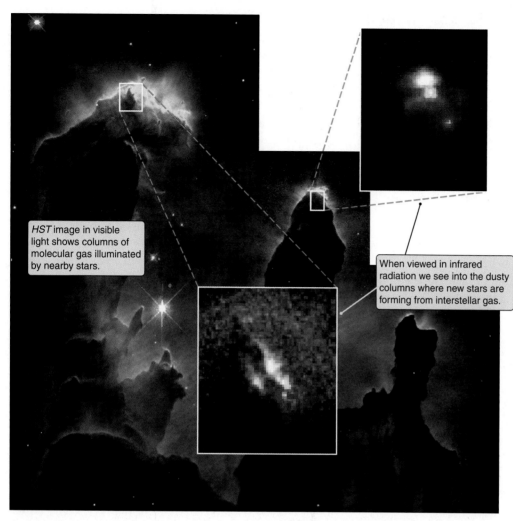

HST image in visible light shows columns of molecular gas illuminated by nearby stars.

When viewed in infrared radiation we see into the dusty columns where new stars are forming from interstellar gas.

Figure 14.15 *This* HST *image of the Eagle Nebula shows dense columns of molecular gas and dust at the edge of an H II region. Infrared images of the same field, also taken with the* HST, *show young stars forming within these columns.*

the protostar is always in balance, with the inward force of gravity matched by the outward force of pressure (see **Figure 14.16b**). But its balance, too, is changing with time. Just like sand that has fallen onto the spring balance, material that has fallen onto the protostar compresses the protostar more and more, and the protostar evolves.

As material continues to fall onto the protostar, it adds weight to the outer layers of the protostar. This growing weight compresses the material in the interior of the protostar, and as it is compressed, the interior of the protostar grows hotter. (Compressing a gas always causes it to heat up, and letting a gas expand always causes it to cool down. The cooling systems in your air-conditioner and refrigerator work by compressing gas to make it hot, then letting this hot gas cool so that when it expands again it gets really cold.) If the interior of the protostar is now denser and hotter, that also means that the pressure is higher—just enough higher to balance the increased weight of the material above it. Balance is always maintained.

Even after material stops falling on the protostar, the protostar *still* goes on evolving in much the same way. Earlier we saw that the protostar is radiating away many times as much energy as the Sun. The source of the energy being radiated away is the thermal energy trapped in the interior of the protostar—the same ther-

... but that balance is constantly changing.

The protostar radiates away thermal energy and shrinks.

consider a more everyday example. **Figure 14.16(a)** shows a simple spring balance, which works on the principle that the more you compress a spring, the harder the spring pushes back. If an object is placed on the balance, it compresses the spring until the downward force of the weight of the object is exactly balanced by the upward force of the spring. The weight of the object is measured by seeing at what point the pull of gravity and the push of the spring are equal. The situation is analogous to our protostar, but with the pressure of the gas in the protostar taking the place of the spring. (The analogy is a good one, because compressing the gas in the protostar causes its pressure to increase, just as compressing the spring causes the spring to push back harder.)

Let's now take our spring balance, and slowly pour sand on it. At any point, the downward weight of the sand is balanced by the upward force of the spring. But as the weight of the sand increases, the spring is slowly compressed. The spring and the weight of the sand are always in balance, but this balance is *changing with time* as more and more sand is added. In just the same way,

mal energy responsible for supporting the protostar against the forces of gravity. So, as thermal energy leaks out of the protostar, gravity gains the upper hand, and the protostar slowly contracts. As the protostar shrinks, the forces of gravity become greater (because the parts of the protostar are closer together—remember the inverse square law of gravity). Meanwhile, as the protostar shrinks, its interior is compressed, and so it grows hotter. As density and temperature climb, so does the pressure—*just enough to continue to counteract the growing force of gravity.* Balance between gravity and pressure is always maintained. This process continues, with the star becoming smaller and smaller and its interior growing hotter and hotter, until the center of the star is finally hot enough for fusion of hydrogen into helium to begin.

Something strange is going on here. When an object radiates energy, it normally gets cooler. When you turn off the electric coil on your stove, the coil cools as it loses the thermal energy within it. Yet we just concluded that as a protostar radiates thermal energy away, it actually grows hotter. How can that be? Once again, the answer lies with the concept of conservation of energy. As the protostar contracts, every part of the protostar is slowly falling toward its center, which means that the protostar is *losing gravitational energy.* This gravitational energy has to show up in some other form, and it does so as thermal energy. Some of this thermal energy is radiated away, but not all of it, so the interior of the star grows hotter.

This chain of events is shown in a nutshell in **Figure 14.17.** The protostar radiates away thermal energy, and as it does so, it loses pressure support. Deprived of pres-

Conversion of gravitational energy heats the shrinking protostar.

Figure 14.16 (a) *A spring balance comes to rest at the point where the downward force of gravity is matched by the upward force of the compressed spring. As sand is added, the location of this balance point shifts. (b) Similarly, the structure of a protostar is determined by a balance between pressure and gravity. Like the spring balance, the structure of the protostar constantly shifts as the protostar radiates energy away and additional material falls onto its surface.*

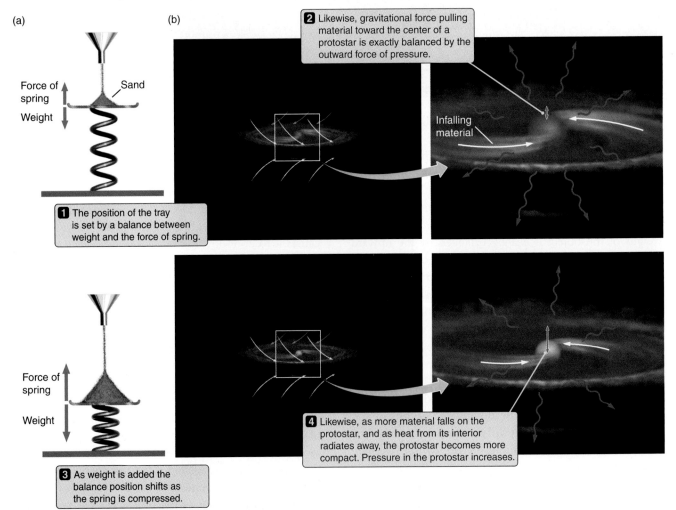

(a)

(b)

Force of spring

Weight

Sand

1 The position of the tray is set by a balance between weight and the force of spring.

2 Likewise, gravitational force pulling material toward the center of a protostar is exactly balanced by the outward force of pressure.

Infalling material

Force of spring

Weight

3 As weight is added the balance position shifts as the spring is compressed.

4 Likewise, as more material falls on the protostar, and as heat from its interior radiates away, the protostar becomes more compact. Pressure in the protostar increases.

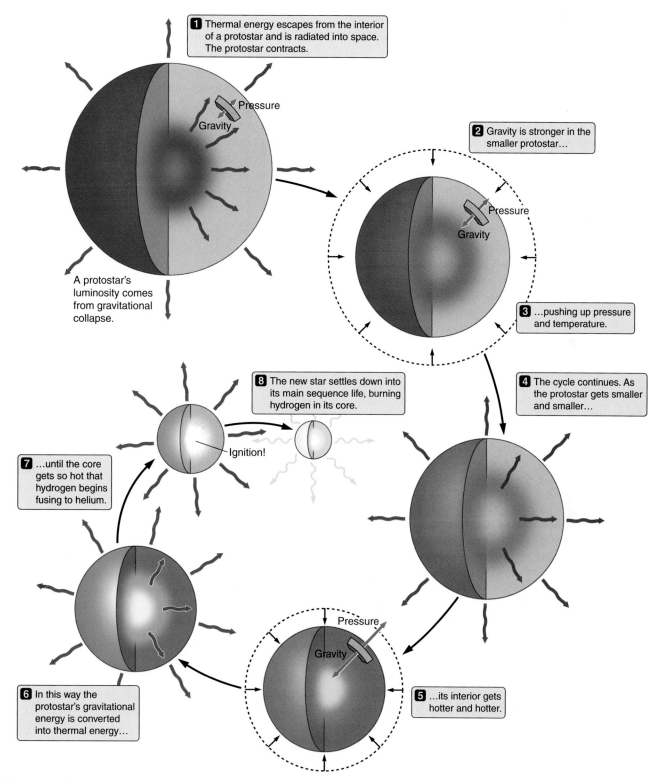

Figure 14.17 *As a protostar radiates away energy, it loses pressure support in its interior and contracts. This contraction drives the interior pressure up. The counterintuitive result is that radiating energy away causes the interior of the protostar to grow hotter and hotter until nuclear reactions begin in its interior.*

Fusion begins and the protostar becomes a star.

energy, driving up the temperature in the protostar's

sure support, the protostar contracts. As the protostar contracts, gravitational energy is converted to thermal energy, driving up the temperature in the protostar's

interior. If the protostar is massive enough, then the temperature in the interior of the protostar will eventually become hot enough for nuclear fusion to begin. Here is the point at which the transition from protostar to star takes place. The distinction between the two is that a *protostar* draws its energy from gravitational collapse, while a *star* draws its energy from thermonuclear reactions in its interior.

As noted, it is the protostar's mass that determines whether it will ever become a star. As the protostar slowly collapses, the temperature at its center gets higher and higher. If the protostar is more massive than about 0.08 $M_\odot$, the temperature in its core will eventually reach the 10-million-kelvin mark, and fusion of hydrogen into helium will begin. When this happens, the newly born star will once again adjust its structure until it is radiating energy away from its surface at just the rate energy is being liberated in its interior. As it does so, it "settles" onto the main sequence of the H-R diagram, where it will spend the majority of its life. If the protostar has a mass of less than 0.08 $M_\odot$, it will never reach the point where nuclear burning takes place. Such a failed star, the first examples of which were discovered in the closing years of the 20th century, is called a **brown dwarf**. A brown dwarf is neither star nor planet, but something in between. It forms the same way a star forms, yet in many respects it is like a giant Jupiter. Brown dwarf candidates are seen as especially cool M stars, and as members of the recently-introduced L and T spectral types discussed in Chapter 12. A brown dwarf never burns hydrogen in its core, but instead glows by continuously cannibalizing its own gravitational energy. As the years pass, the brown dwarf gets progressively smaller and fainter.

At least 0.08 $M_\odot$ is needed to be a star.

Brown dwarfs are failed stars.

EVOLVING STARS AND PROTOSTARS FOLLOW "EVOLUTIONARY TRACKS" ON THE H-R DIAGRAM

Within the evolving protostar it is convection rather than radiation that carries energy outward, keeping the protostar's interior well-stirred. Interestingly, while the interior of the protostar grows hotter and hotter as it contracts, its *surface* stays at about the same temperature through most of this phase of its evolution. This distinction is important. The surface temperature of a star or protostar is *not* the same as the temperature deep in its interior. For example, we found that the temperature of the surface of the Sun is

The protostar's surface temperature does not change much.

5,860 K, while the temperature of its interior is millions of kelvins. As the protostar contracts, the temperature deep within the star grows hotter and hotter, but the temperature of the star's *surface* remains nearly constant.

In the 1960s the Japanese theoretician Chushiro Hayashi explained why this is so. Hayashi pointed out that the atmospheres of stars and protostars contain a natural thermostat: the H^- ion. (An H^- or "H minus" ion is a hydrogen atom that has acquired an extra electron.) The amount of H^- in the atmosphere of a protostar is highly sensitive to the temperature at the surface of the protostar. The cooler the atmosphere of a star, the more slowly atoms and electrons are moving, and the easier it is for a hydrogen atom to hold on to an extra electron. As a result, the cooler the atmosphere of the star, the more H^- there is.

The H^- ion, in turn, helps to control how much energy is radiated away by a star or protostar. The more H^- there is in the atmosphere of a star or protostar, the more opaque the atmosphere is, and the more effectively the thermal energy of the protostar is trapped in its interior. Imagine that the surface of the protostar is too cool. Too cool means that extra H^- forms in the atmosphere, making the atmosphere of the protostar more opaque. This traps more of the radiation that is trying to escape, and the trapped energy heats up the star. As the temperature climbs, H^- ions are destroyed (that is, changed to neutral H atoms). Now imagine the other possibility, that the protostar is too hot. Then H^- in its atmosphere will be destroyed, so the atmosphere will become more transparent, allowing radiation to more freely escape from the interior. Since the protostar cannot hold on to enough of its energy to stay warm, the surface cools. In either case—too cold or too hot—H^- is formed or destroyed until the star's atmosphere once again traps *just* the right amount of escaping radiation.

H^- is a stellar thermostat.

The H^- ion is basically doing the same job that you do with your bed covers at night. If you get too cold, you pile on extra covers to trap your body's thermal energy and keep you warm (more H^- ions). If you get too hot, you kick some covers off so you will cool off (fewer H^- ions). It is a shame that we have no such "automatic cover" to maintain our body temperature at night as effectively as the H^- ion controls the surface temperature of a protostar.

The amount of H^- in the atmosphere of the protostar keeps the surface temperature of the protostar somewhere between about 3,000 and 5,000 K, depending on the mass and age of the protostar. Because the temperature of the star is not changing much, the amount of energy per unit time (power) radiated away by each square meter of the surface of the star does not change much either. Recall Stefan's law from Chapter 4, which says that the amount of power radiated by each square meter of the star's surface is determined by its temperature. But as the star shrinks, the area of its surface shrinks as well. There are fewer square meters of surface to radiate, so the luminosity of the protostar drops. As viewed from the outside, the protostar stays nearly the same temperature and color, but gradually gets fainter as it evolves toward its eventual life as a main sequence star.

In Chapter 12 we introduced the H-R diagram, and used the H-R diagram to begin to understand how the properties of stars differ. For the next several chapters we will also use the H-R diagram to keep track of how stars change as they evolve through their lifetimes.

> **Evolutionary tracks are plotted on the H-R diagram.**

The path across the H-R diagram that a star follows as it goes through the different stages of its life is called the star's **evolutionary track.** The particular path that an evolving protostar follows as it approaches the main sequence is called its **Hayashi track.** The protostar is brighter than it will be as a true star on the main sequence, and so a protostar's Hayashi track is located above the main sequence on the H-R diagram. Since the surface temperature of the protostar stays nearly constant as the protostar contracts, the protostar's Hayashi track prior to the beginning of hydrogen burning is an almost vertical line on the H-R diagram. **Figure 14.18** shows the pre–main sequence evolutionary tracks of stars of several different masses. Astronomers say that an evolving protostar "descends the Hayashi track."

NOT ALL STARS ARE CREATED EQUAL

In Chapter 12 we found that stars can have a wide range of masses, varying from less than a 10th the mass of the Sun up to perhaps 100 times the mass of the Sun. What determines how massive a star will be? The answer to this question is unclear, and is a topic of a great deal of ongoing research. One obvious possibility is that a

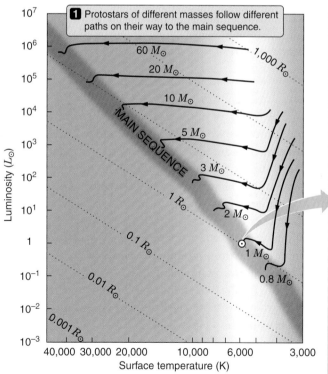

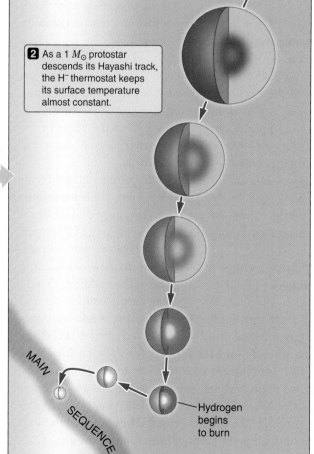

Figure 14.18 *The evolution of pre–main sequence stars can be followed on the H-R diagram. Protostars in the upper right portion of the diagram are large and cool. The roughly vertical, constant-temperature parts of the evolutionary tracks of low-mass protostars are referred to as "Hayashi tracks."*

forming star grows until it uses up all of the gas around it: it becomes no larger simply because it has run out of material. While this explanation would be easy to understand, it does not match our observations of how star formation actually takes place. At this point in our story, the contracting protostar is at the center of a collapsing molecular cloud core, which in turn is a denser-than-average region inside a molecular cloud whose total mass may be hundreds of thousands of times greater than the mass of the Sun. Observations indicate that under most circumstances, star formation is a very inefficient process. Only a small fraction of the material in a molecular cloud—perhaps a few percent—ends up as part of the stars forming within it. Something must prevent most of the material in a molecular cloud from ever actually falling onto protostars. There are a number of ideas as to what this something might be. One intriguing possibility is that forming stars control their own masses.

Star formation is inefficient.

When observing young stellar objects, it is common to find clear signs that material is falling onto the central protostar and accretion disk, as discussed earlier. However, it is also common to observe powerful flows of material moving *away* from forming stars at the same time that material is being accreted *onto* the stars. How can this be? **Figure 14.19** shows how this happens. Material falls onto the accretion disk, and moves inward toward the equator of the star, while at the same time material is blown away from the protostar and disk in the two opposite directions away from the plane of the disk. The resulting stream of material away from the protostar is called a **bipolar outflow.**

Mass flows simultaneously onto and away from the protostar.

Some bipolar outflows from young stellar objects are slow and fairly disordered, but others produce remarkable "jets" of material that moves away from the central protostar and disk at velocities of hundreds of kilometers per second. These jets flow out into the interstellar medium where they heat, compress, and push away surrounding interstellar gas. Knots of glowing gas accelerated by jets are referred to as **Herbig-Haro objects** (or **HH objects** for short), named after the two astronomers who first identified them, and associated them with star formation. **Figure 14.20** shows a *Hubble Space Telescope* image of the first HH objects discovered, HH1-2. These two HH objects are the two sides of the bipolar outflow from a single source.

Protostars drive powerful bipolar outflows.

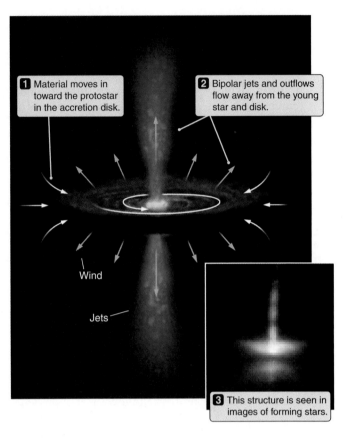

Figure 14.19 *Material falls onto an accretion disk around a protostar, then moves inward, eventually falling onto the star. In the process, some of this material is driven away in powerful jets that stream perpendicular to the disk.*

The origin of outflows from protostars is not as well understood as we would like, but current models suggest that they are the result of magnetic interactions between the protostar and the disk. The interior of a protostar on its Hayashi track is convective. Great cells of hot gas are rising from the interior of the star, while other cells of cooler gas are falling toward the center. This convection, coupled with the protostar's rapid rotation, can lead to the formation of a dynamo, similar to the dynamo that drives the Sun's magnetic field. The dynamo in the center of a protostar would be much more powerful than the Sun's dynamo, however. The protostar's resulting strong magnetic field might cause the protostar to begin blowing a powerful wind. It might also act something like the blade in a blender, tearing at the inner edge of the accretion disk and flinging material off into interstellar space.

Protostars may be powerful dynamos.

Powerful protostellar winds, jets, and other outflows could disrupt the cloud core and accretion disk from

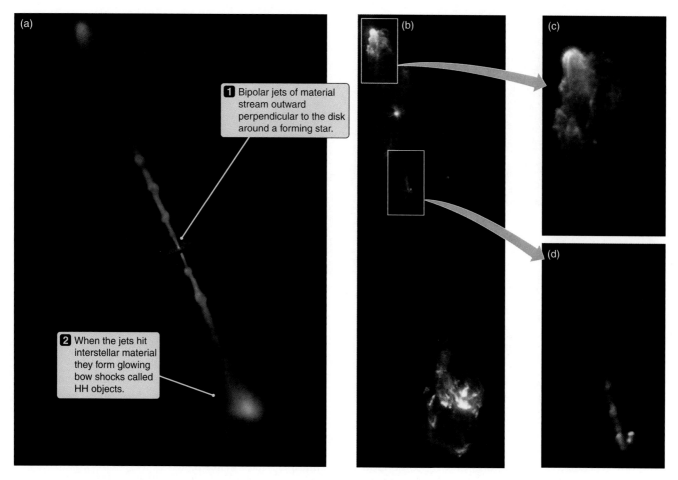

1 Bipolar jets of material stream outward perpendicular to the disk around a forming star.

2 When the jets hit interstellar material they form glowing bow shocks called HH objects.

Figure 14.20 (a) *Jets from protostars slam into surrounding interstellar gas, heating the gas and causing it to glow.* (b) *This* HST *image shows the bow shocks formed at the ends of a bipolar jet from a protostar. Enlargements of the bow shock* (c) *and jet* (d) *are shown at right. Only one side of the jet itself can be seen because the other side is hidden behind the dark cloud in which the star is embedded.*

which the protostar formed, shutting down the flow of material onto the protostar. Up until the time that the protostar wind begins, the protostar is enshrouded in the dusty molecular cloud core from which it was born. As the wind from the protostar disperses this obscuring envelope, we get our first direct, visible–light view of the protostar: the protostar is "revealed." Once the contracting protostar makes its appearance, it is referred to as a **T Tauri star.** This name comes from the star labeled T in the constellation Taurus. The star T Tauri was the first recognized member of this class of objects.

Astronomers have long known that stars are often found together in closely knit collections called

Stars in clusters form together.

star clusters. Figure 14.21 shows one such star cluster, called the Pleiades, or Seven Sisters. If you have normal eyesight (or a pair of binoculars), on a clear winter

night in the Northern Hemisphere you can see the brightest of the stars in this cluster as a tight bunch in the constellation Taurus. Star clusters provided astronomers with their first clues that many stars of all different masses can form together at the same place and at about the same time. When we look around the Galaxy at large, we see a hodgepodge collection of stars, some that are very old, and others very young. If these were the only stars we had to study, it would be extremely difficult to learn much about how stars evolve. Star clusters, on the other hand, are large collections of stars that all formed *at the same time, in the same place, and from the same material*. They provide us with tailor-made samples to study star formation.

Even though the few brightest and most-massive stars in a cluster dominate what we see, the vast majority of stars in a cluster are much lower mass stars like our Sun. In fact, some star-forming regions do not seem to form any especially massive stars at all. There is

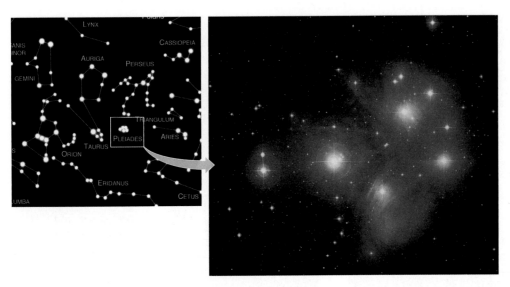

Figure 14.21 *If you have good eyes (or a pair of binoculars), on a clear winter night you can see the brightest of the stars in this cluster—a tight bunch in the constellation Taurus called the Pleiades, or Seven Sisters. The diffuse blue light around the stars is starlight scattered by interstellar dust.*

great interest among astronomers in how and why molecular clouds subdivide themselves into low- and high-mass stars. The details of this division—specifically, what fraction of newly formed stars will be of what masses—are crucial if we are to use observations of the stars around us today to untangle the history of star formation in our Galaxy. Unfortunately, we are still far from a detailed understanding of why some cloud cores become 1 $M_\odot$ stars while others become 10 $M_\odot$ stars.

Following a cloud-core collapse, the subsequent evolution of a protostar is determined largely by its mass. Calculations suggest that a star with the mass of our Sun probably takes about 10 million years or so to descend the Hayashi track and become a star on the main sequence. If we look at the entire history, including the collapse and fragmentation of the molecular cloud itself, the total time is probably more like 30 million years. More-massive stars go through this process much faster. A 10 $M_\odot$ star might go from the stage of being a molecular cloud core to burning hydrogen in its interior in only 100,000 years. A 100 $M_\odot$ star might make the journey in less than 10,000 years. By comparison, a 0.1 $M_\odot$ star might take 100 million years to finally reach the main sequence.

Star formation may take millions of years.

The 30 million years or so that it took for our Sun to form is a long time, but it is a tiny fraction of the 10 billion years which the Sun will spend steadily fusing hydrogen into helium as a main sequence star. It is no wonder that so few of the stars that we see in the sky are such young objects. But every star that we see was young *at one time,* including our own Sun.

Not surprisingly, many questions about star formation remain. One question is, "How must we modify our story to account for the formation of binary stars or other multiple star systems?" When we observe the sky, we find that about half of the stars we see are part of multiple star systems. At what point during star formation is it determined that a collapsing cloud core will form several stars instead of just one? Some models suggest that this split may happen early on in the process, during the fragmentation and collapse of the molecular cloud. The advantage of these ideas is that they provide a natural way of dealing with much of the angular momentum of the cloud core; it goes into the orbital angular momentum of the stars about each other. Other models suggest that additional stars may form from the accretion disk around an initially single protostar.

Often multiple stars form.

The picture of star formation presented here is remarkably complete, considering that we have never visited a protostar or watched a star form. Instead, astronomers have observed many different stars at different stages in their formation and evolution, then used their knowledge of physical laws to tie these observations together into a coherent, consistent description of how, why, and where stars form. This two-pronged attack—using observations to see what things exist in the Universe, and using physics to understand how they work and the relationships between them—is how all astronomy (and in a certain sense, all physical science) works.

SEEING THE FOREST THROUGH THE TREES

We began our discussion of the Solar System in Chapter 5 by describing its formation. Nearly five billion years ago, a vast interstellar cloud collapsed under its own weight to form a swirling disk of gas and dust. Within that disk, small objects stuck together to become parts of larger objects, culminating in the formation of the planets and other solid bodies of our Solar System. We now can see that earlier story—and in some sense our entire discussion of the Sun, Earth, and Solar System—as a "sidebar" to a larger story, the thread of which we picked up again on this leg of our journey. The tenuous expanse of gas and dust that fills the vast reaches of interstellar space sets the stage for the ongoing formation of generations upon generations of stars and planetary systems, of which ours is but one among thousands of billions of billions in the Universe.

The interstellar medium is a difficult topic, even for experts in the field. The time that we have for our journey is limited. To go beyond the basic description of the interstellar medium here would be a long and arduous excursion. It would also quickly move onto more speculative ground. For now, we suggest that you think of the interstellar medium as you might a complex and subtle ecosystem. It is one thing to catalog the inhabitants of an ecosystem, but it is quite another to understand the interrelationships and dependencies among those organisms well enough to say that you know how the ecosystem works. In like fashion, astronomers at the turn of the 21st century have a fairly complete picture of what makes up the interstellar medium, but do not fully understand the complex interplay between the phases of the interstellar medium and the stars that are born and die there. Even so, we have learned some important lessons, and have developed some important tools. For example, 21-cm radiation from ubiquitous interstellar hydrogen—radiation that easily penetrates the dust that obscures our view of the Universe at visible wavelengths—gives us the tool to see out to the far reaches of our Galaxy and map out our home in the Universe.

Fortunately, our understanding of stars is far more complete. In fact, the workings of stars represent such a nicely posed physics problem that we would probably include them on our journey even if this were a pure physics text rather than an astronomy text. Many of the physical principles employed in our discussion of star formation and the evolution of protostars in this chapter should have sounded familiar. The balance between gravity and pressure that gives a protostar its structure is the same balance at work within the Sun. The chain of physical reasoning that we used to understand the collapse of a protostar (gravitational energy that is converted to thermal energy and then radiated away) applies almost unmodified to understanding the ongoing (albeit slow) collapse of Jupiter.

Looking forward in our journey, the physical insight that we have developed will continue to serve us well. Stars are temporal objects. They are born from interstellar gas, they shine by the light of nuclear fires deep within their hearts, and when they exhaust their fuel, they die. The changing balance of the protostar is only the first chapter in a process of evolution that continues throughout the star's life. We found it convenient to follow the changes taking place within an evolving protostar by tracking its progress across the face of the H-R diagram. In so doing we discovered the best way that astronomers have found to draw a road map of a star's life. That is the road map that will be our guide for the next stretch of our journey.

STUDENT QUESTIONS

THINKING ABOUT THE CONCEPTS

1. Clouds in our sky are composed of water condensed from atmospheric water vapor. How does the process that forms our clouds compare with the process that forms interstellar hydrogen clouds from the interstellar medium?

2. The interstellar medium is approximately 99% gas and 1% dust. Why is it the dust and not the gas that blocks our visible-light view of the galactic center?

3. Using your own experience of walking or riding in a dense fog, explain how the existence of large amounts of interstellar dust influenced the views held by 19th-century astronomers regarding the structure of our Galaxy.

4. Explain how the development of infrared astronomy has made it possible for astronomers to study the detailed processes of star formation.

5. If you placed your hand in boiling water (100°C) for even one second, you would get a very serious burn. If you placed your hand in a hot oven (200°C) for a second or two, you would hardly feel the heat. Explain why this is so, and how it

relates to million-Kelvin regions of the interstellar medium.

6. Molecular clouds are much more massive than the stars they form, and many stars can form from a single large cloud. What does this imply about the environment in which our Sun and Solar System were born?

7. Stellar radiation can convert atomic hydrogen (H I) to ionized hydrogen (H II).
   a. Why does a B8 main sequence star ionize far more interstellar hydrogen in its vicinity than a K0 giant of the same luminosity?
   b. What properties of a star are important in determining whether it can ionize large amounts of nearby interstellar hydrogen?

8. When a hydrogen atom is ionized, a single particle becomes two particles.
   a. Identify the two particles.
   b. If both particles have the same kinetic energy, which moves faster?

9. Explain how the important discovery of 21-cm radio emission has allowed us to detect interstellar clouds of neutral hydrogen (H I), even when large amounts of interstellar dust are in the way.

10. Molecular hydrogen is very difficult to detect from the ground, but we can easily detect carbon monoxide (CO) by its 2.6-cm microwave emission. Describe how observations of CO might help astronomers infer the amounts and distribution of molecular hydrogen within giant molecular clouds?

11. You can think of a brown dwarf as a failed star, that is, one lacking sufficient mass for nuclear reactions to begin. What similarities and differences do you see between a brown dwarf and a giant planet such as Jupiter? Would you classify a brown dwarf as a super-giant planet? Explain your answer.

APPLYING THE CONCEPTS

12. The mass of a proton is 1,850 times the mass of an electron. If a proton and an electron have the same kinetic energy, K.E. = $\frac{1}{2} mv^2$, how many times greater is the velocity of the electron than that of the proton?

13. If a typical hydrogen atom in a collapsing molecular cloud core starts at a distance of 10,000 AU (1.5×10^{12} km) from the core's center and falls inward at an average velocity of 1.5 km/s, how many years does it take to reach the newly forming protostar? Assume that a year is 3×10^7 s.

14. From Table 12.1 we can see that the ratio of hydrogen atoms (H) relative to carbon atoms (C) in the Sun's atmosphere is approximately 2,400 to 1. It would be reasonable to assume that this ratio also applies to molecular clouds. If 2.6-cm radio observations indicate 100 $M_\odot$ of carbon monoxide (CO) in a giant molecular cloud, what is the implied mass of molecular hydrogen (H_2) in the cloud? (Carbon represents $\frac{3}{7}$ of the mass of a CO molecule.)

15. Neutral hydrogen emits radiation at a radio wavelength of 21 cm when an atom drops from a higher energy magnetic state to a lower energy magnetic state. On average, each atom remains in the higher energy state for 11 million years (3.5×10^{14} s).
    a. What is the probability that any given atom will make the transition in 1 second?
    b. If there are 6×10^{59} atoms of neutral hydrogen in a 500 $M_\odot$ cloud, how many photons of 21-cm radiation will the cloud emit each second?
    c. How does this compare with the 1.8×10^{45} photons emitted each second by a solar-type star?

16. The Sun took 30 million years to evolve from a collapsing cloud core to a star, with 10 million of those years spent on the Hayashi track. It will spend a total of 10 billion years on the main sequence. Suppose we were to compress the Sun's main sequence lifetime into just a single year.
    a. How long would the total collapse phase last?
    b. How long would it spend on the Hayashi track?

It is said an Eastern monarch once charged his wise men to invent him a sentence to be ever in view, and which should be true and appropriate in all times and situations. They presented him the words: "And this, too, shall pass away."

ABRAHAM LINCOLN (1809–1865),
SEPTEMBER 30, 1859

STARS IN THE SLOW LANE

15.1 THIS, TOO, SHALL PASS AWAY

If you go outside tonight and look up, you will not see the same sky that you would have seen a year ago. The stars themselves probably have not changed, and the Moon might even be in the same place and same phase. Nonetheless, what you see will be radically different, because *the way you see it* has changed. A year ago, the sky was probably full of points of light, called stars, that were "somewhere off in space." Now you see an expanse of distant suns, each farther away than the mind can easily comprehend, and each shining with a devastating brilliance powered by a nuclear inferno deep in its heart. When you look at the dark splotches that mar the Milky Way's eerie glow, you see vast clouds of interstellar gas and dust. In your mind's eye you look inside these clouds at the new suns and new solar systems being born there, and imagine the birth of our own planet and the star we orbit, 4.6 billion years ago. The sky is like a page from a book. To those who cannot read, written words are so many hen scratches, but to those who can, thoughts and words and ideas spring off the page to touch our hearts and stimulate our minds and imaginations. The sky is a book worth reading, as is the rest of nature. At this stage in our journey we are well on our way to becoming literate.

KEY CONCEPTS

Within its core, the Sun fuses over 4 billion kilograms of hydrogen to helium each second, and while the Sun may seem immortal by human standards, eventually it will run out of fuel. When it does, some 5 billion years from now, the Sun's time on the main sequence will come to an end. As we examine what happens when a low-mass star like the Sun nears the end of its life, we will find that:

* The more massive a star, the shorter its lifetime;
* We can follow the post–main sequence evolution of stars by tracing their paths on the H-R diagram;
* When the Sun runs out of fuel at its center, it will grow into a larger, more luminous red giant star;
* Red giants, and some other evolved stars, are shaped by dense "degenerate" cores in which atoms have been crushed by gravity;
* Evolving low-mass stars go through a series of stages, eventually coming to burn helium to carbon, and building up a core of carbon ash;
* In the end, a low-mass star will eject its outer layers, possibly forming a planetary nebula, and leaving behind a tiny degenerate white dwarf; and
* Low-mass stars in binary systems may experience more exciting fates as novae or supernovae.

So far on our journey, we have seen interstellar clouds collapse under the force of gravity to form immense protostars surrounded by swirling disks of gas and dust. We have watched as dust that is left behind in the disk accumulates into planets and moons, asteroids and comets. We have followed along as protostars continue their collapse until the nuclear fuel in their cores ignites, and we have recognized a grand pattern—the main sequence—in the stars which those protostars go on to become. We have even looked inside one of these stars, our Sun, and come to appreciate the battle between gravity and pressure that gives our Sun its structure. Our understanding of that battle provides our understanding of all stars along the main sequence. A star's mass sets the strength of its gravity, and the need to balance that gravity in turn sets the pressure in the star's interior. The more massive the star, the higher the pressure needed to hold it up, and the more rapidly the star must burn its nuclear fuel to support its own weight. From luminous O stars on the hot end of the main sequence to faint M stars on the cool end, mass is the fundamental and overriding property that makes a main sequence star what it is.

We have seen stars born, and seen how they live. Now the time has come to watch them die. Like all main sequence stars, the Sun gets its energy by converting hydrogen to helium in its core. This is what *defines* the main sequence: being on the main sequence *means* that a star is burning hydrogen in its core. But a star cannot remain a main sequence star forever.

Stars eventually exhaust their nuclear fuel.

Hydrogen at the core of a star is a consumable resource. Any star eventually exhausts this resource—it "runs out of gas"—and when it does, its structure begins to change dramatically. Just as the balance between pressure and gravity within a protostar constantly changes as it descends the Hayashi track toward the main sequence, new balances must also constantly be found as a star evolves beyond the main sequence, until at last no balance is possible at all.

The mass and composition of a star control the star's life on the main sequence, and they remain at center stage in the closing acts of the star's story as well. The

Mass and composition determine a star's fate.

evolutionary course followed by each of the hundred billion or so stars that make up the Milky Way Galaxy (and each of the stars in every other galaxy throughout the Universe) is locked in place when the star forms, determined foremost by the seemingly incidental fact of the amount of mass incorporated into the star at the time of its birth, and secondarily by the chemical composition of the material from which it formed.

Each star is unique. Relatively minor differences in the masses and chemical compositions of two stars can sometimes result in significant, and possibly even dramatic, differences in their fates. The course followed by a star with a mass of 1.1 $M_\odot$ is *not* identical to the fate of a star with a mass of 0.9 $M_\odot$. Despite that fact, stars can be divided roughly into two broad categories

Low-mass and high-mass stars evolve differently.

whose members evolve in qualitatively different ways. Massive, luminous O and B stars follow a fundamentally different course than cooler, fainter, less massive stars found toward the lower right end of the main sequence. These stars, which have masses less than about 3 $M_\odot$, are referred to as **low-mass stars,** and are typified by our Sun. In this chapter we begin our discussion of stellar evolution by examining the stages through which low-mass stars progress, and what better place to start than by asking what fate awaits our own Sun?

15.2 THE LIFE AND TIMES OF A MAIN SEQUENCE STAR

In Chapter 13 we learned that the structure of the Sun is determined by a balance between the inward force of gravity and the outward force of the pressure. The pressure within the Sun is in turn maintained by energy released by nuclear fusion in the heart of the star. If you were to add mass to the Sun, the weight of material

The balance between pressure and gravity controls temperature and density in stars.

pushing down on the inner regions of the star would increase. Gravity would gain an advantage. The inner parts of the Sun would be compressed by the added weight, driving up the temperature and density there. This increase in temperature and density would in turn cause the nuclear reactions occurring there to run faster.

We have followed this chain of reasoning before, but it is so crucial to what is to come that it is worth reviewing here. Increasing the temperature and pressure at the center of a star means several things. It means that more atomic nuclei are packed together into a smaller volume. You know that you are far more likely to bump into another person while strolling through a crowded shopping mall than through an empty park. Similarly, packing atomic nuclei more tightly

Increasing temperature and pressure speeds up nuclear burning.

together increases the likelihood that they will run into each other and fuse. Higher density means more frequent collisions between atomic nuclei, and more collisions means faster burning. Higher temperature also drives up the rate of nuclear reactions: Higher temperature means that atomic nuclei are moving faster, so they are more likely to bump into each other. More important, higher temperature means that atomic nuclei collide with each other more *violently*, making it more likely they will overcome the electric repulsion which pushes the positively charged nuclei apart. As a result of the combined effects of temperature and density, modest increases in the pressure within a star can sometimes lead to dramatic increases in the amount of energy released by nuclear burning.

Here is the key to understanding why the main sequence is primarily a sequence of masses, with low-mass stars on the faint end and

> **A higher mass main sequence star *must* be hotter and more luminous.**

high-mass stars on the luminous end. More mass means stronger gravity, stronger gravity means higher temperature and pressure in the star's interior, higher temperature and pressure mean faster nuclear reactions, and faster nuclear reactions mean a more luminous star. If the Sun were more massive, it would lead to a different balance between gravity and pressure—a balance in which the Sun would burn its nuclear fuel more rapidly, and so would be more luminous. In other words, if the Sun were more massive, it would be located at a different position on the H-R diagram, farther up and to the left on the main sequence. Mass determines the structure of a star and its place on the main sequence. This, in a nutshell, is the heart of our understanding of stellar structure.

HIGHER MASS MEANS A SHORTER LIFETIME

A main sequence star can only live for so long. But how long is "long"? The question "How long can a main sequence star continue to burn hydrogen in its core?" is much like the question "How long can you drive your car before it runs out of gas?" In the case of the car, the answer depends in part on how much gas your tank holds. The larger the gas tank, the more fuel you have, and so the longer your motor might run. But the answer also depends on the size and efficiency of your motor. An eight-cylinder, four-barrel-carburetor, supercharged 409-cubic-inch gas-guzzler drinks fuel a lot faster than a motor scooter. The amount of time your motor will run is determined by a competition between these two effects. The larger motor might run out of gas first, even if it is attached to a much larger tank.

The competition between these two effects—tank size and motor size—is most readily expressed as a ratio. How long your motor runs is given by the amount of gas in the tank, divided by how quickly the motor uses it:

$$\begin{matrix} \text{Lifetime of} \\ \text{tank of gas} \\ \text{(hours)} \end{matrix} = \frac{\begin{matrix}\text{amount of fuel}\\\text{(gallons)}\end{matrix}}{\begin{matrix}\text{rate fuel is used}\\\text{(gallons/hour)}\end{matrix}}$$

If you have a 15-gallon tank and your motor is burning fuel at a rate of 3 gallons each hour, then your motor will use up all of the gas in just

$$\frac{15 \text{ gallons}}{3 \text{ gallons/hour}} = 5 \text{ hours.}$$

The same principle works for main sequence stars. The "amount of fuel" is determined by the mass of the star. The more massive the star, the more hydrogen there is available to power nuclear burning. The "rate at which fuel is used" is

> **How quickly a star runs out of fuel depends on its mass and luminosity.**

measured by the luminosity of the star. Main sequence stars are "in balance," so energy is radiated into space from the surface of the star at the same rate at which energy is being generated in its core. (This balance between energy generation and luminosity remains true at almost every stage of a star's evolution.) If one main sequence star has twice the luminosity of another, then it must be burning hydrogen at twice the rate as well.

An expression for the lifetime of the star looks very similar to our previous expression for the time it takes for your car to run out of fuel:

$$\begin{matrix}\text{Lifetime}\\\text{of star}\end{matrix} = \frac{\begin{matrix}\text{amount of fuel}\\(\propto \text{ mass of star})\end{matrix}}{\begin{matrix}\text{rate fuel is used}\\(\propto \text{luminosity of star})\end{matrix}}$$

If we use what we know about how much hydrogen must be converted into helium each second to produce a given amount of energy, as well as the fraction of its hydrogen that a star burns, this equation says that the length of time a star spends on the main sequence is given approximately by

$$\tau_{MS} = (1.0 \times 10^{10}) \times \frac{M\ (M_\odot)}{L\ (L_\odot)} \text{ years.}$$

By definition, the Sun has a mass M equal to $1\ M_\odot$ and a luminosity L equal to $1\ L_\odot$. Plugging these numbers into this equation tells us that our Sun will use up the hydrogen in its core after spending 10 billion years on the main sequence. This number, written τ_{MS} in the equation above, is called the **main sequence lifetime** of the star in question, in this case our Sun.

Now compare this number with the lifetime of a more massive star. The relationship between mass and luminosity of stars is very sensitive. Relatively small differences in the masses of stars result in large differences in their main sequence luminosities. From **Table 15.1,** or by referring back to Figure 12.16, you can see that a main sequence O5 star has a mass 60 times the mass of the Sun. But a 60 $M_\odot$ main sequence star is *not* just 60 times as luminous as the Sun, it is 794,000 times as luminous! Putting in $M = 60\ M_\odot$ and $L = 794,000\ L_\odot$ means that

$$\tau_{\text{MS}}\ (\text{O5 star}) = (1.0 \times 10^{10})\ \frac{60}{794,000}$$
$$= 8 \times 10^5\ \text{years.}$$

So instead of the 10 billion year lifespan of the Sun, an O5 star has a main sequence lifetime of *less than 1 million years!* Several generations of O stars have lived and died in the time that hominids have walked the surface of Earth. Even though the 60 $M_\odot$ star starts out with 60 times as much fuel as the Sun, it burns that fuel so much faster that it uses it up in less than a ten-thousandth the time.

Low-mass stars live much longer than high-mass stars.

The formula above is a rule of thumb for calculating stellar lifetimes. Better values, such as those in Table 15.1,

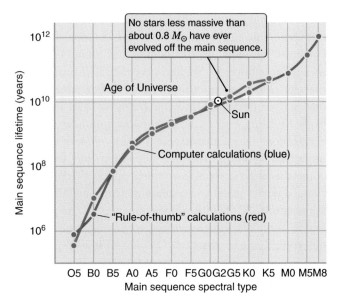

Figure 15.1 *The main sequence lifetimes of stars.*

are based on detailed computer models of stars at each mass. Even so, the values that you get by applying this rule of thumb are very close to the model values. (**Figure 15.1** gives a comparison between the two sets of numbers.) The rule of thumb also lets us see the reasons behind the differences in lifetimes of stars of different masses. While high-mass stars live fast and die young, low-mass stars live their lives in the slow lane.

THE STRUCTURE OF A STAR CHANGES AS IT USES ITS FUEL

To say that a main sequence star is stable is not to say that it does not continually change. At the time the Sun formed, about 90% of the atoms in the Sun were hydrogen atoms. Since that time the Sun has produced its energy by converting hydro-gen into helium via the

TABLE 15.1

MAIN SEQUENCE LIFETIMES
Lifetimes marked with an asterisk (★) are based on the "rule of thumb" discussed in the text. The other lifetimes are based on computer models.

Spectral Type	Mass ($M_\odot$)	Luminosity ($L_\odot$)	Main Sequence Lifetime (years)
O5	60	794,000	3.6×10^5
B0	17.5	52,500	1.0×10^7
B5	5.9	832	7.2×10^7
A0	2.9	54	3.9×10^8
A5	2.0	14	1.1×10^9
F0	1.6	6.5	2.1×10^9
F5	1.3	3.2	3.5×10^9
G0	1.05	1.5	8.3×10^9
G2 (Our Sun)	1.0	1.0	1.0×10^{10}
G5	0.92	0.8	1.5×10^{10}
K0	0.79	0.4	3.7×10^{10}
K5	0.67	0.15	5.3×10^{10}
M0	0.51	0.08	7.9×10^{10}★
M5	0.21	0.011	2.9×10^{11}★
M8	0.19	0.0012	1.1×10^{12}★

proton–proton chain. As the composition of a star changes, so must its structure. When we discussed the collapse of a protostar toward the main sequence in Chapter 14, we considered the idea of a "changing balance" between gravity and pressure. The protostar was always in balance, but as it radiated away thermal energy, this balance was constantly changing, shifting toward that of a smaller and denser object. The same concept applies here. As the main sequence star uses the fuel in its core, its structure must continually shift in response to the changing core composition. At any given point in its lifetime, a main sequence star like the Sun is in balance, but the balance in the Sun today is slightly different from the balance in the Sun 4 billion years ago, and slightly different from the balance it will be 4 billion years from now. Between the time the Sun was born and the time it will leave the main sequence, its luminosity will roughly double, with most of this change occurring during the last billion years of its life on the main sequence. Stars evolve even as they "sit" on the main sequence, although this evolution is slow and modest in comparison with the events that follow.

The balance in an evolving main sequence star must change.

What happens to the helium that is produced in the core of a low-mass main sequence star like our Sun? We might imagine that it begins to burn, with helium atoms fusing to form heavier elements, but such is not the case. For fusion to occur, atomic nuclei must be slammed together with enough energy to overcome the electric repulsion between them. Getting two helium nuclei close enough to fuse involves two protons in one nucleus pushing against two protons in the other nucleus, which is four times the repulsive force of a hydrogen nucleus pushing on a hydrogen nucleus. At the temperature at the center of a low-mass main sequence star, collisions are not energetic enough to overcome the electric repulsion between helium nuclei.[1] As hydrogen burns in the core of a low-mass star, the resulting helium collects there, building up like the nonburning ash in the bottom of a fireplace.

Helium ash builds up in the core of a main sequence star.

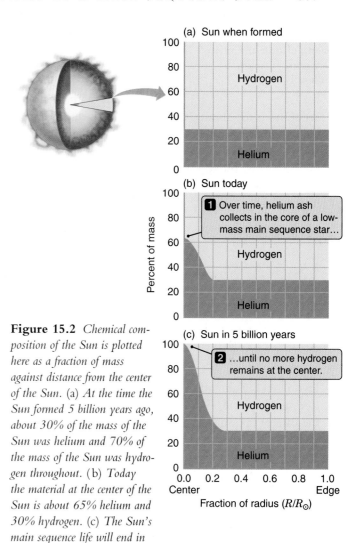

Figure 15.2 *Chemical composition of the Sun is plotted here as a fraction of mass against distance from the center of the Sun.* (a) *At the time the Sun formed 5 billion years ago, about 30% of the mass of the Sun was helium and 70% of the mass of the Sun was hydrogen throughout.* (b) *Today the material at the center of the Sun is about 65% helium and 30% hydrogen.* (c) *The Sun's main sequence life will end in about 5 billion years, when all of the hydrogen at the center of the Sun is gone.*

HELIUM ASH BUILDS UP IN THE CENTER OF THE STAR

Helium does not build up at the same rate throughout the interior of a star. In Chapter 6 we found that the temperature and pressure within Earth *must* be highest at the center of the planet. No other configuration makes sense. Exactly the same arguments apply equally well to stars. The fact that the temperature and pressure is highest at the center of a main sequence star means that hydrogen burns most rapidly there as well. As a result, nonburing helium "ash" accumulates most rapidly at the center of the star.

If we could cut a star open and watch as it evolves, we would see the chemical composition of the star changing most rapidly at its center, and less rapidly as we move outward through the star. **Figure 15.2** shows how the chemical composition inside a star like the

---

[1] An astute student may be puzzled, remembering from Chapter 13 that one step of the proton–proton chain is the fusion of two ³He nuclei, which have as strong an electric repulsion as two ⁴He nuclei do. The answer is that the strong nuclear force interaction between the two ³He nuclei is much more powerful than the strong force interaction between two ⁴He nuclei, so two ³He nuclei do not have to get as close together as do two ⁴He nuclei in order to fuse.

The composition of a main sequence star changes continuously.

Sun changes throughout its main sequence lifetime. When the Sun formed, it had a uniform composition throughout, with hydrogen accounting for about 70% of the mass in the Sun and helium accounting for most of the remaining 30%. With time, the helium fraction in the center of the Sun climbed. Today, roughly 5 billion years later, only about 35% of the mass at the center of the Sun is hydrogen.

15.3 A STAR RUNS OUT OF HYDROGEN AND LEAVES THE MAIN SEQUENCE

Hydrogen burning in the core of a star cannot continue forever. Eventually—about 5 billion years from now in the case of the Sun—a star exhausts all of the hydrogen fuel available at its center. At this point the innermost core of the star is composed entirely of helium ash. As thermal energy leaks out of the helium core into the surrounding layers of the star, no more energy is generated within the core to replace it. The balance that has maintained the structure of the star throughout its life is now broken. The star's life on the main sequence has come to an end.

THE HELIUM CORE IS DEGENERATE

Throughout your life you have experienced forms of matter that are mostly *empty space.* Just as the Solar System is mostly empty, except for the tiny bit of space occupied by the Sun and the planets, an atom is mostly empty except for the tiny bit of space occupied by the nucleus and the electrons. The same is true for the matter within the Sun. At the enormous temperatures within the Sun the electrons have almost all been stripped away from their atoms by energetic collisions. (In other words, the gas is completely ionized.) So the gas inside the Sun is a mixture of electrons and atomic nuclei all flying about freely. Even so, the gas that makes up the Sun is still mostly empty space, with the electrons and atomic nuclei filling only a tiny fraction of the volume.

Normal matter is mostly empty space.

When a low-mass star like the Sun exhausts the hydrogen at its center, the situation changes. As gravity begins to win the shoving match with pressure, the helium core is crushed to an ever smaller size and an ever greater density, but there is a limit to how dense the core can get. The rules of quantum mechanics (the same rules that say that atoms can only have certain discrete amounts of energy and that light comes in packets called photons) limit the number of electrons that can be packed into a given volume of space at a given pressure.[2] As the matter in the core of the star is compressed further and further, it finally bumps up against this limit. The space occupied by the core of the star is no longer mostly empty, but is now effectively "filled" with electrons that are smashed tightly together. Matter at the center of the star is now so dense that a single cubic centimeter of this material can have a mass of a metric ton (1,000 kg) or more. Matter which has been compressed to this point is said to be **electron degenerate.**

The crushed helium core is electron degenerate.

HYDROGEN BURNS IN A SHELL SURROUNDING A CORE OF HELIUM ASH

Once a low-mass star exhausts the hydrogen at its center, nuclear burning may end there, but the story is very different outside the core. The layers surrounding the degenerate core still contain hydrogen, and this hydrogen continues to burn. Astronomers speak of **hydrogen shell burning,** since the hydrogen is burning in a shell surrounding a core of helium, like eating the flesh of a fruit around its pit.

The electron-degenerate core of a star has a number of fascinating properties. For example, as more and more helium ash piles up on the degenerate core, the core *shrinks* in size. It does so because the added mass increases the strength of gravity and therefore the weight bearing down on the core, which means that the electrons can be smashed together into a smaller volume. The presence of the degenerate core triggers a chain of events that will dominate the evolution of our 1 $M_\odot$ star for the

Adding mass to a degenerate core makes it shrink.

---

[2]If you have taken high school or college chemistry, you may remember the *Pauli exclusion principle,* which limits the number of electrons that can go into a single orbital in an atom. This principle also limits the number of electrons that can be packed into the states available in the center of a star.

next 50 million years. Follow along as we step through the chain of cause and effect that takes center stage in the post–main sequence evolution of a low-mass star.

1. From our discussion in Chapter 9 (see Figure 9.3) we know that just outside the star's degenerate core the gravitational acceleration g_{core} is given by

$$g_{core} = \frac{GM_{core}}{r^2_{core}},$$

where M_{core} is the mass of the helium core and r_{core} is its radius. As the helium core grows, its larger mass (bigger M_{core}) and its shrinking size (smaller r_{core}) *both* cause the strength of gravity at the surface of the core to increase. As more helium is added to the core, the strength of gravity at its surface increases dramatically.

2. Increasing the strength of gravity around the core increases the weight of the overlying material pushing down on the hydrogen-burning shell surrounding the core.

3. This increase in weight must be balanced by an increase in pressure in the inner parts of the star. In particular, the pressure in the hydrogen-burning shell must increase.

4. Increasing the pressure in the hydrogen-burning shell drives up the rate of the nuclear reactions occurring in the shell.

5. Faster nuclear reactions means that more energy is released, so the luminosity of the star *increases*.

This is a very counterintuitive result! We might have imagined that, when a star like the Sun uses up the nuclear fuel at its center, it would grow fainter. Yet just the opposite happens.

Running out of fuel makes the star grow more luminous. A degenerate core means stronger gravity, stronger gravity means higher pressure, higher pressure means faster nuclear burning, and faster nuclear burning means a more luminous star. When the low-mass star "runs out of gas" at its center, it responds by getting more luminous!

TRACKING THE EVOLUTION OF THE STAR ON THE H-R DIAGRAM

The changes that occur in the heart of a star with a degenerate helium core are reflected in changes in the overall structure of the star. With time, the mass of the degenerate helium core grows as more and more hydrogen is converted into helium ash in the surrounding shell. And as the mass of the degenerate helium core grows, so does the rate of energy generation in the surrounding hydrogen-burning shell. This increase in energy generation heats the overlying

A bloated luminous giant surrounds a tiny degenerate core.

layers of the star, causing them to expand. The star becomes a bloated, luminous giant. As illustrated in **Figure 15.3,** the internal structure of the star is now fundamentally different from when the star was on the main sequence. The giant can grow to have a luminosity hundreds of times the luminosity of the Sun, and a radius over 50 $R_\odot$. Yet at the same time, the core of the giant star is far more compact than that of the Sun, with much of the star's mass becoming concentrated into a volume which is only a few times the size of Earth.

From our vantage point outside the star, we are not privy to the changes taking place deep within its interior. All we can see directly is that the star becomes larger and more luminous, and perhaps surprisingly, *cooler and redder* as well. The enormous expanse of the star's surface allows it to cool very efficiently. Even though its *interior* grows hotter and its luminosity higher, the *surface* temperature of the star actually begins to drop.

Just as we used the H-R diagram to follow the changes in a protostar on its way to the main sequence, the H-R diagram is a handy device for keeping track of the changing luminosity and temperature of the star as it evolves away from the main sequence (see **Figure 15.4**). As soon as the star exhausts the hydrogen in its core, it leaves the main sequence and begins to move upward and to the right on the H-R diagram, growing more luminous but cooler. We refer to such a star, which is somewhat brighter and larger than it was on the main sequence, as a subgiant star. As the subgiant continues to evolve, it grows larger and cooler, but after a time its progress to the right on the

An evolving low-mass star moves up and to the right on the H-R diagram.

H-R diagram hits a roadblock, the H$^-$ thermostat. When the temperature of the subgiant star has dropped by about 1,000 K relative to its temperature on the main sequence, H$^-$ ions start to form in great abundance in its atmosphere. We have encountered formation of H$^-$ ions before. In our discussion of protostars in Chapter 14, it was the H$^-$ ion in the protostar's atmosphere that acted as a thermostat, regulating the star's temperature. The H$^-$ ion serves exactly the same role here, regulating how much radiation can escape from the star and preventing it from becoming any cooler.

Since the star can cool no further, it begins to move almost vertically upward on the H-R diagram, growing

Figure 15.3 *The structure of a star near the top of the red giant branch is compared with the structure of the Sun. Left panels compare the size of the Sun with the size of the red giant. Right panels compare the size and structure of the Sun with the core of the red giant. The panels at right are blown up by about 50 times compared to panels on the left.*

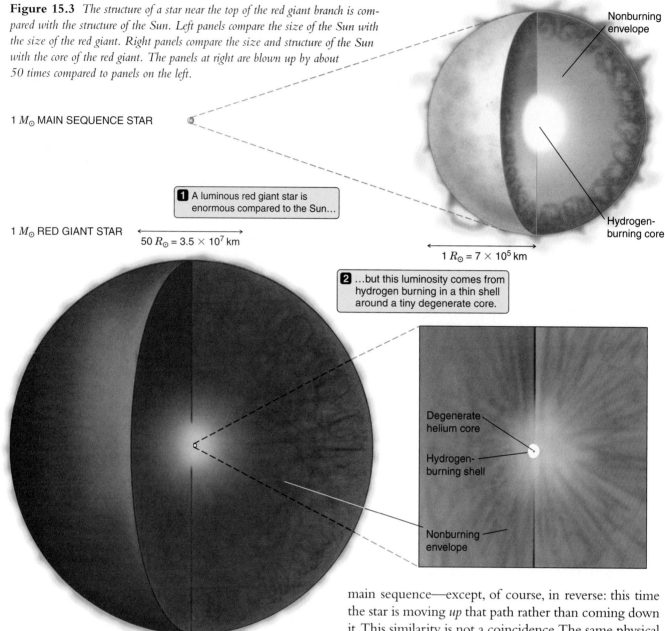

1 $M_\odot$ MAIN SEQUENCE STAR

Nonburning envelope

Hydrogen-burning core

1 $R_\odot$ = 7 × 10⁵ km

1 $M_\odot$ RED GIANT STAR

1 A luminous red giant star is enormous compared to the Sun...

50 $R_\odot$ = 3.5 × 10⁷ km

2 ...but this luminosity comes from hydrogen burning in a thin shell around a tiny degenerate core.

Degenerate helium core

Hydrogen-burning shell

Nonburning envelope

larger and more luminous, but remaining about the same temperature. The star has become a **red giant**—an obvious name for a star which is now both redder in color and larger in size than it was on the main sequence. We can think of the path that a star follows on the H-R diagram as it leaves the main sequence as being a tree "branch" growing out of the "trunk" of the main sequence. Astronomers refer to these tracks as the **subgiant branch** and the **red giant branch** of the H-R diagram. Interestingly, the path that a red giant follows on the H-R diagram closely parallels the path that the collapsing protostar followed on its way toward the

The red giant climbs the Hayashi track.

main sequence—except, of course, in reverse: this time the star is moving *up* that path rather than coming down it. This similarity is not a coincidence. The same physical processes (such as the H⁻ "thermostat") that give rise to the Hayashi track followed by a collapsing protostar also control the relationship between luminosity, size, and surface temperature in an expanding red giant.

As the star leaves the main sequence, the changes in its structure occur sluggishly at first, but then pick up steam as the star moves up the red giant branch faster and faster. It takes 200 million years or so for a star like the Sun to go from the main sequence to the top of the red giant branch. Roughly the first half of this period is spent on the subgiant branch, as the star's luminosity increases to about 10 $L_\odot$. During the second half of this time, the star's luminosity skyrockets

The star accelerates up the red giant branch.

from 10 $L_\odot$ to almost 1,000 $L_\odot$. The evolution of the star, illustrated in **Figure 15.5,** is reminiscent of the growth of a snowball rolling downhill. The larger the snowball becomes, the faster it grows, and the faster it grows, the larger it becomes. "Growth" and "size" feed off each other, and what began as a bit of snow at the top of the mountain becomes an avalanche by the time it reaches the bottom.

The analogy between the evolution of a red giant star and the growth of a snowball is actually a pretty good one. The helium core of the star grows in mass (but not in radius!) as hydrogen is converted to helium in the hydrogen-burning shell. The increasing mass of the ever-more-compact helium core drives up the force of gravity in the heart of the star. Stronger gravity means higher pressure, and higher pressure means faster nuclear burning in the shell. But faster nuclear reactions in the shell mean that hydrogen is being converted into helium more quickly, so the mass of the core grows more rapidly. We have come full circle in a cycle that feeds on itself. Increasing core mass leads to ever faster burning in the shell, and the faster hydrogen is burned in the shell, the faster the core mass grows. As a result, the star's luminosity climbs at an ever faster rate.

15.4 HELIUM BEGINS TO BURN IN THE DEGENERATE CORE

The growth of the red giant cannot continue forever, and we find ourselves once again asking the crucial question for understanding the evolution of stars: "What will be the next thing to give?" The answer lies in another of the unusual properties of the degenerate helium core, and we next turn our attention there.

THE ATOMIC NUCLEI IN THE CORE FORM A "GAS WITHIN A GAS"

The core of the red giant star is electron degenerate, which means that as many *electrons* are packed into that space as the rules of quantum mechanics allow at that pressure. However, *atomic nuclei* in the core are still able

Nuclei move freely in the electron-degenerate core.

Figure 15.4 *The evolution of a red giant star on the H-R diagram. The structure of a red giant star consists of a degenerate core of helium ash surrounded by a hydrogen-burning shell. As the star moves up the red giant branch, it comes close to retracing the Hayashi track that it followed when it was a protostar collapsing toward the main sequence.*

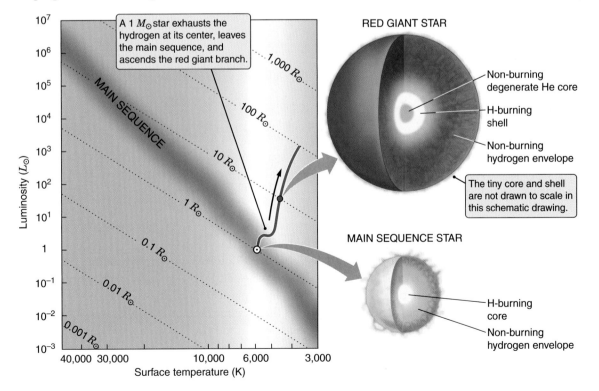

Figure 15.5 *As a star moves up the red giant branch, burning hydrogen to helium in a shell surrounding a degenerate helium core, its evolution feeds on itself. The luminosity of the star grows faster and faster.*

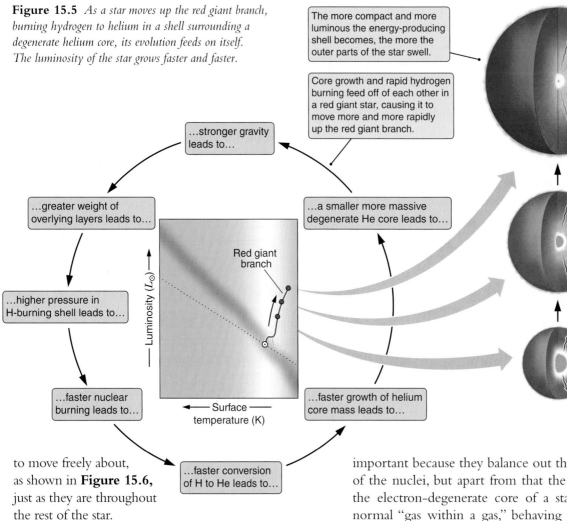

to move freely about, as shown in **Figure 15.6,** just as they are throughout the rest of the star.

"Wait a minute!" you say. "That last statement is nonsense for at least two different reasons. First, if the electrons are packed as tightly as possible into the core of the star, then surely the atomic nuclei are packed as tightly as possible into the core as well. After all, we normally think of atomic nuclei as being larger than electrons." But the world behaves in strange ways at the quantum mechanical level. The rules of quantum mechanics say that when packed together tightly, less massive particles like electrons effectively "take up more space" than more massive particles like atomic nuclei. As the core of the star is compressed, electrons become degenerate much sooner than the atomic nuclei do.

"Fine," you say, "but how can the atomic nuclei go moving freely about within a core that is wall-to-wall electrons?" Actually, this is no problem at all, because the laws of quantum mechanics place few restrictions on electrons and atomic nuclei occupying the same physical space. As far as the atomic nuclei are concerned, the electron-degenerate core of the star is still mostly empty space. The nuclei are free to go flying about through the sea of degenerate electrons almost as if the electrons were not there. The negative charges of the electrons are important because they balance out the positive charges of the nuclei, but apart from that the atomic nuclei in the electron-degenerate core of a star are a perfectly normal "gas within a gas," behaving just as the (electrically neutral) atoms and molecules in the air around you do.

HELIUM BURNING AND THE TRIPLE-ALPHA PROCESS

As the star evolves up the red giant branch, its helium core not only grows smaller and more massive, but hotter as well. This is partly because of the gravitational energy released as the core shrinks (just as the protostar's core grew hotter as it collapsed), and in part due to the energy released by the ever faster pace of hydrogen burning in the surrounding shell. The climbing temperature of the core means that the thermal motions of the atomic nuclei in the core become more and more energetic. Eventually, at a temperature of around 10^8 K (a hundred million kelvins), the collisions between helium nuclei in the core become energetic enough to overcome the electric repulsion between them. Helium

The core of the evolving red giant grows hotter.

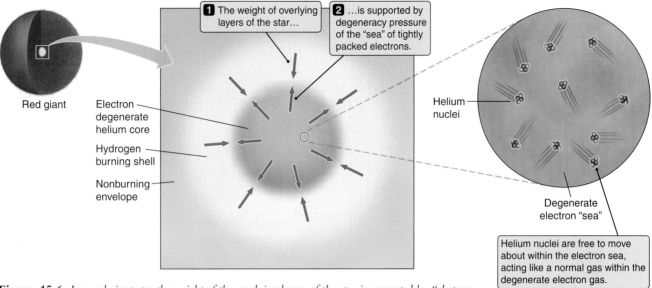

Figure 15.6 *In a red giant star the weight of the overlying layers of the star is supported by "electron degeneracy pressure" in the core arising from the fact that electrons are packed together as tightly as allowed by quantum mechanics. Even so, atomic nuclei in the core of the star are able to move freely about within the sea of degenerate electrons, and so behave as a normal gas.*

nuclei are slammed together hard enough for the strong nuclear force to act, and helium burning begins.

Helium burns in a two-stage process, referred to as the **triple-alpha process,** illustrated in **Figure 15.7.** First, two helium nuclei (^{4}He) fuse to form a beryllium-8 nucleus (^{8}Be) consisting of four protons and four neutrons. The beryllium-8 nucleus is extremely unstable. Left on its own, it would break apart again after only about a trillionth (10^{-12}) of a second. But if, in that short time, it collides with another ^{4}He nucleus, then the two nuclei will fuse into a stable nucleus of carbon-12 (^{12}C) consisting of six protons and six neutrons. The triple-alpha process takes its name from the fact that it involves the fusion of three ^{4}He nuclei, which are traditionally referred to as **alpha particles.**

Helium burns via the triple-alpha process.

THE HELIUM CORE IGNITES IN A HELIUM FLASH

The next phase of the star's evolution is shown in **Figure 15.8.** Degenerate material is a very good conductor of thermal energy, so any differences in temperature within the core are quickly evened out. As a result, the degenerate core of a red giant star is at almost exactly the same temperature throughout. When helium burning be-

The degenerate helium core ignites.

gins at the center of the core, the energy released quickly heats the entire core. Within a few minutes, the entire core is burning helium into carbon by the triple-alpha process.

In a normal gas like the air around you, the pressure of the gas comes from the random thermal motions of the atoms. Increasing the temperature of such a gas means that the motions of the atoms become more

Figure 15.7 *The triple-alpha processes: Two ^{4}He nuclei fuse to form an unstable ^{8}Be nucleus. If this nucleus collides with another ^{4}He nucleus before it breaks apart, the two will fuse to form a stable nucleus of carbon-12 (^{12}C). The energy produced is carried off both by the motion of the ^{12}C nucleus and by a high-energy gamma ray emitted in the second step of the process.*

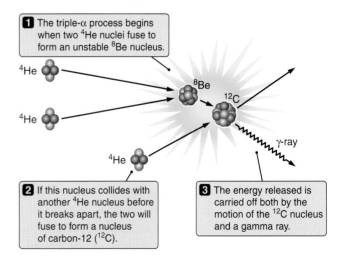

1 The triple-α process begins when two ^{4}He nuclei fuse to form an unstable ^{8}Be nucleus.

2 If this nucleus collides with another ^{4}He nucleus before it breaks apart, the two will fuse to form a nucleus of carbon-12 (^{12}C).

3 The energy released is carried off both by the motion of the ^{12}C nucleus and a gamma ray.

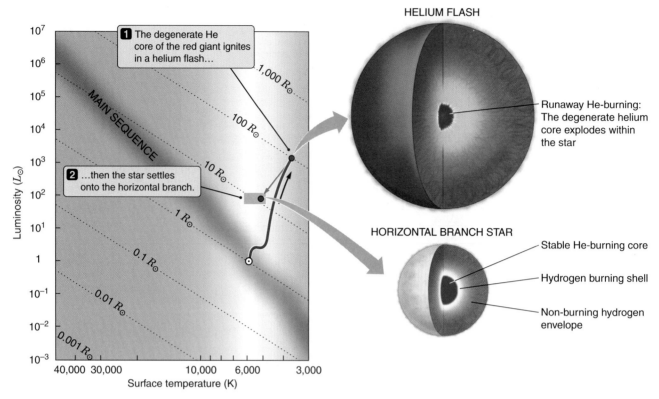

Figure 15.8 *When the core temperature of the red giant reaches about 10^8 K, He begins to burn explosively in the degenerate core, leading to the helium flash. After a few hours the core of the star begins to inflate, ending the helium flash. Over about the next 100,000 years (a relatively short time), the star settles onto the horizontal branch, where it burns He in its core and H in a surrounding shell.*

energetic, so the pressure of the gas increases. If the helium core of a red giant star were a normal gas, then the increase in temperature that accompanies the onset of helium burning would lead to an increase in pressure. The core of the star would expand; the temperature, density, and pressure would decrease; nuclear reactions would slow; and the star would settle down into a new balance between gravity and pressure. These are exactly the sorts of changes that are steadily occurring within the cores of main sequence stars like the Sun, as the structure of the star steadily and smoothly shifts in response to the changing composition in the star's core.

However, the degenerate core of a red giant is *not* a normal gas. In a certain sense, the degenerate core is more like a rock than a normal gas. A rock is hard to crush because of how the atoms within it are packed together. Heating a rock does not cause it to swell up like a hot-air balloon, and cooling a rock does not cause it to deflate. Similarly, the pressure that prevents the degenerate core of a white

Higher temperature does not cause the degenerate core to expand.

dwarf from collapsing comes from how tightly the electrons in the core are packed together. Heating the degenerate core of a red giant does not change the number of electrons that can be packed into its volume, and so heating the core does not change the pressure by much. And if the pressure does not increase, then there is nothing to cause the core to expand.

Take another run at that to be sure it clicks. When helium begins to burn in the degenerate core, the *temperature* of the core goes up, but the *pressure* does not. *So the onset of helium burning in the degenerate core of a red giant does not cause the core to expand!* Yet even though the higher temperature does not change the pressure, it *does* cause the helium nuclei to be slammed together with more frequency and greater force, so the nuclear reactions become more vigorous. The process begins to snowball again. More vigorous reactions mean higher temperature, and higher temperature means even more vigorous reactions. Thermonuclear burning in the degenerate core runs away with itself, wildly out of control, as increas-

The helium flash is a thermonuclear runaway—an explosion within the star.

ing temperature and increasing reaction rates feed each other. This is the **helium flash.**

It is difficult to imagine just how dramatic the thermonuclear runaway that takes place during the helium flash is. Helium burning begins at a temperature of about 100 million K. By the time the temperature has climbed by a mere 10%, to 110 million K, the rate of helium burning has increased to 40 times what it was at 100 million K. By the time the temperature in the core reaches 200 million K, it is burning helium 460 *million* times faster than it was at 100 million K! As the temperature in the core grows higher and higher, the thermal motions of the electrons and nuclei become more energetic, and the pressure due to these thermal motions becomes greater and greater. Within seconds of ignition the thermal pressure increases to the point that it is no longer smaller than the degeneracy pressure, at which point the core literally explodes. We do not see the explosion, however, because it is contained within the star. The energy released in this runaway thermonuclear explosion lifts the overlying layers of the star, and as the core expands, the electrons are able to spread out. The drama is over within a few hours. The expanded helium-burning core is no longer degenerate, and the star is on its way toward a new equilibrium.

Following the helium flash, the star once again does something counterintuitive. You might imagine that helium burning in the core would cause the star to grow more luminous, but it does not. The tremendous energy released during the helium flash goes into fighting gravity and puffing up the core. After the helium flash the core (which is no longer degenerate) is much larger, so the acceleration of gravity within it and the surrounding shell is much weaker. (Again, $g = GM/r^2$, so a larger core radius means smaller values of g.) Weaker gravity means less weight pushing down on the core and the shell, which means lower pressure. Lower pressure in turn means that the nuclear reactions slow down. The net result is that following the helium flash, core helium burning keeps the core of the star puffed up, and the star becomes *less* luminous than it was as a red giant.

The star's luminosity drops following the helium flash.

The star spends the next 100,000 years or so settling into a stable structure in which helium burns to carbon in a normal, nondegenerate core, while hydrogen burns to helium in a surrounding shell. The star is now about a hundredth as luminous as it was at the time of the helium flash. The lower luminosity means that the outer layers of the star are not as puffed up as they were as a red giant. The star shrinks, and as it does so its surface temperature climbs. (This is just the reverse of the sequence of events which caused the red giant to become larger and redder as it grew more luminous.) At this point in their evolution, low-mass stars with chemical compositions similar to the Sun will bunch up on the H-R diagram just to the left of the red giant branch. Stars which contain much less iron than the Sun tend to distribute themselves away from the red giant branch, along a nearly horizontal line on the H-R diagram. This stage of stellar evolution takes its name from this horizontal band. The star is now referred to as a **horizontal branch star.**

Horizontal branch stars burn helium in the core and hydrogen in a shell.

15.5 THE LOW-MASS STAR ENTERS THE LAST STAGES OF ITS EVOLUTION

The evolution of a solar-type star from the main sequence to the helium flash and horizontal branch is fairly well understood. Just as our understanding of the interior of the Sun comes from computer models of the physical conditions within our local star, our understanding of the evolution of the red giant comes from computer models that look at the changes in structure as the star's degenerate helium core grows. These models show that any star with a mass of about 1 $M_\odot$ will follow the march from main sequence to helium flash, then drop down onto the horizontal branch. However, when we try to use computer models to push our understanding to what happens next, the road that we follow gets a bit trickier. We already noted that differences in chemical composition between stars significantly affect where they fall on the horizontal branch. From this point on, small changes in the properties of the star—in mass, in chemical composition, in the strength of the star's magnetic field, or even in the rate at which the star is rotating—can lead to qualitative differences in how the star evolves. The further we go in time beyond the helium flash, the more possible divergent evolutionary paths are open to a low-mass star.

With this caution in mind, we continue our story of the evolution of a 1 $M_\odot$ star with solar composition, following what we currently believe to be the most likely sequence of events awaiting our Sun.

15.1
LOW

THE STAR MOVES UP THE ASYMPTOTIC GIANT BRANCH

The structure of a horizontal branch star is much like the structure of a main sequence star in many respects. The biggest difference is that now instead of burning hydrogen into helium in a stable, nondegenerate core, the horizontal branch star is burning helium into carbon in a stable, nondegenerate core. (The other difference, of course, is that hydrogen is continuing to burn in a shell surrounding the core.) The behavior of a star on the horizontal branch is remarkably similar to that of a star on the main sequence. It is this similarity that offers the key to understanding what comes next. In fact, the next stage in the star's evolution might be summarized by the old song lyric, "Second verse, same as the first." If you review our account of what happened as the star evolved off the main sequence, and if you replace "hydrogen" with "helium" and "helium" with "carbon," you will have a pretty good description of how the star evolves as it leaves the horizontal branch.

Leaving the horizontal branch resembles leaving the main sequence.

The star's life on the horizontal branch is, however, much shorter than its life on the main sequence. For one thing, there is now less fuel to burn in its core. In addition, the star is more luminous, and so must be consuming fuel more rapidly. Finally, helium is a much less efficient nuclear fuel than hydrogen. Even so, for 50 million years the horizontal branch star remains stable, burning helium into carbon in its core and hydrogen to helium in a shell.

The temperature at the center of a horizontal branch star is not high enough for carbon to burn, so carbon ash builds up in the heart of the star, just as helium ash accumulates at the center of a main sequence star. When the horizontal branch star has burned all of the helium at its core, gravity once again begins to win. The nonburning carbon ash core is crushed by the weight of the layers of the star above it, until once again the electrons in the core are packed together as tightly as the laws of quantum mechanics allow at its pressure. The carbon core is now electron degenerate, with physical properties much like those of the degenerate helium core at the center of a red giant.

The star forms a degenerate carbon core as it leaves the horizontal branch.

The small, dense electron-degenerate carbon core drives up the strength of gravity in the inner parts of the star, which in turn drives up the pressure, which speeds up the nuclear reactions, which causes the degenerate core to grow more rapidly . . . we have heard this story before. The internal changes occurring within the star are similar to the changes that took place at the end of the star's main sequence lifetime, and the path the star follows as it leaves the horizontal branch echoes that earlier phase of evolution as well. Just as the star accelerated up the red giant branch as its degenerate helium core grew, the star now leaves the horizontal branch and once again begins to grow larger, redder, and more luminous as its degenerate carbon core grows. The path that the star follows on the H-R diagram **(Figure 15.9)** closely parallels the path it followed as a red giant, getting closer to the red giant branch as the star grows more luminous. That is why this phase of evolution is referred to as the **asymptotic giant branch** of the H-R diagram. An **AGB star** burns helium and hydrogen in nested concentric shells surrounding a degenerate carbon core.

GIANT STARS LOSE MASS

Building on our analogy between AGB stars and red giants, you might imagine that the next step in the evolution of an AGB star should be a "carbon flash," when carbon burning begins in the star's degenerate core. Yet a carbon flash never happens. Before the temperature in the carbon core becomes high enough for carbon to burn, the star loses its grip on itself, and expels its outer layers into interstellar space.

Red giant and AGB stars are huge objects. The AGB star into which a 1 $M_\odot$ main sequence star evolves can grow to a radius hundreds of times the radius of our Sun. When our Sun becomes an AGB star, its outer layers will swell to the point that they engulf the orbits of the inner planets, possibly including Earth. (Do not lose much sleep over this eventuality, however. It will be 5 billion years before Earth is engulfed by the burgeoning Sun, and well before then, Earth will be well toasted by the thousandfold increase in the Sun's luminosity.) When a star expands to such a size, its hold on its outer layers becomes tenuous indeed.

A giant's grip on its outer layers is tenuous.

Once again an understanding of what happens with the star rests with Newton's universal law of gravitation. Recalling that $g = GM/r^2$, the acceleration due to gravity (g) at the surface of a star with a mass of 1 $M_\odot$ and a radius of 100 $R_\odot$ is only 1/10,000 as strong as the gravity at the surface of the Sun. It takes very little in the way of a kick from below to push material near the sur-

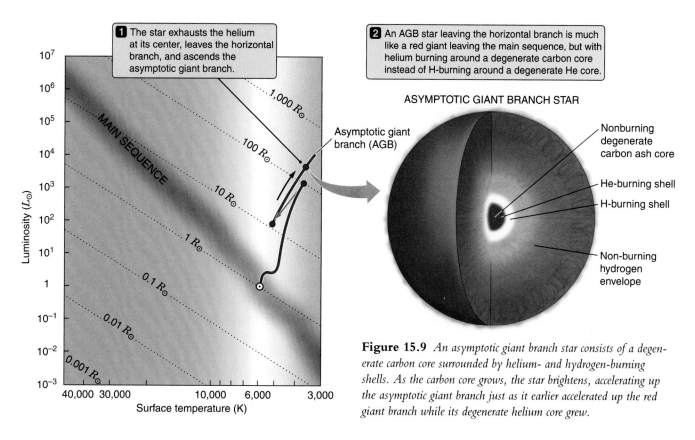

1 The star exhausts the helium at its center, leaves the horizontal branch, and ascends the asymptotic giant branch.

2 An AGB star leaving the horizontal branch is much like a red giant leaving the main sequence, but with helium burning around a degenerate carbon core instead of H-burning around a degenerate He core.

ASYMPTOTIC GIANT BRANCH STAR

Asymptotic giant branch (AGB)

Nonburning degenerate carbon ash core

He-burning shell

H-burning shell

Non-burning hydrogen envelope

Figure 15.9 *An asymptotic giant branch star consists of a degenerate carbon core surrounded by helium- and hydrogen-burning shells. As the carbon core grows, the star brightens, accelerating up the asymptotic giant branch just as it earlier accelerated up the red giant branch while its degenerate helium core grew.*

Red giants and AGB stars lose mass from their outer layers.

face of such a giant star over the edge, driving it completely away from the star. The process of **stellar mass loss** begins when the star is still on the red giant branch, and by the time a 1 $M_\odot$ main sequence star reaches the horizontal branch, it may have lost 10% to 20% of its total mass. As the star ascends the asymptotic giant branch, it loses another 20% or so of its mass. By the time it is well up on this branch, a star that began as a 1 $M_\odot$ star probably has a mass less than about 0.7 $M_\odot$. Mass loss on the asymptotic giant branch can be spurred on by a lack of stability in the star's interior. The extreme sensitivity of the triple-alpha process to temperature can lead to episodes of rapid energy release, which can provide the extra kick needed to expel material from the star's outer layers. It also means that stars that are initially quite similar can behave very differently when they reach this stage in their evolution.

THE POST-AGB STAR CAUSES A PLANETARY NEBULA TO GLOW

Toward the end of an AGB star's life, mass loss itself becomes a runaway process. When a star loses a bit of its outer layers, it reduces the weight pushing down on

the underlying layers of the star. Without this weight holding them down, the outer layers of the star puff up even larger than they were before. The star, which is now both less massive and larger, is even less tightly bound by gravity, so even less energy is needed to push outer layers away from the star. The situation is a bit like taking the lid off of a pressure cooker. Mass loss leads to weaker gravity, which leads to faster mass loss, which leads to weaker gravity. . . . When the end comes, much of the remaining mass of the star is ejected into space, typically at speeds of 20 to 30 km/s.

After ejection of its outer layers, all that is left of the low-mass star itself is a tiny, very hot electron-degenerate carbon core, surrounded by a thin envelope in which hydrogen and helium are still burning. This star is now somewhat less luminous than when it was at the top of the asymptotic giant branch, but it is still much more luminous than a horizontal branch star. The remaining hydrogen and helium in the star rapidly burn to carbon, and as more and more of the mass of the star ends up in the carbon core, the star itself shrinks and becomes hotter and hotter. Over the course of only 30,000 years or so following the beginning of runaway mass loss, the star moves very rapidly from

A hot degenerate core is left behind.

Figure 15.10 *At the end of the AGB star's life, it ejects most of the star's mass in a planetary nebula, becoming a post-AGB star and finally leaving behind nothing but the star's degenerate core, which appears on the H-R diagram as a cooling white dwarf star.*

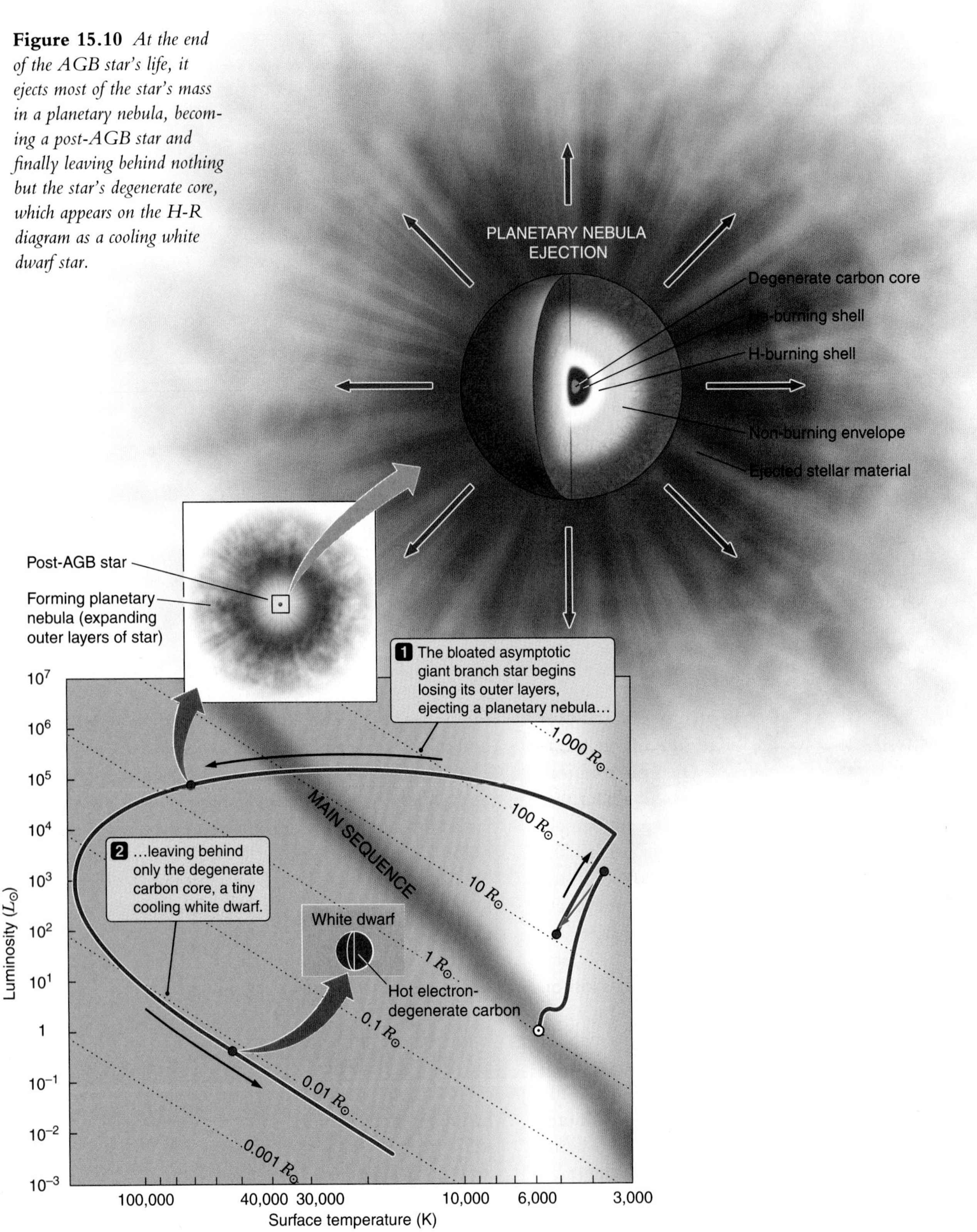
PLANETARY NEBULA EJECTION
Degenerate carbon core
He-burning shell
H-burning shell
Non-burning envelope
Ejected stellar material
Post-AGB star
Forming planetary nebula (expanding outer layers of star)
1 The bloated asymptotic giant branch star begins losing its outer layers, ejecting a planetary nebula...
2 ...leaving behind only the degenerate carbon core, a tiny cooling white dwarf.
White dwarf
Hot electron-degenerate carbon
MAIN SEQUENCE
1,000 R⊙
100 R⊙
10 R⊙
1 R⊙
0.1 R⊙
0.01 R⊙
0.001 R⊙
Luminosity (L⊙)
10^7
10^6
10^5
10^4
10^3
10^2
10^1
1
10^-1
10^-2
10^-3
100,000
40,000 30,000
10,000
6,000
3,000
Surface temperature (K)

right to left across the top of the H-R diagram as shown in **Figure 15.10.**

The surface temperature of the star may eventually reach 100,000 K or hotter. Wien's law says that at such temperatures most of the light from the star is in the hard (high-energy) UV part of the spectrum:

$$\lambda_{\text{peak}} = \frac{2{,}900\ \mu\text{m K}}{100{,}000\ \text{K}} = 0.029\ \mu\text{m}$$

The intense UV light from what remains of the star heats and ionizes the expanding shell of gas which was recently ejected by the star. The ultraviolet light causes

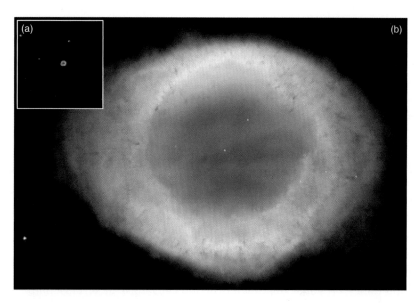

Figure 15.11 *At the end of its life, a low-mass star ejects its outer layers, and may form a "planetary nebula" consisting of an expanding shell of gas surrounding the white-hot remnant of the star.* (a) *This picture of a famous planetary nebula called the Ring Nebula, as it appears through a small amateur telescope, shows why astronomers thought these objects looked like planets.* (b) *However, an HST image of the ring shows the remarkable and complex structure of this expanding shell of gas.*

the shell of gas to glow in the same way that UV light from an O star causes an H II region to glow.

If conditions are right, the mass ejected by the AGB star will pile up in a dense, expanding shell. If you were to look at such a shell through a telescope, you would see it as a round or perhaps oblong patch of light surrounding the remains of the AGB star. At first glance you might almost confuse the glowing shell with the appearance of a planet **(Figure 15.11a),** which is why such objects are referred to as **planetary nebulae.** But there is nothing truly "planetary" about a planetary nebula, as is apparent in **Figure 15.11(b).** Instead, it is the remaining outer layers of a star, ejected into space as a dying gasp at the end of the star's ascent of the asymptotic giant branch.

A planetary nebula may form.

Rather than being simple spherical shells of the sort we might naively predict would surround a nice spherical star, planetary nebulae show a dazzling range of appearances, earning them names like Owl Nebula, Clown Nebula, Cat's Eye Nebula, and Dumbbell. This extraordinary menagerie, illustrated in **Figure 15.12,** serves to rub astronomers' noses in the complexity of the late stages of a star's evolution, telling of eras when mass loss from the star was slower or faster, and of times when mass was primarily ejected from the star's equator or its poles. The gas in the planetary nebula also contains chemical elements that were produced by nuclear burning in the star offering us our first look at the processes responsible for the chemical evolution of the Universe (see Connections 15.1). The planetary nebula is visible for 50,000 years or so before the gas ejected by the star disperses so far that the nebula is too faint to be seen.

Figure 15.12 *Planetary nebulae are not all simple spherical shells around their parent stars. These HST images of planetary nebulae show a wealth of structure resulting from the complex processes by which low-mass stars eject their outer layers.*

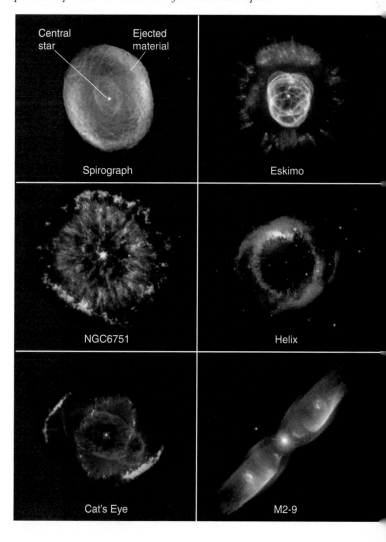

SEEDING THE UNIVERSE WITH NEW CHEMICAL ELEMENTS

One of the grand themes of 20th- and 21st-century astronomy is the origin of the chemical elements. When we discuss the properties of the early Universe, we will see that the only chemical elements that were formed in abundance when the Universe was young were hydrogen and helium, along with trace amounts of lithium, beryllium, and boron. As we mentioned in Chapter 1, the chemical elements that make up the world around you were formed in the interiors of stars. You are made of stardust.

Astronomers studying red giant and AGB stars see direct evidence that chemical elements formed deep within the interiors of low-mass stars find their way to the "outside" world. In Chapter 13 we learned that if the change in temperature within a star becomes too steep, convection will begin. Like the boiling motion in a pot of water, convection is a process of *mixing*. Material from the inner parts of the star is carried upward and mixed with material in the outer parts of the star. As a star ascends first the red giant and later the asymptotic giant branch of the H-R diagram, the core of the star grows hotter and hotter, and the temperature gradient within the star grows steeper. Convection spreads through more and more of the star. Under certain circumstances convection can spread so deep into a star that it is able to dredge up chemical elements formed by nuclear burning within the star and carry them to the star's surface. Many AGB stars, known as **carbon stars,** show an overabundance of carbon and other by-products of nuclear burning in their spectra. This extra carbon originated in the star's helium-burning shell, then was carried to its surface by convection.

Mass loss from giant stars carries the chemical elements enriching the stars' outer layers off into interstellar space. Planetary nebulae—the ejected atmospheres of low-mass stars—often show an overabundance of elements such as carbon, nitrogen, and oxygen, which are by-products of nuclear burning. Once this chemically enriched material leaves the star, it mixes with the interstellar gas there, enriching the chemical diversity of the Universe. Novae and Type I supernovae, discussed at the end of this chapter, also do their part. The chemical burning that occurs in these explosive events goes far beyond the formation of elements such as carbon. Novae and Type I supernovae help to seed the Universe with much more massive atoms such as iron and nickel.

As we continue to discuss the evolution of stars and the Universe in the chapters to come, we will learn more about the origin of the chemical elements. Today when scientists study the patterns of the abundances of different chemical elements on Earth and in our Solar System, they recognize the patterns they see. Those patterns are clear signatures of nuclear reactions in stars.

THE STAR BECOMES A WHITE DWARF

Within 50,000 years or so, a post-AGB star burns all of the fuel remaining on its surface, leaving nothing behind but a cinder—a nonburning ball of carbon with a mass of about 0.6 $M_\odot$. In the process the star plummets down the left side of the H-R diagram, becoming smaller and fainter. Within a few thousand years the burnt-out core shrinks to about the size of Earth, at which point it has become fully degenerate and so can shrink no further.

The white dwarf is hot but tiny. The degenerate stellar cinder is now referred to as a **white dwarf.** The white dwarf continues to radiate energy away into space, and as it does so it cools, just as the filament of a lightbulb cools when the switch is turned off. Since the white dwarf is electron degenerate, its size does not change much as it cools, so it moves down and to the right on the H-R diagram, following a line of constant radius. Even though the white dwarf may remain very hot for ten million years or so, its luminosity may now only be 1/1,000 that of a main sequence star like our Sun. Many white dwarfs are known, but all are much too faint to be seen without the aid of a telescope. Yet when you look at Sirius, you are also looking at a white dwarf. The brightest star in Earth's sky has a faint white dwarf as a binary companion.

Here we have the fate of our Sun, some 5 billion years from now. It will become a white dwarf which, bereft of all sources of nuclear fuel, will shine ever more feebly as it continues to cool, radiating its thermal energy away into space. It is amazing to recall that this superdense ball—with a density of a ton per

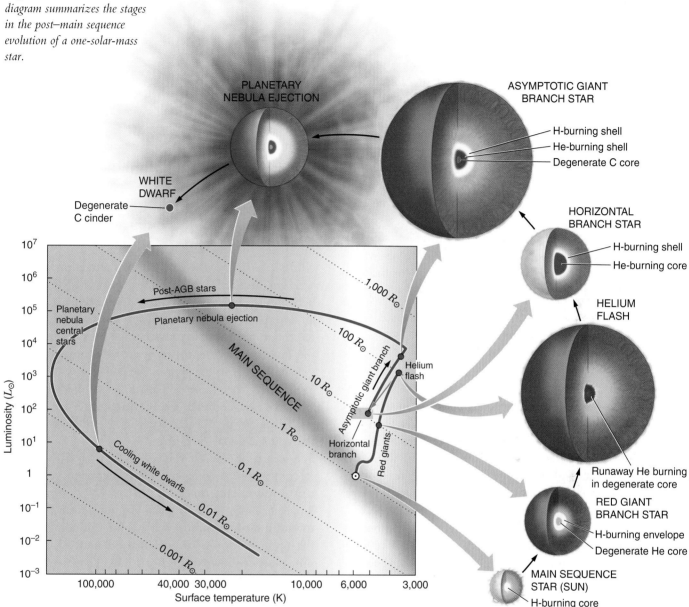

Figure 15.13 *This H-R diagram summarizes the stages in the post–main sequence evolution of a one-solar-mass star.*

teaspoonful—actually began its life billions of years earlier as a cloud of interstellar gas billions of times more tenuous than the vacuum in the best vacuum chamber on Earth.

Figure 15.13 recaps the evolution of a solar-type $1\ M_\odot$ main sequence star through to its final existence as a $0.6\ M_\odot$ white dwarf. We point out once again that what we have described is *representative* of the fate of low-mass stars. While all low-mass stars form white dwarfs at the end points of their evolution, the exact path a low-mass star follows from core hydrogen burning on the main sequence to white dwarf depends on many de-

> **Details differ, but all low-mass stars become white dwarfs.**

tails particular to the star. Some stars less massive than the Sun may become white dwarfs composed largely of helium rather than carbon. Temperatures in the cores of evolved $2\ M_\odot$ to $3\ M_\odot$ stars are high enough to allow additional nuclear reactions to occur, leading to the formation of somewhat more massive white dwarfs composed of materials such as oxygen, neon, and magnesium. Differences in chemical composition of a star can also lead to dramatic differences in its post–main sequence evolution. Finally, we note that studies of the evolution of stars whose main sequence masses are less than about $0.8\ M_\odot$ or so are pure theory. These stars are able to burn hydrogen in their cores for longer than the age of the Universe. No star with a mass lower than that has ever evolved beyond the main sequence, except in the virtual world of a computer's calculations.

In our discussion of stellar evolution we have focused on what happens after a star leaves the main sequence. It is important, however, that we remind ourselves that the spectacle of a red giant or AGB star is ephemeral. The Sun will travel the path from subgiant to white dwarf in less than a 10th of the time that it spends on the main sequence steadily burning hydrogen to helium in its core. Stars spend most of their luminous lifetimes on the main sequence, which is why most of the stars that we see in the sky *are* main sequence stars. (Were it otherwise, the "main sequence" would not be "main" at all!) Yet in the end it is white dwarfs that will carry the day. Too faint to be noticed as we look at the sky on a clear summer evening, white dwarfs still constitute the final resting place for the vast majority of stars that have been or ever will be formed.

15.6 MANY STARS EVOLVE AS PAIRS

Possibly the most significant complication in our picture of the evolution of low-mass stars arises from the fact that many stars in the sky are members of binary systems. Binary stars were quite useful to us when we were uncovering the properties of stars. We used them as a sort of celestial bathroom scale, applying Kepler's laws to the orbits of two stars about each other to determine their masses. While the members of a binary pair are on the main sequence, they probably have little effect on each other. However, when one of the stars evolves off the main sequence, it can swell to hundreds of times its original size, and its outer layers can literally cross the line between what belongs to the evolved star and what belongs to its companion. We will shortly see the consequences of this.

MASS FLOWS FROM AN EVOLVING STAR ONTO ITS COMPANION

Think for a moment about what would happen if you were to travel in a spacecraft from Earth toward the Moon. When you are still near Earth, the force of Earth's gravity is far stronger than that of the Moon. There is no question that you "belong" to Earth, just as there is no question that you "belong" to Earth as you sit in your easy chair, reading this book. Yet as you move away from Earth and closer to the Moon, this terrestrial hegemony becomes less secure: the gravitational attraction of Earth weakens, and the gravitational attraction of the Moon becomes stronger. There comes a point when you reach an interplanetary no-man's-land where neither body has the upper hand. If you continue beyond this point, the lunar gravity begins to assert itself, until you find yourself firmly in the grip of the Moon.

Gravitational "ownership" between two bodies depends on distance and mass.

Exactly the same situation exists between two stars. Gas near each star clearly "belongs" to that star. But what happens if, as a star swells up, its outer layers cross that gravitational dividing line separating the star from its companion? Any material that crosses this line no longer belongs to the first star, but instead can be pulled toward the companion. A star reaches this point when it fills up its portion of an imaginary figure-8-shaped volume of space shown in **Figure 15.14**. These regions surrounding the two stars are referred to as the **Roche lobes** of the system (see Connections 15.2). Once the star has expanded to fill its Roche lobe, material begins to pour through the "neck" of the figure 8 and fall

An evolving star loses mass to its companion.

15.2 LOBES

CONNECTIONS 15.2

ROCHE LIMITS AND ROCHE LOBES

You may recall the name *Roche* from our earlier discussion of the tidal disruption of moons and comets leading to the formation of planetary rings. The Roche limit determines how close a small object might come to a planet before being torn apart by the planet's tides. In this chapter we discuss Roche lobes in connection with the transfer of mass from

one member of a binary star system to the other. The two problems are actually similar. Mass transfer between stars in a binary can be thought of as the "tidal stripping" of one star by the other. The physical principles at work in the two situations are much the same, and Edouard Roche is responsible for early calculations of both.

toward the other star. Astronomers refer to this exchange of material between the two stars as **mass transfer.**

EVOLUTION OF A BINARY SYSTEM

The best way to understand how mass transfer affects the evolution of stars in a binary system is to apply what we have learned from the evolution of single low-mass stars. Figure 15.14(a) shows a binary system consisting of two low-mass stars. We will rather unimaginatively call the more massive of the two stars star 1, and the less massive of the two star 2. This is an ordinary binary system, and each of these stars is an ordinary main sequence star for most of the system's lifetime. However, when one of the stars begins to evolve, things start to get interesting.

The more massive a main sequence star is, the more rapidly it evolves. Therefore star 1, which is more massive, will be the first to use up the hydrogen at its center and begin to evolve off the main sequence (Figure 15.14b). If the two stars are close enough together, star 1 will eventually grow to fill its Roche lobe, and the overflow from star 1 will start falling onto star 2 (Figure 15.14c). A number of interesting things can happen at this point. For example, the transfer of mass between the two stars can result in a sort of "drag" that causes the orbits of the two stars to shrink, bringing the stars closer together and further enhancing mass loss. The two stars can even reach the point where they are effectively two cores sharing the same extended envelope.

The more massive star evolves first.

Despite these complexities, star 2 probably remains a basically normal main sequence star throughout this process, burning hydrogen in its core. However, it does not remain the *same* main sequence star that it started out to be. The mass of star 2 increases, courtesy of the largesse of its companion. As it does so, the structure of star 2 must change to accommodate its new status as a higher-mass star. If we could watch how star 2 moves on the H-R diagram during this period, we would see it move up and to the left along the main sequence, becoming larger, hotter, and more luminous.

The second (smaller) star grows and changes.

While star 2 gains from the interaction, star 1 suffers. Star 1 never gets to experience the full glory of being an isolated red giant or AGB star, because star 1 can never grow larger than its Roche lobe. The presence of star 2 prevents star 1 from swelling

The first star becomes a white dwarf orbiting a main sequence star.

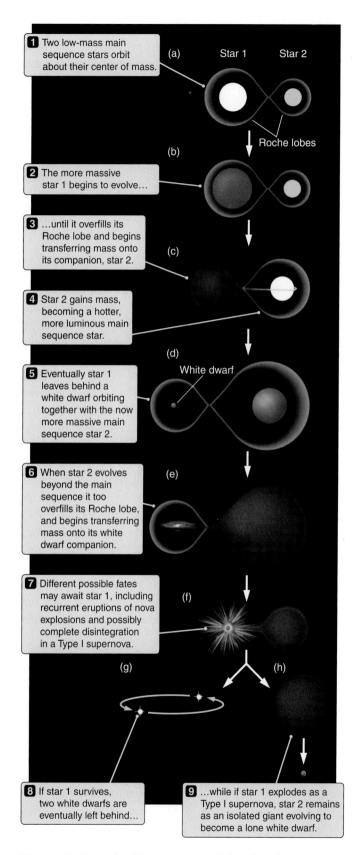

1 Two low-mass main sequence stars orbit about their center of mass.

(a)   Star 1   Star 2

Roche lobes

(b)

2 The more massive star 1 begins to evolve...

3 ...until it overfills its Roche lobe and begins transferring mass onto its companion, star 2.

(c)

4 Star 2 gains mass, becoming a hotter, more luminous main sequence star.

(d)   White dwarf

5 Eventually star 1 leaves behind a white dwarf orbiting together with the now more massive main sequence star 2.

6 When star 2 evolves beyond the main sequence it too overfills its Roche lobe, and begins transferring mass onto its white dwarf companion.

(e)

7 Different possible fates may await star 1, including recurrent eruptions of nova explosions and possibly complete disintegration in a Type I supernova.

(f)

(g)   (h)

8 If star 1 survives, two white dwarfs are eventually left behind...

9 ...while if star 1 explodes as a Type I supernova, star 2 remains as an isolated giant evolving to become a lone white dwarf.

Figure 15.14 *A close binary system consisting of two low-mass stars passes through a sequence of stages as the stars evolve and mass is transferred back and forth. The evolution of the binary system can lead to novae or even to a Type I supernova that destroys the initially more massive star 1.*

to the size of the behemoths that populate the top of the H-R diagram. Yet star 1 continues to evolve, spending its allotted time burning helium in its core on the horizontal branch, proceeding through a stage of helium shell burning, and finally losing its outer layers and leaving a white dwarf behind. Figure 15.14(d) shows the binary system after star 1 has completed its evolution. All that remains of star 1 is a white dwarf, orbiting about its bolstered main sequence companion, star 2.

FIREWORKS OCCUR WHEN THE SECOND STAR EVOLVES

The stage is now set for a sequence of events that can lead to one of nature's most spectacular displays. Figure 15.14(e) picks up the evolution of the binary system as star 2 begins to evolve off of the main sequence. Like star 1 before it, star 2 grows to fill its Roche lobe, and as it does, material from star 2 begins to pour through the "neck" connecting the Roche lobes of the two stars. However, this time the mass is not being added to a normal star, but is drawn toward the tiny white dwarf left behind by star 1. Because the white dwarf is so small, the orbit of the infalling material generally causes it to miss. Instead of landing directly on the white dwarf, the infalling mass forms an **accretion disk** around the white dwarf, similar in some ways to the accretion disk that forms around a protostar. As in the process of star formation, the accretion disk serves as a way station for material destined to find its way onto the white dwarf, but which starts out with too much angular momentum to hit the white dwarf directly.

Next, the second star evolves, transferring mass onto the white dwarf.

We have already seen that a white dwarf has a mass comparable to that of the Sun but is closer to the size of Earth. Large mass and small radius mean strong gravity. In Chapter 5 we saw how the gravitational energy of material falling toward a forming protostar is converted into heat when the material hits the accretion disk around the protostar. The same principle applies here, but with a vengeance. A kilogram of material falling from space onto the surface of a white dwarf releases a hundred times more energy than a kilogram of material falling from the outer Solar System onto the surface of the Sun. The material streaming toward the white dwarf in our binary system falls down into an incredibly deep gravitational "well." All of this energy has to go somewhere, and it goes into thermal energy. The

Material falling on the white dwarf is heated.

spot where the stream of material from star 2 hits the accretion disk can be heated to millions of kelvins, where it glows in the far ultraviolet and X-ray parts of the electromagnetic spectrum.

Eventually the infalling material accumulates on the surface of the white dwarf **(Figure 15.15a),** where it is compressed by the enormous gravitational pull of the white dwarf to a density close to that of the white dwarf itself. As more and more material builds up on the surface of the white dwarf, the white dwarf shrinks (just as the core of the red giant shrinks as it grows more massive). The density gets greater and greater, and at the same time, the release of gravitational energy drives the temperature of the white dwarf higher and higher. Now recall that the infalling material is from the outer, unburnt layers of star 2, so it is mostly composed of hydrogen. We have a dangerous situation here. Hydrogen, the best nuclear fuel around, is being compressed to higher and higher densities and heated to higher and higher and higher temperatures on the surface of the white dwarf.

A degenerate hydrogen layer accumulates on the white dwarf.

The mental picture you should have at this point is of gasoline pooling on the floor of a match factory. Once the temperature at the base of the layer of hydrogen reaches about 10 million K, hydrogen begins to burn. But this is not the contained hydrogen burning taking place in the center of the Sun. Rather, this is explosive hydrogen burning in a degenerate gas. Energy released by hydrogen burning drives up the temperature, and the higher temperature drives up the rate of hydrogen burning. This runaway thermonuclear reaction is much like the runaway helium burning that takes place during the helium flash, except now there are no overlying layers of a star to keep things contained. The result is a tremendous explosion—a **nova**—that blows part of the layer covering the white dwarf out into space at speeds of thousands of kilometers per second (see Figure 15.14f and **Figure 15.15b**).

Runaway burning of hydrogen on a white dwarf causes a nova.

About 50 novae are thought to occur in the Milky Way Galaxy each year, but because our view is restricted by interstellar extinction, we can only see a few of these, typically two or three each year. Novae get bright in a hurry—typically reaching their peak brightness in only a few hours—and for a brief time can be almost half a million times more luminous than the Sun. While the brightness of a nova sharply declines in the weeks following the outburst, it can sometimes still be seen for years. During this time the glow from the expanding cloud of material ejected by the explosion is powered by the decay of radioactive isotopes created in the explosion.

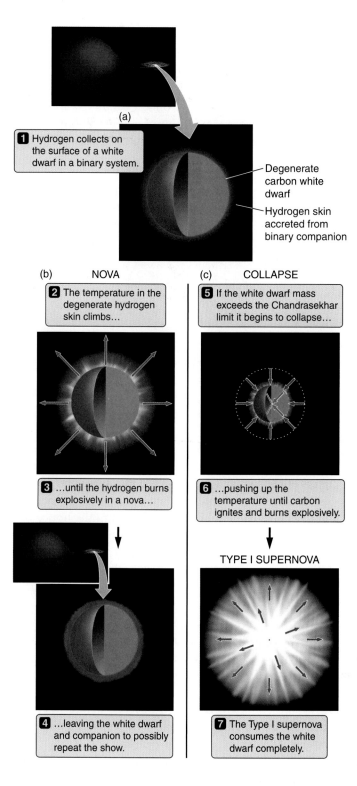

(a)

1 Hydrogen collects on the surface of a white dwarf in a binary system.

Degenerate carbon white dwarf

Hydrogen skin accreted from binary companion

(b) NOVA

2 The temperature in the degenerate hydrogen skin climbs…

3 …until the hydrogen burns explosively in a nova…

4 …leaving the white dwarf and companion to possibly repeat the show.

(c) COLLAPSE

5 If the white dwarf mass exceeds the Chandrasekhar limit it begins to collapse…

6 …pushing up the temperature until carbon ignites and burns explosively.

TYPE I SUPERNOVA

7 The Type I supernova consumes the white dwarf completely.

Figure 15.15 *In a binary system in which mass is transferred onto a white dwarf, a skin of hydrogen builds up on the surface of the degenerate white dwarf. If hydrogen burning ignites on the surface of the white dwarf, the result is a nova. If enough hydrogen accumulates to force the white dwarf to begin to collapse, carbon ignites and the result is a Type I supernova.*

The explosion of a nova does *not* destroy the underlying white dwarf star. In fact, much of the material that had built up on the white dwarf can remain behind after the explosion. The nova leaves the binary system in much the same configuration it was in previously— the configuration shown in Figure 15.15(a)—with material from star 2 still pouring onto the white dwarf. This cycle can repeat itself many times, with material building up and igniting over and over again on the surface of the white dwarf. In most cases outbursts are probably separated by thousands of years, and so most novae have been seen only once in historical times. However, some novae are known to be recurrent, erupting over and over again every decade or so.

> **Novae can be recurrent.**

A STELLAR CATACLYSM MAY AWAIT THE WHITE DWARF IN A BINARY SYSTEM

It is possible that eventually star 2 will simply go on to form a white dwarf, leaving behind a binary system consisting of two white dwarfs, as in Figure 15.14(g). There is another possibility, however. Novae may be spectacular events, but they pale in comparison with the fate that may eventually await the white dwarf in a binary system. Through millions of years of mass transfer from star 2 onto the white dwarf, and possibly through countless nova outbursts, the mass of the white dwarf slowly increases. As it does, the white dwarf shrinks in size, and the grip of self-gravity on this already extreme object gets tighter and tighter. The mass of a white dwarf can only be pushed so far. There comes a point when even the pressure supplied by degenerate electrons is no longer enough to balance gravity. This precipice occurs at a mass of about 1.4 $M_\odot$, a value referred to as the **Chandrasekhar limit.** It is impossible for a star to increase in mass above 1.4 $M_\odot$ and remain a white dwarf. Raise its mass above the Chandrasekhar limit, and it will begin to collapse, to become even more extreme forms of matter.

> **The mass of a white dwarf cannot exceed 1.4 $M_\odot$.**

As the white dwarf in our binary system approaches this limit, it grows smaller and smaller and hotter and hotter. As the star crosses the Chandrasekhar limit it begins to collapse, and the conversion of gravitational energy into thermal energy quickly drives the temperature in the white dwarf past the 6×10^8 K needed to slam carbon nuclei together with enough force to overcome their electric repulsion and fuse them. Once again, the onset of nuclear burning in a degenerate gas leads to a thermonuclear runaway, with ever higher temperatures and ever faster nuclear burning feeding off each other. But while the thermonuclear runaway in a nova involves only a thin shell of material on the surface of the white dwarf, runaway carbon burning involves the entire white dwarf. Within about a second, the entire white dwarf is consumed in the resulting conflagration. *In this single instant 100 times more energy is liberated than will be given off by the Sun over the entire course of its 10 billion years on the main sequence.* Runaway fusion reactions convert a large fraction of the mass of the star into elements such as iron and nickel, and the explosion blasts the shards of the white dwarf into space at top speeds in excess of 20,000 km/s.

The carbon white dwarf can ignite in a Type I supernova.

The explosion completely destroys star 1, leaving star 2 behind as a lone giant to continue its evolution toward a white dwarf (Figure 15.14h).

Explosive carbon burning in a white dwarf is a leading theory used to explain colossal events called **Type I supernovae.** Type I supernovae[3] occur in a galaxy the size of the Milky Way about once a century. For a brief time they can shine with a luminosity 10 billion times that of our Sun, possibly outshining the galaxy itself.

---

[3]Strictly speaking, we should refer here to Type Ia supernovae. Technically, a "Type I" supernova is a supernova that shows no emission lines from hydrogen. This includes some variants of the supernovae from massive stars that will be discussed in the next chapter. We hope the purist will forgive us the simplification of referring to all supernovae from white dwarfs as Type I and all supernovae from massive stars as Type II.

SEEING THE FOREST THROUGH THE TREES

Stars are born of the tenuous matter of interstellar space, then settle down onto the main sequence as stable and steady cosmic citizens. But the Universe is a place of constant change and evolution. Stars consume the fuel they were born with, and when that fuel is gone, they must die.

In the world of humans it is wealth and power and birthright that determine our lifestyle. In the lives of stars, birthright is spelled "mass." Mass determines the weight bearing down on the interior of a star, which in turn controls the rate of the nuclear reactions taking place there. High-mass stars are spectacular but ephemeral, burning out the fuel in their cores in only a few million years. Our Sun is typical instead of the much longer-lived low-mass stars. Yet eventually even these steady performers must exhaust the hydrogen at their centers.

In this chapter we have followed the evolution of a low-mass star like our Sun as it evolves beyond the main sequence, passing through a fascinating series of evolutionary stages. Some of these stages are dramatic and violent, such as the runaway thermonuclear reactions of the helium flash or the ejection of a planetary nebula that carries the chemical legacy of nuclear burning deep within the dying AGB star. Other phases are far more quiescent, such as the star's time on the horizontal branch burning helium to carbon in its core and hydrogen to helium in a surrounding shell. Eventually even an undistinguished star like our Sun may reach a luminosity rivaling all but the most luminous main sequence O stars, and swell to a size that would swallow much of our inner Solar System. Curiously, such a behemoth is supported by frenzied nuclear reactions in a region surrounding the tiniest of cores—a superdense inert cinder crushed by the weight of the overlying star to a size little larger than Earth. At a density of a thousand kilograms per cubic centimeter, the electron-degenerate core defies our normal notions of matter.

All of these stages are signposts along the star's inexorable march toward its eventual fate when the outer layers of the star have been ejected into space and nothing remains but a tiny cooling cinder—a degenerate white dwarf that will eventually fade from the notice of the rest of the Universe. Yet if it happens to orbit about a binary companion, the white dwarf may live again. Unburnt nuclear fuel pouring onto the surface of the white dwarf may trigger the episodic nuclear outbursts of novae, or even a cataclysmic Type I supernova.

We have never witnessed the life cycle of a low-mass star from birth to death, nor will we. Not even a hundred million human lifetimes would be enough to see the saga through from beginning to end. Yet we are confident of most of what we have reported here, just as we are confident of our knowledge of the internal structure of the Sun and Earth. Physics works, and the closing decades of the 20th century saw steady improvements in the quality of our calculations of the structure and evolution of stars. And while we cannot witness the life cycle of a single star, when we look at the sky we find many examples of red giants and AGB stars and white dwarfs and each of the other members of this evolutionary menagerie. Later on our journey we will also see how clusters of stars that form together provide snapshots that allow us to test our theories of stellar evolution in exquisite detail.

What we have seen is representative of the fate awaiting the vast majority of stars in the Universe. In the next chapter we turn our attention to the evolution of the small fraction of stars that fall into the category of high-mass stars. There we will find that the extreme and counterintuitive behavior of white dwarfs presages a whole zoo of bizarre and exotic objects. Where low-mass stars bent our notions about matter, high-mass stars will shatter those notions, leading us to consider forms of matter so extreme that they cease to be matter at all.

STUDENT QUESTIONS

THINKING ABOUT THE CONCEPTS

1. Why do most nearby stars turn out to be low-mass, low-luminosity stars?

2. The Sun is slowly becoming more luminous as it ages on the main sequence. What effect will this have on terrestrial life?

3. What is the primary reason that the most massive stars have the shortest lifetimes? (The answer can be expressed in just a few words.)

4. Consider a hypothetical star similar to our Sun being one of the first stars that formed when the Universe was still very young. Would you expect it to be surrounded by planets? Explain your answer.

5. Suppose you were an astronomer making a survey of the observable stars in our Galaxy. What would be your chances of seeing a star undergoing the helium flash? Explain your answer.

6. The intersection of the Roche Lobes in a binary system is the equilibrium point between the two stars where the gravitational attraction from both stars is equally strong and opposite in direction. Is this an example of stable or unstable equilibrium? Explain.

7. We typically say that the mass of a newly formed star determines its destiny from birth to death, but there is a special case in which this is not true. Explain the circumstances under which this is not true.

8. Suppose Jupiter were not a planet, but instead were a G5 main sequence star with a mass of 0.8 $M_\odot$.
   a. What effect, if any, do you think this would have on terrestrial life?
   b. How would this affect the Sun as it came to the end of its life?

9. In Latin, *nova* means new. Novae, as we now know, are not "new" stars. Explain how novae might have gotten their name.

10. T Corona Borealis is a well-known recurrent nova.
    a. Is it a single star or a binary system? Explain.
    b. What mechanism causes a nova to flare up?
    c. How can a nova flare-up happen more than once?

APPLYING THE CONCEPTS

11. After leaving the main sequence, low-mass stars "seed" the interstellar medium with heavy elements such as carbon, nitrogen, and oxygen. Assume that the Universe is approximately 1.3×10^{10} (13 billion) years old and that the Sun is 4.6 billion years old. Refer to Table 15.1 for the masses and main sequence lifetimes of stars. How many generations of each type of star preceded the birth of our Sun and Solar System?

12. Assume there are twice as many F0 stars as A0 stars, and that each type contributes heavy elements to the interstellar medium in approximate proportion to its main sequence mass. What was the relative contribution from the two types of star in the time before the Sun was born?

13. A white dwarf has a density of approximately 10^9 kg/m³. Earth has an average density of 5,500 kg/m³ and a diameter of 12,700 km. If Earth were compressed to the same density as a white dwarf, how large would it be?

14. The escape velocity ($v_{esc} = \sqrt{2GM_\odot/r_\odot}$) from the Sun's surface today is 618 km/s.
    a. What will the escape velocity be when the Sun becomes a red giant with a radius 50 times greater and a mass only 0.9 times that of today?
    b. What will it be when the Sun becomes an AGB star with a radius 200 times greater and a mass only 0.7 times that of today?
    c. How would these changes in escape velocity affect mass loss from the surface of the Sun as a red giant, and later as an AGB star?

15. Each form of energy generation in stars depends on temperature.
    a. Hydrogen fusion (proton–proton chain) at 10^7 K increases with an increase in temperature at a rate of T^5. If you raise the temperature of the hydrogen-burning core by 10%, how much does the hydrogen-fusion energy increase?
    b. Helium fusion (triple-alpha) at 10^8 K increases with an increase in temperature at a rate of T^{38}. If you raise the temperature of the helium-burning core by 10%, how much does the helium-fusion energy increase?

16

*It does not do to leave a live dragon
out of your calculations,
if you live near him.*

J. R. R. TOLKIEN (1892–1973)
THE HOBBIT

LIVE FAST, DIE YOUNG

16.1 THERE BE DRAGONS. . .

So far in our discussion of the lives of stars we have concentrated on what happens with low-mass stars like our Sun. These are stars that steadily burn hydrogen to helium in their cores and produce a fairly constant output of energy for billions of years—long enough for our Earth to cool from a molten ball of rock, and for life to evolve to the point of contemplating its place in the Universe. High-mass stars—stars with masses greater than about 8 $M_\odot$—are very different beasts. These are objects which burn with luminosities of thousands or even millions of times that of our Sun, and which squander their generous allotments of nuclear fuel in less time than mammalians have walked on Earth.

The difference between the ways in which high-mass stars and low-mass stars evolve lies in the same relationship between gravity, pressure, and the rate of nuclear burning that determines the balance in a star like our Sun. The litany should be familiar by now. More mass means stronger gravity and more weight bearing down on the inner parts of the star. This means higher pressure, higher pressure means faster reaction rates, and faster reaction rates mean greater luminosity. There are many differences in the evolution of low- and high-mass stars, but in the end, all of these trace back to the greater gravitational force bearing down on the interior of a high-mass star.

KEY CONCEPTS

Our Sun is but a pale ember compared with the brilliance of more massive stars. These stars consume their fuel in a cosmic blink of the eye, then in their death dazzle us with brilliant fireworks, seed the Universe with the elements of life, and force us to once again alter our concept of space and time. As our exploration of stars draws to a close, we will discover:

* How the structure of high-mass stars differs from that of low-mass stars;
* The stages of nuclear burning that evolving high-mass stars experience as they form progressively more massive chemical elements;
* How H-R diagrams allow us to measure the ages of stars and test our theories of stellar evolution;
* The death of massive stars in Type II supernovae;
* The origin of the most massive chemical elements;
* Degenerate neutron stars the size of small cities, with masses several times that of the Sun;
* A menagerie of bizarre objects, including pulsars and X-ray binary stars;
* That gravity is a consequence of the way that mass distorts the very shape of space-time; and
* The bottomless pits in space-time that we call black holes.

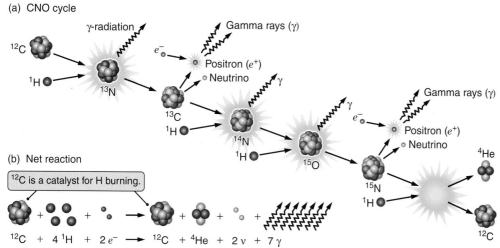

(a) CNO cycle

(b) Net reaction

^{12}C is a catalyst for H burning.

^{12}C + 4 ^{1}H + 2 e^- ⟶ ^{12}C + ^{4}He + 2 ν + 7 γ

Figure 16.1 *In high-mass stars, carbon serves as a catalyst for fusion of hydrogen to helium. This process is called the carbon–nitrogen–oxygen (or CNO) cycle.*

16.2 HIGH-MASS STARS FOLLOW THEIR OWN PATH

Recall from Chapter 13 that the first step in the hydrogen-burning proton–proton chain is the collision and fusion of two protons. Protons have only a single positive charge, so they are the easiest atomic nuclei to force close enough together to fuse. This is a great advantage for low-mass stars, where hydrogen burns at temperatures as low as a few million kelvins. However, the proton–proton chain suffers from a large disadvantage as well. The "affinity" between two protons is not very great. Even when two protons are slammed together hard enough for the strong nuclear force to act on them, the likelihood that they will fuse is low. The lack of vigor of this first step in the proton–proton chain limits how rapidly the proton–proton chain can run.

At the much higher temperatures at the center of a high-mass star additional nuclear reactions become possible. In particular, hydrogen nuclei are able to interact with the nuclei of more massive elements such as carbon. It takes a lot of energy to get past the electric barrier that separates a carbon nucleus, with its six protons, from a hydrogen nucleus. But if this barrier can be overcome, the strong nuclear force interaction between the carbon and the hydrogen is much more favorable than the interaction between two hydrogen nuclei. This reaction, in which a ^{12}C nucleus and a proton combine to form a ^{13}N nucleus consisting of seven protons and six neutrons, is the first reaction in a chain of events that is referred to as the **carbon–nitrogen–oxygen (CNO) cycle.** The remaining steps in the CNO cycle are shown in **Figure 16.1.** Note that carbon is not consumed by the CNO cycle but instead is a catalyst. The ^{12}C nucleus that started the cycle is there again at the end of the cycle, but now instead of the four hydrogen nuclei that took part in the chain of reactions, we have a helium nucleus. **Figure 16.2** shows that the CNO cycle is far more efficient than the proton–proton chain in stars more massive than 1.5 $M_\odot$.

The CNO cycle burns hydrogen in massive stars.

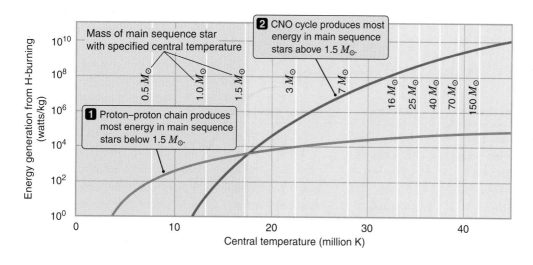

Figure 16.2 *Plots of the rate of energy generation as a function of temperature for the proton–proton chain and the CNO cycle. At the higher central temperatures of stars more massive than 1.5 M$_\odot$, it is the CNO cycle that more efficiently fuses hydrogen into helium.*

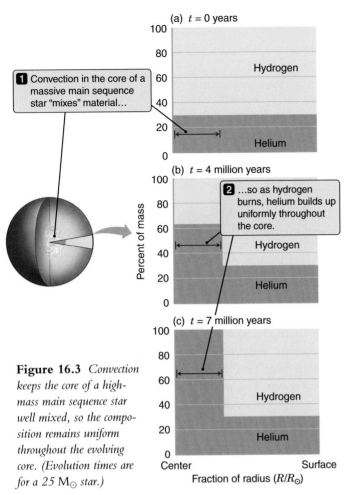

(a) *t* = 0 years

1 Convection in the core of a massive main sequence star "mixes" material...

(b) *t* = 4 million years

2 ...so as hydrogen burns, helium builds up uniformly throughout the core.

(c) *t* = 7 million years

Percent of mass

Center   Surface
Fraction of radius (*R/R*⊙)

Figure 16.3 *Convection keeps the core of a high-mass main sequence star well mixed, so the composition remains uniform throughout the evolving core. (Evolution times are for a 25 M⊙ star.)*

a boiling pot. Compare **Figure 16.3** with the similar figure in Chapter 15 (Figure 15.2). Rather than building up from the center outward, helium ash is spread uniformly throughout the core of a high-mass star as the star consumes its hydrogen.

THE HIGH-MASS STAR LEAVES THE MAIN SEQUENCE

These differences between high- and low-mass stars might seem a bit esoteric, but as the star's life on the main sequence comes to an end, the visible differences in the structure and evolution of the star become far more pronounced. As the high-mass star runs out of hydrogen in its core, the weight of the overlying star compresses the core, just as it did in a low-mass star. Yet long before the core of the high-mass star becomes electron degenerate, the pressure and temperature in the core reach the 10^8 K point needed for helium burning to begin. There will be no growing degenerate core in the high-mass star. There will be no accelerating ascent up the red giant and asymptotic giant branches on the H-R diagram. Instead, the star makes a fairly smooth transition from hydrogen burning to helium burning. The overall structure of the star responds to the changes taking place in its interior, but its luminosity will change relatively little. When a low-mass star leaves the main sequence, the path that it follows on the H-R diagram is largely vertical, going to higher and higher luminosities. As the

> **No degenerate core develops as a high-mass star leaves the main sequence.**

> **High-mass stars "skip" the red giant phase.**

16.1
HIGH

high-mass star leaves the main sequence, on the other hand, it grows in size while its surface temperature falls, so it moves mostly off to the right on the H-R diagram **(Figure 16.4).** The massive star now has the same structure as a low-mass horizontal branch star, burning helium in its core and hydrogen in a surrounding shell, but it has skipped the red giant phase of low-mass star evolution.

The different ways that hydrogen burning takes place in high- and low-mass stars are reflected in the different structures of their cores. The temperature gradient in the core of a high-mass star is so steep that convection sets in within the core itself, "stirring" the core like the water in

> **Convection stirs the core of a high-mass star.**

TABLE 16.1

BURNING STAGES IN HIGH-MASS STARS

Core Burning Stage	9$M_⊙$ Star	25$M_⊙$ Star	Typical Temperatures
H burning	20 million years	7 million years	$(3–10) \times 10^7$ K
He burning	2 million years	700,000 years	$(1–7.5) \times 10^8$ K
C burning	380 years	600 years	$(0.8–1.4) \times 10^9$ K
Ne burning	1.1 years	1 year	$(1.4–1.7) \times 10^9$ K
O burning	8 months	6 months	$(1.8–2.8) \times 10^9$ K
Si burning	4 days	1 day	$(2.8–4) \times 10^9$ K

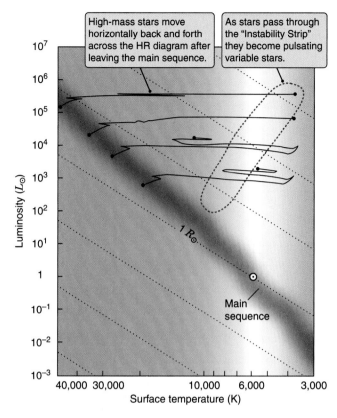

Figure 16.4 *When massive stars leave the main sequence they move horizontally across the H-R diagram.*

The next stage in the evolution of a high-mass star has no analog in low-mass stars. When the high-mass star exhausts the helium in its core, its core again begins to collapse, but this time as the core collapses it reaches temperatures of 8×10^8 K or higher, and carbon begins to burn (see **Table 16.1**). Carbon burning produces a number of more massive elements, including sodium, neon, and magnesium. The star now consists of a carbon-burning core surrounded by a helium-burning shell, which in turn is surrounded by a hydrogen-burning shell. The

Progressively more massive elements burn.

sequence does not end here. When carbon is exhausted as a nuclear fuel at the center of the star, neon burning picks up the slack, and when neon is exhausted, oxygen begins to burn. The structure of the evolving high-mass star, shown in **Figure 16.5,** is reminiscent of an onion, with layer surrounding layer surrounding layer. As we move inward toward the center of the star, we pass through a layer of hydrogen burning, then a layer of helium burning, then a layer of carbon burning, and so on. Finally we reach the most advanced stages of nuclear burning in the core of the star.

NOT ALL STARS ARE STABLE

When we discussed the structure of a low-mass star in Chapter 15, we asserted that a single stable solution always exists. Given one solar mass with a mix of 70% hydrogen and 30% helium, that mass *will* form a stable star with the structure of our Sun. As we "what-if" the Sun, considering the consequences of forcing the Sun

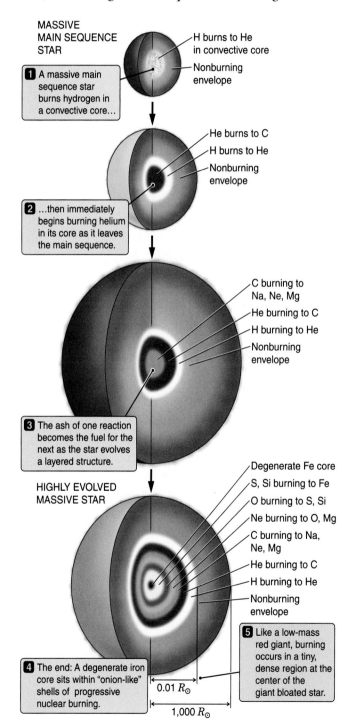

Figure 16.5 *As a high-mass star evolves, it builds up a layered structure like an onion, with progressively more advanced stages of nuclear burning found deeper and deeper within the star.*

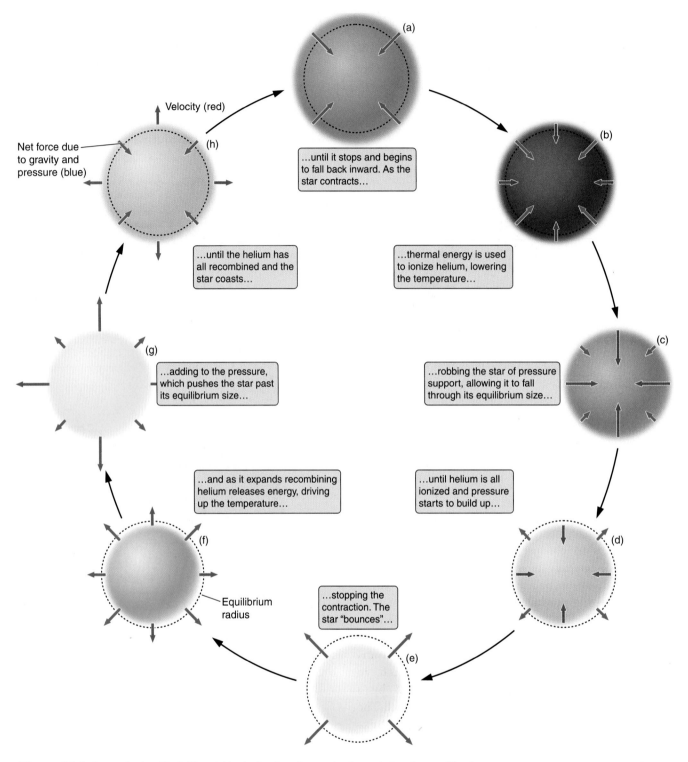

Figure 16.6 *In a pulsating Cepheid variable, ionization of atoms in the star's interior acts like the valves in a steam engine, alternately allowing the surface of the star to fall inward, then pushing it back out again. (Color changes shown here are greatly exaggerated.)*

to be larger or smaller than it is, we find that changes in the star would cause it to settle back into this unique balance between the inward pull of gravity and the outward push of pressure. This remains true for high-mass stars on the main sequence as well. However, it is *untrue* for many evolved stars. As a star undergoes post–main sequence evolution, it may make one or more passes through a region of the H-R diagram known as the **instability strip,** shown in Figure 16.4. Rather than finding a steady balance, stars in the in-

Evolved stars passing through the instability strip pulsate.

stability strip pulsate, alternately growing larger and smaller.

Stars lying within the instability strip of the H-R diagram are heat engines, powering their pulsation by tapping into the thermal energy flowing outward through them. Changes in the ionization state of the gas within the star alternately trap thermal energy, forcing the pressure in the star up and causing it to expand, and then release this energy, allowing the star to contract again. These changes serve the role of the valves in an engine: they rob the star of pressure support at one part of its cycle, allowing it to shrink too far, and then pump the pressure in the star back up at a different point in its cycle, puffing up the star to a larger size than it can support. The process is illustrated in **Figure 16.6.** Rather than settling at a constant radius, like the Sun, the star alternately expands and shrinks, moving out and in like the piston of a steam engine. The pulsations in the outer parts of the star have very little effect on the nuclear burning in the star's interior. However, the pulsations do affect the light escaping from the star. Both the luminosity and color of the star change as the star expands and shrinks. The star is at its brightest and bluest while it expands through its equilibrium size, and at its faintest and reddest while it falls back inward. Such stars are referred to as **pulsating variable stars.**

The highest-mass and most luminous pulsating variable stars are named **Cepheid variables,** after Delta Cephei, the first recognized member of this class. A Cepheid variable completes one cycle of its pulsation in anywhere from about 1 to 100 days, depending on its luminosity. The more luminous the star, the longer it takes it to complete its cycle. This **period–luminosity relationship** for Cepheid variables was first discovered experimentally by **Henrietta Leavitt** in 1912, and is the basis for the use of Cepheid variables as indicators of the distances to galaxies beyond our own.

Cepheid variables are not the only type of variable star. The horizontal branch of the evolutionary tracks of low-mass stars (see Figure 15.8) may pass through the instability strip as well. These unstable horizontal branch stars are known as **RR Lyrae variables** after their prototype. They pulsate by the same mechanism as Cepheid variables, but are typically hundreds of times less luminous. The instability strip also intersects the main sequence around spectral type A, and many A stars do show significant variability (although it is much less pronounced than the variability in Cepheids or RR Lyrae variables). A number of other kinds of variable stars are seen elsewhere in the H-R diagram, each driven by its own variety of "engine."

> There are many kinds of variable stars.

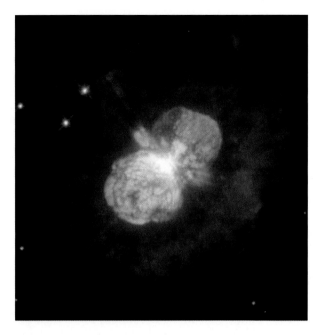

Figure 16.7 *This* Hubble Space Telescope *image shows an expanding cloud of dusty material ejected by the luminous blue variable star Eta Carinae. The star itself, which is largely hidden by the surrounding dust, has a luminosity of 3 million times that of the Sun, and a mass probably in excess of 100* M$_\odot$.

As low-mass stars evolve beyond the main sequence, they expel significant amounts of mass back into interstellar space. For a high-mass star, violent mass loss is a fact of life throughout its existence. Even while on the main sequence, massive O and B stars drive tenuous winds away from them at velocities that can reach 3,000 km/s. These winds are pushed outward by the pressure of the radiation from the star. We do not normally think of light as "pushing" on something, but the pressure of the intense radiation at the surface of a massive star is enough to overcome the star's gravity and drive away material in its outermost layers. Main sequence O and B stars lose mass at rates ranging from about 10^{-7} to 10^{-5} M$_\odot$ of material per year, with the greatest mass loss occurring in the most massive stars. These numbers may sound tiny, but added up over millions of years they mean that mass loss plays a prominent role in the evolution of high-mass stars. O stars with masses of 20 M$_\odot$ or more may lose around 20% of their mass while on the main sequence, and possibly more than 50% of their mass over their entire lifetime. Even an 8 M$_\odot$ star may lose 5% to 10% of its mass. An extreme example of mass loss in a massive star is seen in **Eta Carinae (Figure 16.7),** a 100 M$_\odot$ star with a lumi-

> Massive stars have high-velocity winds.

> Eta Carinae is an extreme case.

nosity of 3 million Suns. Currently Eta Carinae is losing mass at a rate of about $10^{-3}M_\odot$ per year (or 1 $M_\odot$ every 1,000 years). However, during a 19th-century eruption, when Eta Carinae became the second-brightest star in the sky, its mass loss must have reached the amazing level of 0.1 $M_\odot$ per year, shedding 2 $M_\odot$ of material over the course of only 20 years.

16.3 HIGH-MASS STARS GO OUT WITH A BANG

A low-mass star approaches the end of its life relatively slowly and gently as the outer parts of the star are ejected into a planetary nebula, leaving behind the degenerate core of the star. In stark contrast, for a high-mass star the end comes suddenly and amid considerable fury. An evolving high-mass star builds up its onion-like structure (Figure 16.5) as nuclear burning in its interior proceeds to more and more advanced stages. Hydrogen burns to helium, helium burns to carbon and oxygen, carbon burns to magnesium, oxygen burns to sulfur and silicon, then silicon and sulfur burn to iron. There are many different types of nuclear reactions that occur up until this point, forming almost all of the different stable isotopes of elements less massive than iron. But the essential point is this: *with iron, the chain of nuclear fusion stops.*

Iron is the most massive element formed by fusion.

Why does hydrogen burn while iron does not? More generally, what must be true for a material to serve as fuel? Gasoline burns because when it is combined with oxygen, energy is released. This extra energy pushes up the temperature and causes the chemical reaction between gasoline and oxygen to go faster. The reaction is *self-sustaining*, which means that the reaction itself is the source of thermal energy needed to cause the reaction to go. The same is true of nuclear fusion reactions in the interiors of stars. When four hydrogen atoms are combined to form a helium atom, the resulting helium atom has less energy than the four hydrogen atoms did separately. This difference in energy gets converted to thermal energy, which maintains the temperature of the gas at the high levels needed to sustain the reaction.

Burning must release energy.

How much energy is available from a nuclear reaction of combining, say, three helium nuclei to form a ^{12}C nucleus (the triple-alpha process)? We can answer the question in steps. First, how much energy would it

take to break the three helium nuclei into their constituent six neutrons and six protons? Next, how much energy would be released if these six protons and neutrons were combined to form a ^{12}C nucleus? The net energy produced by the reaction is just the difference between these two amounts.

The energy that it would take to break an atomic nucleus up into its constituent parts is called the **binding energy** of the nucleus. The net energy released by a nuclear reaction is the difference between the binding energy of the products and the binding energy of the reactants:

Different nuclei have different binding energies.

$$\frac{\text{Net}}{\text{energy}} = \frac{\text{binding energy}}{\text{of products}} - \frac{\text{binding energy}}{\text{of reactants}}$$

Figure 16.8 shows a plot of the nuclear binding energy per kilogram for a number of different atomic nuclei.

It is worth taking a brief digression to see how this works in practice. Using these values to work through the above example of the triple-alpha process, we find that the binding energy of a helium nucleus is 4.534×10^{-12} joules, or 6.822×10^{14} J per kilogram of helium. The binding energy of a ^{12}C nucleus is 1.477×10^{-11} J, or 7.410×10^{14} J per kilogram of carbon. To find the amount of energy available from fusing a kilogram of helium nuclei into carbon, take the difference between the two numbers:

$$\begin{aligned}\frac{\text{Net energy from}}{\text{burning 1 kg He}} &= \frac{\text{binding energy}}{\text{of C formed}} - \frac{\text{binding energy}}{\text{of He burned}} \\ &= 7.410 \times 10^{14}\text{ J} - 6.822 \times 10^{14}\text{ J} \\ &= 5.84 \times 10^{13}\text{ J}\end{aligned}$$

So helium is a good nuclear fuel. We can see this directly by looking at Figure 16.8. Moving from helium to carbon on the plot increases the binding energy of each nucleon, so fusing helium to carbon releases energy. But what if we instead try to fuse iron to form more massive elements? Since iron is at the peak of the binding-energy curve, the products of iron burning will have *less* binding energy than the reactants—going from iron to more massive elements takes us down on the binding energy curve in Figure 16.8—so the net energy in the reaction will be *negative*. Rather than producing energy, fusion of iron *uses* energy. Iron does not burn.[1]

Iron does not burn.

---

[1] Iron does not burn in a nuclear sense, but iron does burn *chemically*. Combining iron and oxygen increases the *chemical* binding energy of the atoms involved in the reaction. Of course, chemical burning plays no role in the interior of a star.

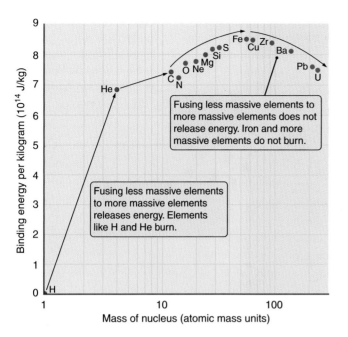

Figure 16.8 *The nuclear binding energy of a kilogram of material is plotted against the mass of the atomic nuclei. This is the energy that it would take to break a kilogram of the material apart into protons and neutrons. Energy is released by nuclear fusion only if the binding energy of the products is greater than the binding energy of the reactants.*

THE FINAL DAYS IN THE LIFE OF A MASSIVE STAR

The nuclear reactions following hydrogen burning are energetically much less favorable than conversion of hydrogen to helium. A look at Figure 16.8 shows that conversion of 1 kg of helium to carbon produces less than $\frac{1}{10}$ as much energy as conversion of 1 kg of hydrogen to helium. In order to support the star against gravity, this less efficient nuclear fuel must be consumed more rapidly. While conversion of hydrogen into helium can provide the energy needed to support the high-mass star against the force of gravity for millions of years, helium burning is able to support the star for only a few hundred thousand years.

Following helium burning, the nature of the balance within the star becomes qualitatively different.

Stages of burning after helium burning are progressively shorter-lived.

There is almost as much energy available from burning a kilogram of carbon, neon, oxygen, or silicon as there is from burning a kilogram of helium. We might expect, then, that each of these stages would last about as long as helium burning. Yet when we look at Table 16.1, we see that the star proceeds from helium burning through to the end of its life in a cosmic blink of the eye. So where is the catch?

You can think of the balance in a star as something like trying to keep a leaky balloon inflated. The larger the leak, the more rapidly you have to pump air into the balloon. A star that is burning hydrogen or helium is like a balloon with a slow leak. At the temperatures generated by hydrogen or helium burning, energy leaks out of the interior of a star primarily by radiation and/or convection. Neither of these processes is very efficient because the outer layers of the star act like a thick warm blanket. Nuclear fuel need only burn at a relatively modest rate to support the weight of the outer layers of the star while keeping up with the energy escaping outward. Beginning with carbon burning, this balance shifts in a dramatic and fundamental way. Rather than being carried by radiation and convection, energy now begins to escape from the core primarily in the form of neutrinos produced by the many nuclear reactions occurring there. When this happens, the floodgates are opened. Like air pouring out through a huge rent in the side of a balloon, neutrinos produced in the interior of the star stream through the overlying layers of the star as if they were not even there, carrying the energy from the stellar interior out into space. Keeping the balloon inflated becomes difficult indeed. As heat energy pours out of the interior of the star, the outer layers of

Burning proceeds at a breakneck pace to replace energy lost in neutrinos . . .

the star push inward, driving up the density and temperature, and forcing nuclear reactions to run at the furious rate necessary to replace the energy escaping in the form of neutrinos.

Once this process of **neutrino cooling** becomes significant, the star begins evolving much more rapidly. Carbon burning is only able to support the star for something less than about a thousand years. Oxygen burning holds the star up for only about a year. Once a massive star reaches the point that it begins to burn silicon, it is within a few days of the end. The luminosity of a silicon-burning star in electromagnetic radiation may be little greater than it was when the star was burning helium in its core. But if our eyes could sense neutrinos, we would see that a star burning silicon is actually giving off about 200 million times more energy per second than its former self!

Following silicon burning, we come to the end of the line. Once a star forms its iron core, no

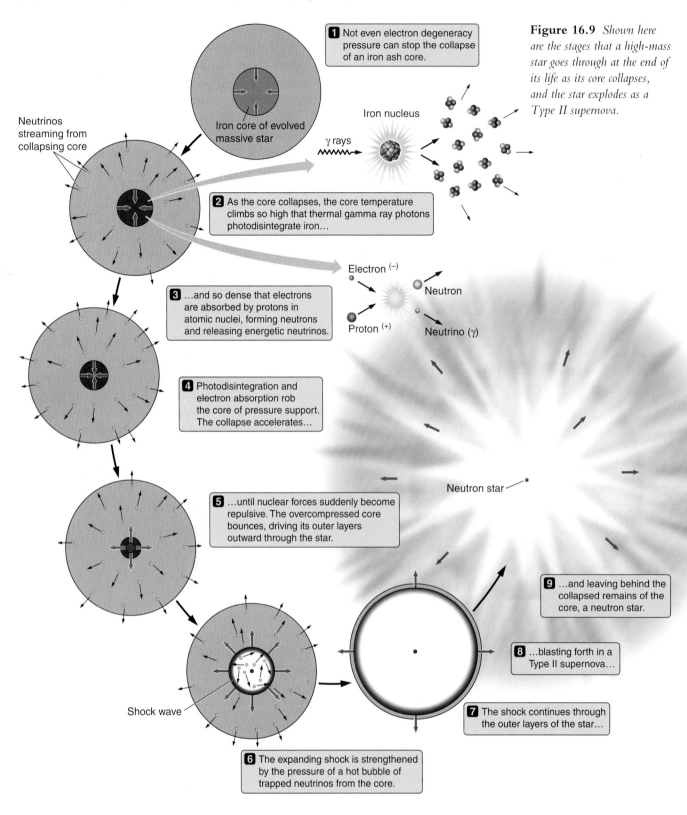

1 Not even electron degeneracy pressure can stop the collapse of an iron ash core.

Iron core of evolved massive star

Iron nucleus

γ rays

Neutrinos streaming from collapsing core

2 As the core collapses, the core temperature climbs so high that thermal gamma ray photons photodisintegrate iron…

3 …and so dense that electrons are absorbed by protons in atomic nuclei, forming neutrons and releasing energetic neutrinos.

Electron (−)

Neutron

Proton (+)

Neutrino (γ)

4 Photodisintegration and electron absorption rob the core of pressure support. The collapse accelerates…

5 …until nuclear forces suddenly become repulsive. The overcompressed core bounces, driving its outer layers outward through the star.

Neutron star

9 …and leaving behind the collapsed remains of the core, a neutron star.

8 …blasting forth in a Type II supernova…

7 The shock continues through the outer layers of the star…

Shock wave

6 The expanding shock is strengthened by the pressure of a hot bubble of trapped neutrinos from the core.

Figure 16.9 *Shown here are the stages that a high-mass star goes through at the end of its life as its core collapses, and the star explodes as a Type II supernova.*

. . . until no energy source remains to balance gravity.

source of nuclear energy remains to replenish the energy that is being lost in neutrinos. The high-mass star's life as a balancing act between gravity and a controlled thermonuclear furnace is over. Gravity will have its way.

THE CORE COLLAPSES AND THE STAR EXPLODES

Throughout the story of stellar evolution, we have seen that, while gravity can be held off for a time—perhaps a very long time—it is infinitely patient, and in the end generally wins the battle. Bereft of support from thermonuclear fusion, the iron core of the massive star begins to collapse. Refer to **Figure 16.9** as we follow the dramatic sequence of events that comes next.

The early stages of the collapse of the iron ash core of an evolved massive star are much the same as in the collapse of a nonburning core in a low-mass star. As the core collapses, its density and temperature skyrocket, and the force of gravity becomes even stronger. The gas in the core is once again compressed beyond the realm of normal matter, becoming electron degenerate when it reaches about the size of Earth. Unlike the electron-degenerate core of a low-mass red giant, however, the weight bearing down on the interior of the iron ash core is too great to be held up by electron degeneracy. This time gravity is too strong even for the quantum mechanical rules that limit how tightly electrons can be packed together. As the collapse continues, the core reaches temperatures of 10 billion K (10^{10} K) and higher, while the density exceeds 10 metric tons per cubic centimeter, or 10 times the density of an electron-degenerate white dwarf.

Density and temperature skyrocket as the iron core collapses.

These phenomenal temperatures and pressures trigger fundamental changes in the makeup of the core. The laws describing thermal radiation say that at these temperatures the nucleus of the star will be awash in extremely energetic thermal radiation. This radiation is so energetic that thermal gamma ray photons are produced with enough energy to break iron nuclei apart into helium nuclei. This process, known as **photodisintegration,** literally begins undoing the results of nuclear fusion—a task that uses up a tremendous amount of the thermal energy of the core. At the same time, the density of the core is so great that electrons are squeezed into atomic nuclei, where they combine with protons to produce neutron-rich isotopes in the core of the star. This process uses up thermal energy as well, robbing the core of even more of its pressure support. All the while, neutrinos continue to pour out of the core of the dying star. These events take place over the course of only about a second! The collapse of the core accelerates, reaching velocities of 70,000 km/s, or almost a quarter of the speed of light, on its inward fall.

Thermal energy is lost through photodisintegration and neutrino cooling.

The next hurdle standing in the way of the collapsing core is the force that holds atomic nuclei together. As material in the collapsing core reaches and exceeds the density of an atomic nucleus, the strong

The collapsing core bounces.

nuclear force actually becomes repulsive. Computer models say that about half of the collapsing core suddenly slows its inward fall. The remaining half slams into the innermost part of the star at a significant fraction of the speed of light and "bounces," sending a tremendous shock wave back out through the star.

Under the extreme conditions in the center of the star, neutrinos are being produced in copious abundance. Over the next second or so, almost a fifth of the mass of the material in the core is converted into neutrinos. Most of these neutrinos pour outward through the star, but at the phenomenal densities found in the collapsing core of the massive star, not even neutrinos pass with complete freedom. A few tenths of a percent of the energy of the neutrinos streaming out of the core of the dying star are trapped by the dense material behind the expanding shock wave. The energy of these trapped neutrinos drives the pressure and temperature in this region ever higher, inflating a bubble of extremely hot gas and intense radiation around the core of the star. The pressure of this bubble adds to the strength of the shock wave moving outward through the star. Within about a minute the shock wave has pushed its way out through the helium shell within the star. Within a few hours it reaches the surface of the star itself, heating the stellar surface to 500,000 K and blasting

The star explodes in a Type II supernova.

material outward at velocities of up to about 30,000 km/s. Our evolved massive star has exploded in an event referred to as a **Type II supernova.**

16.4 THE SPECTACLE AND LEGACY OF SUPERNOVAE

A Type II supernova is comparable in its spectacle to a Type I supernova. For a brief time, the Type II supernova can shine with the light of 100 billion suns, yet the energy that comes out of the supernova in the form of light is but a tiny fraction of the energy of the explosion itself. A hundred times more energy comes out as the kinetic energy of the outer parts of the star that are blasted into space by the explosion. There is about 10^{47} J of kinetic energy in the ejected material, or enough energy to accelerate the mass of our Sun to a speed of

Tremendous energy is released in the explosion.

EXCURSIONS 16.1

STARS ARE UNIQUE

The distinction made in this book between low-mass and high-mass stars is a useful and convenient way to think about stellar evolution. After all, there is a fundamental and qualitative difference between stars that end their lives quietly as dying cinders, and those that die in spectacular explosions. Even so, this distinction is an oversimplification. In reality, each star is unique, with its own individual mass and composition, and its own particular circumstances.

Stars with masses between about 3 $M_\odot$ and 8 $M_\odot$ exist in something of a gray area between the low-mass stars discussed in Chapter 15 and the high-mass stars in this chapter. While they are on the main sequence, these **intermediate-mass stars** burn hydrogen via the CNO cycle like massive stars. These stars also leave the main sequence as massive stars do, burning helium in their cores immediately after hydrogen is exhausted, and skipping the red giant and helium flash phases of low-mass star evolution. However, at the completion of the He-burning phase, the temperature at the center in intermediate-mass star is too low for carbon to burn. From this point on, the star evolves more like low-mass star, ascending the asymptotic giant branch burning helium and hydrogen in shells around a degenerate core, then ejecting its outer layers and leaving behind a white dwarf. Yet even at the end, the star carries the signature of its early "high-mass" years. The chemical composition of planetary nebulas and white dwarfs left behind by intermediate-mass stars can be quite distinct from those of truly low-mass stars.

10,000 km/s. It is the kinetic energy of the ejecta from supernovae of both Type I and Type II that is responsible for heating the hottest phases of the interstellar medium and pushing the clouds in the interstellar medium around. Yet even this amount of energy is a pittance in comparison with the energy carried away from the supernova explosion by neutrinos—an amount of energy a hundred times larger yet!

One of the most important astronomical events in the last part of the 20th century was the explosion of a massive star in the small companion galaxy to the Milky Way known as the Large Magellanic Cloud. Even at a distance of 160,000 light-years, supernova 1987A was so bright that it dazzled sky gazers in the Southern Hemisphere (see **Figure 16.10**). While astronomers working in all parts of the electromagnetic spectrum scrambled to point their telescopes at the new supernova, solar astronomers had already quietly and unknowingly captured one of the true scientific prizes of SN1987A. Solar neutrino telescopes recorded a burst of neutrinos passing through Earth—neutrinos which originated not in the Sun but in the tremendous stellar explosion that occurred beyond the bounds of our Galaxy itself. The detection of neutrinos from SN1987A provided us with a rare and crucial glimpse of the very heart of a massive star at the moment of its death, confirming a fundamental prediction of our theories about the collapse of the core and its effects.

THE ENERGETIC AND CHEMICAL LEGACY OF SUPERNOVAE

Type II supernova explosions leave a rich and varied legacy to the Universe. We have already mentioned their energetic legacy: the energy from these explosions is responsible for driving much of the dynamics of the interstellar medium. When we look at the sky, we see huge expanding bubbles of million-kelvin gas, like that in **Figure 16.11(a)**, aglow in X-rays, and driving visible shock waves **(Figure 16.11b)** into the surrounding interstellar medium. These bubbles are the still-powerful blast waves of supernova explosions that took place thousands of years ago. In fact, in many cases supernova explosions are thought to be responsible for compressing nearby clouds (like the one in **Figure 16.11c**) enough to trigger their collapse toward the formation of new generations of stars.

Figure 16.10 *SN1987A was a supernova that exploded in a small companion galaxy of the Milky Way called the Large Magellanic Cloud (LMC). These images show the LMC before the explosion, then while the supernova was near its peak.*

Perhaps even more important to us is the chemical legacy left behind by supernova explosions. Later in our journey we will turn our attention to the earliest moments after the birth of the Universe, and find that the only chemical elements that formed at that time were the least massive elements—hydrogen, helium, and trace amounts of lithium, beryllium, and boron. All of the rest of the chemical elements, including a large fraction of the atoms of which we are made, were formed in the hearts of stars, then returned to the interstellar medium. This process of progressive chemical enrichment of the Universe is called **nucleosynthesis.** We introduced the idea of nucleosynthesis in our discussion of planetary nebulae in Chapter 15, but while mass loss from both low- and high-mass stars enriches the interstellar medium with massive elements formed in their interiors, Type I and II supernovae are the true champions of nucleosynthesis. A look at Figure 16.5 should offer the first clue as to why this is so. Whereas low-mass stars are able to form elements only as massive as carbon and oxygen, nuclear burning in high-mass stars produces elements as massive as iron. Yet this is not the whole story. A look at a table of elements that occur in nature shows that there are many elements, up through uranium, that are far more massive than iron. If iron is the most massive element that can be formed by nuclear burning, then where do these more massive elements come from?

> **Supernovae eject newly formed massive elements into interstellar space.**

Figure 16.11 *The Cygnus Loop is a supernova remnant—an expanding interstellar blast wave caused by the explosion of a massive star.* (a) *Gas in the interior with a temperature of millions of kelvins glows in X-rays, while* (b) *visible light comes from locations where the expanding blast wave pushes through denser gas in the interstellar medium.* (c) *A* Hubble Space Telescope *image of a location where the blast wave is hitting an interstellar cloud.*

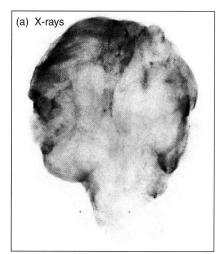
(a) X-rays

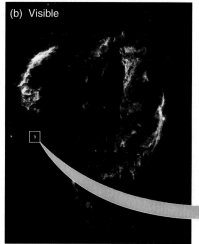
(b) Visible

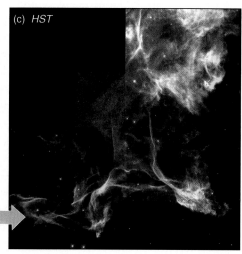
(c) *HST*

Answering this question takes us back to the reason why high temperatures are necessary for fusion to occur. Under normal circumstances electric repulsion keeps positively charged atomic nuclei far apart. Extreme temperatures are needed to slam nuclei together hard enough to overcome this electric repulsion.

Neutrons are not repelled by a nucleus's electric charge.

Free neutrons, on the other hand, are not subject to these ground rules: they have no net electric charge, so there is no electric repulsion to prevent them from simply running into an atomic nucleus, regardless of how many protons that nucleus contains. Under normal circumstances in nature, free neutrons are very rare. However in the interiors of evolved stars, a number of nuclear reactions produce free neutrons, and under some circumstances—including those shortly before and during a Type II supernova—free neutrons can be produced in very large numbers. Free neutrons are easily captured by atomic nuclei and later decay to become protons. In this way, elements with higher and higher mass are formed. The most massive element that is stable enough to last for long times is uranium. As discussed in Foundations 16.1, stellar nucleosynthesis correctly predicts the patterns in the abundances of atoms of different kinds found in the very earth beneath our feet.

We have seen the fate of the outer parts of the star which are blasted back into interstellar space by the explosion of a Type II supernova. But what remains of the core that was left behind? Picking up our story where we left off, the material at the center of the massive star has collapsed to the point where it has about the same density of matter as the nucleus of an atom. As long as the mass of the core left behind by the explosion is no more than about 3 $M_\odot$, this collapse will be halted by quantum mechanical rules similar to those responsible for holding up a white dwarf. Only now, instead of electrons it is neutrons that are forced together as tightly as the rules of quantum mechanics allow. The neutron-degenerate core left behind by the explosion of a Type II supernova is referred to as a **neutron star.** It has a radius of perhaps 10 km, making it roughly the

The neutron-degenerate core is left behind.

FOUNDATIONS 16.1

THE CHEMICAL COMPOSITION OF THE UNIVERSE: COMPARISON OF OBSERVATION AND THEORY

Calculations of nucleosynthesis in stars make clear predictions about which nuclei should be formed in abundance and which should not. These same patterns are found in measurements of the abundances of nuclei on Earth, in meteorite material, and in the atmospheres of stars. While some of these patterns are subtle, others can be appreciated by comparing what we have learned about nucleosynthesis in this book with a plot of the relative abundance of elements found in nature **(Figure 16.12)**. Notice first of all that less-massive elements are far more abundant than more-massive elements—a consequence of the way more-massive elements are progressively built up from less-massive elements. An exception to this is the dip in the abundances of the light elements lithium, beryllium, and boron. These light elements are easily destroyed by nuclear burning, but their production is mostly bypassed by the main reactions involved in burning of H and He. Conversely, carbon, nitrogen, and oxygen are big winners in the CNO cycle of hydrogen burning and the triple-alpha process of helium burning, and their high abundances reflect this fact. The spike in the abundances of the "iron-peak"

elements is evidence of the nucleosynthetic processes including those in Type I and Type II supernovae that favor these most tightly bound of all nuclei. Even the sawtooth pattern in the abundances of even- and odd-numbered elements can be understood as a consequence of the process of stellar nucleosynthesis. (To complete our discussion of nucleosynthesis, we will need to consider the formation of elements that took place, not in stars, but at the time of the formation of the Universe itself. We will take up this discussion in Chapter 20.)

Once again we find connections and insights in places that prior to our study of the Universe we might never have imagined to look. Our understanding of the processes at work within the interiors of dying stars is being confirmed by analyzing the chemical composition of Earth under our feet. At the same time, we can now make a pretty good stab at saying what kinds of stars are responsible for forming the atoms that make up our own bodies. Our growing understanding of the chemical evolution of the Universe and our connection to it represents one of the triumphs of modern astronomy.

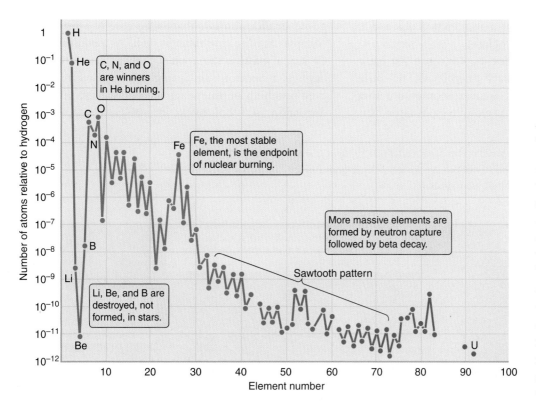

Figure 16.12 *The relative abundances of different elements on Earth are plotted against the mass of the nucleus. This pattern can be understood as a result of the process of nucleosynthesis in stars.*

C, N, and O are winners in He burning.

Fe, the most stable element, is the endpoint of nuclear burning.

Li, Be, and B are destroyed, not formed, in stars.

More massive elements are formed by neutron capture followed by beta decay.

Sawtooth pattern

other exotic objects. If the massive star responsible for the formation of a neutron star is part of a binary system, then the neutron star will be left with a binary companion. Such a system might remind you of the white dwarf binary systems responsible for novae and Type I supernovae discussed in Chapter 15. As the second star in such a binary system evolves and overfills its Roche lobe, matter plummets down the deep gravitational well of the tiny but massive neutron star. This matter slams into the accretion disk around the neutron star with enough energy to heat the disk to temperatures of millions of kelvins, and the accretion disk glows brightly in X-rays. Such an

Neutron stars form X-ray binaries.

object, illustrated in **Figure 16.13,** is known as an **X-ray binary.** Many fascinating phenomena can occur in X-ray binaries, including the formation of powerful jets that blast away from the neutron

size of a small city, but into that volume is packed a mass more than 1.4 times that of our Sun. At a density of around a billion metric tons per cubic centimeter, the neutron star is a billion times more dense than a white dwarf and a thousand trillion (10^{15}) times more dense than water! That density is roughly the same as we would get by crushing the entire Earth down to an object the size of a football stadium.

As if neutron stars were not extraordinary enough in their own right, they also form the heart of a number of

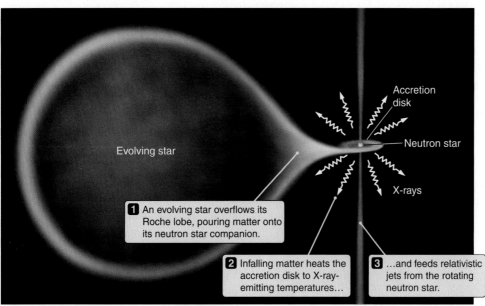

1 An evolving star overflows its Roche lobe, pouring matter onto its neutron star companion.

2 Infalling matter heats the accretion disk to X-ray-emitting temperatures…

3 …and feeds relativistic jets from the rotating neutron star.

Evolving star

Accretion disk

Neutron star

X-rays

Figure 16.13 *X-ray binaries are systems consisting of a white dwarf, neutron star, or black hole and a normal evolving star. As the evolving star overflows its Roche lobe, mass falls toward the collapsed object. The gravitational well of the collapsed object is so deep that when the material hits the accretion disk, it is heated to such high temperatures that it radiates away most of its energy as X-rays.*

star in directions perpendicular to its accretion disk and at speeds approaching the speed of light.

Besides its phenomenal density, a neutron star has a number of other extraordinary properties. The same principle of conservation of angular momentum that requires a collapsing molecular cloud to spin faster as it grows smaller, also says that as the core of a massive star collapses, it must spin faster as well. As a main sequence O star, a massive star rotates perhaps once every few days. As a neutron star, it might instead rotate tens or even hundreds of times each second. We also saw in our discussion of a collapsing interstellar cloud how the

Figure 16.14 *As a highly magnetized neutron star rotates rapidly, light is given off, much like the beams from a rotating lighthouse lamp. From our perspective, as these beams sweep past us, the star will appear to pulse on and off, earning it the name* pulsar.

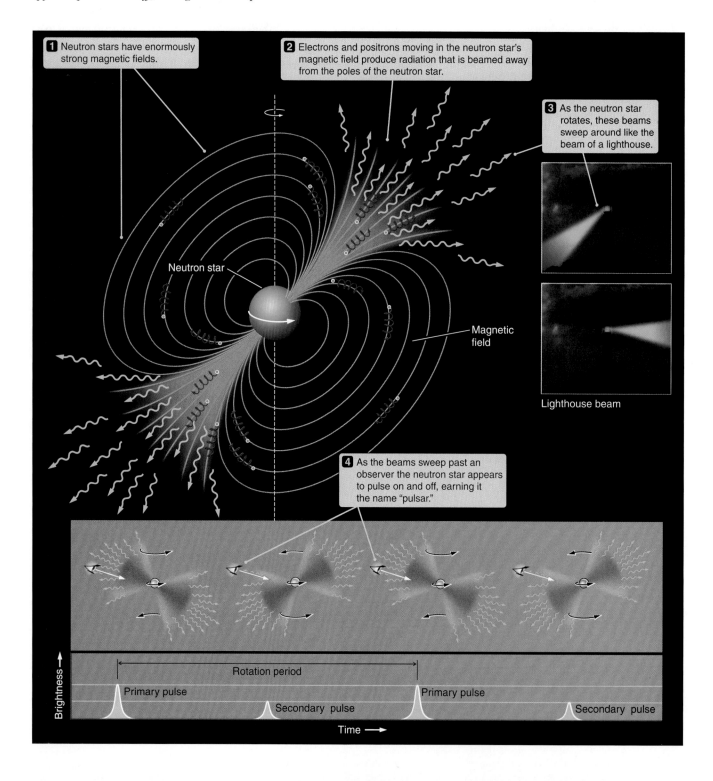

1 Neutron stars have enormously strong magnetic fields.

2 Electrons and positrons moving in the neutron star's magnetic field produce radiation that is beamed away from the poles of the neutron star.

3 As the neutron star rotates, these beams sweep around like the beam of a lighthouse.

Neutron star

Magnetic field

Lighthouse beam

4 As the beams sweep past an observer the neutron star appears to pulse on and off, earning it the name "pulsar."

Brightness

Rotation period

Primary pulse

Secondary pulse

Primary pulse

Secondary pulse

Time →

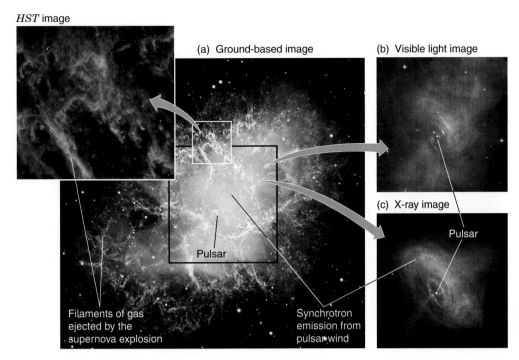

HST image

(a) Ground-based image

Pulsar

Filaments of gas ejected by the supernova explosion

Synchrotron emission from pulsar wind

(b) Visible light image

(c) X-ray image

Pulsar

Pulsar

Figure 16.15 *The Crab Nebula is the remnant of a supernova explosion witnessed by Chinese astronomers in A.D. 1054. (a) The object we see today is an expanding cloud of "shrapnel" from that earlier cataclysm. The spinning pulsar at the heart of the Crab Nebula sends off a "wind" of electrons and positrons moving at close to the speed of light. The synchrotron radiation from these is shown (b) in visible light and (c) in X-rays.*

magnetic field in the cloud is carried along and concentrated by the collapse. This phenomenon also occurs in the collapse of the star, amplifying the magnetic field to values that are trillions of times greater than the magnetic field at Earth's surface. A neutron star has a magnetosphere, just as Earth and several other planets do, except that the neutron star's magnetosphere is unimaginably stronger, and is whipped around many times a second by the spinning neutron star.

Energetic subatomic particles such as electrons and positrons move along the magnetic field lines of the neutron star, and are "funneled" by the field toward the magnetic poles of the system. Conditions there produce intense electromagnetic radiation which is beamed out in the directions away from the magnetic poles of the neutron star as in **Figure 16.14.** As the neutron star rotates, these beams of radiation sweep through space much like the rotating beams of a lighthouse. When we are located in the paths of these beams, then we see what a ship coming into harbor at night would see looking at a lighthouse. The neutron star appears to flash on and off with a regular period equal to the period of rotation of the star (or half the rotation period, if we see both beams).

Rapidly pulsing objects were first discovered by observers working at radio wavelengths in 1967. These objects, which blinked like regularly ticking clocks, puzzled astronomers. One of the early tongue-in-cheek names given to these objects was *LGMs*, which stood for "little green men." Today these objects are

Pulsars are rapidly spinning magnetized neutron stars.

referred to by the less flamboyant but more accurately descriptive term **pulsar.** At the time of this writing well over a thousand pulsars are known, and more are being discovered all the time.

THE CRAB NEBULA—REMAINS OF A STELLAR CATACLYSM

In A.D. 1054, Chinese astronomers recorded the presence of a "guest star" in the part of the sky that we call the constellation Taurus. The new star was so bright that it could be seen during the daytime for three weeks, and did not fade from visibility altogether for a period of many months. On the basis of the Chinese description of the changing brightness and color of the object, we can say today that the guest star of 1054 was a fairly typical Type II supernova. When we look at this spot in the sky today, we see an expanding cloud of debris from this explosion—an extraordinary object called the **Crab Nebula (Figure 16.15).**

The Crab Nebula consists of several components. Images of the Crab Nebula taken in the light of nebular emission lines show filaments of glowing gas. Doppler shift measurements of these filaments reveal a pattern much like that seen

An expanding cloud of debris carries the Crab's chemical legacy.

in planetary nebulae—the hallmark of an expanding shell. But where planetary nebulae are expanding at 20 to 30 km/s, the shell of the Crab is expanding at closer to 1,500 km/s. Studies of the spectra of these filaments show that they contain anomalously high abundances of helium and other more massive chemical elements—the products of the nucleosynthesis that took place in the supernova and its progenitor star.

There is a pulsar at the center of the Crab Nebula. This was the first pulsar to be seen at visible wavelengths as it flashed on and off 30 times a second. Actually the Crab pulsar flashes 60 times a second with a main pulse associated with one of the lighthouse beams, then a fainter secondary pulse associated with the other lighthouse beam. Today the Crab pulsar has been observed in all parts of the electromagnetic spectrum, from radio waves through to X-rays and even higher-energy gamma rays.

As the Crab pulsar spins 30 times a second, it whips its powerful magnetosphere around with it. At a distance from the pulsar about equal to the radius of the Moon, material in the magnetosphere must move at almost the speed of light to keep up with this rotation. Like a tremendous slingshot, the rotating pulsar magnetosphere flings elementary particles—probably mostly electrons and antimatter electrons called positrons—away from the neutron star in a powerful wind moving at close to the speed of light. Material from this wind fills the space between the pulsar and the expanding shell. The Crab Nebula is almost like a big balloon, but instead of being filled with hot air, it is filled with a mix of relativistic particles and strong magnetic fields—an environment more like what we might find in a physicist's particle accelerator than what we normally think of as interstellar space. Images of the Crab Nebula (Figure 16.15b and c) show this bizarre bubble as an eerie glow. This glow is synchrotron radiation from the relativistic electrons and positrons as they spiral around the magnetic field in the Crab.

The pulsar powers the Crab's eerie synchrotron glow.

16.5 STAR CLUSTERS ARE SNAPSHOTS OF STELLAR EVOLUTION

Over the course of this and the previous chapter, we have told a remarkable tale about the evolution of stars of different masses, presenting the story as a well-corroborated theory. In other words, we have presented the story of stellar evolution as fact. Yet even the most massive stars take hundreds of thousands of years to evolve, which is far longer than the handful of decades that we have spent studying their ways. Upon what testable predictions of our theories of stellar evolution do we base such solid claims of knowledge?

As we learned in Chapter 14, when an interstellar cloud collapses, it fragments into pieces, forming not one star but many stars of different masses. We see many such **star clusters** around us today, containing anywhere from a few dozen to millions of stars. The fact that all of the stars in a cluster formed together at very nearly the same time means that clusters are snapshots of stellar evolution. A look at a cluster that is 10 million years old shows us what stars of all different masses evolve into during the first 10 million years after they are formed. A look at a cluster 10 *billion* years after it formed shows us what has become of stars of different masses after 10 billion years have passed.

Stars in clusters formed together at about the same time.

This basic result—the fact that high-mass stars in a cluster evolve more rapidly than low-mass stars that formed at the same time—provides the key to our knowledge of stellar evolution. **Figure 16.16** shows simulations of the H-R diagram of a cluster of 40,000 stars as it would appear at several different ages. In Figure 16.16(a) stars of all masses are located on the zero age main sequence, showing where they begin their lives as main sequence stars. The increasing masses of stars along the main sequence are indicated. We would never expect to see a cluster H-R diagram that looks like the one in Figure 16.16(a), however, for the simple reason that the stars in a cluster do not all reach the main sequence at exactly the same time. Star formation in a molecular cloud is spread out over several million years, and it takes a considerable time for lower mass stars to contract to reach the main sequence. The H-R diagram of a very young cluster normally shows many lower-mass stars located well above the main sequence, still descending their Hayashi tracks.

The more massive a star is, the shorter its life on the main sequence will be. After only 4 million years (Figure 16.16(b), all stars with masses greater than about 20 $M_\odot$ have evolved off of the main sequence and are now spread out across the top of the H-R diagram. The most massive stars have already disappeared from the H-R diagram entirely, having vanished in supernovae.

As time goes on, lower and lower mass stars evolve off the main sequence. As in Figure 16.16(c), by the time the cluster is 10 million years old, only stars with masses less than about 15 $M_\odot$ remain on the main sequence. The cluster H-R diagram looks as if we grabbed the top of the band of cluster stars and gradually peeled it away from its original location along the main sequence. The location of the most massive star that is still on the main sequence is called the **main sequence turnoff.** As the cluster ages, the main sequence turnoff moves farther and farther down the main sequence to stars of lower and lower mass.

The main sequence turnoff shifts to lower-mass stars as a cluster grows older.

As a cluster ages (Figure 16.16d and e) we see more than the motion of the main sequence turnoff to lower and lower masses; we see the details of all stages of stellar evolution. By the time the star cluster is 10 *billion* years old (Figure 16.16f), stars with masses of only 1 $M_\odot$ are beginning to pull away from the main sequence. Stars slightly more massive than this are seen as giant stars of various types. In this cluster the horizontal branch appears as a knot of stars sometimes referred to as the "red clump." Note how few giant stars are present in any of the cluster H-R diagrams. The giant, horizontal, and asymptotic giant branch phases in the evolution of low- and intermediate-mass stars pass so quickly in comparison with a star's main sequence lifetime that even though the cluster started with 40,000 stars, only a handful of stars are seen in these phases of evolution at any given time. Similarly, it takes a newly formed white dwarf only a few tens of millions of years to cool to the point that it disappears off of the bottom of these figures. Even though the majority of evolved stars in an old cluster are white dwarfs, all but a few of these stars will have cooled and faded into obscurity at any given time.

The cluster H-R diagrams in Figure 16.16 are theoretical calculations of what clusters of different ages *should* look like. The crucial point is that this is what H-R diagrams of real clusters *must* look like if our theories of stellar evolution are correct. Fortunately, this is also what H-R diagrams of real clusters *do* look like.

Real cluster H-R diagrams match predictions of stellar evolution models.

Figure 16.16 *H-R diagrams of star clusters are snapshots of stellar evolution. These are simulated H-R diagrams of a cluster of 40,000 stars of solar composition seen at different times following the birth of the cluster. Note the progression of the main sequence turnoff to lower and lower masses.*

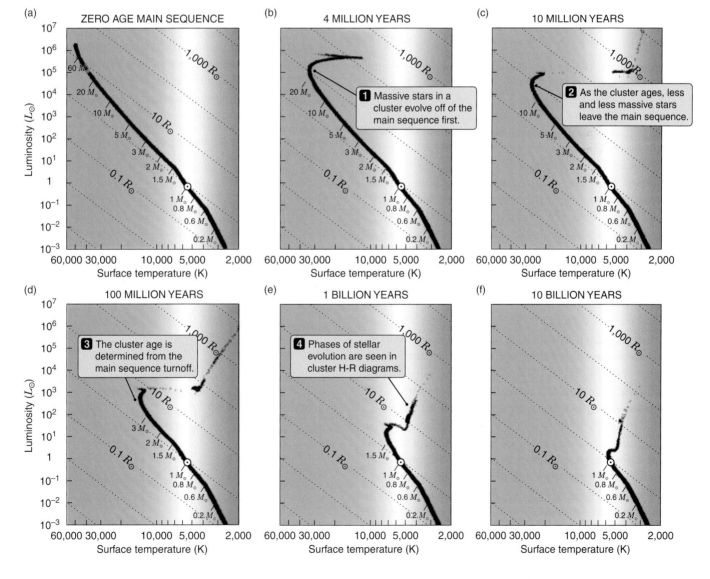

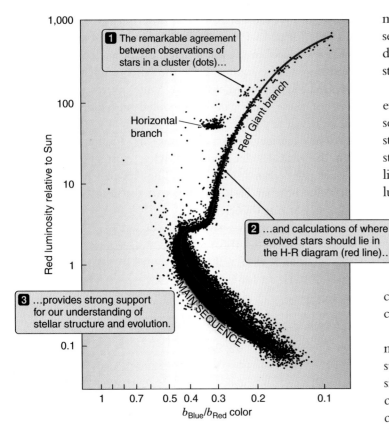

Figure 16.17 *The observed H-R diagram of the cluster 47 Tucanae agrees remarkably well with the theoretical calculation (solid line) of the H-R diagram of a 12-billion-year-old cluster.*

masses. Detailed comparisons between models and observed cluster H-R diagrams must account for the abundances of massive elements in the atmospheres of cluster stars, as well as the cluster age.)

We can apply our understanding of stellar evolution even when groups of stars are so far away that we cannot see each star individually. While many fewer high-mass stars form in a cluster than low-mass stars, higher-mass stars are *far* more luminous than their lower mass siblings. Likewise giant, evolved low-mass stars are far more luminous than less massive stars that remain on the main sequence. As a result, the most massive, most luminous stars present will dominate the light from a star cluster. If the cluster is young, then most of the light we see will usually come from luminous hot, blue stars. If the cluster is old, then the light from the cluster will have the color of red giants and relatively cool, low-mass stars.

As always, there are caveats to such general statements. Young clusters may contain very luminous red supergiants, while stars with lower abundances of massive elements in their atmospheres often look significantly bluer than their more chemically enriched counterparts. Even so, we can usually get some idea about the properties of a group of stars by looking at its overall color. We will put this knowledge to good use in the chapters that follow as we turn our attention to the

Figure 16.17 shows the observed H-R diagram for the cluster 47 Tucanae, along with a theoretical calculation of the H-R diagram for a 12-billion-year-old cluster. The quality of the agreement speaks for itself. The fact that observed star clusters have H-R diagrams that agree so well with the predictions of models is strong support for our theories of stellar evolution.

Armed with confidence in the reliability of our ideas about stellar evolution, we can turn cluster evolution models into a powerful tool for studying the history of star formation. When we observe a star cluster, the location of the main sequence turnoff immediately tells us the age of the cluster. If shown any of the cluster H-R diagrams in Figure 16.16, we should have no difficulty estimating the age of the cluster. **Figure 16.18** traces the observed H-R diagrams for several real star clusters. Once we know what to look for, the difference between young and old clusters is obvious. NGC 2362 is clearly a young cluster. Its complement of massive, young stars shows it to be only a few million years old. In contrast, 47 Tucanae (Figure 16.17) has a main sequence turnoff of around 0.85 $M_\odot$, signifying a cluster age of around 12 billion years. (We should note that the evolution of stars depends on their chemical composition as well as their

Figure 16.18 *H-R diagrams for clusters having a range of different ages. The ages associated with the different main sequence turnoffs are indicated.*

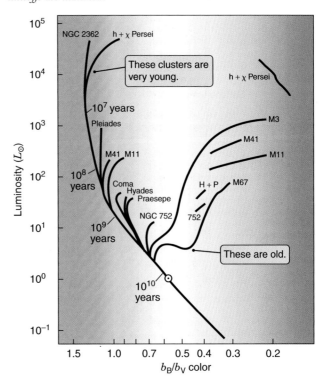

Colors reveal the ages of stellar populations.

large collections of stars called galaxies. A group of stars with similar ages and other characteristics is referred to as a **stellar population.** If a galaxy or a part of a galaxy is especially blue in color, it often signifies that the galaxy contains a young stellar population that still includes hot, luminous, blue stars that must have formed recently. In contrast, if a galaxy or part of a galaxy is red in color, we expect that it is composed primarily of an old stellar population.

16.6 GRAVITY IS A DISTORTION OF SPACE-TIME

Having discussed the evolution of clusters of stars of all masses, we now return to the stellar remnants left behind by the evolution of the most massive stars. You might imagine that, surely, the neutron star represents the final extreme of stellar evolution, but there is one more step a star can take. The physics of a neutron star is much like the physics of a white dwarf, except that it is neutrons rather than electrons that are degenerate. A white dwarf can have a mass of no more than about 1.4 $M_\odot$. This is the Chandrasekhar limit, discussed in Chapter 15. If the mass of the object exceeds this limit, then gravity will be able to overcome electron degeneracy, and the white dwarf will begin collapsing again.

Just as there is a Chandrasekhar limit for white dwarfs, there is a Chandrasekhar limit for the mass of neutron stars as well. If the mass of a neutron star exceeds about 3 $M_\odot$, then gravity will begin to win out over matter once again.

If a neutron star's mass exceeds 3 $M_\odot$, it will collapse to a black hole.

The neutron star grows smaller, and gravity becomes stronger and stronger at an ever accelerating pace. However, this time there is no force in nature powerful enough to prevent gravity's final victory. The collapsing object quickly crosses a threshold where the escape velocity from its surface exceeds the speed of light, and not even light can escape its gravity. From this point on, nothing can escape from the collapsing object and find its way back into the Universe of which it was once a part. The object is now a **black hole.** A black hole will form if the stellar core left behind by a Type II supernova exceeds about 3 $M_\odot$. Alternatively, a neutron star will collapse to become a black hole if it accretes enough matter from a binary companion to put it over the 3 $M_\odot$ limit. Regardless of how it formed, any collapsed object with a mass greater than 3 $M_\odot$ must be a black hole.

FREE FALL IS THE SAME AS FREE FLOAT

In Chapter 4, our exploration of special relativity began with the observation that the speed of light is always the same (regardless of the motion of an observer), and ended by shattering our everyday notions of space and time. Now as we confront the properties of black holes—indeed, of all massive objects in the Universe—our concepts of space and time will be pulled even further from the comfortable absolutes of Newtonian physics. First we learned that what we comfortably

Mass warps the fabric of space-time.

called (three-dimensional) space and time are actually just a result of our particular, limited perspective on a four-dimensional space-time that is different for each observer. Now we discover that this four-dimensional space-time is warped and distorted by the masses it contains. One of the consequences of this deformation is the gravity that holds you to Earth. This realization, called the **general theory of relativity,** is one of Albert Einstein's great contributions to science.

A crucial clue to the fundamental connection between gravity and space-time has been with us since Chapter 3, where we found that the *inertial mass* of an object—the mass appearing in Newton's $F = ma$—is *exactly* the same as the object's gravitational mass. Another clue is that, left on their own, any two objects at the same location and moving with the same velocity will follow the same path through space-time, regardless of their masses.

Inertial mass and gravitational mass are the same.

The space shuttle astronaut falls around Earth, moving in lockstep with the space shuttle itself. A feather dropped by an Apollo astronaut standing on the Moon falls toward the surface of the Moon at exactly the same rate as a dropped hammer. In some sense, rather than thinking of gravity as a "force" that "acts on" objects, it is more accurate to think of it as a consequence of the path through space-time that objects will follow in the absence of other forces. *Gravitation is the result of the shape of the space-time terrain through which objects move.*

The essence of special relativity is that any inertial reference frame is as good as any other. There is no experiment that you can do to distinguish between sitting in an enclosed spaceship floating stationary in deep space **(Figure 16.19a)** and sitting in an enclosed spaceship traveling through our Galaxy at 0.99999 times the speed of light **(Figure 16.19b).** You cannot tell any difference between these two cases—they do not *feel* any different—because there *is* no difference between them. Each of these reference frames is an equally valid inertial reference frame. As long as nothing acts to *change* the motion—that is, nothing pushes on either spaceship—the laws of physics are exactly the same inside both spacecraft.

General relativity begins by applying this same idea to an astronaut inside the space shuttle orbiting Earth, as shown in **Figure 16.19(c).** So long as we restrict our attention to a small enough volume of space and a short enough period of time that we can ignore changes in the strength and direction of gravity from place to place, our astronaut again has no way to tell the difference between being inside the space shuttle as it falls around Earth and being inside a spaceship coasting through interstellar space. Close your eyes and jump off a diving board. For the brief time that you are falling freely through Earth's gravitational field, the sensation you feel is exactly the same as the sensation that you would feel adrift in the abyss of interstellar space!

A freely falling object defines an inertial reference frame.

The implications of this result are somewhat startling. Even though its velocity is constantly changing as it falls, *the inside of a space shuttle orbiting Earth is as good an inertial frame of reference as that of an object drifting along a straight line through interstellar space.* This principle—which can be simply stated as "free fall is the same as free float"—is called the **equivalence principle.**

The equivalence principle says that a falling object is simply following its "natural" path through space-time—it is going where its inertia carries it—every bit as much

Falling objects follow straight lines through curved space-time.

as an object that drifts along a straight line at a constant speed through deep space. The natural path that an object will follow through space-time in the absence of other forces is referred to as the object's "world line" or **geodesic.** In the absence of a gravitational field, the geodesic of an object is a straight line—hence Newton's statement of inertia that, unless acted on by an external force, an object will move at a constant velocity in a constant direction. However, in the presence of mass the shape of space-time becomes distorted, so an object's geodesic becomes curved.

The equivalence principle gives us a way of understanding why gravitational mass and inertial mass are one and the same thing. When we discussed Newton's law,

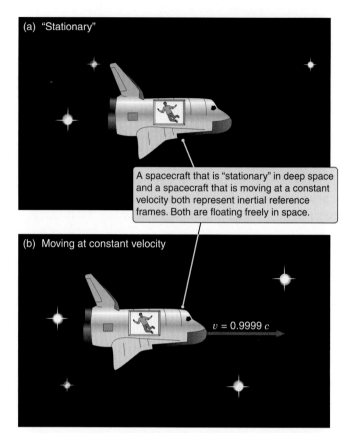

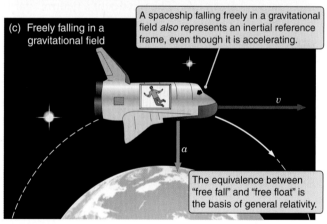

Figure 16.19 *Special relativity says that there is no difference between (a) a reference frame which is floating "stationary" in space and (b) one that is moving through the galaxy at constant velocity. General relativity adds that there is no difference between these inertial reference frames and (c) an inertial reference frame that is falling freely in a gravitational field. "Free fall" is the same as "free float," as far as the laws of physics are concerned.*

$F = ma$, we found it useful to state it instead as $a = F/m$. When you apply a force F to an object you move it away from its natural path, and its inertial mass m tells you how strongly it resists the change. General relativity says that when you are standing still on the surface of Earth, your natural path through space-time—your geodesic—is actually a path falling inward toward the center of Earth. In the absence of any external forces, this is what you would do. Of course, the surface of Earth gets in your way. Put another way, the surface of Earth exerts an external force on your feet, and that force causes you to accelerate continuously away from your natural path through space-time.

This leads to a different, equally valid, way of stating the equivalence principle, and another thought experiment, as shown in **Figure 16.20,** demonstrates the point. Imagine you are in a box inside a rocket ship that is accelerating through deep space at a rate of 9.8 m/s^2 in the direction of the arrow. The floor of the box exerts enough of a force on you to overcome your inertia and cause you to accelerate at 9.8 m/s^2, so you feel as though you are being pushed into the floor of the box. Now imagine instead that you are sitting in a closed box on the surface of Earth. Again the floor of the box exerts enough upward force on you to overcome your inertia, causing you to accelerate at 9.8 m/s^2. You feel as though you are being pushed into the floor of the

> **Being stationary in a gravitational field is the same as being in an accelerated reference frame.**

box. According to the equivalence principle, *the two cases are identical.* There is no difference between sitting in an armchair in a rocket ship traveling through deep space with an acceleration of 9.8 m/s^2, and sitting in an armchair on the surface of Earth reading this book. In the first case, the force of the rocket ship is pushing you away from your "floating" straight-line geodesic through space-time. In the second case, Earth's surface is pushing you away from your curved "falling" geodesic through a space-time that has been distorted by the mass of Earth. An acceleration is an acceleration, regardless of whether you are being accelerated off a straight-line geodesic through deep space or being accelerated off a "falling" geodesic in the gravitational field of Earth. And in all cases, it is the same mass—the mass that gives an object inertia—that resists the change. Gravitational mass and inertial mass are the same thing!

There is an important caveat to the equivalence principle. In an accelerated reference frame such as an accelerating rocket ship, the *same* acceleration is experienced *everywhere.* In contrast, the curvature of space by a massive object changes from place to place. Tides are one result of changes in the curvature of space from one place to another. A more careful statement of the equivalence principle is that the effects of gravity and acceleration are indistinguishable *locally*—that is, so long as we restrict our attention to small enough volumes of space that changes in gravity can be ignored.

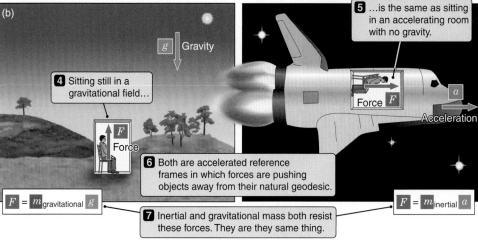

Figure 16.20 (a) *According to the equivalence principle, an object falling freely in a gravitational field is in an inertial reference frame, while* (b) *an object at rest in a gravitational field is in an accelerated frame of reference.*

SPACE-TIME AS A RUBBER SHEET

General relativity is a *geometrical* theory. It describes how mass distorts the *geometry* of space-time. You can get a sense for how mass distorts space-time by imagining the surface of a tightly stretched rubber sheet. The rubber sheet is flat. If you roll a marble across the sheet, it will roll in a straight line.

Mass distorts the geometry of space-time . . .

All of the Euclidean geometry that you learned in high school applies on the surface of the sheet as well: The angles in a triangle on the sheet add up to 180°. Right triangles obey the Pythagorean theorem. If you draw a circle on the sheet, you will find that the circumference of the circle is equal to 2π times its radius. If you draw a line on this sheet and a point off to one side, there will be exactly one line which passes through that point but never intersects the first line.

But now think about what happens if you place a bowling ball in the middle of the rubber sheet as in **Figure 16.21.** The surface of the sheet will be stretched and distorted. Now if you roll a marble across the sheet, its path will dip and curve (Figure 16.21a).

. . . much as a bowling ball distorts a rubber sheet . . .

You might even find that you can roll the marble so that it moves around and around the bowling ball, like a planet orbiting about the Sun. Next, you might revisit the relationship between the radius of a circle and its circumference. If you draw a circle around the bowling ball, measure the distance around that circle, and then compare that distance with the distance from the circle to its center along the surface of the sheet (Figure 16.21b), you will find that the circumference of the circle is less than $2\pi r$. Finally, you might try to draw a triangle on the surface of the sheet, connecting three points around the bowling ball with the straightest and shortest lines you can draw on the surface of the sheet, as in Figure 16.21(c). If you do this and then look at the sheet from above, you will be amazed to see that, rather than adding to 180°, the angles in this new triangle sum to more than 180°. The surface of the sheet is no longer flat, and Euclid's geometry no longer applies. (That is why Euclid's geometry is called "plane geometry.")

Mass has an effect on the fabric of space-time that is analogous to the effect of the bowling ball on the fabric of the rubber sheet. The bowling ball stretches

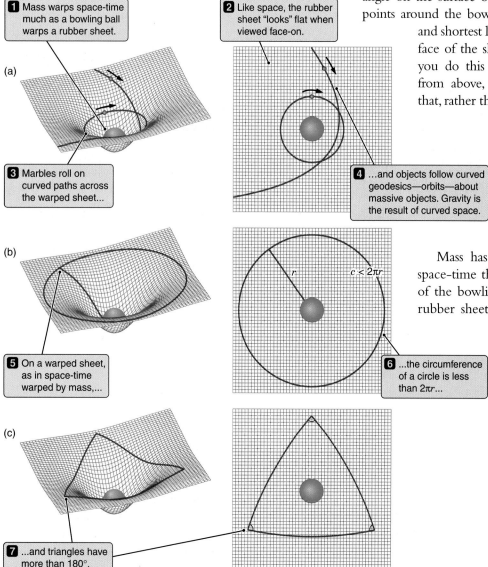

1 Mass warps space-time much as a bowling ball warps a rubber sheet.

2 Like space, the rubber sheet "looks" flat when viewed face-on.

(a)

3 Marbles roll on curved paths across the warped sheet...

4 ...and objects follow curved geodesics—orbits—about massive objects. Gravity is the result of curved space.

(b)

5 On a warped sheet, as in space-time warped by mass,...

r $c < 2\pi r$

6 ...the circumference of a circle is less than $2\pi r$...

(c)

7 ...and triangles have more than 180°.

Figure 16.21 *Mass warps the geometry of space-time in much the same way that a bowling ball warps the surface of a stretched rubber sheet. This distortion of space-time has many consequences, such as: (a) objects follow curved paths or geodesics through curved space-time, (b) the circumference of a circle around a massive object is less than 2π times the radius of the circle, and (c) angles in triangles need not sum to exactly 180°.*

CONNECTIONS 16.1

GRAVITATION: WHEN ONE PHYSICAL LAW SUPPLANTS ANOTHER

If you have been reading this chapter closely, you might feel some justifiable annoyance with this business of gravity. Throughout the book up until this point we have described gravity as a force that obeys Newton's universal law of gravitation: $F = Gm_1m_2/r^2$. Now we suddenly introduce the ideas of general relativity, and in the process ask you to totally change the way you think about gravity. So which is the real deal? If general relativity is right, does that not imply that Newton's formulation of gravity is wrong? And if so, then why have we continued to use Newton's law?

The answers to these questions go to the heart of how science progresses and how our conception of the Universe evolves. Under most circumstances there is virtually no difference between the predictions made using general relativity and the predictions made using Newton's law. So long as a gravitational field is not too strong, then Newton's law of gravitation is a very close *approximation* to the results of a calculation using general relativity. The meaning of "too strong" in this context is relative. For example, in most ways the enormous gravitational field near the core of a massive main sequence star would be considered "weak." Had we used a general relativistic formulation of gravity to compute the structure of a main sequence star, it would have made virtually no difference in the results of the calculation. Similarly, even though space-time is curved by gravity, this curvature near Earth is very slight, so over small regions it can be ignored entirely. The flat Euclidean geometry you learned in high school is a very good "local" approximation even to a curved space-time—and is a lot easier to use. This is exactly the kind of approximation we use when, despite the curvature of Earth, we navigate about a city using a flat road map.

This is not the first time we have run into the idea that the physics of our everyday experience is only an approximation to the more general rules governing the behavior of matter and energy. Newton's laws of motion are one of the great triumphs of the human intellect, and still form the basis of several years of study for students of physics. Yet we now know that Newton's laws of motion are actually *approximations* to the more generally applicable rules of special relativity and quantum mechanics. In fact, we can *derive* Newton's laws from special relativity and quantum mechanics by making the "everyday" assumptions that speeds of objects are much less than the speed of light, and that the sizes of objects are much larger than the subatomic particles that atoms are made from. We use Newton's laws of motion and gravitation most of the time because they are far easier to apply than the relativistically and quantum mechanically "correct" laws, and because any inaccuracies we introduce by using the Newtonian approximations are usually far too tiny to matter. It is only in conditions very different from those of our everyday lives (such as the behavior of an electron in an atom, or the gravitational field of a black hole), or in special cases when very high accuracy is needed (such as the precise timing used by the Global Positioning System satellite network), that the more general laws must be used.

This is a general feature of new scientific theories. If a new theory is to replace an earlier, highly successful, scientific theory, the new theory must normally "hold the old theory within it"—it must be able to reproduce the successes of the earlier theory—just as general relativity holds within it the successful Newtonian description of gravity that we have used throughout the book.

the sheet, changing the distances between any two points on the surface of the sheet. (Think of the deep depression in the rubber sheet as a well. We will frequently use this language.) Similarly, mass distorts the shape of space-time, changing the "distance" between any two locations or events in that space-time.

. . . changing the distances between points.

It is easy to understand how the two-dimensional surface of the sheet is distorted by the bowling ball because we can visualize how the sheet is stretched through a third spatial dimension. It is much more difficult for us to "see" in our mind's eye what a curved four-dimensional space-time would "look like." Once again we have run into a limitation in how our brains are wired. Yet there are experiments that we can do, much

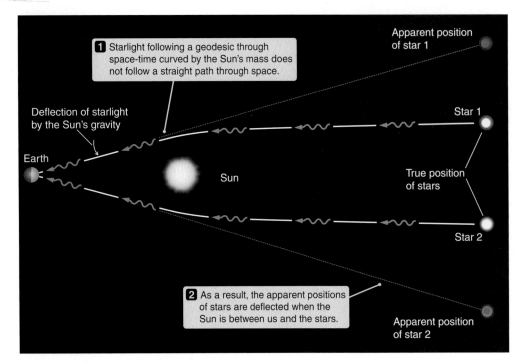

1 Starlight following a geodesic through space-time curved by the Sun's mass does not follow a straight path through space.

Deflection of starlight by the Sun's gravity

Earth

Sun

Apparent position of star 1

Star 1

True position of stars

Star 2

2 As a result, the apparent positions of stars are deflected when the Sun is between us and the stars.

Apparent position of star 2

Figure 16.22 *Measurements obtained by Sir Arthur Eddington during the total solar eclipse of 1919 found that the gravity of the Sun bends the light from distant stars by the amount predicted by Einstein's general theory of relativity. This is an example of gravitational lensing. Note that the "triangle" formed by Earth and the two stars contains more than 180°, just like the triangle in Figure 16.21(c).*

like those done on the surface of the rubber sheet, which demonstrate that the geometry of our four-dimensional space-time is distorted much like the rubber sheet.

Before going any further in our discussion of general relativity, however, it is important to point out that general relativity does not mean that Newton's law of gravitation is "wrong." See Connections 16.1 for a discussion of what happens when one physical law supplants another.

THE OBSERVABLE CONSEQUENCES OF GENERAL RELATIVITY

There are many observable consequences of curved space-time. You can imagine, at least in principle, stretching a rope all the way around the circumference of Earth's orbit about the Sun, and then comparing the length of that rope with the length of a rope taken from the orbit of Earth to the center of the Sun. Having taken geometry in high school, you might expect to find that the circumference of Earth's orbit is equal to 2π times the radius of Earth's orbit, just like a circle drawn on a flat piece of paper. However, if you could carry this experiment out, you would find instead that *the rope around the circumference of Earth's orbit was shorter than 2π times the length of the rope stretched from Earth to the center of the Sun*—just like the circumference of the circle on the stretched rubber sheet is less than 2π times the radius

The consequences of curved space-time include precession of Mercury's orbit . . .

of the circle. It is not practical to stretch a rope from Earth to the Sun. But we *can* do an experiment that is conceptually much the same thing. The long axis of Mercury's elliptical orbit about the Sun is slowly *precessing*. That is, the axis of Mercury's orbit is slowly changing its direction. Even after allowing for the perturbations caused by the gravity of the other planets, this precession is not predicted by Newton's inverse square law of gravity, working in a flat Euclidean space. However, it is *just* what is predicted for the path of a planet that is dipping in and out of the stretched-out non-Euclidean fabric of space-time that has been warped by the Sun.

The real-life equivalent of the triangle with more than 180° is probably easier to understand. A straight line in space is *defined* by the path followed by a beam of light. This is the shortest distance between any two points. A beam of light moving through the distorted space-time around a massive object is bent by gravity, just as the lines in Figure 16.21(c) are bent by the curvature of the sheet. This phenomenon is called **gravitational lensing** because the curvature of space-time bends the path of light much like the glass lens in a pair of eyeglasses.

. . . and gravitational lensing.

The first measurement of gravitational lensing came during the solar eclipse of 1919. Prior to the eclipse, Sir Arthur Eddington measured the positions of a number of stars in the part of the sky where the eclipse would occur. Eddington then repeated his measurement during the solar eclipse and found that the apparent positions of the stars had been deflected outward by the presence of the Sun. The light from the stars followed a bent path through the curved space-time around the Sun, causing the stars to appear farther apart in Eddington's measurement (see **Figure 16.22**). During the eclipse the triangle formed by Earth and the two stars contained more than 180°—just like the triangle on the surface of our rubber sheet. The results of Eddington's measurements were just as predicted by Einstein's theory. Eddington's result was the first experimental test of a prediction of general relativity, and is considered to be one of the landmark experiments of 20th-century physics. More recently, gravitational lensing has been used to search for unseen massive objects adrift in space. These objects do not give off light, but their gravity can distort the light from background stars that they happen to pass in front of.

General relatively also affects space-time in ways that have no direct comparison with a rubber sheet, for mass distorts not only the geometry of space, but the geometry of time as well. The deeper we descend into the gravitational field of a massive object, the more slowly our clocks appear to run from the perspective of a distant observer. This effect is called **general relativistic time dilation.** To understand one consequence of general relativistic time dilation, suppose a light is attached to a clock sitting on the surface of a neutron star. The light is timed so that it flashes once a second. However, since time near the surface of the star is dilated—stretched out—to an observer far from the neutron star it seems that the light is pulsing more slowly than once a second. The frequency of the flashing is lowered. Now suppose that we have an emission line source on the surface of the neutron star. Since time is running slowly on the surface of the neutron star, at least from our distant perspective, the light that reaches us will have a lower frequency as well. The light from the source will be seen at a longer, redder, wavelength than the wavelength at which it was emitted.

Time runs more slowly near massive objects.

This phenomenon, shown in **Figure 16.23,** is called the **gravitational redshift** because the wavelengths of light from objects deep within a gravitational well are shifted to longer wavelengths. Gravitational redshift is similar in its effect to the Doppler redshift we saw earlier. In fact, there is no way to tell the difference between light that is red-shifted by gravity and the Doppler-shifted light from an object moving away from us. Astronomers often describe the gravitational redshift of an object as an "equivalent velocity." The gravitational redshift of lines formed on the surface of the Sun is equivalent to a Doppler shift of 0.6 km/s. The gravitational redshift of light from the surface of a white dwarf is equivalent to a Doppler shift of about 50 km/s. The gravitational redshift from the surface of a neutron star is equivalent to a Doppler shift of about a 10th the speed of light. Sometimes astronomers get

Objects deep in a gravitational well appear red-shifted.

sloppy and talk about the gravitational redshift as if it truly were a Doppler shift. We might say, for example, that the "gravitational redshift of the surface of a particular white dwarf is 57.1 km/s." However, this does not mean that the surface of the white dwarf is moving away from us at 57.1 km/s. It means that time is running slowly enough on the surface of the white dwarf that the light reaching us from the white dwarf *looks like* it is coming from an object moving away from us at 57.1 km/s.

Bringing this a bit closer to home, a clock on the top of Mount Everest gains about 80 nanoseconds a day compared with a clock at sea level. The difference between an object on the surface of Earth and an object in orbit is much greater. A Global Positioning System (GPS) receiver uses the results of sophisticated calculations of the effects of general relativistic time dilation to help you accurately find your position on the surface of Earth. Even after allowing for slowing due to special relativity, the clocks on the satellites that make up the GPS run faster than clocks on the surface of Earth. If your GPS receiver did not correct for this and other effects of general relativity, then the position that it would report would be in error by up to half a kilometer. The fact that the GPS system works is actually a strong experimental confirmation of a number of predictions of general relativity, including general relativistic time dilation.

We could easily fill the rest of this book with fascinating tales about general relativity. It is pretty heady stuff if you think about it—discussing the fabric of the Universe itself as though it were a substance in a test tube, there to be poked and prodded (see Connections 16.2). One final phenomenon that we should mention before moving on is the phenomenon of **gravity waves.** If you thump the surface of our rubber sheet, waves will move away from where you thump it, something like ripples spreading out over the surface of a pond. Similarly, the equations of general relativity predict that if you "thump" the fabric of space-time, then ripples in space-time will move outward at the speed of light. These gravity waves are like electromagnetic waves in some respects. Accelerating an electrically charged particle gives rise to an electromagnetic wave. Accelerating a massive object gives rise to gravity waves.

Gravitational waves travel through the fabric of space-time.

Gravity waves have never been observed in a laboratory, but there is strong circumstantial evidence for their reality. In 1974 astronomers discovered a binary system consisting of two neutron stars, one of which is an observable pulsar. By using the pulsar as a precise clock, astronomers are able to very accurately measure the orbits of the stars. The stars themselves are 2.8 solar

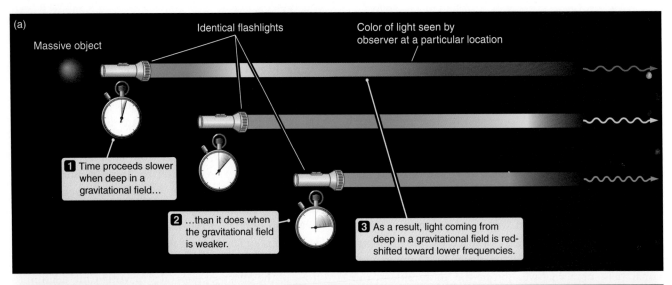

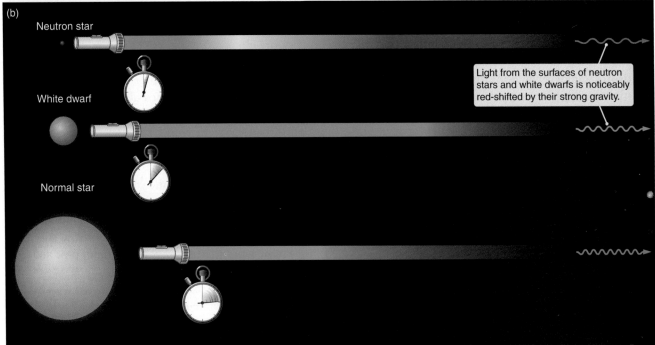

Figure 16.23 *Time passes more slowly near massive objects because of the curvature of space-time. As a result, to a distant observer, light from near a massive object will have a lower frequency and longer wavelength. (a) The closer to the object the source of radiation is, or (b) the more massive and compact the object is, the greater the gravitational redshift.*

radii apart, but their orbits are very gradually decaying, which means that they are losing energy *somewhere*. Calculations show that the energy being lost by the system is just what general relativity predicts the system should be losing in the form of gravity waves. With a new form of radiation to observe, the opening of the 21st century saw physicists working to construct a new kind of "telescope" to look for gravity waves—traveling distortions in space-time—from extreme events such as supernovae or the formation of black holes.

BACK TO BLACK HOLES

We began our digression into general relativity when we encountered black holes, and we return to the nature of black holes now. When we placed an object on the surface of our rubber sheet, it caused a funnel-shaped distortion that is analogous to the distortion of space-time by a mass. Now imagine such a funnel-

Black holes are bottomless pits in space-time.

shaped distortion in the rubber sheet that is *infinitely deep*—a funnel that keeps getting narrower and narrower as we go deeper and deeper, but which has no bottom. This is the rubber-sheet analog to a black hole. A black hole is a place where the mathematics describing the shape of space-time fails in the same way that the mathematical expression $1/x$ fails when $x = 0$. Such a mathematical anomaly is called a **singularity.** Black holes are singularities in space-time (see **Figure 16.24**).

A black hole has only three properties—mass, electric charge, and angular momentum. The amount of mass that falls into a black hole determines the extent of its distortion of space-time. The electric charge of a black hole is the net electric charge of the matter that fell into it. The angular momentum of a black hole causes the space-time around the black hole to be twisted around, much like the water around an eddy in a river. Apart from these three properties, all information about the material which fell into the black hole is lost. Nothing of its former composition, structure, or history survives.

16.3
HOLES

We can never actually "see" the singularity at the center of a black hole. The closer an object is to a black hole, the greater is its escape velocity (that is, the speed it would have to move to escape from the gravity of the black hole). There is a radial distance from the black hole at which the escape velocity reaches the speed of light. This point of no return, beyond which even light is trapped by the black hole, is called its **event horizon.** The radius of the event horizon of a black hole is proportional to the black hole's mass. A black

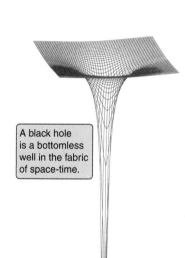

A black hole is a bottomless well in the fabric of space-time.

Figure 16.24 *A black hole is a singularity in the curvature of space-time. It is a gravitational well with no bottom.*

1 As an explorer approaches a stellar black hole, tides rip her apart...

2 ...and her signals to us are red-shifted by the black hole's gravity.

Figure 16.25 *A trip into a black hole.*

hole with a mass of 1 $M_\odot$ has a radius of about 3 km. A 2 $M_\odot$ black hole has a radius of about 6 km. The radius of a 3 $M_\odot$ black hole is about 9 km. A black hole with the mass equivalent to that of Earth would have a radius of only about a centimeter—the mass of our planet compressed into a volume equal to that of a pecan!

The event horizon is the boundary of no return.

Let us consider what would happen if an adventurer were willing to journey into a black hole **(Figure 16.25).** From our perspective outside the black hole, we would see our adventurer fall toward the event horizon, but as she did her watch would run more and more slowly, and her progress toward the event horizon would get slower and slower as well. Like Hercules in Zeno's famous paradox, even though our adventurer got closer and closer to the event horizon, she would never quite make it. The event horizon is where the gravitational redshift becomes infinite, and where clocks stop altogether. Yet our adventurer's own experience would be very different. From her standpoint there would be nothing special about the event horizon at all. She would fall past the event horizon and on, deeper into the black hole's gravitational well. However, she would now have entered a region of space-time that was cut off from the rest of the Universe. The event horizon is like a one-way door: once our adventurer

CONNECTIONS 16.2

GENERAL RELATIVITY AND THE STRUCTURE OF THE UNIVERSE

Our discussion of gravity as the warping of space-time by the mass it contains was motivated by the events that accompany the collapse of a massive star. Were it only for black holes, it might not have been appropriate to devote so much effort in this chapter to a discussion of general relativity. However, this is not the last time that we will run across general relativity. Many of the same physical processes that we have seen at work here also shape events in the larger Universe. In our discussion of galaxies, for example, we will find that some galaxies contain supermassive black holes at their centers—objects with masses millions of times that of our Sun, which grow by consuming entire stars and interstellar clouds. Material falling into such monsters is a leading candidate to explain the powerful radiation from *quasars*—beacons at the edge of the observable Universe—which we will explore in greater detail in Chapter 17. We will also find that just as the Sun bends the path of light passing near it, as observed by Eddington during the total eclipse of 1919, the curved space-time around distant galaxies can act as a lens, magnifying and distorting the appearance of even more distant objects behind them.

The grandest application of general relativity will come as we consider the history and fate of the Universe itself. We live in an expanding Universe—one that is the result of a singular event that took place around 14 billion years ago. This event, called the *Big Bang,* was not the result of mass coming into existence *within* the space-time of the Universe. Rather, *it was the coming into existence of the space-time of the Universe itself.* When we say that the Universe is expanding, we do not mean that the stars and the galaxies in the Universe are flying away from each other through space. Instead, it is the space-time of the Universe itself that is expanding with time. Just as the space-time around a black hole has a shape, so does the space-time of the Universe, and that shape is determined at least in part by the mass that our Universe contains. What was the Universe like when it was very young? Will the Universe expand forever, or is there enough mass in the Universe to warp space-time into a closed shape that will eventually collapse back in on itself? General relativity provides astronomers with the tools that they need to address such basic questions about the nature of existence itself.

---

has passed through, she can never again pass back into the larger Universe of which she was once a part.

Actually, we have overlooked a rather crucial fact. Our intrepid explorer would have been torn to shreds long before she reached the black hole. Near the event horizon of a 3 $M_\odot$ black hole the difference in gravitational acceleration between our explorer's feet and her head—the tidal "force" pulling her apart—is about a billion times her weight on the surface of Earth!

"SEEING" BLACK HOLES

In 1974 the British physicist **Stephen Hawking** (1942–) realized that black holes should actually be *sources* of radiation. In the ordinary vacuum of empty space, quantum theory says that particles and their antiparticle "mates" spontaneously spring into existence and then quickly annihilate each other and

Black holes glow . . .

disappear. These particle pairs typically live for less than about 10^{-21} second, but their effects are seen in sensitive measurements of atomic transitions. If such a pair of **virtual particles** comes into existence near the event horizon of a very small black hole, as shown in **Figure 16.26,** then one of the particles might wind up falling into the black hole while the other particle is able to escape. Some of the gravitational energy of the black hole will have been used up in making one of the pair of virtual particles real. When all of the esoteric physics is taken into account, Hawking was able to show that a black hole should actually emit a Planck spectrum, and that the effective temperature of this spectrum would increase as the black hole became smaller. While this phenomenon, called **Hawking radiation,** is of considerable interest to physicists and astronomers, in a practical sense it is usually negligible. A 3 $M_\odot$ black hole should emit radiation at a whopping temperature of only 2×10^{-8} K,

. . . but incredibly faintly.

which means that black hole should radiate with a power of 1.6×10^{-29} watts—very feeble indeed!

Hawking radiation is hardly a useful way to "see" a black hole. For all intents and purposes, black holes remain true to their name. Nonetheless, by the end of the 20th century, astronomers had found strong circumstantial evidence for black holes in two very different kinds of systems. In Chapter 17 we will find that there is strong evidence for supermassive black holes in the very centers of galaxies, but the first and perhaps strongest evidence for black holes comes from X-ray binary stars in our own Galaxy. In 1972 astronomers did not yet have a good understanding of the X-ray emission from stars. It was in that year that the recently-launched *Uhuru* X-ray satellite made a puzzling discovery. The brightest X-ray source in the constellation Cygnus was found to be rapidly flickering. We now know that the brightness of the X-ray emission from this object, called **Cygnus X-1**, can change in as little as 0.01 second. For reasons we will cover in more detail in our discussion of quasars in Chapter 17, this means that the source of the X-rays must be smaller than the distance light travels in 0.01 second, or

Black holes are located through the effect of their gravity.

3,000 km. Thus, the source of X-rays in Cygnus X-1 must be smaller than Earth!

When astronomers began to study this object in other parts of the electromagnetic spectrum, Cygnus X-1 was identified with both a radio star and with an already-cataloged optical star called HD226868. The spectrum of HD226868 shows that it is a normal B0 supergiant star with a mass of about 30 $M_\odot$. Such a star is far too cool to explain the X-ray emission from Cygnus X-1. But HD226868 was also discovered to be part of a binary system. The wavelengths of absorption lines in the spectrum of HD226868 are Doppler-shifted back and forth with a period of 5.6 days. Using the same techniques we used to measure the masses of stars in Chapter 12 (namely, analyzing the orbits of binaries), astronomers found that the mass of the unseen compact companion of HD226868 must be *at least* 6 $M_\odot$. (Only a lower limit can be determined because the tilt of the orbit of the binary is not known.) The companion to HD226868 is too compact to be a normal star, yet it is much more massive than the Chandrasekhar limit for a white dwarf or a neutron star. According to our understanding of the laws of physics, such an object can only be a black hole.

Figure 16.26 *In the vacuum, particles and antiparticles are constantly being created and then annihilating each other. However, near the event horizon of a black hole, one particle may cross the horizon before it recombines with its partner. The remaining particle leaves the black hole as Hawking Radiation.*

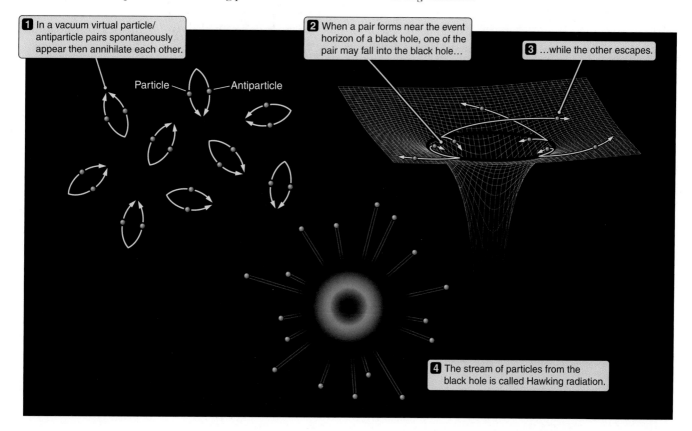

1 In a vacuum virtual particle/antiparticle pairs spontaneously appear then annihilate each other.

2 When a pair forms near the event horizon of a black hole, one of the pair may fall into the black hole...

3 ...while the other escapes.

Particle — Antiparticle

4 The stream of particles from the black hole is called Hawking radiation.

Since 1972 a number of other good candidates for stellar mass black holes have been discovered. One such object is a rapidly varying X-ray source in our companion galaxy, the Large Magellanic Cloud. Called LMC X-3, this X-ray source orbits a B3 main sequence star every 1.7 days, and the data show that the compact source must have a mass of at least 9 $M_\odot$! Although the evidence that these systems contain black holes is circumstantial, the arguments that lead to this conclusion seem airtight. With more than a half dozen compelling examples of such objects on the books, the evidence is in. Black holes, once regarded as nothing more than a bizarre quirk of the mathematics describing gravitation and space-time, exist in nature!

SEEING THE FOREST THROUGH THE TREES

The story of low-mass stars like our Sun is by and large a story of longevity and stability. No one has ever seen a star less massive than about 0.8 $M_\odot$ evolve off the main sequence because all the time in the Universe has literally not been enough for this to happen even once. What a contrast on this leg of our journey to look instead at stars that inhabit the high-mass end of the family of stars. Rather than stability, these stars offer spectacle. Living for only a short while, they blaze with the light of thousands or even millions of suns. Such stars are few and far between, but the role they play in the life of the Universe (and in our understanding of the Universe) is significant far beyond their numbers. Indeed, when we move on to the study of galaxies and the Universe, massive stars such as Cepheid variables will provide the signposts used to gauge the scale of our Universe.

Contemplating the lives of massive stars has also taken us even further afield into the extremes to which matter, space, and time can be molded. White dwarfs may have seemed incomprehensibly bizarre when we encountered them in our discussion of low-mass stars. Objects with densities of a ton per teaspoonful are so far beyond our everyday experience as to be all but unimaginable. Now we have encountered neutron stars—objects a billion times more dense yet. The matter in a neutron star is to the electron-degenerate matter in a white dwarf as the matter in a white dwarf is to the air in a summer breeze. But even the neutron star is but a way station on our journey to the limits of matter and mind. As we contemplate massive stars, we also come face to face with the most extreme object of all—if *object* is even an appropriate term for the bottomless pit in the fabric of the Universe that we call a black hole.

This leg of our journey has not only taken us into the realm of tortured space-time, but has also shown us the answers to immediate questions of our own existence. We have watched as high-mass stars begin their lives by tapping the same reservoir of nuclear fuel as their low-mass counterparts, fusing hydrogen into helium and helium into elements such as carbon. But the alchemy of high-mass stars does not end with a cinder of carbon and oxygen. Instead it goes on, fusing massive elements into more massive elements, and those elements into more massive elements yet.

And finally, we have seen the spectacle of a Type II supernova. As gravity has its final victory, pulling the inner parts of the massive star down into a neutron star or black hole, the outer layers of the star blaze forth, casting into the Universe the seeds of future generations of planets and, in all likelihood, life.

We now come to the end of this leg of our journey, having filled out the family album of stars, following them from their origins in the vast reaches of interstellar space to their final places of rest. But that is not to say that we are done with stars—far from it. All that we have seen has happened on the stage of galaxies, and the Universe is the grand hall in which that stage resides. Knowledge of stars and the interstellar medium is the astronomers' starting point as we endeavor to tease from galaxies and the Universe their secrets. At the same time, knowledge of the theater will give us further insight into the players on the stage.

And so we again head outward on the last major segment of our journey: an investigation that will begin with the properties of galaxies, and end with a consideration of the origin and nature of the Universe itself.

STUDENT QUESTIONS

THINKING ABOUT THE CONCEPTS

1. Of the four forces in nature (strong nuclear, electromagnetic, weak nuclear and gravity), gravity is by far the weakest. Why then is gravity such a dominant force in stellar evolution? *Note:* Although not explicitly discussed in this text, the weak nuclear force is involved in certain decay processes within the nucleus.

2. Explain the differences between the way that hydrogen is converted to helium in a low-mass star (proton–proton chain) and in a high-mass star (CNO cycle). What is the catalyst in the CNO cycle and how does it take part in the reaction?

3. Cepheids are highly-luminous variable stars, in which the period of variability is directly related to luminosity. Explain why Cepheids are good indicators for determining stellar distances that lie beyond the limits of accurate parallax measurements.

4. Why does the accretion disk around a neutron star have so much more energy than the accretion disk around a white dwarf, even though both stars have approximately the same mass?

5. What are the two reasons why each post-helium-burning cycle for high-mass stars (carbon, neon, oxygen, silicon, and sulfur) becomes shorter than the preceding cycle?

6. An experienced astronomer can take one look at the H-R diagram of a star cluster and immediately estimate its age. How is this possible?

7. What do we mean by the binding energy of an atomic nucleus? How does this measurement help us to calculate the energy given off in nuclear fusion reactions?

8. List and explain two important ways that supernovae influence the formation and evolution of new stars.

9. An astronomer sees a redshift in the spectrum of an object. With no other information available, can he determine whether this is an extremely dense object (gravitational redshift) or one that is receding from us (Doppler redshift)? Explain your answer.

10. If you could watch a star falling into a black hole, how would the color of the star change as it approaches the event horizon?

APPLYING THE CONCEPTS

11. In 1841 the 150 $M_\odot$ star, Eta Carinae, was losing mass at the rate of 0.1 $M_\odot$ per year. Let's put that into perspective.
    a. The mass of the Sun is 2×10^{30} kg. How much mass (in kg) did Eta Carinae lose each minute?
    b. The mass of the Moon is 7.35×10^{22} kg. How does Eta Carinae's mass loss per minute compare with the mass of the Moon?

12. If the Crab Nebula has been expanding at an average velocity of 3,000 km/s since A.D. 1054, what was its average radius in the year 2002? (There are approximately 3×10^7 seconds in a year.)

13. In our Galaxy there are about 50,000 stars of average mass (0.5 $M_\odot$) for every main sequence star of mass 20 $M_\odot$. But stars with 20 $M_\odot$ are about 10^4 times more luminous than the Sun, and 0.5 $M_\odot$ stars are only 0.08 times as luminous as the Sun.
    a. How much more luminous is a single massive star than the total luminosity of the 50,000 less massive stars?
    b. How much mass is in the lower-mass stars compared to the single high-mass star?
    c. What does this tell you about which stars contain the most mass in the Galaxy and which stars produce the most light?

14. The Moon has a mass equal to 3.74×10^{-8} $M_\odot$. Suppose the Moon suddenly collapsed into a black hole.
    a. What would be the radius of the event horizon (the "point of no return") around the black hole Moon?
    b. What affect would this have on tides raised by the Moon on Earth? Explain.
    c. Do you think this event would or would not generate gravity waves? Explain.

15. The approximate relationship between the luminosity and the period of Cepheid variables is L_{star} ($L_\odot$ units) = 335 P (days). Delta Cephei has a cycle period of 5.4 days and a parallax 0.0033 arcseconds. A more distant Cepheid variable appears 1/1,000 as bright as Delta Cephei and has a period of 54 days.
    a. How far away (in parsecs, or pc) is the more distant Cepheid variable?
    b. Could the distance of the more distant Cepheid variable be measured by parallax? Explain.

GALAXIES, THE UNIVERSE, AND COSMOLOGY

. . . it may not be amiss to point out some other very remarkable Nebulae which cannot well be less, but are probably much larger than our own system; and being also extended, the inhabitants of the planets that attend the stars which compose them must likewise perceive the same phenomena. For which reason they may also be called milky ways. . . .

SIR WILLIAM HERSCHEL (1738–1822)

GALAXIES

17.1 TWENTIETH-CENTURY ASTRONOMERS DISCOVERED THE UNIVERSE OF GALAXIES

We have come a long way on our journey of discovery, building an understanding of stars and the planetary systems that surround them. Yet, cosmically speaking, everything we have come across so far is in our own backyard. A deep-space image of a piece of "blank" sky like the *Hubble Space Telescope* image shown in **Figure 17.1** reveals myriad faint smudges of light filling the gaps among a sparse smattering of nearby stars. It has long been known that the sky contains faint, misty patches of light. These objects were originally referred to as "nebulae" because of their nebulous appearance. Prior to the 1780s, only about 100 of these smudges of light had been found by eye in telescopes. In 1784, the French comet hunter **Charles Messier** (1730–1817) published a catalog of 103 nebulous objects, mostly as a warning to other comet hunters not to waste their time on these objects. Yet twenty years later, courtesy of the remarkable observations of William Herschel and his sister Caroline, that number jumped to 2,500. From this time on, astronomers were aware of systematic differ-

KEY CONCEPTS

As humans it is difficult for us to comprehend the scale of even a single star such as our Sun, yet stars in the Universe are as grains of sand on all the world's beaches. Stars are not spread uniformly through space, but are instead grouped into what Kant referred to as "island universes," and what we refer to today as "galaxies." As we look beyond stars to the galaxies of which they are a part, we will find that:

✴ Galaxies are classified into different morphological types that are given their shapes by the properties of the orbits of the stars they contain;

✴ The arms of spiral galaxies, which form whenever the disk of a spiral galaxy is disturbed, are sites of star formation;

✴ Stars and gas account for only a small fraction of the mass of a galaxy; galaxies are mostly composed of an unknown form of "dark matter";

✴ Most, and perhaps all, large galaxies have supermassive black holes at their centers; and

✴ When these supermassive black holes are fed, as during encounters between galaxies, they may blaze forth with the light of thousands of normal galaxies coming from an active galactic nucleus no larger than our own Solar System.

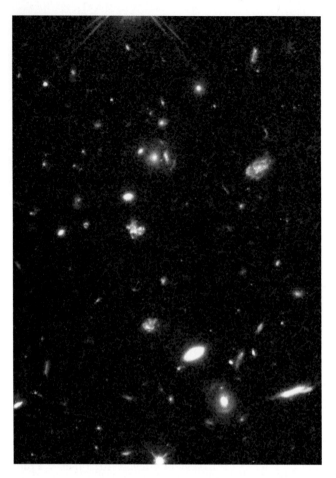

Figure 17.1 *A deep image of a section of "blank" sky made with the* Hubble Space Telescope. *When we look hard enough, the sky seems literally to be covered with faint galaxies.*

that the technological tools became available to turn philosophical musing into scientific knowledge.

Today we know that Kant was correct. Our Milky Way is only one of Kant's island universes, which were renamed **galaxies** to reflect this change in understanding. (The term **Universe** is now used to refer to the full expanse of space and all that it contains.) While most diffuse nebulae are nearby clouds of gas and dust in our own Milky Way, the elliptical and spiral nebulae are instead galaxies located far beyond the bounds of our local Galaxy. Each of the tiny smudges in Figure 17.1 is such a galaxy. The Universe contains billions upon billions of galaxies— possibly more galaxies than

> There are hundreds of billions of galaxies in the Universe.

stars in our own Galaxy! Each galaxy is a collection ranging from millions to hundreds of billions of stars rivaling our own cosmic home. Most of these galaxies are located at such astonishing distances from us that they appear too small and faint to see with any but the most powerful telescopes.

As has often been the case in astronomy, the nature of galaxies was, at its heart, a question about size and distance. Early attempts to understand the size of the Milky Way were confounded by interstellar dust,

> The nature of galaxies was a question of size and distance.

which blocks the passage of visible light, limiting our view. Early astronomers did not know of the existence or consequences of dust, and assumed that what they could see in visible light was all that there is. Unable to see past this obscuring shroud, they concluded that we live in a system of stars some 6,000 light-years across. It was not until the mid-1910s that **Harlow Shapley** (1885–1972), of the Harvard College Observatory, found that our Galaxy is over 300,000 light-years in size.

In an interesting historical twist, the same insight that brought Shapley to the correct conclusion about the size of the Milky Way also led him to an erroneous conclusion about the nature of the spiral and elliptical nebulae. In 1920, Shapley met Lick Observatory's

> The Great Debate focused attention on what nebulae are.

H. D. Curtis (1872–1942) in Washington, D.C., to publicly debate these issues. Historians call this meeting astronomy's Great Debate. When the Great Debate was held, Curtis defended the earlier, smaller model of our Galaxy, but the tide against that picture was already turning. However, the question about the nature of what we now call **spiral galaxies** was still wide open. In Shapley's opinion, his far-larger Milky Way was ample to encompass everything in the Universe. (Having

ences in the appearance of nebulae. While some of the Herschels' nebulae looked diffuse and amorphous, most were round or elliptical or resembled spiraling whirlpools. These distinctions formed the original three categories—diffuse, elliptical, and spiral—used to classify nebulae.

Discovering the existence of nebulae was one thing, but uncovering their true nature was quite another. Speculations about the nature of these objects abounded for the next 140 years. Prior to the 1920s, many astronomers thought that the sum of existence—the Universe—consisted solely of the swarm of stars to which our Sun belongs. It was even suggested that spiral nebulae might be planetary systems in various stages of formation. The great 18th-century philosopher **Immanuel Kant** (1724–1804) had a very different idea. He speculated that nebulae were instead "island universes"—realms of existence separate from our own. Herschel himself shared this belief, but realized that no telescope of his day would ever be able to resolve the issue. It was not until the first third of the 20th century

worked to show that the Galaxy is 50 times larger than previously thought, Shapley balked at the idea that the whole Universe was hugely larger still.) Curtis, on the other hand, favored the idea that the spiral nebulae were really galaxies separate from our own, and that the Universe was, indeed, far larger than our own Galaxy.

Unlike questions of politics and law, scientific questions are not resolved by the rhetorical skills of partisans. Instead they are settled by the results of well-crafted and carefully conducted experiments and observations. However, scientific debates *do* help bring issues into sharper focus, leading scientists to concentrate their attention and efforts on key questions. The reason we call that 1920s meeting the Great Debate is because it clearly marked the final steps that would lead to a correct understanding of nebulae. The 1920 debate set the stage and pointed the direction for the work of **Edwin P. Hubble** (1889– 1953), whose name was to become forever entwined with our modern understanding of the Universe.

Using the newly finished 100-inch telescope on Mt. Wilson, high above the then-small city of Los Angeles, Hubble was able

The nondebater Hubble settled the Great Debate.

to see some individual stars in the Andromeda Nebula and other nearby nebulae. He went on to discover that some of these stars were variable stars, recognizable as the same type of Cepheid variable stars studied by Henrietta Leavitt, but far

Figure 17.2 *Galaxies are like a handful of coins when thrown in the air. We see some face-on, some edge-on, and most somewhere in between.*

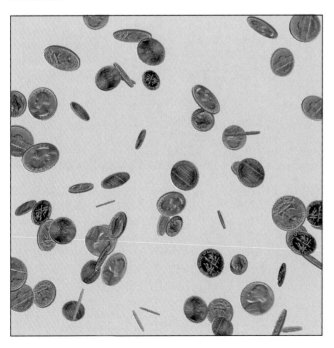

fainter in appearance. Using the period-luminosity relation for Cepheid variable stars discussed in Chapter 16, Hubble turned his observations of these stars into measurements of the distances to these objects. The results showed that these nebulae are far more distant than even Shapley's measurement of the size of our Galaxy. No doubt remained: spiral and elliptical nebulae are really galaxies in their own right, similar in size to our own Galaxy but located at truly immense distances. Hubble may not have shared the stage during the Great Debate, but when he spoke through his results, those earlier questions were answered once and for all. Shapley's Milky Way, itself vast beyond comprehension, is but a speck adrift in a Universe full of galaxies.

17.2 GALAXIES ARE CLASSIFIED BASED ON THEIR APPEARANCE

Imagine what you would see if you were to take a handful of coins and throw them into the air as shown in **Figure 17.2.** You know that all of these objects are very much the same: dimes, pennies, nickels, quarters— all flat and circular. When you look at the objects falling through the air, however, they all do not appear to be the same. Some coins you see face on, appearing circular. Some coins are instead seen edge-on, appearing as nothing but thin lines. Most coins are seen from an angle between these two extremes, and appear with various degrees of *ellipticity*.

In principle, we can learn a lot about the properties of coins by looking at a picture like Figure 17.2. If we begin by assuming that coins have some particular three-dimensional shape, we can predict what we should see in Figure 17.2. We could then compare our prediction with what is actually observed. In this example, we would probably have little trouble convincing ourselves that coins must be disks. As astronomers, we play exactly this game in our efforts to discover the true three-dimensional shape of galaxies.

A quick look at a group of galaxies, like that in **Figure 17.3,** shows that galaxies come in a wide range of sizes and shapes. The first step in understanding galaxies came by sorting these different shapes into categories. The classifications we use today date back to the 1930s, when

We classify galaxies according to Hubble's classification scheme.

Edwin Hubble devised a scheme much like that shown

Figure 17.3 *A* Hubble Space Telescope *image of a small group of galaxies called Hickson Compact Group 87. This image shows something of the range of shapes and sizes found among galaxies.*

in **Figure 17.4.** On the bottom (or "handle") of this **tuning fork diagram** sit oval-shaped objects called **elliptical galaxies.** On the two "tines" of the fork are the spirals (as well as a class we will describe shortly, the *S0 galaxies*). Galaxies that fall into none of these classes are called **irregular galaxies.** Originally, Hubble thought that his tuning fork diagram might do for galaxies what the H-R diagram did for stars. This hope turned out to be incorrect, but his classification scheme did succeed in bringing order to the study of these objects.

STELLAR MOTIONS GIVE GALAXIES THEIR SHAPES

Galaxies are not solid objects like coins, but collections of stars, gas, and dust orbiting about under the influence of the galaxy's overall gravitational field. In any region in an elliptical galaxy, some stars are falling in while others are climbing out—in fact, there are stars moving in all possible directions. Unlike planets, which move on simple elliptical orbits about the Sun, stars in an elliptical galaxy follow orbits with a wide range of different shapes, as shown in **Figure 17.5.** These orbits are more complex than the orbits of planets because the gravitational field within an elliptical galaxy does not come from a single, central object.

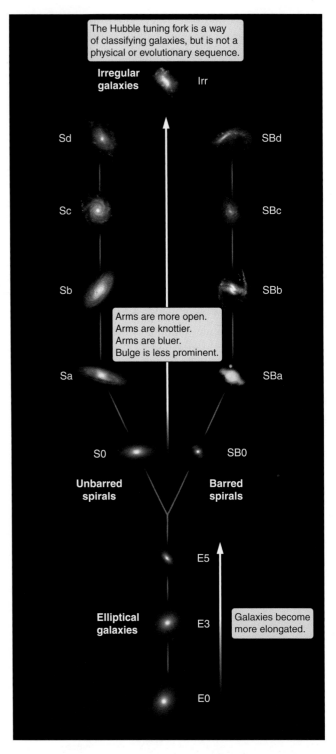

Figure 17.4 *Tuning fork diagram showing Edwin Hubble's scheme for classifying galaxies based on their appearance. Elliptical galaxies form the "handle" of a tuning fork, shown here at the bottom of the diagram. Unbarred and barred spiral and S0 galaxies lie along the left and right tines of the fork, respectively. Irregular galaxies are not placed on the tuning fork.*

All of these stellar orbits taken together are what give an elliptical galaxy its shape. The faster the stars are moving, the more spread out the galaxy is. (After all, if the stars were not moving at all, they would all clump together at the center of the galaxy.) If the stars in an elliptical galaxy are moving in random directions, then the galaxy will be spherical in shape. However, if stars tend to be moving faster in one direction than in others, then the galaxy will be more spread out in that direction, giving it an elongated shape. Hubble noted that some elliptical galaxies (those on the bottom of the tuning fork handle—see Figure 17.4) are round while others (those on the top of the handle) are elongated. One of the most frustrating problems faced by astronomers studying elliptical galaxies, however, is that their appearance in the sky does not necessarily tell us their true shape. For example, a galaxy might actually be shaped like an American football, but if we happen to see it end-on, it will look instead like a baseball.

The collective orbits of all its stars give an elliptical galaxy its shape.

The orbits of stars in spiral galaxies are quite different from those of stars in elliptical galaxies. The components of a spiral galaxy are shown in **Figure 17.6.** The defining feature of a spiral galaxy is a flattened, rotating disk. Like the planets of our Solar System, most of the stars in the disk of a spiral galaxy follow nearly circular orbits in the same direction

Spiral galaxies have a rotating disk and a central bulge.

TABLE 17.1

THE HUBBLE SEQUENCE OF GALAXIES
A morphological classification scheme based on the gross properties of galaxies.

Category/Criteria	Sequence			
	Abbreviation	Range of Features		
Ellipticals	E0	Rounder		
	E1	↑		
Mostly bulge	E2			
Old, red stellar population	E3	↓		
Smooth appearing	E4			
	E5	Flatter		
S-zeros (unbarred/barred)	S0/SB0	Smooth disk and bulge		
Bulge and disk with no arms				
Bulge and disk contain mostly old, red stars				
Spirals (unbarred/barred)	Sa/SBa	More bulge	Tightly wound arms	Smooth arms
Bulge and disk with arms	Sb/SBb	↑	↑	↑
Bulge has old, red stars	Sc/SBc			
Disk has both old, red stars	Sd/SBd	↓	↓	↓
and young, blue stars		Little bulge	Open arms	Knotty arms
Spirals (S) have roundish bulges				
Barred spirals (SB) have elongated or barred bulges				
Irregulars	Irr			
No arms				
No bulge				
Some old stars, but mostly young stars, giving a knotty appearance				

about the center of the galaxy. Spiral galaxies also contain the spiral arms that give these galaxies their name. In addition to disks and arms, spiral galaxies have central **bulges,** which look like elliptical galaxies. This similarity is more than appearance: like elliptical galaxies, the bulges of spiral galaxies get their shapes from the range of orbits of the stars of which they are made.

Hubble noticed that the bulges of some spiral galaxies are bar shaped. He called these **barred spirals,** and placed them along the right tine of the tuning fork in Figure 17.4. Spirals that lack a bar-like bulge lie along the left tine. Placement of spiral galaxies vertically along the tines of the fork is based on the prominence of the central bulge and how tightly the spiral arms are wound. The criteria Hubble used to classify galaxies are summarized in **Table 17.1.**

There is another big difference between spiral and elliptical galaxies. Most spiral galaxies contain large amounts of molecular gas and dust concentrated in the midplanes of their disks. Just as the dust in the disk of our own Galaxy can be seen on a clear summer night as a dark band slicing the Milky Way in two (see Chap-

ter 14, Figure 14.1), the dust in an edge-on spiral galaxy appears as a dark, obscuring band running down the midplane of the disk **(Figure 17.7).** The

Spirals and ellipticals differ in the gas they contain.

molecular gas that accompanies the dust can also be seen in radio observations of spiral galaxies. In contrast, elliptical galaxies contain large amounts of very hot gas seen primarily by observing the X-rays it emits. The difference in shape between elliptical and spiral galaxies offers some insight into why the gas in ellipticals is hot, while spirals contain a large amount of cold dense gas. Just as gas settles into a disk around a forming star, so too does conservation of angular momentum cause cold gas to settle into the disk of a spiral galaxy. In contrast, the only place in an elliptical galaxy where cold gas could collect is at the center. However, the density of stars in elliptical galaxies is so high that Type I supernovae continually reheat this gas, preventing most of it from cooling off and forming cold clouds.

Hubble recognized that the distinction between spiral and elliptical galaxies is not always clear-cut. Some galaxies seem to be a cross between the two types, having stellar disks but no spiral arms. Hubble called these **S0 galaxies,** and placed them near the junction of his tuning fork. Today the lines between these different categories are even less distinct. Based on better observations, it now seems that most, if not all, elliptical galaxies contain small rotating disks.

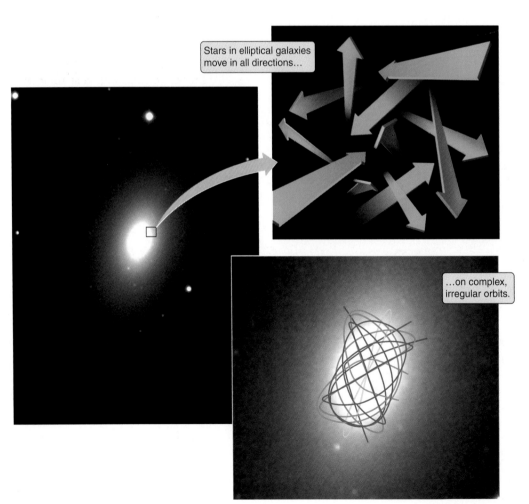

Stars in elliptical galaxies move in all directions...

...on complex, irregular orbits.

Figure 17.5 *Elliptical galaxies take their shape from the orbits of the stars they contain. Shown here is a sample of stellar orbits in an elliptical galaxy superposed on the galaxy itself.*

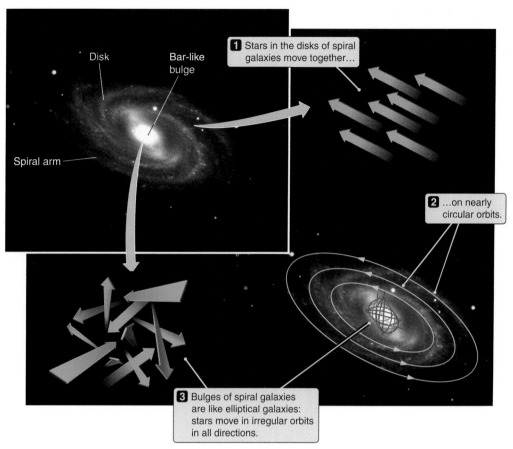

Disk

Bar-like bulge

Spiral arm

1 Stars in the disks of spiral galaxies move together...

2 ...on nearly circular orbits.

3 Bulges of spiral galaxies are like elliptical galaxies: stars move in irregular orbits in all directions.

Figure 17.6 *The components of a barred spiral galaxy. The orbits of stars in the rotating disk and the ellipse-like bulge are indicated.*

OTHER DIFFERENCES AMONG GALAXIES

Stars form from dense clouds of cold molecular gas. Spiral galaxies contain large amounts of such gas, while ellipticals do not. From these two observations we would predict that many spiral galaxies are actively forming stars today, while star formation is quite rare in ellipticals. One consequence of this prediction is that the disks of spiral galaxies should contain both young and old stars, while elliptical galaxies should contain only a much older population of stars. These predictions have proven correct. The differences in stellar population show up as differences in the colors of the two types of galaxies. The disks of spiral galaxies tend to be fairly blue in color, reflecting the fact that their light is dominated by the luminous, young, massive, hot stars that they contain. In contrast, elliptical galaxies are much redder in color, reflecting the fact that they only contain an older population of lower-mass stars.

Turning this around, the colors of spiral and elliptical galaxies tell us a great deal about their star forma-

Stars form in spiral galaxies but not in elliptical galaxies.

tion histories. The red colors of elliptical galaxies tell us that we are looking at an old stellar population, and that there has been little or no star formation for quite some time. The blue colors of the disks of spiral galaxies, on the other hand, tell us that we are looking at regions of ongoing star formation, where massive young stars are being born. Even though *most* of the stars in a spiral disk are old, the massive young stars are so luminous that their blue light dominates what we see. When it comes to star formation, most irregular galaxies are more like spirals than ellipticals. Some irregular galaxies are currently forming stars at prodigious rates, given their relatively small sizes.

Galaxies range in luminosity from around a million solar luminosities up to a million million solar lumi-

Figure 17.7 *The dust in the plane of an edge-on spiral galaxy is seen as a dark obscuring band in the midplane of the galaxy. Compare this image with Figure 14.1, which shows the dust in the plane of the Milky Way.*

nosities (10^6–10^{12} $L_\odot$) and in size from around 3,000 light-years up to hundreds of thousands of light-years. There is no strict size difference between elliptical and spiral galaxies. While it is true that the most luminous elliptical galaxies are much larger than the most luminous spiral galaxies, there is considerable overlap in the range of sizes and luminosities among all Hubble types **(Figure 17.8).**

Galaxies come in a wide range of sizes for all types.

Earlier in our journey, we found that mass is the single most important parameter in determining the properties and evolution of a star. In contrast, differences in mass and size do not lead to such obvious differences in the appearance of galaxies. While slight differences in color and concentration exist between large and small galaxies, these differences are subtle. Even when a smaller, nearby spiral galaxy is seen next to a larger, distant spiral **(Figure 17.9),** it can be difficult to tell which is which. Still, astronomers prefer to call galaxies that are of relatively low luminosity (less than 1 billion solar luminosities) **dwarf galaxies** and galaxies more luminous than this **giant galaxies.**

17.3 STARS FORM IN THE SPIRAL ARMS OF A GALAXY'S DISK

Spiral galaxies take their name from the spiral arms they contain, but what is a spiral arm? From pictures of spiral galaxies outside of our own, we might have

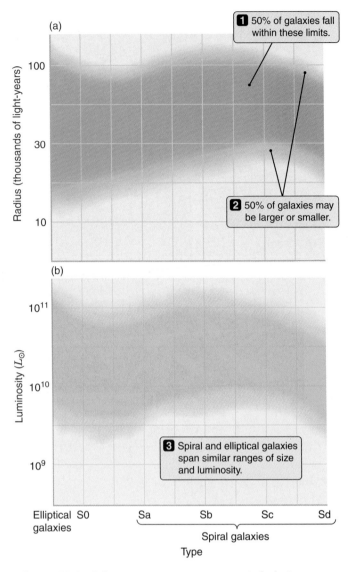

Figure 17.8 *Galaxies span an enormous range in both size (a) and luminosity (b). Note the great overlap in luminosity and size among galaxies of different Hubble types.*

Figure 17.9 *The mass or size of a galaxy does not determine its appearance. Here a larger, more distant galaxy (left) looks much the same as a smaller, closer galaxy.*

guessed that stars in the disk of a spiral galaxy are concentrated in the spiral arms. This turns out *not* to be the case. **Figure 17.10** shows images of the same spiral galaxy taken in ultraviolet light (17.10a) and in red light (17.10b). Notice that while the spiral arms are very prominent in the UV image, they are much less prominent when viewed in red light. If we carefully trace the actual numbers of stars rather than just their brightness, we find that while stars are slightly concentrated in spiral arms, this concentration is not strong—certainly not strong enough to account for the prominence of the spiral arms we see. In fact, the concentration of stars in the disks of spiral galaxies varies quite smoothly as it decreases outward from the center of the disk to the edge of the galaxy.

> **Stars in a spiral galaxy are not strongly concentrated in the spiral arms.**

Spiral arms look so prominent when viewed in blue or UV light because that is where the young, massive, luminous stars are concentrated. In other words, what *is* strongly concentrated in the arms of spiral galaxies is ongoing star formation. H II regions (17.10c), molecular clouds, associations of O and B stars, and other structures that we have learned to associate with star formation are all found predominantly in the spiral arms of galaxies.

We can say a lot about what spiral arms must be like just by applying what we already know about star formation. Stars form when dense interstellar clouds become so massive and concentrated that they begin to collapse under the force of their own gravity. If stars

> **Gas, dust, and young stars are concentrated in spiral arms.**

form in spiral arms, then spiral arms must be places where clouds of interstellar gas pile up and are compressed. Such is indeed the case. There are many ways to trace the presence of gas in the spiral arms of galaxies. Pictures of face-on spiral galaxies, like that in **Figure 17.11(a)**, show dark lanes where clouds of dust block starlight. These lanes provide one of the best tracers of spiral arms. Spiral arms also show up in other tracers of concentrations of gas, such as 21-cm radiation from neutral hydrogen or radio emission from carbon monoxide **(Figure 17.11b)**.

WHEN A GALAXY'S DISK IS "KICKED," SPIRAL STRUCTURE FORMS

Spiral arms are concentrations of gas where stars form, but why do spiral arms exist at all? Part of the answer is that *any* disturbance in the disk of a spiral galaxy will naturally be made into a spiral pattern by the disk's rotation. Material closer to the center takes less time to

Figure 17.10 *Images of the spiral galaxy UGC 12343, taken in (a) ultraviolet light, (b) red light, and (c) emission from glowing interstellar clouds of ionized hydrogen gas. Note that the spiral arms are most prominent in the image in ultraviolet light, which is dominated by young hot stars, and in emission from interstellar clouds that are ionized by the radiation from young hot stars. The spiral arms are much less prominent in the image taken in red light, in which the smooth underlying disk of old stars accounts for more of what we see.*

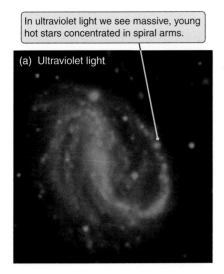

In ultraviolet light we see massive, young hot stars concentrated in spiral arms.

(a) Ultraviolet light

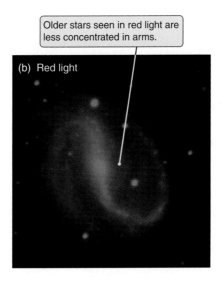

Older stars seen in red light are less concentrated in arms.

(b) Red light

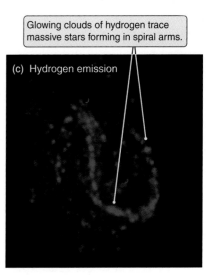

Glowing clouds of hydrogen trace massive stars forming in spiral arms.

(c) Hydrogen emission

Dust lane

Spiral arms are traced by concentrations of gas and dust, seen here as dark absorbing lanes...

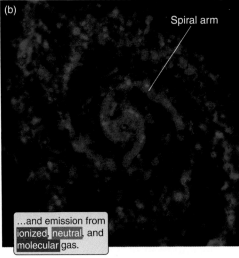

Spiral arm

...and emission from ionized, neutral, and molecular gas.

Figure 17.11 *Two images of a face-on spiral galaxy showing the spiral arms.* (a) *An image in visible light, which also shows dust absorption.* (b) *21-cm emission, which shows the distribution of neutral interstellar hydrogen CO emission from cold molecular clouds and Hα emission from ionized gas.*

complete a revolution around the galaxy than material farther out in the galaxy. **Figure 17.12** illustrates the point. We begin with a single linear arm through the center of a model galaxy, then watch what happens as the model galaxy rotates. In the time that it takes for the inner part of the galaxy to complete several rotations, the outer parts of the galaxy may not have completed even a single revolution. In the process, the originally straight arms are slowly made into the spiral structure shown.

Rotation in a disk galaxy naturally produces spiral structure.

The way disk galaxies rotate implies that any "kicks" to a galaxy's disk will naturally lead to spiral structure in that disk. One way that a galaxy can be kicked is by gravitational interactions with other galaxies. We will see much more of these interactions later on in the chapter.

Gravitational interactions and star formation are processes that kick disks.

Another spur to the formation of spiral structure comes from the process of star formation itself. Regions of star formation dump considerable energy into their surroundings in the form of UV radiation, stellar winds, and supernova explosions. All of this energy drives up the pressure in the region, compressing clouds of gas and triggering more star formation. Since many massive stars typically form in the same region at about the same time, their combined mass outflows and supernova explosions will occur one after another in the same region of space over the course of only a few million years. The result can be large, expanding bubbles of hot

17.1 Spiral

gas that sweep out cavities in the interstellar medium and concentrate the swept-up gas into dense star-forming clouds, much like the snow that piles up in front of a snowplow. In this way, star formation can actually propagate through the disk of a galaxy. The resulting strings of star-forming regions can then be swept into spiral structures by rotation.

Just giving a galaxy disk a single kick (whether through gravitational interactions or star formation) will not produce a stable spiral-arm pattern. For the same reason that spiral arms form at all, the spiral arms produced from a "one-shot" kick will wind themselves up completely in two or three rotations of a disk, then disappear. Some types of kicks are repetitive, however, and so are capable of sustaining spiral structure indefinitely. Some of these kicks come from within a galaxy itself. If the bulge in the center of a spiral galaxy is not spherically symmetric, then the bulge will produce a gravitational disturbance in the disk. As the disk rotates through this disturbance, it is subjected to the kick needed to trigger star formation and the formation of spiral structure.

Many galaxies show clear evidence of a relationship between the shapes of their bulges and the structure of their spiral arms. Barred spirals, for example, have a characteristic two-armed spiral pattern that is tied to

Regular disturbances lead to two-armed spirals.

the elongated bulge (see Figure 17.4). Even the bulges of galaxies that are not obviously barred may be non-spherical enough to contribute to the formation of two-armed spiral structure. Smaller galaxies in orbit about larger galaxies can also give rise to a periodic kick, triggering the same sort of two-armed structure.

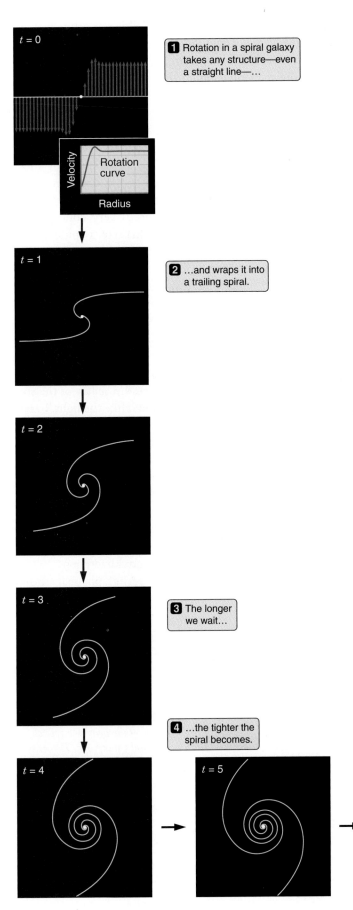

1 Rotation in a spiral galaxy takes any structure—even a straight line—…

2 …and wraps it into a trailing spiral.

3 The longer we wait…

4 …the tighter the spiral becomes.

Regular disturbances in the disks of spiral galaxies are called **spiral density waves.** We call them density waves because they are regions of greater mass density and increased pressure in the interstellar medium that move around a disk in the pattern of a two-armed spiral. Spiral density waves act like the spiral-shaped blade in a blender. Models of how spiral density waves form tell us that this spiral-shaped two-armed wave pattern does not necessarily rotate in the same direction or at the same rate as the rest of the galaxy. This means that as material in the disk orbits about, it passes through the spiral density waves.

The motions of stars are little affected as they pass through the spiral density wave. However, for gas it is a very different story. Consider what happens when you turn on the tap in your kitchen sink. The water hits the bottom of the sink and spreads out in a thin, rapidly moving layer. A few inches out, depending on how much water is flowing, there is a sudden increase in the depth of the water called a "hydrostatic jump." Spiral arms in **Spiral density waves compress gas, triggering star formation.** galaxies work in much the same way. Gas flows into the spiral density wave and piles up like water in a hydrostatic jump. Stars form in the resulting compressed gas. Massive stars have such short lives that they never get the chance to drift far from the spiral arms where they were born, and so that is where we see them. Less massive stars, on the other hand, have plenty of time to move away from their places of birth, forming a smooth underlying disk.

17.4 GALAXIES ARE MOSTLY DARK MATTER

As we found earlier, while mass may play the dominant role in determining the properties of stars, the same is not true for galaxies. Even so, efforts to measure the masses of galaxies during the last decades of the 20th cen- **We can estimate galaxy mass from what we see.**

Figure 17.12 *The rotation of a spiral galaxy will naturally take even an originally linear structure and wrap it into a progressively tighter spiral as time goes by.*

tury led to some of the most remarkable and surprising findings in the history of astronomy. To understand this work, we first need to ask how to go about measuring the mass of a galaxy. One way is to look at the amount of light it gives off. We can use the spectrum of starlight from a galaxy to determine the types of stars the galaxy contains. We then use our knowledge of stellar evolution to turn the luminosity of the galaxy into an estimate of the total mass in stars. Finally, our knowledge of the physics of radiation from interstellar gas at X-ray, infrared, and radio wavelengths allows us to estimate the mass of these other components. Together, the stars, gas, and dust in a galaxy are called **luminous matter,** or simply **normal matter,** because this matter emits electromagnetic radiation.

Looking at the light from a galaxy is not a sure way to determine its mass, however. Imagine, for example, if we were to replace the Sun with a black hole of the same mass. We would see no light coming from the black hole—we would not include its mass in our estimate based on the luminosity of starlight from our galaxy—and yet the planets would continue on their orbits, moving under the influence of its gravity. At every point in this book we have relied on measuring the effect of gravity on motion as the one sure way of determining the masses of objects. This time will be no different.

Gravity and Kepler's laws are the sure way to measure galaxy mass.

The disks of spiral galaxies are rotating, which means that the stars in those disks are following orbits that are much like the Keplerian orbits of planets around their parent stars and binary stars around each other. To measure the mass of a spiral galaxy, all we need do is apply Kepler's laws, just as we did for those other systems.

We might begin our study of the rotation of spiral galaxies with the hypothesis that mass in a galaxy is distributed in the same way as the light. That is, we could begin by assuming that the luminous mass in these galaxies is all the mass that there is. On the basis of this hypothesis, we would make a prediction. The light of all galaxies, including spiral galaxies, is very concentrated toward their centers. If all the mass of a spiral galaxy were contained in its centrally concentrated stars, gas, and dust, we would predict that the orbital velocities of the gas and stars in the disk should behave like the orbital velocities of the planets in our Solar System. We should see fast orbital velocities close in and slower orbital velocities farther out.

Figure 17.13 shows the prediction of how the orbital velocities of material in the disk of a spiral galaxy should change with distance from the center of the spiral, assuming that mass is distributed like the light. To test this prediction, we use the Doppler effect to measure orbital motions. There are several ways to do this. We might measure the velocities of stars from observations of absorption lines in their spectra, or we might measure the velocities of interstellar gas using emission lines such as Hα emission or 21-cm emission from neutral hydrogen.

A graph that shows how orbital velocity in a galaxy varies with distance from the galaxy's center is called a **rota-**

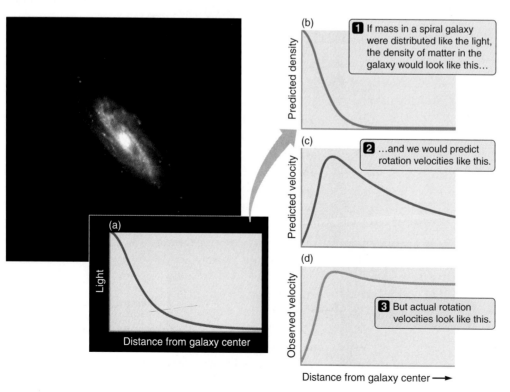

(b)
Predicted density

❶ If mass in a spiral galaxy were distributed like the light, the density of matter in the galaxy would look like this...

(c)
Predicted velocity

❷ ...and we would predict rotation velocities like this.

(d)
Observed velocity

❸ But actual rotation velocities look like this.

(a)
Light
Distance from galaxy center

Distance from galaxy center ⟶

Figure 17.13 (a) *The profile of the light in a typical spiral galaxy.* (b) *The mass density of stars and gas located at a given distance from the galaxy's center. If stars and gas accounted for all of the mass of the galaxy then the galaxy's rotation curve should be as shown in (c), but real galaxies have observed rotation curves more like that shown in (d).*

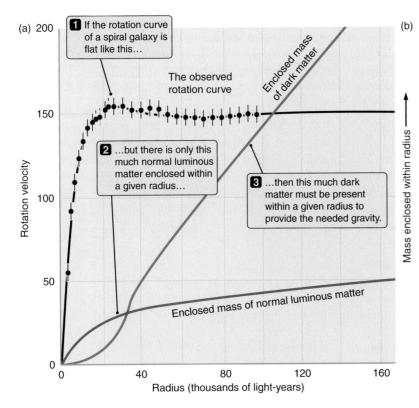

(a)

1 If the rotation curve of a spiral galaxy is flat like this…

The observed rotation curve

Enclosed mass of dark matter

2 …but there is only this much normal luminous matter enclosed within a given radius…

3 …then this much dark matter must be present within a given radius to provide the needed gravity.

Enclosed mass of normal luminous matter

Rotation velocity

Radius (thousands of light-years)

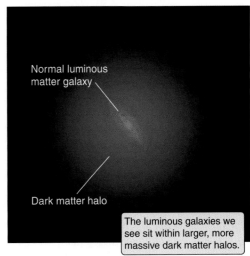

(b)

Normal luminous matter galaxy

Dark matter halo

Mass enclosed within radius

The luminous galaxies we see sit within larger, more massive dark matter halos.

Figure 17.14 *The flat rotation curve of the spiral galaxy NGC 3198, along with the total mass within a given radius that can be accounted for by stars and gas, and the extra "dark" mass needed to explain the rotation curve. In addition to matter we can see, galaxies must be surrounded by halos containing a large amount of dark matter.*

tion curve. Astronomers had to wait until the mid-1970s before telescope instrumentation permitted such rotation curves to be measured reliably outside the inner, bright regions of galaxies. Much to their surprise, when the observations were made, the prediction that mass in galaxies is distributed like the light from galaxies was falsified! Rather than finding rotation curves that fall off to lower and lower velocities in the outer parts of spiral galaxies, as predicted, observations instead showed that spiral galaxy rotation velocities remain about the same out to the most distant measured parts of the galaxies. As shown in Figure 17.13(d), the rotation curves of spiral galaxies appear level, or "flat," in their outer parts. These are naturally referred to as **flat rotation curves.** Observations of 21-cm radiation from neutral hydrogen even show that the rotation curves remain flat well outside the extent of the visible disks. This discovery came as a shock. It meant that long-held ideas about the distribution of mass in galaxies were wrong.

The rotation curve of a spiral galaxy allows us to directly determine how the mass in that galaxy is distributed. All we need to do is apply Kepler's laws to these rotation curves and ask, How much mass must be present inside a given radius to account for the orbital velocity we measure at that radius? (Recall from Chapter 9 that only the mass inside a given radius contributes to the net

Rotation curves of spiral galaxies are flat.

gravitational force felt by an object.) **Figure 17.14** shows the result of such a calculation. In addition to the centrally concentrated luminous matter, there must be a second component to these galaxies consisting of matter that does not show up in our census of stars, gas, and dust. This material, which reveals itself only by the influence of its gravity, is called **dark matter.**

We see the presence of dark matter in spiral galaxies because it is distributed differently from the starlight. While both the starlight and the dark matter are concentrated toward the center of a galaxy, this is less true for dark matter than for luminous matter. The rotation curves of the inner parts of spiral galaxies match fairly well what would be predicted by their luminous matter, indicating that normal luminous matter dominates the inner part of spiral galaxies. Within the part of a galaxy that can be seen in visual light, the mix of dark and luminous matter is about half and half. However, rotation curves measured using 21-cm radiation from neutral hydrogen show that the outer parts of spiral galaxies are mostly dark matter. It is currently estimated that up to 95% of the total mass in a spiral galaxy consists of a greatly extended **dark matter halo,** far larger than the visible spiral portion of the galaxy located at its center. This is a startling statement. A spiral galaxy illuminates only the inner part of a

Most of the mass in spiral galaxies is dark matter.

much larger distribution of mass that is dominated by some type of matter we cannot see!

We measure the luminous matter in elliptical galaxies the same way we measure the luminous matter in spiral galaxies. However, elliptical galaxies do not rotate, so we need a different approach to measure their masses. Our earlier discussion of planetary atmospheres in Chapter 7 provides just the tool we need. A planet's ability to hold on to its atmosphere depends on its mass. In like fashion, an elliptical galaxy's ability to hold on to its hot, X-ray-emitting gas depends on its mass. When the masses of elliptical galaxies are inferred from X-ray images, such as the one in **Figure 17.15,** we discover the same thing that we found from the rotation curves of spirals. Elliptical galaxies contain up to 20 times as much mass as can be accounted for by their stars and gas alone, so they too are dominated by dark matter. As with spirals, the luminous matter in ellipticals is more centrally concentrated than is the dark matter.

Ellipticals' dark matter enables them to hold on to their hot gases.

The transition from the inner part of galaxies (where luminous matter dominates) to the outer parts of galaxies (which are dominated by dark matter) is remarkably smooth. This seamless combination of dark and luminous matter is an important piece of evidence that any successful theory of galaxy formation must explain.

So what is this dark matter of which galaxies are mostly made? We do not yet know. A number of suggestions have been made over the years, ranging from space filled with Jupiter-like objects, to

The nature of dark matter is not known.

swarms of black holes, to copious numbers of white dwarf stars, to exotic unknown elementary particles. (The last of these is currently the favored explanation.) If galaxies are mostly formed of dark matter, then have we been wasting our time paying so much attention to the small fraction of mass tied up in stars and the interstellar medium? No. For one thing, stars and gas are the only parts of galaxies that we can see directly. Stars are also of undeniable importance from our human perspective. Stars formed the atoms in our bodies, and are the kernels around which planetary systems form. The star we call our Sun supplies the energy that makes life on Earth possible. The attention we have paid to stars is well placed indeed!

17.5 THERE IS A BEAST AT THE CENTERS OF GALAXIES

Galaxies are remarkable objects, each shining with the light of hundreds of billions of stars. However, galaxies themselves pale in comparison with the most brilliant beacons of all—**quasars.** *Quasar* is short for "quasi-stellar radio source," so named because they were first observed as unresolved points at radio wavelengths. Quasars are phenomenally powerful, pouring forth the luminosity of a trillion to a thousand trillion (10^{12}–10^{15}) Suns! Quasars are objects of the distant Universe. The nearest

Quasars are phenomenally luminous.

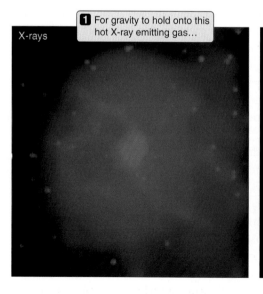

1 For gravity to hold onto this hot X-ray emitting gas…

X-rays

2 …an elliptical galaxy must be much more massive than the combination of all stars we see.

Visible light

3 Elliptical galaxies are mostly dark matter.

Figure 17.15 *The X-ray emission from hot gas around an elliptical galaxy extends well beyond the region of visible light from stars.*

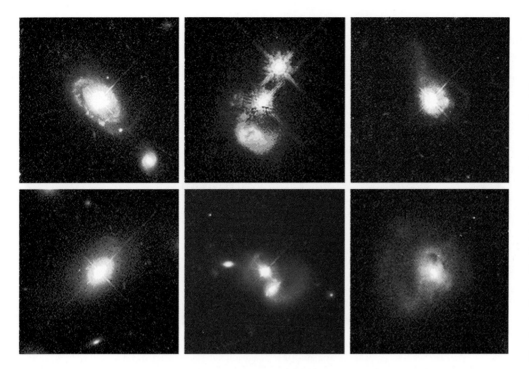

Figure 17.16 HST *images of the environments around quasars. Quasars are found in the centers of galaxies. Those galaxies often show evidence of interactions with other galaxies.*

like Seyfert nuclei, however, AGNs in elliptical galaxies are usually most prominent in the radio portion of the electromagnetic spectrum, earning them the name **radio galaxies.** Radio galaxies, and their distant, extremely luminous cousins the quasars, are often the sources of thin jets that extend outward millions of light-years from the galaxy, powering twin lobes of radio emission, such as those seen in **Figure 17.17.**

What could power such phenomenal cosmic beacons as quasars, Seyfert galaxies, and radio galaxies? Clues lie in the light they emit. Much of the light from AGNs is synchrotron radiation. This is the same type of radiation

AGNs emit synchrotron radiation.

that we first encountered coming from Jupiter's magnetosphere, and later saw again in such extreme environments as the Crab Nebula. Recall from Chapter 8 that synchrotron radiation comes from relativistic charged particles spiraling around the direction of a magnetic field (see Figure 8.15). The fact that AGNs accelerate large amounts of material to close to the speed of light indicates that they are very violent objects indeed. In addition to synchrotron radiation, the spectra of many quasars and Seyfert nuclei show emission lines that are smeared out by the Doppler effect across a wide range of wavelengths. This implies that gas in AGNs is swirling around the centers of these galaxies at speeds of thousands or even tens of thousands of kilometers per second.

quasar to us is approximately 1 billion light-years away. There are literally billions of galaxies that are closer to us than the nearest quasar. Because light travels at a finite speed, the distance to an object also tells us the amount of time that has passed since the light from that object left its source. The fact that the nearest quasar is seen as it existed a billion years ago tells us that quasars are quite rare in the Universe today. Quasars were once much more common. The discovery of quasars in the distant and therefore earlier Universe provided one of the first pieces of evidence that the Universe has evolved over time.

Quasars are not isolated beacons, but instead are centers of violent activity in the hearts of large galaxies, as seen in **Figure 17.16.** Quasars are now recognized as only the most extreme form of activity that can occur in the nuclei of galaxies. In today's Universe, approximately 3% of galaxies contain brilliant points of light in their centers that may outshine all of the stars in the galaxies that host them. Together, quasars and their less luminous but still active cousins are called **active galactic nuclei,** or simply **AGNs.**

Seyfert galaxies (named after **Carl Seyfert** 1911–1960, who discovered them in 1943) are spiral galaxies that contain AGNs discernable in visible light at their centers. The luminosity of a typical Seyfert nucleus can be 10 billion to 100 billion $L_\odot$, comparable to the luminosity of the rest of the galaxy as a whole. The luminosities of AGNs found in elliptical galaxies are similar to those of Seyfert nuclei (10 billion to 100 billion $L_\odot$). Un-

There are several types of active galactic nuclei.

AGNS ARE AS SMALL AS THE SOLAR SYSTEM

The enormous radiated power and mechanical energy of AGNs are mindboggling on their own, but they are made even more spectacular by the fact that all this

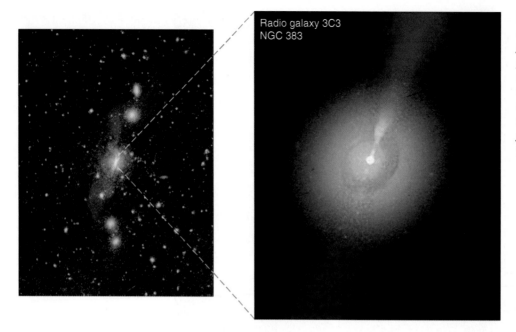

Radio galaxy 3C3
NGC 383

Figure 17.17 *Radio emission from a double-lobed radio galaxy (shown in red) is superposed over an image of visible starlight from the galaxy (shown in blue). The lobes, powered by a relativistic jet streaming outward from the nucleus of the galaxy, are over 3 million light-years from the galaxy.*

power emerges from a region that can be no larger than a light-day or so across. That is comparable in size to our own Solar System! How can we even make such a claim? Quasars and other AGNs appear only as unresolved points of light even in our most powerful telescopes. What information is there in the light we receive from AGNs that tells us they are such compact objects? For an answer we turn not to the sky but to the halftime show at a local football game.

Figure 17.18 illustrates the unalterable bane of the existence of every marching band director. When a band is all together in a tight formation at the center of the field, the notes you hear in the stand are clear and crisp. The band plays together beautifully. But as the band spreads out across the field, its sound begins to get mushy. This is not because the marchers are poor musicians. Rather, it is a consequence of the fact that sound travels at a finite speed. The speed of sound on a cold, dry December day is around 330 m/s. At this speed, it takes sound approximately a third of a second to travel from one end of the football field to the other. Even if every musician on the field played a note at exactly the same instant in response to the director's cue, in the stands you would hear the musicians close to you first but would have to wait longer for the sound from the far end of the field to arrive.

If the band is spread from one end of the field to the other, then the beginning of a note will be smeared out over about a third of a second, or the difference in sound travel time from the near side of the field to the

> **A marching band cannot play a clean note.**

far side. If the band were spread out over two football fields, it would take about two-thirds of a second for the sound from the most distant musicians to arrive at your ear. If our marching band were spread out over a kilometer, then it would take roughly three seconds—the time it takes sound to travel a kilometer—for us to hear a sharply played note start and stop. Even with our eyes closed, it would be easy to tell whether the band was in a tight group or spread out across the field.

Exactly the same principle works for AGNs, only here we are working with the speed of light, not the speed of sound. Quasars and other AGNs can change their brightness dramatically over the course of only a day or two. The AGN powerhouse must therefore be no more than a light-day or so across because, if the powerhouse were larger, what we see could not possibly change in a day or two. Here is the image that should come to mind when you think of quasars: *the light of 10,000 galaxies pouring out of a region of space that would come close to fitting within the orbit of Pluto!*

> **AGNs vary rapidly and so must be relatively small.**

SUPERMASSIVE BLACK HOLES AND ACCRETION DISKS RUN AMOK

When AGNs were first discovered, a variety of ideas were put forward to explain them. However, as their tiny sizes and incredible energy densities became clear,

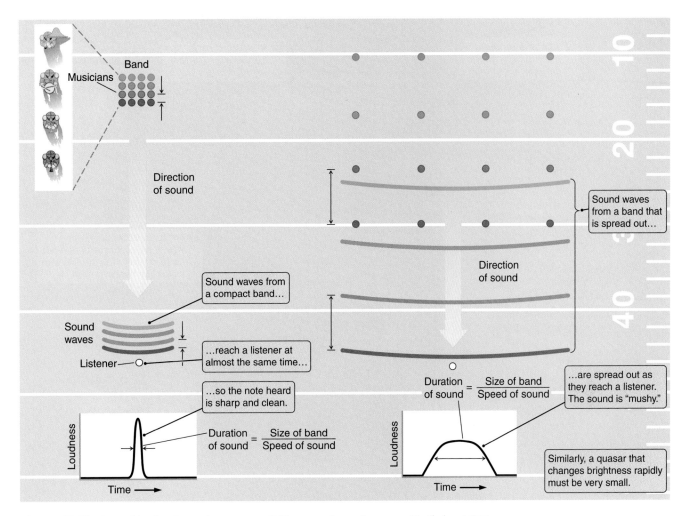

Figure 17.18 *A marching band spread out across a field cannot play a clean note. Similarly, AGNs must be very small to explain their rapid variability.*

only one answer seemed to make sense. AGNs are powered by violent accretion disks surrounding **supermassive black holes** with masses from thousands to tens of billions of solar masses.

We have run across accretion disks on a number of occasions during our journey. Accretion disks surround young stars, providing the raw material for solar systems. Accretion disks around white dwarfs, fueled by material torn from their bloated evolving companions, lead to novae and Type I supernovae. Accretion disks around neutron stars and star-sized black holes a few kilometers across are seen as X-ray binary stars. Take this mental picture and scale it up to a black hole with a mass of a billion solar masses and a radius comparable in size to the orbit of Uranus. Rather than small amounts of material being siphoned off a star, imagine an accretion disk fed by substantial fractions of entire galaxies. *That* is an active galactic nucleus.

AGNs are powered by accretion onto supermassive black holes.

This basic picture of a supermassive black hole surrounded by an accretion disk has been developed into a more complete physical description called the **unified model of AGNs.** The unified model takes its name from the fact that it attempts to unify our understanding of all AGNs—quasars, Seyfert galaxies, and radio galaxies—within the same framework of understanding.

Figure 17.19(a) shows a diagram of the various components of the unified model of AGNs. In the unified model, a supermassive black hole is surrounded by an accretion disk. Much farther out lies a large torus (i.e., donut) of gas and dust consisting of material that is feeding the central engine. Each of the different components of the unified model accounts for different observed properties of AGNs.

In our discussion of star formation, we learned how gravitational energy is converted to thermal energy as material moves inward toward the growing protostar. As material moves inward toward a supermassive black

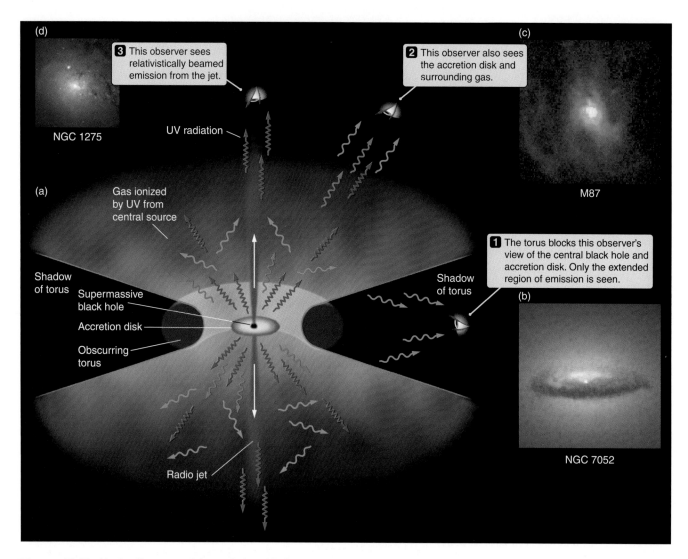

Figure 17.19 (a) *An illustration of the unified model of active galactic nuclei. The appearance of the object changes when the disk and torus are viewed (b) from the edge, (c) at higher inclination, and (d) close to face-on. Viewing angle, along with the mass of the central black hole and the rate at which it is being fed, is thought to determine the properties of an AGN that we see.*

hole, conversion of gravitational energy heats the disk to hundreds of thousands of kelvins, causing it to glow brightly in visible and ultraviolet light. Conversion of gravitational energy to thermal energy as material falls onto the accretion disk is also a source of X-rays, UV radiation, and other energetic emission. When we discussed the Sun, we marveled at the efficiency of fusion, which converts 0.7% of the mass of hydrogen into energy. In contrast, upwards of 50% of the mass of infalling material around a supermassive black hole is converted to luminous energy. The rest of that mass is pulled into the black hole itself, causing it to grow even more massive.

AGN accretion disks are phenomenally efficient at converting mass to energy.

The interaction of the accretion disk with the black hole gives rise to powerful radio jets—superpowerful analogs to the Herbig-Haro jets formed by accretion disks around young stellar objects. Throughout, twisted magnetic fields accelerate charged particles such as electrons and protons to relativistic speeds, accounting for the observed synchrotron emission. Gas in the accretion disk or in nearby clouds orbiting the central engine at high speeds gives off emission lines that are smeared out by the Doppler effect into the broad lines seen in AGN spectra.

The outer torus plays a somewhat different role in the unified model. Located far from the inner turmoil of the accretion disk, and far larger than the central engine, some of the outer torus is ionized by UV light from the AGN much as an H II region is ionized by the UV light from O stars. This provides one possible source for H II–region-like emission lines seen in

The outer torus determines what we can see.

many AGNs. And of course, the outer torus *must* be there in some form if material is continuing to fall inward toward the accretion disk. Yet the most important conceptual role of the outer torus is that it allows a single model to explain many of the differences observed among various AGNs. The outer torus obscures our view of the central engine in different ways depending on the angle from which we happen to see it. By invoking variations in the viewing angle, the mass of the black hole, and the rate at which it is being fed, the unified model of AGNs can account for a wide range of AGN properties.

WHAT AGN WE SEE DEPENDS ON OUR PERSPECTIVE

When we view the unified model edge-on, we see emission lines from the surrounding torus and other surrounding gas. We can also sometimes see the torus in absorption against the background of the galaxy. **Figure 17.19(b)** is a *Hubble Space Telescope* image of the inner part of the galaxy NGC4261, showing what appears to be just such a shadow. When viewed from this orientation, we cannot see the accretion disk itself, so we do not expect to see the Doppler-smeared lines that

Viewed edge-on, an AGN's accretion disk is not visible.

originate closer to the supermassive black hole. However, if jets are present in the AGN, we should be able to see these emerging from the center of the galaxy.

If we look at the inner accretion disk somewhat more face-on, then we can see over the edge of the torus, giving us a more direct look at the accretion disk and the black hole. In this case we should see more of the syn-

When viewed more face-on, we see the accretion disk.

chrotron emission from the region around the black hole and the Doppler-broadened lines produced in and around the accretion disk. **Figure 17.19(c)** shows a *Hubble Space Telescope* image of one such object called M87 (object 87 in Messier's catalog). M87 is a source of powerful jets that carry on for millions of light-years, but originate in the tiny engine at the heart of the galaxy **(Figure 17.20)**. Spectra of the disk at the center of this galaxy **(Figure 17.21)** show the rapid rotation of material around a central black hole with a mass of 3 billion $M_\odot$.

The material in an AGN jet travels very close to the speed of light. As a result, what we see is strongly influenced by relativistic effects. One of these is an extreme form of the Doppler effect called **relativistic beam-**

Relativistic effects influence what we see.

ing. Matter traveling at close to the speed of light con-

centrates any radiation it emits into a tight beam pointed in the direction in which it is moving. As a result of relativistic beaming, an AGN jet coming toward us will look much brighter than its twin moving away from us. The unified model clearly predicts that AGN jets should be two-sided, but most of the time what we actually see appear to be one-sided jets. This is because we see only the portion of the jet that is being beamed toward us. The emission from the other portion of the jet always exists. We infer this from the fact that the radio lobes of radio galaxies are always two-sided. We do not see the jet moving away from us because it is beaming its radiation in the other direction.

In rare cases we happen to see the accretion disk in a quasar or radio galaxy almost directly face-on. When this happens, relativistic beaming dominates what we see. Rather than emission lines and other light coming from hot gas in the accretion disk, we are blinded by the bright glare of jet emission beamed directly at us **(Figure 17.19d)**.

Relativistic beaming is not the only thing that makes appearances deceiving when looking down the barrel of an AGN jet. The material in an AGN jet is moving so close to the speed of light that the radiation it emits is barely able to outrun its source. As observers, we see all of the light emitted by the jet over thousands of years arrive at our telescopes over the course of only a few years. Time appears to be compressed. From our perspective, the jet seems to travel great distances in brief periods of time. In extreme cases, such as the jet in M87 **(Figure 17.22)**, features in the jet *appear* to be moving across the sky faster than the speed of light. We stress "appears" because this phenomenon, referred to as **superluminal motion,** is an optical illusion. Despite the name, nothing in these jets is actually traveling through space faster than the speed of light. Einstein's theory of special relativity remains safe.

It is worth noting that we have used the same galaxy, M87, as an example of several of the phenomena we have discussed. The presence of many different phenomena predicted by the unified model in the same object is important support for the view that the model truly is *unified*.

NORMAL GALAXIES AND AGNs—A QUESTION OF FEEDING THE BEAST

The unified model of AGNs has been around in some form since the 1980s. This model was constructed to explain existing observations of AGN, and has been constantly modified to account for new and better

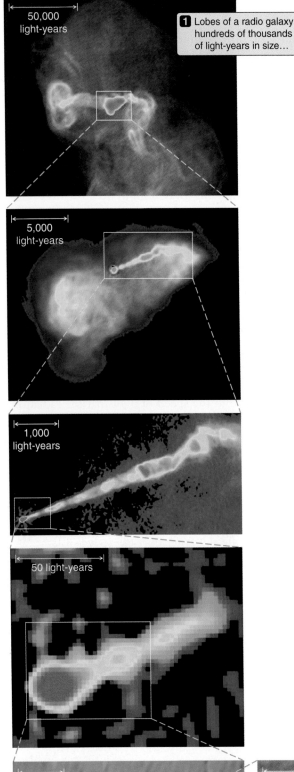

1 Lobes of a radio galaxy hundreds of thousands of light-years in size...

data. It is comforting that the unified model of AGNs has been able to keep up. No observations have come along that have forced us to abandon the model altogether. On the other hand,

Consistency with its own assumptions does not make the unified model true.

the model was *invented* to explain observations of AGNs. The unified model cannot *help* but be consistent with those observations. However, scientific knowledge is not based on whether a theory is consistent with its own assumptions, but on experimental confirmation of a theory's testable predictions. What testable predictions does the unified model make, and how have those predictions fared in the light of new observational tests?

The essential elements of the unified model are a central engine (an accretion disk surrounding a supermassive black hole) and a source of fuel (gas and stars flowing onto the accretion disk). We commonly speak of the infall of material onto the accretion disk as "feeding the beast." If we were to shut off this infall—if we were to stop feeding the beast—the supermassive black hole would remain, absent all the fireworks. Without a source of matter falling onto the black hole, an AGN would no longer be active. If we were to look at such an object, we should see a normal (not active) galaxy with a supermassive black hole sitting in its center.

Should such objects be common? As noted above, only about 3% of present-day galaxies contain AGNs. However, when we look at more distant galaxies (and therefore look back in time), that fraction is much larger. Our observations show that when the Universe was younger, there were many more AGNs than there are today. If the unified model of AGNs is correct, then all of

The unified model predicts that normal galaxies contain supermassive black holes.

the supermassive black holes that powered those long-

Figure 17.20 *The visible jet from the galaxy M87 originates in a tiny volume at the heart of the galaxy, but extends over hundreds of thousands of light-years.*

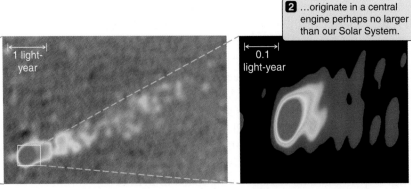

2 ...originate in a central engine perhaps no larger than our Solar System.

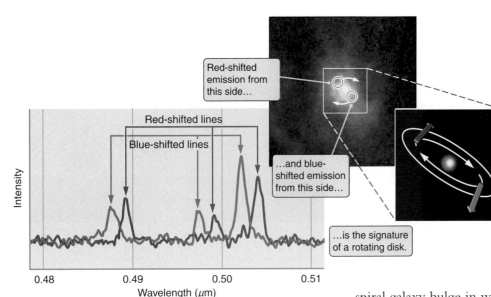

Red-shifted lines

Blue-shifted lines

Intensity

Wavelength (μm)

0.48   0.49   0.50   0.51

Red-shifted emission from this side…

…and blue-shifted emission from this side…

…is the signature of a rotating disk.

Figure 17.21 *An* HST *measurement of the rotation velocities of gas near the center of the nearby radio galaxy M87. The high velocities observed provide direct evidence of rotation about a supermassive black hole at this galaxy's center.*

dead AGNs should still be around. If we combine what we know of the number of AGNs in the past with ideas about how long a given galaxy remains in an active phase, we are led by the unified model to predict that many—perhaps even *most*—normal galaxies today contain supermassive black holes!

This is a somewhat startling prediction. Quiescent collections of stars, gas, and dust like our own Milky Way should have slumbering beasts at their centers. It is a bit like suggesting that all of our dignified, soft-spoken grandmothers were once members of biker gangs. Yet here is a prediction of the unified model that can be tested.

If supermassive black holes are present in the centers of normal galaxies, they should reveal themselves in a number of ways. For one thing, the presence of such a concentration of mass at the center of a galaxy should draw surrounding stars close to it. The central region of such a galaxy would be much brighter than could be explained if stars alone were responsible for the gravitational field in the inner part of the galaxy. Stars feeling the gravitational pull of a supermassive

Supermassive black holes have been discovered in the nuclei of many nearby galaxies.

black hole in the center of a galaxy should also orbit at very high velocities. We should therefore see large Doppler shifts in the light from stars near the centers of normal galaxies. Evidence of this sort has now been found in every normal galaxy with a substantial bulge in which a careful search has been conducted. The masses inferred for these black holes range from 10,000 $M_\odot$ (for a "small" black hole) to 5 billion $M_\odot$ (a "gargantuan" one)! The mass of the supermassive black hole seems to be related to the mass of the elliptical galaxy or

spiral galaxy bulge in which it is found. At the beginning of the 21st century, we have come to accept that all large galaxies probably contain supermassive black holes. These observations confirm the most fundamental prediction of the unified model. They also tell us something remarkable about the structure and history of normal galaxies.

Figure 17.22 *Because the jet material in M87 is moving toward us at relativistic speeds, we see an optical illusion: the radio blobs appear to be moving apart at more than six times the speed of light.*

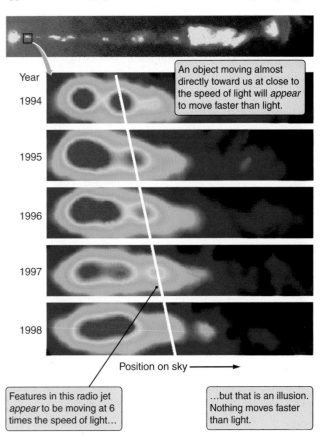

Year

1994

1995

1996

1997

1998

An object moving almost directly toward us at close to the speed of light will *appear* to move faster than light.

Position on sky

Features in this radio jet *appear* to be moving at 6 times the speed of light…

…but that is an illusion. Nothing moves faster than light.

Apparently, the only difference between a normal galaxy and an active galaxy is whether the supermassive black hole at its center is being fed at the time we see that galaxy. The fact that only 3% of present-day galaxies have AGNs does not indicate which galaxies have the potential for AGN activity. Rather, it indicates which galaxy centers are being lit up at the moment. If we were to drop a large amount of gas and dust directly into the center of any large galaxy, this material would fall inward toward the central black hole, forming an accretion disk and a surrounding torus. The predicted result of this process is that the nucleus of this galaxy would change into an AGN.

Figure 17.23 *These tidally interacting galaxies show severe distortions, including stars and gas drawn into long tidal tails.*

MERGERS AND INTERACTIONS MAKE THE DIFFERENCE

Galaxies do not exist in isolation. Even our own Milky Way has several neighbors. In Chapter 9 we discussed tidal interactions between galaxies. **Figure 17.23** shows the havoc that such interactions can cause, pulling interacting galaxies into distorted shapes in which stars and gas are drawn out into sweeping arcs and tidal tails (also see Figure 9.13). Sometimes when galaxies interact, they pass each other by and go their separate ways. Such interactions can trigger the formation of spiral structure, and they can also slam clouds of interstellar gas together at high speeds, triggering additional star formation. Tidally distorted or interacting galaxies often contain regions of vigorous ongoing star formation. Sometimes when two galaxies interact, they merge to form a single, larger galaxy. This turns out to be an important part of the process of galaxy formation, as we will learn in Chapter 20.

Interactions and mergers must have been much more prevalent in the past to account for the many large galaxies that we see today. The prevalence of interacting galaxies when the Universe was younger explains the large number of AGNs that existed in the past. Computer models show that galaxy–galaxy interactions can cause gas located tens of thousands of light-years from the center of a galaxy to fall inward toward the center, where it can provide fuel for an AGN. During mergers, a significant fraction of a cannibalized galaxy might wind up being fed to the beast. *HST* images of quasars, such as those in Figure 17.16,

> **Galaxies do not exist in isolation.**

> **Galaxy–galaxy interactions fuel AGN activity.**

often show that quasar host galaxies are tidally distorted, or are surrounded by other visible matter that is probably still falling into the galaxies. The most violent forms of AGN activity were likely most common in the early Universe, because that was the time when galaxies were forming, and large amounts of matter were constantly being drawn in by the gravity of newly formed galaxies. This process is still at work today. Galaxies that show evidence of recent interactions with other galaxies are far more likely than average to house AGNs in their centers.

There are still many puzzles. For example, the unified model has not been developed to the point that it predicts how long an outburst of AGN activity will last, or how often galaxies will undergo episodes of AGN activity. To answer these questions the unified model of AGNs will have to be combined with much better models of galaxy formation and evolution than we currently have. There are also some observations that the unified model of AGNs does not yet explain. For instance, the unified model does not account for why one quasar can be a very powerful radio source while another, identical in all other respects, is radio-quiet, even when observed with the most sensitive of radio telescopes.

Our understanding of AGNs is far from complete. Even so, the unified model has had many successes—enough for us to say with confidence that any large galaxy, including our own, is only a chance encounter away from becoming an AGN. How different our own sky might be if, a few tens of millions or hundreds of millions of years from now, our descendants look toward the center of our Galaxy and see the brilliant light of a powerful Seyfert nucleus blazing forth.

SEEING THE FOREST THROUGH THE TREES

A century is but a blink of an eye compared with the age of Earth, or even the age of our species. Even so, a century can seem like forever when viewed from the perspective of the changes that century brings. Such is the story of 20th-century astronomy. A hundred years ago, humankind knew virtually nothing of the true answers to the most basic questions we might ask about the Universe. Virtually all that we know of these matters we have learned within living memory. The discovery and study of galaxies is one of the major threads running throughout the story of that century.

From the time that Copernicus first dislodged Earth from the center of creation, each new discovery has pushed us further and further from the human-centered conceptions that shaped our view of the world for millennia (and persists in many ways to this day). The discovery of the almost unthinkable size of our own Milky Way and the realization that the Universe contains countless more galaxies comparable to our own were huge steps in humanity's expanding awareness of the Universe. The work of Shapley, Hubble, and others did far more than merely answer a few arcane and esoteric scientific questions. These giants of 20th-century science tore away some of the last vestiges of a curtain that had hidden the reality of existence from the human mind. These scientists forever changed the way we must view everything, including ourselves.

The scale and distance of galaxies may have shattered preconceptions about the extent of the Universe, but galaxies themselves seemed at first to be composed of familiar objects. The zoo grew far larger, but the animals themselves seemed the same. Stars in distant galaxies shine in accordance with the same laws that govern the Sun. Galaxies are held together by the same force of gravity that set the course of the cannonball in Newton's famous thought experiment. But as our understanding grew and new observations were made, observations of galaxies forced astronomers to push our knowledge of physics in extreme and unexpected ways.

When we first learned of black holes, we were faced with the thought of several solar masses compressed into a region only a few kilometers across, and had to confront ideas of general relativity that turn our everyday notions of space and time inside out. Now we discover that such star-sized black holes are the merest of specks compared with the monsters residing in the centers of large galaxies. We have witnessed the consequences when these behemoths, with masses as great as billions of times that of the Sun, are fed with gas supplied by collisions between galaxies. Radio observations of the sky show us relativistic jets stretching across millions of light-years of intergalactic space. Yet the source of such a jet is tiny. The light of a quasar outshines a thousand galaxies—a beacon that can be seen from the very edge of the Universe—yet originates from a region no larger than our Solar System. As we continue our journey we will find that such a monster lurks at the heart of our own, mundane Milky Way, waiting (perhaps forever) for its next meal.

Active galaxies stir the imagination, but one of the most startling and fundamental results of our study of galaxies comes from the simple application of Newton's derivation of Kepler's laws to the motions of stars in galaxies. We have used this technique over and over on our journey to measure the mass of planets and stars. When we apply this comfortable tool to galaxies, however, the results are shocking. The matter we see in the stars, gas, and dust is but the tip of a much larger iceberg, the rest of which is composed of a substance known only as dark matter. For the most part, the Universe is made up of we know not what.

The 20th century saw dramatic progress in building a physical understanding of the formation, evolution, and death of stars. We can use the laws of physics to pull back the layers of a star, peer into the heart of a supernova, or run the clock forward and see our Sun's ultimate fate. No such understanding exists for galaxies. This is nothing to apologize for. Remember how little time has passed since galaxies were even recognized for what they are. It is important to remember that the study of galaxies is still in its infancy. Many galactic astronomers still spend their lives comparing and contrasting morphological traits, much as Linnaeus sorted living things into kingdoms without real understanding of how those kingdoms arose or what they signified. It is unlikely there will ever be a simple scheme that does for galaxies what the H-R diagram did for stars. On the other hand, astronomers *are* beginning to better understand what the "ecology" within galaxies is like and to piece together something of how they form and evolve. In the next chapter, we take a step down this road by looking in more detail at the galaxy we know best— our Milky Way.

STUDENT QUESTIONS

THINKING ABOUT THE CONCEPTS

1. Name and describe some objects that appear so dissimilar when viewed from different angles that you might think from these different views that they are actually different objects.

2. What are the principal differences between spiral and elliptical galaxies?

3. Explain why star formation in spiral galaxies takes place mostly in the spiral arms.

4. What determines the shape of an elliptical galaxy?

5. Some galaxies have regions that are relatively blue in color, and others have regions that appear redder. Aside from color, what can you say about the differences between these regions?

6. What distinguishes a "normal" galaxy from one we call "active"?

7. What evidence do we have that most galaxies are composed largely of dark matter?

8. The nearest quasar is about a billion light years away. Why do we not see any that are closer?

9. Describe the conditions at the centers of galaxies that contain AGNs.

10. It is likely that most galaxies contain supermassive black holes, yet in many galaxies there is no obvious evidence for their existence. Why do some black holes reveal their presence while others do not?

APPLYING THE CONCEPTS

11. Assume that there are one trillion (10^{12}) galaxies in the Universe, that the average galaxy has a mass equivalent to 100 billion (10^{11}) average stars, and that an average star has a mass of 10^{30} kg.
    a. Ignoring dark matter, how much mass (in units of kg) does the Universe contain?
    b. If the mass of an average particle of normal matter is 10^{-27} kg, how many particlces are there in the entire Universe?

12. A disk star has an orbital period of 3×10^8 years at an average distance of 2.4×10^4 light-years (which equals 1.5×10^9 AU) from the center of its galaxy. Remember that Newton's explanation of Kepler's Third Law has mass $(M_\odot) = [A \text{ (AU)}]^3/[P(\text{yr})]^2$.
    a. What is the mass of the galaxy, in units of solar masses, that lies inside the orbit of this star?
    b. If luminous energy implies 2.5×10^9 solar masses of visible matter inside the star's orbit, what is the ratio of dark matter to visible matter?

13. A quasar has the same brightness as a foreground galaxy that happens to be 6 million light-years distant. If the quasar is one million times more luminous than the galaxy, what is the distance of the quasar?

14. You read in the newspaper that astronomers have discovered a "new" cosmological object that appears to be flickering with a period of 83 minutes. Having read *21st Century Astronomy*, you are able to quickly estimate the maximum size of this object. How large can it be?

15. Assume the Sun is located 27,000 light-years (2.6×10^{17} km) from the center of the Milky Way Galaxy, and is moving along a circular orbit at a speed of 220 km/s. How long does it take our Solar System to make one complete circuit around our Galaxy?

16. A solar-type star (mass = 2×10^{30} kg), accompanied by its retinue of planets, approaches a supermassive black hole. As it crosses the event horizon, half of its mass falls into the black hole, while the other half is completely converted to luminous energy.
    a. As it signals its demise in a burst of electromagnetic radiation, how much energy (in units of joules) does the dying solar system send out to the rest of the Universe?
    b. This is likely how quasars emit energy. If a luminous quasar has a luminosity of 2×10^{41} joules/second, how many solar masses per year does this quasar consume to maintain its average energy output?

O Milky Way, sister in whiteness
To Canaan's rivers and the bright
Bodies of lovers drowned,
Can we follow toilsomely
Your path to other nebulae?

GUILLAUME APOLLINAIRE (1880–1918)

THE MILKY WAY— A NORMAL SPIRAL GALAXY

18.1 WE LOOK UP AND SEE OUR GALAXY

We live in a Universe full of galaxies of many sizes and types, visible in our most powerful telescopes all the way to the edge of the observable Universe. Yet when we go outside at night and look up, it is not this Universe of galaxies that we see. Rather, the night sky is filled with a single galaxy—our home, the galaxy we call the Milky Way. With what we know of galaxies from the previous chapter, there is a great deal that the appearance of the night sky can tell us about our local galaxy. **Figure 18.1(a)** shows a complete picture of what the sky looks like as seen from Earth, while **Figure 18.1(b)** shows an image of an edge-on spiral galaxy. Comparison of the two can leave little doubt that we live in the disk of such a spiral. The flattened disk of the Milky Way is obvious when we look at the sky, once we appreciate what we are looking at. We can even see dark bands where clouds of interstellar gas and dust obscure much of the central plane of our Galaxy. This single look tells us a great deal about our Galaxy, but there is much that it does not tell us. There are many things that we know about galaxies only from our experience with the Milky Way. We see it from a

We live in a barred spiral galaxy called the Milky Way.

KEY CONCEPTS

Of the hundreds of billions of galaxies in the Universe, the one that means the most to us is our cosmic home, the Milky Way. The Milky Way may be just another galaxy, but it is the only galaxy that we can study at close range. As we focus our attention on our Galaxy we will learn:

✳ How variable stars in globular clusters are used as standard candles, allowing us to measure the size of the Milky Way;

✳ How Doppler-shifted radio emission from gas throughout the rotating disk of the Milky Way allows us to map the Galaxy's structure;

✳ That the Milky Way is a typical giant barred spiral galaxy, and like all such galaxies, it is mostly composed of dark matter;

✳ How the chemical composition of the Milky Way has evolved with time;

✳ What differences in age and chemical composition of groups of stars tell us about the history of star formation in our Galaxy;

✳ About the environment within the disk of the Milky Way, and the halo of stars, globular clusters, and dark matter that surrounds our Galaxy; and

✳ About the black hole at the center of our Galaxy.

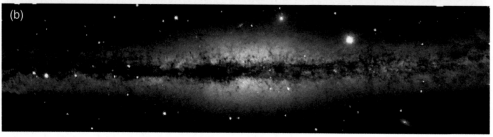

Figure 18.1 (a) *The night sky as viewed from Earth.* (b) *The edge-on spiral galaxy NGC891, whose disk greatly resembles the Milky Way.*

much closer perspective than we do any other galaxy. We see it from the inside!

At the same time, there are real disadvantages in the perspective that we have on the Milky Way Galaxy. Buried within the disk of the Milky Way, we lack a bird's-eye view of our galactic home. The situation is further complicated by the dust in the surrounding interstellar medium that limits our view. **Figure 18.2** shows a model of the structure of the Milky Way proposed in the early 1800s by William Herschel. Herschel did not know about the interstellar extinction of starlight, the properties of stars, or the relationship between our Galaxy and the "spiral nebulae" that he studied throughout his life. His model, which bears little resemblance to our modern understanding of the Milky Way, was based simply on counting the number and brightness of stars seen in different directions.

Dust obscures our view of our own Galaxy.

Early models of our Galaxy are of interest today mostly as historical curiosities. Even so, today's astronomers must cope with the same suite of difficulties that stood in the way of 19th- and early-20th-century astronomers. What type of spiral galaxy do we live in? Are the spiral arms prominent? Is the bulge barlike, and is it large or small relative to the disk? The questions that are so easy to ask and answer for distant galaxies are far more difficult to address for our own Galaxy. Yet at this stage in our journey, we have already learned far more about our Galaxy than any other galaxy. In fact, everything that we have discussed in this book up until the previous chapter falls in the category of "things we know about the Milky Way." Stars, planets, and the interstellar medium—almost all that we know about these we have learned within the context of our own galactic home. It is now time to make this home the focus of our study. It is time to merge the perspective of our look outward at the Universe of other galaxies with our knowledge of our locality to better understand our Milky Way as a spiral galaxy.

18.2 MEASURING THE MILKY WAY

One of the more difficult practical issues faced by astronomers is determining the distances to objects in the sky. In Chapter 12 we learned how geometrical parallax is used to measure the distances to nearby stars, but this only got us out to distances of a few hundred light-years. In the previous chapter we mentioned that Harlow Shapley determined the size of the Milky Way and the effect his measurement had on astronomy. We also mentioned Hubble's discovery of Cepheid variables in nearby galaxies, and discussed galaxies that are many millions or even billions of light-years distant. Yet for the most part the important question of how we actually *know* these distances was temporarily swept under the rug. The difficult question of measuring distances in the Universe will be a theme in much of the rest of our journey, and it is an issue that must now come to the foreground.

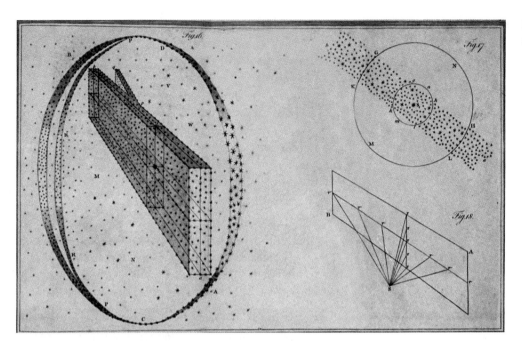

Figure 18.2 *Two models of the Milky Way proposed by William Herschel in the late 1700s and early 1800s.*

held together by gravity. Many clusters can be seen through small telescopes. At first glance they look something like small elliptical galaxies, and the analogy is not a bad one. The motions of stars within a globular cluster are much like the motions of stars within an elliptical galaxy. However, globular clusters are quite different from elliptical galaxies in size and in concentration of stars. There are more than 150 cataloged globular clusters in our Galaxy (and likely many more, as dust in the disk of our Galaxy may hide them from view). The known globular clusters have luminosities ranging from a low of 400 $L_\odot$ to a high of around 1 million $L_\odot$. A typical

> **Globular clusters are very luminous and easy to identify.**

globular cluster consists of 500,000 stars packed into a volume of space with a radius of only 15 light-years. In contrast, we find only about 50 stars within that distance from our own Sun. Globular clusters are therefore much denser concentrations of stars than occur on average throughout our Galaxy. On the other hand, our Galaxy has over 100,000 times as many stars as a typical globular cluster.

About a quarter of the globular clusters in the Milky Way reside in or near the disk of our Galaxy. The rest of them occupy a large volume of space surrounding the disk and bulge, referred to as the **halo** of the Milky Way. Globular clus-

> **Globular clusters are seen to great distances within our Galaxy.**

ters offer two distinct advantages that make them relatively easy to study. First, they are very luminous, and so can be easily seen to great distances. Second, because many globular clusters lie outside the disk of the Milky Way, we can see them at great distances without encountering much absorption due to the obscuring dust within the disk. However, simply being able to see globular clusters does not necessarily put us any closer

Globular Clusters and the Size of the Milky Way

The key to discovering the size of the Milky Way Galaxy turned out to be the presence of **globular clusters** in our Galaxy. A globular cluster, such as the one in **Figure 18.3**, is a large spheroidal group of stars

Figure 18.3 *A Hubble Space Telescope image of the globular cluster M80.*

to measuring their distances. To do that, we need to look at the properties of the stars they contain.

This is not the first time we have been confronted with measuring the distance to remote objects. If we know the luminosity of a star and can measure its brightness, then we can use the inverse square law of radiation to determine its distance. We discussed the inverse square law at length in Chapter 4. Rearranging that law for our present purpose, we have

$$d = \sqrt{\frac{L_{star}}{4\pi b_{star}}}.$$

In words, the distance of a star is proportional to the square root of the ratio of its luminosity (L_{star}) to its brightness (b_{star}) as seen from Earth.

Some types of stars are especially useful for determining distances. These are stars that are both very luminous (so that they can be seen at great distances) and have *known* luminosities. Such stars are referred to as **standard candles**.

The story of measuring distances to remote objects has largely involved the search for better, more luminous standard candles.

Do globular clusters contain good standard candles? **Figure 18.4** shows the H-R diagram for the stars in globular cluster M92. The main sequence turnoff in this cluster's H-R diagram occurs for stars with masses of around 0.8 $M_\odot$, which (for the low abundance of heavy elements in the stars in this globular cluster) corresponds to a main sequence lifetime of around 13 billion years. The age of this globular cluster is similar to those of many others, making them the oldest objects known in our Galaxy or in any nearby galaxy. Globular clusters must have formed when the Universe and our Galaxy were very young. Compared to globular cluster stars, our Sun at 5 billion years old is a relatively recent member of our Galaxy.

In Chapter 16 we found that there is a region in the H-R diagram, called the instability strip, in which stars pulsate. As they do so, their luminosity changes. In an old cluster like a globular cluster, the horizontal branch of the H-R diagram crosses the instability strip. Recall that horizontal branch stars that lie in the instability strip are variable stars called *RR Lyrae stars*. RR Lyrae stars in globular clusters are easy to spot because they are very luminous (horizontal branch stars are giant stars) and because their periodic changes in brightness signal their identity. As with Cepheid variables, the time it takes for an RR Lyrae star to undergo one pulsation is related to

the star's luminosity. Shapley used Henrietta Leavitt's determination of this period-luminosity relationship to determine the luminosities of RR Lyrae stars in globular clusters, then used the inverse square law of radiation to combine these luminosities with measured brightnesses to determine the distances to globular clusters. Shapley then cross-checked himself by noting that more distant clusters (as measured

Figure 18.4 *The globular cluster M92 and an H-R diagram of the stars it contains. (In this instance the color plotted is based on the ratio of visible to infrared light.) The main sequence turnoff at 0.8 $M_\odot$ indicates that the cluster is about 13 billion years old.*

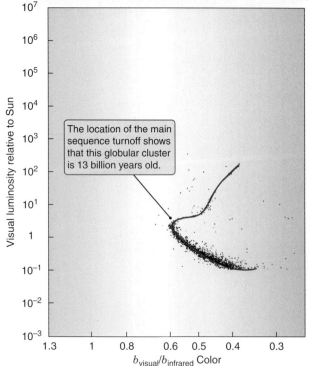

The location of the main sequence turnoff shows that this globular cluster is 13 billion years old.

EXCURSIONS 18.1

NIGHTFALL

The famous science and science fiction writer **Isaac Asimov** (1920–1992) imagined what might happen to a civilization on a planet orbiting within a system of six stars, located in the heart of a giant globular cluster. His story *Nightfall* has become one of the more famous works of science fiction. On Asimov's fictional planet Kalgash, at least one of its six stars is almost always above the horizon. "Nightfall" occurs on Kalgash only once every 2,000 years. The story tells of the great madness that afflicts the inhabitants on this one night, as they recoil in fear from a sky filled with hundreds of thousands of bright stars.

Asimov's story leads us to consider how our perspectives as human beings are formed by the circumstances in which we live. Yet we need not go to science fiction to ask these kinds of questions. Our perspective on the Universe changes with time. At

the moment we are traversing an open, rather dust-free part of the Milky Way. On a moonless night we see a dark sky and gaze with our telescopes into a Universe of galaxies, but that will not always be the case. Just "down the road," astronomically speaking, are dark interstellar clouds and star-forming regions filled with glowing gas and obscuring dust through which the Sun and its entourage of planets will occasionally pass. How different would our view of the Universe be if, instead of a dark sky, we looked up each night and saw a sky filled with a soft green glow, punctuated by a few points of intense light? How much different would our history be if at some point during the rise of our civilization we had suddenly emerged over the course of just a few years from within a molecular cloud and gotten our first look at the larger Universe?

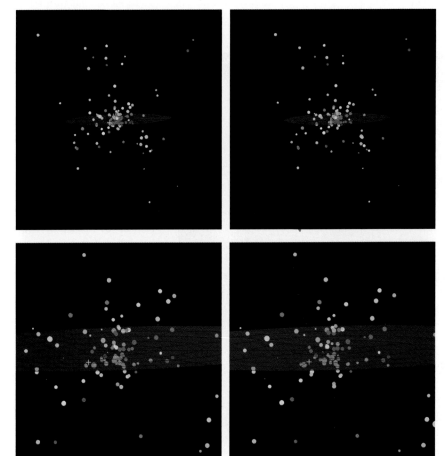

with his standard candle) also tended to appear smaller on the sky, as expected.

Knowing both the distances to globular clusters and where they appear in the sky, Shapley was able to make a three-dimensional map of the distribution of globular clusters in space. This map showed that globular clusters occupy a roughly spherical region of space with a radius of about 300,000 light-years! These globular clusters trace out the halo of the Milky Way Galaxy, as shown in **Figure 18.5.**

The globular clusters around the Milky Way are moving about under the gravitational influence of the Galaxy, just as stars in an elliptical galaxy move about within the gravitational influence of that kind of galaxy. Symmetry therefore requires that the center of the dis-

Figure 18.5 *Stereoscopic views of the distribution of globular clusters in the Milky Way Galaxy. Colors represent the average colors of stars in each cluster. The green cross shows the location of the Sun. (See Figure 12.1 for viewing instructions.)*

tribution of globular clusters coincide with the gravitational center of the Galaxy itself. What Shapley realized was that, since he could determine the distance to the center of this distribution, he had determined the size of the Milky Way itself.

Figure 18.6 identifies the disk, bulge, and halo of the Milky Way. Modern determinations indicate that the Sun is located about 27,000 light-years from the center of the Galaxy, or roughly halfway out toward the edge of the disk. Armed with strong evidence, Shapley confidently presented the then-new, greatly expanded Galaxy in 1915. It took all of human history up to 1610 to go from an Earth-centered Universe to a Sun-centered Solar System. It took 305 years to go from a small Galaxy to a large Galaxy. Yet it was barely a decade later that Edwin Hubble, using Cepheid stars and the period-luminosity relationship, proved that Shapley's enormous Milky Way is but one of billions of galaxies in the Universe.

18.1 MWG

ROTATION OF THE GALAXY IS MEASURED USING 21-cm RADIATION

Clearly there is much we would like to know about the Milky Way, apart from its size and the fact that it is a spiral galaxy. However, as noted many times, there is a problem: we live inside the dusty disk of the Galaxy, which badly obscures the visible light view of our own Galaxy. If you go out at night and look in the direction of the center of our galaxy (located in the constellation Sagittarius), instead of a bright spot you will instead see a dark lane of dusty clouds. To probe the structure of our Galaxy, we must therefore use long-wavelength infrared and radio radiation that can penetrate the disk without being affected much by dust. The most powerful tool for this work is the same 21-cm line from neutral interstellar hydrogen that we used in the previous chapter to measure the rotation of other galaxies.

Dust greatly obscures our view of the center of our Galaxy.

Figure 18.7 is a map (another name for a plot or graph) of the velocities of interstellar hydrogen measured from 21-cm radiation, shown as a function of the direction in which we are looking. Looking in the region around the center of the Galaxy we see that, on one side, hydrogen clouds are moving toward us, while on the other side, clouds are moving away from us. This is a pattern we have seen before—it is the pattern of the rotation velocity of gas in a disk. The only difference is that instead of looking at it from outside, we see our own Galaxy's rotation curve from a vantage point located within—and rotating with—the Galaxy. In other directions the velocities that we see are complicated by our moving vantage point within the disk, and so are more difficult to interpret at a glance. Even so, observed velocities of neutral hydrogen allow us to measure our Galaxy's rotation curve, and even determine the structure present through-out the disk of our Galaxy.

21-cm Doppler velocities show that we live inside a rotating galaxy.

18.2 MEASURE

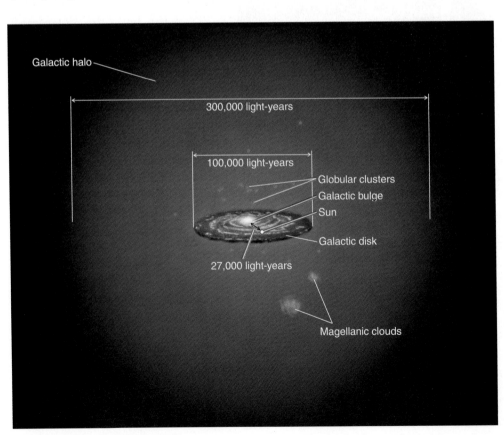

Figure 18.6 *A diagram of the disk, bulge, and halo of the Milky Way Galaxy showing the location of the Sun.*

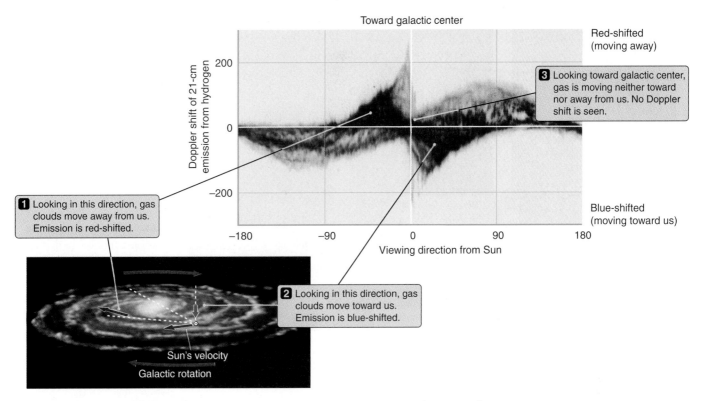

Figure 18.7 *The Doppler velocities measured from observations of 21-cm emission from interstellar clouds of neutral hydrogen change as we look around in the plane of our Galaxy. Notice that we see the clear signature of a rotating disk when looking on either side of the center of the Galaxy.*

Figure 18.8 shows a reconstruction of the Milky Way based on 21-cm data and other observations. Our Galaxy has a modest bulge, which studies of the distribution and motions of stars toward the center of our Galaxy suggest is barlike. Spiral arms sweep through the Galaxy's disk, just like the arms we saw in external spiral galaxies. If we put all of the available information together, we conclude that the Milky Way is a middle-of-the-road giant barred spiral. From outside, our Galaxy probably looks much like the galaxy shown in **Figure 18.9,** and would be placed about halfway along the right tine of the Hubble tuning fork diagram (see Figure 17.4).

Figure 18.8 *The Milky Way Galaxy as reconstructed from 21-cm radiation from neutral hydrogen and other observations.*

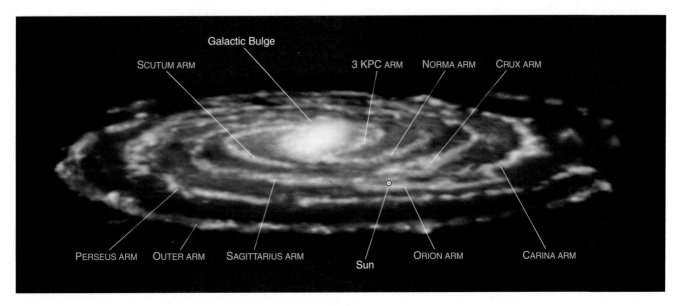

Irr

Sd   SBd

Sc   SBc

1 The Milky Way would lie here on the Hubble tuning fork...

Sb   SBb

Sa   SBa

S0   SB0

2 ...and look much like this SBbc galaxy.

E5

E3

E0

Figure 18.9 *From the outside, the Milky Way would look much like this barred spiral galaxy, NGC6744.*

THE MILKY WAY IS MOSTLY DARK MATTER

Figure 18.10 shows the rotation curve of the Milky Way as inferred primarily from 21-cm observations. The outermost point in the rotation curve, at a distance of roughly 160,000 light-years from the center of the Galaxy, is obtained from the orbital motion of the nearby dwarf galaxy called the Large Magellanic

> **Our Galaxy has a flat rotation curve.**

Cloud. As we have come to expect for spiral galaxies, the Milky Way has a fairly flat rotation curve.

EXCURSIONS 18.2

SEARCHING FOR DARK MATTER IN THE HALO

There is a very clever way to search for dark matter within the halo of our own Galaxy, if the dark matter consists of compact objects such as low-mass stars, planets, white dwarfs, neutron stars, or black holes. We refer to such dark matter candidates as **MACHOs,** which stands for "massive compact halo objects."[1] If the dark matter in our halo consists of MACHOs, there would have to be a lot of these objects, and they would each exert gravitational force but not emit much light.

18.3 MACHOs

How might we detect MACHOs? Because of their gravity, MACHOs can gravitationally deflect light according to Einstein's general theory of

relativity. If we are observing a distant star and if a MACHO were to pass between us and the star, the star's light would be deflected and perhaps amplified by the intervening MACHO as it passed across our line of sight, as illustrated in **Figure 18.11(a).**

We would be remarkably lucky if such an event occurred just as we were observing a single distant star. However, astronomers have monitored the stars in the Large and Small Magellanic Clouds (two of the small satellite galaxies of the Milky Way Galaxy), observing tens of millions of stars for several years. They found a number of examples of events of the sort shown in **Figure 18.11(b),** but not nearly enough to account for the amount of dark matter in the halo of our Galaxy. Thus it was concluded that the dark matter in our Galaxy is probably *not* primarily composed of MACHOs.

[1]MACHOs are not to be confused with another dark matter candidate called "weakly interacting massive particles," or WIMPs. Who says astronomers have no sense of humor?

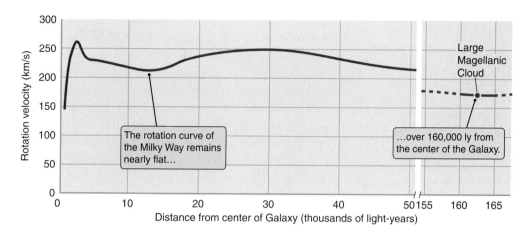

Figure 18.10 *A plot showing rotation velocity versus distance from the center for the Milky Way. The most distant point comes from measurements of the orbit of the Large Magellanic Cloud.*

The most important result from our analysis of galactic rotation curves in the previous chapter was the use of such curves to measure the distribution of mass within galaxies. By applying these techniques to the rotation curve of the Milky Way, we infer that the Galaxy's mass must be about 6×10^{11} $M_\odot$. However, if we instead estimate the mass of the Milky Way by measuring its infrared luminosity (using infrared to see through the dust), we find a much lower value. As are other spiral galaxies, the Milky Way is mostly dark matter. The spatial distribution of dark and normal matter within the Milky Way is also much like what we have seen in other galaxies. The inner part of our Galaxy is dominated by visible matter. Its outer parts, at least to a distance of 150,000 light-years from the center of our Galaxy, are dominated by dark matter.

Figure 18.11 (a) *The light from a distant star is affected by a compact object crossing our line of sight. Because gravity affects all wavelengths equally, such "lensing events" should look the same in all colors.* (b) *The observed light curves of a real star experiencing a lensing event.*

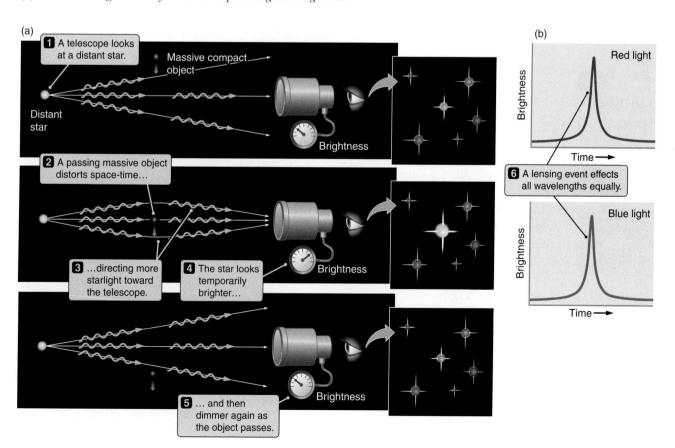

18.3 STUDYING THE MILKY WAY GALAXY UP CLOSE AND PERSONAL

Earlier in the chapter, we concentrated on the disadvantages of our perspective that make it hard to see the Milky Way as a galaxy. However, our perspective also has advantages. For example, we can study the stellar content of the Milky Way at very close range, star by star, looking at subtle aspects of the populations of stars that give us direct clues about how spiral galaxies like ours form. It is much more difficult to glean such information by observing other spiral galaxies.

As with other galaxies, the shapes of the different parts of our Galaxy are determined by the shapes of the orbits of the stars they contain. The stars in the disk rotate about the center of the Galaxy, just as do the gas and dust in the disk. The stars in the halo move in orbits similar to those of stars in elliptical galaxies. The bar shape of the bulge of our Galaxy is defined primarily by stars and gas moving in highly elongated orbits up and down the long axis of the bar.

The Sun is a middle-aged disk star, located mostly among stars that are middle-aged as well. Yet, also near the Sun are stars that are part of the halo, passing through the disk in their orbits. As a result, we can study stars both in the disk and in the halo just by studying the ages, chemical abundances, and motions of stars near our Sun.

STARS OF DIFFERENT AGE AND CHEMICAL COMPOSITION

The most fundamental categories into which populations of stars can be grouped are based on their ages and the abundances of massive elements found in their atmospheres. Conveniently, some stars come prepackaged into groups that split up along just these two lines. There are two different varieties of star clusters in the Milky Way. Globular clusters—the densely packed collections of millions of stars studied by Shapley—are mostly halo objects. With ages of up to 13 billion years, they are the oldest objects known. In contrast, **open clusters,** like the one in **Figure 18.12(a)**, are much less tightly bound collections of a few tens to a few thousands of stars that are found orbiting in the disk of our Galaxy. As with globular clusters, the stars in

Globular and open clusters differ in age, location, and chemical abundance.

an open cluster all formed in the same region at about the same time. When we study the H-R diagrams of open clusters **(Figure 18.12b)**, we find a wide range of ages. Some open clusters contain the very youngest stars known. Other open clusters contain stars that are somewhat older than the Sun. There is no overlap in age between open clusters and globular clusters, however. Even the youngest globular clusters are several billion years older than the oldest open clusters.

The differences in ages between globular and open clusters immediately tell us something interesting about

Figure 18.12 (a) *The open star cluster NGC6530, located 5,200 light-years away, in the disk of our Galaxy.* (b) *The H-R diagram of stars in this cluster shows that it has an age of a few million years or less.*

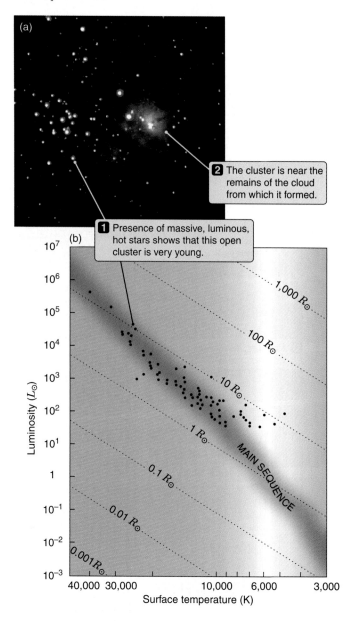

2 The cluster is near the remains of the cloud from which it formed.

1 Presence of massive, luminous, hot stars shows that this open cluster is very young.

the history of star formation in our Galaxy. Stars in the halo formed first, but this epoch of star formation did not last long. No young globular clusters are seen. In the disk of the Galaxy, star formation seems not to have gotten started until later, but it has been continuing ever since. The process that formed the stars in the massive, compact globular clusters must have also been much different from the more sedate process responsible for creating stars in the less massive, more scattered open clusters.

It is obvious why we would be so interested in grouping stars according to their ages, but why would we place emphasis on the chemical composition of stars? We have discussed the chemical evolution of the

Massive element abundances record the cumulative history of star formation.

Universe on a number of occasions. When the Universe was very young, only the least massive of elements existed. All elements more massive than boron must have formed by nucleosynthesis in stars. For this reason, the abundance of massive elements in the interstellar medium provides a record of the cumulative amount of star formation that has taken place up until the present time. Gas that shows large abundances of massive elements must have gone through a great deal of stellar processing, while gas with low abundances of massive elements is more pristine.

In turn, the abundance of massive elements in the atmospheres of a star provides a snapshot of the chemical composition of the interstellar medium *at the time that star formed!* (In main sequence stars, material from the core does not mix with material in the atmosphere, so the abundances of chemical elements inferred from the spectra of a star are the same as the abundances in the interstellar gas from which the star formed.) As illustrated in **Figure 18.13,** the chemical composition of a star's atmosphere is a record of the cumulative amount of star formation up until the time that star formed.

If our ideas about the chemical evolution of the Universe are correct, we would expect to see large differences in massive-element abundances between glob-

Younger stars typically have higher massive-element abundances than older stars.

ular and open clusters. Stars in globular clusters, being among the earliest stars to form, should contain only very small amounts of massive elements. That is exactly what we see. Some globular cluster stars contain only 0.5% as much of these massive elements as does our Sun. This relationship between age and abundances of massive elements is seen throughout much of the Galaxy. The chemical evolu-

tion of the Milky Way has continued within the disk, as generation after generation of disk stars has further enriched the interstellar medium with the products of their nucleosynthesis. Within the disk, younger stars typically have higher abundances of massive elements than do older stars. Similarly, older stars in the outer parts of our Galaxy's bulge have lower massive element abundances than young stars in the disk.

We can even see differences in abundances of massive elements from place to place within the Galaxy that are related to the rate of star formation in different regions. Star formation is generally more active in the inner part of the Milky

Chemical abundances vary from place to place within the Galaxy.

Way than in the outer parts. This is a result of the denser concentrations of gas that are found in the inner Galaxy. If this has continued throughout the history of our Galaxy, we might predict massive elements to be more abundant in the inner parts of our Galaxy than in the outer parts. Observations of chemical abundances in the interstellar medium, based both on interstellar absorption lines in the spectra of stars and on emission lines in glowing H II regions, confirm this prediction. As expected, there is a smooth decline in abundances of massive elements from the inner to the outer parts of the disk. Similar trends are often seen in other galaxies. These trends can be seen in stars as well. Relatively old stars near the very center of a galaxy often have massive-element abundances that are greater than those of young stars in the outer parts of the disk.

Our basic idea about higher massive-element abundances following the more prodigious star formation in the inner Galaxy seems correct, but as always, the full picture is not this simple. The chemical composition of the interstellar medium at any location depends on a wealth of factors. New material falling into the Galaxy might affect interstellar chemical abundances. Chemical elements produced in the inner disk might be blasted into the halo in great fountains powered by the energy of massive stars, only to fall back onto the disk elsewhere. Past interactions with other galaxies might have stirred the Milky Way's interstellar medium, and mixed gas from those other galaxies in with our own. At this moment, a small companion of the Milky Way, called the Sagittarius Dwarf, is plowing through the disk of the Galaxy on the other side of the bulge. How chemical abundances vary from place to place within the Milky Way and other galaxies and what these variations tell us about the history of star formation and nucleosynthesis are active topics of research at the beginning of the 21st century.

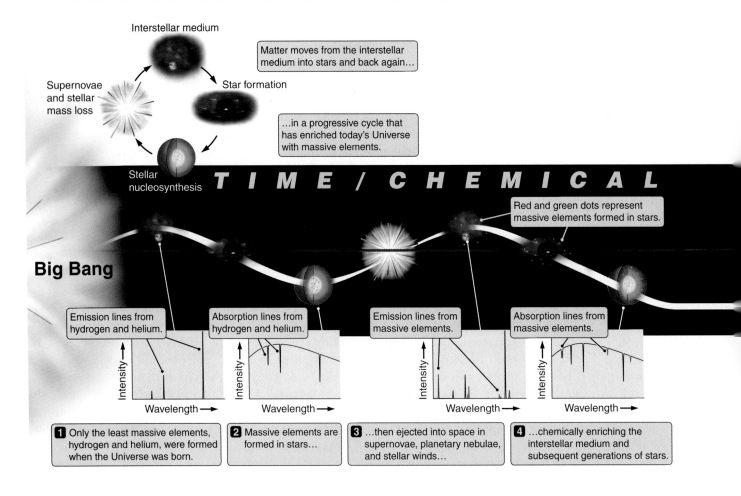

Interstellar medium

Matter moves from the interstellar medium into stars and back again...

Star formation

Supernovae and stellar mass loss

...in a progressive cycle that has enriched today's Universe with massive elements.

Stellar nucleosynthesis

T I M E / C H E M I C A L

Red and green dots represent massive elements formed in stars.

Big Bang

Emission lines from hydrogen and helium.

Absorption lines from hydrogen and helium.

Emission lines from massive elements.

Absorption lines from massive elements.

Intensity / Wavelength →

1 Only the least massive elements, hydrogen and helium, were formed when the Universe was born.

2 Massive elements are formed in stars...

3 ...then ejected into space in supernovae, planetary nebulae, and stellar winds...

4 ...chemically enriching the interstellar medium and subsequent generations of stars.

While the details are complex, there are several clear and important lessons to be learned from patterns in massive-element abundances in the Galaxy. The first is that even the very oldest globular cluster stars contain *some* amount of massive chemical elements. The implication is clear: Globular cluster stars were not the first stars in our Galaxy to form. *There must have been at least one generation of massive stars that lived and died, ejecting newly synthesized massive elements into space, before even the oldest globular clusters formed.* Further, every star less massive than about $0.8\ M_\odot$ that ever formed is still around today. Even so, we find *no* disk stars with exceptionally low massive-element abundances. We would have found these stars by now if they existed. The gas that wound up in the plane of the Milky Way must have seen a significant amount of star formation *before* it settled into the disk of the Galaxy.

Generations of stars must have formed even before globular clusters did.

We have placed great emphasis on variations in chemical abundances from place to place. These tell us a lot about the history of our Galaxy and a lot about the origin of the material that we are made from. It is important to remember, however, that even a chemically "rich" star like the Sun, which is made of gas processed through approximately 9 billion years of previous generations of stars, is still composed of less than 2% massive elements. Luminous matter in the Universe is still dominated by hydrogen and helium formed long before the first stars.

A CROSS SECTION THROUGH THE DISK

The youngest stars in our Galaxy are most strongly concentrated in the plane of our Galaxy, defining a disk about 400 light-years thick (but over 100,000 light-years across—thin indeed). The older population of disk stars, distinguishable by lower abundances of massive elements, has a much "thicker" distribution (about 4,000 light-years thick). **Figure 18.14** illustrates how the population of stars changes with distance from the galactic plane. Not too surprisingly, the youngest stars are concentrated closest to the plane of the Galaxy, for the simple reason that this is where the molecular clouds are. Older stars make up the thicker parts of the

There are thin and thick parts of our disk.

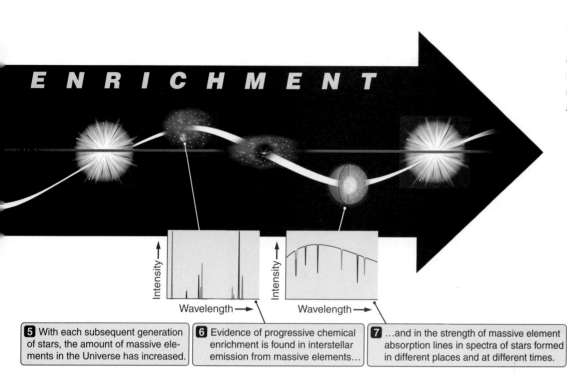

ENRICHMENT

5 With each subsequent generation of stars, the amount of massive elements in the Universe has increased.

6 Evidence of progressive chemical enrichment is found in interstellar emission from massive elements…

7 …and in the strength of massive element absorption lines in spectra of stars formed in different places and at different times.

Figure 18.13 *As subsequent generations of stars formed, lived, and died, they enriched the interstellar medium with massive elements—the products of stellar nucleosynthesis. The chemical evolution of our Galaxy and other galaxies can be traced in many ways, including by the strength of interstellar emission lines and stellar absorption lines.*

disk. These stars formed in the midplane of the disk long ago, but have since been kicked up out of the plane of the Galaxy, primarily by gravitational interactions with massive molecular clouds.

When we discussed the formation of accretion disks in Chapters 5 and 14, we found that a rotating cloud of gas could do nothing other than collapse into a thin disk. This was a consequence of the fact that gas falling from one direction ran into gas falling from the other direction. Clouds of gas cannot pass through each other, so the gas had no choice but to settle into a disk. The same thing applies to clouds of gas that are pulled by gravity toward the midplane of the disk of a spiral galaxy. While stars are free to pass back and forth from one side of the disk to the other, cold, dense clouds of interstellar gas settle down into the central plane of the disk.

Gas tends to concentrate into thin sheets.

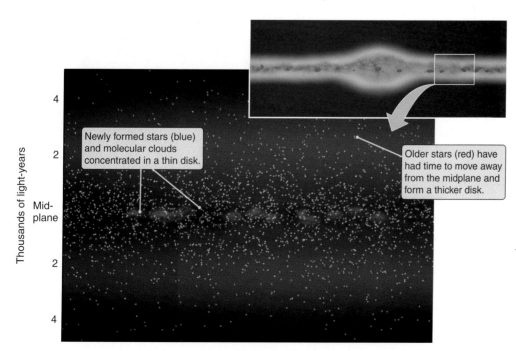

Newly formed stars (blue) and molecular clouds concentrated in a thin disk.

Older stars (red) have had time to move away from the midplane and form a thicker disk.

Figure 18.14 *A vertical profile of the disk of the Milky Way Galaxy. Gas and young stars are concentrated in a thin layer in the center of the disk, but older populations of stars define ever thickening portions of the disk.*

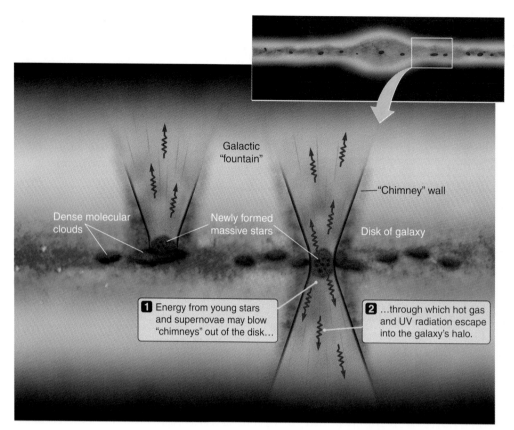

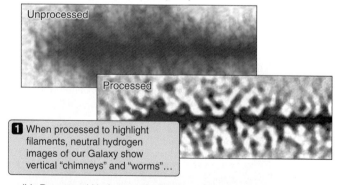

Figure 18.15 *In the "galactic fountain" model of the disk of a spiral galaxy, gas is pushed away from the plane of the galaxy by energy released by young stars and supernovae, then falls back onto the disk.*

These clouds are seen as the concentrated dust lanes that slice the disks of spiral galaxies (see Figure 18.1b) like a layer of bologna in a very thin, flat sandwich. This thin lane slicing through the mid-plane of the disk is both where new stars form and where new stars are found.

We mentioned earlier that energy from regions of star formation can impose interesting structure on the interstellar medium, clearing out large regions of gas in the disk of a galaxy. Many massive stars forming in the same region can blow "chimneys" out through the disk of the galaxy via a combination of supernova explosions and strong stellar winds. If enough massive stars are formed together, sufficient energy may be deposited to blast holes all the way through the plane of the galaxy. In the process, dense interstellar gas can be thrown high above the plane of the galaxy **(Figure 18.15)**. Maps of the 21-cm emission from neutral hydrogen in our Galaxy and visible light images of hydrogen emission from some edge-on external galaxies **(Figure 18.16)** show a wealth of vertical structure in the interstellar medium of disk galaxies. These vertical structures are often interpreted as the "walls" of these chimneys.

Figure 18.16 *Observations of emission (a) from neutral hydrogen in our own Galaxy, and (b) ionized hydrogen in other galaxies such as NGC891, show "chimneys" extending up out of the disks of these galaxies. These chimneys are understood to be the walls of cavities blown by energy released from young stars and supernovae.*

(a) 21-cm image of part of Milky Way disk

1 When processed to highlight filaments, neutral hydrogen images of our Galaxy show vertical "chimneys" and "worms"...

(b) Processed Hα image of edge-on spiral galaxy

2 ...much like those seen in the disks of other galaxies.

THE HALO IS MORE THAN GLOBULAR CLUSTERS

Earlier, when discussing the halo of our Galaxy, we concentrated our attention on its globular clusters. These are very important to astronomy, in part because they tell us a great deal about the history of star formation in the halo. Yet globular clusters account for only about 1% of the total mass of stars in the halo. As halo stars fall through the disk of the Milky Way, some pass close to the Sun, giving us a sample of the halo that we can study at close range.

Some halo stars pass close to the Sun.

Since most of the stars near our Sun are, like the Sun, disk stars, you might be wondering how we can tell them apart from nearby halo stars. They are distinguishable in two ways. First, most halo stars have much lower abundances of massive elements than disk stars. Second, halo stars appear to be whizzing by us at high velocities. Actually, it is often not the halo stars that are moving fast, but we who are moving fast relative to them. This is because halo stars do not rotate about the center of the Galaxy in the same way as disk stars do, but rather are in the same kind of odd-shaped orbits as stars in elliptical galaxies. In contrast, the disk stars near the Sun are moving, mostly together, in 220 km/s rotation about the center of our Galaxy. Just as it is easy to tell the difference between a person sitting beside you on a bus and the people who (in your frame of reference) are whizzing by outside the window, it is easy to tell the difference between disk stars that share our motion and halo stars that, to us, are just passing through.

Reflecting once again the bias of the observer, astronomers call the halo stars **high-velocity stars,** even though it is usually the disk stars that are moving faster relative to the Galaxy as a whole! Enough halo stars are near us that we can measure their distances and map out their orbits, giving us a very detailed look at the kinds of orbits that stars in a halo (or in an elliptical galaxy) have. The picture we get is one of confusion. About half the halo stars are orbiting in the same direction as the disk rotation, while the other half are moving in the *opposite* direction. The motions of the halo stars near us indicate that, in their full orbits, these halo stars fill a volume of space similar to that occupied by the globular clusters in the halo. Halo stars and globular clusters both paint the same picture of the halo of our Galaxy.

MAGNETIC FIELDS AND COSMIC RAYS FILL THE GALAXY

Almost all that we know about the Galaxy we learn from analyzing the electromagnetic radiation it emits—almost, but not quite all. An important component of the Galaxy's disk, whose energy we can measure directly on Earth, is **cosmic rays.** Despite their name, cosmic rays are not a form of electromagnetic radiation. Rather, they are charged particles moving at close to the speed of light. Cosmic rays are continually hitting Earth.

Most cosmic ray particles are protons, but some are nuclei of helium, carbon, and all of the other elements produced by nucleosynthesis. What makes cosmic rays especially fascinating is that they span an enormous range in particle energy. We can observe the lowest-energy cosmic rays using interplanetary spacecraft. These cosmic rays have energies as low as about 10^{-11} joules, which corresponds to the energy of a proton moving at a velocity of a few tenths the speed of light. In contrast, the most-energetic cosmic rays are 10 trillion (10^{13}) times more energetic than the lowest-energy cosmic rays. These high-energy cosmic rays are known from the showers of elementary particles that they cause when crashing through Earth's atmosphere.

It is thought that cosmic rays are accelerated to these incredible energies in the shock waves produced in supernova explosions. The highest-energy cosmic rays are more difficult to explain. These are as much as a billion times more energetic than any particle ever produced in a particle accelerator on Earth. These cosmic rays have energies up to 100 J, corresponding to the energy of a proton moving at a velocity that is 99.999 999 999 999 999 999 999 9% of the speed of light. A better way to visualize this is to realize that if you were to drop your copy of *21st Century Astronomy* (this irreplaceable textbook) from the ceiling, the energy it would have when it hit the floor is about the same as the energy concentrated into a *single* high-energy cosmic ray proton!

Some cosmic rays have incredibly high energies.

The interstellar medium of the Milky Way is laced with remarkably strong magnetic fields that are wound up and strengthened by the rotation of the Galaxy's disk. When we discussed Earth's magnetosphere, we saw that charged particles are unable to move freely across the direction of a magnetic field. The other

Cosmic rays are trapped by the Milky Way's magnetic field.

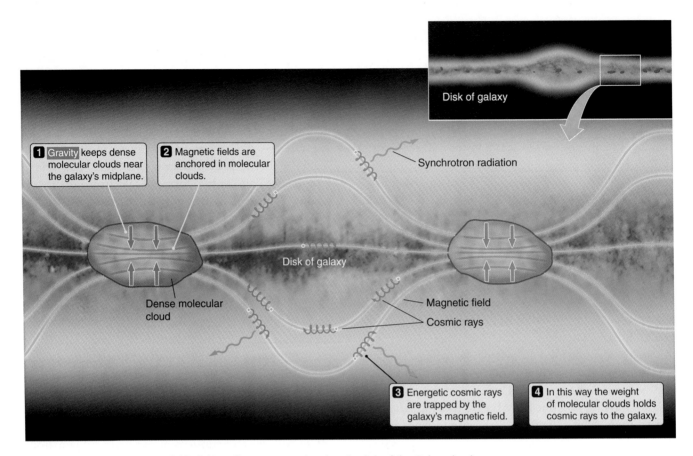

Figure 18.17 *The magnetic field of the Milky Way is anchored to the disk of the Galaxy by the weight of interstellar clouds. The magnetic field in turn traps the Galaxy's cosmic rays, much as a planetary magnetosphere traps charged particles.*

side of this coin is that magnetic fields cannot freely escape from a cloud of gas containing charged particles. Earlier we mentioned that dense clouds of interstellar gas are concentrated by gravity in the midplane of the Milky Way. The weight of these clouds anchors the Galaxy's magnetic field to the disk, as shown in **Figure 18.17.**

The disk of our Galaxy glows from synchrotron radiation produced by cosmic rays (mostly electrons) looping around the direction of the Galaxy's magnetic field. Such synchrotron emission is seen in the disks of other spiral galaxies as well, telling us that they, too, have magnetic fields and populations of energetic cosmic rays. Even so, the very highest energy cosmic rays are moving much too fast to be confined to our Galaxy. Any such cosmic rays formed in the Milky Way soon stream away from the Galaxy into intergalactic space. It is also quite possible that some fraction of the energetic cosmic rays reaching Earth have had their origins in energetic events outside our Galaxy.

The total energy of all of the cosmic rays in the galactic disk can be estimated from the energy of the cosmic rays reaching Earth. The strength of the interstellar magnetic field can be measured in a variety of ways, including the effect that it has on the properties of radio waves passing through the interstellar medium. These measurements indicate that in our Galaxy, the magnetic field energy and the cosmic ray energy are about equal to each other. Both are comparable to the energy present in other energetic components of the Galaxy including the motions of interstellar gas and the total energy of electromagnetic radiation within the Galaxy. Magnetic fields and cosmic rays are *not* bit players in the production we call the Milky Way.

THE MILKY WAY HOSTS A SUPERMASSIVE BLACK HOLE

In the previous chapter we discussed evidence that most galaxies with bulges have supermassive black holes at their centers. The Milky Way Galaxy is no exception. Observations of our Galaxy's rotation curve show rapid rotation velocities very close to its center. This is the same sort of evi-

The center of our Galaxy shows rapid rotation and AGN-like emission.

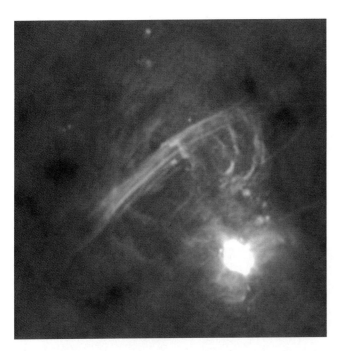

Figure 18.18 *Radio observations of the center of the Milky Way showing strong synchrotron emission. The galactic center is also a source of strong X-rays and gamma rays, evidence of the "beast" at the center of our own Galaxy.*

dence that led us to surmise the presence of supermassive black holes in other galaxies. We estimate the black hole at the center of our own Galaxy is a relative lightweight, having a mass of "only" $2.6 \times 10^6 \, M_\odot$.

Radio observations of the inner part of our Galaxy **(Figure 18.18)** show synchrotron radiation from fascinating wisps and loops of material throughout the region. These are reminiscent of the synchrotron emission seen from AGNs, albeit at far lower levels. Without a source of large amounts of material to feed the beast that resides at the center of our Galaxy, the Milky Way is quiescent at the moment. However, like steam rising from the cauldron of a dormant volcano, the synchrotron emission from the inner Milky Way is a reminder that our Galaxy was likely "active" at some time in the past, and, like an episodic volcano, could become active again in the future.

18.4 THE MILKY WAY OFFERS CLUES ABOUT HOW GALAXIES FORM

One of the fundamental goals of stellar astronomy is to understand the life cycle of stars, including how stars form from clouds of interstellar gas. In Chapter 14 we were able to tell a fairly complete story of this process,

at least as it occurs today, and tie this story strongly to observations of our galactic neighborhood. Galactic astronomy has the same basic goal. That is, astronomers would like very much to have a complete and well-tested theory of how our Galaxy formed. Unfortunately, such a complete theory is not yet in hand. Even so, what we have seen so far offers us many clues.

Important among these clues are the properties of globular clusters and high-velocity stars in the halo of the Galaxy. For reasons discussed earlier, these objects must have been among the first stars formed. The fact that they are not concentrated in the disk or bulge of the Galaxy says that they formed from clouds of gas well before those clouds had settled into the Galaxy's disk. The observations that globular clusters are very old and that the youngest cluster is older than the oldest disk stars agree with this surmise. The presence of small amounts of massive elements in the atmospheres of halo stars also tells us that there must have been at least one generation of stars that lived and died *before* the formation of the halo stars we see today. We have yet to find any stars from that first generation in our Galaxy today.

> **At least one generation of stars had to exist before today's halo stars were formed.**

Based on these and other clues, we have come to understand that our Galaxy must have formed when the gas within a large "clump" of dark matter collapsed into a large number of small protogalaxies. Some of these smaller clumps are still around today in the form of small dwarf galaxies near our own, including the Sagittarius Dwarf and the Large and Small Magellanic Clouds, shown in **Figure 18.19.** The remainder of these protogalaxies merged to form the barred spiral galaxy we call the Milky Way. In this process, stars were formed in the halo, in the bulge, and in the disk. The first stars to form ended up in the halo—many in globular clusters, but many not. Gas that settled into the disk of the Milky Way quickly formed several generations of stars. In this process, a seed somehow formed, which quickly grew into the supermassive black hole at the center of the Galaxy. The details of this process are sketchy, but some calculations indicate that so much mass was concentrated in this small region that almost any sequence of events would have led to the formation of a massive black hole.

> **Our Galaxy formed from the mergers of many smaller protogalaxies.**

We need to be careful not to get ahead of ourselves. The Milky Way offers many clues about the way galaxies form, but much of what we know of the process comes from looking beyond our local system.

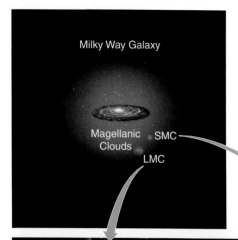

Figure 18.19 *The Large and Small Magellanic Clouds were named for Ferdinand Magellan, who headed the first European expedition to venture far enough into the Southern Hemisphere to see these two dwarf companions of the Milky Way.*

Images of distant galaxies (which we see as they existed billions of years ago), as well as observations of the glow left behind by the formation of the Universe itself, provide equally important pieces of the puzzle. We will have to leave the story of galaxy formation as we currently understand it unfinished, for now, as we turn our attention to the immensely larger structure of the Universe as a whole.

SEEING THE FOREST THROUGH THE TREES

Many books of puzzles contain optical illusions in which familiar patterns lie hidden in jumbles of lines and shapes. When looking at such a picture, at first we see nothing but confusion. Then in a flash of insight we suddenly see the pattern! Once the pattern is recognized, our perception shifts: a moment ago all we saw was a mess, but now when we look at the picture, the "totally obvious" pattern jumps off the page at us. Such is the case with observing our Galaxy. Go outside on a dark, moonless summer night, away from the city lights, and stare at the sky. In addition to the dazzling jewels of individual stars that fill your view, you will also see a faint, patchy river of light running from horizon to horizon. This river—this "Milky Way"—has been known since earliest prehistory. Even so, its nature was unknown until a remarkably short time ago. As recently as the early 20th century, it was unclear how this river of light fit into models of our stellar system based on counts of individual stars. Yet once the answer is known, boom! Stretched across our sky lies an unmistakable spiral galaxy, as viewed edge-on from within the disk. There it is for all to see—how could we ever have been so blind as to miss the obvious?

Our conclusion from this flash of insight is confirmed by the results of countless observational studies. Whether we look at the distribution of globular clusters, 21-cm observations of the motions of hydrogen gas, or any number of other tracers of structure, the answer is always the same. The Milky Way is a giant "plain vanilla," dark-matter-dominated, barred spiral galaxy, with a supermassive black hole at its center, just like countless others. The Sun and Earth are located in a relatively open and clear region of interstellar space located part way out in the Galaxy's disk.

Our vantage point within the Milky Way is hardly the best from which to answer questions about global properties of our Galaxy. If ever there was a problem in astronomy of seeing the forest for the trees, this is it. For example, we can determine the morphological classification of the Andromeda Galaxy by glancing at a single picture. In contrast, obtaining a reliable and complete classification of our own Milky Way is a task that has taken decades of detailed study. The exact answer is still not known for certain.

On the other hand, the Milky Way gives us the ultimate insider's view. For example, stars in the Milky Way are not just part of a large blur, but are instead individuals that can be studied as such. Here is where we learn about the stars in globular and open clusters. These collections of stars not only tell us about the Milky Way as a galaxy, but also provide the data we use to test and refine our ideas about the way stars everywhere evolve. In the Milky Way we see stars in the bulge, stars in the thin and thicker portions of the disk, and stars in the halo that are falling through the disk near us. We measure their properties—luminosity, size, temperature, and abundance of the chemical elements of which they are made—and we learn what each variety of stars in a spiral galaxy is like. Spiral arms in the Milky Way are not just bright swirls against the background of a spiral disk. Instead, they are collections of young stars, old stars, and clouds of gas and dust that can be studied at close range. We know that stars form in the arms of other spiral galaxies because we see the brightest of those stars, along with the generally blue color of the arms. We know that stars form in the spiral arms of the Milky Way because that is where we see dense molecular cloud

cores containing individual young stars surrounded by their accretion disks.

So yes, it is difficult to study the Milky Way as a galaxy, but it is worth the investment. We are guided in our study of the grand patterns within the Milky Way by what we know of other galaxies; conversely, in the Milky Way we can learn the details of processes that are at work shaping all of those other islands of activity spread out across the expanse of the Universe. These details are beginning to help us clarify our understanding of some of the "big" questions about the way galaxies form and evolve.

Our location in the Milky Way may not give us the best perspective from which to study our Galaxy, but it is far from the worst. The Fates have been kind. Since the Sun and Earth formed, they have made numerous trips around the Galaxy, passing back and forth through the disk and encountering a wide range of environments. What if our species had become aware at a moment when Earth was in the depths of a molecular cloud, and the sky held nothing of interest other than the Sun and planets? That would bring a whole new and very literal meaning to the notion of the "dark ages."

In some sense, this portion of our journey has been about context. We have learned to see stars within the context of the history of the Milky Way. We have seen star-forming molecular clouds within the context of the sweep of spiral arms. We have learned to see the Milky Way within the context of its similarities with other galaxies. On the next leg of our journey, we will pose the ultimate question of context as we learn to see galaxies, including our Milky Way, within the context of the Universe itself.

STUDENT QUESTIONS

THINKING ABOUT THE CONCEPTS

1. Explain the observational evidence that shows we live in a spiral galaxy, not an elliptical galaxy.
2. What can you see for yourself in the night sky (unaided by a telescope) that suggests that we live in a spiral galaxy? Explain your answer.
3. What do astronomers mean by a "standard candle?"
4. Why are the youngest stars concentrated close to the plane of our Galaxy?
5. Stars in many globular clusters tend to have a lower abundance of massive elements than stars in the disk of our Galaxy. Explain why this is so.
6. How do we know that the stars in globular clusters are the oldest stars in our Galaxy?

7. To observers in Earth's Southern Hemisphere, the Large and Small Magellanic Clouds look like detached pieces of the Milky Way. What are these "clouds" and why is it not surprising that they look so much like pieces of the Milky Way?
8. Why do we need to use 21-cm radio observations to probe the structure of our Galaxy?
9. Discuss reasons for the differences between globular clusters and open clusters, with respect to their relative ages, abundance of massive elements, numbers of stars, and distribution throughout the Galaxy.
10. Describe how our environment would be different and how our skies might appear if our Sun and Solar System were located:

a. Near the center of the Galaxy.

b. Near the center of a large globular cluster.

c. Near the center of a large, dense molecular cloud.

APPLYING THE CONCEPTS

11. The Sun completes one trip around the center of the Galaxy in approximately 230 million years. How many times has our Solar System made the circuit since its formation 4.6 billion years ago?

12. The Sun is located about 27,000 light-years from the center of the Galaxy, but the disk probably extends another 30,000 light-years farther out from the center.

   a. Assuming a truly flat rotation curve, how long would it take a globular cluster located near the edge of the disk to complete one trip around the center of the Galaxy?

   b. How many times has that globular cluster made the circuit since its formation 13 billion years ago?

13. Parallax measurements of the variable star RR Lyrae indicate that it is located 750 light-years from the Sun. A similar star observed in a globular cluster located far above the galactic plane has the same period of variability, but appears 160,000 times fainter than RR Lyrae.

   a. How far from the Sun is this globular cluster?

   b. What does your answer to (a) tell you about the size of the Galaxy's halo compared to the size of its disk?

14. The flat rotation curve indicates that the total mass of our Galaxy is 6×10^{11} $M_\odot$. However, electromagnetic radiation associated with normal matter suggests a total mass of only 3×10^{10} $M_\odot$. Given this information, calculate the fraction of our Galaxy's mass that is made up of dark matter.

15. A cosmic ray proton is traveling at nearly the speed of light (3×10^8 m/s).

   a. Using Einstein's familiar relationship between mass and energy ($E = mc^2$), show how much energy (in joules) the cosmic ray proton would have if m is based only on the proton's rest mass (1.7×10^{-27} kg).

   b. The actual measured energy of the cosmic ray proton, however, is 100 J. What then is the relativistic mass of the cosmic ray proton?

   c. How much greater is the relativistic mass of this cosmic ray proton than the mass of a proton at rest?

A man said to the universe:
"Sir, I exist!"
"However," replied the universe,
"The fact has not created in me
A sense of obligation."

STEPHEN CRANE (1871–1900)

OUR EXPANDING UNIVERSE

19.1 THE COSMOLOGICAL PRINCIPLE SHAPES OUR VIEW OF THE UNIVERSE

As we look back on our journey of the mind—and on the history of our species—it is remarkable how far we have come. Beginning with Copernicus's early realization that Earth is not the center of all things, we have shared the insights of Galileo and Newton and their intellectual heirs as they tore down the conceptual barriers separating terrestrial existence from that of the heavens. Casting aside the aura of mysticism and magic, we can now go out and look at the night sky and begin to see it for what it is. Earth sits within an enormous spiral galaxy consisting of hundreds of billions of stars. This galaxy in turn is but one of hundreds of billions of galaxies that fill a Universe vastly larger than our ancestors might have imagined. Through it all, the cosmological principle has been at the center of our conceptual understanding. No progress has been possible without the enabling realization that the rules that apply to one part of our Universe apply everywhere.

The same rules apply everywhere.

The time has come to put the cosmological principle to work in its namesake field—the field of **cosmology.**

KEY CONCEPTS

Just as stars make up the structure of our own Galaxy, so too do galaxies make up the structure of the Universe itself. Observations of the motions of galaxies led to one of the most remarkable discoveries in the history of our species: that the Universe is expanding! As we investigate this discovery we will learn:

* About the discovery of Hubble's Law relating the redshift of a galaxy to its distance;
* That galaxies are not flying apart through space, but rather "ride along" as space itself expands in accordance with the general theory of relativity;
* How Hubble's Law is used to map the Universe in space, and look back in time;
* That space-time itself was born roughly 13 to 14 billion years ago in an event called the Big Bang;
* That each of the major predictions of Big Bang theory, from the glow of the cosmic background radiation to the types of atoms that fill the Universe, is confirmed by observation;
* How the history, shape, and fate of the Universe are determined by the amount of mass it contains, and by a poorly understood cosmological constant; and
* What observations of the cosmic background tell us about the properties of the young Universe.

The science of cosmology is the study of the Universe itself, including its structure, history, origins, and fate. As we stressed before, the cosmological principle is not an article of faith. Rather, it is a testable scientific theory. An important prediction of the cosmological principle is that the conclusions we reach about our Universe should be more or less the same, regardless of whether we live in the Milky Way or in a galaxy billions of light-years away at the limits of the observable Universe. In other words, we predict that, if the cosmological principle is correct, then our Universe will be **homogeneous.**

According to one popular dictionary, the word *homogeneous* means "having at all points the same composition and properties." Clearly the Universe is not truly homogeneous. The conditions we encounter at the surface of Earth are very different from those we would encounter in deep space or the heart of the Sun. (Even homogenized milk varies at the molecular level from place to place.) When we speak as cosmologists of homogeneity of the Universe, we mean instead that stars and galaxies in our part of the Universe are much the same, and behave in the same manner, as stars and galaxies in remote corners of the Universe. We also mean that stars and galaxies everywhere are distributed in space in much the same way as they are in our cosmic neighborhood, and that observers in those galaxies see the same properties for our Universe that we do. In other words, when we speak of a homogeneous Universe, we are using the term *homogeneity* in a very "broad brush" sense.

It is not easy to verify the prediction of homogeneity directly. We do not have the luxury of traveling from our Galaxy to a galaxy in the remote Universe to see whether conditions are the same. However, we can observe light arriving from the distant Universe and see in what ways features look the same or different. For example, we can look at the way galaxies are distributed in space **(Figure 19.1)** and ask whether that distribution is homogeneous.

Figure 19.1 *Homogeneity and isotropy in four universes.* (a) *The distribution of galaxies is uniform, so this universe is both homogeneous and isotropic.* (b) *The density of galaxies is decreasing in one direction, so this universe is neither homogeneous nor isotropic.* (c) *The bands of galaxies define a unique direction, making this universe anisotropic.* (d) *The distribution of galaxies is uniform, but galaxies move along only one direction, so this universe, too, is anisotropic.*

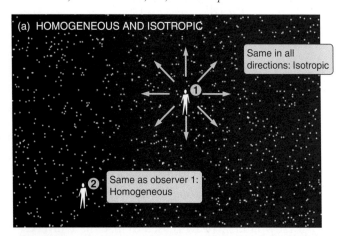

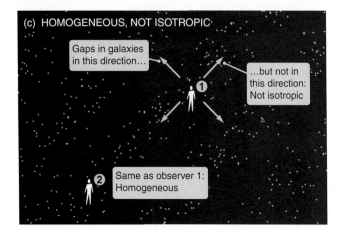

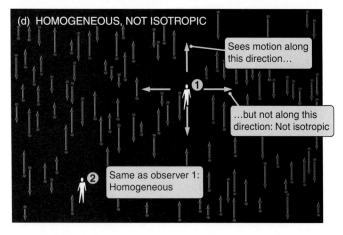

In addition to predicting that the Universe is homogeneous, the cosmological principle requires that all observers (including us) have the same impression of the Universe, regardless of the *direction* in which they are looking. This prediction of the cosmological principle is much easier to test directly than homogeneity. For example, if galaxies were lined up in rows—a violation of the cosmological principle (Figure 19.1c)—we would get very different impressions depending on the direction in which we looked. If something is the same in all directions, then it is **isotropic.** In most instances isotropy goes hand in hand with homogeneity, and the cosmological principle requires them both.

Observers everywhere should see the same Universe.

The isotropy and homogeneity of the distribution of galaxies in the Universe are predictions of the cosmological principle that we can go to the telescope and test directly. As you read this, you can benefit from the experience of decades of astronomers who have used the world's most powerful telescopes to test these predictions. These observations show that the properties of the Universe are basically the same, regardless of the direction in which we look. If we look on very large scales, the Universe appears homogeneous as well. Had these observations turned out otherwise, they would have proven the cosmological principle false, but the cosmological principle again withstood the test. As has been true over and over again on our journey, more of the predictions of the cosmological principle are found to be correct.

The distribution of galaxies is isotropic and homogeneous, as predicted.

19.2 WE LIVE IN AN EXPANDING UNIVERSE

One of the great pioneers of cosmology was Edwin Hubble, who is shown on the cover of a 1948 edition of *Time* magazine in **Figure 19.2.** In the 1920s, Hubble and his coworkers were studying the properties of a large collection of galaxies. In addition to taking images, a colleague of Hubble's named **Vesto Slipher** (1875–1969) was using the facilities at Lowell Observatory to obtain spectra of those galaxies. Part of what Slipher's spectra showed was no surprise at all. Galaxy spectra look like the spectra of ensembles of stars but with a bit of glowing interstellar gas mixed in for good measure. The surprise, however, was that the

Emission from distant galaxies is red-shifted to longer wavelengths.

emission and absorption lines in the spectra of galaxies that Slipher observed were seldom seen at the same wavelengths at which these features appeared in laboratory-generated spectra. The lines were almost always shifted to longer wavelengths, as shown in **Figure 19.3.**

Slipher characterized the observed shifts in galaxy spectra as **redshifts** since almost all galaxies have spectral lines shifted to longer (in other words, redder) wavelengths. The wavelength at which a line is observed in an object that is stationary relative to the observer is called the rest wavelength of the line, written λ_{rest}. The redshift of a galaxy, written z, is defined as the difference between the observed wavelength, $\lambda_{observed}$, and the rest wavelength, divided by the rest wavelength:

$$z = \frac{\lambda_{observed} - \lambda_{rest}}{\lambda_{rest}}$$

Note that the redshift of a galaxy is the same, regardless of the wavelength of the line used to measure it. Also note that the right side of this expression is the same as in the equation for the Doppler shift that we studied in Chapter 4, with z equal to the speed of an object moving away from us (v_r) divided by the speed of light. Fol-

Figure 19.2 *Edwin Hubble, as he appeared on the cover of the February 9, 1948, issue of* Time *magazine.*

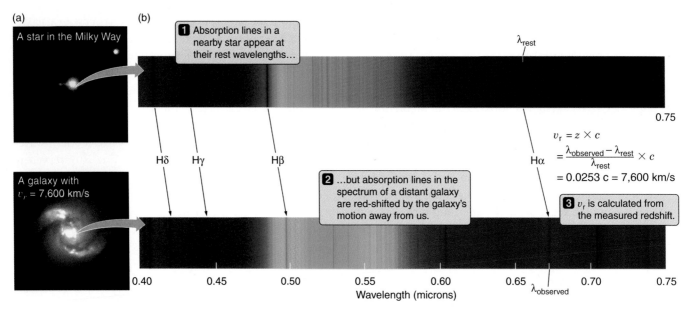

Figure 19.3 (a) *A distant galaxy seen together with a star in our Galaxy.* (b) *The spectrum of each, shown on the same scale. Note that lines in the galaxy spectrum are red-shifted to longer wavelengths.*

lowing this reasoning, Hubble interpreted Slipher's redshifts as Doppler shifts. By setting $z = v_r/c$, he concluded that almost all of the galaxies in the Universe are moving away from the Milky Way.

When Hubble combined these measurements of galaxy recession velocities with his own estimates of the distances to these galaxies, he made one of the greatest discoveries in the history of astronomy. Hubble found that distant galaxies are moving away from us more rapidly than nearby galaxies. Specifically, *the velocity at which a galaxy is moving away from us is proportional to the distance of that galaxy.* This simple relationship between distance and recession velocity has become known as **Hubble's law.**

> **Hubble's law: a galaxy's recession velocity is proportional to its distance.**

When talking about the distances of remote galaxies in the Universe, we need a suitable yardstick. The yardstick we choose is the **mega-light-year (Mly),** equal to a million light-years. According to Hubble's law, a galaxy located 100 Mly away is, on average, moving away from us at twice the speed of a galaxy at a distance of 50 Mly.

We usually write Hubble's law as

$$v_r = H_0 \times d_G,$$

where d_G is the distance to the galaxy and v_r is the recession velocity of the galaxy. H_0 is a constant of proportionality and is called the **Hubble constant.** When thinking of Hubble's

> **Hubble's constant relates a galaxy's recession velocity to its distance.**

law, just remember that if you double the distance d_G you also double the recession velocity v_r.

ANY OBSERVER SEES THE SAME HUBBLE EXPANSION

Hubble's law is a remarkable observation about the Universe that has far-reaching implications. For one thing, Hubble's law helps us test the prediction that the Universe is homogeneous and isotropic. When we look at galaxies in one direction in the sky, we find they obey the same Hubble law as galaxies observed in other directions in the sky. However, while Hubble's law corroborates the prediction that our view of the Universe is isotropic, the law appears at first glance to contradict the prediction of the cosmological principle that the Universe is homogeneous. Hubble's law might seem to imply that we are sitting in a very special place—at the *center* of a tremendous explosion, with everything else in the Universe streaming away from us. However, this initial impression is incorrect. *Hubble's law actually says that we are sitting in a uniformly expanding Universe and that the expansion looks*

> **Hubble's law says the Universe is expanding uniformly.**

the same, regardless of the location of the galaxy from which we view it! To understand why this is the case, we now turn to a useful toy model that you can build for yourself with materials you can probably find in your desk.

Figure 19.4 shows a long rubber band with paper clips attached along its length. If you stretch the rubber

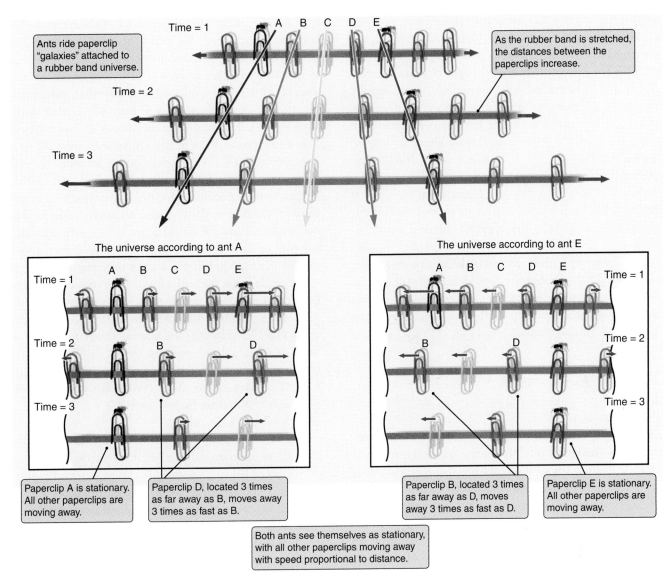

Figure 19.4 *Imagine a stretched rubber band with paper clips evenly spaced along its length. An ant riding on paper clip A will observe a Hubble law along the rubber band, with paper clip C moving away twice as fast as paper clip B. Similarly, an ant riding on paper clip E will see a Hubble law, with paper clip C moving away twice as fast as paper clip D. Any ant would see the same Hubble law, regardless of the paper clip it was riding.*

band, the paper clips, which represent galaxies in an expanding Universe, get farther and farther apart. Imagine what this expansion would look like if you were an ant riding on paper clip A. As the rubber band is stretched, you notice that all of the paper clips are moving away from you. When you look at paper clip B, which is the next paper clip over, you see it moving away slowly. When you look at paper clip C, located twice as far down the line from you as B, you see it moving away twice as fast as B. By the time you work your way down the line to paper clip E (located four times as far away as paper clip B), you see it moving away four times as fast as B. In other words, from the perspective of an

ant riding on paper clip A, all of the other paper clips on the rubber band are moving away with a speed that is proportional to their distance. The paper clips located along the rubber band obey a Hubble law.

This is a handy result, but the key insight comes from realizing that there is nothing special about paper clip A. If instead you were riding on paper clip E, then it would have been paper clip D that was moving away slowly, and paper clip A that was moving away four times as fast. Repeat this experiment for *any* paper clip along the rubber band, and you will arrive at the same result: the speed at which other paper clips are moving away from you is proportional to their distance. Bringing our cosmological terminology to bear, the stretching rubber band is homogeneous. We would see the *same* Hubble law, regardless of the paper clip that we choose as our vantage point.

The observation that nearby paper clips move away slowly and distant paper clips move away more rapidly does not say that we are at the center of anything. Instead, it says that the rubber band is being stretched uniformly along its length.

Hubble's law describes a homogeneous Universe.

In like fashion, *Hubble's law for galaxies does not mean that our Galaxy is at the center of an expanding Universe. Hubble's law means that the Universe is expanding uniformly.* Any observer viewing our Universe from any galaxy will see nearby galaxies moving away slowly and more distant galaxies moving away more rapidly. Any observer would find that the same Hubble law applies from their vantage point as applies from our vantage point on Earth. The expansion of the Universe is homogeneous.

WE MUST BUILD A DISTANCE LADDER TO MEASURE H_0

The discovery of Hubble's law for galaxies marked a fundamental change in our perception of the Universe. It also marked the beginning of a quest that has taken center stage in astronomy for seven decades. The form of Hubble's law—as a proportionality between recession velocity and distance—tells us our Universe is expanding. But to know the present *rate* of that expansion, we need a good value for the Hubble constant H_0.

How shall we go about measuring a value for the Hubble constant? In principle, the answer is straightforward. If we can measure the redshifts and distances of a number of galaxies, and plot velocity versus distance, then H_0 will be the slope of the resulting line. Yet while measuring the redshifts of galaxies is easy, given a large enough telescope, measuring the actual distances to galaxies is much more difficult.

The difficulty in measuring the Hubble constant comes from the fact that we must measure the distances not only to nearby galaxies, but to galaxies that are very far away. To see why, think about the motion of the water in a river. All of the water in a flowing river moves downstream, but even very uniform and steady rivers contain eddies and cross currents that disturb the uniform flow. If you want to get a good overall picture of river flow, you need to look at a large portion of the river, not just the motion of a single leaf or two drifting downstream.

Peculiar velocities make it hard to measure H_0.

Likewise, there are eddies and cross currents in the motions of galaxies that make up our Universe. The overall motion of galaxies in accord with Hubble's law is often referred to as *Hubble flow*. Departures from a smooth Hubble flow, referred to as **peculiar velocities,** are the result of gravitational interactions that cause galaxies to fall toward their neighbors or toward large concentrations of mass in the Universe. If we look only at nearby galaxies, their motions due to Hubble's law will be small, so most of the motion that we see will instead be due to their peculiar velocities. If we want to measure the Hubble flow itself to obtain a reliable value for H_0, we need to study galaxies that are far enough away that the part of their velocities due to Hubble flow is much greater than any peculiar velocities they might have.

Peculiar velocities of galaxies are typically a few hundred kilometers per second, so to measure H_0 we need to accurately measure the distances to galaxies with Hubble velocities of several thousand km/s or more (that is, galaxies with redshift z greater than about 0.01). The distances to such galaxies—150 million light-years and beyond—are far

Measuring H_0 requires measuring distances to remote galaxies.

too great to measure using the same techniques that allowed us to measure the distances of objects in our own Galaxy. Instead, we must find new standard candles luminous enough to be seen at great distances. This is done in a series of steps, referred to as the **distance ladder.**

The distance ladder begins with stellar parallax (see Chapter 12), which allows us to measure distances to nearby stars and trace the position of the main sequence. By using the main sequence itself as a standard candle, we can measure the distance to very nearby galaxies such as the Magellanic Clouds, shown in Figure 18.19. The Magellanic Clouds, located 160,000 light-years away, contain many Cepheid variable stars, allowing us to improve the calibration of these standard candles. Cepheid variables in turn allow us to accurately measure distances to galaxies as far out as about 100 million light-years, but even this is not enough for us to measure a reliable value for the Hubble constant. In that volume of space, however, there are many galaxies that we can scour for yet more powerful standard candles. Among the best of these are Type I supernovae.[1]

Recall that Type I supernovae are thought to occur when gas flows from an evolved star onto its white dwarf companion, pushing the white dwarf over the

---

[1]Formally, a specialist in the field would refer to these as Type Ia supernovae, as noted previously in Chapter 15.

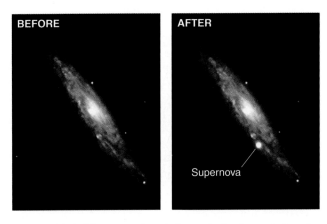

Figure 19.5 *The same galaxy before and after the explosion of a Type I supernova. Type I supernovae are extremely luminous standard candles.*

Chandrasekhar limit for the mass of an electron-degenerate object. When this happens, the overburdened white dwarf first begins to collapse, and then explodes. Because all Type I supernovae occur in white dwarfs of the same mass, we might expect all such explosions to have about the same luminosity. This is borne out by observations of Type I supernovae in galaxies with known distances. With a peak luminosity

that outshines a hundred billion suns **(Figure 19.5)**, Type I supernovae can be seen and measured using current telescopes reaching much of the way to the edge of the observable Universe.

> **Type I supernovae are very luminous standard candles.**

It is often said that a chain is no stronger than its weakest link. When constructing the cosmic distance ladder the situation is even worse, for the final ladder is no better than the *combination* of the weaknesses in each of its rungs. For the last few decades of the 20th century, astronomers working to measure H_0 were largely split into two camps. One group favored a Hubble constant of around 18 km/(s Mly), while a second camp viewed the data as supporting a value of around 35 km/(s Mly). A virtual war raged between these two groups for years, until *HST* observations in the mid-1990s converged on an intermediate value of about 22 km/(s Mly).[2]

Figure 19.6 plots the measured recession velocities of galaxies against their measured distances. Notice how well the Universe follows Hubble's law. It is thought that current measurements of the Hubble constant are

> **Current measurements give H_0 at 23 km/(s Mly).**

accurate to within about 10%. That is, astronomers working in the field are fairly confident that the true value lies between about 20 and 25 km/(s Mly). Even so, those many steps in the cosmic distance ladder remain, and efforts to test and refine our knowledge of the Hubble constant will doubtless continue for years to come.

---

[2]Astronomers normally express H_0 in units of km/s per million parsecs rather than km/s per million light-years. In these units, 22 km/(s Mly) corresponds to 72 km/(s Mpc).

(a)

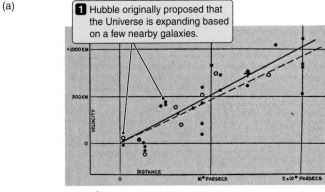

1 Hubble originally proposed that the Universe is expanding based on a few nearby galaxies.

(b)

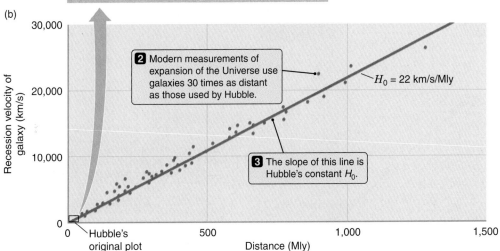

2 Modern measurements of expansion of the Universe use galaxies 30 times as distant as those used by Hubble.

$H_0 = 22$ km/s/Mly

3 The slope of this line is Hubble's constant H_0.

Hubble's original plot

Recession velocity of galaxy (km/s)

Distance (Mly)

Figure 19.6 (a) *Hubble's original figure illustrating that more distant galaxies are receding faster than less distant galaxies.* (b) *Modern data on galaxies up to 30 times farther away than those studied by Hubble show that recession velocity is proportional to distance.*

HUBBLE'S LAW MAPS THE UNIVERSE IN SPACE AND TIME

So why all the fuss about Hubble's law and the Hubble constant? What do these tell us about the Universe? First, Hubble's law provides us with a practical tool for measuring distances to remote objects. Once we know the value of H_0, Hubble's law allows us to turn a straightforward measurement of the redshift of a (relatively nearby) galaxy into knowledge of its distance. For example, a galaxy with a redshift of 0.1 is 1.4 billion light-years away; a galaxy with a redshift of 0.2 is twice that distance away. In short, once we know H_0, Hubble's law makes the once-difficult task of measuring distances in the Universe relatively easy, providing us with a tool to literally map the structure of the Universe. We will put this tool to good use in the next chapter when we turn our attention to the large-scale structure of the Universe.

> **Redshift tells us a galaxy's distance.**

Hubble's law does more than place galaxies in space. It also places galaxies in time. Light travels at a huge but finite speed. When we look at the Sun, we see it as it existed $8\frac{1}{3}$ minutes ago. When we look at Alpha Centauri, the nearest stellar system beyond the Sun, we see it as it existed 4.3 years ago. When we look at the center of our Galaxy, the picture that we see is 27,000 years old. When looking at a distant object, we speak of its **look-back time**—the time it has taken for the light from that object to reach our telescope. As we look into the distant Universe, look-back times become very great indeed. The distance to a galaxy where $z = 0.1$ is 1.4 billion light-years (assuming $H_0 = 22$ km/(s Mly)), so the look-back time to that galaxy is 1.4 billion years. The look-back time to a galaxy where $z = 0.2$ is 2.7 billion years. As we look at objects with greater and greater redshifts, we are seeing increasingly younger versions of our Universe.

> **When observing the Universe, elsewhere is "elsewhen."**

19.3 THE UNIVERSE BEGAN IN THE BIG BANG

Hubble's law provides a very powerful and practical tool for mapping the distribution of galaxies throughout the Universe. Yet the most significant aspect of Hubble's law is what it tells us about the structure of the Universe itself. We know that all galaxies in the Universe are moving away from each other, but if we could run the movie backward in time, we would find the galaxies getting closer and closer together.

> **Galaxies used to be closer together.**

Figure 19.7 shows two galaxies located 100 Mly ($d_G = 9.5 \times 10^{20}$ km) away from each other. If these two galaxies are flying apart from each other, then at some point in the past, they must have been together in the same place at the same time. According to Hubble's

Figure 19.7 *The origin of the Hubble time. Assuming the velocities of galaxies remain constant, the time needed for any pair of galaxies to reach their current locations is the same, regardless of their separation.*

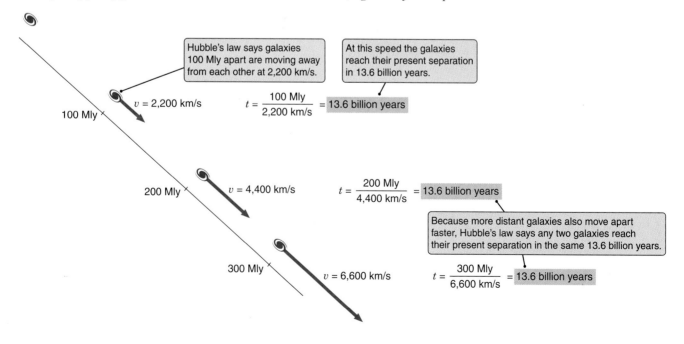

Hubble's law says galaxies 100 Mly apart are moving away from each other at 2,200 km/s.

At this speed the galaxies reach their present separation in 13.6 billion years.

100 Mly

$v = 2,200$ km/s $\qquad t = \dfrac{100 \text{ Mly}}{2,200 \text{ km/s}} = $ 13.6 billion years

200 Mly

$v = 4,400$ km/s $\qquad t = \dfrac{200 \text{ Mly}}{4,400 \text{ km/s}} = $ 13.6 billion years

Because more distant galaxies also move apart faster, Hubble's law says any two galaxies reach their present separation in the same 13.6 billion years.

300 Mly

$v = 6,600$ km/s $\qquad t = \dfrac{300 \text{ Mly}}{6,600 \text{ km/s}} = $ 13.6 billion years

law, the distance between these two galaxies is increasing at the rate $v_r = H_0 \times d_G = 2{,}200$ km/s, assuming that $H_0 = 22$ km/(s Mly). Traveling at this speed, the two galaxies would have taken about 13.6 billion years to travel the 100 Mly that separates them. [Time = distance/speed = $(9.5 \times 10^{20}$ km)/(2,200 km/s) = 4.32 $\times 10^{17}$ s.] In other words, *if* expansion of the Universe has been constant, two galaxies that today are 100 Mly apart started out at the same place 13.6 billion years ago.

Now we do the same calculation with two galaxies that are twice as far (200 Mly) apart (see Figure 19.7).

Hubble's law shows galaxies have been traveling apart for 13 billion years.

These two galaxies are twice as far apart, but the distance between them is increasing twice as rapidly: $v_r = H_0 \times d_G = 4{,}400$ km/s. We again calculate time equals distance divided by speed (twice the distance divided by twice the speed) to find that these galaxies *also* took 13.6 billion years to reach their current locations. We also find that galaxies beginning 300 Mly apart also were in the same place 13.6 billion years ago. We could do this calculation again and again for any pair of galaxies in the Universe today. The farther apart the two galaxies are, the faster they are moving. The thing that is the *same* for all galaxies is the *time* that it took them to get to where they are today.

A look at the math below makes this clear. Velocity equals Hubble's constant times distance, so when we calculate a time by saying "time equals distance divided by velocity," the distance factors on top and bottom cancel out. Writing it out as an equation, we get

$1/H_0$ is one measure of the age of the Universe.

$$\text{Time} = \frac{\text{distance}}{\text{velocity}} = \frac{\text{distance}}{H_0 \times \text{distance}} = \frac{1}{H_0}.$$

The startling implications of this result are illustrated in **Figure 19.8**. About 6.8 billion years ago, when the Universe was half its present age, all of the galaxies in the Universe were separated from each other by half their present distances. Twelve billion years ago, all of the galaxies in the Universe must have been separated from each other by about a tenth of their present distance. Assuming that galaxies have been moving apart at the same speed that we see today, then 13.6 billion years ago (a time equal to $1/H_0$), *all the stars and galaxies that*

Expansion started in a Big Bang.

Figure 19.8 *Looking backward in time, the distance between any two galaxies is smaller and smaller, until all matter in the Universe is concentrated together at the same point, the Big Bang.*

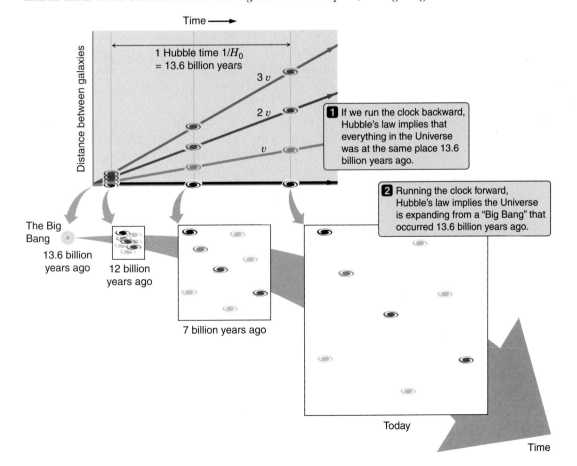

1 If we run the clock backward, Hubble's law implies that everything in the Universe was at the same place 13.6 billion years ago.

2 Running the clock forward, Hubble's law implies the Universe is expanding from a "Big Bang" that occurred 13.6 billion years ago.

The Big Bang 13.6 billion years ago

12 billion years ago

7 billion years ago

Today

make up today's Universe must have been concentrated together at the same location! The value of 1 divided by the Hubble constant is referred to as the **Hubble time.** If our line of reasoning is correct, then today's Universe is hurtling outward from a tremendous explosion that took place approximately 13 to 14 billion years ago. This colossal event, which marked the beginning of our Universe, is referred to as the **Big Bang.**

The idea of the Big Bang greatly troubled many astronomers in the early and middle years of the 20th century. Several different suggestions were put forward to explain the observed fact of Hubble expansion without resorting to the idea that the Universe came into existence in an extraordinarily dense fireball billions of years ago. However, as more and more observations have come in, and more discoveries about the structure

The Big Bang theory is scientific fact.

of the Universe have been made, the Big Bang theory has only grown stronger. Today there are few astronomers working seriously in this field who doubt whether the Big Bang took place. Virtually all the major predictions of the Big Bang theory (expansion of the Universe being but one of them) have proven to be correct. As we will see, the Big Bang theory for the origin of our Universe is now such a well-corroborated theory that most astronomers would probably say it has crossed into the realm of scientific fact.

The implications of Hubble's law are striking. This single discovery forever changed our conception of the origin, history, and possible future of the Universe in which we live. At the same time, Hubble's law has pointed to many new questions about the Universe. To address them, we next need to consider exactly what we mean by the term *expanding Universe.*

GALAXIES ARE NOT FLYING APART THROUGH SPACE

The mental picture that you have of the expanding Universe at this point in our discussion is probably

The Big Bang happened everywhere.

one of a cloud of debris from an explosion flying outward through space. One of the first questions students usually ask about the Big Bang is, "Where did the explosion take place?" The answer to this question, amazing as it seems, is that the explosion took place *everywhere.* Wherever you are in the Universe today, you

Space itself is expanding.

are sitting at the site of the Big Bang. The reason for this is that galaxies are not flying apart through space at all. Rather, it is *space itself* that is expanding, carrying the stars and galaxies that populate the Universe along with it.

This may seem an incredible notion, but we have already dealt with the basic ideas that allow us to understand the expansion of space. In our discussion of neutron stars and black holes, back in Chapter 16, we encountered Einstein's general theory of relativity. General relativity says that space is distorted by the presence of mass, and that the consequence of this distortion is gravity. For example, the mass of the Sun, like any object, distorts the geometry of space-time around it, so that Earth, coasting along in its inertial frame of reference, follows a curved path around the Sun. We illustrated this with the analogy of a ball placed on a stretched rubber sheet, showing how the ball distorted the surface of the sheet.

There are other ways to distort the surface of a rubber sheet as well. Imagine a number of coins placed on a rubber sheet, as shown in **Figure 19.9.** Suppose we grab the edges of the sheet and begin pulling them outward. As the rubber sheet stretches, each coin remains at the same location on the surface of the sheet, but the distances between the coins increase. Two coins sitting close to each other move apart only slowly, while coins farther apart move away from each other more rapidly. In other words, the distances and relative motions of the coins on the surface of a rubber sheet would obey a Hubble-type relationship as the sheet is stretched.

The Universe is like an expanding rubber sheet.

This is what is happening in the Universe, with galaxies taking the place of the coins and space itself taking the place of the rubber sheet. Obviously, in the case of coins on a rubber sheet, there is a limit to how far we can stretch the sheet before it breaks. With space and the real Universe, there is no such limit. The fabric of space can, in principle, go on expanding forever. Hubble's law is the observational consequence of the fact that the space making up the Universe is expanding.

19.1 EXPAND

EXPANSION IS DESCRIBED WITH A SCALE FACTOR

When astronomers discuss the expansion of the Universe, they talk in terms of the **scale factor** of the Universe. To understand this concept, we return to our

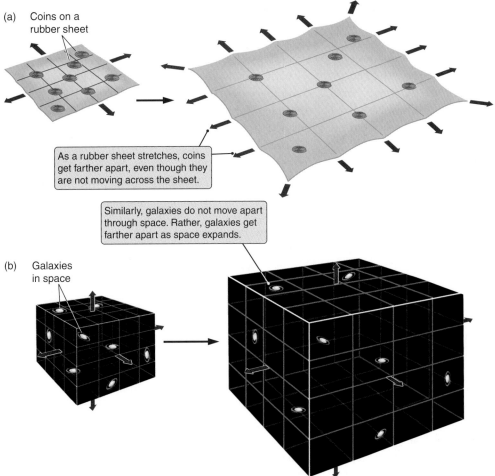

(a) Coins on a rubber sheet

As a rubber sheet stretches, coins get farther apart, even though they are not moving across the sheet.

Similarly, galaxies do not move apart through space. Rather, galaxies get farther apart as space expands.

(b) Galaxies in space

Figure 19.9 (a) *As a rubber sheet is stretched, coins on its surface get farther apart even though they are not moving with respect to the sheet itself. Any coin on the surface of the sheet will observe a Hubble law in every direction.* (b) *In analogous fashion, galaxies in an expanding universe are not flying apart through space. Rather, it is space itself that is stretching.*

analogy of the rubber sheet. Suppose we draw a ruler on the surface of the sheet, placing a tick mark every centimeter, as in **Figure 19.10(a).** If we want to know the distance between two points on the sheet, all we need to do is count the marks between the two points and multiply by 1 cm per tick mark.

But as the sheet is stretched, the distance between the tick marks does not remain 1 cm. When the sheet is stretched to 150% of the size that it had when we drew our ruler, each tick mark is separated from its neighbors by one and one half times their original distance, or 1.5 cm. If we wanted to know the distance between two points, we could still count the marks, but we would have to *scale up* the distance in tick marks by 1.5 to find the distance in centimeters. The scale factor of the sheet is now 1.5. And when the sheet is twice the size that it was when we drew the ruler **(Figure 19.10b),** each mark would correspond to 2 cm of actual distance; the scale factor of the sheet would now be 2. The scale factor tells us the size of the sheet relative to its size at the time when we drew our ruler. The scale factor also tells us how much the distance between points on the sheet has changed.

> **Expansion of the Universe is measured with a scale factor.**

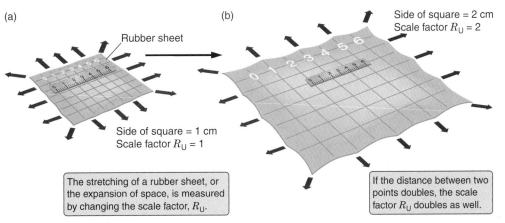

(a) Rubber sheet

Side of square = 1 cm
Scale factor R_U = 1

The stretching of a rubber sheet, or the expansion of space, is measured by changing the scale factor, R_U.

(b) Side of square = 2 cm
Scale factor R_U = 2

If the distance between two points doubles, the scale factor R_U doubles as well.

Figure 19.10 *Tick marks are drawn on a rubber sheet at a distance of 1 cm apart* (a). *As the sheet is stretched, the tick marks get farther apart. When the spacing between the tick marks is 2 cm, or twice the original value, we say that the scale factor of the sheet,* R, *has doubled* (b). *A similar scale factor,* R_U, *is used to describe the expansion of the Universe.*

We can apply this same idea to the Universe. Suppose we choose today to lay out a "cosmic ruler" on the fabric of space, placing an imaginary tick mark every 10 Mly. We define the scale factor of the Universe at this time to be 1. In the past, when the Universe was smaller, distances between the points in space marked by our cosmic ruler would have been less than 10 Mly apart. The scale factor of that younger, smaller Universe would have been less than 1, compared with today. In the future, as the Universe continues to expand, the distances between the tick marks on our cosmic ruler will grow to more than 10 Mly, and the scale factor of the Universe will be greater than 1. In this way we can use

The scale factor R_U increases as the Universe expands.

the scale factor, usually written as R_U, to keep track of the changing scale of the Universe.

It is important to remember, when thinking about the expanding Universe, that the laws of physics are themselves unchanged by the changing scale factor. For example, when we stretched out the rubber sheet, we did not change the properties of the coins on its surface. In like fashion, as the Universe expands, the properties of atoms, stars, and galaxies do not change.

Expansion does not affect local physics—stars, atoms, or anything else.

We return to the question of locating the center of expansion. Looking back in time, the scale factor of the Universe gets smaller and smaller, approaching zero as

FOUNDATIONS 19.1

WHEN REDSHIFT EXCEEDS ONE

In our discussion of Doppler shift, we found that $(\lambda_{observed} - \lambda_{rest})/\lambda_{rest}$ is equal to the velocity of an object away from us, divided by the speed of light. In this chapter we see how Edwin Hubble used this result to interpret the observed redshifts of galaxies as evidence that galaxies throughout the Universe are moving away from us. Einstein's special theory of relativity says that nothing can move faster than the speed of light. Hubble's initial assumption was that redshifts are due to the Doppler effect. The resulting relation $z = v_r/c$ would then seem to imply that no object can have a redshift z greater than 1. Yet that is not the case. Astronomers routinely observe redshifts significantly in excess of 1. As of this writing, the most distant objects known have redshifts of over 6! What is wrong here? It is worth taking a brief digression to consider how redshifts can exceed 1.

The first thing to note is that, to arrive at the expression for the Doppler effect—$v_r/c = (\lambda_{observed} - \lambda_{rest})/\lambda_{rest}$—we have to *assume* that v_r is much less than c. If that were not the case—if v_r were close to c—then we would have to take into account more than just the fact that the waves from an object are stretched out by the object's motion away from us. We would also have to consider relativistic effects, including the fact that moving clocks run slowly. When combining these effects, we would find that, as the speed of an object approaches the speed of light, its redshift becomes arbitrarily large **(Figure 19.11)**.

A second source of redshift is the *gravitational redshift* discussed in Chapter 16. As light escapes from

deep within a gravitational well, it loses energy, so photons are shifted to longer and longer wavelengths. If the gravitational well is deep enough, then the observed redshift of this radiation can be boundlessly large. In fact, the event horizon of a black hole—that is, the surface around the black hole from which not even light can escape—is where the gravitational redshift becomes infinite.

Cosmological redshift, which is most relevant to this chapter, results from the amount of "stretching" space has undergone during the time the light from its original source has been en route to us. The amount of stretching that has occurred is given by the factor $1 + z$. When we look at a distant galaxy whose redshift $z = 1$, we are seeing the Universe at a time when it was half the size that it is today. When we see light from a galaxy with $z = 2$, we are seeing the Universe when it was $\frac{1}{3}$ of its current size. Nearby, this means that distance and lookback time are proportional to z. As we look back closer and closer to the Big Bang, however, redshift climbs more and more rapidly.

Doppler's original formula is essentially correct—as long as we are looking at shifts due to an object's motion and the velocities we measure are far less than the speed of light. In this case, $v_r/c = (\lambda_{observed} - \lambda_{rest})/\lambda_{rest}$. When we look at the motions of orbiting binary stars or the peculiar velocities of galaxies relative to the Hubble flow, this equation works just fine. But anytime you have a redshift of 1 or greater, you should immediately recognize that you are now in the realm of relativity.

Just after the Big Bang, our observable Universe was very small.

we get closer and closer to the Big Bang. The fabric of space that today spans billions of light-years spanned much smaller distances when the Universe was young. When the Universe was only a day old, all of the space that we see today amounted to a region only a few times the size of our Solar System. When the Universe was a fiftieth of a second old, the vast expanse of space that makes up today's observable Universe (and all the matter in it) occupied a volume only the size of today's Earth. As we continue to approach the Big Bang itself going backward in time, the space that makes up today's observable Universe becomes smaller and smaller—the size of a grapefruit, a marble, an atom, a proton. . . . Every point in the fabric of space that makes up today's Universe was right there at the beginning, a part of that unimaginably tiny, dense Universe that emerged from the Big Bang.

These points bear repetition: Where is the center of the Big Bang? There is no center. The Big Bang did not occur at a specific point in space, because it was space itself that came into existence with the Big Bang. Where did the Big Bang happen? It happened everywhere, including right where you are sitting. This is an important result. If there were a particular point in today's Universe that marked the site of the Big Bang, that would be a very special point indeed. But there is no such point. The Big Bang happened everywhere. A Big Bang universe is homogeneous and isotropic, consistent with the cosmological principle.

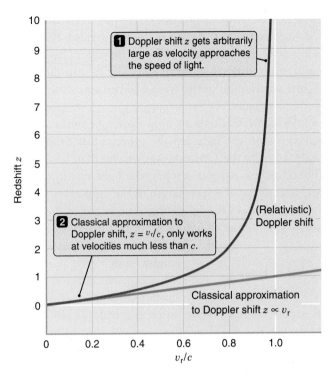

Figure 19.11 *Plot of the redshift z of an object versus its recession velocity* v_r *as a fraction of the speed of light. According to special relativity, as* v_r *approaches* c, *the redshift becomes large without limit.*

Let's return to our rubber-sheet analogy. If we were to draw a series of bands on the rubber sheet to represent the crests of an electromagnetic wave, as in **Figure 19.12,** we could watch what happens to the wave as the sheet is stretched out. By the time the sheet is

Figure 19.12 *Bands drawn on a rubber sheet represent the positions of the crests of an electromagnetic wave in space. As the rubber sheet is stretched—that is, as the Universe expands—the wave crests get farther apart. The light is red-shifted.*

REDSHIFT IS DUE TO THE CHANGING SCALE FACTOR OF THE UNIVERSE

General relativity gives us a powerful tool for interpreting Hubble's great discovery. It also forces us to rethink just what we mean when we talk about the redshift of distant galaxies. While it is true that the distance between galaxies is increasing as a result of the expansion of the Universe, and that we can use the equation for Doppler shifts to measure the redshifts of galaxies, these redshifts are not due to Doppler shifts at all! As light comes toward us from distant galaxies, the scale factor of the space through which the light travels is constantly increasing, and as it does so, the distance between adjacent wave crests increases as well. The light is "stretched out" as the space it travels through expands. (See Foundations 19.1.)

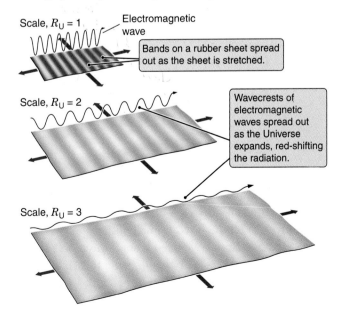

stretched to twice its original size—that is, by the time the scale factor of the sheet is 2—the distance between wavecrests has doubled. When the sheet has been stretched to three times its original size (a scale factor of 3), the wavelength of the wave will be three times what it was originally.

We apply this idea to light coming from a distant galaxy. When the light left the galaxy of its origin, the scale factor of the Universe was smaller than it is today. The Universe expanded while the light was in transit, and as it did so, the wavelength of the light grew longer in proportion to the increasing scale factor of the Universe. The redshift of light from distant galaxies is therefore a direct measure of how much the Universe has expanded since the time the radiation left its source. Redshift measures how much the scale factor of the Universe, R_U, has changed since the light was emitted.

R_U is related to redshift.

If we see radiation with a redshift of 1, then its wavelength is twice as long as when that radiation left its source. When the light left its source, R_U was equal to $\frac{1}{2}$ compared with today. If we see radiation with a redshift of 2, then the wavelength of the radiation is three times its original wavelength. The radiation was emitted when the scale factor of the Universe was $\frac{1}{3}$. This wonderfully direct relationship lets us convert from the observed redshift of a galaxy to knowledge about the size of the Universe at the look-back time to that galaxy. Written as an equation, the scale factor of the Universe we see when looking at a distant galaxy is equal to 1 divided by 1 plus the redshift of the galaxy:

$$R_U(z) = \frac{1}{1+z}.$$

19.4 THE MAJOR PREDICTIONS OF THE BIG BANG THEORY ARE RESOUNDINGLY CONFIRMED

The questions we are grappling with when discussing ideas like the origin of the Universe are some of the most fundamental questions humankind can ask about the Universe. Throughout human history, answers to these same questions have been among the most prized goals of philosophers and theologians. It is quite remarkable that we live in a time when we are finding real, testable answers to these questions by appealing not to divine inspiration or the blind logic of philosophy, but rather to the empirical methods of science. It is essential, then, that we place extraordinary demands on the quality of the evidence that we rely on to support the theory of the Big Bang. What evidence is there, apart from the observed expansion of the Universe itself, that requires us to accept that the Big Bang actually took place?

WE SEE RADIATION LEFT OVER FROM THE BIG BANG

The story of the single most important confirmation of the Big Bang theory begins in the mid-1940s, when **George Gamow** (1904–1968) was thinking, along with his student **Ralph Alpher** (1921–), about the implications of Hubble expansion. When a gas is compressed, it grows hotter. So, reasoned Gamow and Alpher, when the Universe was very young and small, it must have consisted of an extraordinarily hot, dense gas. As with any hot, dense gas, this early Universe would have been awash in the same kind of radiation that we have encountered so many times before on our journey of discovery: radiation from a blackbody, which exhibits a Planck spectrum.

Gamow and Alpher took this idea a step further. As the Universe expanded, they reasoned, this radiation would have been red-shifted to longer and longer wavelengths. Recall Wien's law, which states that the temperature associated with Planck radiation is inversely proportional to the peak wavelength: $T = (2{,}900~\mu\mathrm{m})/\lambda_{\mathrm{peak}}$. Shifting the wavelength of Planck radiation to longer and longer wavelengths is therefore the equivalent of shifting the characteristic temperature of the radiation to lower and lower values. As illustrated in **Figure 19.13,** doubling the wavelength of the photons in a Planck spectrum by doubling the scale factor of the Universe is equivalent to cutting the temperature of the Planck spectrum in half.

On April Fools Day, 1948, Alpher, **Hans Bethe** (1906–), and Gamow published a paper asserting that this radiation should still be visible today, and should have a Planck spectrum with a temperature in the neighborhood of 5 to 10 K. (The paper actually listed as authors not only Alpher and Gamow, but also Hans Bethe, who, though an esteemed physicist in his own right, did not actively par-

Alpher, Bethe, and Gamow predicted the glow from a young, hot Universe in 1948.

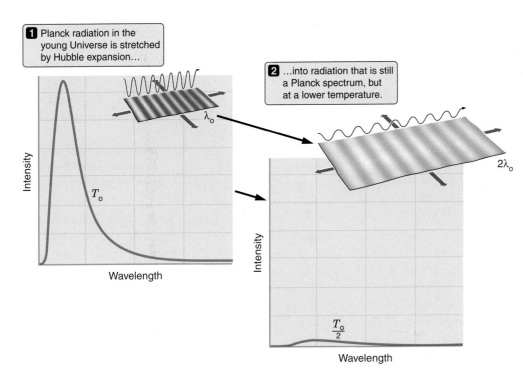

1 Planck radiation in the young Universe is stretched by Hubble expansion...

2 ...into radiation that is still a Planck spectrum, but at a lower temperature.

Figure 19.13 *As the Universe expanded, Planck radiation left over from the hot, young Universe was red-shifted to longer wavelengths. Red-shifting a Planck spectrum is equivalent to lowering its temperature.*

Figure 19.14 *Penzias and Wilson next to the horn telescope with which they discovered the cosmic background radiation.*

ticipate in this research. Gamow, being a renowned jokester, added Bethe's name between his and Alpher's. "Alpher, Bethe, and Gamow" is a play on the first three letters of either the Greek and Hebrew alphabets—an appropriate authorship for a paper making predictions about the very early Universe!)

This prediction languished until the early 1960s, when two physicists from Bell Laboratories, **Arno Penzias** (1933–) and **Robert Wilson** (1936–), were trying to bounce radio signals off the newly launched *ECHO* satellites. This hardly seems much of a feat today, when we routinely use handheld units that communicate directly with satellites. Yet at the time, it pushed radio technology to its limits. Penzias and Wilson needed a very sensitive microwave telescope for their work. Any spurious signals coming from the telescope itself might wash out the faint signals bounced off a satellite. Penzias and Wilson, shown in **Figure 19.14** along with their radio telescope, worked tirelessly to eliminate all possible sources of interference originating from within their instrument. This work included such endless and menial tasks as keeping the telescope free of bird droppings and other extraneous material. Even so, Penzias and Wilson found that no matter how hard they tried to eliminate sources of extraneous noise, they could still detect a faint microwave signal when they pointed the telescope at the sky. Eventually they came to accept that the signal they were detecting was real. The sky faintly glows in microwaves.

In the meantime, Robert Dicke (1916–1997) and his colleagues at Princeton University had also predicted a hot early Universe, arriving independently at the same basic conclusions that Alpher and Gamow had reached two decades earlier. When Dicke and colleagues heard of the signal that Penzias and Wilson had found, they interpreted it as the radiation left behind by the hot early Universe. The strength of the detected signal was consistent with the glow from a blackbody with a temperature of about 3 K, very close to the predicted value. Their results, published in 1965, reported the discovery of the glow left behind by the Big Bang. Penzias and Wilson shared the 1979 Nobel Prize in physics for their remarkable discovery. (It is worth noting that timing can be everything in science. Alpher, who first predicted the existence of a faint glow from the Big Bang, searched unsuccessfully for the signal 10 years before Penzias and Wilson made their discovery. Unfortunately, however, the technology of the late 1940s and early 1950s was simply not up to the task.)

> **Penzias and Wilson discovered the cosmic background radiation.**

This radiation left over from the early Universe is called the **cosmic background radiation (CBR).** Today the cosmic background radiation and the conditions in the early Universe are much better understood than they were in the early 1960s. Origin of the CBR is illustrated in **Figure 19.15.** When the Universe was young, it was hot enough that all of the atoms in the Universe were ions. In our discussion of the structure of the Sun and stars, we found that radiation does not travel well through an ionized plasma. Free electrons in a plasma interact strongly with the radiation, blocking its progress. At this time in the early Universe, the condi-

> **The CBR is thermal radiation that arose when the Universe was hot and ionized.**

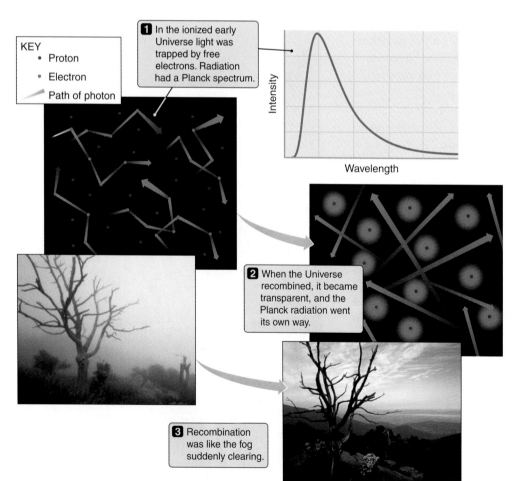

KEY
- Proton
- Electron
- Path of photon

1 In the ionized early Universe light was trapped by free electrons. Radiation had a Planck spectrum.

Intensity / Wavelength

2 When the Universe recombined, it became transparent, and the Planck radiation went its own way.

3 Recombination was like the fog suddenly clearing.

Figure 19.15 *The origin of the cosmic background radiation. Prior to recombination the Universe was like a foggy day. Radiation interacted strongly with free electrons, and so could not travel far. The trapped radiation had a Planck spectrum. When the Universe recombined, the fog cleared, and this radiation was free to go on its way.*

tions within the Universe were much like the conditions within a star: the Universe was an opaque blackbody.

As the Universe expanded, the gas filling the Universe cooled. By the time the Universe was about a thousandth of its current size, the temperature had dropped to a few thousand kelvins, so protons and electrons were able to combine to form hydrogen atoms. This event, called the **recombination of the Universe,** occurred when the Universe was several hundred thousand years old.

Hydrogen atoms are much less effective at blocking radiation than free electrons are, so when recombination occurred, the Universe suddenly became transparent to radiation. Since that time, the radiation left behind from the Big Bang has been able to travel largely unimpeded throughout the Universe. At the time of recombina-

> Since recombination, the CBR has traveled freely and cooled by a factor of 1,000.

tion, when the temperature of the Universe was a few thousand kelvins, the wavelength of this radiation peaked at around 1 μm according to Wien's law. As the Universe expanded, this radiation was red-shifted to longer and longer wavelengths. Today the scale of the Universe has increased a thousandfold since recombination, and the peak wavelength of the cosmic background radiation has increased by a thousandfold as well to a value close to 1 mm. The spectrum of the CBR still has the shape of a Planck spectrum, but with a characteristic temperature of 2.73 K—only a thousandth what it was at recombination.

COBE REMOVED ANY REASONABLE DOUBT THAT THE CBR IS REAL

The presence of cosmic background radiation with a Planck spectrum is a very strong prediction of the Big Bang theory. Penzias and Wilson had confirmed that a signal with the correct strength was there, but they could not say for certain whether the signal they saw had the spectral shape of a Planck spectrum. From the remainder of the 1960s to the 1980s, most experiments at different wavelengths supported these same conclusions. Yet it was not until the end of the 1980s that the predictions of Big Bang cosmology for the CBR were put to the ultimate test. The year 1989 saw the launch of a satellite called the *Cosmic Background*

Explorer, or *COBE. COBE* carried onboard instruments capable of making extremely precise measurements of the CBR at many wavelengths, from a few micrometers out to 1 cm. In January 1990, hundreds of astronomers gathered in a large conference room in Washington, D.C., at the winter meeting of the American Astronomical Society to hear the *COBE* team present its first results. Security surrounding the new findings had been tight, so the atmosphere in the room was electric. The tension did not last for long: presentation of a single viewgraph brought the room to its feet in a spontaneous ovation.

The data shown on that viewgraph are reproduced in **Figure 19.16.** The small circles in the figure are the *COBE* measurements of the CBR at different frequencies. The uncertainty in each measurement is far less than the

> *COBE* unambiguously showed the CBR to have a Planck spectrum at 2.73 K.

size of each dot. The line in the figure, which runs perfectly through the data points, is a Planck spectrum with a temperature of 2.73 K. The agreement between theoretical prediction and observation is truly remarkable. The observed spectrum so perfectly matches the one predicted by Big Bang cosmology that there can be no real doubt we are seeing the residual radiation left behind from the primordial fireball of the early Universe.

Figure 19.16 *The spectrum of the cosmic background radiation as measured by the* Cosmic Background Explorer (COBE) *satellite (red dots). The uncertainty in the measurement at each wavelength is much less than the size of a dot. The line running through the data is a Planck spectrum with a temperature of 2.728 K.*

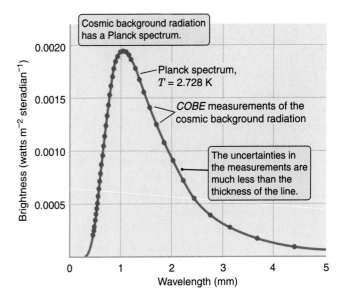

(a)

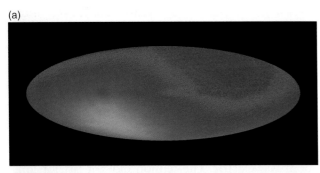

(b)

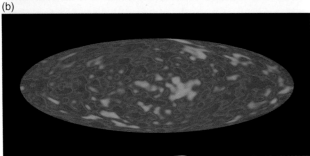

Figure 19.17 (a) *The* COBE *map of the cosmic background radiation. The CBR is slightly hotter (by about 0.003 K) in one direction in the sky than in the other direction. This is due to Earth's motion relative to the CBR.* (b) *The* COBE *map with Earth's motion removed, showing tiny ripples remaining in the CBR.*

THE CBR MEASURES EARTH'S MOTION RELATIVE TO THE UNIVERSE ITSELF

COBE provided us with more than a measurement of the spectrum of the cosmic background radiation. **Figure 19.17(a)** shows a map of the CBR from the entire sky obtained by *COBE*. The different colors in the map correspond to variations in the temperature of the CBR. The differences are not as extreme as the colors might suggest, however. The range in temperature in the map corresponds to a variation of only about 0.1% in the temperature of the CBR. Most of this range of temperature is present because one side of the sky looks slightly warmer than the opposite side of the sky. This difference has nothing to do with the large-scale structure of the Universe itself, but rather is the result of the motion of Earth with respect to the CBR.

Time and again on our journey we have stressed that there is no preferred frame of reference. The laws of physics are the same in *any* inertial reference frame, so none is better than any other. Yet in a certain sense there *is* a preferred frame of reference at any point in the Universe. This is the frame of reference that is at rest with respect to the expansion of the Universe, and in which the CBR is isotropic. The *COBE* map

> **Our motion makes the CBR slightly hotter in the direction we are moving toward and slightly cooler behind us.**

shows that one side of the sky is slightly hotter than the other because our Sun and our Earth are moving at a velocity of 368 km/s in the direction of the constellation of Crater, relative to this cosmic reference frame. Radiation coming from the direction in which we are moving is slightly blue-shifted (shifted to a higher characteristic temperature) by our motion, while radiation coming from the opposite direction is Doppler-shifted toward the red (or cooler temperatures).

If we subtract from the *COBE* map this asymmetry in the CBR caused by the motion of Earth, only slight variations in the CBR remain, as shown in **Figure 19.17(b).** The slight variations seen in this map have an amplitude that is only about 1/100,000 the brightness of the CBR. That means that the brighter parts of this image are only about 1.00001 times brighter than the fainter parts. These slight variations might not seem

> **COBE found tiny variations in the CBR resulting from the formation of structure during early times.**

like much, but they are actually of crucial importance in the history of the Universe. These fluctuations are the result of the gravitational redshift of the cosmic background radiation caused by concentrations of mass in the early Universe. These concentrations later gave rise to galaxies and the rest of the structure that we see in the Universe today. Subsequent observations from the South Pole and from instruments carried aloft by balloons support the *COBE* findings, and new satellites will soon provide a detailed look at these irregularities, telling us even more about the seeds from which galaxies formed.

THE BIG BANG THEORY CORRECTLY PREDICTS THE ABUNDANCE OF THE LEAST MASSIVE ELEMENTS

The next confrontation between the predictions of the Big Bang theory and observations of the Universe came from a very different direction. When the Universe was only a few minutes old, the temperature and density in the Universe

> **Light elements were produced by nuclear reactions a few minutes after the Big Bang.**

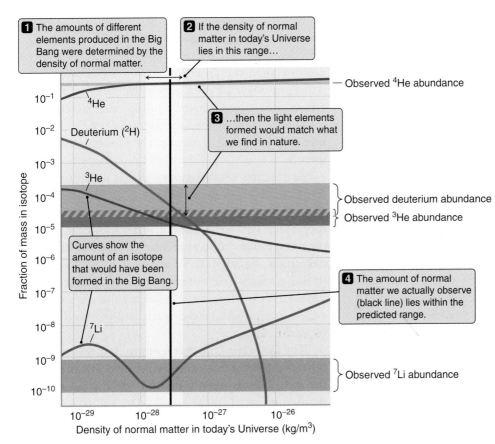

1 The amounts of different elements produced in the Big Bang were determined by the density of normal matter.

2 If the density of normal matter in today's Universe lies in this range...

3 ...then the light elements formed would match what we find in nature.

Observed ⁴He abundance

Curves show the amount of an isotope that would have been formed in the Big Bang.

Observed deuterium abundance
Observed ³He abundance

4 The amount of normal matter we actually observe (black line) lies within the predicted range.

Observed ⁷Li abundance

⁴He
Deuterium (²H)
³He
⁷Li

Fraction of mass in isotope

Density of normal matter in today's Universe (kg/m³)

Figure 19.18 *Calculations of the abundances of the products of Big Bang nucleosynthesis, plotted against the density of normal matter in today's Universe. Big Bang nucleosynthesis correctly predicts the amount of these isotopes found in the Universe today.*

were high enough for nuclear reactions to take place. Just as we can use our knowledge of nuclear physics to calculate the nuclear reactions occurring in the interiors of stars, we can use this knowledge to calculate the nuclear reactions that took place in the early Universe. Collisions between protons in the early Universe built up low-mass nuclei, including deuterium (heavy hydrogen) and isotopes of helium, lithium, beryllium, and boron. The formation of new elements in this early nuclear brew, called **Big Bang nucleosynthesis,** determined the final chemical composition of the matter that emerged from the hot phase of the Big Bang.

We have previously discussed how differences in the abundances of the products of *stellar* nucleosynthesis help us track the chemical evolution of the Universe and the history of star formation. These ideas played an important role in our discussion of the Milky Way in Chapter 18, and will come to the fore again in our later discussion of galaxy formation. Here we focus on the products of *Big Bang* nucleosynthesis.

The amounts of various elements that formed from Big Bang nucleosynthesis depended in detail on the temperature and density of matter in the early Universe. **Figure 19.18** shows the calculated predictions of Big Bang nucleosynthesis

Big Bang theory predicts a Universe that is 24% helium by mass. This is observed.

plotted as a function of the present-day density of normal (luminous) matter in the Universe. The first thing that we see from this figure is that about 24% of the mass of the normal matter formed in the early Universe should have ended up in the form of the very stable isotope ⁴He, regardless of exactly what the density of matter in the Universe was. Indeed, when we look about us in the Universe today, we find that about 24% of the mass of normal matter in the Universe is in the form of ⁴He, in good agreement with the prediction of Big Bang nucleosynthesis.

Unlike helium, the abundances of most of the isotopes formed in the Big Bang depended sensitively on the density of normal matter in the Universe. Beginning with the amounts of isotopes such as deuterium (²H) and ³He found in the Universe (shown as horizontal bands in Figure 19.18) we can ask what the density of normal matter in today's Universe must be for these isotopes to have been formed in the Big Bang. This prediction is shown as the yellow vertical band in the figure. We have talked a great deal about the ways astronomers measure the amount of normal matter in the Universe. These measurements give a value of about 3×10^{-28} kg/m³ for the average density of normal matter in the Universe today. This value, shown as a vertical black line in Figure 19.18, lies well within the predicted range. Once again, the agreement is remarkable. Turning this around, we can begin with an observation of the amount of normal matter in and around galaxies, then use our understanding of the Big Bang to calculate what the chemical composition emerging from the Big

Abundances of other low-mass elements are also consistent with the Big Bang model.

Bang should have been. When we do the answer we get agrees remarkably well with the amounts of these elements we find in nature.

Two other points are worth noting here. The first is that no elements more massive than boron could have been formed in the Big Bang. Reactions such as the triple-alpha process, which forms carbon in the interior of stars, simply would not work under the conditions existing in the early Universe. That is how we know that all the more massive elements in the Universe, including the atoms making up the bulk of our planet and of ourselves, must have formed in subsequent generations of stars. The second point is that the agreement between the observed density of normal matter in the Universe and the abundances of light elements also provides a powerful constraint on the nature of the dark matter dominating the mass in the Universe. Dark matter (see Chapter 17) *cannot* consist of normal matter made up of neutrons and protons; if it did, the density of neutrons and protons in the early Universe would have been much higher, and the resulting abundances of light elements in the Universe would have been much different from what we actually observe.

Heavy elements do not come from the Big Bang but from stellar evolution.

Big Bang nucleosynthesis is inconsistent with dark matter being composed of protons and neutrons.

19.5 THE UNIVERSE HAS A DESTINY AND A SHAPE

We live in an expanding Universe, but will that expansion continue forever? This is clearly one of the great questions of modern cosmology. What is the fate of the Universe? The answer depends in part on the amount of mass the Universe contains. Think back on our discussion of escape velocity in Chapter 3. The fate of a projectile fired straight up from the surface of the Moon depends on its speed. As long as the speed is less than the *escape velocity* from the Moon (2.4 km/s), gravity will eventually stop the rise of the projectile and pull it back to the Moon's surface. However, if the speed of the projectile is greater than the Moon's escape velocity, then gravity will lose. While the projectile will slow, it will never stop. It will escape from the Moon entirely.

Just as the gravity of the Moon pulls on a projectile, slowing its climb, so the gravity of the mass in the Universe acts to slow its expansion. If there is enough mass in the Universe, then gravity will be strong enough to stop the expansion. If that is the case, then the Universe will slow, stop, and eventually collapse in on itself in a catastrophic "Big Crunch." On the other hand, if there is not enough mass, then the expansion of the Universe may slow, but it will never stop. The Universe will expand forever.

Gravity acts to slow the expansion of the Universe.

The escape velocity from a planet is determined by the planet's mass and radius. The "escape velocity" of the Universe is also determined by its mass and size—specifically, its average *density*. If the Universe is denser on average than some particular value, called the **critical mass density,** then we expect gravity will be strong enough to eventually stop and reverse the expansion. If the Universe is less dense than this value, we expect gravity will be too weak, and the Universe will expand forever.

The faster the Universe is expanding, the more mass is needed to turn that expansion around. For that reason, the critical mass density depends on the value of H_0. Assuming that $H_0 = 22$ km/(s Mly), and that gravity is the only thing we have to worry about, the critical density has a value of 8×10^{-27} kg/m^3. Rather than trying to keep track of such awkward numbers, we will instead talk about the *ratio* of the actual density of the Universe divided by its critical density. We call this ratio Ω_{mass} (pronounced "omega sub mass"). Note that Ω_{mass} is dimensionless (has no units).

We can follow the expansion of different possible universes in **Figure 19.19,** which shows a plot of the scale factor R_U versus time for different values of Ω_{mass}. (In these plots we have assumed that gravity alone controls the fate of the Universe, an assumption that we will revisit in the following section.) If Ω_{mass} in a universe controlled by gravity alone is greater than 1, then gravity is strong enough to turn the expansion around. Like a projectile fired from the Moon at less than the escape velocity, the expansion of such a universe will slow and eventually stop, then the universe will fall back in on itself. Conversely, if Ω_{mass} in a possible universe is less than 1, that universe will expand forever. The dividing line, where Ω_{mass} equals 1, corresponds to a universe that expands more and more slowly forever, but never quite turns around.

Ω_{mass} determines the fate of a Universe governed exclusively by gravity.

Until the closing years of the 20th century, most astronomers thought that was all there was to the question of expansion and collapse. Great efforts were focused on carefully measuring the mass of galaxies and assemblages of galaxies. As we have seen time and again,

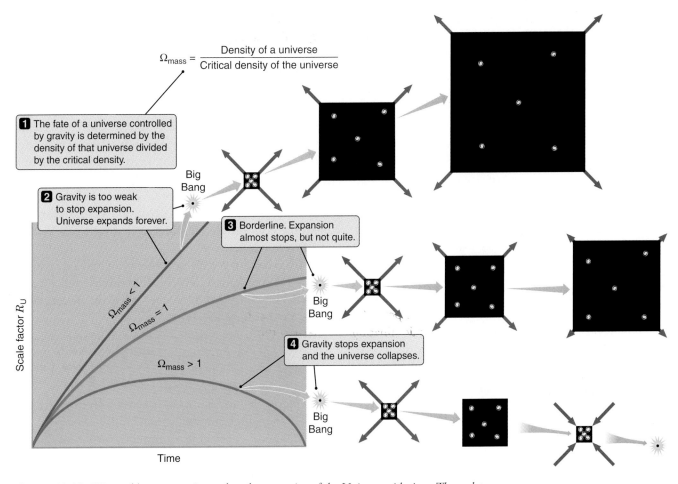

Figure 19.19 *We would expect gravity to slow the expansion of the Universe with time. These plots show the changing scale factor of the Universe as a function of time for three possible scenarios, depending on the density of mass in the Universe. (These plots ignore any cosmological constant.)*

measuring masses of objects in the Universe is tricky. When we calculate Ω_{mass} using just the luminous matter we see in galaxies and groups of galaxies, we get a value for Ω_{mass} of about 0.02. There is about 10 times as much dark matter as normal matter in galaxies, so adding in the dark matter in galaxies pushes the value of Ω_{mass} up to about 0.2. Finally, when we include the mass of dark matter *between* galaxies (a subject we will return to in the next chapter), Ω_{mass} could increase to 0.3 or higher. In other words, by our current accounting, there is perhaps about a third as much mass in the Universe as we would expect it should take to stop its expansion.

REVIVING EINSTEIN'S "GREAT BLUNDER"

As they were closing in on a good value for Ω_{mass}, it seemed to astronomers that our understanding of the expansion of the Universe was almost complete. Then the other shoe dropped.

If the expansion of the Universe has been slowing with time, as we might expect, then when the Universe was young, it must have been expanding more rapidly than it is today. That is a prediction that we can go to telescopes and check. Objects that are very far away (so that we see them as they were long ago) should have larger velocities than we would expect if the Hubble expansion were carrying on at the same pace since the Universe was young.

In the closing years of the 20th century, some groups of astronomers began using tools such as the *Hubble Space Telescope* and the giant Keck telescopes in Hawaii to test this prediction. They measured the brightnesses of standard candles (in particular, Type I supernovae) in very distant galaxies, and compared those brightnesses with expected brightness based on the redshifts of those galaxies. The early findings of these studies have set the astronomical community abuzz. Rather than showing that

There is evidence that the expansion of the Universe is accelerating.

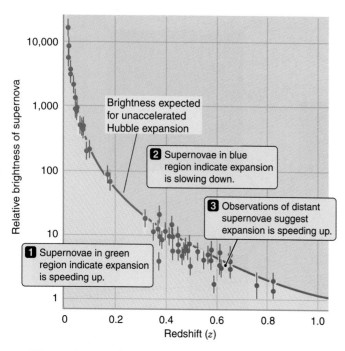

Figure 19.20 *Observed brightness of Type Ia supernovae, plotted as a function of their redshift, z. If Hubble expansion had remained constant throughout the history of the Universe, the observations should fall along the line shown. Surprisingly, the observations generally lie below the line, suggesting that the redshifts are too small for their distance—evidence that the Universe is expanding* faster *today than in the past.*

the expansion of the Universe has slowed down over time, the data indicate that it is *speeding up*!

Figure 19.20 shows the data indicating that the expansion of the Universe is speeding up. It is important to emphasize that these results are preliminary. The observations on which they are based are extremely difficult to carry out, and many different factors have to be taken into account when interpreting the data. Even so, the fact that similar findings have been obtained independently by different groups of astronomers lends credence to the results. As this book goes to press, this question is a hot topic that may be resolved by the time you read this.

How could it be that the rate of expansion of the Universe has *increased* over time? For this to be the case, there would have to be some force other than gravity at work to push the Universe outward. Physicists have some ideas about how such a hypothetical repulsive force might originate. These are related to theories from particle physics concerning the nature of what we call the **vacuum.** Normally we think of a vacuum as "empty space," but it turns out that even empty space has some very interesting physical properties. Discussion of these ideas (which are related to Hawking radiation from black holes, discussed in Chapter 16) will have to

wait until Chapter 20, where we will take up the question of the origin of physical law. For now, we turn our attention to the effects that such a force would have.

The idea of a repulsive force opposing the attractive force of gravity is not new. When Einstein used his newly formulated equations of general relativity to calculate the structure of space-time in the Universe, he was greatly troubled. The equations clearly indicated that any Universe containing mass could not be static, any more than a ball can hang motionless in the air. However, this was over a decade before Hubble did his epic work on the expansion of the Universe, and the conventional wisdom at that time was that the Universe is static.

In order to force his new theory of general relativity to allow for a static Universe, Einstein inserted a "fudge factor" into his equations. He called this fudge factor the **cosmological constant,** which we write as Ω_Λ.[3] The cosmological constant acts as a repulsive force in the equations, opposing gravity and allowing galaxies to remain stationary despite their mutual gravitational attraction. The cosmological constant was Einstein's version of the magician's trick that allows a subject to hang apparently unsupported in midair.

> **Einstein invented a cosmological constant to oppose gravity in a static Universe.**

When Hubble announced his discovery that the Universe is expanding, Einstein realized his mistake. The unmodified equations of general relativity demand that the structure of the Universe be dynamic. Instead of doctoring them with the inclusion of Ω_Λ, Einstein realized that he should have *predicted* that the Universe must either be expanding or contracting with time. What a coup it would have been for his new theory to successfully predict such an amazing and previously unsuspected result. He called the introduction of his fudge factor, the cosmological constant, the "greatest blunder" of his career as a scientist. It is ironic that, with the new results on the brightness of Type I supernovae, "Einstein's greatest blunder" has returned to center stage. The repulsive force represented by the infamous Ω_Λ in Einstein's equations is just what is needed to describe a Universe that is expanding at an ever accelerating rate.

How does the possibility of a non-zero value for Ω_Λ affect the possible fates that await our Universe? If Ω_Λ is not zero, the fate of the Universe is no longer controlled exclusively by Ω_{mass}. If there is something effectively pushing outward from within the Universe,

---

[3]Einstein probably never wrote Ω_Λ, referring instead to the related quantity Λ.

adding to its expansion, then gravity will have a harder time turning the expansion around. In that case, the mass needed to halt the expansion of the Universe would be greater than the critical mass discussed above.

Figure 19.21 shows plots of scale factor versus time, similar to those shown in Figure 19.19, but now we have included the effects of a nonzero cosmological constant. The evolution of a universe that collapses back on itself (one in which $\Omega_{mass} > 1$) is similar, regardless of whether Ω_Λ is zero or not. In contrast, if a universe with a nonzero cosmological constant expands forever, its evolution will look drastically different from any universe in which Ω_Λ is zero. As a universe expands, gravity gets weaker and weaker because the mass is more and more spread out. Unlike gravity, however, the effect of the cosmological constant becomes *greater*. While a universe is young and compact, gravity is strong enough to dominate the effect of the cosmological constant. Unless gravity is able to turn the expansion around, however, the cosmological constant wins

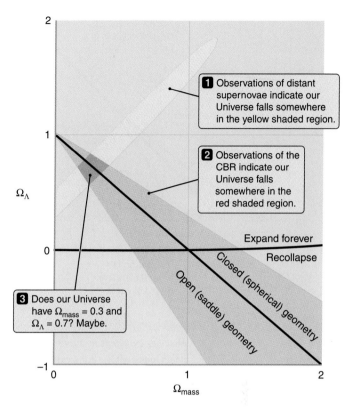

Figure 19.22 *Values of Ω_Λ and Ω_{mass} allowed by current observations from different sources (Type Ia supernovae, measurements of mass in galaxies and clusters, and detailed observations of the structure of the CBR). These sources together suggest that the best current estimate for Ω_Λ is about 0.7, which means the expansion of the Universe is accelerating.*

out in the end, causing the expansion to continue accelerating forever.

So which of these fates awaits our Universe? **Figure 19.22** shows the range of values for Ω_{mass} and Ω_Λ allowed by current observations. Assuming that these observations are correct, the values for Ω_{mass} and Ω_Λ are about 0.3 and 0.7, respectively. It appears that the expansion of our Universe is *already* accelerating under the dominant effect of the cosmological constant.

THE AGE OF THE UNIVERSE

Our values for Ω_{mass} and Ω_Λ not only affect our predictions for the future of the Universe. They also influence our interpretation of the past. **Figure 19.23** shows plots of the scale factor of the Universe versus time. Measurement of the Hubble constant H_0 tells us how fast the Universe is expanding *today*. That is, it tells us the *slope* of the curves in Figure 19.23 *at the current time*. As we saw earlier, if the expansion of the Universe has not changed in time, then the plot of R_U versus time is the straight red line in Figure 19.23. The age of the Universe in this case is equal to the Hubble time,

Figure 19.21 *Plots of scale factor R_U, versus time for cosmologies with and without a cosmological constant, Ω_Λ. If there is enough mass in a universe, gravity may still overcome the cosmological constant and cause that universe to collapse. Any universe without enough mass to eventually collapse will instead end up expanding at an ever-increasing rate.*

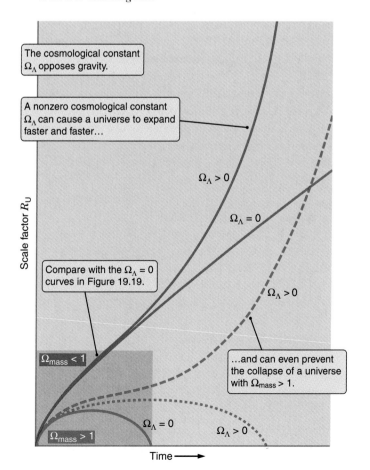

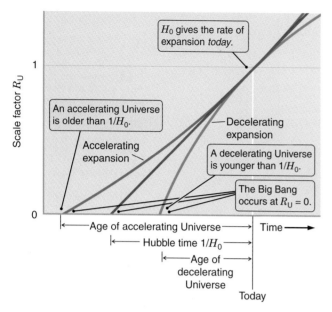

Scale factor R_U

H_0 gives the rate of expansion *today*.

An accelerating Universe is older than $1/H_0$.

Accelerating expansion

Decelerating expansion

A decelerating Universe is younger than $1/H_0$.

The Big Bang occurs at $R_U = 0$.

Age of accelerating Universe — Time →

Hubble time $1/H_0$

Age of decelerating Universe

Today

Figure 19.23 *A plot of the scale factor,* R_U, *versus time for three possible universes. If the Universe has expanded at a constant rate, then its age is equal to the Hubble time,* $1/H_0$. *If the expansion of the Universe has slowed with time, it is younger than a Hubble time. If the expansion of the Universe has sped up with time, then the Universe is older than* $1/H_0$.

$1/H_0$. If the expansion of the Universe has been slowing down with time (the green line in Figure 19.23), then the Universe is actually *younger* than the Hubble time. (The curve crosses $R_U = 0$ at a point more recent than $1/H_0$.) If, on the other hand, the expansion of the Universe has been speeding up with time (blue line), then the true age of the Universe is greater than the Hubble time.

The current measured value for H_0 (22 km/(s Mly)) corresponds to a Hubble time ($1/H_0$) of about 13.6 billion years. If expansion of the Universe has slowed over

time, the Universe is actually younger than 13.6 billion years. This would be a problem if the measured ages of globular clusters—13 billion years—is correct. (Globular clusters clearly cannot be older than the Universe that contains them!) On the other hand, if the expansion of the Universe has sped up with time, as suggested by the observations of Type I supernovae, then the Universe is around 15 billion years old—comfortably older than measurements of the ages of globular clusters.

THE UNIVERSE HAS A SHAPE

By this time, the idea that space-time is described by general relativity is getting to be a comfortable friend. Space is a "rubber sheet" that has stretched outward from the Big Bang. In Chapter 16 we saw that the rubber sheet of space is also curved by the presence of mass. In that chapter we saw how the shape of space around a massive object can be detected through changes in simple geometrical relationships, such as the ratio of the circumference of a circle to its radius, or the sum of the angles in a triangle. If the mass of a star, planet, or black hole causes a distortion in the shape of space, then should not the mass of all of the galaxies and dark matter in the Universe also distort the shape of the Universe *as a whole?* The answer is yes.

There are three basic shapes that our Universe might have. Which shape actually describes the Universe is determined by the sum of Ω_{mass} and Ω_Λ. Continuing with the rubber-sheet analogy, the first possibility, corresponding to $\Omega_{mass} + \Omega_\Lambda = 1$, is that we live in a **flat universe.** A flat universe is described overall by the rules of Euclidean geometry. As shown in **Figure 19.24(a)**, circles in a flat universe have a circumference of 2π times their radius, and triangles contain angles whose sum is 180°. A flat universe stretches on forever.

Figure 19.24 *Two-dimensional representations of the possible geometries that space can have in a Universe.* (a) *In a flat universe, Euclidean geometry holds. Triangles have angles that sum to 180°, and the circumference of a circle equals 2π times the radius. In (b), an open universe, or (c), a closed universe, these relationships are no longer correct over very large distances.*

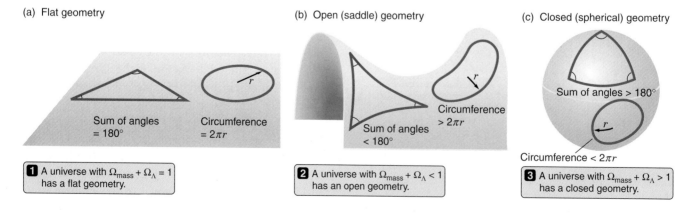

(a) Flat geometry

Sum of angles = 180°

Circumference = $2\pi r$

1 A universe with $\Omega_{mass} + \Omega_\Lambda = 1$ has a flat geometry.

(b) Open (saddle) geometry

Sum of angles < 180°

Circumference > $2\pi r$

2 A universe with $\Omega_{mass} + \Omega_\Lambda < 1$ has an open geometry.

(c) Closed (spherical) geometry

Sum of angles > 180°

Circumference < $2\pi r$

3 A universe with $\Omega_{mass} + \Omega_\Lambda > 1$ has a closed geometry.

The second possibility is that the geometry of the Universe is shaped something like the surface of a saddle **(Figure 19.24b).** This type of universe, in which $\Omega_{mass} + \Omega_\Lambda < 1$, is also infinite, is referred to as an **open universe.** In this type of universe, the circumference of a circle is greater than 2π times its radius, and triangles contain less than 180°.

The final possibility, in which $\Omega_{mass} + \Omega_\Lambda > 1$, is a universe with a geometry shaped like the surface of a sphere **(Figure 19.24c).** The geometric relationships on a sphere are similar to those in the vicinity of a massive object, as discussed in Chapter 16. The circumferences of circles on a sphere are less than 2π times their radii, and triangles contain more than 180°. This possibility is called a **closed universe,** because space is finite and closes back on itself.

Again we face the question, Which of these shapes describes the Universe in which we live? The measurements are difficult. Even so, as seen in Figure 19.22, $\Omega_{mass} + \Omega_\Lambda$ is close to 1 (0.3 + 0.7), meaning that our Universe is very nearly flat.

19.6 PROBLEMS LEAD TO NEW UNDERSTANDING

It is remarkable that Big Bang cosmology makes so many correct predictions about the properties of the Universe in which we live. A century ago, astronomers were struggling just to get a handle on the size of the Universe. Today we have a comprehensive theory that ties together many diverse facts about nature, ranging from the constancy of the speed of light, to the proper-

ties of gravity, to the motions of galaxies, to the origins of the very atoms of which we are made. The case for the Big Bang is compelling. Even so, as our knowledge of the expansion of the Universe has grown and our observations of the cosmic background radiation have improved, a number of puzzles have arisen. The solution to these puzzles has forced us to consider some remarkable ideas about how our Universe expanded when it was very young.

THE UNIVERSE IS MUCH TOO FLAT

The first problem that we run into when observing our Universe is that the Universe is too flat. In fact, the Universe is much too close to being exactly flat for this to have happened by chance. To see why this is a problem, imagine a model universe in which $\Omega_\Lambda = 0$. (This is a reasonable approximation for the very early days of *any* possible universe. When a universe is very young, it is also very dense, and Ω_{mass} is all that matters.) As our model universe expands out of its own version of the Big Bang, its density falls. At the same time, because the universe is expanding more slowly, the critical density needed to eventually stop the expansion falls as well. If Ω_{mass} is *exactly* 1, the decline in the actual density and the decline in the critical density go hand in hand: the ratio between the two, Ω_{mass}, remains 1 for all time, as shown by the middle curve in **Figure 19.25.** A universe that starts out perfectly flat *stays* perfectly flat.

On the other hand, a universe that does *not* start out perfectly flat has a very different fate. If a universe

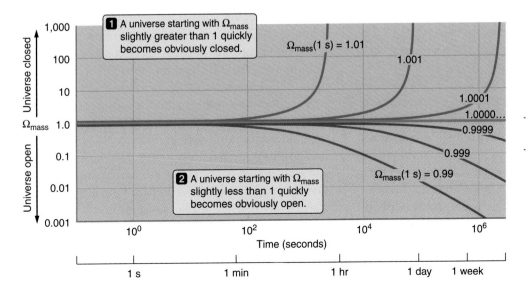

Figure 19.25 *Universes in which Ω_{mass} has slightly different values at an age of 1 second. Notice that even tiny differences from $\Omega_{mass} = 1$ are rapidly amplified as the model universe expands. To have a value of Ω so close to 1 today, our Universe must have started out with a value of Ω exquisitely close to 1. (The value of Ω_Λ does not effect the calculations shown.)*

started out with Ω_{mass} even slightly greater than 1, its expansion would slow more rapidly than that of the flat universe, meaning that less and less density would be required to stop the expansion. At the same time, the actual density would be falling less rapidly than in the flat universe. This disparity between the actual density of the universe and the critical density of the universe would increase, causing the ratio between the two, Ω_{mass}, to skyrocket. This is shown by the curves that climb toward the top of Figure 19.25. A universe that starts out even *slightly* closed rapidly becomes *obviously* closed.

> **The curvature of an expanding universe rapidly increases.**

Conversely, if a universe started with Ω_{mass} even a tiny bit less than 1, the expansion would slow *less* rapidly than in a flat universe. As time passed, more and more mass would be required for gravity to stop the too-rapidly-expanding universe. At the same time, the actual density of the universe would be dropping faster than in a flat universe. In this case, Ω_{mass} (the ratio between the actual density and the critical density) would plummet, leading to the curves that dive toward the bottom of Figure 19.25.

Adding Ω_Λ to the picture makes the math a bit more complex, but it does not change the basic results. Try balancing a razor blade on its edge. If the blade is tipped just a tiny bit one direction, it quickly falls that way. If the blade is tipped just a tiny bit the other direction, it quickly falls in the other direction instead. By all rights, we would expect our Universe to be either *obviously* open or *obviously* closed analogous to the tipped razor blade. We find ourselves instead in a Universe in which $\Omega_{mass} + \Omega_\Lambda$ is so close to 1 that we have difficulty telling which way the razor blade is tipped at all! Discovering that $\Omega_{mass} + \Omega_\Lambda$ is extremely close to 1 after 13 billion years or more is like balancing a razor blade on its edge and coming back 10 years later to find that it still has not tipped over!

> **The Universe is too flat for it to be chance.**

For the present-day value of $\Omega_{mass} + \Omega_\Lambda$ to be as close to 1 as it is, $\Omega_{mass} + \Omega_\Lambda$ could not have differed from 1 by more than 1 part in 100,000 when the Universe was 2,000 years old. When the Universe was 1 second old, it had to be flat to 1 part in 10 billion. At even earlier times, it had to be much flatter still. This is simply too special a situation to be due to chance—a fact referred to in cosmology as the **flatness problem.** *Something* about the early Universe must have *forced* $\Omega_{mass} + \Omega_\Lambda$ to have a value that was incredibly close to 1.

THE COSMIC BACKGROUND RADIATION IS MUCH TOO SMOOTH

The second problem faced by our cosmological models is that the cosmic background radiation is uncomfortably smooth. Following the discovery of the CBR in the 1960s, many observers turned their attention to mapping this background glow. At first they were reassured as result after result showed that the temperature of the CBR was remarkably constant, regardless of where one looked in the sky. Yet over time this strong confirmation of Big Bang cosmology turned instead into a puzzle that challenged our view of the early Universe. Once we remove our motion relative to the CBR from the picture, the CBR is not just smooth, it is *too* smooth.

Why should we expect the CBR to be less uniform than it is? To understand the answer, we need to shift our attention from the very large to the very small. In Chapter 4 we discovered the bizarre world of quantum mechanics that shapes the world of atoms, light, and elementary particles. When the Universe was extremely young, it was so small that quantum mechanical effects played a role in shaping the structure of the Universe as a whole. In particular, the early Universe was subject to the quantum mechanical uncertainty principle. The uncertainty principle says that as we look at a system at smaller and smaller scales, the properties of that system become less and less well determined. This applies whether we are talking about the properties of an electron in an atomic orbital, or about our entire Universe at the time when it would fit within the size of an atom.

> **We expect nonuniformities in the early Universe.**

As this is being written, your guide on this journey is sitting on the beach looking out across the ocean. Looking off into the distance, the surface of the ocean is smooth and flat. The horizon looks almost like a geometrical line. Yet the apparent smoothness of the ocean as a whole hides the tumultuous structure present at smaller scales, where waves and ripples upon waves fluctuate dramatically from place to place. In similar fashion, quantum mechanics says that as we look at smaller and smaller scales in the Universe, conditions *must* fluctuate in unpredictable ways. In particular, quantum mechanics says that the smaller the Universe we consider (that is, the earlier in the history of our Universe that we go), the more dramatic those fluctuations become. When the Universe was young, it could *not* have been smooth. There must have been dramatic variations (that is, "ripples") in the density and temperature of the Universe from place to place.

If the Universe had expanded slowly, those ripples would have smoothed themselves out. But the Universe expanded much too rapidly for this. Different parts of the Universe could not have "communicated" with each other (telling each other to smooth out the ripples) rapidly enough to smooth these ripples out. So, when we look at the Universe today, we should see the fingerprint of those early ripples imprinted on the cosmic background radiation—but we do not. The fact that the CBR is so smooth is referred to as the **horizon problem** in cosmology, because different parts of the Universe are too much like other parts of the Universe that should have been "over their horizon" and beyond the reach of any signals that might have smoothed out the early quantum fluctuations.

The CBR is smoother than the early Universe should have been.

"INFLATION" SOLVES THE PROBLEMS

In the early 1980s, **Alan Guth** (1947–) offered a solution to the flatness and horizon problems of cosmology. Guth suggested that for a brief time the young Universe underwent a period of **inflation** during which the Universe itself expanded at a rate *far* in excess of the speed of light. Like so many things we have seen on our journey, the numbers used to describe inflation are far beyond our ability to grasp intuitively. Between ages of about 10^{-35} s and 10^{-33} s, the scale factor R_U of the Universe increased by a factor of at least 10^{30} and perhaps much more. In that incomprehensibly brief instant, the size of the observable Universe grew from 10 trillionths the size of the nucleus of an atom, to a region about 3 meters across. That is like a grain of very fine sand growing to the size of today's *entire Universe*—all in a billionth the time that it takes light to cross the nucleus of an atom! (At this point you may well be saying to yourself, "Wait a minute! What happened to all that business about nothing traveling faster than the speed of light?" Inflation does not violate the rule that no signal can travel *through* space at greater than the speed of light. During inflation space *itself* expanded so rapidly

Inflation was a period of intense expansion of the Universe.

that the distances between points in space increased at greater than the speed of light.)

To understand how inflation solves the flatness and horizon problems of cosmology, imagine that you are an ant living in the two-dimensional universe defined by the surface of a golf ball, as shown in **Figure 19.26**. Your universe would have two very apparent characteristics. First, it would be obviously curved. If you were to walk around the circumference of a circle in your two-dimensional universe and then measure the radius of the circle, you would find the circumference to be less than 2π times the radius. If you were to draw a triangle in your universe, the sum of its angles would be greater than 180°. The second obvious characteristic would be the dimples, approximately a millimeter deep, on the surface of the golf ball.

Now imagine how this would change if your golf ball universe suddenly grew to the size of Earth. (Granted, this change is nowhere near comparable to the inflation experienced by the real Universe, but you can get the idea.) First, the curvature of your Universe would no longer be apparent. An ant walking along the surface of Earth would be hard pressed to tell that Earth is not flat. The circumference of a circle would be 2π times its radius, and there would be 180° in a triangle. In fact, it took us most of our history as a species to realize that Earth really is round. In the case of inflationary cosmology, the Universe after inflation will be extraordinarily flat (that is, with $\Omega_{mass} + \Omega_\Lambda$ extraordinarily close to 1), *regardless* of what the geometry of the Universe was before inflation. Because the Universe was inflated by a factor of at least 10^{30}, $\Omega_{mass} + \Omega_\Lambda$ immediately after inflation must have been 1 to within 1 part in 10^{30}, which is flat enough for $\Omega_{mass} + \Omega_\Lambda$ to remain close to 1 today. Today's Universe is not flat by chance. It is flat because *any* universe that underwent inflation would emerge with a value for $\Omega_{mass} + \Omega_\Lambda$ that was within a gnat's eyelash of 1.

Inflating the size of a sphere makes it seem flatter.

So much for the flatness problem. What about the horizon problem? When our golf ball universe inflated to the size of Earth, the dimples that covered the surface of the golf ball were stretched out as well. Instead of being a millimeter or so deep and a few millimeters across, these dimples now are only an atom deep but are hundreds of kilometers across. Again, our ant would be hard pressed to tell that any dimple was there at all. In the case of the real Universe, inflation took the large fluctuations in conditions caused by quan-

Huge expansion also smoothes out inhomogeneities.

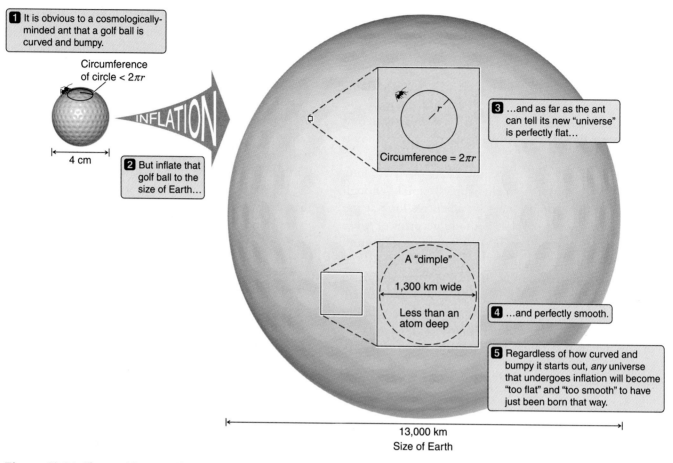

Figure 19.26 *If a round lumpy golf ball were suddenly inflated to the size of Earth, it would seem extraordinarily flat and smooth to an ant on its surface. Similarly, following inflation, any universe would be both extremely flat and extremely smooth, regardless of the exact geometry and irregularities it started out with.*

tum uncertainty in the preinflationary Universe and stretched them out so much that they are unmeasurable in today's postinflationary Universe. The slight irregularities that we see in the CBR are the faint ghosts of quantum fluctuations that occurred as the Universe inflated.

An early era of inflation in the history of the Universe offers a handy way of solving the horizon and flatness problems, but why would the real Universe have the temerity to do such a thing? It seems quite remarkable that the Universe should undergo a period during which it expanded at such an "astronomical" rate. But there are many remarkable things about the Universe. It is remarkable that space-time has a shape.

It is remarkable that light is both a wave and a particle. It is remarkable that galaxies and stars and planets exist at all.

The Universe is pretty amazing.

We live in a Universe of structure. Expansion, inflation, and the physical laws that govern everything are all part of this structure, as is the distribution of stars and galaxies that we see around us today. The fundamental question we face at this stage of our journey is *not* simply, "Why did the Universe undergo inflation?" but rather, "How did structure in the Universe arise?" It is to this question that we now turn.

SEEING THE FOREST THROUGH THE TREES

The journey that we have taken in *21st Century Astronomy* has involved many amazing discoveries. We have peered into the heart of the Sun, watched as planetary systems formed, and witnessed the violent death throes of massive stars. We have seen our ideas of cause and effect shattered by the quantum mechanical world of atoms, and have come to accept that space and time are joined in a four-dimensional fabric of space-time that is bent and even torn by the presence of mass. Yet even when witnessed within the context of such a remarkable journey of discovery, what we see when we look at the structure of the Universe itself is truly mind-boggling.

You never know when you get up in the morning if this will be one of those rare days when you happen upon some extraordinary experience that leaves you looking at the world in a different way. So it was for Edwin Hubble. Hubble, who also was the first to prove the true nature of galaxies, was certainly at the right place at the right time in his career. He was doing as scientists do, using the cutting-edge tools of his day and following his instincts about what questions might turn out to have interesting answers. His investigation into the basic properties of the newly recognized class of objects called galaxies took an amazing and unexpected turn. With one graph, plotting the redshift of galaxies against his estimates of their distances, he forever changed our view of the Universe. What must it have felt like to be Edwin Hubble, looking at his data and realizing for the first time what the implications of those data were? What would it mean to be the first human to glimpse the true history of the Universe?

We live in a Universe in which the space-time described by general relativity has been expanding outward for perhaps 14 or 15 billion years, an expansion that can be traced back to a single moment when space-time and all that it contains came into existence. The Big Bang is a theory, but one of the lessons we have learned along the way is that a well-tested and well-corroborated scientific theory is the closest humans can ever come to certain knowledge. The predictions of the Big Bang have been borne out time and again, in everything from observations of the cosmic background radiation, to the predictions the theory makes about the abundances of chemical elements, to the fact of the expansion of the Universe itself. A full accounting of the experimentally verified predictions of the Big Bang theory could easily fill a huge book. By any reasonable definition of the term, the Big Bang is a fact.

Again and again during our exploration we have stopped to marvel at how vast the Universe is. We have pondered the thousands of stars in the night sky that can be seen with the naked eye, and realized they are but the tiniest fraction of the hundreds of billions of stars that make up our Milky Way. Then we faced the task of comprehending a Universe in which our Milky Way is itself but one of hundreds of billions of galaxies spread out across a Universe so large that even light takes nearly 30 billion years to cross it. Now we are forced to confront the idea that at one time, everything we see in that vast Universe, including space itself, was contained in a volume that was as small relative to the nucleus of an atom as we are as compared with the enormity of today's Universe. All we need to do to be reminded of the reality of that hot, dense early Universe is look at the faint microwave glow that fills the sky. Looking in the other direction in time, we have considered the fate of the Universe, and found it equally startling. While the jury is still out, the best models of the Universe as of the beginning of the 21st century tell of a future in which the expansion of the Universe continues forever at an ever accelerating pace.

Seldom have novelists, theologians, or philosophers dared dream of anything so extraordinary as the Universe of modern cosmology. Yet we arrive at our conclusion not by fantasy or conjecture, but by following a path laid down by hard-won observation and well-tested physical law.

In this chapter we have set the stage. We have discovered the shape of the canvas on which our existence is painted. Yet our world is not defined by space-time alone. The trees and planets and stars are still there for us to see. We are part of a Universe that is filled with structure. How did that structure arise? What are the acts in the grand play that continues on the cosmic stage? The time has come to tell the story from the beginning—to start as close to the Big Bang as our physics will take us—and run the clock forward, watching as the structure and complexity of today's Universe emerges.

Hold on to your hats. We have found the rabbit hole. Now we shall see how deep it goes.

STUDENT QUESTIONS

THINKING ABOUT THE CONCEPTS

1. Why are we not able to use the measured radial velocities of galaxies in the Local Group to evaluate the Hubble constant (H_0)?

2. Imagine you are standing in the middle of a dense fog. Would you describe your environment as isotropic? Would you describe it as homogeneous? Explain your answers.

3. Knowing that you are studying astronomy, a curious friend asks where the center of the Universe is located. You smile and answer "right here and everywhere." Give a detailed explanation for why you would give this answer to your friend.

4. The general relationship between radial velocity (v_r) and redshift (z) is $v_r = cz$. This simple relationship fails, however, for very distant galaxies with large redshifts. Explain why.

5. Why is Earth not expanding together with the rest of the Universe?

6. As astronomers extend their distance ladder beyond 100 Mly, they change their standard candle from Cepheid variable stars to Type I supernovae. Explain why this is necessary.

7. Very distant galaxies have redshifts indicating recession velocities of 100,000 km/s or more, yet their true speeds are probably no more than a few hundred kilometers per second. Explain.

8. Describe the observational evidence suggesting that Einstein's "cosmological constant" (a repulsive force) may be needed to explain the historical expansion of the Universe.

9. If you could accurately measure a triangle and a circle drawn on the surface of Earth, you would find that the sum of the triangle's interior angles is more than 180 degrees and the circle's circumference is less than $2\pi r$. This is contrary to what you probably learned about triangles and circles in your introductory geometry class. Explain why this is so, and how it relates to a "spherical Universe."

10. During the period of inflation, the Universe may have briefly expanded at 10^{25} (ten trillion trillion) times the speed of light. Why did this not violate Einstein's special theory of relativity, which says that neither matter nor communication can travel faster than the speed of light?

APPLYING THE CONCEPTS

11. Hubble time ($1/H_0$) represents the age of a universe that has been expanding at a constant rate since the Big Bang. Assuming a H_0 value of 23 km/s/Mly and a constant rate of expansion, calculate the age of the Universe in years. Note that one year = 3.16×10^7 s and one light-year = 9.46×10^{12} km.

12. One of the most distant known quasars has a redshift $z = 5.82$, and a recession velocity of 287,000 km/s or about 96% of the speed of light.
   a. If $H_0 = 23$ km/s/Mly and the rate of expansion of the Universe is constant, how far away is this quasar?
   b. Assuming a Hubble time of 13 billion years, how old was the Universe at the look-back time of this quasar?
   c. What was the scale factor (R_U) of the Universe at that time?

13. The spectrum of a distant galaxy shows the Hα line of hydrogen ($\lambda_{rest} = 0.65628$ microns) at a wavelength of 0.98442 microns. Assume that $H_0 = 23$ km/s/Mly.
   a. What is the redshift (z) of this galaxy?
   b. What is its recession velocity (in km/s)?
   c. What is the distance of the galaxy (in Mly)?

14. The rest wavelength of the Lyα line of hydrogen is in the extreme ultraviolet region of the spectrum at 0.1216 microns.
   a. What would be the wavelength of this line in the spectrum of a quasar with a redshift $z = 5.82$?
   b. In what region of the spectrum would this red-shifted line be located?

15. Suppose we observe two galaxies, one at a distance 35 Mly with a radial velocity of 580 km/s, and another at a distance of 1,100 Mly with radial velocity of 25,400 km/s.
   a. Calculate the Hubble constant for each of these two observations.
   b. Which of the two calculations would you consider to be more trustworthy? Why?
   c. Estimate the peculiar velocity of the closer galaxy.
   d. If the more distant galaxy had this same peculiar velocity, how would that change your calculated value of the Hubble constant?

16. The Universe today has an average density $\rho = 3 \times 10^{-28}$ kg/m^3. What was the scale factor and age of the Universe when its average density was about the same as Earth's atmosphere at sea level ($\rho = 1.3$ kg/m^3)?

There is a theory which states that if ever anyone discovers exactly what the Universe is for and why it is here, it will instantly disappear and be replaced by something even more bizarre and inexplicable.

There is another theory which states that this has already happened.

DOUGLAS ADAMS (1952–2001)
HITCHHIKER'S GUIDE TO THE GALAXY

THE ORIGIN OF STRUCTURE

20.1 WHENCE STRUCTURE?

Throughout the ages, there have been as many different ideas about the origin of the Universe as there have been cultural traditions. Thoughts about what the Universe was like once upon a time have been part of the mythologies and traditions of all great civilizations. We live at a remarkable moment in history, when the nature of the early Universe has moved from philosophical speculation to the realm of scientific fact. All that we have to do to see the early Universe is point our microwave telescopes at the sky. The glow of the cosmic background radiation is there for anyone with the appropriate technology to see.

The early Universe was an extraordinary place—an expanding fireball that was far more uniform than the blue of the bluest sky on the clearest day. How different that Universe was from the Universe we see about us today! Today's Universe is a universe of stars and galaxies, viewed from the surface of a planet with oceans and mountains and uncountable species of living things. The contrast between these two realities cries out for explanation. How did we get from there to here? We come to one of the most philosophically intriguing questions in the history of human thought: what is the origin of structure?

KEY CONCEPTS

The Universe that emerged from the Big Bang was incredibly uniform, wholly unlike today's Universe of galaxies, stars, and planets. As our journey comes to its end, we tackle face-on the question that has been with us all along: Where does structure in the Universe come from? Here we find that complex structure is a natural, unavoidable consequence of the action of physical law in an evolving Universe:

* Matter and the fundamental forces of nature "froze out" of the uniformity of the expanding and cooling Universe moments after the Big Bang;
* Galaxies formed as slight ripples in the dark matter emerging from the Big Bang collapsed under the force of gravity, pulling in normal matter as well;
* Galaxies were drawn together by gravity to form the large groupings of galaxies we see today;
* Much lies ahead, for Earth, and for the Universe;
* Like planets, stars, and galaxies, life is another form of structure that evolved through the action of the physical processes that shape the Universe—the Universe may teem with life; and
* Evolution of structure through the action of physical law is *the* lynchpin that ties all of modern science together, and brings sense to what we see.

20.2 GALAXIES FORM GROUPS, CLUSTERS, AND LARGER STRUCTURES

Just as stars and clouds of glowing gas show us the structure of our Galaxy, it is the distribution of galaxies themselves that shows us the structure of our Universe. And just as gravity holds galaxies together, giving them their shape, it is gravity that shapes the Universe itself. No galaxy exists in utter isolation. The vast majority of galaxies are parts of gravitationally bound collections of galaxies. The smallest and most common of these are called **galaxy groups.** A galaxy group is an irregular collection of a few tens of galaxies, most of them dwarf galaxies. Our Milky Way is a member of the **Local Group,** which consists of two giant spirals (the Milky Way and the Andromeda Galaxy), along with close to 30 smaller dwarf galaxies in a volume of space about 4 Mly in diameter. Close to 98% of the galaxy mass in the Local Group resides in just its two giant galaxies.

Galaxies collect together into gravitationally bound groups and clusters.

Larger systems of galaxies, which can consist of hundreds of galaxies and often have a more regular structure than groups, are called **galaxy clusters.** Galaxy clusters are larger than groups, typically occupying a volume of space 10 to 15 Mly across, but clusters and groups are similar in composition. Giant galaxies dominate the mass of the far more numerous dwarf galaxies. One interesting difference among galaxy groups and clusters is that while spiral galaxies are common in most systems, elliptical galaxies are common in only about a quarter of groups and clusters. The Virgo cluster **(Figure 20.1a)**, located 53 Mly from Earth, is an example of the first type of cluster, while the Coma cluster **(Figure 20.1b)** is an example of a cluster that is dominated by giant elliptical and S0 galaxies.

Clusters and groups of galaxies themselves bunch together to form enormous **superclusters,** which contain tens of thousands or even hundreds of thousands of galaxies, and span regions of space typically more than 100 Mly in size. Our Local Group is part of the Virgo supercluster, which also includes the Virgo cluster.

Hubble's law is a powerful tool for mapping the distribution of galaxies in space. Using Hubble's law, all that we need to determine the distance to a galaxy is a single spectrum from which we can measure the galaxy's redshift. While it is easy in principle to measure the redshift of a galaxy, in practice it can be a time-consuming business. The first redshifts were measured from spectra recorded on photographic plates. Exposures of several hours were required to capture the feeble signal, and results rolled in at the breakneck pace of one or two redshifts per night of painstaking observing. As of 1975, redshifts had been measured for only about 1,000 of the hundred billion or so galaxies we can see. Since that time, there has been a happy marriage of large telescopes, new instruments (such as CCD detectors and

Galaxy redshift surveys measure distances to large numbers of galaxies.

Figure 20.1 (a) *The Virgo cluster contains a large fraction of spiral galaxies.* (b) *The larger, richer Coma cluster is dominated by elliptical galaxies.*

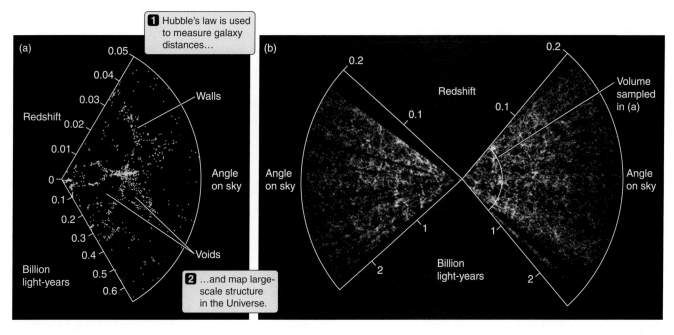

Figure 20.2 *Redshift surveys use Hubble's law to map the Universe. (a) The Harvard-Smithsonian Center for Astrophysics redshift survey, called "A Slice of the Universe," was the first to show that clusters and superclusters of galaxies are part of even larger scale structures. (b) The 2dF Galaxy Redshift Survey shows similar structures at even larger distances.*

spectrographs capable of observing many galaxies at once), and powerful computers that can process large amounts of data in rapid, automated fashion. By 1990 the redshifts of over 10,000 galaxies had been measured, and by the time you read this book, that number will be approaching 1 million galaxy redshifts.

The first large redshift survey was conducted by the Harvard Center for Astrophysics, which in 1986 presented the astronomical community with a "Slice of the Universe." **Figure 20.2(a)** shows what this slice looked like. The observations show that the clusters and superclusters, rather than being scattered randomly through space, are linked together in an intricate network of "filaments" and "walls." These concentrations of galaxies in turn surround large voids, regions of space that are largely devoid of galaxies. Clusters and superclusters are located within the walls and filaments. This structure is not peculiar to the "nearby" Universe. Subsequent surveys have looked at much larger volumes of space. **Figure 20.2(b)** shows the results of one such survey conducted using the Anglo-Australian Telescope in Siding Spring, Australia. Out as far as our observatories can currently measure, the Universe has a porous structure reminiscent of a sponge. Together, galaxies and

Large-scale structure fills the Universe.

the larger groupings in which they are found are referred to as **large-scale structure.**

20.3 GRAVITY FORMS LARGE-SCALE STRUCTURE

Large-scale structure cries out for explanation. Clearly it is the consequence of well-defined physical processes, but what processes? A number of ideas have been proposed. For example, early on it was suggested that voids were the result of huge expanding blast waves from tremendous explosions that might have occurred in the early Universe. The correct answer has turned out to be less fanciful, but far more satisfying. Large scale structure is the fingerprint of our old friend gravity.

At the close of the previous chapter, we learned that the slight ripples in the glow of the cosmic background radiation probably result from quantum mechanical variations that imprinted structure on the early Universe at the time of inflation. These variations provided the seeds from which galaxies and collections of galaxies grew.

Ripples in the Universe were the seeds of structure.

In our discussion of star formation in Chapter 14 we learned about gravitational instabilities. As illustrated in Figure 14.14, we begin with a molecular cloud with clumps inside it, and gravity then causes those clumps to collapse faster than their surroundings. If conditions are right, then what began as relatively minor variations in the density of a cloud are turned by gravity into stars. If you replace "molecular cloud" with "Universe," and "star" with "galaxy," you will have a good starting point for understanding the way galaxies formed from slight inhomogeneities in the distribution of matter following recombination.

It is one thing to say that galaxies and larger structures formed from gravitational instabilities that begin with slight irregularities in the early Universe. It is quite another to turn this statement into a real scientific theory with testable predictions. To make that step, we return to the same basic technique we have used to look into the centers of planets and stars, and to answer many other questions about things we cannot observe directly. We take the ideas we want to test, combine them with the laws of physics, construct a model, and then compare the predictions of that model with observations of the Universe.

Models show how gravity turns early seeds into large-scale structure.

To build a model of the formation of large-scale structure, we have to begin by making three key choices. First we have to decide what universe we are going to model. That is, what values of Ω_{mass} and Ω_{Λ} (see Chapter 19) are we going to assume? These are important in part because they determine how rapidly the Universe expands, and therefore how difficult it is for gravity to overcome this expansion in some region. The more rapidly a universe is expanding, or the less mass it contains, the more difficult it will be for gravity to pull material together into galaxies and larger scale structures.

The second thing we need to know to construct our model is what the early bumps in the density of the Universe looked like. How large and how concentrated were these early bumps? There are several ways to approach this question. One is to make use of observations of variations in the CBR. As this book is being written, new observations are just beginning to give us a better picture of what structure in the CBR looks like in detail. Upcoming space missions such as *MAP (Microwave Anisotropy Probe)* and *Planck* promise to provide better information yet.

A different way to approach this question is to look at models of inflation, which make predictions about the structure that will emerge following this episode of rapid expansion. These are among the predictions of the inflation model that will be tested in the years to come. These are especially important predictions to test because they serve to tie together the large-scale structure of today's Universe with our most basic ideas about what the Universe was like in the briefest instant after the Big Bang. While we still await final answers about the details of structure, we do know enough to say that the early Universe was more irregular on smaller (galaxy-sized) spatial scales than it was on the scales of the clusters, superclusters, filaments, and voids seen in Figure 20.2. An immediate consequence of this fact is that smaller structures (such as galaxies and even subgalactic clumps) formed first, while larger structures took more time to form. The idea of small structure forming first and larger structure forming later is referred to as **hierarchical clustering.** Hierarchical clustering has become one of the most important themes in our growing understanding of how structure in the Universe formed.

Smaller structures form first. Larger structures form later.

The third thing that we need to know if we are to model the formation of large-scale structure is a list of the types and amounts of ingredients the early Universe was made of. Specifically, we need to know the balance between radiation, normal matter, and dark matter. We also need to make some choice about the nature of the dark matter that we use in our model. More about this below.

Once we have made these three choices—the shape of the universe we are modeling, the way matter is distributed, and the nature and mixture of different forms of matter we are using—the rest is physics and calculations: we simply need to "turn the crank," as the expression goes. The question then becomes, What choices lead to a model universe that most resembles the real Universe in which we live?

GALAXIES FORMED BECAUSE OF DARK MATTER

Observations of nonuniformity in the CBR made with *COBE* show variations of about 1 part in 100,000. Models clearly show that such tiny variations at the time of recombination (when the Universe was about half a million years old) are far too slight to explain the structure that we see in today's Universe. Gravity is simply not strong enough to grow galaxies and clusters of galaxies from such poor "seeds." These models indicate that, for ripples in the

The ripples we see in the CBR are not adequate seeds for galaxy formation.

density of the Universe to collapse to form today's galaxies, the density of those ripples must have been at least 0.2% greater than the average density of the Universe at the time of recombination.

If normal luminous matter in the early Universe had been clumped at this level, then the variations in the CBR today would be at least 30 times larger than what is observed. At first glance this discrepancy might seem to be a horrible problem with our understanding of the origin of structure in the Universe, but it has turned out instead to be a crucial result that ties together a number of pieces of the puzzle. The theme of this story is "dark matter."

Dark matter first appeared on our journey back in Chapter 17 as an ad hoc construction—an annoyance, really—used by astronomers to explain the oddly flat rotation curves of spiral galaxies. When we turn our attention to clusters of galaxies, we find that dark matter dominates normal matter on those scales as well. Now we find that irregularities in normal matter in the early Universe were too slight to form galaxies. So where are we going to turn for an answer to the question of how structure in the Universe *did* form? You guessed it: dark matter.

In our discussion of primordial nucleosynthesis, we learned that dark matter cannot be made of normal matter consisting of neutrons, protons, and electrons. If it were, then it would have affected the formation of chemical elements in the early Universe. The abundances of several isotopes of the least massive elements would be quite different from what we find in nature. Dark matter must be something else—something that has no electric charge (so it does not interact with electromagnetic radiation) and that interacts only feebly with normal matter. Clumps of such dark matter in the early Universe would not interact with radiation or normal matter, so they would not be seen directly when looking at the CBR. This unseen dark matter saves the day for modeling the formation of galaxies and clusters of galaxies.

Dark matter provides the seeds for structure.

What is so different about the behavior of dark matter and normal matter in the early Universe? First, pressure waves and radiation did a very good job of smoothing out ripples in the distribution of normal matter in the early Universe. Feebly interacting dark matter would be immune to these processes, so clumps of dark matter survived long after bumps in the normal matter had been smoothed out, as illustrated in **Figure 20.3**. Second, the dark matter in these clumps does not glow, so we do not see it directly when we look at the cosmic background radiation. While these clumps of dark matter do cause slight gravitational redshifts in the light coming from normal matter, the resulting variations fit well with current observations of the CBR. Thus, clumps of dark matter could have (and, according to our theories of the very early Universe, *should* have) existed in the early

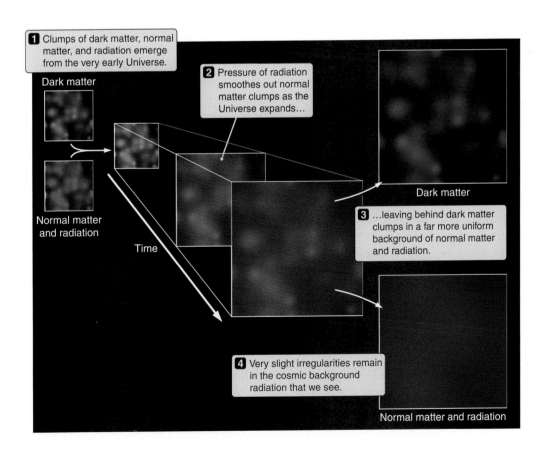

1 Clumps of dark matter, normal matter, and radiation emerge from the very early Universe.

Dark matter

Normal matter and radiation

Time

2 Pressure of radiation smoothes out normal matter clumps as the Universe expands…

Dark matter

3 …leaving behind dark matter clumps in a far more uniform background of normal matter and radiation.

4 Very slight irregularities remain in the cosmic background radiation that we see.

Normal matter and radiation

Figure 20.3 *Radiation pressure and other processes in the early Universe smoothed out variations in normal matter, but irregularities in the dark matter survived to become the seeds of galaxy formation.*

Universe, even though we are unable to see evidence of them directly.

Just because we cannot see clumps of dark matter in the early Universe does not mean that they had no effect. In the same way that dark matter dominates the gravitational fields of today's galaxies and clusters, so too did the mass of dark matter control the growth of gravitational instabilities in the early Universe.

THERE ARE TWO CLASSES OF DARK MATTER

Here is the story of galaxy formation in a nutshell. Dark matter in the early Universe was much more strongly clumped than normal matter. Within a few million years after recombination, these dark matter clumps pulled in the surrounding normal matter. Later, gravitational instabilities caused these clumps to collapse. The normal matter in the clumps went on to form visible galaxies. This story seems plausible enough, but the details of how this happened depend greatly on the properties of dark matter itself. Even though we do not yet know exactly what dark matter in the Universe is made of, we can talk about two broad classes of dark matter based on how it behaves.

One possibility is that dark matter consists of feebly interacting particles that are moving about relatively slowly, like the slowly moving atoms and molecules in a cold gas. For obvious reasons, this type of dark matter is called **cold dark matter.** There are several candidates for cold dark matter. It is possible that cold dark matter consists of tiny black holes that might have been produced in the early Universe. Few physicists and cosmologists favor this idea. Most think instead that cold dark matter consists of some unknown elementary particle. One candidate is the **axion,** a hypothetical particle first proposed to explain some observed properties of neutrons. Even though axions would have very low mass, they would have been produced in great abundance in the Big Bang. Another candidate is the **photino,** an elementary particle related to the photon. Some theories of particle physics predict that these particles exist and have a mass of around 10,000 times that of the proton. Our state of knowledge might change very quickly in the near future: photinos might be detected in current-day particle accelerators, and experiments are under way to search for axions and photinos that are trapped in the dark matter halo of our Galaxy.

Cold dark matter consists of relatively massive, slowly moving particles.

If the first class of dark matter is called cold dark matter, then the other class of dark matter is called— are you ready?—**hot dark matter.** Hot dark matter consists of particles that are moving very rapidly. Neutrinos are one example of hot dark matter that we know exists. We have seen that neutrinos interact with matter so feebly that they are able to flow freely outward from the center of the Sun, passing through the dense overlying layers of matter as if they were not there. There is no question that the Universe is filled with neutrinos. Calculations indicate that around 300 million of these cosmic relics of the Big Bang fill each cubic meter of space. Preliminary measurements of the neutrino mass obtained at the Super-Kamiokande detector in Japan and the Sudbury Neutrino Observatory in Canada suggest that neutrinos might account for as much as 5% of the mass of the Universe. While this percentage is not enough to account for all of the dark matter in the Universe, it may be enough to have had an effect on the formation of structure.

Hot dark matter consists of less massive, rapidly moving particles.

The different effects of cold and hot dark matter on structure formation have to do with differences in the way the two types of dark matter cluster in a gravitational field. The faster an object is moving, the harder it is for gravity to hold on to it. This is the same basic idea that we used to explain why hydrogen and helium are able to escape Earth's atmosphere while molecules such as oxygen and nitrogen are not. It is also the same idea we used to explain why the hot gas in elliptical galaxies extends far beyond their visible boundaries. Slow-moving particles are more easily corralled by gravity than fast-moving particles, so particles of cold dark matter clump together more easily into galaxy-sized structures than do particles of hot dark matter. The result of the calculations of galaxy formation is clear: to account for the formation of today's galaxies, we need cold dark matter.

Cold dark matter forms galaxies.

A GALAXY FORMS WITHIN A COLLAPSING CLUMP OF DARK MATTER

The way models of galaxy formation work can best be seen by following the events predicted by the models step by step. Let us consider the case of a universe made up of 90% cold dark matter and 10% ordinary matter, clumped together in a manner consistent with observations of the cosmic background radiation.

Figure 20.4(a) shows the state of affairs of one clump of dark matter at the time of recombination. The dark matter is less uniformly distributed than normal matter, but overall the distribution of matter is still remarkably uniform. By a few million years after recombination **(Figure 20.4b)**, the universe of our model calculation has expanded several-fold. However, the clump of dark matter is not expanding as rapidly as its surroundings because its self-gravity has slowed its expansion down. The clump now stands out more with respect to its surroundings. Another change that has taken place is that the gravity of the dark matter clump has begun to pull in normal matter as well. By the stage shown in **Figure 20.4(c),** normal matter is clumped in much the same way as dark matter.

A ball thrown in the air will slow, stop, and then fall back to Earth. In like fashion, by the time the universe is around a billion years old the clump of dark matter has reached its maximum size and is beginning to collapse (Figure 20.4c). The collapse of the dark matter clump stops when the clump is about half its maximum size, however, because the particles making up the cold dark matter are moving too rapidly to be pulled in any closer **(Figure 20.4d).** The clump of cold dark matter is now given its shape by the orbits of particles, in the same way that an elliptical galaxy is given its shape by the orbits of the stars it contains.

Clumps of dark matter can only collapse so far.

Unlike dark matter (which cannot emit radiation), the normal matter in the clump is able to radiate away energy and cool. As the gas cools, it loses pressure, and as it loses pressure, it collapses. Models show that small concentrations of normal matter within the dark matter clump collapse under their own gravity to form clumps of normal matter that range from the sizes of globular clusters to the sizes of dwarf galaxies. These clumps of normal matter then fall inward toward the center of the dark matter clump, as shown in **Figure 20.4(e).** We discussed a similar chain of events in Chapter 14, for the collapse of a molecular cloud core on its way to becoming a star. It is important to note that, according to models, gas in our Universe is only able to cool quickly enough to fall in toward the center of the dark matter clump if the clump has a mass of 10^8 to 10^{12} $M_\odot$. This is just the range of masses of observed galaxies! This agreement between theory and observations is an important success of the cold dark matter theory of galaxy formation.

Normal matter cools and falls toward the center of the dark matter halo.

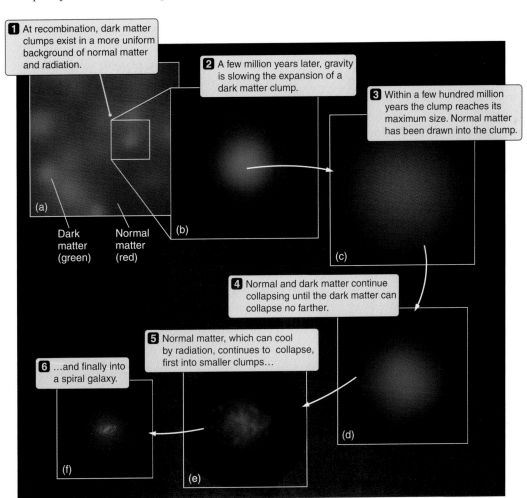

20.2 CLUMPS

1 At recombination, dark matter clumps exist in a more uniform background of normal matter and radiation.

2 A few million years later, gravity is slowing the expansion of a dark matter clump.

3 Within a few hundred million years the clump reaches its maximum size. Normal matter has been drawn into the clump.

4 Normal and dark matter continue collapsing until the dark matter can collapse no farther.

5 Normal matter, which can cool by radiation, continues to collapse, first into smaller clumps…

6 …and finally into a spiral galaxy.

Dark matter (green)   Normal matter (red)

(a) (b) (c) (d) (e) (f)

Figure 20.4 *Stages in the formation of a spiral galaxy from the collapse of a clump of cold dark matter.*

CONNECTIONS 20.1

Parallels Between Galaxy and Star Formation

As you read about galaxy formation, you might find it enlightening to think back to our discussion of star formation in Chapter 14. Both processes involve the gravitational collapse of vast clouds to form denser, more concentrated structures. To help you with the comparison, we list here a few of the similarities and differences between the two:

Gravitational Instability In both star and galaxy formation, the collapse begins with a gravitational instability. Regions only slightly denser than their surroundings are pulled together by their own self-gravity. As the matter in these regions becomes more compact, gravity becomes stronger, and the collapse process snowballs. One key difference is that for a galaxy to form, the dark matter clump must collapse rapidly enough to counteract the overall expansion of the Universe itself.

Fragmentation In both cases, the original cloud separates into smaller pieces as a result of the gravitational instability. However, the order of fragmentation differs between star and galaxy formation. In molecular clouds, large regions begin to collapse first, then fragment further to form individual stars. In contrast to this "top–down" process, galaxy formation is "bottom–up": Smaller structures collapse first, and then merge together to form galaxies and eventually assemblages of galaxies.

Compression, Heating, and Thermal Support As an interstellar molecular cloud collapses, its temperature climbs and the pressure in the cloud increases. The higher pressure would eventually be enough to prevent further collapse except that the cloud core is able to radiate away thermal energy. That is the bright infrared radia-

tion that allows us to see star-forming cores. Compare this with galaxy formation: As a dark matter clump collapses, its temperature also climbs, and it too reaches a point where there is a balance between gravity and the thermal motions of the dark matter. However, dark matter is *not* able to radiate away energy, so once this balance is reached, the collapse of the dark matter is over. Only the normal matter within the cloud of dark matter is able to radiate away thermal energy and continue collapsing. That is why normal matter collapses to form galaxies, while the dark matter remains in much larger dark matter halos.

As galaxies form, dark matter remains in extended halos. Dark matter is too hot to settle into galaxy disks. It is far too hot to become concentrated into even smaller structures such as molecular clouds, or to take part in the formation of stars. Dark matter may be the dominant form of matter in the Universe, and determine the structure of galaxies, but dark matter can never collapse enough to play a role in the processes that shape stars, planets, or the interstellar medium.

Angular Momentum and the Formation of Disks Conservation of angular momentum is responsible for the formation of disk galaxies, just as conservation of angular momentum is responsible for the formation of the accretion disks around young stars. The Milky Way and the Solar System are both flat for the same reasons.

The End Product Once a stellar accretion disk forms, most of the matter moves inward and is collected into a star. In contrast, much of the matter in a spiral galaxy remains in the disk.

As discussed in Connections 20.1, there are many similarities between the collapse of a molecular cloud to form stars and the collapse of a clump in the early Universe to form galaxies. Another process that we saw at work in star formation also comes into play at this point. The clumps of matter collapsing to form galaxies do not exist in isolation. They have been tugged on by the gravity of neighboring clumps and have been pushed

around by the pressure waves that ran through the young Universe, smoothing out its structure. As a result, each protogalactic clump has a little bit of rotation when it begins its collapse. As normal matter falls inward toward the center of the dark matter clump, this rotation forces much of the gas to settle into a rotating disk **(Figure 20.4f),** just as the collapsing cloud that was to become our Sun settled first into an accretion disk. And just as the accretion disk around the Sun provided the raw materials for the planets of our Solar System, the disk formed by the collapse of each protogalactic clump becomes the disk of a spiral galaxy.

SEARCHING FOR SIGNS OF DARK MATTER

Dark matter has come to play a very important role in our understanding of the Universe. We might reasonably be uncomfortable having our models of galaxy formation rely so heavily on something that we cannot see directly. Fortunately, dark matter shows itself in a variety of ways. The flat rotation curves of spiral galaxies, for example, are compelling evidence for the existence of extended halos comprised of dark matter. Similarly, the ability of elliptical galaxies to hold on to hot gas convincingly demonstrates that they contain far more mass than can be accounted for by their stars alone.

Dark matter reveals its presence in several ways.

We can use a similar approach to look for evidence of dark matter on larger scales. Galaxy clusters are filled with extremely hot (10 million to 100 million K) gas, which occupies the space between galaxies **(Figure 20.5)**. Even though this gas is *extremely* tenuous, the volume of space that it occupies is enormous; the mass of this hot gas can be up to five times the mass of all the stars in that cluster. X-ray spectra show that this gas contains significant amounts of massive elements that must have formed in stars. This chemically enriched gas has either been blown out of galaxies in winds driven by the energy of massive stars, or stripped from galaxies during encounters with neighboring galaxies. Were it not for the gravity of the dark matter filling the volume of the cluster, this hot gas would have dispersed long ago.

We can also look at the motions of the cluster's galaxies (which behave like very large "atoms" in this context) and again ask, How strong must gravity be to hold this cluster together? Again, the answer is that clusters must be ten to twenty times more massive than the normal matter they contain.

A third way to look for dark matter relies on the predictions of Einstein's general theory of relativity, which states that mass distorts the geometry of space-time, causing even light to bend near a massive object. In particular, light from a distant object is bent by the gravity of a galaxy or cluster of galaxies, so that images of the distant object can be seen magnified on either side of the intervening galaxy or cluster. The result is a **gravitational lens (Figure 20.6a)**.

We have already encountered gravitational lensing in our discussion of MACHOs in Excursions 18.2, where we found that lenses can make background objects appear brighter. When considering the more complex three-dimensional geometry

Gravitational lenses are one way to measure the masses of galaxy clusters.

of gravitational lenses, we find that lenses can show multiple images of background objects, and that these magnified images are often drawn out into an arclike appearance. The greater the gravitational lensing, the greater the mass that must be in the cluster.

(a) X-rays

1 For gravity to hold onto this hot X-ray emitting gas…

1 Mly

(b) Visible light

2 …this galaxy cluster must be more massive than the galaxies we see.

3 Galaxy clusters are mostly dark matter.

Figure 20.5 (a) *An x-ray image of a cluster of galaxies shows that the cluster is full of hot tenuous gas.* (b) *An image of the same field of view taken in visible light. (The two brightest objects are foregound stars.)*

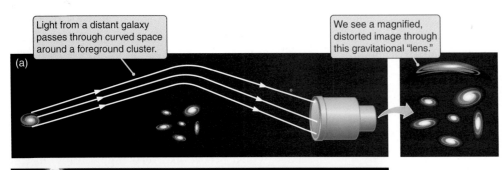

Light from a distant galaxy passes through curved space around a foreground cluster.

We see a magnified, distorted image through this gravitational "lens."

(a)

Figure 20.6 (a) *The geometry of a gravitational lens: a mass can gravitationally focus the light from a distant object, thereby magnifying and distorting the image.* (b) *A Hubble* Space Telescope *image of the cluster Abell 2218 showing many gravitationally lensed galaxies, seen as arcs.*

(b)

Foreground cluster galaxies

Gravitationally lensed background galaxies

substructure within the original galaxy-sized dark matter clump, that substructure would itself have collapsed into subgalaxy-sized objects before the galaxy as a whole finished forming. This gives us a way to understand the existence of such objects as the globular star clusters and dwarf companion galaxies near the Milky Way and other galaxies. These objects formed inside the larger, cold dark matter clump when the luminous matter in the clump was still settling toward what would become the disk of the Galaxy.

It is also likely that the same larger clump often produced more than one galaxy, and that these galaxies later interacted or even merged. When two spiral (or protospiral) galaxies merged, all sorts of commotion were likely. The tidal interactions between the galaxies and the collision between gas clouds in the galaxies probably triggered many regions of star formation throughout the combined system.

The merging of galaxies also answers another puzzle. We have seen how spiral galaxies could form in the early Universe, but what about the formation of elliptical galaxies? Ellipticals are now thought to result from the merger of two or more spiral galaxies. **Figure 20.7**

Figure 20.6(b) shows an image of a galaxy cluster that is acting as a gravitational lens for a number of background galaxies. Analysis of such images allows us to determine the mass of the lensing cluster.

Regardless of how we measure the masses of galaxy clusters—by looking at the motions of their galaxies, by

Dark matter dominates the mass of groups and clusters.

measuring their hot gas, or by using them as gravitational lenses—the results are the same. By mass, galaxy clusters, like the galaxies they are made of, are dominated by the dark matter they contain.

Ellipticals form from the mergers of spirals.

shows a computer simulation of such a merger. If the merging galaxies were not originally spinning in the same direction, then the resulting merged galaxy loses its disklike character. The dark matter halos of the galaxies merge, and the stars eventually settle down into the bloblike shape of an elliptical galaxy. As we might expect on the basis of this picture, elliptical galaxies are known to be more common in dense clusters where mergers are likely to have been more frequent.

MERGERS PLAY A LARGE ROLE IN GALAXY FORMATION

The picture of galaxy formation given in the previous sections is much cleaner and more idealized than what happened in reality. In our

Galaxy formation was messy business.

hierarchical clustering picture, smaller structures collapsed first. If there was

Such events also played a major role in feeding supermassive black holes, which themselves must have formed very early on in the collapse of galaxies and protogalaxies. The same mergers and interactions that formed the giant galaxies that we see today also provided the fuel to power the quasars and other AGNs that were common in the early Universe.

With galaxies crashing together, supermassive black holes forming, quasars flooding the young Universe with intense radiation and powerful jets, and star formation running amok, galaxy formation in the young Universe must have been a violent, messy business. This conclusion has clear consequences for what we should expect to see when looking back to a time in the history of the Universe when galaxies were actively forming. Rather than the well-formed spirals and ellipticals that dominate today's Universe, the early Universe should have contained many clumpy, irregular objects. As expected, images of the young Universe such as the *Hubble Space Telescope* image shown in Figure 17.1 are filled with lumpy, distorted objects that are still in the process of settling in.

We see the messy process of galaxy formation in the early Universe.

When we look at the bright, active galaxies that populate the early Universe, it is important to remember that what we actually see is only the tip of the iceberg. The luminous galaxies are small in comparison with the dark matter clumps that surround them. These clumps are still found in today's Universe as the invisible dark matter halos responsible for the flat rotation curves of spiral galaxies.

OBSERVATIONS HELP FILL GAPS IN THE MODELS

We would like to be able to turn our models of collapsing cold dark matter clumps into detailed models of what galaxies in the young Universe should look like, but we have not yet reached such a level of sophistication. One major problem is that we lack a good understanding of how stars form in young galaxies. We know that most of the normal matter in a galaxy winds up as stars, but when and where do those stars form, and how long does it take? While we can say a great deal about how individual stars form in our Galaxy, as we did in Chapter 14, we do not yet have theories or models that predict such important pieces of the puzzle as what fractions of stars should have what masses. Nor do we have a clear understanding of the differences between star formation in the early Universe and the star formation going on around us today. In a Universe devoid of massive elements there were no dense, dusty molecular clouds from which stars might form, and no spiral disks in which such clouds might congregate. Instead, the first stars must have formed from the collapse of clouds of hydrogen and helium gas within the overall clump of matter of which they were a part.

Star formation on a galactic scale is not well understood.

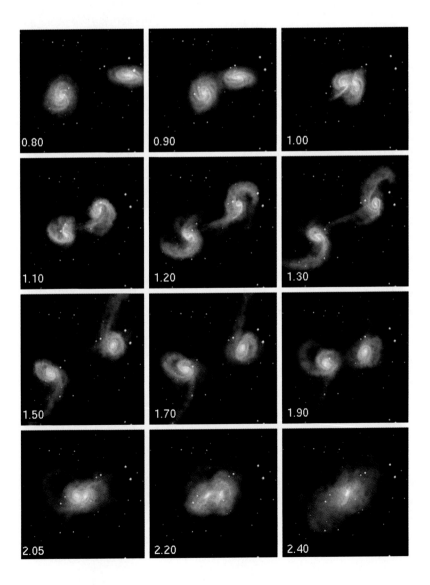

Figure 20.7 *In this computer simulation, two spiral galaxies merge to form an elliptical galaxy. Numbers indicate time since the interaction began.*

While a detailed understanding of star formation in a forming galaxy is in the future, we can still make a number of factual statements about when and where stars form in a collapsing galaxy. These clues come from the study of our own Galaxy. The fact that the atmospheres of even the oldest halo stars in the Milky Way contain some amount of massive elements tells us that a significant amount of star formation must have taken place *very* early on during the collapse of the Milky Way. If we

Star formation began very early on in collapsing galaxies.

could closely inspect a collapsing protogalactic clump in the stages depicted in Figure 20.4(d,e), we would expect to see stars already forming and supernovae exploding from place to place in the collapsing halo. The halo stars that we see today must have formed while the Milky Way was still collapsing, before the gas they formed from coalesced into the disk. In contrast, disk stars (which all have relatively high abundances of massive elements) formed from gas that was enriched by early generations of halo stars before it settled into the Galaxy's disk.

Thus, while we cannot say exactly how stars formed in the early Universe, we can say that star formation must have been going on throughout the process of galaxy formation. The pattern of stellar ages and massive element abundances evident in our Galaxy today fits naturally with our current models in which galaxies formed from collapsing clumps of cold dark matter.

Uncertainties in our understanding of star formation complicate quantitative comparisons between models of galaxy formation and observations of the early Universe. Fortunately, though, when we talk about structure on scales much larger than galaxies, star formation becomes less of an issue. To understand the formation of clusters and larger structures, we only need worry about the way that galaxy-sized clumps of matter themselves fall together under the force of gravity.

Figure 20.8 shows the results of one computer simulation that follows the motions of millions of clumps of dark matter as they fall through space under their mu-

Clusters, filaments, and voids form after galaxies form.

tual gravitational attraction. The scale of this simulation is so large that we cannot follow the outcome of individual galaxies, but instead are looking at the overall distribution of dark matter. The simulation shows that the smaller-scale structures develop first. Over the first few billion years, dark matter falls together into structures comparable in size to today's clusters of galaxies. Only later do the spongelike filaments, walls, and voids become well defined.

The similarities between the results of these models and observations of large-scale structure, such as those in

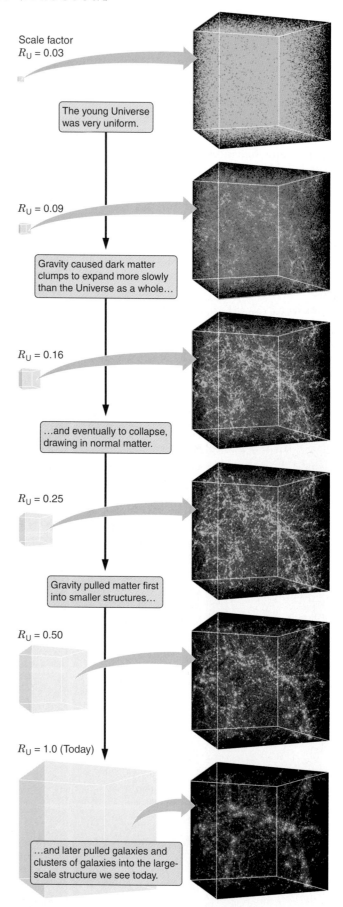

Scale factor
$R_U = 0.03$

The young Universe was very uniform.

$R_U = 0.09$

Gravity caused dark matter clumps to expand more slowly than the Universe as a whole...

$R_U = 0.16$

...and eventually to collapse, drawing in normal matter.

$R_U = 0.25$

Gravity pulled matter first into smaller structures...

$R_U = 0.50$

$R_U = 1.0$ (Today)

...and later pulled galaxies and clusters of galaxies into the large-scale structure we see today.

Figure 20.8 *A computer simulation of the formation of very large-scale structure in a universe filled with cold dark matter.*

Figure 20.2, are quite remarkable. Not any model will do, however. Only model universes with certain combinations of shape, mass, nature of ripples, type of dark matter, and values for the cosmological constant will produce structure similar to what we actually see. It is a very important result that models beginning with assumptions consistent with our knowledge of the early Universe wind up predicting the formation of large-scale structure similar to what we see in today's Universe.

Peculiar Velocities Trace the Distribution of Mass

In our comparison between models and observations, there is an important caveat. The models show where the

The distribution of light may not be the same as the distribution of mass.

mass is, while our observations show where the *light* is. We have already noted that our understanding of star formation in the early Universe is incomplete at best. What if a dark matter clump formed and collapsed, but for some reason the normal matter associated with that clump did not produce stars? Such clumps could still contain a significant amount of mass, and even affect the geometry of the Universe, but would themselves not be seen in images of the sky.

There is a way to get at the question of the overall distribution of dark matter more directly. If the clumping of galaxies is a result of the action of gravity, then we should be able to look at how galaxies are moving and see them falling together. In fact, measurements of the way galaxies fall together should allow us to infer the gravitational field they are experiencing, and hence the overall distribution of dark matter.

In the previous chapter we learned that the cosmic background radiation is blue-shifted in one direction in space and red-shifted in the opposite direction, telling us of our peculiar velocity relative to the CBR. Peculiar velocities of galaxies other than our own are difficult to measure. They require us to accurately determine the distances to galaxies using standard candles, and then use those distances along with Hubble's law to predict what the redshifts of those galaxies should be. Comparison of observed redshifts with redshifts predicted on the basis of Hubble expansion then tells us how fast these galaxies are moving with respect to the cosmic background radiation (at least along our line of sight).

Figure 20.9(a) shows one such map of the peculiar velocities inferred for galaxies in our part of space. If we use observations of peculiar velocities to map out concentrations of mass in this part of the Universe, we get a picture that looks like two mountains surrounded by foothills and valleys, as shown in **Figure 20.9(b).** Unfortunately, these two "mountains" of mass, which exert the greatest gravitational pulls on our own Galaxy, are located in regions of the sky that are heavily obscured by the Milky Way's dusty disk. The largest pull that we are

Peculiar velocities tell us how dark matter is distributed.

experiencing comes from a region dubbed the Great Attractor, so called because of its gravitational tug.

The picture painted in this chapter of the formation of large-scale structure is almost certainly correct in its broad outlines. Even so, uncertainties in star formation, our understanding of the exact nature of dark matter, and our knowledge of the shape of the Universe limit the amount of detail in any comparisons we can make between models and observations. As the 21st century opens, this is a field of very active research. Several pro-

Our understanding of galaxy formation is rapidly improving.

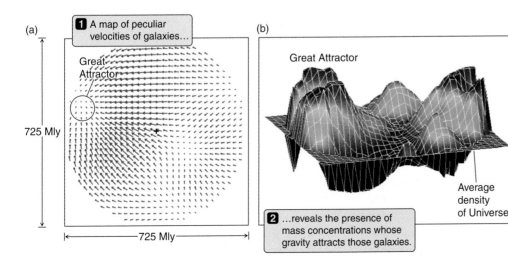

Figure 20.9 (a) *A map of peculiar velocities of galaxies in our neighborhood. Arrows show the velocities of material lying in the plane of our own local supercluster. Motions seem to be converging onto the position of the Great Attractor.* (b) *A map of the mass distribution inferred from observations of peculiar velocities.*

jects are currently collecting large amounts of data on the distribution and redshifts of galaxies, and a number of balloon and spacecraft experiments should dramatically improve our knowledge of the structure in the CBR. Larger and more sophisticated models are being built all the time, while studies of star-forming regions in our Galaxy and in galaxies with lower massive element abundances are continuing to improve our understanding of this important process. There is also hope that new theories and laboratory experiments might identify the particles that make up dark matter. A more complete understanding of the formation of galaxies and large-scale structure in the Universe will not likely come in a single "eureka" moment, but will instead arrive in bits and pieces throughout the first part of the 21st century as a result of advances in all of these different areas.

20.4 THE EARLIEST MOMENTS

It is easy to look at galaxies and stars and say, "Ah ha, structure!" but there is more to the origin of structure than simply the clumping of matter. The forces that govern the behavior of matter and energy in the Universe are themselves a kind of structure. There are four fundamental forces in nature, and everything in the Universe is a result of their action. Chemistry and light are products of the *electromagnetic force* acting between protons and electrons in atoms and molecules. The energy produced in fusion reactions in the heart of the Sun comes from the *strong nuclear force* that binds together the protons and neutrons in the nuclei of atoms. Beta decay of nuclei, in which a neutron decays into a proton, an electron, and an antineutrino, is governed by the **weak nuclear force.** Finally, there is our old friend *gravity*, which has played such a major role at every point along our journey. How these forces—these physical laws—came into being is part of the history of the Universe as well.

There are four fundamental forces in nature.

MAY THE FORCES BE WITH YOU

In Chapter 4 we spoke of electromagnetism using the electric and magnetic fields, but we also spoke of the quantum mechanical description of light as a stream of particles called photons. Since there is only one reality, both of these descriptions of electromagnetism have to

coexist. The branch of physics that deals with this reconciliation is called **quantum electrodynamics, or QED.**

QED treats charged particles almost as if they were baseball players engaged in an endless game of catch. As the baseball players throw and catch baseballs, they experience forces. Similarly, in QED, charged particles "throw" and "catch" an endless stream of virtual photons, as illustrated in **Figure 20.10.** Earlier we grappled with the idea that quantum mechanics is a science of possibilities rather than certainties. QED describes the electromagnetic interaction between two charged particles by averaging over all of the possible ways that the particles could throw photons back and forth. The result is a force that, over large scales, acts like the classical electric and magnetic fields described by Maxwell's equations. Physicists speak of the electromagnetic force

In QED, photons carry the electromagnetic force.

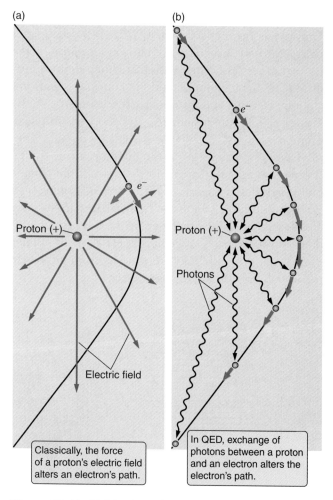

(a) Classically, the force of a proton's electric field alters an electron's path.

(b) In QED, exchange of photons between a proton and an electron alters the electron's path.

Figure 20.10 (a) *The classical view of an electron being deflected from its course by the electric field from a proton.* (b) *According to quantum electrodynamics, the interaction is properly viewed as an ongoing exchange of photons between the two particles.*

being "mediated by the exchange of photons." As is always the case with quantum mechanics, the world described by QED is hard to picture. Even so, QED is one of the most accurate, well-tested, and precise branches of physics. As of this writing, not even the tiniest measurable difference between the predictions of the theory and the outcome of an actual experiment has been found.

The central idea of QED—forces mediated by the exchange of carrier particles—provides a template for understanding two of the other three fundamental forces in nature. The electromagnetic and weak nuclear forces have been combined into a single theory called **electroweak theory.** This theory predicts the existence of three particles—labeled the W^+, W^-, and Z^0—which mediate the weak nuclear force. The essential predictions of electroweak theory were confirmed in the 1980s when these particles were identified in laboratory experiments.

> **The weak nuclear and electromagnetic forces combine in electroweak theory.**

The strong nuclear force is described by a third theory, called **quantum chromodynamics, or QCD.** In this theory, particles such as protons and neutrons are composed of more-fundamental building blocks, called **quarks,** that are bound together by the exchange of another type of carrier particles dubbed **gluons.** Together, electroweak theory and QCD are referred to as the **standard model** of particle physics. A deeper investigation of the standard model must await another journey. Here we leave the discussion by pointing out that the standard model is able to explain all the currently observed properties of matter, and has made many predictions that were subsequently confirmed by laboratory experiments.

> **Electroweak theory + QCD = the standard model of particle physics.**

A UNIVERSE OF PARTICLES AND ANTIPARTICLES

For modern theories of particle physics to make any sense, every type of particle in nature must also have an alter ego—an *antiparticle*—that is the opposite of that particle in every way described by quantum mechanics. For the electron, there is the antielectron, otherwise known as the positron. For the proton there is the antiproton, for the neutron the antineutron, and so on down the list. One fascinating property of these

> **Particle–antiparticle pairs form and annihilate.**

particle–antiparticle pairs is that if you bring such a pair together, the two particles will annihilate each other.

When a particle–antiparticle pair annihilates, the mass of the two particles is converted into energy in accord with Einstein's special theory of relativity ($E = mc^2$). For example, in **Figure 20.11(a)** an electron and positron annihilate each other, and the energy is carried away by a pair of photons. (This is the idea behind *Star Trek's* "antimatter" engines.) When we run time backwards (as we are allowed to do in particle physics), we see two high-energy photons colliding with each other, as in **Figure 20.11(b)**, creating in their place an electron–positron pair. This is an example of how an energetic event can create a particle and its corresponding antiparticle—a process called **pair creation.**

In principle, *any* type of particle and its antiparticle can be created in this way. The only limitation comes when there is not enough energy available to supply the rest mass of the particles being created. A specific example helps to show how this works. An electron (or positron) has a mass of $m_e = 9.11 \times 10^{-31}$ kg, which corresponds to an energy ($E = m_e c^2$) of 8.20×10^{-14} joules. If two gamma ray photons with a combined energy greater than 16.40×10^{-14} J ($= 2 \times m_e c^2$) collide, then the two photons may disappear and leave an electron–positron pair behind in their place. If the photons have more than the necessary energy, then the extra energy goes into the kinetic energy of the two newly formed particles.

Now we apply this idea to a hot Universe awash in a bath of Planck radiation. Using Wien's law from Chapter 4 [$\lambda_{peak} = (2,900\ \mu m\ K)/T$] and the expression for photon energy ($E = hc/\lambda$), we can show that when the Universe was

> **The early Universe was filled with photons, electrons, and positrons.**

Figure 20.11 (a) *An electron and positron annihilate, creating two gamma ray photons that carry away the energy of the particles.* (b) *In the reverse process, pair creation, two gamma ray photons collide to create an electron–positron pair.*

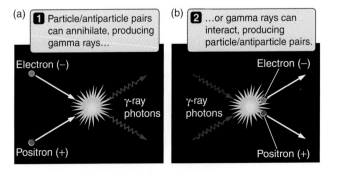

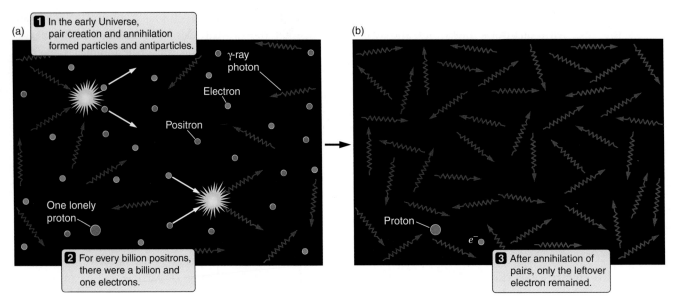

Figure 20.12 (a) *A swarm of electrons, positrons, and photons in the very early Universe. For every billion positrons, there were really a billion and one electrons.* (b) *After these particles annihilated, only the one electron was left.*

less than about 100 seconds old and had a temperature greater than a billion kelvins, it was filled with photons that had enough energy to create electron–positron pairs. Under these conditions, photons were constantly colliding, creating electron–positron pairs, and electron–positron pairs were constantly annihilating each other, creating pairs of gamma ray photons. The whole process reached an equilibrium, determined strictly by temperature, in which pair creation and pair annihilation exactly balanced each other. Rather than being filled only with a swarm of photons, at this time the Universe was filled with a swarm of photons, electrons, and positrons, as illustrated in **Figure 20.12(a).**

THE FRONTIERS OF PHYSICS

Physics has given us the tools to understand all the structures we have seen so far on our journey. There are still many gaps in what we know, but we are confident that everything from planets to stars to galaxies can be understood by applying our current knowledge of the four fundamental forces. At the turn of the 21st century, most of astronomy is a struggle with the complexity of the Universe, rather than a shortcoming of our understanding of the fundamental laws governing matter, energy, and space-time. However, as we push further back toward the Big Bang itself, the nature of the game changes. We need new physical theories.

In Foundations 9.1 we explored the power of symmetry—the idea that we can learn a lot about nature just by thinking about the way one part of some-

thing matches up with another. There we were talking about the gravitational forces that hold planets together, but other kinds of symmetry exist as well. In the process of pair creation there is a symmetry between matter and antimatter. For every particle created, its antiparticle is created as well.

As the Universe cooled, there was no longer enough energy to support the creation of particle pairs, so the swarm of particles and antiparticles that filled the early Universe annihilated each other and were not replaced. When this happened, every electron should have been annihilated by a positron. Every proton should have been annihilated by an antiproton. This was almost the case, but not quite. For every electron in the Universe today, there were a billion and one elec-

For every billion positrons there was a billion and one electrons.

trons in the early Universe, but only a billion positrons. This one part in a billion excess of electrons over positrons meant that when electron–positron pairs were through annihilating each other, some electrons were left over—enough to account for all the electrons in all the atoms in the Universe today **(Figure 20.12b).**

If the standard model of particle physics were a complete description of nature, then the one-part-in-a-billion imbalance between matter and antimatter would not have been there in the early Universe. The symmetry between matter and antimatter would have been complete. No matter at all would have survived into today's Universe, and we would not exist. The fact that you are reading this page demonstrates that something more needs to be added to the model. According to current ideas, the symmetry between matter and anti-matter may be broken in a theory that joins the electroweak and strong nuclear forces together in much the same way that electroweak theory unified our understanding of electromagnetism and the weak nuclear force. Such a theory, which combines three of the four fundamental forces into a single grand, unified force, is referred to as a **grand unified theory,** or **GUT.**

GUTs unify the strong and electroweak interactions.

Many possible grand unified theories exist, and make many predictions about the world. The problem is that the particles that mediate GUTs are so massive that it takes enormous amounts of energy to bring them into existence—roughly a trillion times as much energy as can be achieved in today's particle accelerators! Even so, some predictions of GUTs are testable with current technology. For example, GUTs predict that protons should be unstable particles that, given enough time, will decay into other types of elementary particles. This is a *very* slow process. Over the course of your life, GUTs predict that there is about a 1% chance that *one* of the 10^{28} or so protons in your body will decay. As of this writing, proton decay has yet to be observed. As we speak, however, large arrays of detectors are peering into huge tanks of water, waiting to see the signature of such an event. Perhaps the newspaper sitting in the rack at your local news stand holds a story heralding the confirmation of this central prediction of GUTs.

GUTs predict that even the proton will decay.

The particles that mediate GUTs may be beyond the reach of today's high-energy physics labs, but when the Universe was *very* young (younger than about 10^{-35} second) and *very* hot (hotter than about 10^{27} K), there was enough energy available for these particles to be freely created. During this time, the distinction between the electromagnetic, weak, and strong nuclear forces had not yet come into being. There was only the one grand, unified force. Welcome to the era of GUTs.

A GUT ruled in the very young Universe.

During the era of GUTs, the entire Universe was less than a trillionth the size of a single proton. This may seem virtually incomprehensible, yet the basic ideas needed to understand this Universe are in place. However, as we move backward in time, there is one threshold we have yet to cross. The story of advances in our understanding of physical law has been a story of unification—of the electromagnetic and weak forces into the electroweak theory, and then of these and the strong nuclear force into a GUT. But this program is incomplete. How does gravity fit into this scheme?

General relativity provides a beautifully successful description of gravity that correctly predicts the orbits of planets, describes the ultimate collapse of stars, and even allows us to calculate the structure of the Universe. Yet general relativity's description of gravity "looks" very different from our theories of the other three forces. Rather than talking about the exchange of photons or gluons or other carrier particles, general relativity talks instead about the smooth, continuous canvas of space-time upon which events are painted.

Gravity does not fit into the GUT picture.

We might be tempted to say, "Oh well. Gravity works one way and the other forces work another way, and that is how the Universe happens to be." In practice that is exactly what we do when we call on quantum mechanics to tell us about the properties of atoms, then use relativity to describe the passage of time or the expansion of the Universe. Even the era of GUTs is described perfectly well by treating gravity as a separate force. However, as we push back even closer to the moment of the Big Bang, this happy coexistence between relativity and quantum mechanics turns instead into a brutal confrontation.

TOWARD A THEORY OF EVERYTHING

When the Universe was younger than about 10^{-42} second old, the density of the Universe was incomprehensibly high. The Universe was so small that 10^{60} Universes would have fit into the volume of a single proton! Under these extreme conditions, quantum mechanical fluctuations in the matter and radiation making up the Universe involved immense amounts of mass—so much mass that quantum fluctuations made mincemeat out of space-time. Rather than a smooth sheet, space-time was a quantum mechanical froth. General relativity fails to describe this early Universe in ways that are reminiscent of the failure of Newtonian mechanics to describe the structure of atoms. An electron in an atom must be thought of in terms of probabilities rather than certainties. Similarly, there is no unique history for the earliest moments after the Big Bang. This era in the history of the Universe is referred to as the **Planck era,** signifying that the structure of the Universe during this period itself can only be understood using the ideas of quantum mechanics.

In the Planck era, the whole Universe was a quantum-mechanical froth.

The conflict between the continuous and the discrete—between general relativity and quantum mechanics—brings us to the current limits of human knowledge. The physics that we know can take us back to a time when the Universe was a millionth of a trillionth of a trillionth of a trillionth of a second old, but to push back any further we need something new. We need a theory that combines general relativity and quantum mechanics into a single theoretical framework unifying all four of the fundametal forces. Here we have reached the holy grail of modern physics. To understand the earliest moments of the Universe, we need a **theory of everything (TOE).**

A successful theory of everything would do more than unify general relativity with quantum mechanics.

Superstring theory is a possible theory of everything.

It would tell us which of the possible GUTs is correct, and would provide an answer for the nature of dark matter. A successful theory of everything would also necessarily answer several outstanding issues in cosmology, including the how, when, and why of inflation, and the physics underlying the possible cosmological constant adding to the expansion of the Universe. Physicists are currently grappling with what a TOE might look like. The current contender for the title is **superstring theory.** Here elementary particles are viewed not as points but as tiny loops called strings. A guitar string vibrates in one way to play an F, another way to play a G, and yet another way to play an A. According to superstring theory, different types of elementary particles are like different "notes" played by vibrating loops of string.

Superstring theory in principle provides a way to reconcile general relativity and quantum mechanics, but there is a price to pay for this success. To make superstring theory work, we have to imagine that these tiny loops of string are vibrating in a Universe with 10 spatial dimensions! (Adding time to the list would make our Universe 11-dimensional.) How can that be, when we clearly only experience three spatial dimensions? While the three spatial dimensions that we know spread out across the vastness of our Universe, the other seven dimensions predicted by string theory wrap tightly around themselves **(Figure 20.13),** extending no farther today than they did a brief instant after the Big Bang.

Superstring theory predicts 11 dimensions, but we experience only four of them.

To better appreciate how this bizarre notion works, imagine what it would be like to live in a three-dimensional universe (like the one we experience), in which one of those dimensions extended for only a tiny distance. Living in such a universe would be like living within a thin sheet of paper that extended on for billions of light-years in two directions, but was far smaller than an atom in the third. In such a universe we would easily be aware of length and width—we could move in those directions at will. In contrast, we would have no freedom to move in the third dimension at all, and might not even recognize that the third dimension exists. Perhaps our only inkling of the true nature of space would come from the fact that in order to explain the results of particle physics experiments, we would have to assume that particles extended into a third, unseen dimension. In like fashion, if superstring theory is correct, we see three spatial dimensions extending on possibly forever, but are unaware of the fact that each point in our three-dimensional space actually has a tiny but finite extent in seven other dimensions at the same time!

Superstring theory is only a pale shadow of the sort of well-tested theories that we have made use of throughout this book. In some respects, superstring theory is no more than a promising idea providing

The Big Bang was the ultimate particle accelerator.

direction to theorists searching for a TOE. It is worth noting that we will probably never be able to build particle accelerators that allow us to directly search for the most fundamental particles predicted by a TOE. The energies required are simply too high. Fortunately, however, nature has provided us with the ultimate particle accelerator—the Big Bang itself! The structure of the Universe that we see around us today is the observable result of that grand experiment.

To look at them, particle physics and cosmology would seem to have almost nothing in common. Parti-

Figure 20.13 *It is difficult to visualize what seven spatial dimensions wrapped up into structures far smaller than the nucleus of an atom would be like. Here such geometries are projected onto the two-dimensional plane of the paper.*

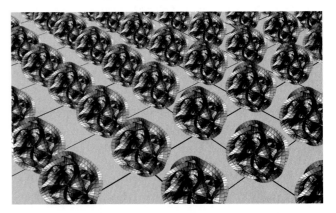

cle physics is the study of the quantum mechanical world that exists on the tiniest scales imaginable, while cosmology is the study of the changing structure of a Universe that extends for billions of light-years, and probably much farther. Yet the last quarter of the 20th century saw the boundary between these two fields fade and eventually disappear, as cosmologists and particle physicists came to realize that the structure of the Universe and the fundamental nature of matter are two sides of the same scientific coin.

One of the most intriguing questions that a successful TOE may answer is whether the Universe could have been different. Do we live in only one of many possible universes, each with different physical laws, or are the physical laws that govern our Universe the *only* consistent set of physical laws that could exist? We do not know the answer yet, but possibly within your lifetime we will.

Is ours the only possible universe?

ORDER "FROZE OUT" OF THE COOLING UNIVERSE

We started our discussion of the fundamental forces of nature by pointing out that these forces themselves represent a type of structure in the Universe. Just as galaxies and stars are structure that condensed out of the uniformity of the Big Bang, so too are the four fundamental forces structure that condensed out of the uniformity of the theory of everything.

Figure 20.14 illustrates the origin of structure in the evolving Universe. The first 10^{-43} second after the Big Bang is described by the TOE, when the physics of elementary particles and the physics of space-time were one and the same. As the Universe expanded and cooled, gravity parted ways with the forces described by the GUT. Space-time took on the properties described by general relativity. Inflation may also have been taking place at this time.

20.3 EVOLVE

As the Universe continued to expand and its temperature fell further, less and less energy was available for the creation of particle–antiparticle pairs. When the particles responsible for GUT interactions could no longer form, the strong force split off from the electroweak force. One might speak of this transition as the strong force "freezing out" of the GUT, because this and other similar transitions are reminiscent of the phase change that occurs as

The four fundamental forces separated as the early Universe cooled.

water changes to ice, and molecules become more constrained in their motions. Somewhere along the line, as the unity of the original TOE was lost, the symmetry between matter and antimatter was broken. As a result, the Universe ended up with more matter than antimatter.

The next big change took place when the particles responsible for unifying the electromagnetic and weak nuclear forces froze out, leaving these two forces independent of each other. The four fundamental forces of nature that govern today's Universe had now come into their own. The temperature of the Universe had fallen to a chilly 10^{16} K, and a ten-trillionth of a second had ticked off the cosmic clock. It was a full minute or two before the Universe cooled to the billion-kelvin mark, below which not even pairs of electrons and positrons could form.

While the Universe was now too cool to form additional particles and their antiparticles, it was still hot enough for the thermal motions of protons to overcome the electric barriers between them, allowing nuclear reactions to take place. These reactions formed the least massive elements, including helium, lithium, beryllium, and boron, but could not create more massive elements.

Atomic nuclei and atoms formed.

Big Bang nucleosynthesis came to an end by the time the Universe was 5 minutes old, and the temperature of the Universe had dropped below about 300 million K. The density of the Universe at this point had fallen to only about a tenth that of water. Normal matter in the Universe now consisted of atomic nuclei and electrons, awash in a bath of radiation. So the Universe remained for the next several hundred thousand years, until finally the temperature dropped so far that electrons were able to combine with atomic nuclei to form neutral atoms. We have encountered this event before. This was the era of recombination, which we see directly when we look at the cosmic background radiation.

20.5 LIFE IS ANOTHER FORM OF STRUCTURE

As we near the end of our journey, let's briefly appreciate the distance we have covered. We have followed the origin of structure in the Universe from the earliest instants after the Big Bang, when the fundamental forces of nature came to be, through to formation of the galaxies and other large-scale structure that is

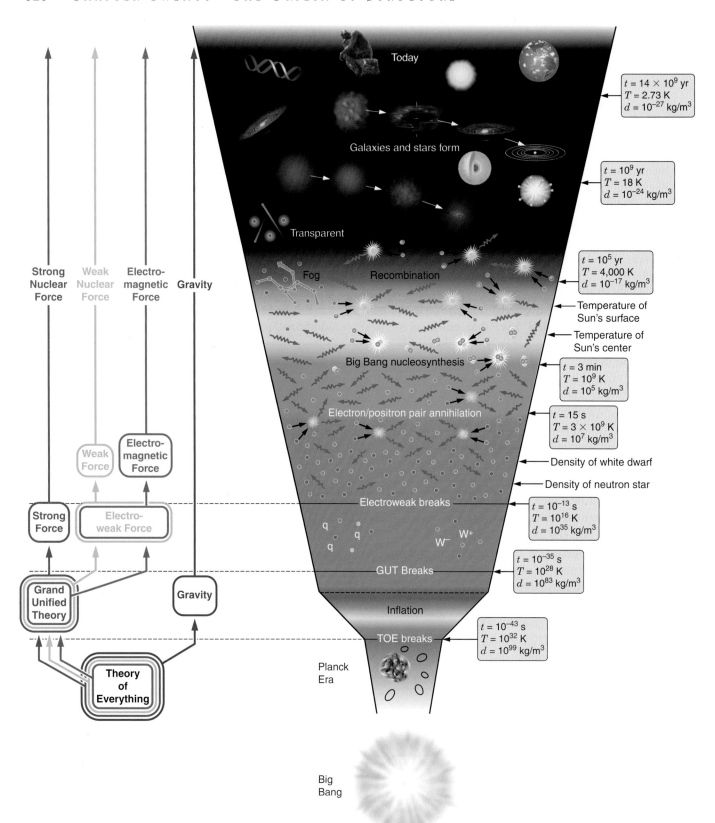

Figure 20.14 *Eras in the evolution of the Universe. As the Universe expanded and cooled following the Big Bang, it went through a series of phases determined by what types of particles could be created freely at that temperature. Later, the structure of the Universe was set by the gravitational collapse of material to form galaxies and stars, and by the chemistry made possible by elements formed in stars.*

Figure 20.15 *The search for our origins is one of the grand themes of modern science.*

visible today. Earlier on our journey, we watched as stars (including our Sun) formed from clouds of gas and dust within these galaxies, and planets (including our Earth) formed around those stars. We learned of the geological processes that shaped early Earth into the planet we see around us today. In short, we have traced the origin of structure from the instant the Universe came into existence up until the modern day **(Figure 20.15).** To help put this into perspective, Excursions 20.1 shows how the major events in the history of the Universe would fit into a single 24-hour day that began with the Big Bang and ended today. Yet there is one piece that we have left out. While the focus of our journey is 21st century *astronomy,* no discussion of how structure evolved in the Universe would be complete without some consideration of the origin of the particular type of structure we refer to as "life."

EVOLUTION IS UNAVOIDABLE

Imagine that just once during the first few hundred million years after the formation of Earth, a single molecule formed by chance somewhere in Earth's oceans. That molecule had a very special property: chemical reactions between that molecule and other molecules in the surrounding water resulted in that molecule making a copy of itself. Now there were two such molecules. Being duplicates of the original, chemical reactions would produce copies of each of these molecules as well, making four. Four be-

To create life, only one self-replicating molecule needed to form by chance.

came eight, eight became 16, 16 became 32, . . . and so on. By the time the original molecule had copied itself just 100 times, over a *million trillion trillion* (10^{30}) of these molecules existed. That is 100 million times more of these molecules than there are stars in the observable Universe!

Chemical reactions are never perfect. Sometimes when a copy is made, it is not an exact duplicate of its predecessor. The imperfection in the copy of the original is called a **mutation.** Most of the time such an error would be devastating, leading to a molecule that could no longer reproduce at all, but occasionally a mutation would actually help. It would lead to a molecule that was *more* successful in duplicating itself than the original. Even if imperfections in the copying process cropped up only once every 100,000 times a molecule reproduced itself, and even if only 1 out of 100,000 of these errors turned out to be beneficial, that still meant that after only 100 generations there were a hundred million trillion (10^{20}) errors that, by dumb luck, improved on the original design. Copies of each of these improved molecules would inherit the change. These molecules would have happened upon a form of **heredity,** the ability of one generation of structure to pass on its characteristics to future generations.

Success breeds success.

In this way, as these molecules continued to interact with their surroundings and make copies of themselves, they split into many different varieties. Eventually, these descendants of our original molecule became so numerous that the building blocks they needed in order to reproduce became scarce. Faced with this scarcity of resources, varieties of molecules that were more successful than others in reproducing themselves became more numerous. Varieties that could break down other varieties of self-replicating molecules and use them as raw material were especially successful in this world of limited resources. Competition and predation had entered the picture. After a few generations, certain varieties of molecules came to dominate the mix, while less successful varieties became less and less common. This competition, in which better-adapted molecules thrive and less well-adapted molecules die out, is referred to as **natural selection.**

Four billion years is a long time—enough time for the combined effects of heredity and natural selection to shape the descendants of that early self-copying molecule into a huge variety of complex, competitive, successful structures. Geological processes on Earth, such as sedimentation, have preserved a fossil record of the history of these structures, as illustrated in

EXCURSIONS 20.1

FOREVER IN A DAY

We have seen that the events that ultimately led to our existence here as inhabitants of planet Earth stretched out over many billions of years, intervals that even astronomers have difficulty visualizing. We can sometimes get a better grasp of such enormous spans of time by compressing them into much shorter intervals with which we have day-to-day experience. Try then to imagine the age of the Universe and those important events associated with our origins as if they were all taking place within a single day. Our cosmic day begins at the stroke of midnight:

12:00:00 A.M.   The embryonic Universe is a mixed broth, minute specks of matter suspended in a vast soup of radiation. The entire Universe exists only as an extraordinarily hot bath of photons and a zoo of elementary particles.

12:00:02 A.M.   It is now just 2 seconds after midnight. All of the early eras in the history of the Universe have passed. The fundamental forces have frozen out, Big Bang nucleosynthesis has formed the Universe's original complement of atomic nuclei, and things have cooled down enough for atomic nuclei to combine with electrons to produce neutral atoms. Both normal and dark matter are now available to create galaxies and stars, but that process will take some time.

1:30 A.M.   The first quasars and galaxies appear. At some point—we are not sure exactly when—our own Galaxy becomes visible as star formation begins. Throughout the cosmic day, stars will continue to form. The more massive stars each go through their brief life cycles in only 5 to 10 seconds of our imaginary 24-hour clock. Each massive star shines briefly, creates its heavy elements, and then disburses this material throughout interstellar space as it dies in a violent supernova explosion. Stars similar to our Sun go through less dramatic life cycles, each lasting about 16 cosmic hours. Stars with masses less than 0.8 $M_\odot$. will last for several tens of cosmic hours, and will survive beyond the end of our cosmic day.

3:35 P.M.   Our Solar System forms out of a giant cloud of gas and dust. Collapse of the cloud's pro-

tostellar core, followed by the appearance of the Sun and the planets—including Earth—all take place within the span of a single cosmic minute.

3:40 P.M.   A Mars-size planetesimal crashes into Earth, forming the Moon.

5:00 P.M.   The first primitive life appears on Earth. It evolves into the simplest life forms, unicellular organisms such as bacteria, cyanobacteria, and archaebacteria.

8:40 P.M.   More complex single-cell organisms appear, making it possible for multicellular life to develop.

11:00 P.M.   Multicellular organisms become abundant. This paves the way for still larger and more complex life-forms.

11:40 P.M.   The first dinosaurs make their appearance. Various small animals appear as well, but they remain subdued by larger life-forms.

11:52:48 P.M.   A large comet or asteroid crashes into Mexico's Yucatan Peninsula. Seventy percent of all species on Earth (including the dinosaurs) suddenly vanish. In the minutes that follow, the mammals, being more adaptable in the changed environment, gain prominence.

11:59:35 P.M.   Our earliest human ancestors finally appear on the plains of Africa just 25 seconds before the end of our cosmic day.

11:59:59.8 P.M.   Modern humans now arrive with only a fifth of a second to spare, a fraction of a heartbeat before the day's end.

12:00:00 A.M.   Just as our cosmic day draws to a close, will a worldwide catastrophe occur? Will 21st-century humans permanently scar the face of the planet, driving many of its life-forms into oblivion by polluting the atmosphere and poisoning the land, the rivers, and the oceans? Or, will 21st-century humans finally break free of their gravitational bondage to their planetary home, and begin to claim the rest of the Solar System as their own? In comparison to all that came before, what will happen in the next century will occur in a blur—less than a blink of the eye.

CONNECTIONS 20.2

LIFE, THE UNIVERSE, AND EVERYTHING

Popular discussions of the origin of life almost always wind up struggling with the question of how it is that complex, highly ordered structures such as living things could have emerged from a simpler, more disordered past. Place a drop of ink in a glass of water and watch what happens. The ink spreads out, diffusing through the water, until the only sign that the ink is there is the fact that the water is a different color. The order represented by the discrete drop of ink naturally fades away as the ink spreads out through the water. Yet no matter how long we watch, we will never see that drop of ink spontaneously reassemble itself.

Physicists discuss the degree of order of a system using the concept of **entropy,** which is a measure of the number of different ways a system could be rearranged and still appear the same. A neatly ordered system (such as the drop of ink and the glass of clear water) has low entropy, while a disordered system (the glass of inky water, which can be stirred or turned and still look the same) has higher entropy. The **second law of thermodynamics** says that, left on its own, an isolated system will always move toward higher entropy—that is, from order toward disorder. This is commonsense. As time goes by, we are more likely to find a system in a state that is, well, more likely.

In light of the inescapable march toward disorder dictated by the second law, how can ordered structure emerge spontaneously? Creationist claims that the origin of life flies in the face of the second law of thermodynamics are about as common as reports of Elvis sightings in supermarket tabloids. Yet such claims make a crucial mistake, by focusing attention on one player, while ignoring the rest of the game.

We have all seen what happens when we set a glass of ice water out on a hot, humid day **(Figure 20.16).** Water vapor from the surrounding air condenses into drops of liquid water on the surface of the cold glass. This is amazing! We have just watched as ordered structure (drops of water) spontaneously emerged from disorder (water vapor in the air)! It is almost as if we saw the drop of ink reassemble itself. This simple, everyday event appears to violate the second law . . . but it does not.

To understand why, we need to step back and look at more than just the drops of water on the glass. When the drops condensed, they released a small amount of thermal energy that slightly heated both the glass and the surrounding air. Heating something increases its entropy. The decrease in entropy due to the formation of the drops is more than made up for by the increase in the entropy of their surroundings. Ordered structure spontaneously emerged, but *overall,* disorder increased.

It is important to note that energy can be used to reduce entropy at one location while increasing entropy somewhere else. An air-conditioner is a good example. Electric energy produced at a power plant is used to pump thermal energy from inside a home and dump that thermal energy outside. When an air-conditioner is run in reverse to provide heating, it is referred to as a "heat pump," but "entropy pump" might be a more accurate description. As you sit in your armchair on a summer day, all that you immediately notice is that when the air-conditioner comes on, the temperature drops. Entropy decreases inside your house. If you look at the system as a whole, on the other hand, including the heating of air by the outside coils and the entropy produced by the burning of coal or natural gas during the production of electricity, then you see that turning on the A/C causes an overall *increase* in entropy.

This idea is especially important when considering the origin of life. A living thing, whether an amoeba or a human being, represents a huge local increase in order. However, there is no violation of the second law involved. In our everyday lives, the food we eat gives us the energy we need to stave off the relentless advance of entropy. Viewed even more broadly, the evolution of life on Earth was powered primarily by energy striking Earth in the form of sunlight. A local increase in order on Earth (such as you) is "paid for" in the end by the much greater decrease in order that accompanies thermonuclear fusion in the heart of the Sun. Order emerges in localized regions within a system, but the second law is obeyed overall.

The unifying theme of this chapter and, in many ways, this entire book, has been understanding the origin of structure. The answer we have found is clear: whether discussing the freezing out of matter and the fundamental forces in the young Universe; the gravitational collapse of stars, planets, and galaxies; the evolution of life; or water beading up on the outside of a cold glass—ordered structure *does* emerge spontaneously as an unavoidable consequence of the action of physical law.

Figure 20.16 *On a humid day, water condenses into droplets on the surface of a cold glass. The second law of thermodynamics does not mean that ordered structure cannot spontaneously emerge.*

Figure 20.17. Among these descendants are "structures" capable of thinking about their own existence and unraveling the mysteries of the stars.

The molecules of DNA (deoxyribonucleic acid) that make up the chromosomes in the nuclei of the cells throughout your body are direct descendants of those early self-duplicating molecules that flourished in the oceans of a young Earth. While the game played by the molecules of DNA in your body is far more elaborate than the game played by those early molecules in Earth's oceans, the fundamental rules remain the same. We now realize that this process is inevitable: any system that combines the elements of heredity, mutation, and natural selection *must* and *will* evolve.

Evolution is inevitable.

In *The Selfish Gene* (1976), Richard Dawkins points out that, when viewed from a purely utilitarian perspective, a human being is a machine whose "purpose" is to produce copies of its genetic material. The remarkable thing about humanity is that our intelligence (itself a very powerful tool for survival in a competitive world) has also allowed us to develop more noble pursuits (for example, astronomy) than those dictated by our basic biochemical imperative.

ARE WE ALONE?

The story of the evolution of life cannot be separated from the story of 21st-century astronomy. We know that we live in a Universe full of stars, and that systems of planets orbit many and probably most of those stars. We also know that there is nothing mysterious about the origin of life or the processes that cause it to evolve. In fact, as discussed in Connections 20.2, the evolution of life on Earth is but one of the many examples that we have encountered of the emergence of structure in an evolving Universe. This leads naturally to one of the

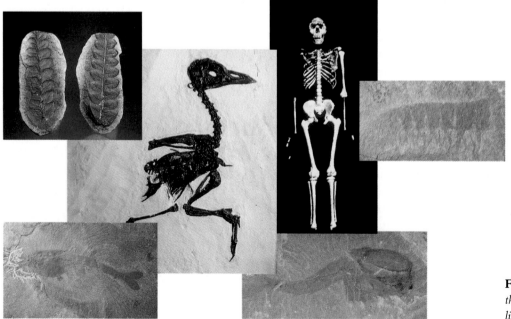

Figure 20.17 *Fossils record the history of the evolution of life on Earth.*

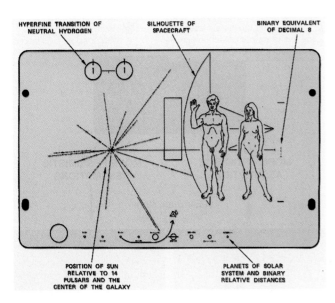

HYPERFINE TRANSITION OF NEUTRAL HYDROGEN

SILHOUETTE OF SPACECRAFT

BINARY EQUIVALENT OF DECIMAL 8

POSITION OF SUN RELATIVE TO 14 PULSARS AND THE CENTER OF THE GALAXY

PLANETS OF SOLAR SYSTEM AND BINARY RELATIVE DISTANCES

Figure 20.18 *The plaque included with the* Pioneer 11 *probe, which left the Solar System to travel through the millennia in interstellar space.*

more profound questions that we ask about the Universe. Has life arisen elsewhere, and is it intelligent? Are we alone?

Humans have already made preliminary efforts to "reach out and touch someone." The *Pioneer 11* spacecraft, which will probably spend eternity drifting through interstellar space, carries the plaque shown in **Figure 20.18** describing ourselves to any future interstellar traveler who might happen to find it. This may

We have sent messages to the stars.

not be the most efficient way to make contact with the Universe, but it is a significant gesture nonetheless. A somewhat more practical effort was made in 1974, when the kilometer-wide dish of the Arecibo radio telescope was used to beam the message shown in **Figure 20.19** toward the star cluster M13. (If someone from M13 answers, we will know in 48,000 years!)

The first serious effort to search for intelligent extraterrestrial life was made by astronomer Frank Drake in 1960. Drake used what was then astronomy's most powerful radio telescope to listen for signals from two nearby stars. Although his search revealed nothing unusual, it prompted him to develop an equation that still bears his name. The **Drake equation** is a prescription for calculating the likelihood that intelligent civilizations exist beyond our own Solar System.

In its basic form, the Drake equation is a prescription for calculating $\mathcal{N}$, the number of technologically advanced civilizations in our Galaxy today:

$$\mathcal{N} = \mathcal{T} f_{\mathrm{p}} n_{\mathrm{pm}} f_{\ell} f_{\mathrm{c}} \mathscr{L}$$

The six factors on the right side of the equation relate to the conditions that must be met for a civilization to exist:

The Drake equation estimates the number of advanced civilizations in the Milky Way.

* $\mathcal{T}$ is the total number of stars in our Galaxy. We know this to be several hundred billion stars.
* f_{p} is the fraction of stars that form planetary systems. Questions remain, but we do know that planets form as a natural by-product of star formation, and that many other stars are known to have planets. For this calculation we will assume that f_{p} is between 0.5 and 1.
* n_{pm} is the average number of planets and moons in each planetary system capable of supporting life. If such planets are like Earth, they will have to be of the terrestrial type and orbit within a "life zone," where temperatures are neither too hot nor too cold. If our own Solar System is a good example, there would be one or two in each planetary system. (In addition to Earth, Mars appears capable of supporting life, whether or not it presently exists there.) While we now know that many planetary systems do not resemble our own, it is possible that moons orbiting giant planets near stars could also harbor life. For purposes of calculation, we put the value of n_{pm} between 0.1 and 2.
* f_{ℓ} is the fraction of planets and moons capable of supporting life on which life actually arises. Remember that just a single self-replicating molecule may be enough to get the ball rolling. Many biochemists now believe that if the right chemical and environmental conditions are present, then life *will* develop. If they are correct, then f_{ℓ} is close to 1. We will use a range for f_{ℓ} of 0.01 to 1.
* f_{c} is the fraction of those planets harboring life that eventually develop technologically advanced civilizations. With only one example of a technological civilization to work with, f_{c} is hard to estimate. Intelligence is certainly the kind of survival trait that might often be strongly favored by natural selection. On the other hand, on Earth it took 4 billion years—half the expected lifetime of our star—to evolve tool-building intelligence. The correct value for f_{c} might be close to 1, or it might be closer to 1 in 1,000.
* $\mathscr{L}$ is the likelihood that any such civilization exists *today*. This is certainly the most difficult factor of all to estimate, because it depends on the long-term stability of advanced civilizations. We have had a technological civilization on Earth for around 100 years, and during that time have de-

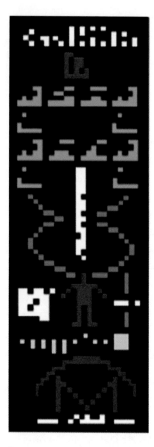

Figure 20.19 *The message we beamed toward the star cluster M13. A reply may be forthcoming in 48,000 years.*

veloped, deployed, and used weapons with the potential to eradicate our civilization and render Earth hostile to life for many years to come. We have also so degraded our planet's ecosystem that many respectable biologists and climatologists wonder if we are not nearing the brink. Do all technological civilizations destroy themselves within a thousand years? A thousand years is only one ten-millionth of the lifetime of our star. On the other hand, if most technological civilizations learn to use their technology for survival rather than self-destruction, perhaps they instead will survive for hundreds of millions of years. For our calculation, we will use a range for $\mathcal{L}$ of between 10^{-7} (for civilizations that live 1,000 years) and 10^{-2} (for civilizations that live 100 million years).

As illustrated in **Figure 20.20,** the conclusions we draw based on the Drake equation depend a great deal on the assumptions we make. Using the most pessimistic of our estimates above, the Drake equation sets the likelihood of finding a technological civilization in our Galaxy at 1%, or 1 chance in 100. If this is correct, then we are in all likelihood the *only* technological civilization in the Milky Way. In fact, this would suggest that only 1 in 100 galaxies would contain a technological civilization *at all.* Such a Universe would still be full of intelligent life. With a hundred billion galaxies in the observable Universe, even these pessimistic assumptions would mean that there are a billion technological civilizations out there somewhere. On the other hand, we would have to go a *very* long way (30 Mly or so) to find our nearest neighbor.

At the other extreme, what if we take the most optimistic numbers, assuming that intelligent life arises

The nearest advanced civilization may be as far away as 30 Mly . . .

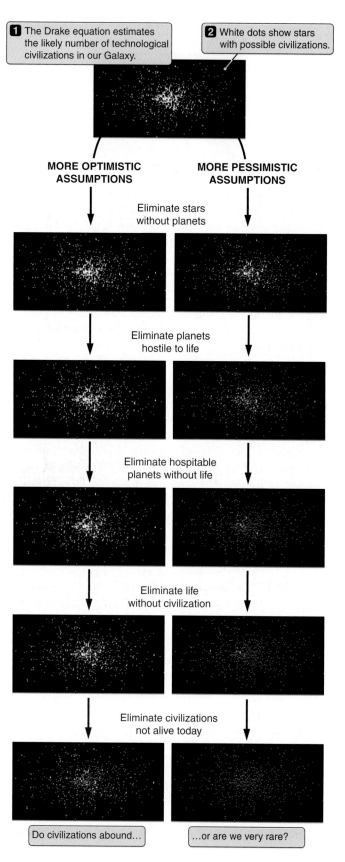

1 The Drake equation estimates the likely number of technological civilizations in our Galaxy.

2 White dots show stars with possible civilizations.

MORE OPTIMISTIC ASSUMPTIONS   **MORE PESSIMISTIC ASSUMPTIONS**

Eliminate stars without planets

Eliminate planets hostile to life

Eliminate hospitable planets without life

Eliminate life without civilization

Eliminate civilizations not alive today

Do civilizations abound...   ...or are we very rare?

Figure 20.20 *Simulation of the Drake equation for optimistic and pessimistic assumptions about the factors affecting the prevalence of intelligent life in the Universe.*

. . . or as near as 40 light-years.

and survives everywhere it gets the chance? The Drake equation then says that there should be 40 million technological civilizations in our Galaxy alone! In this case, our nearest neighbor may be "only" 40 or 50 light-years away. If that civilization is listening to the Universe with its own radio telescopes, hoping to answer the question of life in the Universe for themselves, then, as we speak, they may be sitting back to watch an episode of *I Love Lucy*.

It is fascinating to speculate about the implications of the Drake equation. If there are civilizations around every corner, cosmically speaking, then why have we not heard from them? Perhaps they are not interested in talking to the new kid on the block, or perhaps the fact that we know of no other civilizations simply means that there are none nearby.

If we did run across another technologically advanced civilization, what would it be like? Looking

Any civilization we discover will almost certainly be advanced.

back at the Drake equation, it is highly unlikely that we have neighbors nearby unless civilizations typically live for many thousands or even millions of years. In this case, any civilization that we encountered would almost certainly have been around for much longer than we. Having survived that long, would its members have learned the value of peace, or would they have developed strategies for controlling pesky neighbors? Movie theaters and the science fiction shelves of libraries and bookstores are filled with amusing and thoughtful stories that explore what life in the Universe might be like **(Figure 20.21).** For the moment, we set aside such speculation to look at the question as scientists. Given this fascinating question, how might we go about finding a real answer?

THE SEARCH FOR SIGNS OF INTELLIGENT LIFE IN THE UNIVERSE[1]

One question we may be able to answer using observations of our own Solar System is, What is the likelihood of life originating at all? When the *Viking* landers failed

Life may exist elsewhere in the Solar System.

to discover life on Mars in 1976, hopes faded for the presence of other life on worlds orbiting the Sun. Since that time, however, optimism has been renewed. A

---

[1]With apologies to Jane Wagner and Lily Tomlin.

better understanding of the history of Mars indicates that at one time, the planet was wet and warm, and many scientists believe that fossil life or even living microbes may be buried under the planet's surface.

Even more exciting are discoveries in the outer Solar System. There is strong evidence that life on Earth may have originated deep under the oceans, where geothermal vents **(Figure 20.22)** provided the thermal and chemical energy needed for life to gain a toehold. If we are correct that geological activity churns the floor of a deep ocean on Jupiter's moon Europa, for example, then this could be an excellent place to find life. One other example is all we need. If life arose independently *twice* in the same planetary system, then we would have no choice but to conclude that f_ℓ in the Drake equation is close to 1, and that life is ubiquitous throughout the Universe.

Another way to search for intelligent life is to simply turn our ear to the sky and listen. Drake's original project of listening for radio signals from intelligent life around two nearby stars has grown over the years into a much more elaborate program that is referred to as the Search for Extraterrestrial Intelligence, or **SETI.** Scientists from around the world have thought carefully about what

SETI listens for radio signals from other civilizations.

strategies might be useful for finding life in the Universe. Most of these have focused on the idea of using radio telescopes to listen for signals from space that bear an unambiguous signature of intelligent design. Some have listened intently at certain "magic" frequencies, such as the frequency of the interstellar 21-cm line from hydrogen gas. The assumption behind this technique is that if a civilization wanted to be heard, they would tune their broadcasts to a channel that astronomers throughout the Galaxy should be listening to. More recent searches have made use of advances in technology to

Figure 20.21 *The classic 1951 film* The Day the Earth Stood Still *portrayed intelligent extraterrestrials who had no interest in our internal affairs, but promised destruction of Earth if we carried our violent ways into space.*

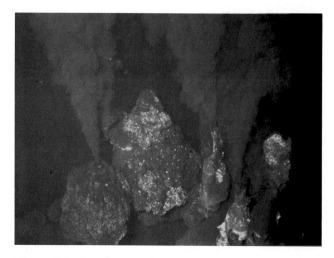

Figure 20.22 *Life on Earth may have arisen near ocean geothermal vents like the one at* left. *Similar environments might exist elsewhere in the Solar System. Life around vents* (right) *is powered by geothermal rather than solar energy.*

record as broad a range of radio signals from space as possible, and then use computers to search these databases for types of regularity in the signals that might suggest that they are intelligent in origin.

Unlike much astronomical research, SETI is funded mostly through private rather than governmental means, and SETI researchers have been very ingenious at finding ways to accomplish as much as possible with limited resources. One especially clever idea, coming out of the SETI Institute in Mountain View, California, uses the underutilized computing power of thousands of personal computers around the world to analyze their data. SETI screen saver programs installed on desktops around the country download radio observations from the SETI Institute over the network, analyze these data while the computers' owners are off living their lives, and then report the results of their searches back to the Institute. It is fun to think that the first sign of intelligent life in the Universe might be found by a computer sitting on a table in the corner of your living room!

There are a number of SETI projects that are on the drawing board. One is called the Allen Telescope Array (ATA; see **Figure 20.23**), named after the cofounder of Microsoft, Paul Allen, who is providing much of the financing for the project. The ATA will consist of a "farm" of hundreds of small, inexpensive radio dishes like those used to capture signals from orbiting communication satellites. The ATA, a joint venture between the SETI Institute and the University of California, will use sensitive modern receiver technology to search the sky 24 hours a day, seven days a week, for signs of intelligent life. Just as your brain can sort out sounds coming from different directions,

There will be profound implications if we find life elsewhere.

this array of radio telescopes will be able to determine the direction a signal is coming from, allowing it to listen to many stars at the same time. Over several years' time, the ATA is expected to survey as many as a million stars, hoping to find a civilization that has sent a signal in our direction. If reality is anything like the more optimistic of the assumptions we used in evaluating the Drake equation, this project will stand a good chance of success.

If SETI finds even *one* nearby civilization in our Galaxy, then the message will be clear. If we find a *second* technological civilization in our own small corner of the Galaxy, then it means that the Universe as a whole must literally be teeming with intelligent life. SETI may not be in the mainstream of astronomy, and the likelihood of its success may be difficult to predict, but its potential payoff is enormous. Few discoveries would do more to change our understanding of ourselves than certain knowledge that we are not alone.

Figure 20.23 *When complete, the Allen Telescope Array will listen for evidence of intelligent life from as many as a million stellar systems.*

20.6 THE FUTURE, NEAR AND FAR

We have used our understanding of physics and cosmology to look back through time and watch as structure formed throughout the Universe. Now, as our journey nears its end, we look toward the future and contemplate the fate that awaits Earth, humanity, and the Universe as a whole. Beginning close to home, around 5 billion years from now, the Sun will end its long period of relative stability. Shedding its identity as the passive, benevolent star that has nurtured life on Earth for nearly 4 billion years, the Sun will expand to become a red giant and later an AGB star, swelling to hundreds of times its present size. The giant planets, orbiting outside the extended red giant atmosphere, should survive the Sun's cranky old age in some form. Even so, they will suffer the blistering radiation from a Sun grown thousands of times more luminous than it is today.

The terrestrial planets will not be so lucky. Some and perhaps all of the worlds of the inner Solar System will be engulfed by the expanding Sun. Just as an artificial satellite is slowed by drag in Earth's tenuous outer atmosphere and eventually falls to the ground in a dazzling streak of white-hot light, so too will a terrestrial planet caught in the Sun's atmosphere be consumed by the burgeoning star. If this is Earth's fate, our home world will leave no trace other than a slight increase in the amount of massive elements in the Sun's atmosphere. As the Sun loses more and more of its atmosphere in an AGB wind, our atoms may be expelled back into the reaches of interstellar space from which they came.

Far future Earth will be consumed by the Sun or left as an icy cinder.

Another fate is possible, however. As the red giant Sun loses mass in a powerful wind, its gravitational grasp on the planets will weaken, and the orbits of both the inner and outer planets will spiral outward. If Earth moves out far enough, it may survive as a seared cinder, orbiting the white dwarf that the Sun will become. Barely larger than Earth and with its nuclear fuel exhausted, the white dwarf Sun will slowly cool, eventually becoming a cold, inert sphere of degenerate carbon, orbited by what remains of its retinue of planets. The ultimate outcome for our Earth—consumed in the heart of the Sun or left behind as a frigid burnt rock orbiting a long-dead white dwarf—is not yet certain. In either case Earth's status as a garden spot will be at an end.

THE FATE OF LIFE ON EARTH

It is far from certain that the descendants of today's humanity will survive to see the death throes of the Sun. Some of the threats that await us come from beyond Earth. For the remainder of the Sun's life, the terrestrial planets, including the Earth, will continue to be bombarded by asteroids and comets. Perhaps a hundred or more of these impacts will involve kilometer-sized objects, capable of causing the kind of devastation that eradicated the dinosaurs 65 million years ago. While these events may create new surface scars, they will have little effect on the integrity of Earth itself. Earth's geological record is filled with such events, and each time it happens, life manages to recover and reorganize.

It seems likely, then, that some form of life will survive to see the Sun begin its march toward becoming a red giant. On the other hand, individual species do not necessarily fare so well when faced with cosmic cataclysm. If the descendants of humankind survive, it will be because we chose to become players in the game by changing the odds of such planetwide biological upheavals. We are rapidly developing technology that could allow us to detect most threatening asteroids and modify their orbits well before they can strike Earth. Comets are more difficult to guard against because long-period comets appear from the outer Solar System with little warning. To offer protection, defense capabilities would have to be in place, ready to be used on very short notice. We have been slow to take such threats seriously. While impacts from kilometer-sized objects are infrequent, objects a few tens of meters in size, carrying the punch of a several-megaton bomb, strike Earth about once every 100 years. Perhaps an explosion like the 1908 Tunguska blast occurring over New York or Paris would be enough to convince us that such precautions are worthwhile.

To survive, humanity must learn to manage the threat of impacts.

We might protect ourselves from the fate of the dinosaurs, but in the long run the descendants of humanity will either leave this world or die out. Planetary systems surround other stars, and all that we know tells us that many other Earth-like planets should exist throughout our Galaxy. Colonizing other planets is currently the stuff of science fiction, but if our descendants are ultimately to survive the death of our home planet, off-Earth colonization must become science fact at some point in the future.

While humankind may soon be capable of protecting Earth from life-threatening comet and asteroid impacts, in other ways we are our own worst enemy. We are poisoning the atmosphere, the water, and the land that form the habitat for all terrestrial life. As our population grows unchecked, we are occupying more and more of Earth's land and consuming more and more of its resources, while sending thousands of species of plants and animals to their extinction each year. At the same time, human activities are dramatically affecting the balances of atmospheric gases. The climate and ecosystem of Earth constitute a finely balanced, complex system capable of exhibiting chaotic behavior. The fossil record shows that Earth has undergone sudden and dramatic climatic changes in response to even minor perturbations. Such drastic changes in the overall balance of nature would be certain to have consequences for our own survival. When politics is added to the mix, even more immediate dangers await. For the first time in human history, we possess the means to unleash nuclear or biological disasters that could threaten the very survival of our species. In the end, the fate of humanity will depend more than anything on whether we accept stewardship of ourselves and of our planet.

The Future of Our Expanding Universe

So much for our planet. What about the Universe itself? As we have seen, if the mass of the Universe is large enough and the cosmological constant small enough, gravity will win in the end. Hubble expansion will eventually reverse, and the Universe will collapse back into a state resembling that of the young, hot Universe. Galaxies, stars, planets, molecules, atoms, and subatomic particles—all might cease to exist as matter is replaced by pure energy. Perhaps from such a state, a new Universe would emerge. At the dawn of the 21st century, however, few cosmologists see such a "Big Crunch" in the future of the Universe. It appears that our Universe will expand forever, perhaps even at an ever accelerating pace. Does this mean the Universe will go on without end? Yes, but not in the form that we see today.

In 1997 Fred Adams and Gregory Laughlin of the University of Michigan published their calculations of the great eras, past and future, in the history of the Universe. During the first era—the first 500,000 years after the Big Bang and before recombination—the Universe

was a swarm of radiation and elementary particles. Today we live during the second era, the Era of Stars, but this too will end. Some 100 trillion (10^{14}) years from now, the last molecular cloud will collapse to

form stars, and a mere 10 trillion years later the least massive of these stars will evolve to form white dwarfs.

Following the Era of Stars, most of the normal matter in the Universe will be locked up in degenerate stellar objects: brown dwarfs, white dwarfs, and neutron stars. During this Era of Degeneracy, the occasional star will still flare up as ancient substellar brown dwarfs collide, merging to form low-mass stars that burn out in a short trillion years or so. However, the main source of energy during this era will come from the decay of protons and neutrons and the annihilation of particles of dark matter. Even these processes will eventually run out of fuel. In 10^{39} years, white dwarfs will have been destroyed by proton decay, and neutron stars will have been destroyed by the beta decay of neutrons.

As the Era of Degeneracy comes to an end, the only significant concentrations of mass left will be black holes. These will range from black holes with the masses of single stars to greedy monsters that grew during the Era of Degeneracy to have masses as large as those of galaxy clusters. During the period that follows, the Era of Black Holes, these black holes will slowly evaporate into elementary particles through the emission of Hawking radiation. A black hole with a mass of a few solar masses will evaporate into elementary particles in 10^{65} years, while galaxy-sized black holes will evaporate in around 10^{98} years. By the time the Universe reaches an age of 10^{100} years, even the largest of the black holes will be gone. A Universe vastly larger than ours will contain little but photons with colossal wavelengths, neutrinos, electrons, positrons, and other waste products of black hole evaporation. The Dark Era will have arrived, as the Universe continues to expand forever into the long, cold, dark night of eternity. This may be the final victory of entropy—the **heat death** of the Universe.

From our perspective, the Universe of the far distant future sounds like an extremely dull and lifeless place, but will that necessarily be so? Imagine for a moment that intelligent life evolved amid the swarm of free quarks and gluons that filled the Universe immediately after inflation. Such organisms would have been far smaller than today's atoms, and perhaps would have lived out their lives in 10^{-40} second or so.

To such organisms, the Universe—all three meters of it—would have seemed incomprehensibly vast. These creatures and their entire civilization would have had to evolve, live, and die out in a millionth of a trillionth of a trillionth of the time that it takes for a single synapse in our brains to fire. If such creatures ever calculated the conditions in *today's* Universe, they would have recoiled in horror. They would have imagined a frozen time when the temperature of the Universe was only a thousandth of a trillionth of a trillionth of what they knew—a time when most matter had ceased to exist altogether, and the tiny fraction that remained was spread out over a volume of space 10^{75} times greater than that of their Universe. In short, such creatures would have looked forward and seen *our* Universe as the frozen, desolate future. It seems doubtful they could have foreseen the existence of stars, galaxies, planets, and intelligent creatures for whom a single thought took longer than a billion trillion times the entire history of the Universe they knew.

Now turn the tables, and think *forward* to a time when the Universe is 10^{50} times older than it is today. Who can say that there will not be life then as well? Perhaps they will be organisms of magnetic fields and tenuous electron plasmas, spread across countless trillions of light-years of space, whose lives unfold over untold aeons of time. For such organisms, if they ever exist, it will be *we* who are the impossible creatures, alive for the briefest of instants, still immersed in the momentary fireball of the Big Bang.

SEEING THE FOREST THROUGH THE TREES

As our journey nears its end, we have reached high ground, a vantage point from which we can look back and survey the terrain that we have covered. From this perspective it becomes clear that from the beginning, our journey has been guided by a single underlying quest: to understand how we came to be. Whether talking about the formation of galaxies, the origin and evolution of stars, the history of our Solar System, the geology of our own world, the evolution of life, the changes that shaped the Universe itself during its earliest moments, or the eras of the far future, 21st-century astronomy (the science, as well as the book) is organized around the desire to better understand the origin of structure in the Universe.

The last century has seen remarkable strides toward this goal, and along the way a pattern has emerged. Look back on the quotation opening Chapter 15. What statement remains true in all circumstances? "And this, too, shall pass away." Structure is ephemeral. The Universe is not about destination. The Universe is a place of process and change. The things that "matter" are the things that happen along the way!

When you see your breath fog up on a cold winter day, think about elementary particles freezing out of the early Universe, giving rise not only to matter but to the fundamental forces of nature. As you watch the clouds build before a thunderstorm, think of galaxies coalescing within halos of dark matter. Glance at a crystalline snowflake as it lands on the sleeve of your coat, and in your mind's eye see a star that condensed out of clouds of interstellar gas and dust. The origin of structure is everywhere around us, and is no less remarkable for its familiarity.

As it is for galaxies and stars and planets, so too it is for ourselves. A recurring theme on our journey of discovery has been the systematic dismantling of conceptual walls that in our minds separated us from the larger Universe. Humanity does not stand outside the processes that shape the Universe. We are instead one more variety of the structure to which the Universe has given birth—a way station on a long road of evolving structure stretching back 14 billion years. Few single words are capable of eliciting as much emotional reaction from some people as the word *evolution*. Yet the public controversy surrounding evolution cannot change the scientific standing of this theory as one of the best tested and most successful theories in all of science. As stated by Stephen J. Gould, "The theory of evolution is in about as much trouble as the theory that Earth revolves around the Sun." By any reasonable standards of scientific knowledge, evolution is a fact.

Opponents of evolution often act as if evolution were a tiny piece of science that can simply be cast aside without doing violence to the rest. Having traveled the journey of *21st Century Astronomy*, it should be clear how absurd such a claim is. Modern astronomy would simply cease to be were it stripped of our understanding of the origin and evolution of galaxies, stars, planets, and every other component of the Universe that we observe. Everything that we see says that the Universe formed 14 billion years ago, not 6,000. Cosmology is the ultimate evolutionary science—the science of the origin and evolution of the Universe itself. During the 20th century, geology became the science of the evolution of the surface of Earth, while planetary science applied our understanding of terrestrial geology to understanding the evolution of other planets and their moons. The most fundamental questions in physics concern the origin and evolution of physical laws. So too is the case in modern biology, which simply makes no sense until it is organized around the theme of evolution by natural selection.

Evolution is anything but a scientific appendix that can be harmlessly removed. It is the backbone and central nervous system of modern science. In *Darwin's*

Dangerous Idea (1995), Daniel Dennett speaks of the concept of evolution as "universal acid: it eats through just about every traditional concept, and leaves in its wake a revolutionized world-view." As the 21st century gets under way, evolution of structure is *the* unifying theme that ties the breadth of modern science together into a beautiful, powerful, comprehensive whole. If we tried to pull the thread of evolution out of this tapestry, the whole cloth would unravel before our eyes.

STUDENT QUESTIONS

THINKING ABOUT THE CONCEPTS

1. Of the four fundamental forces in nature, which does not depend on electric charge?

2. As clumps of cold dark matter and normal matter collapse, they heat up. When a clump is about half its maximum size, the increased pressure caused by increased thermal motion of particles tends to inhibit further collapse. Yet, normal matter can overcome this effect and continue to collapse while dark matter cannot. Explain the reason for this difference.

3. How do astronomers use the following to measure the amount of dark matter contained in a cluster of galaxies?
   a. Motions of individual members of the cluster.
   b. Extremely hot gas that fills the intergalactic space within the cluster.
   c. Gravitational lensing by the cluster.

4. Imagine that there are galaxies in the Universe composed mostly of dark matter with lesser amounts of non-luminous normal matter. If this were true, how might we learn of the existence of such galaxies?

5. As the sensitivity of our instrumentation increases, we are able to look ever farther into space and, therefore, ever further back in time. When we reach the era of recombination, however, we run into a wall and can see no further back in time. Explain why.

6. What are the basic differences between a Grand Unified Theory (GUT) and the Theory of Everything (TOE)?

7. Suppose you could view the early Universe at a time when galaxies were first forming. How would it be different from the Universe we see today?

8. If we should eventually find life on Europa, what would this tell you about the probability of finding life on worlds around other stars?

9. A few scientists believe we may be the only advanced life in the Galaxy today. If this were indeed the case, which factors in the Drake equation would have to be extremely small?

10. The second law of thermodynamics says that the entropy (a measure of disorder) of the Universe is always increasing. Yet living organisms exist by creating order from disorder. Why does this not violate the second law of thermodynamics?

APPLYING THE CONCEPTS

11. The proton and antiproton each have the same mass, $m_p = 1.67 \times 10^{-27}$ kg. What is the energy (in joules) of each of the two gamma rays created in a proton-antiproton annihilation?

12. Suppose you brought together a gram of ordinary-matter hydrogen atoms (each composed of a proton and an electron) and a gram of antimatter hydrogen atoms (each composed of an antiproton and a positron). Keeping in mind that two grams is less than the mass of a dime:
    a. Calculate how much energy (J) would be released as the ordinary-matter and antimatter hydrogen atoms annihilated one another.
    b. Compare this with the energy released by a one-megaton hydrogen bomb (1.6×10^{14} J).

13. Excursion 20.1 ("Forever in a Day") takes events spread out over enormous intervals of time and compresses them into the more comprehensible interval of a single 24-hour day. Make your own "Life in a Day" by compressing all the important events of your lifetime into a single day, starting with your birth at the stroke of midnight and continuing to the present at the end of the day.

14. Consider an organism Beta that, because of a genetic mutation, has a 5% greater probability of survival than its non-mutated form, Alpha. Alpha has only a 95% probability (p_r) of reproducing itself compared to Beta. After n generations, Alpha's population within the species would be $S_p = (p_r)^n$ compared to Beta's. Calculate Alpha's relative population after 100 generations. (You may need a scientific calculator or help from your instructor to evaluate the quantity 0.95^{100}.)

15. To fully appreciate the power of heredity, mutation, and natural selection, consider Alpha's relative population (from Question 14) after 5,000 generations in a case where Beta has a mere 0.1% survivability advantage over Alpha.

Sometimes the light's all shining on me
Other times I can barely see
Lately it occurs to me
What a long strange trip it's been.

ROBERT HUNTER (1938–)

Epilog: We Are Stardust in Human Form

The Long and Winding Road

Go out at night and look at the stars, and feel the same sense of wonder and awe that our kind has always felt at the sight. Take it in, be amazed at the majestic canopy overhead, and let your imagination roam, just as our ancestors have for thousands of years. As you do, drift back down the road we have followed over the course of our journey. Reflect on all that we have come to know about the Universe, and on how much more magnificent the heavens are than our ancestors ever could have imagined. Our journey has been more than a description of the Universe and what it contains. It has been a travelog of the struggle and triumph of the human mind and spirit. Ever since humans first recognized that the patterns shaping our lives are echoed in the sky, we have searched for the threads connecting us to the cosmos, and have sought to understand our place in it. We have the privilege of being among the first generations of humans to find those threads, and to learn real answers to those age-old questions.

We have no way to count how many different stories have been told about the heavens, but we do know that for thousands of years most of those stories shared a common foundation. One cornerstone was the belief that Earth occupies a special place in the scheme of things. In our minds, we were at the center of Creation, fixed and immovable, and all that we saw was present only to give meaning to our existence. A second cornerstone of this traditional worldview was the belief that the heavens are fundamentally different from Earth. To our ancestors our world was the realm of the ordinary and mundane—a terrestrial existence built from Aristotle's earth, wind, fire, and water. In contrast, the heavens had their own separate reality. There we saw a realm of gods and angels, of mysticism and magic, of the perfect and unchanging fifth element.

So it remained for thousands of years until, at the dawn of the Renaissance, a Polish monk dared to challenge the wisdom of the ages and to think the unthinkable. Reviving a notion that had been discarded long before by the Greeks, Nicholas Copernicus allowed himself to imagine that perhaps it was the motion of Earth, rather than the motions of the Sun and stars, that shaped the passage of the days and the years. In so doing, he not only conceptually dislodged Earth from its moorings, but he also broke the shackles that had for so long constrained the human mind. What began as a crack in the foundation of our preconceptions would in the end turn that old view of the world to rubble. In its place we would construct an edifice of knowledge that has given us dominion over our world, and carried our thoughts to the edges of the Universe.

Copernicus was one of a succession of great minds who could not rest without first picking at the loose threads of the ideas in which they had been raised to believe. Water runs into the cracks in a slab of granite and freezes, expanding and pushing the cracks open, exposing the flaws in the rock. As the seasons come and go, the imposing boulder stands no chance in the face of this persistent onslaught. In like fashion, the persistent questioning and probing of great minds would eventually shatter the reign of enforced ignorance and entrenched authority. We have met a few of these great minds on our journey; there were many others. They were of different nationalities, different upbringings, different dispositions and beliefs, but their work shared a common theme: the answers to questions about the world come not from the pronouncements of authority or the prejudices taught in childhood, but from observing nature itself and thinking carefully, honestly, and openly about what we see.

In Newton's famous thought experiment, a ball fired from an imaginary cannon moves rapidly enough that it falls around the world in a circle. Such a "cannon" could not be built with the technology available in Newton's day. That would have to await the launch of *Sputnik* hundreds of years later. Yet while Newton could not make his cannon reality, he did not have to. All he had to do was look at the sky and watch as the Moon traced out its monthly path. Newton realized that the force holding the Moon in its orbit about Earth is the same force that gives us weight and guides the path of a ball thrown into the air. In fact, all Newton had to do to see his cannonball was look out across the English countryside, for all of us ride Newton's cannonball as the force of the Sun's gravity holds Earth in its yearly orbit.

This thought experiment was an important step in Newton's work. Eventually it led him to invent calculus, which he used to calculate the motions of the planets, making predictions that were confirmed by Kepler's empirical laws. The philosophical and scientific significance of Newton's thought experiment goes far deeper, however, for it signifies the final collapse of the barriers that humankind had placed between Earth and the heavens. With Newton's insight we came to see the heavens as part of the world around us—made of the same substance and shaped by the same physical laws. It is ironic that for knowledge to progress, we had to turn the early cornerstones of our thinking upside down. The modern foundation of our understanding of the Universe, the cosmological principle, is the literal negation of those early beliefs: Not only is Earth *not* the center of the Universe, but Earth occupies no special

place in the Universe *at all*. The heavens are *not* "the other," but are instead "the same." The heavens are knowable by going into terrestrial laboratories and learning about the nature of matter and energy and radiation, then applying this knowledge to careful observations of a Universe of stars, planets, and galaxies that is governed by physical law.

In our journey we have followed the trail of discovery that grew from this profound change in our understanding. We have watched as stars and planets formed, as stars lived out their lives and died, and as galaxies coalesced out of the primordial fireball of the Big Bang. We have followed our physics back to the very briefest of instants after the event that brought space and time into existence, and have seen the hints of theories that may in our lifetimes carry us the final step. There is no doubting the wonder of what we have seen. Yet for you there is another aspect to our journey that, in a practical sense, should be even more significant. While learning *about* the Universe, you have also come to better appreciate *how we know* those things, and in so doing have found a powerful, workable definition for what it means "to know."

According to Richard Paul, director of the Foundation for Critical Thinking in Dillon Beach, California, the three most common standards that people apply to knowledge even today are "it is true because I believe it," "it is true because we believe it," and "it is true because I want to believe it." Of course, none of these has anything to do with what really *is* true. To learn about the Universe and our world, we have had to set aside these notions, which blur the line between reality and fantasy, and replace them with a tough, unforgiving, and very different standard: "It is *provisionally* true because we have worked very hard to show that it is false, but so far have failed." This standard alone places reality itself in front of our parochial ideas and beliefs. It puts what *is* true ahead of what we would *like* to be true. Only by testing the falsifiable predictions of our theories about the world have we learned to push aside the comfortable notions that for so long prevented us from truly seeing our world and our Universe.

WE ARE STARDUST IN HUMAN FORM

As witnessed by the obelisks of Stonehenge or the ruins of a Mayan pyramid, humans have always built temples to the stars. We still build temples to the stars today. They are seen as an array of radio telescopes spread across the high desert of New Mexico, or a city of domes atop the summit of a dormant Hawaiian volcano, or a satellite telescope carried into orbit and subsequently repaired by space shuttle astronauts, or a tiny rover crawling across the surface of Mars. These modern temples are the legacy of insights by Copernicus, Kepler, Galileo, Newton, Einstein, Hubble, and countless others.

The discoveries that pour forth from these modern-day temples stretch the mind and stir the imagination. We have walked on the Moon, and have come to see the planets not as points of light in the sky, but as worlds as rich and complex as our own. We have looked at the remnants of stars that exploded long ago, and peered into eerie columns of glowing interstellar gas within which new stars are being born. We have gazed back in time at galaxies forming when the Universe was young, and have even learned to recognize the birth of the Universe itself in the faint glow of the cosmic background radiation. As we contemplate those wonders, the words from act I, scene V of Shakespeare's *Hamlet* seem almost frighteningly appropriate. As Hamlet faces the challenges and revelations brought by the ghost of his father, Horatio cries out:

"O day and night, but this is wondrous strange!"

To this comes Hamlet's immortal reply:

And therefore as a stranger give it welcome.
There are more things in heaven and earth, Horatio,
Than are dreamt of in your philosophy.

"There are more things in heaven and earth, Horatio, than are dreamt of in your philosophy,"—here is a message to shout back through the ages. At the start of our journey we asked the most basic of questions about what we see in the sky—"How big is it?"—and the answers were enough to expose the comedy of humanity's ancient conceits. Traveling at the speed of a modern jetliner, it would take us over 5 million years to cross the distance to even the nearest star beyond our Sun. Even so, we live in a Galaxy containing hundreds of billions of such stars, which are themselves outnumbered by other galaxies filling a Universe that may stretch on forever. Using the speed of light as our yardstick, we have come to realize that Earth—the world of our birth and the stage on which all of human history has been played—is to the expanse of the observable Universe as a single snap of our fingers is to the aeons that have transpired since time itself came into existence, roughly 14 billion years ago.

As we stare at the images that have come to symbolize 21st-century astronomy and consider what they show, it is easy to understand why some people recoil from these insights. "Wouldn't it be nice," they say, "if we could just go back to imagining that Earth is only 6,000 years old, and that humanity occupies a special place at the pinnacle of Creation?" Indeed, if this were where our story ended—with the fact of our seeming insignificance in the Universe—we might *all* long for an excuse to retreat into ignorance. Fortunately, this is not where our story ends. Rather, this is where our true journey of discovery begins. While modern science may have shattered our egotistical views about our exalted place in the scheme of things, it has also offered us a wonderful new appreciation and understanding of ourselves to fill that void.

When we look at distant galaxies, the light we see is produced by stars like our Sun. Each of those stars formed when a cloud of interstellar gas and dust collapsed under the same force of gravity that guided Newton's cannonball. As each of those clouds collapsed, it spun faster and faster, obeying the same laws of motion that accelerate the spin of an Olympic skater as she pulls her arms and leg ever more tightly to her body. This spin prevented those clouds from collapsing directly into stars, forcing them instead to settle into flat rotating disks. We see such disks today when we look at the newest generation of stars. Inside those disks, grains of dust stick together to make larger grains which stick together to make still larger grains—the beginning of a bottom-up process that culminates with planetesimals crashing together to make worlds. Our Earth is one such world. We have come to view our Sun, Earth, and Solar System as products of natural processes still going on around us today. As we watch new generations of stars form and search for the planets that surround them, we are witnessing a replay of the birth of our own world, 4.6 billion years ago. We have found our roots in the stars.

Go out at sunset on an evening when a waxing crescent Moon hangs low above the western horizon, and several planets stretch out across the darkening sky. The plane of the ecliptic is there in front of you, and your mind's eye might even envision the flat, rotating accretion disk from which our Solar System formed. Once you realize what you are looking at, the cradle of our world and ourselves hangs there in the night sky for all to see.

Looked at in this way, the sunset takes on a whole new significance. It will never be the same. Yet even the majesty of the planets spread across the sky fails to capture the intimacy of our connection with the Universe. In the most basic sense, the question "What are we?" is easily answered. Humans and all other terrestrial life are an organized assemblage of various organic molecules, most of which are very complex. Counting by the numbers of atoms in our bodies, we are approximately 60% hydrogen, 26% oxygen, 11% carbon, 2% nitrogen, with a small fraction of metals and other heavy elements mixed in. Over the course of our journey we have witnessed the history of those atoms. A very long time ago—roughly 14 billion years—something wonderful happened. The Universe came into being, and time began. From an infinitesimally small volume of concentrated energy, the Universe expanded, growing in size at the speed of light. Within the first few minutes, particles of solid matter, including protons and electrons, condensed out of this dense, primordial ball of energy. Nuclear reactions caused some of the protons to fuse into other light nuclei. Several hundred thousand years later, when the Universe had cooled to a temperature of a few thousand kelvins, those nuclei combined with electrons to form atoms. Of those atoms, roughly 90% were hydrogen atoms and 10% were helium atoms. There were traces of lithium, beryllium, and boron as well, but that was all that existed in the way of normal luminous matter as the Universe expanded past the threshold of recombination.

The hydrogen atoms in our bodies date back to this early time in the history of the Universe, but what of the rest? Having taken our journey of discovery, you know the answers. As the Universe emerged from the Big Bang, clumps of dark matter began to collapse under the force of gravity, pulling normal matter along with it. Within these collapsing protogalaxies the first generations of stars formed—nuclear furnaces powered by the fusion of those original hydrogen atoms into increasingly massive elements. Carbon, oxygen, silicon, sulfur, all the way up to iron and nickel were formed in those stellar infernos. As those first generations of stars ended their lives, they blasted this nuclear ash back into the reaches of interstellar space. Nucleosynthesis did not end with fusion, however. In the extreme environments of supernovae, free neutrons were captured by the products of fusion, building more-massive elements still. Atoms of copper, zinc, tin, silver, and gold, all the way up to the most massive naturally occurring element—uranium—were formed and expelled into space. Here were the chemical elements to fill the periodic table, and to build the compounds of life. As early protogalaxies merged to form early galaxies, more generations of stars continued to enrich the Universe with the fruits of their alchemy. As early galaxies settled into the well ordered ellipticals and spirals of today's Universe, still more generations of stars came and went, adding to the chemical richness of the Universe. When our Solar System appeared on the scene 4.6 billion years ago, it formed from interstellar material that carried the chemical building blocks of planets and of life—atoms produced both in the Big Bang, and in the hearts of generations of stars that had lived and died during the 9 billion years that had transpired since the Universe began. "What are we, and how did we get here?" We are stardust in human form.

THE FUTURE ARRIVES EVERY DAY

While traveling the highways and back roads of 21st-century astronomy, we have come to see our world and ourselves in a very different light. Even so, nothing we have seen has changed the most basic circumstances of our day-to-day existence. Earth remains our world, our home. The hopes and dreams we humans feel are no less real today than they were a thousand years ago. In 1968, Stanley Kubrick and Arthur C. Clarke collaborated to make the film *2001: A Space Odyssey*. This provocative piece of speculative fiction captured the imagination of a generation, and the year 2001 came to signify the future. That future is now here, yet little of our modern-day life is recognizable in those cinematic prognostications, dating from only three decades ago. It is ironic that while we can forecast the future of the Sun with great accuracy, and can even calculate the fate of the Universe itself, our vision of our own future is so much less certain.

These words were first drafted on the afternoon of December 31, 2000. At that moment in the life of this coauthor, roughly half the surface of Earth remained in the 20th century, while the other half of the world had witnessed the beginning not only of a new century, but also of a new millennium. As the dividing line between the two swept across Europe and out into the Atlantic Ocean, it was just another moment in the long dance of Earth as it spins on its axis and falls around the Sun. The Sun took no notice of the moment as it continued its orbit around the center of the Milky Way Galaxy, which itself is but a speck in an expanding Universe. Yet while that December afternoon had no objective physical significance, it was loaded with symbolic meaning for our species. The year of Kubrick and Clarke's mind-bending tale was at hand. Though the details of their vision have turned out to be incorrect, there is no question but that we live at an amazing moment in history—a moment when science has for the first time allowed us to truly see the Universe beyond ourselves, and offers the promise of showing us the universe within ourselves, as well.

There is no denying the fact that we live in an evolving Universe—a place of ongoing and unending change that continues to shape humanity as certainly as it shapes the cosmos itself. The grim prospects mentioned in the closing sections of Chapter 20 are real. Saying that they do not exist or choosing to push them from our minds will not make them go away. Yet at the same time, modern medicine, food production, transportation, and a thousand other technologies that push back the ancient scourges of humanity are equally real. Differences among peoples remain, but modern communication offers the hope of spreading understanding. Meanwhile, a picture of Earthrise above the lunar horizon taken by the *Apollo 8* astronauts **(Figure 1)** hangs forever as part of the human experience and perspective, showing us as nothing else could that Earth is a tiny fragile island to be cherished. Whether we like it or not, humanity is a single, interrelated, interdependent global village that will face the future together—or not at all.

At the dawn of the 21st century, science has shown us the wonders of the Universe, and at the same time has given us the knowledge and power to shape our world and choose our future. The future of humanity may depend entirely on how well we treat Earth and ourselves over the few decades and centuries ahead. If 21st-century astronomy does nothing else, it forces us to change our perspective. As we study the laws that govern the workings of atoms, we learn something about the conditions of our own lives: the future is not yet written. As we use our telescopes to stare at galaxies 10 billion light-years away, collecting light from stars that died billions of years before our Sun was even born, manifest destiny seems a pretty silly concept. There are no guarantees that things will work out in the end for one tiny world or for the species to which it gave birth. If we choose to destroy our world, either through direct action or simple neglect, so be it. The Universe as a whole will carry on in sublime indifference to our fate.

This is not, however, a message of despair. Instead, it is a message of hope, responsibility, and maturity. We have it in our power to make our Earth a paradise, or to leave our children's children to cope with a world choking from our shortsighted excess. The choices are ours, whether we want them or not. Borrowing from the book of Genesis, we have truly tasted of the tree of knowledge. As we stand at the beginning of the 21st century, we face a future filled with choices, but one choice we are not allowed is refusing to acknowledge responsibility for our own destiny.

As you read these words, this coauthor's moment of introspection at the threshold between the second and third millennia will have passed. The remaining months needed to turn a draft manuscript into the volume that you hold are, for you, in the past. Even so, the moment when you read these words is not so different from that New Year's Eve of the year A.D. 2000. While the symbolism of your moment may not be as palpable as that of an instant when the world passes a major milestone in humanity's accounting of events, your moment is a milestone nonetheless. It is one of a stream of moments that define your life as the possibilities of the future cross, inexorably, into the unchangeable reality of the past. That moment is there for you to use as you will. The course of the future is yours to shape.

With that thought, we come to the end of our journey. We, the authors, hope that this journey has helped to open your eyes to the wonder of the world and the Universe around you. Even more, we hope this journey has given you pause to reflect on who we are as humans, and on our place in the larger reality in which we find ourselves. Finally, we hope this journey has changed the shape of the way you think, not only about the sights you see in the night sky, but also about the events of your daily life. If any or all of these hopes are fulfilled, then the journey will have been worth taking, worthy of your time and thought, and of ours.

Figure 1 *This image of Earth taken from lunar orbit by the* Apollo 8 *astronauts forever changed our understanding of our Earth and ourselves.*

GLOSSARY

A

aberration of starlight The apparent displacement in the position of a star due to the finite speed of light and Earth's orbital motion around the Sun.

absolute zero The temperature at which thermal motions cease. The lowest possible temperature. Zero on the kelvin temperature scale.

absorption The capture of electromagnetic radiation by matter.

absorption line An intensity minimum in a spectrum due to absorption of electromagnetic radiation at a specific wavelength determined by the energy levels of an atom or molecule.

acceleration The rate at which the speed and/or direction of an object's motion is changing.

accretion The process by which smaller bodies coalesce to form larger ones.

accretion disk A flat, rotating disk of gas and dust surrounding a condensed mass, such as a young stellar object, a forming planet, or a collapsed star in a binary pair.

achondrite Stony meteorite that does not contain chondrules.

active comet Comet nucleus that approaches close enough to the Sun to show signs of activity, such as the production of a coma and tail.

active galactic nucleus (AGN) A highly luminous, compact galactic nucleus whose luminosity may exceed that of the rest of the galaxy.

AGB star A star on the asymptotic giant branch.

AGN See *active galactic nucleus*.

albedo The fraction of electromagnetic radiation incident on a surface that is reflected by the surface.

alpha particle An He^4 nucleus consisting of two protons and two neutrons. Alpha particles are given off in the type of radioactive decay referred to as alpha decay, hence their name.

Alpher, Ralph (1921–) The physicist who worked with George Gamow to predict the existence of the cosmic background radiation.

amplitude In a wave, the maximum excursion from equilibrium. For example, in a water wave the amplitude is the vertical distance from crests to the undisturbed water level.

Amors Family of asteroids with orbits that cross the orbit of Mars but not that of Earth.

angular momentum A conserved property of a rotating or revolving system whose value depends on the velocity and distribution of its mass.

annular eclipse A solar eclipse that occurs when the apparent diameter of the Moon is less than that of the Sun, leaving a visible ring of light (annulus) surrounding the dark disk of the Moon.

Antarctic Circle The circle on Earth with latitude 66.5° south, marking the limit of the Antarctic region.

antimatter Matter made from antiparticles.

antiparticle An elementary particle of antimatter identical in mass but opposite in charge and all other properties to its corresponding ordinary matter particle.

aphelion (pl. **aphelia**) The point in a solar orbit that is farthest from the Sun.

Apollos A group of asteroids whose orbits cross the orbits of both Earth and Mars.

arcminute See *minute of arc*.

arcsecond See *second of arc*.

Arctic Circle The circle on Earth with a latitude of 66.5° north marking the limit of the Arctic region.

asteroid A primitive rocky or metallic body (planetesimal) that has survived planetary accretion. Asteroids are parent bodies of meteoroids.

asteroid belt The region between the orbits of Mars and Jupiter that contains most of the asteroids in our Solar System.

astronomical seeing A measurement of the degree to which Earth's atmosphere degrades the resolution of a telescope's view of astronomical objects.

astronomical unit (AU) The average distance from the Sun to Earth, approximately 150 million kilometers.

astronomy The scientific study of planets, stars, galaxies, and the Universe as a whole.

astrophysics The application of physical laws to the understanding of planets, stars, galaxies, and the Universe as a whole.

asymptotic giant branch (AGB) In the H-R diagram, a separate "branch" that goes from the horizontal branch toward higher luminosities and lower temperatures, asymptotically approaching and then rising above the red giant branch.

Atens A group of asteroids whose orbits cross Earth's orbit, but not the orbit of Mars.

atmosphere The gravitationally bound, outer gaseous envelope surrounding a planet, moon, or star.

atmospheric greenhouse effect A warming of planetary surfaces produced by atmospheric gases that transmit optical solar radiation but partially trap infrared radiation. Compare with *greenhouse effect*.

atmospheric windows Regions of the electromagnetic spectrum in which radiation is able to penetrate a planet's atmosphere.

atom The smallest piece of an element that retains the properties of that element. Each atom is composed of a nucleus (neutrons and protons) surrounded by a cloud of electrons.

aurorae Emission in the upper atmosphere of a planet from atoms that have been excited by collisions with energetic particles from the planet's magnetosphere.

autumnal equinox (a) The point where the ecliptic crosses the celestial equator, with the Sun moving from north to south. (b) The day, around September 23rd, that the Sun appears at this location, which is the first day of autumn (fall) in the Northern Hemisphere.

axion A hypothetical elementary particle first proposed to explain certain properties of the neutron and now considered a candidate for cold dark matter.

B

backlighting Illumination from behind a subject as seen by an observer. Fine material such as human hair and dust in planetary rings stands out best when viewed under backlighting conditions.

bar A unit of pressure. One bar is equivalent to 10^5 newtons per square meter, approximately equal to Earth's atmospheric pressure at sea level.

barred spiral Spiral galaxy whose bulge has an elongated, barlike shape.

basalt Gray to black volcanic rock, rich in iron and magnesium.

beta decay The decay of a neutron into a proton by emission of an electron (beta ray) and an antineutrino, or the decay of a proton into a neutron by emission of a positron and a neutrino.

Bethe, Hans (1906–) The physicist who won the Nobel Prize for his work on the source of stellar energy.

Big Bang The event that occurred about 14 billion years ago that marks the beginning of time and the Universe.

Big Bang nucleosynthesis The formation of low-mass nuclei (H, He, Li, Be, B) during the first few minutes following the Big Bang.

binary star System in which two stars are in gravitationally bound orbits about their common center of mass.

binding energy The energy required to separate an atomic nucleus into its component protons and neutrons.

bipolar outflow Material streaming away in opposite directions from either side of the accretion disk of a young star.

blackbody An object that absorbs and re-emits all electromagnetic energy it receives.

blackbody spectrum See *Planck spectrum*.

black hole An object so dense its escape velocity exceeds the speed of light. A *singularity* in space-time.

blueshift The Doppler shift toward shorter wavelengths of light from an approaching object. Compare with *redshift*.

bolide A very bright, exploding meteor.

Boltzmann's constant (k) The constant that relates the temperature of a gas to the average kinetic energy of the molecules of the gas.

bound orbit A closed orbit in which the velocity is less than the escape velocity.

Brahe, Tycho (1546–1601) The foremost observer before the invention of the telescope. Tycho provided the data that permitted Kepler to deduce the motions of the planets around the Sun.

brightness The apparent intensity of light from a luminous object. Brightness depends both on the luminosity of a source and its distance. Units at the detector: watts per square meter (W/m^2).

brown dwarf A "failed" star that is not massive enough to cause hydrogen fusion in its core.

bulge The central region of a spiral galaxy that is similar in appearance to a small elliptical galaxy.

b_B/b_V color Measurement of the color of an object on the basis of the ratio of its brightness in blue light to its brightness in "visual" (or yellow-green) light.

C

Cannon, Annie Jump (1863–1941) The American astronomer who systematically classified the spectral types of stars in the Henry Draper Catalog, the first catalog to list stellar spectral types for 250,000 stars.

carbonaceous chondrite Primitive stony meteorites that contain chondrules and are rich in carbon and volatile materials.

carbon–nitrogen–oxygen cycle See *CNO cycle*.

carbon star Cool *red giant* or AGB star that has an excess of carbon in its atmosphere.

Cassini, Jean-Dominique (1625–1712) The Italian astronomer (nationalized French in 1673) who discovered four satellites of Saturn as well as the division within Saturn's ring that bears his name.

Cassini Division The largest gap in Saturn's rings, discovered by Jean-Dominique Cassini in 1675.

catalyst Atomic and molecular structure that permits or encourages chemical and nuclear reactions, but does not change its own chemical or nuclear properties.

CBR See *cosmic background radiation*.

CCD See *charge-coupled device*.

celestial equator The imaginary great circle that is the projection of Earth's equator onto the celestial sphere.

celestial sphere An imaginary sphere with celestial objects on its inner surface and Earth at its center. The celestial sphere has no physical existence but is a convenient tool for picturing the directions in which celestial objects are seen from the surface of Earth.

Celsius (C) The arbitrary temperature scale with 0°C at the freezing point of water and 100°C at the boiling point of water at sea level. Defined by Anders Celsius (1701–1744). Also known as the Centigrade scale.

center of mass The location within an isolated system in which we may regard the entire mass of the system as being concentrated. The point in any isolated system that moves according to Newton's first law of motion.

centripetal acceleration The acceleration of an object directed toward the center of curvature of its motion.

centripetal force A force directed toward the center of curvature of an object's curved path.

Cepheid variable An evolved high-mass star with an atmosphere that is pulsating, leading to variability in the star's luminosity and color.

Chandrasekhar limit The upper limit on the mass of an object supported by electron degeneracy pressure—approximately 1.4 $M_\odot$.

chaos Behavior in complex, interrelated systems in which tiny differences in the initial configuration of a system result in dramatic differences in the system's later evolution.

charge-coupled device (CCD) A common type of solid-state detector of electromagnetic radiation that transforms the intensity of light directly into electric signals.

chondrite Stony meteorite containing chondrules.

chondrule Spherical inclusion of rapidly cooled melt found inside some meteorites.

chromosphere The region in the Sun's atmosphere located between the photosphere and the corona.

circular velocity The orbital velocity needed to keep an object moving in a circular orbit.

circumpolar That part of the sky, near either celestial pole, that can always be seen above the horizon from a specific location on Earth.

classical mechanics The science of applying Newton's laws to the motion of objects.

climate The state of an atmosphere averaged over an extended time.

closed universe A finite universe with a curved spatial structure such that the angles of a triangle always exceed 180°.

CNO cycle (carbon–nitrogen–oxygen cycle) One of the ways in which hydrogen burning (the fusion of four hydrogen atoms to form one helium atom) can take place.

cold dark matter Dark matter particles that move slowly enough to be gravitationally bound even in the smallest galaxies.

coma (pl. **comae**) The nearly spherical cloud of gas and dust surrounding the nucleus of an active comet.

comet A complex object consisting of a small solid, icy nucleus, an atmospheric halo, and a tail of gas and dust.

comet nucleus (pl. **nuclei**) A primitive planetesimal, composed of ices and *refractory materials* that has survived planetary accretion. The "heart" of a comet, containing nearly the entire mass of the comet. A "dirty snowball."

comparative planetology The study of planets through comparison of their chemical and physical properties.

complex system An interrelated system capable of exhibiting chaotic behavior. See also *chaos*.

conservation of energy The conservation law stating that in an isolated, closed system, the total energy does not change.

conservation law Physical law stating that the amount of some physical quantity (such as energy or angular momentum) of an isolated system does not change over time.

constant of proportionality The multiplicative factor by which one quantity is related to another.

constellation Imaginary image formed by patterns of stars; any of 88 defined areas on the celestial sphere used by astronomers to locate celestial objects.

continental drift The slow motion (centimeters per year) of Earth's continents relative to each other and to Earth's mantle.

continuous radiation Electromagnetic radiation with intensity that varies smoothly over some range of wavelengths.

convection The transport of thermal energy from the lower (hotter) to the higher (cooler) layers of a fluid by motions within the fluid driven by variations in buoyancy.

convective zone A region within a star in which energy is transported outward by convection.

Copernicus, Nicholas (1473–1543) The Polish monk who first proposed the model of a Sun-centered Solar System to the Western world.

core (a) The innermost region of a planetary interior. (b) The innermost part of a star.

Coriolis effect The apparent displacement of objects in a direction perpendicular to their true motion as viewed from a rotating frame of reference. On a rotating planet, different latitudes rotating at different speeds cause this effect.

corona The hot, outermost part of the Sun's atmosphere.

coronal hole Lower-density region in the solar corona containing "open" magnetic field lines along which coronal material is free to stream into interplanetary space.

coronal mass ejection An eruption on the Sun that ejects hot gas and energetic particles at much higher speeds than are typical in the solar wind.

cosmic background radiation (CBR) Isotropic microwave radiation from the celestial sphere having a 2.73 K Planck spectrum. The CBR is understood as residual radiation from the Big Bang.

cosmic ray Very fast moving particle (mostly atomic nuclei) that resides in the disk of our Galaxy.

cosmological constant A constant introduced into general relativity by Einstein that characterizes an extra, repulsive force in the Universe due to the vacuum of space itself.

cosmological principle The (testable) assumption that the same physical laws that apply here and now also apply everywhere and at all times, and that there are no special locations or directions in the Universe.

cosmological redshift (z) The redshift that results from the expansion of the Universe, rather than from the motions of galaxies or gravity (see *gravitational redshift*).

cosmology Study of the large-scale structure and evolution of the Universe as a whole.

Crab Nebula The remnant of the Type II supernova explosion witnessed by Chinese astronomers in A.D. 1054.

crescent Phases of the Moon, Mercury or Venus in which the object appears less than half-illuminated by the Sun.

Cretaceous period   The interval in Earth's history from 146 million to 65 million years ago.

Cretaceous–Tertiary (K-T) boundary   The boundary between the Cretaceous and Tertiary periods in Earth's history; corresponds to the time of the impact of an asteroid or comet and the extinction of the dinosaurs.

critical mass density   The value of the mass density of the Universe which, ignoring any cosmological constant, is just barely capable of halting expansion of the Universe.

crust   The relatively thin, outermost, hard layer of a planet chemically distinct from the interior.

C-type asteroid   An asteroid made of material that has largely been unmodified since the formation of the Solar System; the most primitive type of asteroid.

Curtis, Heber D. (1872–1942)   The Lick Observatory astronomer who argued, in the Great Debate of 1920, that galaxies are island universes.

cyclonic motion   The rotation of a weather system resulting from the Coriolis effect as air moves toward a region of low atmospheric pressure.

Cygnus X-1   A binary X-ray source and probable black hole.

D

dark matter   The matter existing in galaxies and in groups and clusters of galaxies that does not emit or absorb electromagnetic radiation; thought to comprise most of the mass in the Universe.

dark matter halo   The centrally condensed, greatly extended dark matter component of a galaxy that contains up to 95% of the galaxy's mass.

daughter product   An element resulting from radioactive decay of a more massive parent element.

decay   The process of a radioactive nucleus changing into its daughter element, or an atom or molecule dropping from a higher energy state to a lower energy state.

density   The measure of an object's mass per unit of volume. Units: kg/m^3.

differential rotation   Rotation of different parts of a system at different rates.

differentiation   The process by which materials of higher density sink toward the center of a molten or fluid planetary interior.

diffraction   The spreading of a wave after it passes through an opening or past the edge of an object.

diffraction limit   The limit of a telescope's angular resolution caused by diffraction.

distance ladder   A sequence of techniques for measuring cosmic distances: each method is calibrated using the results from other methods that have been applied to closer objects.

Doppler effect   The change in wavelength of sound or light due to the relative motion of the source toward or away from the observer.

Doppler redshift   See *redshift*.

Doppler shift   The change in observed wavelength in a wave from a source moving toward or away from an observer.

Drake equation   A prescription for estimating the number of intelligent civilizations existing elsewhere; named after Frank Drake (1930–).

dust tail   A type of comet tail consisting of dust particles that are pushed away from the comet's head by radiation pressure from the Sun.

dwarf galaxy   A small galaxy with a luminosity ranging from 1 million to 1 billion solar luminosities.

dynamic equilibrium   A situation in which the configuration of a system does not change even though the components of that system are in motion.

dynamo   A device that converts mechanical energy into electric currents and magnetic fields.

E

eccentricity (e)   A measure of the departure of an ellipse from circularity; the ratio of the distance between the two foci of an ellipse to its major axis.

eclipse season   Times during the year when the line of nodes is sufficiently close to the Sun for eclipses to occur.

eclipsing binary   A binary system in which the orbital plane is oriented such that the two stars appear to pass in front of one another as seen from Earth.

ecliptic   (a) The apparent annual path of the Sun against the background of stars. (b) The projection of Earth's orbital plane onto the celestial sphere.

Einstein, Albert (1879–1955)   The eminent 20th-century physicist whose theories on relativity and the properties of light changed the way we view the physical universe.

ejecta   (a) Material thrown outward by the impact of an asteroid or comet on a planetary surface, leaving a crater behind. (b) Material thrown outward by a stellar explosion.

electric field   A field that is able to exert a force on a charged object, whether at rest or moving.

electric force   The force exerted on a charged particle by an electric field.

electromagnetic force   The force, including both electric and magnetic forces, that acts on electrically charged particles. One of four fundamental forces of nature. The force mediated by photons.

electromagnetic radiation   A traveling disturbance in the electric and magnetic fields caused by accelerating electric charges. In quantum mechanics, a stream of photons. Light.

electromagnetic spectrum   The spectrum made up of all possible frequencies or wavelengths of electromagnetic radiation, ranging from gamma rays through radio waves and including the portion our eyes can use.

electromagnetic wave   Wave consisting of oscillations in the electric field strength and the magnetic field strength.

electron (e^-)   A fundamental particle having a negative charge of 1.6×10^{-19} coulomb, a rest mass of 9.1×10^{-31} kg, and mass-equivalent energy of 8×10^{-14} joules.

electron degenerate   Describes the state of material compressed to the point where electron density reaches the limit imposed by the rules of quantum mechanics.

electroweak theory   The quantum theory that combines descriptions of both the electromagnetic force and the weak nuclear force.

element   One of 92 naturally occurring substances (such as hydrogen, oxygen and uranium) and more than 20 human-made ones (such as plutonium). Each element is chemically defined by the specific number of protons in the nuclei of its atoms.

ellipse   A conic section produced by the intersection of a plane with a cone by passing the plane through the cone at some angle to the axis other than 0° or 90°.

elliptical galaxy   Galaxy with a circular to elliptical outline on the sky containing almost no disk and a population of old stars (abbreviated E galaxy).

emission   The release of electromagnetic energy when an atom, molecule, or particle drops from a higher energy state to a lower energy state.

emission line   An intensity peak in a spectrum due to sharply-defined emission of electromagnetic radiation in a narrow range of wavelengths.

empirical   Based primarily on observations and experimental data. Descriptive rather than theoretical.

energy   The conserved quantity that gives objects and systems the ability to do work. Units: joules (J).

energy transport   The transport of energy from one location to another. In stars, energy transport is mostly carried out by radiation or convection.

entropy   A measure of the disorder of a system related to the number of ways a system can be rearranged without affecting its appearance.

equator   The great circle on the surface of a body midway between its poles. The equatorial plane passes through the center of the body and is perpendicular to its rotation axis.

equilibrium   The state of an object in which physical processes balance each other so that its properties or conditions remain constant.

equinox ("equal night")   (a) One of two positions on the ecliptic where it intersects the celestial equator. (b) Either of the two times of year at which the Sun is at one of these two positions. At this time, night and day are of the same length everywhere on Earth. See also *autumnal equinox, vernal equinox.*

equivalence principle   The principle stating that there is no difference between a reference frame that is freely floating through space and one that is freely falling within a gravitational field.

escape velocity   The minimum velocity needed for an object to achieve a parabolic trajectory and thus permanently leave the gravitational grasp of another mass.

Eta Carinae   An extremely massive luminous star that underwent an episode of extreme mass loss during the 1840s.

event   A particular location in space-time.

event horizon   The effective "surface" of a black hole. Nothing inside this surface—not even light—can escape from a black hole.

evolutionary track   The path that a star follows across the H-R diagram as it evolves through its lifetime.

excited state   An energy level of a particular atom, molecule, or particle, higher than its ground state.

F

Fahrenheit (F)   The arbitrary temperature scale with 32°F at the melting point of water and 212°F at the boiling point of water at sea level. Defined by Gabriel Fahrenheit (1686–1736).

fault   A fracture in the crust of a planet or moon along which blocks of material can slide.

filter   An instrument element that transmits a limited wavelength range of electromagnetic radiation. For the optical range such elements are typically made of different kinds of glass and take on the hue of the light they transmit.

first quarter Moon   The phase of the Moon as seen from Earth in which only the western half of the Moon is illuminated by the Sun; occurs about a week after a new Moon.

fissure   A fracture in the planetary lithosphere from which magma emerges.

flatness problem   The surprising result that the sum of Ω_{mass} plus Ω_Λ is so close to unity in the present-day Universe; equivalent to saying that it is surprising the Universe is so flat.

flat rotation curve   Rotation curve of a spiral galaxy in which rotation rates do not decline in the outer part of the galaxy, but remain relatively constant to the outermost points.

flat universe   An infinite universe whose spatial structure obeys Euclidean geometry, such that the sum of the angles of a triangle always equals 180°.

flux   The total amount of energy passing through each square meter of a surface each second. Units: watts per square meter (W/m^2).

flux tube   Tubelike structures in a plasma, such as the atmosphere of the Sun, or the interaction of Io and Jupiter, that are "bundled" together by magnetic fields.

focal length   The optical distance between a telescope's objective lens or primary mirror and the plane (called the focal plane) upon which the light from a distant object is focused.

focus (pl. foci)   (a) One of two points that define an ellipse. (b) A point in the focal plane of a telescope.

fold   Location where rock is bent or warped.

Foucault, Jean-Berrnard-Leon   (1819–1868)   The French physicist who was the first to demonstrate that Earth rotates using the periodic motions of a pendulum. A pendulum that demonstrates Earth's rotation is called a Foucault pendulum in his honor.

Foucault pendulum   A pendulum used to demonstrate the rotation of Earth.

frame of reference   A frame against which an observer measures positions, motions, etc.

free fall The motion of an object when the only force acting on it is gravity.

frequency The number of times per second that a periodic process occurs. Unit: hertz (Hz), 1/s.

full Moon The phase of the Moon as seen from Earth in which the near side of the Moon is fully illuminated by the Sun; occurs about two weeks after a new Moon.

G

galaxy A gravitationally bound system that consists of stars and star clusters, gas, dust, and dark matter; typically greater than 1,000 light-years across; and recognizable as a discrete, single object.

galaxy cluster A large, gravitationally bound collection of galaxies containing hundreds to thousands of members; typically 10 to 15 Mly across.

galaxy group A small, gravitationally bound collection of galaxies containing from several to one hundred members; typically 4 to 6 Mly across.

Galileo Galilei (1564–1642) The first to use a telescope to make useful discoveries about the heavens, and develop the concept of inertia.

gamma ray Electromagnetic radiation with higher frequency, higher photon energy, and shorter wavelength than all other types of electromagnetic radiation.

Gamow, George (1904–1968) Physicist who, with his student Ralph Alpher, first predicted the existence of the *cosmic background radiation.*

gas giant A giant planet formed mostly of hydrogen and helium.

general relativistic time dilation The verified prediction that time passes more slowly in a gravitational field than in the absence of a gravitational field. See *time dilation.*

general relativity See *general theory of relativity.*

general theory of relativity Einstein's theory explaining gravity as the distortion of space-time by massive objects; also known simply as *general relativity.* This theory deals with all types of motion; for contrast, see *special relativity.*

geodesic The path an object will follow through space-time in the absence of external forces.

giant galaxy Galaxy with luminosity greater than about a billion solar luminosities.

giant molecular cloud An interstellar cloud composed primarily of molecular gas and dust, having hundreds of thousands of solar masses.

giant planet One of the largest planets in the Solar System, typically 10 times the size and many times the mass of the terrestrial planets and lacking a solid surface.

gibbous Phases of the Moon, Mercury or Venus in which the object appears more than half-illuminated by the Sun.

global circulation The overall, planetwide circulation pattern of a planet's atmosphere.

globular cluster A spherically symmetric, highly condensed cluster of stars, containing tens of thousands to a million members.

gluon The particle that carries (or, equivalently, mediates) interactions due to the strong nuclear force.

gradation The leveling of a planet's surface through weathering, erosion, transpiration, and deposition of rock debris by water, wind, and gravity.

grand unified theory (GUT) A unified quantum theory that combines the *strong nuclear, weak nuclear,* and *electromagnetic forces* but does not include gravity.

granite Rock that is cooled from magma and is relatively rich in silicon and oxygen.

gravitational lens A massive object that gravitationally focuses the light of a more distant object, to produce multiple brighter, magnified, possibly distorted images.

gravitational lensing The bending of light by gravity.

gravitational potential energy The potential energy that an object has solely due to its position within a gravitational field.

gravitational redshift The shifting to longer wavelengths of the wavelength of radiation from an object deep within a gravitational well.

gravity (a) The mutually attractive force between massive objects. (b) An effect arising from the bending of space-time by massive objects. (c) One of four fundamental forces of nature.

gravity wave Wave in the fabric of space-time emitted by accelerating masses.

great circle Any circle on a sphere that has as its center the center of the sphere. The celestial equator, the meridian, and the ecliptic are all great circles on the sphere of the sky, as is any circle drawn through the zenith.

Great Red Spot (GRS) The giant, oval, brick-red anticyclone seen in Jupiter's southern hemisphere.

greenhouse effect The solar heating of air in an enclosed space, such as a closed building or car, resulting primarily from the inability of the hot air to escape. Compare with *atmospheric greenhouse effect.*

greenhouse molecule One of a group of atmospheric molecules such as carbon dioxide that are transparent to visible radiation but absorb infrared radiation.

ground state The lowest-possible energy state for a system or part of a system, such as an atom, molecule, or particle.

GUT See *grand unified theory.*

Guth, Alan (1947–) The physicist who first proposed that the Universe underwent an early, rapid *inflation,* which resolves both the flatness and horizon problems.

H

Hadley circulation A simplified, and therefore uncommon, atmospheric global circulation that carries thermal energy directly from the equatorial to the polar regions of a planet.

half-life The time it takes half a sample of a particular radioactive element to decay to a daughter element.

halo The spherically symmetric, low-density distribution of stars and dark matter that defines the outermost regions of a galaxy.

harmonic law Another name for *Kepler's third law* of planetary motion.

Hawking, Stephen (1942–) The British physicist/astrophysicist who has done much to further our understanding of the Universe and the phenomena surrounding black holes.

Hawking radiation Radiation from a black hole.

Hayashi track The path that a protostar follows on the H-R diagram as it contracts toward the main sequence.

head The part of a comet that includes both the nucleus and the inner part of the coma.

heat death The possible eventual fate of an open universe, in which entropy has triumphed and all energy- and structure-producing processes have come to an end.

heavy element See *massive element.*

helioseismology The use of solar oscillations to study the interior of the Sun.

helium flash The runaway explosive burning of helium in the degenerate helium core of a *red giant* star.

Herbig-Haro (HH) object Glowing, rapidly moving knot of gas and dust that is excited by bipolar outflows in very young stars.

heredity The process by which one generation passes on its characteristics to future generations.

hertz (Hz) A unit of frequency equivalent to cycles per second.

Hertz, Heinrich (1857–1894) The 19th-century physicist who supplied the first experimental data confirming Maxwell's predictions about electromagnetic radiation.

Hertzsprung, Einar (1873–1967) The Danish astronomer who, along with Henry Norris Russell, defined the main diagnostic tool for studying the properties of stars, in which we graph stellar luminosity versus stellar spectral type (or temperature, or color). The Hertzsprung-Russell (or H-R) diagram is so-named in his honor.

Hertzsprung-Russell diagram See *H-R diagram.*

HH object See *Herbig-Haro object.*

hierarchical clustering The "bottom up" process of forming large-scale structure. Small-scale structure first produces groups of galaxies, which in turn form clusters, which then form superclusters.

high-velocity star Star belonging to the halo found near the Sun, distinguished from disk stars by moving far faster and often in the direction opposite to the rotation of the disk and its stars.

homogeneous In cosmology, describes a universe in which observers in any location would observe the same properties.

horizon The boundary that separates the sky from the ground.

horizon problem The puzzling observation that the cosmic background radiation is so uniform in all directions, despite the fact that widely separated regions should have been "over the horizon" from each other in the early Universe.

horizontal branch A region on the H-R diagram defined by stars burning helium to carbon in a stable core.

hot dark matter Particles of dark matter that move so fast that gravity cannot confine them to the volume occupied by a galaxy's normal luminous matter.

hot spot A place where hot plumes of mantle material rise near the surface of the planet.

H-R diagram The Hertzsprung-Russell diagram, which is a plot of the luminosities versus the surface temperatures of stars. The evolving properties of stars are plotted as tracks across the H-R diagram.

HST *Hubble Space Telescope*

H II region Region of interstellar gas that has been ionized by UV radiation from nearby hot, massive stars.

Hubble, Edwin P. (1889–1953) The astronomer who discovered the true nature of galaxies and that the Universe is expanding. The *Hubble Space Telescope* is named after him.

Hubble constant (H_0) The constant of proportionality relating the recession velocities of galaxies to their distances. See also *Hubble time.*

Hubble's law The law stating that the speed at which a galaxy is moving away from us is proportional to the distance of that galaxy.

Hubble time An estimate of the age of the Universe from the inverse of the Hubble constant, $1/H_0$.

Huygens, Christiaan (1619–1695) Among the founders of mechanics and optics; the first to recognize that Saturn has rings.

hydrogen burning The release of energy from the nuclear fusion of four hydrogen atoms into a single helium atom.

hydrogen shell burning Fusion of hydrogen in a shell surrounding a stellar core that may be either degenerate or fusing more massive elements.

hydrosphere That portion of Earth that is largely liquid water.

hydrostatic equilibrium The condition in which the weight bearing down at some point within an object is balanced by the pressure within the object.

hypothesis A well-thought-out idea, based on scientific principles and knowledge, that makes testable predictions.

hypothetical Proposed but not yet tested.

I

ice The solid form of a volatile material. Sometimes the volatile material itself, regardless of its form.

ice giant A giant planet formed mostly of the liquid form of volatile substances.

ideal gas A gas in which all collisions between its individual atoms or molecules are like collisions between billiard balls or marbles.

ideal gas law The relationship between pressure (P), number density of particles (n), and temperature (T) expressed as $P = nkT$, where k is Boltzmann's constant.

igneous activity The formation and action of molten rock or magma.

impact crater The scar of the impact left on a solid planetary or moon surface by collision with another object.

inert gas   Gaseous element that combines with other elements only under conditions of extreme temperature and pressure. For example, helium, neon, and argon.

inertia   The tendency for objects to retain their state of motion.

inertial frame of reference   A reference frame that is not accelerating. In general relativity, a reference frame that is falling freely in a gravitational field.

inflation   An extremely brief phase of ultra-rapid expansion of the very early Universe. Following inflation, the standard Big Bang models of expansion apply.

infrared radiation   Electromagnetic radiation occurring in the spectral region between that of visible light and microwaves.

instability strip   A region of the H-R diagram containing stars that pulsate with a periodic variation in luminosity.

intensity   Amount per second per unit area. In the case of electromagnetic radiation: W/m^2.

intercloud gas   Low-density regions of the interstellar medium that fill the space between interstellar clouds.

intermediate-mass star   Star with mass between about $3\ M_\odot$ and $8\ M_\odot$.

interstellar cloud   Discrete, high-density region of the interstellar medium made up mostly of atomic or molecular hydrogen and dust.

interstellar dust   Small particles or grains (0.01 to 10 μm) of matter, primarily carbon and silicates, distributed throughout interstellar space.

interstellar extinction   The dimming of visible and ultraviolet light by interstellar dust.

interstellar medium   The gas and dust that fills the space between the stars within a galaxy.

inverse square law   The rule that a quantity or effect diminishes with the square of the distance from the source.

ionosphere   A layer high in Earth's atmosphere in which most of the atoms are ionized by solar radiation.

ion tail   A type of comet tail consisting of ionized gas. Particles in the ion tail are pushed directly away from the comet's head in the antisolar direction at high speeds by the solar wind.

iron meteorite   Metallic meteorite composed mostly of iron-nickel alloys.

irregular galaxy   Galaxy without regular or symmetric appearance.

irregular moon   Moon that has been captured by a planet. Some revolve in the opposite direction from the rotation of the planet, and many are in distant, unstable orbits.

isotope   Form of the same element with differing numbers of neutrons.

isotropic   In cosmology, a universe whose properties observers find to be the same in all directions.

J

joule (J)   Unit of energy or work. 1 J = 1 newton meter.

K

Kant, Immanuel (1724–1804)   The 18th-century philosopher who hypothesized that the nebulae seen by Messier and Herschel were "island universes," separate from our own.

Kelvin (K)   The temperature scale using Celsius-sized degrees, but with 0 K defined as *absolute zero* instead of the melting point of water. Defined by William Thompson, better known as Lord Kelvin (1824–1907).

Kepler, Johannes (1571–1630)   The discoverer of the elliptical shape of planetary orbits, and how orbital period is related to average distance from the Sun.

Kepler's first law   The law stating that planets move in orbits of elliptical shapes with the Sun at one focus.

Kepler's laws   The three rules of planetary motion inferred by Johannes Kepler from the data acquired by Tycho Brahe.

Kepler's second law   The law stating that a line drawn from the Sun to a planet Sun sweeps out equal areas in equal times as the planet orbits about the Sun. Also called the *Law of Equal Areas*.

Kepler's third law   The relationship between the period of a planet's orbit and its distance from the Sun. Also called the *harmonic law*.

kinetic energy   The energy of an object due to its motions. K.E. = $1/2\ mv^2$. Units: joules (J).

Kirkwood gap   Gap in the main asteroid belt related to orbital resonances with Jupiter.

Kuiper Belt   A disk-shaped population of comet nuclei extending from Neptune's orbit to perhaps several thousand AU from the Sun.

Kuiper Belt objects (KBOs)   The icy planetesimals (comet nuclei) that orbit within the Kuiper Belt.

L

Lagrangian equilibrium point   One of five points of equilibrium in a system consisting of two massive objects in near-circular orbit around a common center of mass. Only two (L_4 and L_5) represent stable equilibrium. A third smaller body located at one of the five points will move in lock-step with the center of mass of the larger bodies.

lambda peak (λ_{peak})   The wavelength at which the thermal radiation from an object is most intense. Literally, the peak of the Planck spectrum as specified by Wien's law.

large-scale structure   Observable aggregates on the largest scales in the Universe, including galaxy groups, clusters, and superclusters.

latitude   The angular distance north (+) or south (−) from the equatorial plane of a nearly spherical body.

law of conservation of angular momentum   The physical law stating that a system's total angular momentum remains constant, unless acted upon by an external torque.

Law of Equal Areas   See *Kepler's second law.*

law of gravitation   See *universal law of gravitation.*

Leavitt, Henrietta (1868–1921) The American astronomer who discovered that periods of variation of Cepheid stars are related to their luminosities, by studying the variable stars in the Large and Small Magellanic Clouds.

Leonids A November meteor shower associated with the dust debris left by comet Tempel-Tuttle.

libration The apparent wobble of an orbiting body that is tidally locked to its companion (such as Earth's Moon) resulting from the fact that its orbit is elliptical rather than circular.

light All electromagnetic radiation, which comprises the entire *electromagnetic spectrum*.

light-year (ly) The distance light travels in one year—about 9 trillion kilometers.

limb darkening The darker appearance caused by increased atmospheric absorption near the limb of a planet or star.

line of nodes (a) A line defined by the intersection of two orbital planes. (b) The line defined by the intersection of Earth's equatorial plane and the plane of the ecliptic.

lithosphere The solid, brittle part of Earth (or any planet or moon), including the crust and the upper part of the mantle.

lithospheric plate Independent piece of Earth's lithosphere capable of moving independently. See *continental drift, plate tectonics*.

Local Group The small group of galaxies of which the Milky Way is a member.

longitudinal wave A wave that oscillates parallel to the direction of the wave's propagation.

long-period comet Comet with an orbital period greater than 200 years.

look-back time The time that it has taken the light from an astronomical object to reach Earth.

low-mass star A star with a main sequence mass of less than about 3 $M_\odot$.

luminosity The total flux emitted by an object. Unit: watts (W).

luminosity-temperature-radius relationship A relationship among these three properties of stars indicating that if any two are known, the third can be calculated.

luminous matter Matter in galaxies—including stars, gas, and dust—that emits electromagnetic radiation.

lunar eclipse An eclipse that occurs when the Moon is partially or entirely in Earth's shadow.

lunar tide The component of Earth's tides due to the Moon.

M

MACHO "Massive compact halo object," such as brown dwarfs, white dwarfs, and black holes, which are candidates for dark matter.

magma Molten rock often containing dissolved gases and solid minerals.

magnetic field A field that is able to exert a force on a moving electric charge. See *electromagnetic force*.

magnetic force A force associated with, or caused by, the relative motion of charges.

magnetosphere The region surrounding a planet filled with relatively intense magnetic fields and plasmas.

magnitude A system used by astronomers to describe the brightness or luminosity of stars. The brighter the star, the smaller its magnitude.

main asteroid belt See *asteroid belt*.

main sequence The strip on the H-R diagram where most stars are found. Main sequence stars are fusing hydrogen to helium in their cores.

main sequence lifetime The amount of time a star spends on the main sequence, fusing hydrogen into helium in its core.

main sequence turnoff The location on the H-R diagram of a single-aged stellar population (such as a star cluster) where stars have just evolved off the main sequence. The position of the main sequence turnoff is determined by the age of the stellar population.

mantle The solid portion of a rocky planet that lies between the crust and the core.

mare (pl. **maria**) A dark region on the Moon, composed of basaltic lava flows.

mass (a) Inertial mass is the property of matter that resists changes in motion. (b) Gravitational mass is the property of matter defined by its attractive force on other objects. According to general relativity the two are equivalent.

massive element (a) In astronomy, refers to all elements more massive than helium. (b) In other sciences (and sometimes also in astronomy), refers to the most massive elements in the periodic table, such as uranium and plutonium.

mass transfer The transfer of mass from one member of a binary star system to its companion. Mass transfer occurs when one of the stars evolves to the point that it overfills its Roche lobe, so that its outer layers are pulled toward its binary companion.

mathematics The science and language of patterns; the language of science. Mathematics provides the tools used by scientists to understand, describe, and predict the patterns found in nature.

matter Objects made of particles that have mass, such as protons, neutrons, and electrons; anything that occupies space and has mass.

Maunder minimum The period 1645–1715, when there were few sunspots.

Maxwell, James Clerk (1831–1879) The 19th-century physicist who summarized all of classical electromagnetism, including electromagnetic radiation, with a set of four equations.

mega-light-year (Mly) A unit of distance equal to 1 million light-years.

meridian The imaginary arc in the sky running from the horizon at due north through the zenith to the horizon at due south. The meridian divides the observer's sky into eastern and western halves.

mesosphere A portion of Earth's atmosphere that lies above the stratosphere.

Messier, Charles (1730–1817) An early French comet hunter; published the first catalog of bright, diffuse celestial objects to prevent their being confused with comets.

Messier's Catalog   The catalog made by Charles Messier in the late 18th century in which he identify 103 bright, diffuse-appearing objects visible from France, all of them either star clusters, planetary nebulae, H II regions, or galaxies; later expanded to 110 objects.

meteor   The incandescent trail produced by a small piece of interplanetary debris as it travels through the atmosphere at very high speeds.

meteorite   A meteoroid that survives to reach a planet's surface.

meteoroid   A small cometary or asteroidal fragment ranging in size from 100 μm to 100 m. When entering a planetary atmosphere, the meteoroid creates a *meteor,* which is an atmospheric phenomenon.

meteor shower   A larger than normal display of meteors, occurring when Earth passes through the orbit of a disintegrating comet, sweeping up its debris.

micrometer (μm) 10^{-6} meter; a unit of length used for measuring the wavelength of visible light. Also called *micron.*

micron   See *micrometer.*

microwave radiation   Electromagnetic radiation occurring in the spectral region between that of infrared radiation and radio waves.

minute of arc (')   A unit for measuring angles. A minute of arc is 1/60 of a degree of arc. Also called *arcminute.*

Mly   See *mega light-year.*

modern physics   A term usually used to refer to those physical principles, including relativity and quantum mechanics, developed since Maxwell's equations were published.

molecular cloud   Interstellar cloud composed primarily of molecular hydrogen.

molecular cloud core   Dense clump within a molecular cloud that forms as the cloud collapses and fragments. Protostars form from molecular cloud cores.

moon   A less massive satellite orbiting a more massive object. Moons are found around both planets and asteroids.

M-type asteroid   An asteroid once part of the metallic core of a larger, differentiated body that has since been broken into pieces; mostly made of iron and nickel.

mutation   An imperfect reproduction of self-replicating material.

N

natural selection   The process by which forms of structure, ranging from molecules to whole organisms, that are best adapted to their environment become more common than less well adapted forms.

nature   The word frequently used by scientists to denote the physical Universe. Note the difference from other common uses.

neap tide   An especially weak tide that occurs around the time of the first or third quarter Moon when the gravitational forces of the Moon and the Sun on Earth are at right angles to each other, thus producing the least-pronounced tides.

near-Earth asteroid   Asteroid whose orbit brings it close to the orbit of Earth.

neutrino   A very low-mass, electrically neutral particle emitted during beta decays. (Neutrinos interact with matter only very feebly and so can penetrate through great quantities of matter.)

neutrino cooling   The process in which thermal energy is carried out of the center of a star by neutrinos rather than by electromagnetic radiation or convection.

neutron (n^0)   A subatomic particle having no net electric charge, and a rest mass and rest energy nearly equal to that of the proton.

neutron star   The neutron-degenerate remnant left behind by a Type II supernova.

new Moon   The phase of the Moon that occurs when the Moon is between Earth and the Sun, and we see only the side of the Moon not being illuminated by the Sun.

Newton, Sir Isaac   (1642–1727)   Discoverer and codifier of the laws describing the motion of objects and of the law of gravity; in many respects, the founding father of modern science.

Newton's first law of motion   The law stating that an object will remain at rest or will continue moving along a straight line at a constant speed until an unbalanced force acts on it.

Newton's laws   See *Newton's first law of motion, Newton's second law of motion* and *Newton's third law of motion.*

Newton's second law of motion   The law stating that if an unbalanced force acts on a body, the body will have an acceleration proportional to the unbalanced force and inversely proportional to the object's mass. The acceleration will be in the direction of the unbalanced force.

Newton's third law of motion   The law stating that for every force, there is an equal and opposite force.

nonthermal radiation   Any form of electromagnetic radiation, such as *synchrotron radiation*, that does not originate from thermal energy.

normal matter   Matter in galaxies that emits and absorbs electromagnetic radiation. Compare with *dark matter.*

north celestial pole   The northward projection of Earth's rotation axis onto the celestial sphere.

North Pole   The point in the Northern Hemisphere where Earth's rotation axis intersects the surface of Earth.

north star   See *Polaris.*

nova (pl. novae)   A stellar explosion that results from runaway nuclear fusion in a layer of material on the surface of a white dwarf in a binary system.

nuclear fusion   The combination of two less-massive atomic nuclei into a single more-massive atomic nucleus. Release of energy by fusion of low-mass elements is often referred to as "nuclear burning."

nuclear burning   Release of energy by fusion of low-mass elements.

nucleosynthesis   The formation of more massive atomic nuclei from less massive nuclei, either in the Big Bang (Big Bang nucleosynthesis) or in the interiors of stars (stellar nucleosynthesis).

nucleus (a) the dense, central part of an atom. (b) the central core of a galaxy, comet, or other diffuse object.

O

oblateness The flattening of an otherwise spherical planet or star caused by its rapid rotation.

obliquity The inclination of a celestial body's equator to its orbital plane.

observational uncertainty The fact that real measurements are never perfect; all observations are uncertain by some amount.

Occam's razor The principle that the simplest hypothesis is the most likely; named after William of Occam (circa 1285–1349), the medieval cleric to whom the idea is attributed.

Oort Cloud A spherical distribution of comet nuclei stretching from the Kuiper Belt to more than 50,000 AU from the Sun.

opacity A measure of how effectively a material blocks the radiation going through it.

open cluster Group of a few tens to a few thousands of stars that formed together in the disk of a spiral galaxy. Also called *galactic clusters*.

open universe An infinite universe with a negatively curved spatial structure (much like the surface of a saddle) such that the sum of the angles of a triangle is always less than 180°.

orbit The path taken by one object moving around another object under the influence of their mutual gravitational or electric attraction.

orbital resonance A situation in which the orbital periods of two objects are related by a ratio of small integers.

P

pair creation The production of a particle–antiparticle pair from a source of electromagnetic energy.

paleomagnetism The record of Earth's magnetic field as preserved in rocks.

palimpsest Flat, circular feature in icy lithospheres, believed to be a scar of an ancient impact.

parallax The apparent shift in the position of one object relative to another object, caused by the changing perspective of the observer. In astronomy, the displacement in the apparent position of a nearby star caused by the changing location of Earth in its orbit.

parent element Radioactive element. See *daughter element*.

parsec (pc) The distance to a star with a parallax of 1 arcsecond using a base of 1 AU. One parsec is close to 3.26 light-years.

partial solar eclipse Any solar eclipse in which the observer is outside the path of totality.

peculiar velocity The motion of a galaxy relative to the overall expansion of the Universe.

pendulum A mass supported by a string, wire, or rod that is free to swing back and forth in a gravitational field.

penumbra (pl. **penumbrae**) (a) The outer part of a shadow, where the source of light is only partially blocked. (b) The region surrounding the umbra of a sunspot. The penumbra is cooler and darker than the surrounding surface of the Sun but not as cool or dark as the umbra of the sunspot.

penumbral lunar eclipse A lunar eclipse in which the Moon passes through the penumbra of Earth's shadow.

Penzias, Arno (1933–) The Bell Laboratories physicist who, together with Robert Wilson, in the early 1960s, discovered the cosmic background radiation, for which they won the Nobel Prize in physics.

perihelion (pl. **perihelia**) The closest point to the Sun on a solar orbit.

period The time it takes for a regularly repetitive process to complete one cycle.

period–luminosity relationship The relationship between the period of variability of a pulsating variable star, such as a Cepheid or RR Lyrae variable, and the luminosity of the star. Longer-period Cepheid or RR Lyrae variables are more luminous than their shorter-period cousins.

Perseids A prominent August meteor shower associated with the dust debris left by comet Swift-Tuttle.

phase One of the various appearances of the sunlit surface of the Moon or a planet caused by the change in viewing location of Earth relative both to the Sun and the object.

photino One of the leading candidates for the elementary particle composing cold dark matter.

photodisintegration The splitting of iron into helium by gamma rays.

photodissociation The breaking apart of molecules into smaller fragments or individual atoms by the action of photons.

photon A discrete unit or particle of electromagnetic radiation; a *quantum of light*. The energy of a photon is equal to Planck's constant (h) times the frequency (f) of its electromagnetic radiation: $E_{photon} = h \times f$. The particle that mediates the electromagnetic force.

photosphere The apparent surface of the Sun as seen in visible light.

physical laws Broad statements that predict some aspect of how the physical Universe behaves and that are supported by many empirical tests.

physics The scientific study of the fundamental principles that govern the behavior of matter and energy, and the application of those principles to understanding and predicting natural phenomena.

Planck, Max (1858–1947) The person first to postulate the quantum nature of electromagnetic radiation.

Planck era The early time, just after the Big Bang, when the Universe as a whole must be described quantum mechanically.

Planck's constant (h) The constant of proportionality between the energy of a photon and the frequency of the photon. This constant defines how much energy a single photon of a given frequency or wavelength has. Value: $h = 6.63 \times 10^{-34}$ joule-second.

Planck spectrum The spectrum of electromagnetic energy emitted by a blackbody per unit area per second, which is determined only by the temperature of the object. Also called *blackbody spectrum*.

planetary nebula   Expanding shell of material ejected by a dying AGB star. A planetary nebula glows from flourescence caused by intense ultraviolet light coming from the hot, stellar remnant at its center.

planetary system   A system of planets and other smaller objects in orbit around a star.

planet   Large body that orbits the Sun or other star that shines only by light reflected from the Sun or star.

planetesimal   Primitive body of rock and ice, 100 m or more in diameter, that combines with others to form planets.

plasma   A gas composed largely of charged particles but which also may include some neutral atoms.

plate tectonics   The geological theory concerning the motions of lithospheric plates, which in turn provides the theoretical basis for continental drift.

Polaris   The star that currently is the closest bright star to the north celestial pole.

positron   A positively charged subatomic particle; the antiparticle of the electron.

power   The rate at which work is done or at which energy is delivered. Unit: watts or joule/second.

precession of the equinoxes   The slow change in orientation between the ecliptic plane and the celestial equator caused by the wobbling of Earth's axis.

pressure   Force per unit area. Units: newtons per square meter (N/m^2) or bars.

primary atmosphere   An atmosphere, composed mostly hydrogen and helium, that forms at the same time as a planet.

primary wave   Longitudinal seismic wave in which the oscillations involve compression and decompression parallel to the direction of travel (i.e., a pressure wave).

principle   A general idea or sense about how the Universe is that guides us in constructing new scientific theories. Principles can be testable theories.

prograde   (a) Describes rotational or orbital motion of a moon that is in the same sense as the planet it orbits. (b) Describes the counterclockwise orbital motion of the planets as seen from above Earth's orbital plane.

prominence   Archlike projection above the solar photosphere often associated with sunspots.

proportional   Two things are proportional if their ratio is a constant.

proton (p or p^+)   A fundamental particle having a positive electric charge of 1.6×10^{-19} coulomb, a mass of 1.67×10^{-27} kg, and a rest energy of 1.5×10^{-10} joules.

proton–proton chain   One of the ways in which hydrogen burning—the fusion of four hydrogen atoms to form a helium atom—can take place. This is the most important path for hydrogen burning in low-mass stars such as the Sun.

protoplanetary disk   The remains of the accretion disk around a young star from which a planetary system may form.

protostar   A young stellar object that derives its luminosity from the conversion of gravitational energy to thermal energy, rather than from nuclear reactions in its core.

protostellar disk   The accretion disk that forms as an interstellar cloud collapses on its way to becoming a star.

pulsar   A rapidly rotating neutron star that beams radiation into space in two searchlight-like beams. To a distant observer, the star appears to flash on and off, earning it its name.

pulsating variable star   Variable star that undergoes periodic radial pulsations.

Q

QCD   See *quantum chromodynamics*.

QED   See *quantum electrodynamics*.

quantized   Describes a quantity that exists as discrete, irreducible units.

quantum chromodynamics (QCD)   The quantum mechanical theory describing the strong nuclear force and its mediation by gluons.

quantum electrodynamics (QED)   The quantum theory that describes the electromagnetic force and its mediation by photons.

quantum mechanics   The branch of physics that deals with the quantized and probabilistic behavior of atoms and subatomic particles.

quantum of light   The discrete particle of light we call a *photon*.

quark   The building block of protons and neutrons.

quasar   The most luminous of the active galactic nuclei (AGNs), seen only at great distances from our Galaxy; shortened from "quasi-stellar radio" source.

Quaternary period   The period in Earth's history from 1.8 million years ago through today.

R

radial velocity   The component of velocity that is directed toward or away from the observer.

radiant   The direction in the sky from which the meteors in a meteor shower seem to come.

radiation belt   A toroidal ring of high-energy particles surrounding a planet.

radiative transfer   The transport of energy from one location to another by electromagnetic radiation.

radiative zone   A region in the interior of a star through which energy is transported by radiation.

radio galaxy   Elliptical galaxy having very strong emission (10^{35} to 10^{38} watts) in the radio part of the electromagnetic spectrum.

radiometric dating   Use of radioactive decay to measure the ages of materials such as minerals.

radio wave Electromagnetic radiation in the extreme long-wavelength region of spectrum, beyond the region of microwaves.

recombination An event early in the evolution of the Universe in which hydrogen and helium nuclei combined with electrons to form neutral atoms. The removal of electrons caused the Universe to become transparent to electromagnetic radiation.

reddening The effect that stars and other objects, when viewed through interstellar dust, appear redder than they actually are. Reddening is caused by blue light being more strongly absorbed and scattered than red light.

red giant A low-mass star that has evolved beyond the main sequence and is now fusing hydrogen in a shell surrounding a degenerate helium core.

red giant branch A region on the H-R diagram defined by low-mass stars evolving from the main sequence toward the horizontal branch.

redshift The shifting of the wavelength of light to longer wavelengths by any of several effects including Doppler shifts, gravitational redshift, or cosmological redshift. Compare with *blueshift*.

reflecting telescope Telescope that uses mirrors for collecting and focusing incoming electromagnetic radiation to form an image in their focal planes. The size of a reflecting telescope is defined by the diameter of the first mirror from which incoming light is reflected (called the primary mirror).

refracting telescope Telescope that uses objective lenses to collect and focus light.

refractory material Material that remains solid at high temperatures. Compare with *volatile material*.

regular moon Moon that formed together with the planet it orbits.

relative humidity The amount of water vapor held by a volume of air at a given temperature compared (stated as %) to the total amount of water that could be held by the same volume of air at the same temperature.

relativistic Pertaining to physical processes that take place in systems traveling at nearly the speed of light or located in the vicinity of very strong gravitational fields.

relativistic beaming The effect created when material moving at nearly the speed of light beams the radiation it emits in the direction of its motion.

remote sensing The use of images, spectra, radar, or other techniques to measure the properties of an object from a distance.

rest wavelength The wavelength of light we see coming from an object at rest with respect to the observer.

retrograde (a) Describes rotation or orbital motion of a moon that is in the opposite sense to the rotation of the planet it orbits. (b) Describes the clockwise orbital motion of an object as seen from above Earth's orbital plane.

ring An aggregation of small particles orbiting a planet or star. The rings of the four giant planets of the Solar System are composed variously of dust, organic materials, and ices.

ring arc Discontinuous region of higher density within an otherwise continuous, narrow ring.

ringlet Narrow subdivision within larger features called rings.

Rømer, Ole (1644–1719) The Danish astronomer who first estimated the speed of light by timing eclipses of Jupiter's moons when Earth was at different distances from Jupiter.

Roche, Edouard A. (1820–1883) The French astronomer who systematized the effects of tidal stress on the structure of moons and planets.

Roche limit The distance at which a planet's tidal forces exceed the self-gravity of a smaller object, such as a moon, asteroid or comet, causing the object to break apart.

Roche lobe The hourglass or figure eight shaped volume of space surrounding two stars, which constrains material that is gravitationally bound by one or the other.

Roentgen, W. C. (1845–1923) The discoverer of X-rays, for which he won the first Nobel Prize in physics in 1901.

rotation curve A graph showing how the orbital velocity of stars and gas in a galaxy changes with radial distance from the galaxy's center.

RR Lyrae variable Variable giant star whose regularly timed pulsations are good predictors of its luminosity; are used for distance measurements to globular clusters.

Russell, Henry Norris (1877–1957) American astronomer who, together with Einar Hertzsprung, defined the main diagnostic tool for studying the properties of stars. The Hertzsprung-Russell (or H-R) diagram is so-named in his honor, as well as is the Vogt-Russell theorem.

S

satellite (a) An object in orbit about a more massive body. (b) A moon.

saturation Atmosphere reaching the point that it can hold no more water (or other substance) in the gas phase.

scale factor (R_U) A dimensionless number proportional to the distance between two points in space. The scale factor increases as the Universe expands.

scattering The random change in the direction of travel of photons, caused by their interactions with molecules or dust particles.

science The search for the rules that govern the behavior of the Universe and the explanation of observed phenomena in terms of those rules.

scientific method The formal procedure—including hypothesis, prediction, and experiment or observation—used to test (attempt to falsify) the validity of scientific hypotheses and theories.

scientific notation The standard expression of numbers with one digit (which can be zero) to the left of the decimal point and multiplied by 10 to the exponent required to give the number its correct value. Example: $2.99 \times 10^8 = 299,000,000$.

secondary atmosphere A planetary atmosphere that formed—as a result of volcanism, comet impacts, or some other process—sometime after the planet formed.

secondary crater Crater formed from ejecta by a primary impact.

secondary wave Transverse seismic wave.

second law of thermodynamics   The law stating that the entropy or disorder of an *isolated* system always increases as the system evolves.

second of arc (″)   A unit used for measuring very small angles. A second of arc is 1/60 of a minute of arc, or 1/3,600 of a degree. Also called *arcsecond*.

seismic wave   Vibration due to earthquakes, large explosions, or impacts on the surface that travels through a planet's interior.

seismometer   Instrument that measures the amplitude and frequency of seismic waves.

self-gravity   The gravitational attraction between all the parts of the same object.

semimajor axis   Half of the longer axis of an ellipse.

SETI   The Search for Extraterrestrial Intelligence project, which uses advanced technology combined with radio telescopes to search for evidence of intelligent life elsewhere in the Universe.

Seyfert, Carl   (1911–1960)   The astronomer who discovered AGNs in spiral galaxies. Seyfert galaxies are named in his honor.

Seyfert galaxy   A type of spiral galaxy with an active galactic nucleus (AGN) at its center; first discovered in 1943 by Carl Seyfert.

Shapley, Harlow   (1885–1972)   The Harvard College Observatory astronomer who first showed that our Galaxy is nearly 300,000 light-years in size.

shepherd moon   Moon that orbits close to rings and gravitationally confines the orbits of the ring particles.

shield volcano   A volcano formed by very fluid lava flowing from a single source, spreading out from that source.

short-period comet   Comet whose orbital period is less than 200 years.

sidereal   The orbital or rotational period of the orbit measured with respect to the stars.

sidereal time   Time based on the true rotation period of Earth. The 24-hour sidereal day used by astronomers is approximately four minutes shorter than a solar day.

silicate   One of the family of minerals composed of silicon and oxygen in combination with other elements.

singularity   (a) The point where a mathematical expression or equation becomes meaningless, such as the denominator of a fraction going to zero. (b) A black hole.

Slipher, Vesto   (1875–1969)   The Lowell Observatory astronomer who first made measurements of the redshifts of galaxies, supplying most of the data used by Edwin Hubble to discover that our Universe is expanding.

solar abundance   The relative amount of an element detected in the atmosphere of the Sun, expressed as the ratio of the number of atoms of that element to the number of hydrogen atoms.

solar eclipse   Blocking of all or part of the Sun by the Moon.

solar flare   Explosive events on the Sun's surface associated with complex sunspot groups and strong magnetic fields.

solar maximum   The time, occurring about every 11 years, when the Sun is at its peak activity, meaning that sunspot activity and related phenomena (such as prominences, flares, and coronal mass ejections) are at their peak.

solar neutrino problem   The observation that only about a third as many neutrinos are observed from the Sun as are predicted by theory.

Solar System   The gravitationally bound system comprised of the Sun, planets, moons, asteroids, comets and their associated gas and dust.

solar tide   The component of Earth's tide due to the Sun's gravity. Also see *tide* and *lunar tide*.

solar time   Time based on the apparent rotation period of Earth. A 24-hour solar day is the average interval between two successive passages of the Sun across the local meridian.

solar wind   The stream of charged particles emitted by the Sun that flow at high speeds through interplanetary space.

solstice   ("sun standing still")   (a) One of the two most northerly and southerly points on the ecliptic. (b) The time of year when the Sun is at one of these two points.

south celestial pole   The southward projection of Earth's rotation axis onto the celestial sphere.

South Pole   The location in the Southern Hemisphere where Earth's rotation axis intersects the surface of Earth.

space-time   The four-dimensional continuum in which we live, and which we experience as three spatial dimensions plus time.

special relativity   The consequences of the speed of light being a constant for nonaccelerating frames of reference; discovered by Albert Einstein. Compare with *general theory of relativity*.

spectral type   A classification system for stars based on the presence and relative strength of absorption lines in their spectra. Spectral type is related to the surface temperature of a star.

spectrograph   A device that spreads out the light from an object into its component wavelengths.

spectroscopic parallax   Use of the spectroscopically determined luminosity and observed brightness of a star to determine the star's distance.

spectroscopy   The study of electromagnetic radiation from an object in terms of its component wavelengths.

spectrum   (a) The intensity of electromagnetic radiation as a function of wavelength. (b) Waves sorted by wavelength.

spherically symmetric   Describes an object whose properties depend only on distance from the object's center, so that the object has the same form viewed from any direction.

spin-orbit resonance   A relationship between the orbital and rotation periods of an object such that the ratio of their periods can be expressed by simple integers.

spiral density wave   A stable spiral-shaped change in the local gravity of a disk that can be produced by periodic gravitational kicks from neighboring galaxies or from nonspherical bulges and bars in spiral galaxies.

spiral galaxy   Hubble type "S" class, with a discernible disk in which large spiral patterns exist.

sporadic meteor   Meteor not associated with a specific meteor shower.

spreading center   A zone from which two tectonic plates diverge.

spring tide An especially strong tide that occurs near the time of a new or full Moon, when lunar tides and solar tides reinforce each other.

stable equilibrium An equilibrium condition in which the system returns to its former condition after a small disturbance. Compare with *unstable equilibrium*.

standard candle An object whose luminosity either is known or can be predicted in a distance-independent way, so its brightness can be used to determine its distance via the inverse square law of radiation.

standard model The theory of particle physics that combines electroweak theory with quantum chromodynamics to describe the structure of known forms of matter.

Star A luminous ball of gas that is held together by gravity. A normal star is powered by nuclear reactions in its interior.

star cluster A group of stars that all formed at the same time and in the same place.

static equilibrium State in which the forces within a system are all in balance so that the system does not change. Compare with *dynamic equilibrium*.

Stefan–Boltzmann constant (σ) The proportionality constant that relates the flux emitted by an object to the fourth power of its absolute temperature.

Stefan's law Gives the amount of electromagnetic energy emitted from the surface of a body, summed over the energies of all photons of all wavelengths emitted, which is proportional to the fourth power of the temperature of the body.

stellar mass loss The loss of mass from the outermost parts of a star's atmosphere during the course of its evolution.

stellar occultation An event in which a planet or other Solar System body moves between the observer and a star, eclipsing the light emitted by that star.

stellar population A group of stars with similar ages, chemical compositions, and dynamical properties.

stereoscopic vision The way an animal's brain combines the different information from its two eyes to perceive the distances to objects around it.

stony-iron meteorite Meteorite consisting of a mixture of silicate minerals and iron-nickel alloys.

stony meteorite Meteorite composed primarily of silicate minerals, similar to those found on Earth.

stratosphere The atmospheric layer immediately above the troposphere. On Earth it extends upward to an altitude of 50 km.

strong nuclear force The attractive short-range force between protons and neutrons that holds atomic nuclei together; one of the four fundamental forces of nature, mediated by the exchange of gluons.

S-type asteroid An asteroid made of material that has been modified from its original state, likely as the outer part of a larger, differentiated body that has since been broken into pieces.

subduction zone A region where two tectonic plates converge, with one plate sliding under the other and being drawn downward into the interior.

subgiant branch A region of the H-R diagram defined by stars that have left the main sequence but have not yet reached the *red giant branch*.

sublimation The process of a solid becoming a gas without first becoming a liquid.

summer solstice The time of year in the Northern Hemisphere (about June 21) when the Sun is at its most northerly distance from the celestial equator.

sungrazer Comet whose perihelion is within a few solar diameters of the surface of the Sun.

sunspot Cooler, transitory region on the solar surface produced when loops of magnetic flux break through the surface of the Sun.

sunspot cycle The approximate 11-year cycle during which sunspot activity increases and then decreases. This is one-half of a full 22-year cycle in which the magnetic polarity of the Sun first reverses, then returns to its original configuration.

supercluster A large conglomeration of galaxy clusters and galaxy groups; typically more than 100 million light-years in size and containing tens of thousands to hundreds of thousands of galaxies.

superluminal motion The appearance (though not the reality) that a jet is moving faster than the speed of light.

supermassive black hole Black hole of 1,000 solar masses or more that resides in the center of a galaxy, and whose gravity powers active galactic nuclei.

supernova A stellar explosion resulting in the release of tremendous amounts of energy, including the high-speed ejection of matter into the interstellar medium. See *Type I supernova* and *Type II supernova*.

superstring theory The theory that conceives of particles as strings in 11 dimensions of space and time; the current contender for a Theory of Everything.

surface brightness The amount of electromagnetic radiation emitted or reflected per unit area.

surface wave Seismic wave that travels on the surface of a planet or moon.

symmetry (a) The property that an object has if the object is unchanged by rotation or reflection about some point, line or plane. (b) In theoretical physics, the correspondence of different aspects of physical laws or systems, such as the symmetry between matter and antimatter.

synchronous rotation The case in which the period of rotation of a body on its axis equals its period of revolution in its orbit around another body. A special type of spin-orbit resonance.

synchrotron radiation Radiation from electrons moving at close to the speed of light as they spiral in a strong magnetic field; named because this kind of radiation was first identified on Earth in particle accelerators called synchrotrons.

synodic A period or time related to apparent changes in celestial objects as seen from a planet. See also *sidereal*.

S0 galaxy Galaxy with a bulge and a disklike spiral, but smooth in appearance like ellipticals.

T

tail   A stream of gas and dust swept away from the coma of a comet by the solar wind and by radiation pressure from the Sun.

tectonism   Deformation of the lithosphere of a planet.

telescope   The basic tool of astronomers. Telescopes collect and concentrate light from distant objects.

temperature   A measure of the average kinetic energy of the atoms or molecules in a gas, solid, or liquid.

terrestrial planet   An Earth-like planet, made of rock and metal and having a solid surface.

Tertiary period   The period in Earth's history from 65 million until 1.8 million years ago.

theoretical   (a) Using mathematics to formalize rules or laws from ideas and/or empirical results. (b) Using rules and laws to explain observed phenomena or predict phenomena not yet seen.

theoretical model   A detailed description of the properties of some object or system in terms of known physical laws or theories. Often a computer calculation of predicted properties based on such a description.

theory   A well-developed idea or group of ideas that are consistent with known physical laws and make testable predictions about the world. A very well-tested theory may be called a physical law, or simply a fact.

Theory of Everything (TOE)   A theory that unifies all four fundamental forces of nature: strong nuclear, weak nuclear, electromagnetic, and gravitational forces.

thermal conduction   The transfer of energy in which the thermal energy of particles is transferred to adjacent particles by collisions or other interactions. Conduction is most important in solids.

thermal energy   The energy that resides in the random motion of atoms, molecules, and particles, by which we measure their temperature.

thermal equilibrium   The state in which the rate at which thermal energy emitted by an object is equal to the rate at which thermal energy is absorbed.

thermal motion   The random motion of atoms, molecules, and particles that gives rise to thermal radiation.

thermal radiation   Electromagnetic radiation resulting from the random motion of the charged particles in every substance.

thermosphere   That portion of Earth's atmosphere between the exosphere and the mesosphere.

third quarter Moon   The phase of the Moon in which the eastern half of the Moon as viewed from Earth is illuminated. The third quarter Moon occurs about one week after the full Moon.

tidal bulge   A distortion of a body resulting from tidal stresses.

tidal locking   Synchronous rotation of an object caused by internal friction as the object rotates through its tidal bulge.

tidal stress   Stress due to differences in the gravitational force of one mass on different parts of another mass.

tide   (a) The deformation of a mass due to differential gravitational effects of one mass on another, because of the extended size of the masses. (b) On Earth, the rise and fall of the oceans as Earth rotates through a tidal bulge caused by the Moon and Sun.

time dilation   The "stretching" of time by relativistic effects.

TOE   See *Theory of Everything*.

topographic relief   The differences in elevation from point to point on a planetary surface.

total lunar eclipse   A lunar eclipse in which the Moon passes through the umbra of Earth's shadow.

total solar eclipse   The type of eclipse that occurs when Earth passes through the umbra of the Moon's shadow so that the disk of the Sun is completely blocked by the Moon.

transverse wave   A wave in which the oscillations are perpendicular to the direction of the wave's propagation.

triple-alpha process   The nuclear fusion reaction that combines three helium nuclei, or "alpha particles," together into a single nucleus of carbon.

Trojan asteroid   One of a group of asteroids orbiting in the L_4 and L_5 Lagrangian points of Jupiter's orbit.

tropical year   The time between one crossing of the vernal equinox and the next. Due to precession of the equinoxes, a tropical year is slightly shorter than the time that it takes for Earth to orbit once about the Sun.

Tropics   The region on Earth between 23.5° south latitude and 23.5° north latitude and in which the Sun appears directly overhead twice during the year.

tropopause   The top of a planet's troposphere.

troposphere   The convection-dominated layer of a planet's atmosphere. On Earth the atmospheric region closest to the ground within which most weather phenomena take place.

T Tauri star   A young stellar object that has dispersed enough of the material surrounding it to be seen in visible light.

tuning fork diagram   The two-pronged diagram showing Hubble's classification of galaxies into ellipticals, S0s, spirals, barred S0s and spirals, and irregular galaxies.

turbulence   The random motion of blobs of gas within a larger cloud of gas.

Type I supernova   A supernova explosion in which no trace of hydrogen is seen in the ejected material. Most supernovae of this type are thought to be the result of runaway carbon burning in a white dwarf star onto which material is being deposited by a binary companion.

Type II supernova   A supernova explosion in which the degenerate core of an evolved massive star suddenly collapses and rebounds.

U

ultraviolet (UV) radiation   Describes electromagnetic radiation with frequencies and photon energies greater than those of visible light but less than those of X-rays, and wavelengths smaller than those of visible light but longer than those of X-rays.

umbra (a) The darkest part of a shadow, where the source of light is completely blocked. (b) The darkest, innermost part of a sunspot.

unbalanced force The non-zero net force acting on a body.

unbound orbit Orbit in which the velocity is greater than the escape velocity.

unified model of AGN A model in which many different types of activity in the nuclei of galaxies are all explained by accretion of matter around a supermassive black hole.

uniform circular motion Motion in a circular path at a constant speed.

universal gravitational constant (G) The constant of proportionality in the universal law of gravity. Value: $G = 6.673 \times 10^{-11}$ Nm^2/kg^2.

universal law of gravitation The gravitational force between any two objects is proportional to the product of their masses and inversely proportional to the square of the distance between them.

Universe All of space and everything contained therein.

unstable equilibrium An equilibrium state in which a small disturbance will cause a system to move away from equilibrium.

V

vacuum A region of space which contains very little matter. However, in quantum mechanics and general relativity, even a perfect vacuum has physical properties.

variable star Star whose luminosity varies. Many periodic variables are found within the instability strip on the H-R diagram.

velocity The rate and direction of change of an object's position with time. Units: m/s, km/hr.

vernal equinox (a) The point where the ecliptic crosses the celestial equator, with the Sun moving from south to north. (b) The day on which the Sun appears at this location; the first day of spring in the Northern Hemisphere (about March 21).

virtual particle A particle which, according to quantum mechanics, comes into momentary existence. According to theory, fundamental forces are mediated by the exchange of virtual particles.

visual binary A binary star in which the two stars can be seen individually from Earth.

Vogt-Russell theorem A theorem that states that the structure of a star is completely determined by its mass and chemical composition.

volatile material Material that remains gaseous at moderate temperature, sometimes referred to as an "ice." Compare with *refractory material*.

volcanic dome Steep sided, dome shaped volcanic feature, produced by viscous lavas.

volcanism The occurrence of volcanic activity on a planet or moon.

vortex (pl. **vortices**) Any circulating fluid system: (a) An atmospheric anticyclone or cyclone. (b) A whirlpool or eddy.

W

waning The changing phases of the Moon as it becomes less fully illuminated between full Moon and new Moon as seen from Earth.

wave A disturbance moving along a surface or passing through a space or a medium.

wavelength The distance between two adjacent points on a wave that have identical characteristics. The distance a wave travels in one period. Unit: m.

waxing The changing phases of the Moon as it becomes more fully illuminated between new Moon and full Moon as seen from Earth.

weak nuclear force The force underlying some forms of radioactivity and certain interactions between subatomic particles. It is responsible for radioactive beta decay, and for the initial proton–proton interactions that lead to nuclear fusion in the Sun and other stars. One of the four fundamental forces of nature.

weight Force equal to the mass of an object times the local acceleration due to gravity, or (in general relativity) the force equal to the mass of an object times the acceleration of the reference frame in which the object is observed.

white dwarf The stellar remnant left at the end of the evolution of a low-mass star. A typical white dwarf has a mass of 0.6 $M_\odot$, a radius about that of Earth, and is made of nonburning, electron-degenerate carbon.

Wien's Law A relationship describing how the peak wavelength, and therefore the color, of electromagnetic radiation from a glowing black body changes with temperature. Also known as Wien's displacement law. See *lambda peak*.

Wilson, Robert (1936–) The Bell Laboratories physicist who, together with Arno Penzias, in the early 1960s discovered the cosmic background radiation for which they won the Nobel Prize in physics.

winter solstice The time of year (about December 22) when the Sun is at its most southerly distance from the celestial equator. The first day of northern winter.

X

X-ray Electromagnetic radiation having frequencies and photon energies greater than those of ultraviolet light but less than those of gamma rays, and wavelengths smaller than those of UV light but longer than those of gamma rays.

X-ray binary A binary system in which mass from an evolving star spills over onto a collapsed companion such as a neutron star or black hole. The material falling in is heated to temperatures where it glows brightly in X-rays.

Y

year The time it takes Earth to make one revolution around the Sun. A solar year is measured from equinox to equinox. A sidereal year, Earth's true orbital period, is measured relative to the stars.

Z

zenith   The point on the celestial sphere located directly overhead an observer.

zodiac   The twelve constellations lying along the plane of the ecliptic.

zodiacal dust   Particles of cometary and asteroidal debris less than 100 μm in size that orbit the inner Solar System close to the plane of the ecliptic. Compare with *meteoroid* and *planetesimal*.

zodiacal light   A band of light in the night sky caused by sunlight reflected by zodiacal dust.

zonal wind   The planetwide circulation of air that moves in directions parallel to the planet's equator.

MATHEMATICAL TOOLS

WORKING WITH PROPORTIONALITIES

Most of the mathematics in *21st Century Astronomy* involves proportionalities—statements about the way that one physical quantity changes when another quantity changes. Here we offer a practical guide to working with proportionalities.

To use a statement of proportionality to compare two objects, begin by turning the proportionality into a ratio. For example, the price of a bag of apples is proportional to the weight of the bag, or

$$\text{price} \propto \text{weight},$$

where the symbol "$\propto$" is pronounced "is proportional to." What this means is that the ratio of the prices of two bags of apples is equal to the ratio of the weights of the two bags:

$$\text{price} \propto \text{weight} \quad \textit{means} \quad \frac{\text{price of A}}{\text{price of B}} = \frac{\text{weight of A}}{\text{weight of B}}$$

Work a specific example: Suppose that bag A weighs 2 pounds, while bag B weighs 1 pound. That of course means that bag A will cost twice as much as bag B. We can turn the proportionality above into the equation:

$$\frac{\text{price of A}}{\text{price of B}} = \frac{\text{weight of A}}{\text{weight of B}} = \frac{2 \text{ lb}}{1 \text{ lb}} = 2,$$

so the price of bag A is 2 times the price of bag B.

Let's work another, more complicated, example. In Chapter 12 we discuss the relationship between the luminosity, brightness, and distance of stars. The luminosity of a star—the total energy the star radiates each second—is proportional to the star's brightness times the square of its distance, or

$$\text{luminosity} \propto \text{brightness} \times \text{distance}^2.$$

What this proportionality means is that if we have two stars, call them A and B, then

$$\frac{\text{luminosity of A}}{\text{luminosity of B}} = \frac{\text{brightness of A}}{\text{brightness of B}} \times \left(\frac{\text{distance to A}}{\text{distance to B}}\right)^2$$

If we use the symbols L, b, and d to represent luminosity, brightness, and distance, respectively, this becomes:

$$\frac{L_A}{L_B} = \frac{b_A}{b_B} \times \left(\frac{d_A}{d_B}\right)^2$$

As an example, suppose that Star A appears twice as bright in the sky as Star B, but Star A is located 10 times as far away as Star B. Compare the luminosities of the two stars. Solution:

$$\text{luminosity} \propto \text{brightness} \times \text{distance}^2,$$

so we write:

$$\frac{\text{luminosity of A}}{\text{luminosity of B}} = \frac{\text{brightness of A}}{\text{brightness of B}} \times \left(\frac{\text{distance to A}}{\text{distance to B}}\right)^2$$

$$= \frac{2}{1} \times \left(\frac{10}{1}\right)^2 = 200.$$

Star A is 200 times as luminous as Star B.

A final note: In our original example, we said that the price of a bag of apples is proportional to the weight of the bag, which allowed us to say that a 2 pound bag of apples costs twice as much as a 1 pound bag of apples. This is a statement about the way the world *works*. A 2 pound bag of apples costs *more* than a 1 pound bag, not *less* than a 1 pound bag. To actually figure how much a bag of apples will cost, we need another piece of information: the price per pound. The price per pound is an example of a *constant of proportionality*. Wrapped up in the price per pound is all sorts of information about the cost of growing apples, the cost of transporting them, how the market is at the moment, and the profit the grocer needs to make, etc. In other words, this constant of proportionality is a statement about the way the world *is*.

Proportionalities help us understand how the world works, and allow us to compare one object to another. Constants of proportionalities allow us to calculate real values for things. In *21st Century Astronomy* it is usually the "understanding"—the proportionality itself—that we care about.

SCIENTIFIC NOTATION

Astronomy is a science both of the very large and the very small. The mass of an electron, for example, is

0.0000000000000000000000000000009109 kg

while the distance to a galaxy far, far away might be around

100,000,000,000,000,000,000,000,000 m.

All it takes is a quick glance at these two numbers to see why astronomers, like most scientists, make heavy use of scientific notation to express numbers.

POWERS OF 10

Our number system is based on powers of 10. Going to the left of the decimal place,

$$10 = 10 \times 1,$$
$$100 = 10 \times 10 \times 1,$$
$$1,000 = 10 \times 10 \times 10 \times 1,$$

and so on. Going to the right of the decimal place:

$$0.1 = \frac{1}{10} \times 1,$$

$$0.01 = \frac{1}{10} \times \frac{1}{10} \times 1,$$

$$0.001 = \frac{1}{10} \times \frac{1}{10} \times \frac{1}{10} \times 1,$$

for as long as we care to continue. In other words, each place to the right or left of the decimal place in a number represents a power of 10. For example, one million can be written,

$$1 \text{ million} = 1,000,000$$
$$= 1 \times 10 \times 10 \times 10 \times 10 \times 10 \times 10.$$

That is to say, 1 million is 1 times six factors of 10. Scientific notation just takes these factors of 10 and combines them in convenient shorthand. Rather than writing out all six factors of 10, instead we combine them in easy shorthand using an exponent:

$$1 \text{ million} = 1 \times 10^6$$

which means one times six factors of 10.

Moving to the right of the decimal place, each step that we take *removes* a power of 10 from the number. One millionth can be written

$$1 \text{ millionth} = 1 \times \frac{1}{10} \times \frac{1}{10} \times \frac{1}{10} \times \frac{1}{10} \times \frac{1}{10} \times \frac{1}{10}.$$

We are removing powers of 10, so we express this using a negative exponent. We write

$$\frac{1}{10} = 10^{-1},$$

and

$$1 \text{ millionth} = 1 \times 10^{-1} \times 10^{-1} \times 10^{-1} \times$$
$$10^{-1} \times 10^{-1} \times 10^{-1}$$
$$= 1 \times 10^{-6}.$$

Returning to our earlier examples, the mass of an electron is 9.109×10^{-31} kg, while the distant galaxy is located 1×10^{26} m away, which are *much* more convenient ways of writing this information. *Notice that the exponent in scientific notation gives you a feeling for the size of a number at a glance.* The exponent of 10 in the electron mass is -31, which tells us instantly that it is a very small number. The exponent of 10 in the distance to a remote galaxy, $+26$, tells us immediately that it is a very large number. This exponent is often called the *order of magnitude* of a number. When you see a number written in scientific notation while reading *21st Century Astronomy* (or elsewhere), just remember to look at the exponent to better understand what the number is telling you.

Scientific notation is also convenient because it makes multiplying and dividing numbers easier. Again, lets look at an example. Two billion times eight thousandths can be written,

$$2,000,000,000 \times 0.008,$$

but it is more convenient to write these two numbers using scientific notation as,

$$(2 \times 10^9) \times (8 \times 10^{-3}).$$

This can be regrouped as,

$$(2 \times 8) \times (10^9 \times 10^{-3})$$

The first part of the problem is just $2 \times 8 = 16$. The more interesting part of the problem is the multiplication in the right hand parentheses. The first number, 10^9, is just shorthand for $10 \times 10 \times 10 \ldots$ nine times. That is, it represents 9 factors of 10. The second number stands for 3 factors of $1/10$, or removing three factors of 10, if you prefer to think of it that way. All together, that makes $9 - 3 = 6$ factors of 10. In other words,

$$10^9 \times 10^{-3} = 10^{9-3} = 10^6.$$

Putting the problem together,

$$(2 \times 10^9) \times (8 \times 10^{-3}) = (2 \times 8) \times (10^9 \times 10^{-3})$$
$$= 16 \times 10^6.$$

By convention, when a number is written in scientific notation, only one digit is placed to the left of the decimal point. In this case, there are 2. However, 16 is 1.6×10, so we can add this additional factor of 10 into the exponent at right, making the final answer:

$$1.6 \times 10^7.$$

Dividing is just the inverse of multiplication. Dividing by 10^3 means removing 3 factors of 10 from a number. Using the number above:

$$(1.6 \times 10^7) \div (2 \times 10^3) = (1.6 \div 2) \times (10^7 \div 10^3)$$
$$= 0.8 \times 10^{7-3}$$
$$= 0.8 \times 10^4.$$

This time we have only a zero to the left of the decimal point. To get the number into proper form, say that $0.8 = 8 \times 10^{-1}$, giving,

$$0.8 \times 10^4 = (8 \times 10^{-1}) \times 10^4 = 8 \times 10^3.$$

Adding and subtracting numbers in scientific notation is somewhat more difficult, because each number has to be written as a value times the *same* power of 10 before they can be added or subtracted. However, almost all modern calculators have scientific notation built in. They keep up with the powers of 10 for you. If you do not have one, you may want to buy one and learn to use it before tackling the mathematical problems in this book. There are more examples on our website that you can use to better learn how to work with scientific notation.

SIGNIFICANT FIGURES

In the example above we actually broke some rules in the interest of explaining how powers of 10 are treated in scientific notation. The rules that we broke involve the *precision* of the numbers we are expressing. In everyday speech we might say, "The store is a kilometer away," by which we probably mean that the store is *roughly* a kilometer away. If it turned out to be 0.8 km, or 1.2 km, it is unlikely that the recipient of our directions would quibble. But when expressing quantities in science, it is extremely important to have a good idea not only of the value of a number, but also how precise that value is.

The most complete way to keep track of the precision of numbers is to actually write down the uncertainty in the number. For example, if we knew that the distance to the store (call it d) is between 0.8 km and 1.2 km, we could write that

$$d = 1.0 \pm 0.2 \text{ km},$$

where the symbol "$\pm$" is pronounced "plus or minus." In this example, d is between $1.0 - 0.2 = 0.8$ km and $1.0 + 0.2 = 1.2$ km. This is an unambiguous statement about the limitations on our knowledge of the value of d, but carrying along the formal errors with every number that we write down would be cumbersome at best. Instead, we keep track of the approximate precision of a number by using *significant figures*.

The convention for significant figures works like this. We assume that the number that we write has been rounded from a number that had one additional digit to the right of the decimal point. If we say that some quantity d, which might represent the distance to the store, is "1.", what we mean is that d is close to 1. It is likely not as small as 0., and it is likely not as large as 2. If we say instead,

$$d = 1.0,$$

then we mean that d is likely not 0.9, and is likely not 1.1 It is roughly 1.0, to the nearest tenth. The greater the number of significant figures, the more precisely the number is being specified. For example, 1.00000 is *not* the same number as 1.00. The first number, 1.00000, represents a value that is probably not as small as 0.99999, and is probably not as large as 1.00001. The second number, 1.00, represents a value that is probably not as small as 0.99, nor as large as 1.01.

When carrying out mathematical operations, significant figures are important. For example, $2.0 \times 1.6 = 3.2$. It does *not* equal 3.20000000000. The product of two numbers cannot be known to any greater accuracy than the numbers themselves! As a general rule, when multiplying and dividing, the answer should have the same number of significant figures as the less precise of the numbers being multiplied or divided. In other words, $2.0 \times 1.602583475 = 3.2$. Since all we know is that the first factor is probably closer to 2.0 than to 1.9 or 2.1, all that we know is that the product is between about 3.0 and 3.4. It is 3.2. It is not 3.205166950 (*even if that is the answer that your calculator gives!*). The rest of the digits to the right of 3.2 just do not mean anything.

When adding and subtracting, the rules are a bit different. If one number has a significant figure with a particular place value, but another number does not, then their sum or difference cannot have a significant figure in that place value. In other words,

$$
\begin{array}{r}
1,045. \\
+1.34567 \\
\hline
1,046.
\end{array}
$$

The answer is 1,046., *not* 1,046.34567. Again, all of the extra digits to the right of the decimal place have no meaning because 1,045. is not known to that accuracy.

There is always a fly in the ointment, and this is no exception. What is the precision of the number 1,000,000? As written, the answer is unclear. Are all of those zeros really significant, or are they just serving as place holders? If we write the number in scientific notation, on the other hand, there is never a question. Instead of 1,000,000 we write 1.0×10^6, for a number that is known to the nearest hundred thousand or so, or we write 1.00000×10^6 for a number that is one million to the nearest 10.

So, our example above would have been more correct had we said,

$$(2.0 \times 10^9) \times (8.0 \times 10^{-3}) = 1.6 \times 10^7.$$

ALGEBRA

There are many branches of mathematics. The branch that tells us about the relationships between quantities is called *algebra*. If you are reading this book, you have almost certainly taken an algebra class, but you are not alone if you feel a little review is in order. Basically, algebra begins by using symbols to represent quantities. For example, we might write the distance you travel in a day as d. As it stands, d has no value. It might be 10,000 miles. It might be 3 feet. It does, however, have *units*—in this case, the units of distance.

The average speed at which you travel is equal to the distance you travel divided by the time that you take. If we use the symbol v to represent your average speed, and the symbol t to represent the time you took, then instead of writing out, "Your average speed is

equal to the distance that you travel divided by the time taken," we can write simply,

$$v = \frac{d}{t}.$$

The meaning of this algebraic expression is exactly the same as the sentence quoted above, but it is much more concise. As it stands, v, d, and t still have no specific values. There are no numbers assigned to them yet. However, this expression tells us what the relationship between those numbers will be when we *do* look at a specific example. For example, if you go 500 km ($d = 500$ km) in 10 hours ($t = 10$ hours), then this expression tells you that your average speed is,

$$v = \frac{d}{t} = \frac{500 \text{ km}}{10 \text{ hours}} = 50 \text{ km/hour}.$$

Notice that the units in this expression act exactly like the numerical values. They are just multiplicative factors. When we say "500 km" what we really mean is "$500 \times$ kilometers." Likewise, 10 hours is "$10 \times$ hours." When we divide the two we find that the units of v are kilometers divided by hours, or km/hr (pronounced kilometers per hour).

We introduced algebra as shorthand for expressing relations between quantities, but it is far more powerful than that. Algebra provides rules for manipulating the symbols used to represent quantities. We begin with a bit of notation for *powers* and *roots*. When we talk about raising a quantity to a power, we mean multiplying the quantity by itself some number of times. For example, if S is a symbol for something (anything), then S^2 (pronounced "S squared" or "S to the second power") means $S \times S$, and S^3 (pronounced "S cubed" or "S to the third power") means $S \times S \times S$. An example is in order. Suppose that S represents the length of the side of a square. The area of the square is given by

$$\text{area} = S \times S = S^2.$$

If $S = 3$ m, then the area of the square is,

$$S^2 = 3 \text{ m} \times 3 \text{ m} = 9 \text{ m}^2,$$

(pronounced 9 square meters). It should be obvious why raising a quantity to the second power is called "squaring" the quantity. We could have done the same thing for the sides of a cube and found that the volume of the cube is

$$\text{volume} = S \times S \times S = S^3.$$

If $S = 3$ m, then the volume of the cube is,

$$S^3 = 3 \text{ m} \times 3 \text{ m} \times 3 \text{ m} = 27 \text{ m}^3,$$

(pronounced 27 cubic meters). Again, it is clear why raising a quantity to the third power is called "cubing" the quantity.

Roots are the reverse of this process. The square root of a quantity is the value that, when squared, gives the original quantity. The square root of 4 is 2, which means that $2 \times 2 = 4$. The square root of 9 is 3, which means that $3 \times 3 = 9$. Similarly, the cube root of a quantity is the value that, when cubed, gives the original quantity. The cube root of 8 is 2, which means that $2 \times 2 \times 2 = 8$. Roots are written with the symbol $\sqrt{}$. For example, we write,

$$\sqrt{9} = 3$$

for the square root of 9 and

$$\sqrt[3]{8} = 2$$

for the cube root of 8. If the volume of a cube is $V = S^3$, we can also write

$$S = \sqrt[3]{V} = \sqrt[3]{S^3}.$$

Roots can also be written as powers. Powers and roots behave exactly the same way that the exponents of 10 did in our discussion of scientific notation. (They had better, since the exponents used in scientific notation are just powers of 10.) For example, if a, n, and m are all algebraic quantities, then

$$a^n \times a^m = a^{n+m}, \qquad \text{and} \qquad \frac{a^n}{a^m} = a^{n-m}.$$

(To see if you understand all this, explain why the square root of a can also be written $a^{\frac{1}{2}}$ and the cubed root of a can be written $a^{\frac{1}{3}}$.)

Some of the rules of algebra are summarized below. These are really no more than the rules of arithmetic, but applied to the symbolic quantities of algebra. The important thing is this: So long as we apply the rules of algebra properly, then the relationships among symbols we arrive at through our algebraic manipulations remain true for the physical quantities those symbols represent.

Here we summarize a few algebraic rules and relationships. In this summary, a, b, c, n, m, r, x and y are all algebraic quantities:

Associative rule:
$$a \times b \times c = (a \times b) \times c = a \times (b \times c)$$

Commutative rule:
$$a \times b = b \times a$$

Distributive rule:
$$a \times (b + c) = (a \times b) + (a \times c)$$

Cross multiplication:

$$\text{If } \frac{a}{b} = \frac{c}{d}, \text{ then } ad = bc.$$

Working with exponents:

$$\frac{1}{a^n} = a^{-n} \qquad a^n a^m = a^{n+m}$$

$$\frac{a^n}{a^m} = a^{n-m} \qquad (a^n)^m = a^{n \times m} \qquad \left(\frac{a}{b}\right)^n = \frac{a^n}{b^n}$$

Equation of a line with slope m and y-intercept b:

$$y = mx + b$$

Equation of a circle with radius r centered at $x = 0$, $y = 0$:

$$x^2 + y^2 = r^2$$

ANGLES AND DISTANCES

The farther away something is, the smaller it appears. This is common sense and everyday experience. In astronomy, where we seldom get to walk up to the object we are studying and measure it with a meter stick, our knowledge about the sizes of things usually depends on knowing the relationship between the size of an object, it's distance, and the angle it covers on the sky.

The natural way to measure angles is using a unit called *radians*. As shown in **Figure A1.1(a),** the size of an angle in radians is just the length of the arc subtending the angle, divided by the radius of the circle. In the figure, the angle $x = S/r$ radians.

Since the circumference of a circle is 2π times the radius, $C = 2\pi r$, a complete circle has an angular measure of $(2\pi r)/r = 2\pi$ radians. In more conventional angular measure, a complete circle is 360°, so we can say that

$$360° = 2\pi \text{ radians,}$$

or that

$$1 \text{ radian} = \frac{360°}{2\pi} = 57.2958°.$$

When talking about stars and galaxies, we often use seconds of arc to measure angles. A degree is broken into 60 minutes of arc, each of which is broken into 60 seconds of arc, so there are 3,600 seconds of arc in a de-

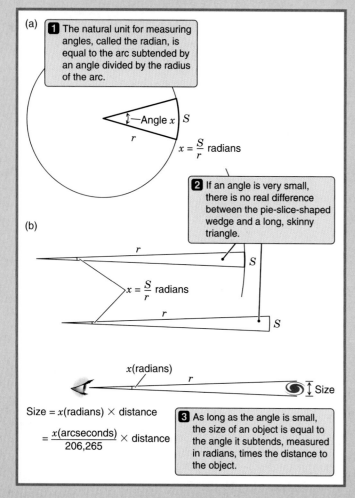

Figure A1.1

gree. That means that

$$3{,}600 \frac{\text{arcseconds}}{\text{degree}} \times 57.2958 \frac{\text{degrees}}{\text{radian}}$$

$$= 206265 \frac{\text{arcseconds}}{\text{radian}}.$$

Now, if the angle is small enough (which it usually is in astronomy), there is precious little difference between the pie slice described above, and a long skinny triangle with a short side of length S (see **Figure A1.1b**). So, if we know the distance d to an object, and we can measure the angular size x of the object, then the size of the object is just

$$Size\ S = x(\text{in radians}) \times d =$$

$$\frac{x(\text{in degrees})}{57.2958\ \text{degree/radian}} \times d$$

$$= \frac{x(\text{in arcseconds})}{206265\ \text{arcseconds/radian}} \times d,$$

which is all we need to turn our knowledge of the angular size and the distance to an object into a measurement of the object's physical size. (As always, our Web site has plenty of examples of such calculations.)

CIRCLES AND SPHERES

To round out our mathematical tools, here are a few useful formulae about circles and spheres. The circle or sphere in each case has a radius r.

$$\text{Circumference}_{\text{circle}} = 2\pi r$$

$$\text{Area}_{\text{circle}} = \pi r^2$$

$$\text{Area}_{\text{sphere}} = 4\pi r^2$$

$$\text{Volume}_{\text{sphere}} = \frac{4}{3}\pi r^3$$

PHYSICAL CONSTANTS AND UNITS

FUNDAMENTAL PHYSICAL CONSTANTS

Constant	Symbol	Value
Speed of light in a vacuum	c	2.99792×10^8 m/s
Universal gravitational constant	G	6.673×10^{-11} N m^2/kg^2
Planck constant	h	6.62607×10^{-34} J s
Electric charge of electron or proton	e	1.60218×10^{-19} C
Boltzmann constant	k	1.38065×10^{-23} J/K
Stefan–Boltzmann constant	σ	5.67040×10^{-8} W/(m^2 K^4)
Mass of electron	m_e	9.10938×10^{-31} kg
Mass of proton	m_p	1.67262×10^{-27} kg

Source: National Institute of Standards and Technology (http://physics.nist.gov).

UNIT PREFIXES

Prefix[a]	Name	Factor[b]
n	nano–	10^{-9}
μ	micro–	10^{-6}
m	milli–	10^{-3}
k	kilo–	10^{3}
M	mega–	10^{6}
G	giga–	10^{9}
T	tera–	10^{12}

These prefixes ([a]), when appended to a unit, change the size of the unit by the factor ([b]) given. For example, 1 km (kilometer) is 10^3 meters.

UNITS AND VALUES

Quantity	Fundamental unit	Values
Length	meter (m)	Radius of Sun ($R_\odot$) = 6.96265×10^8 m Astronomical Unit (AU) = 1.49598×10^{11} m = 149,598,000 km light-year (ly) = 9.4605×10^{15} m = 6.324×10^4 AU 1 parsec (pc) = 3.261 ly = 3.0857×10^{16} m 1 m = 3.281 feet
Volume	meters^3 (m^3)	1 m^3 = 1000 liters = 264.2 gallons
Mass	kilogram (kg)	1 kg = 1000 g Mass of Earth ($M_\oplus$) = 5.9736×10^{24} kg Mass of Sun ($M_\odot$) = 1.9891×10^{30} kg
Time	seconds (s)	Solar day (noon to noon) = 86,400 s Sidereal day (Earth rotation period) = 86,164.1 s Tropical year (equinox to equinox) = 365.24219 days = 3.15569×10^7 s Sidereal year (Earth orbital period) = 365.25636 days = 3.15581×10^7 s
Speed	meters/second (m/s)	1 km/s = 1000 m/s = 3,600 km/h c = 3.00×10^8 m/s = 300,000 km/s
Acceleration	meters/second^2 (m/s^2)	Gravitational acceleration on Earth (g) = 9.78 m/s^2
Energy	joules (J)	1 J = 1 kg m^2/s^2 1 megaton = 4.19×10^{15} J
Power	watt (W)	1 W = 1 J/s Solar luminosity ($L_\odot$) = 3.827×10^{26} W
Force	newton (N)	1 N = 1 kg m/s^2 1 pound (lbs) = 4.448 N
Pressure	newtons/meter^2 (N/m^2)	Atmospheric pressure at sea level = 1.013×10^5 N/m^2 = 1.013 bar
Temperature	kelvins (K)	Absolute zero = 0 K = $-273.15°$C = $-459.67°$F

Sources: National Space Science Data Center (2002); *Observer's Handbook 2002*, Rajiv Gupta (Royal Astronomical Society of Canada, 2001); National Institute of Standards and Technology (2002).

Periodic Table of the Elements

Key

Period number — 1 | H (Number of protons: 1, Symbol: H, Name: Hydrogen, Atomic mass: 1.008)

- Group number
- Number of protons
- Symbol
- Name
- Atomic mass

Legend:
- Alkaline earth metals
- Alkali metals
- Transition metals
- Rare earths
- Metalloids
- Non metals
- Halogens
- Noble gases
- Other metals

1	2	3	4	5	6	7	8	9	10	11	12	13	14	15	16	17	18
1 **H** Hydrogen 1.008																	2 **He** Helium 4.003
3 **Li** Lithium 6.941	4 **Be** Beryllium 9.012											5 **B** Boron 10.81	6 **C** Carbon 12.01	7 **N** Nitrogen 14.01	8 **O** Oxygen 16.00	9 **F** Fluorine 19.00	10 **Ne** Neon 20.18
11 **Na** Sodium 22.99	12 **Mg** Magnesium 24.31											13 **Al** Aluminum 26.98	14 **Si** Silicon 28.09	15 **P** Phosphorus 30.97	16 **S** Sulfur 32.06	17 **Cl** Chlorine 35.45	18 **Ar** Argon 39.95
19 **K** Potassium 39.10	20 **Ca** Calcium 40.08	21 **Sc** Scandium 44.96	22 **Ti** Titanium 47.88	23 **V** Vanadium 50.94	24 **Cr** Chromium 52.00	25 **Mn** Manganese 54.94	26 **Fe** Iron 55.85	27 **Co** Cobalt 58.93	28 **Ni** Nickel 58.69	29 **Cu** Copper 63.55	30 **Zn** Zinc 65.39	31 **Ga** Gallium 69.72	32 **Ge** Germanium 72.64	33 **As** Arsenic 74.92	34 **Se** Selenium 78.96	35 **Br** Bromine 79.90	36 **Kr** Krypton 83.80
37 **Rb** Rubidium 85.47	38 **Sr** Strontium 87.62	39 **Y** Yttrium 88.91	40 **Zr** Zirconium 91.22	41 **Nb** Niobium 92.91	42 **Mo** Molybdenum 95.94	43 **Tc** Technetium (98)	44 **Ru** Ruthenium 101.1	45 **Rh** Rhodium 102.9	46 **Pd** Palladium 106.4	47 **Ag** Silver 107.9	48 **Cd** Cadmium 112.4	49 **In** Indium 114.8	50 **Sn** Tin 118.7	51 **Sb** Antimony 121.8	52 **Te** Tellurium 127.6	53 **I** Iodine 126.9	54 **Xe** Xenon 131.3
55 **Cs** Cesium 132.9	56 **Ba** Barium 137.3	57 **La** Lanthanum 138.9	72 **Hf** Hafnium 178.5	73 **Ta** Tantalum 180.9	74 **W** Tungsten 183.9	75 **Re** Rhenium 186.2	76 **Os** Osmium 190.2	77 **Ir** Iridium 192.2	78 **Pt** Platinum 195.1	79 **Au** Gold 197.0	80 **Hg** Mercury 200.6	81 **Tl** Thallium 204.4	82 **Pb** Lead 207.2	83 **Bi** Bismuth 209.0	84 **Po** Polonium (209)	85 **At** Astatine (210)	86 **Rn** Radon (222)
87 **Fr** Francium (223)	88 **Ra** Radium (226)	89 **Ac** Actinium (227)	104 **Rf** Rutherfordium (261)	105 **Db** Dubnium (262)	106 **Sg** Seaborgium (263)	107 **Bh** Bohrium (262)	108 **Hs** Hassium (265)	109 **Mt** Meitnerium (266)	110 ---- 0	111 ---- 0	112 ---- 0	114 ---- 0			116 ---- 0		118 ---- 0

Lanthanide series 6

58 **Ce** Cerium 140.1	59 **Pr** Praseodymium 140.9	60 **Nd** Neodynium 144.2	61 **Pm** Promethium (145)	62 **Sm** Samarium 150.4	63 **Eu** Europium 152.0	64 **Gd** Gadolium 157.3	65 **Tb** Terbium 158.9	66 **Dy** Dysprosium 162.5	67 **Ho** Holmium 164.9	68 **Er** Erbium 167.3	69 **Tm** Thulium 168.9	70 **Yb** Ytterbium 173.0	71 **Lu** Lutetium 175.0

Actinide series 7

90 **Th** Thorium 232.0	91 **Pa** Protactinium (231)	92 **U** Uranium (238)	93 **Np** Neptunium (237)	94 **Pu** Plutonium (244)	95 **Am** Americium (243)	96 **Cm** Curium (247)	97 **Bk** Berkelium (247)	98 **Cf** Californium (251)	99 **Es** Einsteinium (254)	100 **Fm** Fermium (257)	101 **Md** Mendelevium (256)	102 **No** Nobelium (259)	103 **Lr** Lawrencium (260)

Sources: Los Alamos National Laboratory; National Institute of Standards and Technology.

APPENDIX 4

PROPERTIES OF PLANETS & MOONS

PHYSICAL DATA FOR PLANETS

Planet	Equatorial radius km	Equatorial radius $R/R_\oplus$	Mass kg	Mass $M/M_\oplus$	Average density (relative to water[a])	Sidereal rotation period (days)	Tilt of rotation axis (relative to orbit)	Surface gravity (relative to Earth[b])	Escape velocity (km/s)	Average surface temperature (K)
Mercury	2,440	0.383	3.30×10^{23}	0.055	5.427	58.646	0.01°	0.378	4.3	440 (100–725)[d]
Venus	6,052	0.949	4.87×10^{24}	0.815	5.243	243.021[c]	177.36°	0.907	10.36	737
Earth	6,378	1.000	5.97×10^{24}	1.000	5.515	0.9973	23.45°	1.000	11.19	288 (183–331)[d]
Mars	3,397	0.533	6.42×10^{23}	0.107	3.933	1.026	25.19°	0.377	5.03	210 (133–293)[d]
Jupiter	71,492	11.209	1.90×10^{27}	317.83	1.326	0.4136	3.13°	2.364	59.5	165[e]
Saturn	60,268	9.449	5.68×10^{26}	95.16	0.687	0.4440	26.73°	0.916	35.5	134[e]
Uranus	25,559	4.007	8.68×10^{25}	14.537	1.270	0.7183[c]	97.77°	0.889	21.3	76[e]
Neptune	24,764	3.883	1.02×10^{26}	17.147	1.638	0.6713	28.32°	1.12	23.5	72[e]
Pluto	1,195	0.187	1.25×10^{22}	0.0021	1.750	6.387[c]	122.53°	0.059	1.1	50

[a]The density of water is 1,000 kg/m^3.

[b]The surface gravity of Earth is 9.78 m/s^2.

[c]Venus, Uranus, and Pluto rotate opposite the direction of their orbits. Their north poles are south of their orbital plane.

[d]Where given, values in parentheses give extremes of recorded temperatures.

[e]Temperature where pressure equals 1 bar.

ORBITAL DATA FOR PLANETS

Planet	Mean Distance from Sun (A^a)		Orbital Period (P) (sidereal years)	Eccentricity	Inclination (relative to ecliptic)	Average speed (km/s)
	10^6 km	AU				
Mercury	57.91	0.387	0.2408	0.2056	7.005°	47.87
Venus	108.2	0.723	0.6152	0.0067	3.395°	35.02
Earth	149.6	1.000	1.000	0.0167	0.000°	29.78
Mars	227.9	1.524	1.8809	0.0935	1.850°	24.13
Jupiter	778.6	5.204	11.8618	0.0489	1.304°	13.07
Saturn	1,433.5	9.582	29.4566	0.0565	2.485°	9.69
Uranus	2,872.5	19.201	84.0106	0.0457	0.772°	6.81
Neptune	4,495.1	30.047	164.7856	0.0113	1.769°	5.43
Pluto	5,869.7	39.236	247.6753	0.2444	17.16°	4.72

[a]A is the semimajor axis of the planet's elliptical orbit.

Source: National Space Science Data Center (2002).

PROPERTIES OF SELECTED MOONS[a]

Planet	Moon	Orbital properties		Physical properties		
		P (days)	A (10^3 km)	R (km)	M (10^{20} kg)	Density[b] (water = 1)
Earth (1 moon)	Moon	27.32	384.4	1737.4	735	3.34
Mars (2 moons)	Phobos	0.32	9.38	13.4 × 11.2 × 9.2	0.0001	1.9
	Deimos	1.26	23.46	7.5 × 6.1 × 5.2	0.00002	1.7
Jupiter (28 known moons)	Metis	0.29	127.97	20	0.00096	2.8
	Amalthea	0.50	181.30	131 × 73 × 67	0.0717	2.68
	Io	1.77	421.60	1,821	894	3.53
	Europa	3.55	670.90	1,565	480	2.99
	Ganymede	7.16	1,070	2,634	1,480	1.94
	Callisto	16.69	1,883	2,403	1,080	1.86
	Himalia	250.57	11,480	85	0.0956	3.7
	Pasiphae	736[c]	23,500	18	0.0019	–
	S/1999 J1	767.9[c]	24,250	5	–	–
Saturn (30 known moons)	Pan	0.58	133.58	10	0.00003	–
	Prometheus	0.61	139.35	74 × 50 × 34	0.0027	0.7
	Pandora	0.63	141.70	55 × 44 × 31	0.0022	0.7
	Mimas	0.94	185.52	196	0.38	1.17
	Enceladus	1.37	238.02	250	0.84	1.24
	Tethys	1.89	294.66	530	7.55	1.21
	Dione	2.74	377.40	560	11	1.44
	Rhea	4.52	527.04	765	24.9	1.33
	Titan	15.95	1,222	2,575	1,345.5	1.88
	Hyperion	21.28	1,481	185 × 140 × 113	0.177	1.4
	Iapetus	79.33	3,561	718	18.8	1.02
	Phoebe	550.48[c]	12,952	110	0.04	0.7
	S/2000 S1	1,311[c]	23,100	8	–	–
Uranus (21 known moons)	Cordelia	0.34	49.75	13	–	–
	Miranda	1.41	129.78	236	0.66	1.20
	Ariel	2.52	191.02	579	13.4	1.67
	Umbriel	4.14	266.30	585	11.7	1.40
	Titania	8.71	435.91	789	35.2	1.71
	Oberon	13.46	583.52	761	30.1	1.63
	Setebos	2,235[c]	21,650	10	–	–
Neptune (8 known moons)	Naiad	0.29	48.2	29	–	–
	Larissa	0.55	73.55	104 × 89	–	–
	Proteus	1.12	117.6	218 × 208 × 201	–	–
	Triton	5.88[c]	354.8	1,353	214.7	2.05
	Nereid	360.14	5,513.40	170	0.2	1.0
Pluto (1 moon)	Charon	6.39	19.60	593	19	2.0

[a]Innermost, outermost, largest, and a few other moons for each planet.
[b]The density of water is 1,000 kg/m^3.
[c]Irregular moon (has retrograde orbit).

NEAREST AND BRIGHTEST STARS

STARS WITHIN 15 LIGHT-YEARS OF EARTH

Name[a]	Visual[b] brightness (Sirius = 1,000,000)	Distance (ly)	Spectral type[c]	Visual[b] luminosity (Sun = 1.000)
Sun	1.29×10^{16}	1.58×10^{-5}	G2V	1.000
Alpha Centauri C (Proxima Centauri)	10.3	4.23	M5.5V	0.00006
Alpha Centauri B	76,600	4.40	K1V	0.5
Alpha Centauri A	263,000	4.40	G2V	1.6
Barnard's star (dbl)[d]	40	6.0	M4Ve	0.0004
CN Leonis	1.1	7.8	M5.5	0.00002
BD +36-2147	260	8.3	M2	0.0055
Sirius A	1,000,000	8.6	A1V	23
Sirius B	110	8.6	wdDA	0.0025
BL Ceti A	2.6	8.8	M5.5	0.00006
BL Ceti B	1.7	8.8	M5.5	0.00004
V1216 Sagittarii	19	9.7	M3.5V	0.0005
HH Andromedae	3.2	10.3	M5var	0.0001
Epsilon Eridani (dbl)	8,500	10.5	K2V	0.28
Lacaille 9352	300	10.7	M0.5	0.010
FI Virginis	9.3	10.9	M4	0.0003
EZ Aquarii	3.1	11.1	M5.5	0.0001
61 Cygni A	2,200	11.4	K5V	0.084
61 Cygni B	990	11.4	K7V	0.039
Procyon A	184,000	11.4	F5IV-V	7.4
Procyon B	14	11.4	wdDA	0.00054
Gleise 227-046B	34	11.5	M3.5	0.0014
Gleise 227-046A	69	11.7	M3V	0.0028
Groombridge 34 A	150	11.7	M1	0.0062
Groombridge 34 B	9.5	11.7	M3.5	0.00039
DX Cancri	0.31	11.8	M6	0.00001
Epsilon Indi (dbl)	3,500	11.8	K4.5V	0.15

(Continued)

Stars within 15 Light-years of Earth (continued)

Name[a]	Visual[b] brightness (Sirius = 1,000,000)	Distance (ly)	Spectral type[c]	Visual[b] luminosity (Sun = 1.000)
Tau Ceti	10,500	11.9	G8V	0.45
YZ Ceti	3.8	12.1	M4.5V	0.00017
Luyten's star	30	12.4	M3.5	0.0014
Kapteyn's star	75	12.8	M0	0.0037
AX Microscopii	550	12.9	M1/M2V	0.027
Kruger 60A	38	13.1	M2V	0.0020
Kruger 60B	7.9	13.1	M6V	0.0004
V577 Monocerotis A	9.3	13.4	M4.5V	0.0005
V577 Monocerotis B	0.38	13.4	–	0.00002
CD 25 10553A (dbl)	5.4	13.9	M1	0.0003
Gliese 153-058	24	13.9	M3.5	0.0014
FL Virginis A	1.6	14.2	M5	0.0001
FL Virginis B	1.1	14.2	M7	0.00007
Gliese 267-025	98	14.2	M1.5	0.0060
HIP 15689 (dbl)	3.6	14.4	–	0.0002
Van Maanen 2	2.9	14.4	wdDG	0.0002
TZ Arietis	3.3	14.6	M4.5	0.0002
LHS 288	0.74	14.7	M	0.00005
CD 25 10553B	3.9	14.7	M1.5	0.0003
LP 731-58	0.15	14.8	M6.5	0.00001
Gliese 240-063 (dbl)	57	14.8	M3	0.0037
CD 46 11540	46	14.8	M2.5	0.0030

[a]Stars may carry many names, including common names (such as Sirius), names based on their prominence within a constellation (such as Alpha Canis Majoris, another name for Sirius), or names based on their inclusion in a catalog (such as BD+36−2147). Addition of letters A, B, etc, or superscripts indicate membership in a multiple star system.

[b]Brightness and luminosity in these tables refer only to radiation in "visual" light.

[c]Spectral types such as M3 are discussed in Chapter 12. Other letters or numbers provide additional information. For example, V after the spectral type indicates a main sequence star, while III indicates a giant star.

[d](dbl) means an unresolved double star.

The 52 Brightest Stars in the Sky

Name	Common name	Visual brightness (Sirius = 1,000,000)	Distance (ly)	Spectral type	Visual luminosity (Sun = 1.000)
Sun	Sun	1.29×10^{16}	1.58×10^{-5}	G2V	1.000
Alpha Canis Majoris	Sirius	1,000,000	8.60	A1V	23
Alpha Carinae	Canopus	506,000	310	F0II	15,000
Alpha Bootis	Arcturus	270,000	36.7	K1.5IIIFe-0.5	110
Alpha^1 Centauri	Rigel Kentaurus	263,000	4.40	G2V	1.6
Alpha Lyrae	Vega	254,000	25.3	A0Va	50
Alpha Aurigae	Capella	242,000	42	G5IIIe+G0III	130
Beta Orionis	Rigel	233,000	770	B8Ia:	43,000
Alpha^1 Canis Minoris	Procyon A	184,000	11.4	F5IV-V	7.4
Alpha Eridani	Achernar	171,000	144	B3Vpe	1,100
Alpha Orionis	Betelgeuse	164,000	430	M1-2Ia-Iab	9,300
Beta Centauri	Hadar	149,000	530	B1III	13,000
Alpha Aquilae	Altair	128,000	16.8	A7V	11
Alpha Tauri	Aldebaran	119,000	65	K5+III	160
Alpha Scorpii	Antares	108,000	600	M1.5Iab-Ib+B4Ve	12,000
Alpha Virginis	Spica	106,000	260	B1III-IV+B2V	2,200
Beta Geminorum	Pollux	91,200	33.7	K0IIIb	32
Alpha Piscis Austrinus	Fomalhaut	89,500	25.1	A3V	17
Beta Crucis	Becrux	82,400	350	B0.5III	3,200
Alpha Cygni	Deneb	82,400	3,000	A2Ia	270,000
Alpha^1 Crucis	Acrux A	76,600	320	B0.5IV	2,400
Alpha^2 Centauri	Alpha Centauri B	76,600	4.40	K1V	0.5
Alpha Leonis	Regulus	75,200	77	B7V	140
Epsilon Canis Majoris	Adhara	65,500	430	B2II	3,800
Gamma Crucis	Gacrux	58,100	88	M3.5III	140
Lambda Scopii	Shaula	58,100	700	B2IV+B	8,900
Gamma Orionis	Bellatrix	57,500	240	B2III	1,100
Beta Tauri	El Nath	57,000	131	B7III	300
Beta Carinae	Miaplacidus	55,500	111	A2IV	210
Epsilon Orionis	Alnilam	54,500	1,300	B0Ia	30,000
Alpha^2 Crucis	Acrux B	53,000	800	B1V	11,000
Alpha Gruis	Al Na'ir	52,500	101	B7IV	170
Epsilon Ursae Majoris	Alioth	51,000	81	A0pCr	100
Gamma^2 Velorum	Suhail	50,600	840	WC8+O9I	11,000
Alpha Persei	Mirfak	50,100	590	F5Ib	5,400
Alpha Ursae Majoris	Dubhe	50,100	124	K0IIIa	240
Delta Canis Majoris	Wezen	47,900	1,800	F8Ia	47,500
Epsilon Sagittarii	Kaus Australis	47,400	145	B9.5III	300
Epsilon Carinae	Avior	47,000	630	K3III+B2:V	5,800
Eta Ursae Majoris	Benetnasch	47,000	101	B3V	150
Theta Scorpii	Sargas	46,600	270	F1II	1,100
Beta Aurigae	Menkalinan	45,300	82	A2IV	94
Alpha Trianguli Australis	Atria	44,500	415	K2IIb-IIIa	2,400
Gamma Geminorum	Alhena	44,100	105	A0IV	150
Alpha Pavonis	Peacock	43,700	183	B2IV	450
Delta Velorum	–	42,900	80	A1V	84

(Continued)

THE 52 BRIGHTEST STARS IN THE SKY (CONTINUED)

Name	Common name	Visual brightness (Sirius = 1,000,000)	Distance (ly)	Spectral type	Visual luminosity (Sun = 1.000)
Beta Canis Majoris	Murzim	42,100	500	B1II–III	3,200
Alpha Geminorum	Castor	42,100	52	A1V	35
Alpha Hydrae	Alphard	42,100	177	K3II–III	410
Alpha Arietis	Hamal	41,300	66	K2-IIICa-1	55
Alpha Ursae Minoris	Polaris (Pole Star)	40,600	430	F7:Ib–II	2,300
Sigma Sagittarii	Nunki	40,600	224	B2.5V	630

Sources for Appendix 5: The Hipparcos and Tycho Catalogues, 1997, European Space Agency SP-1200; *Bright Star Catalogue*, 5th Revised Ed. (Hoffleit, 1991), NSSDC Astronomical Data Center; *Burnham's Celestial Handbook* (Dover, 1983).

OBSERVING THE SKY

The purpose of this appendix is to provide enough information to be able to make sense of a star chart or list of astronomical objects, and to find a few objects in the sky.

CELESTIAL COORDINATES

In Chapter 2 we discuss the celestial sphere—the imaginary sphere with Earth at its center upon which celestial objects appear to lie. There are a number of different coordinate systems that are used to specify the positions of objects on the celestial sphere. The simplest of these is the *altitude-azimuth coordinate system*. The altitude-azimuth coordinate system is based on the "map" direction to an object (the object's azimuth, with north = 0°, east = 90°, south = 180°, and west = 270°), combined with how high the object is above the horizon (the object's altitude, with the horizon at 0° and the zenith at 90°). For example, an object that is 10° above the eastern horizon has an altitude of 10° and an azimuth of 90°. An object that is 45° above the horizon in the southwest is at altitude 45°, azimuth 225°.

The altitude-azimuth coordinate system is the simplest way to tell a friend where in the sky to look at the moment, but it is not a very good coordinate system for cataloging the positions of objects. The altitude and azimuth of an object are different for each observer, depending on their position on Earth, and are constantly changing as Earth rotates on its axis. If we need to specify the direction to an object in a way that is the same for everyone, we need a coordinate system that is fixed relative to the celestial sphere. The most common such coordinates are called *celestial coordinates*.

Celestial coordinates are illustrated in **Figure A6.1.** Celestial coordinates are much like the traditional system of latitude and longitude used on the surface of Earth. On Earth, latitude specifies how far you are from Earth's equator, as discussed in Chapter 2. If you are on Earth's equator, your latitude is 0°. If you are at Earth's North Pole, your latitude is 90° north. If you are at Earth's South Pole, your latitude is 90° south.

The latitude-like coordinate on the celestial sphere is called *declination,* often signified with the Greek letter δ (delta). The celestial equator has δ = 0°. The north celestial pole has δ = +90°. The south celestial pole has δ = −90°. (See Chapter 2 if you need to refresh your memory of the celestial equator or celestial poles.) Declination is usually expressed in degrees, minutes of arc, and seconds of arc. For example, Sirius, the brightest star in the sky, has δ = −16° 42′ 58″, meaning that it is located not quite 17° south of the celestial equator.

On Earth, east-west position is specified by longitude. Lines of constant longitude run north-south from one pole to the other. Unlike latitude, for which the equator provides a natural place to call "zero," there is no natural starting point for longitude, so we just have to invent one. By arbitrary convention, the Royal Greenwich Observatory in Greenwich, England is defined to lie at a longitude of 0°. On the celestial sphere, the longitude-like coordinate is called *right ascension,* often signified with the Greek letter α (alpha). Unlike longitude, there *is* a natural point on the celestial sphere to use as the starting point for right ascension—the vernal equinox, or that point at which the ecliptic crosses the celestial equator with the Sun moving from

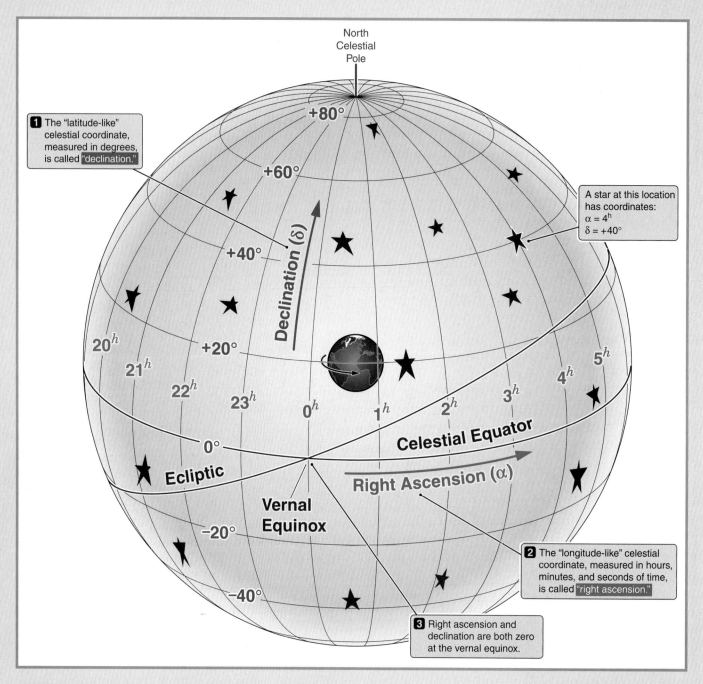

1 The "latitude-like" celestial coordinate, measured in degrees, is called "declination."

A star at this location has coordinates:
$\alpha = 4^h$
$\delta = +40°$

North Celestial Pole

+80°

+60°

+40°

Declination (δ)

+20°

20^h

21^h

22^h

23^h

0^h

1^h

2^h

3^h

4^h

5^h

0°

Ecliptic

Celestial Equator

Right Ascension (α)

Vernal Equinox

−20°

−40°

2 The "longitude-like" celestial coordinate, measured in hours, minutes, and seconds of time, is called "right ascension."

3 Right ascension and declination are both zero at the vernal equinox.

Figure A6.1

the southern sky into the northern sky. The vernal equinox defines the line of right ascension at which $\alpha = 0°$. The autumnal equinox, located on the opposite side of the sky, is at $\alpha = 180°$.

Normally, right ascension is measured in units of time rather than in degrees. It takes Earth 24 hours (of sidereal time) to rotate on its axis, so the celestial sphere is broken into 24 hours of right ascension, with each

hour of right ascension corresponding to 15°. Hours of right ascension are then subdivided into minutes and seconds of time. Right ascension increases going to the east. The right ascension of Sirius, for example, is $\alpha = 06^h\,45^m\,08.9^s$, meaning that Sirius is about 101° (that is, $06^h\,45^m$) east of the vernal equinox. Time is a natural unit for measuring right ascension because time naturally tracks the motion of objects due to Earth's rotation on its axis. If stars on the meridian at a certain time have $\alpha = 6^h$, then an hour later the stars on the meridian will have $\alpha = 7^h$, and an hour after that they will have $\alpha = 8^h$. The *local sidereal time*, or "star time," at your location right now is equal to the right ascension of the stars that are on your meridian at the moment. Because of Earth's motion around the Sun, a sidereal day is about 4 minutes shorter than a solar day, and so local sidereal time constantly gains on solar time. At midnight on September 21, the local sidereal time is 0^h. By midnight on December 21, local sidereal time has advanced to 6^h. On March 21, local sidereal time at midnight is 12^h. Local sidereal time at midnight on June 21 is 18^h.

Putting this all together, right ascension and declination provide a convenient way to specify the location of any object on the celestial sphere. Sirius is located at $\alpha = 06^h\ 45^m\ 08.9^s$, $\delta = -16° 42' 58''$, which means that at midnight on December 21 (local sidereal time = 6^h) you will find Sirius about 45^m east of the meridian, not quite 17° south of the celestial equator.

There is just one final caveat. As we discussed in Chapter 2, the directions of the celestial equator, celestial poles, and vernal equinox are constantly changing as Earth's axis wobbles like the axis of a spinning top. In Chapter 2 we called this 26,000-year wobble the "precession of the equinoxes," meaning that the location of the equinoxes is slowly advancing along the ecliptic. So when we specify the celestial coordinates of an object, we need to specify the date at which the positions of the vernal equinox and celestial poles were measured. By convention, coordinates are usually referred to with the position of the vernal equinox on January 1, 2000. A complete, formal specification of the coordinates of Sirius would then be $\alpha(2000) = 06^h\ 45^m\ 08.9^s$, $\delta(2000) = -16° 42' 58''$, where the "2000" in parenthesis refers to the equinox of the coordinates.

CONSTELLATIONS AND NAMES

While it is certainly possible to specify exactly any location on the surface of Earth by giving its latitude and longitude, it is usually more convenient to use a more descriptive address. We might say, for example, that one of the coauthors of this book works near latitude 37° north, longitude 122° west, but it would probably mean a lot more to you were we to say that George Blumenthal works in Santa Cruz, California.

Just as the surface of Earth is divided into nations and states, the celestial sphere is divided into 88 *constellations*, the names of which are often used to refer to objects within their boundaries (see the star charts in **Figure A6.2**, starting on the next page). The brightest stars within the boundaries of a constellation are referred to using a Greek letter combined with the name of the constellation. For example, the star Sirius is the brightest star in the constellation Canis Major (the Great Dog), and so is referred to as α *Canis Majoris*. The bright red star in the northeastern corner of the constellation of Orion is referred to as α *Orionis*, also known as Betelgeuse. Rigel, the bright blue star in the southwest corner of Orion, is also called β *Orionis*.

It is worth noting that astronomical objects can take on a bewildering range of names. For example, the bright southern star Canopus, also known as α *Carinae* (the brightest star in the constellation of Carina) has no fewer than 34 different names, most of which are about as memorable as "SAO 234480" (number 234,480 in the Smithsonian Astrophysical Observatory catalog of stars).

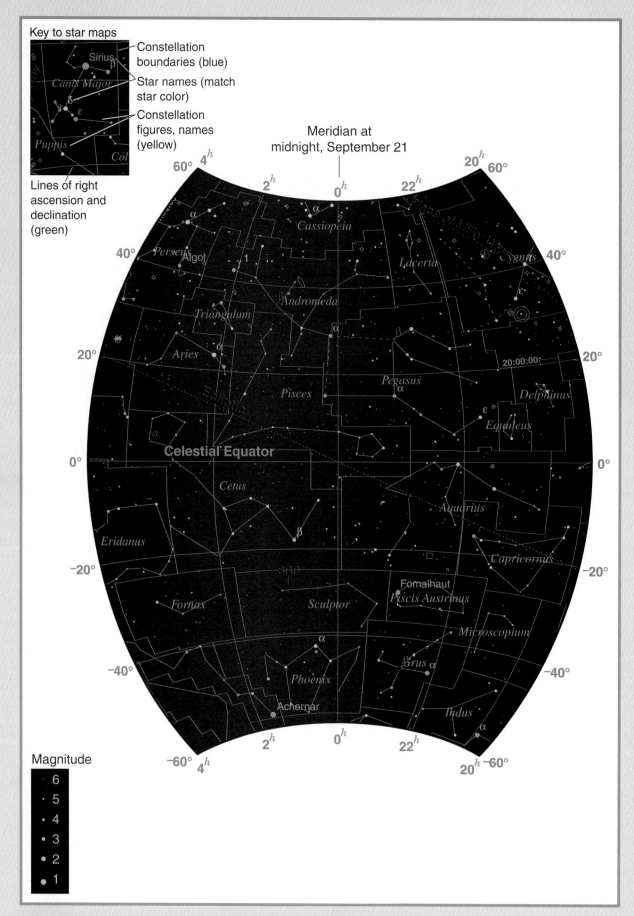

Figure A6.2(a) *The sky from right ascension 20ʰ to 4ʰ and declination −60° to +60°.*

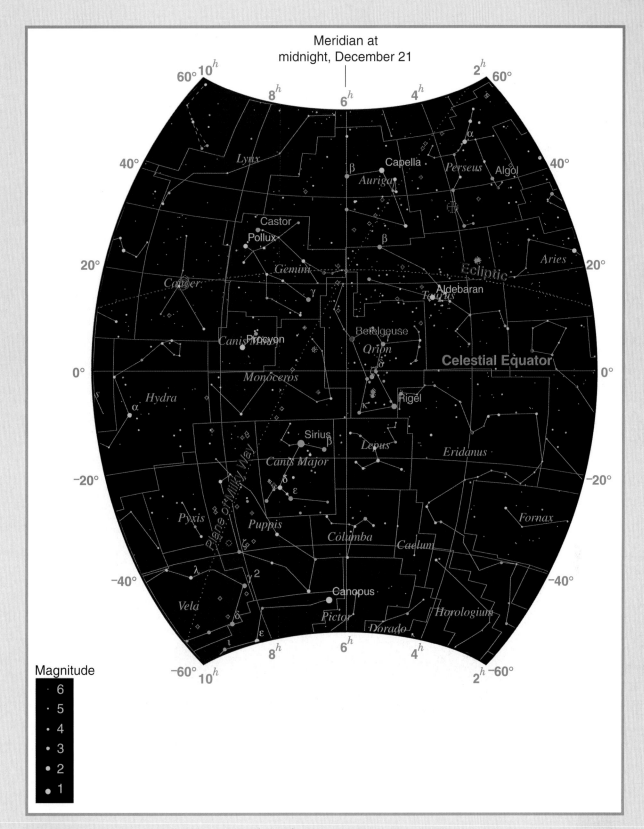

Figure A6.2(b) *The sky from right ascension 2ʰ to 10ʰ and declination −60° to +60°.*

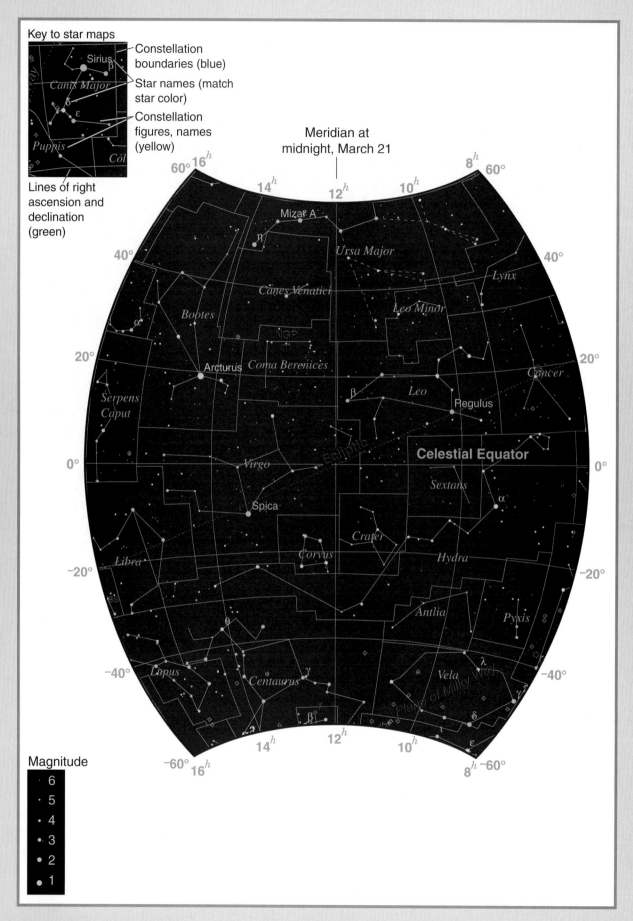

Figure A6.2(c) *The sky from right ascension 8ʰ to 16ʰ and declination −60° to +60°.*

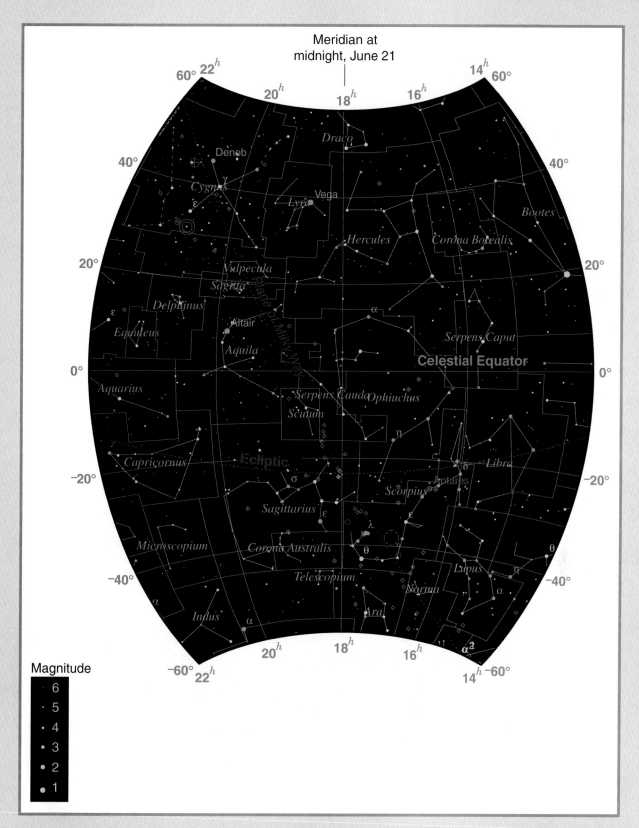

Figure A6.2(d) *The sky from right ascension 14^h to 22^h and declination −60° to +60°.*

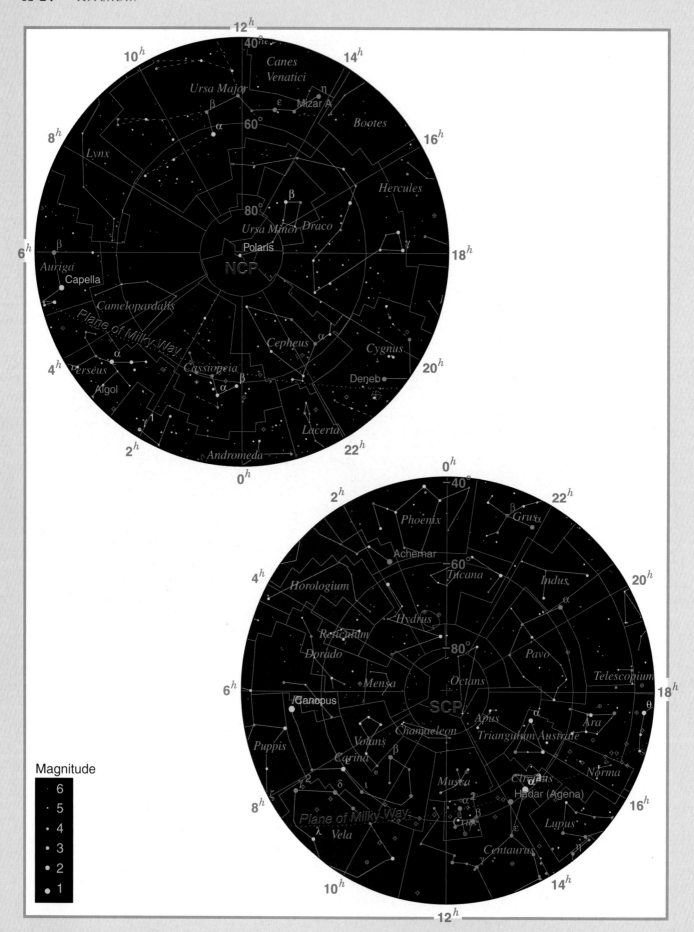

Figure A6.2(e) *The regions of the sky north of declination +40° and south of declination −40°.*

ASTRONOMICAL MAGNITUDES

Throughout the text we refer to the brightness of objects, but when discussing the appearance of an object in the sky, astronomers normally speak instead of the object's *magnitude*. The system of astronomical magnitudes dates back to the Greek astronomer Hipparchus, who when classifying stars ordered them according to their "rank." First rank stars were the brightest stars in the sky, second rank stars were the next brightest, and so on. The faintest stars that could be seen were called sixth rank stars. When methods were devised to actually measure the brightness of stars, it was discovered that first rank stars were typically about 100 times as bright as sixth rank stars, and that the steps from one rank to the next were *logarithmic*. That is to say, a first rank star was typically about 2.51 *times* as bright as a second rank star, while a second rank star was typically about 2.51 *times* as bright as a third rank star.

This rough classification of stellar "rank" was formalized into the modern system of astronomical magnitudes. A difference of five magnitudes between the brightness of two stars (say a star with $m = 6$ and a star with $m = 1$), corresponds to a hundred-fold difference in brightness. Notice that the magnitude scale is backwards—the *greater the magnitude, the fainter the object*.

If five steps in magnitude corresponds to a factor of 100 in brightness, then one step in magnitude must correspond to a factor of $100^{1/5} = 10^{2/5} = 2.512...$ in brightness ($100^{1/5} \times 100^{1/5} \times 100^{1/5} \times 100^{1/5} \times 100^{1/5} = 100$). The relationship between brightness and magnitude is most easily written using common or base-10 logarithms. If star 1 has a brightness of b_1 and star 2 has a brightness of b_2, then the difference in magnitude $m_2 - m_1$ between the two stars is

$$m_2 - m_1 = -2.5 \, log_{10} \frac{b_2}{b_1}.$$

To convert from magnitude differences to brightness ratios, divide by -2.5 (i.e., multiply by -0.4) and raise 10 to the resulting power:

$$\frac{b_2}{b_1} = 10^{-0.4 \times (m_2 - m_1)}$$

The last thing that we need to set the magnitude scale is an object that we define to have a magnitude of zero. Many different magnitude scales exist, but perhaps the most common scale for visual magnitudes defines the star Vega as having a magnitude of 0.

A final note: In Chapters 12 and onward, we used "colors" based on the ratio of the brightness of a star as seen in two different parts of the spectrum. The "b_B/b_V color," for example, was just the ratio of the brightness of a star seen through a blue filter, divided by the brightness of a star seen through a red filter. Normally, astronomers instead discuss the "B-V color" of a star, which is equal to the difference between a star's blue magnitude and its visual magnitude. We can use the expression above for a difference in magnitude to write

$$\mathrm{B} - \mathrm{V} \text{ color} = m_B - m_V = -2.5 \, log_{10} \frac{b_B}{b_V}.$$

So a star with a b_B/b_V color of 1 has a B-V color of 0. A star with a b_B/b_V color of 1.4 has a B-V color of -0.37. Notice that, as with magnitudes, B-V colors are "backwards." The bluer a star the greater its b_B/b_V color, but the lesser its B-V color.

While the system of astronomical magnitudes and colors is actually very convenient in many ways—which is why astronomers continue to use it—it can also be very confusing to new students, especially students for whom math is a foreign language. Having seen something of how astronomical magnitudes work, we think you may agree with our choice to discuss the brightness of stars rather than their magnitudes throughout the text! However, if you do need to use the magnitude system (for example, to read star charts, make sense of a popular astronomy article, or complete a lab assignment), just remember three things and you will probably get by:

1. The greater the magnitude, the *fainter* the object;
2. A magnitude less means about two and a half times brighter; and
3. The brightest stars in the sky have magnitudes less than 1, while the faintest stars that can be seen with the naked eye on a dark night have magnitudes of about 6.

APPENDIX
7

UNIFORM CIRCULAR MOTION AND CIRCULAR ORBITS

In Chapter 3 (see Section 3.5 and Figure 3.15) we discuss the motion of an object moving in a circle at a constant speed. This motion, called uniform circular motion, is the result of centripetal force always acting toward the center of the circle. The key question when thinking about uniform circular motion is, "how hard do I have to pull to keep the object moving in a circle?" Part of the answer to this question is pretty obvious—the more massive an object is, the harder it will be to keep it moving on its circular path. According to Newton's Second Law, $F = ma$, or in this case, the centripetal force equals the mass times the centripetal acceleration. The larger the mass, the greater the force required to keep it moving on its circle.

The centripetal force needed to keep an object moving in constant circular motion also depends on two other quantities: the speed of the object and the size of the circle. The faster an object is moving, the more rapidly it has to change direction to stay on a circle of a given size. The second quantity that influences the needed acceleration is the radius of the circle. The smaller the circle, the greater the pull needed to keep it on track. You can understand this by looking at the motion. A small circle requires a continuous "hard" turn, while a larger circle requires a more gentle change in direction. It takes more force to keep an object moving faster in a smaller circle than it does to keep the same object moving more slowly in a larger circle. (To get a better feeling for how this works, think about the difference between riding in a car that is taking a tight curve at high speed, and a car that is moving slowly around a gentle curve.)

To arrive at the circular velocity and other results discussed in Chapter 3, we need to turn these intuitive ideas about uniform circular motion into a quantitative expression for exactly how much centripetal acceleration is needed to keep an object moving in a circle with radius r at speed v. **Figure A7.1** shows a ball moving around a circle of radius r at a constant speed v at two different times. The centripetal acceleration that is keeping the ball on the circle is a. Remember that the acceleration is always directed toward the center of the circle, while the velocity of the ball is always perpendicular to the acceleration. The ball's velocity and its acceleration are always at right angles to each other. As the object moves around the circle, the direction of motion and the direction of the acceleration change together in lock step.

In the figure we have drawn two triangles. In triangle 1 we show the velocity (speed and direction) at each of the two times. The arrow labeled $\triangle v$ connecting the heads of the two velocity arrows shows how much the velocity changed between time 1 and time 2. This change is the effect of the centripetal acceleration. If we imagine that points 1 and 2 are very close together—so close that the direction of the centripetal acceleration does not change by much between the two—then we can say that the centripetal acceleration is equal to the change in the velocity divided by the time between the two, $\triangle t = t_2 - t_1$. So we have $\triangle v = a \triangle t$.

In triangle 2 we do a similar thing. Here, the arrow labeled $\triangle r$ indicates the change in the position of the ball between time 1 and time 2. Again, if we imagine

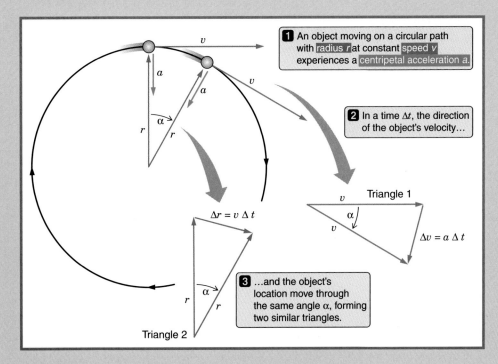

Figure A7.1 *Similar triangles used to find the centripetal force needed to keep an object moving at a constant speed on a circular path.*

that the time between the two points is very short, we can say that $\triangle r$ is equal to the velocity times the time, or $\triangle r = v \triangle t$.

The line between the center of the circle and the ball is always perpendicular to the velocity of the ball. So, if the direction of the ball's velocity changes by an angle α, then the direction of the line between the ball and the center of the circle must also change by the same angle α. In other words, triangles 1 and 2 are *similar triangles*. They have the same *shape*. If the triangles are the same shape it means that the ratio of two sides of triangle 1 must equal the ratio of the two corresponding sides of triangle 2. Using this fact we can write

$$\frac{a\triangle t}{v} = \frac{v\triangle t}{r}.$$

If we divide out the $\triangle t$ from both sides of the equation, then cross multiply, we obtain

$$ar = v^2,$$

which after dividing both sides of the equation by r becomes

$$a_{\text{centripetal}} = \frac{v^2}{r}.$$

We have added the subscript "centripetal" to a to signify that this is the centripetal acceleration needed to keep the object moving in a circle of radius r at speed v. The centripetal force required to keep an object of mass m moving on such a circle is then

$$F_{\text{centripetal}} = ma_{\text{centripetal}} = \frac{mv^2}{r}.$$

CIRCULAR ORBITS

In the case of an object moving in a circular orbit there is no string to hold the ball on its circular path. Instead, this force is provided by gravity.

Think about the case where an object with mass m is in orbit about a much larger object with mass M. The orbit is circular, and the distance between the two objects is given by r. The force needed to keep the smaller object moving at speed v in a circle with radius r is given by the expression above for $F_{\text{centripetal}}$. The force actually provided by gravity (see Chapter 3) is

$$F_{\text{gravity}} = G\frac{Mm}{r^2}.$$

If gravity is responsible for holding the mass in its circular motion, then it had better be the case that $F_{\text{gravity}} = F_{\text{centripetal}}$. That is, if mass m is moving in a circle under the force of gravity, the force provided by gravity *must* be equal to the centripetal force needed to explain that circular motion. Setting the two expressions above for $F_{centripetal}$ and $F_{gravity}$ equal to each other gives

$$\frac{mv^2}{r} = G\frac{Mm}{r^2}.$$

All that remains is a bit of algebra. Dividing out the m on both sides of the equation and multiplying both sides by r gives

$$v^2 = G\frac{M}{r}.$$

Taking the square root of both sides then brings us to the desired result:

$$v_{circular} = \sqrt{\frac{GM}{r}}$$

This is the "circular velocity" that we presented in Chapter 3. It is the velocity at which an object in a circular orbit *must* be moving. If the object were not moving at this velocity, then gravity would not be providing the needed centripetal force, and the object would not move in a circle.

CREDITS

Title Page NASA and The Hubble Heritage Team (STScI). **False Title Page** R. Williams (STScI) and NASA. **Part 1 Opener** STScI. **Part 2 Opener** NASA/JPL. **Part 3 Opener** Jeff Hester and Paul Scowen (Arizona State University), and NASA. **Part 4 Opener** NASA and The Hubble Heritage Team (STScI/AURA).

CHAPTER 1

1.1 Craig Lovell/Corbis. **1.2** Dr. Bernard G. Lindsay, Physics and Astronomy Department, Rice University. **1.3a & b** Neil Ryder Hoos. **1.3c** Owen Franken/Corbis. **1.3d** Image Bank/Gettyone. **1.3e** PhotoDisc. **1.3f & g** American Museum of Natural History, photograph by Denis Finnin and Craig Chesek. **1.3h** NASA/JPL/Caltech. **1.4** Bettmann/Corbis. **1.5 left** STScI/NASA/Arizona State University/Hester/Ressmeyer/Corbis. **1.5 right** Ron Watts/Corbis. **1.6a** Michael J. Tuttle (NASM, NASA). **1.6b** NSSDC. **1.7** NRAO/AUI, James J. Condon, John J. Broderick, and George A. Seielstad. **1.8** © 1969 by The New York Times Co. Reprinted by permission. **1.9 top left** Bettmann/Corbis. **1.9 top right** From *Special Relativity: The M.I.T. Introductory Physics Series* by A. P. French. © 1968, 1966 by Massachusetts Institute of Technology. Used by permission of W. W. Norton & Company, Inc. **1.9 center** Seth Joel/Corbis. **1.9 bottom right** Robbie Jack/Corbis. **1.9 bottom left** Musée d'Orsay, Paris, France. Photo: Réunion des Musées Nationaux/Art Resource, NY. Photograph by Herve Lewandowski. **1.10** © Bettmann/Corbis. **1.11** Four seasons photographs, Jim Schwabel/Index Stock Imagery/PictureQuest. **1.12** NON SEQUITUR © Wiley Miller. Dist. by Universal Press Syndicate. Reprinted with permission. All rights reserved.

CHAPTER 2

2.1 British Library Maps. **2.7 left** Pekka Parviainen/Science Photo Library. **2.7 right** David Nunuk/Science Photo Library. **2.19 top left** AFP/Corbis. **2.19 top right** AP/Wide World Photos. **2.19 bottom right** Lindsay Hebberd/Corbis. **2.19 bottom left** Lynn Goldsmith/Corbis. **2.23** Roger Ressmeyer/Corbis. **2.25a** Reuters New Media Inc./Corbis. **2.25b** Photograph © 2001 by Fred Espenak, courtesy of www.MrEclipse.com.

CHAPTER 4

4.8 Eyewire Collection. **4.10c** Yerkes Observatory photographs: Richard Dreiser. **4.10e** W. M. Keck Observatory/CARA. **4.12 Kitt Peak** NOAO/AURA/NSF. **4.12 Compton** NASA. **4.12 Chandra** Harvard-Smithsonian Center for Astrophysics. **4.12 Fuse** Graphic courtesy Orbital Sciences Corp. **4.12 SIRTF** NASA/JPL/Caltech. **4.12 JCMT** Joint Astronomy Center in Hilo, Hawaii. **4.12 VLA** NRAO VLA Image Gallery. **4.12 Green Bank** The National Radio Astronomy Observatory, Green Bank. **4.12 Arecibo** Dave Finley, Courtesy National Radio Astronomy Observatory and Associated Universities, Inc. **4.21** © Tom Pantages. **4.27** NSSDC/GSFC/NASA.

CHAPTER 5

5.1 STScI photo: Karl Stapelfeldt (JPL). **5.2** Photograph by Pelisson, SaharaMet. **5.9** NSSDC/GSFC/NASA.

CHAPTER 6

6.1 GSFC/NASA. **6.2** U.S. Geological Survey Hawaiian Volcano Observatory. **6.3b** NASA/JSC. **6.4** Montes De Oca & Associates. **6.5** Photograph by D.J. Roddy and K.A. Zeller, USGS, Flagstaff, AZ. **6.6** NASA/JSC. **6.7** NASA/JPL/Caltech. **6.8** © Don Davis. **6.9 top, center & bottom** NASA/JPL/Caltech. **6.12b** Grant Heilman Photography, Inc. **6.13** Photo courtesy of Ron Greeley. **6.18** NASA/JSC. **6.19** NSSDC/NASA. **6.20** NASA Magellan Image JPL P-39225. **6.21** NASA/JSC. **6.22** NASA/JSC. **6.23** NASA/JPL/Caltech. **6.24** NASA/JPL/Caltech. **6.25** NASA/JPL/Malin Space Science Systems. **6.26** NASA/JPL/Caltech. **6.27** NASA/JPL/Caltech, courtesy Alfred McEwen, USGS. **6.27 inset** NASA/JPL/Caltech.

CHAPTER 7

7.1 Venus NASA/NSSDC. **7.1 Earth** NASA/NSSDC. **7.1 Mars** NASA/STScI. **7.8a** Craig Aurness/Corbis. **7.8b** Philip James Corwin/Corbis. **7.10a** Louis A. Frank, The University of Iowa. **7.10b** Raymond Gehman/Corbis. **7.12** Harald Edens, www.weather-photography.com. **7.14** NASA/JPL/Caltech. **7.15** NASA/JPL/Caltech. **7.16a** NASA/JPL/Caltech. **7.16b** NASA/STScI.

CHAPTER 8

8.1 NASA/JPL/Caltech. **8.4** NASA/STScI. **8.5 left** E. Karkoschka (LPL) and NASA. **8.5 right** L. Sromovsky (University of Wisconsin) and NASA. **8.6** CICLOPS/NASA/JPL/University of Arizona. **8.7a** NASA/JPL/Caltech. **8.8** NASA/JPL/Caltech. **8.16** N.M. Schneider, J.T. Trauger, Catalina Observatory. **8.17a** J. Clarke (University of Michigan), NASA. **8.17b** J.T. Trauger, JPL/NASA.

CHAPTER 9

9.7 Christopher Mackay. **9.12** NASA/JPL/Caltech. **9.13** AURA/NASA. **9.14a** Dr. Hal Weaver (Johns Hopkins University) and T. Smith (STScI), and NASA. **9.14b** New Mexico State University Astronomy Department.

CHAPTER 10

10.1 NASA/JPL/Caltech. **10.4** NASA/JPL/Caltech. **10.5** NASA/JPL/Caltech. **10.6** NASA/JPL/Caltech. **10.7** NASA/JPL/Caltech. **10.8** NASA/JPL/Caltech. **10.9** NASA/JPL/Caltech. **10.10a** © 2002 Calvin J. Hamilton. **10.10b** NASA/JPL/Caltech. **10.11** NASA/JPL/Caltech. **10.12** NASA/JPL/Caltech. **10.13** NASA/JPL/Caltech. **10.14** NASA/JPL/Caltech. **10.15** NASA/JPL/Caltech. **10.16** NASA/JPL/Caltech. **10.17** NASA/JPL/Caltech. **10.18** Peter H. Smith and Mark T. Lemmon (University of Arizona) and NASA. **10.19a** NASA/JPL/Caltech. **10.20a** NASA/JPL/Caltech. **10.20b** Courtesy of LUCASFILM LTD, *Star Wars: Episode IV—A New Hope* © 1977 & 1997 Lucasfilm Ltd. & TM, All rights reserved. **10.21** NASA/JPL/Caltech. **10.22a** Dr. R. Albrecht, ESA/ESO Space Telescope European Coordinating Facility/NASA.

CHAPTER 11

11.2 Courtesy of Ron Greeley. **11.5** NASA/JPL/Caltech. **11.6 top:** NASA/Johns Hopkins University Applied Physics Laboratory (JHU/APL). **11.6 bottom** JHU/APL. **11.6 right** JHU/APL. **11.7** NASA/JPL/Caltech. **11.9b** © Terry Acomb. **11.10a** NASA/NSSDC/GSFC. **11.10b** HST/WFPC2 H. Weaver (ARC), NASA. **11.12** Roger Lynds/NOAO/AURA/NSF. **11.15a** NASA/JPL/MPIA. **11.15b** NASA/STScI. **11.15c** R. Evans, J. T. Trauger, H. Hammel and the *HST* Comet Science Team, and NASA. **11.16** Courtesy of the Wolbach Library, Harvard-Smithsonian Center for Astrophysics, Cambridge, MA. **11.17b** The Image Bank © Barrie Rokeach. **11.18** Armagh Observatory. **11.19** NASA.

CHAPTER 13

13.8a Brookhaven National Laboratory. **13.8b & c** Courtesy of Tomasz Barszczak, University of California, Irvine, for the Super-Kamiokande Collaboration. **13.9** NOAO/AURA/NSF. **13.10a** SOHO/ESA/NASA. **13.12** Nigel Sharp, NOAO/NSO/Kitt Peak FTS/AURA/NSF. **13.13a & b** © 2001 by Fred Espenak, courtesy of www.MrEclipse.com. **13.14** NASA/Marshall Space Flight Center. **13.15** SOHO/ESA/NASA. **13.18** SOHO/ESA/NASA. **13.20c & d** SOHO/ESA/NASA. **13.22** SOHO/ESA/NASA. **13.23** SOHO/EIT and SOHO/LASCO (ESA and NASA).

CHAPTER 14

14.1 top Image courtesy of Dr. A. Mellinger. **14.1 bottom** NASA/NSSDC/GSFC Diffuse Infrared Background Experiment (DIRBE) instrument on the Cosmic Background Explorer (COBE). **14.4** NSSDC/GSFC Infrared Astronomical Satellite (IRAS). **14.5** NSSDC/GSFC Roentgen Satellite (ROSAT). **14.6** Courtesy of Ron Reynolds, WHAM project. **14.7** C.R. O'Dell (Rice University), and NASA. **14.8** *HST* WFPC2 Science Team and NASA. **14.9** Courtesy of Dr. Christine Jones. **14.10** Patrick Hartigan, Rice University. **14.15** Jeff Hester and Paul Scowen (Arizona State University), Bradford Smith (University of Hawaii), Roger Thompson (University of Arizona), and NASA. **14.19** Karl Stapelfeldt/JPL/NASA. **14.20** Jeff Hester (Arizona State University), WFPC2 Team, NASA. **14.21** © Anglo-Australian

Observatory/Royal Observatory, Edinburgh, photograph from UK Schmidt plates by David Malin.

CHAPTER 15

Table 15.1 Guy Worthey based on the calculations of A. Bressan, F. Fagotto, G. Bertelli, C. Chiosi, *Astronomy and Astrophysics,* 1993, V100, p. 647. **15.11a** © Minnesota Astronomical Society. **15.11b** NASA and The Hubble Heritage Team (STScI/AURA). **15.12 Spirograph** Dr. Raghvendra Sahai (JPL) and Dr. Arsen R. Hajian (USNO), NASA and The Hubble Heritage Team (STScI/AURA). **15.12 NGC6751** A. Hajian (USNO) et al., Hubble Heritage Team (STScI/AURA), NASA. **15.12 Cat's Eye** J.P. Harrington and K.J. Borkowski (University of Maryland), *HST,* NASA. **15.12 Eskimo** NASA, A. Fruchter and the ERO Team (STScI). **15.12 Helix** Anglo-Australian Telescope, photograph by David Malin. **15.12 M2–9** Bruce Balick (University of Washington), Vincent Icke (Leiden University, The Netherlands), Garrelt Mellema (Stockholm University), and NASA.

CHAPTER 16

16.7 STScI/NASA/Arizona State University/Hester/Ressmeyer/Corbis. **16.10** © Anglo-Australian Observatory, photographs by David Malin. **16.11a** Joachim Trumper, Max-Planck-Institut für extraterrestrische Physik. **16.11b** Courtesy Nancy Levenson and colleagues, © American Astronomical Society. **16.11c** Jeff Hester, Arizona State University, and NASA. **16.15a** © European Southern Observatory, ESO PR Photo 40f199. **16.15a inset** Jeff Hester & Paul Scowen, Arizona State University, and NASA. **16.15b** Jeff Hester, Arizona State University, and NASA. **16.15c** Jeff Hester, Arizona State University, D. Burrows and K. Mori, Pennsylvania State University, and NASA.

CHAPTER 17

17.1 R. Windhorst and S. Pascarelle (Arizona State University) and NASA. **17.3** The Hubble Heritage Team (AURA/STScI/NASA). **17.4** NOAO/AURA/NSF and Z. Frei, Institute of Physics, Eotvos University, Hungary. **17.5** NOAO/AURA/NSF. **17.6** NOAO/AURA/NSF. **17.7** C. Howk (JHU), B. Savage (University of Wisconsin), N.A. Sharp (NOAO)/WIYN/NOAO/NSF. **17.9 left** George Jacoby, Bruce Bohannan, Mark Hanna/NOAO/AURA/NSF. **17.9 right** European Southern Observatory (ESO). **17.10** R. Windhorst and D. Burstein (Arizona State University). **17.11a** Todd Boroson/NOAO/AURA/NSF. **17.11b** Richard Rand, University of New Mexico. **17.13a** Z. Frei, Institute of Physics, Eotvos University, Hungary. **17.14b** Z. Frei, Institute of Physics, Eotvos University, Hungary. **17.15** Images courtesy G. Fabbiano, Harvard-Smithsonian Center for Astrophysics. **17.16** J. Bahcall (Institute for Advanced Study), M. Disney (University of Wales) and NASA. **17.17** NRAO/Alan Bridle et al. **17.19b** Courtesy Bill Keel, University of Alabama, and NASA. **17.19c** Holland Ford, STScI/Johns Hopkins University, and NASA. **17.19d** R.P. van der Marel, STScI, F.C. van den Bosch, University of Washington, and NASA. **17.20** F. Owen, NRAO, with J. Biretta, STScI, and J. Eilek, NMIMT. **17.21** Holland Ford, STScI/Johns Hopkins University; Richard Harms, Applied Research Corp.; Zlatan Tsvetanov,

Arthur Davidsen, and Gerard Kriss at Johns Hopkins University; Ralph Bohlin and George Hartig at STScI; Linda Dressel and Ajay K. Kochhar at Applied Research Corp. in Landover, MD; and Bruce Margon from the University of Washington, Seattle/NASA. **17.22** John Biretta, STScI. **17.23** Halton Arp/Caltech.

CHAPTER 18

18.1a Axel Mellinger. **18.1b** C. Howk and B. Savage (University of Wisconsin); N. Sharp (NOAO)/© WIYN, Inc. **18.2** U.S. Patent Office Library. **18.3** Hubble Heritage Team (AURA/STScI/NASA). **18.4** NOAO/AURA/NSF. **18.5** Courtesy Jeff Hester. **18.7** Courtesy J. Binney. **18.9 left** NOAO/AURA/NSF and Z. Frei, Institute of Physics, Eotvos University, Hungary. **18.9 right** © Anglo-Australian Observatory. **18.12a** The Electronic Universe Project. **18.16a** Carl Heiles, UC Berkeley. **18.16b** C. Howk and B. Savage (University of Wisconsin), N. Sharp (NOAO). **18.18** NRAO. **18.19 LMC and SMC** © Anglo-Australian Observatory, photograph by David Malin.

CHAPTER 19

19.2 TIMEPIX. **19.3 top** Courtesy of Jeff Hester. **19.3 bottom** NASA William C. Keel (University of Alabama, Tuscaloosa). **19.5** WIYN/NOAO/NSF, © WIYN Consortium, Inc. **19.6** From Hubble, E.P.: *The Realm of the Nebulae*. New Haven, CT: Yale University Press 1936. **19.14** Lucent Technologies' Bell Labs. **19.15** © Ann and Rob Simpson. **19.17** NASA COBE Science Team.

CHAPTER 20

20.1a © Anglo-Australian Observatory. **20.1b** Omar Lopez-Cruz and Ian Shelton/NOAO/AURA/NSF. **20.5** AAO/David Malin. **20.6b** NASA, A. Fruchter and the ERO team (STScI, ST-ECF). **20.7** Images produced by Donna Cox © Smithsonian and Motorola Corporation. **20.8** Tolia Klypin, New Mexico State University. **20.9** A. Dekel, Hebrew University, Jerusalem, Israel. **20.13** From *The Elegant Universe* by Brian Greene. © 1999 by Brian R. Greene. Used by permission of W. W. Norton & Company, Inc. **20.15** NASA/JPL/Origins. **20.16** © Jeff J. Daly/ Visuals Unlimited. **20.17 top left** © Mark A. Schneider/Visuals Unlimited. **20.17 bottom left** © Ken Lucas/Visuals Unlimited. **20.17 middle bird** © K. Sandved/Visuals Unlimited. **20.17 right skeleton** © Science VU/National Museum of Kenya/ Visuals Unlimited. **20.17 top right and bottom right** Burgess shale © Ken Lucas/Visuals Unlimited. **20.18** NASA. **20.19** NASA. **20.20** F. Drake (UCSC) et al., Arecibo Observatory (Cornell, NAIC). **20.21** FOXCLIPS. **20.22 left** Dr. Michael Perfit, University of Florida, Robert Embley/NOA. **20.22 right** © Woods Hole Oceanographic Institution, Deep Submergence Operations Group, Dan Fornari. **20.23** Courtesy of Ly Ly.

Epilog NASA/Manned Spacecraft Center/Image # 68-HC-870.

INDEX